AF443482

Nuclei in the Cosmos

Nuclei in the Cosmos

Proceedings of the Second International Symposium on
Nuclear Astrophysics held at Karlsruhe, Germany, 6–10 July,
1992

Edited by F Käppeler and K Wisshak

Sponsored by Kernforschungszentrum Karlsruhe

Institute of Physics Publishing
Bristol and Philadelphia

British Library Cataloguing in Publication Data

A catalogue record for this book is available from the British Library

ISBN 0-7503-0260-7

Library of Congress Cataloging-in-Publication Data are available

Published by IOP Publishing Ltd, a company wholly owned by the Institute of Physics, London
Techno House, Redcliffe Way, Bristol BS1 6NX, England
US Editorial Office: IOP Publishing Inc., The Public Ledger Buildings, Suite 1035, Independence Square, Philadelphia, PA 19106, USA

Printed in the UK by Galliard (Printers) Ltd, Great Yarmouth, Norfolk

Preface

The second International Symposium 'Nuclei in the Cosmos' was held with the intention to gather astronomers, astrophysicists, and nuclear physicists for a thorough discussion of nucleosynthesis, its role in the evolution of the universe and its intriguing possibilities as a diagnostic tool for stellar interiors. With the impressive degree of sophistication achieved over the last decade, the exchange and the collaboration between scientists working on the various aspects of nucleosynthesis becomes increasingly fascinating, but is also necessary for comprehensive investigations.

Another important idea was to combine a review of the topics in nucleosynthesis, provided by a relatively large number of invited talks, with the forefront of research expressed by contributed papers. In this way, young scientists or those interested in joining the astrophysical community were offered a sound basis of information. In total, the symposium attracted 149 scientists from 23 countries.

The motivations outlined were essentially adopted from the meetings which immediately preceded this symposium: 'Quests in Nuclear Astrophysics and Experimental Approaches' organized by T Paradellis *et al* in Crete (1988), and the first International Symposium on 'Nuclei in the Cosmos' organized by H Oberhummer in Baden/Vienna (1990). In order to maintain that structure, we have used the same title, hoping to establish a series of biennial conferences that would possibly continue the recent tradition developed since 1988. Needless to say, we are very glad to see this continuity maintained, at least for the near future, and we are wishing the Torino group the best success for the organization of the next symposium in 1994.

This is also the place to express our gratitude. In the first place, we have to thank Kernforschungszentrum Karlsruhe for its generous sponsorship, in particular Professor W Klose for his continuous support and encouragement. Special thanks are also due to all members of the program committee, M Arnould (Brussels), C A Barnes (Pasadena), C Cesarsky (Saclay), W Hillebrandt (Munich), T Kirsten (Heidelberg), D Lambert (Austin/Texas), B Pagel (Copenhagen), C Rolfs (Bochum), G Schatz (Karlsruhe), and S E Woosley (Santa Cruz), for their ideas and their enthusiasm, by which they have prepared the scientific success of the symposium. We also thank the members of the organizing committee, H Beer, S Jaag, Ch Theis, F Voss and K Wisshak, as well as all students and colleagues at KfK, who helped in so many respects to make the meeting happen. Of course, this holds, in particular, for the friendly and patient assistance of Eva Schröder, Irmgard Antoni, Sabine Burkhardt, Martina Engelmann and Ye Peng.

F Käppeler
Kernforschungszentrum Karlsruhe
September 1992

Contents

Section 1: Astronomical Facts

The processes of nucleosynthesis and chemical evolution as observed in stars: three examples

B   Gustafsson

Astronomical Observatory, Uppsala University, Box 515, S-75120 Uppsala, Sweden

Abstract.   The possibilities and problems of studying nucleosynthesis and chemical evolution in the Galaxy by determining stellar abundances are illustrated in three examples - the abundances in solar-type disk dwarfs with various ages and kinematic properties, the light element abundances in Pop II dwarfs, and the molecular abundances in circumstellar shells around carbon stars.

1.   Introduction

The evolution of the chemical elements in the Galaxy is a truly complex process, involving nuclear reactions in stars and elsewhere, stellar evolution and mass loss, processes in the interstellar medium, and star formation. Also phenomena on larger scales - infall of intergalactic matter, mergers between galaxies, etc. - are probably significant, as are cosmological events for the lightest elements. In spite of considerable progress since the 1950s we are still far from a detailed understanding of this complex interplay between different phenomena. It is, however, encouraging (though not quite astonishing) that high-quality data from one area in this field of research may often be used to promote understanding in another area. Thus, the intimate interrelation between the different areas is not only contributing complexity but also allows firm though often indirect constraints to be established concerning the physical processes involved.

Stellar spectroscopy is an area of increasing significance in this respect. In the last decade a new generation of effective spectrometers, like the ESO CES and the McDonald coudé spectrometer, equipped with linear detectors like Reticons and CCDs, have revolutionized abundance determinations. It is now possible to measure weak stellar absorption lines, so important in abundance analysis since they are not saturated, with a relative accuracy of better than 10%. With contemporary detailed model atmospheres this accuracy may be rather well matched, thus rendering differential abundance determinations almost one order of magnitude more reliable than before. Still more remarkable is that such analyses, of a character that could at the very best be made for a handful of stars two decades ago, may now be made for thousands of stars.

We are still far from a systematic exploitation of these new opportunities. An example of what can be achieved is given below (Sec. 2). This study of nearly 200 disk stars hopefully shows the strength of selecting well defined and large samples of suitable stars in the study of nucleosynthesis. My second example (Sec. 3) deals with the study of the build-up of the lightest elements in the Early Galaxy. Although technical problems are clearly illustrated in this study, the example will show that observing just a few spectral lines just above the noise level in a few stars may also lead to significant new information. The third example (Sec. 4), dealing with attempts to determine the chemical composition of circumstellar shells ejected from carbon stars, should show the significance of failures in abundance determination when these have further physical implications.

2. The production of chemical elements in the Galactic Disk

About 200 stars in the solar neighbourhood were selected from *uvbyβ* photometry such that they fulfilled the following criteria: they were all dwarfs with effective temperatures around the solar one and all in the main-sequence band but at some distance from the Zero-Age Main Sequence so that stellar ages could be determined from that distance when combined with evolutionary tracks. The stars were distributed in metallicity rather evenly in the interval 1/10 - 2 times that of the Sun, i.e. -1.0 < [Fe/H] < 0.3 with the usual logarithmic measure, normalised relative to the solar abundances. Known binaries were excluded. For these stars five different wavelength bands (from 500 nm to 880 nm) were observed with high spectral resolution and high signal/noise. The bands were selected such that a number of weak unblended spectral absorption lines could be found for the following elements: O, Na, Mg, Al, Si, Ca, Ti, Fe, Ni, Y, Zr, Ba and Nd. A sample spectrum, in fact of one of the faintest stars in our sample, for one of the bands is shown in Fig. 1.

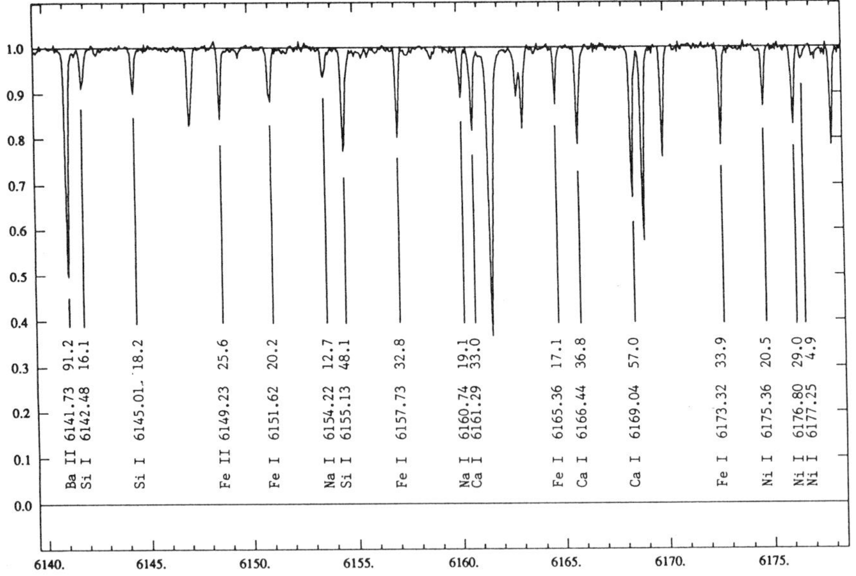

Fig. 1. One spectral region observed for HD 17548, V=8.16. Identifications, wavlengths (in Å) and equivalent widths (in mÅ) are given for each absorption line.

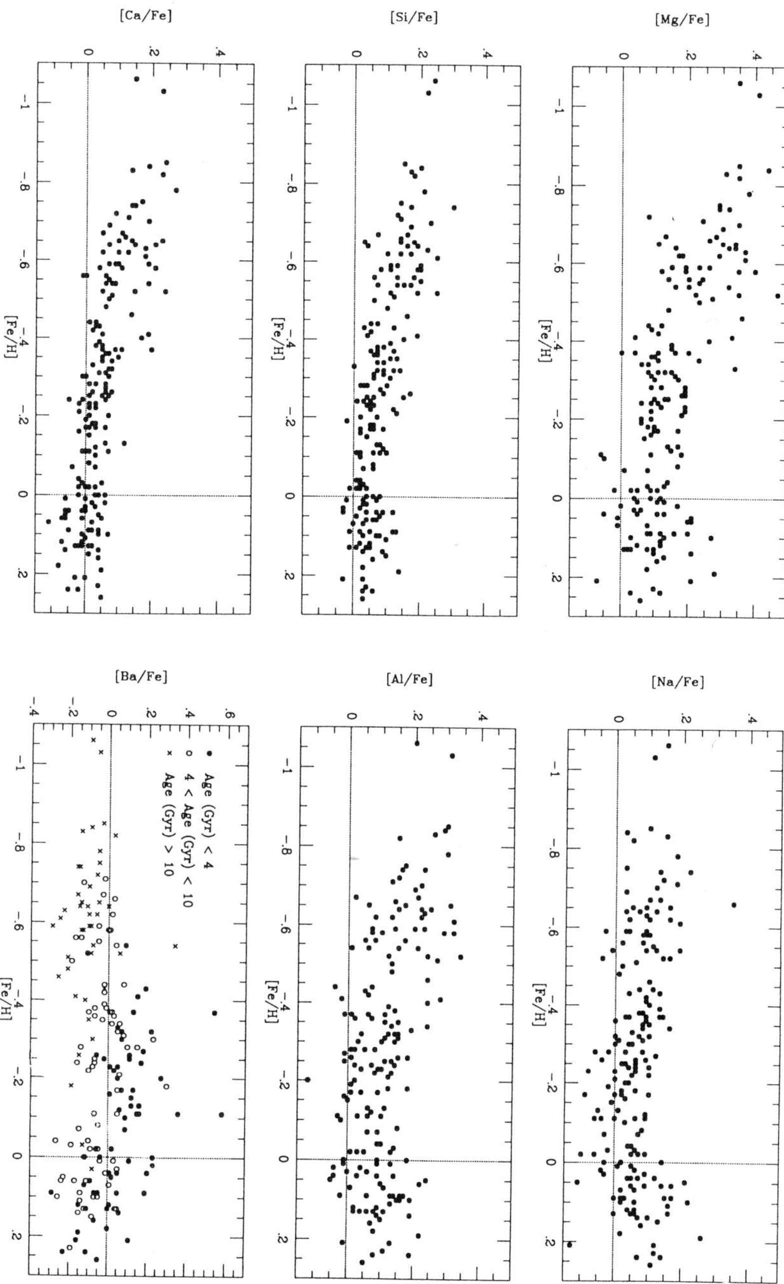

Fig. 2. The logarithmic abundances of disk stars relative to iron for a number of chemical elements vs iron abundance relative to hydrogen. All abundances are with respect to those of the Sun. For Ba, stars of different ages are plotted with different symbols. (From Edvardsson et al. 1992.)

The spectral data were analysed by using a new grid of detailed model atmospheres for which the blanketing effects from millions of weak spectral lines were taken into consideration. The resulting abundances relative to the Sun show a high internal consistency with errors of typically 0.05 dex (about 10%) or even less. In addition to this radial velocities were observed, and good proper motion data and parallaxes were also available such that accurate space motions could be calculated and, using a model of the galactic potential, galactic orbits. This work is presented in full detail in a forthcoming paper (Edvardsson et al. 1992). Here, only some results will be commented on.

The abundances relative to iron are displayed in Fig. 2. It is seen that there is a smooth variation and a rather small scatter around the mean relations. The slopes of these relations, in particular when the lighter elements are compared with iron or nickel, is ususally interpreted as the result of two types of supernovae operating, one (Type II) promptly producing light elements rather early in the history of the Galactic Disk and another (Type Ia) producing most of the heavier elements with some built-in time delay.

One could wonder whether the small scatter in several panels of Fig. 2 is at all significant or only due to observational errors. This is contrary to the relation between [Fe/H] and age which shows a considerable scatter (Fig. 3) which quite certainly is real. To study this further we have subdivided the stellar sample into groups with similar ages and similar $R_m \equiv (R_{min}+R_{max})/2$, R_{min} and R_{max} being the minimum and maximum distances from the Galactic Centre, respectively. These latter quantities are varying with time as a result of inhomogeneities in the gravitational potential, while the variation in R_m is much less (Grenon 1987). Thus, R_m should contain information on the distance from the Galactic

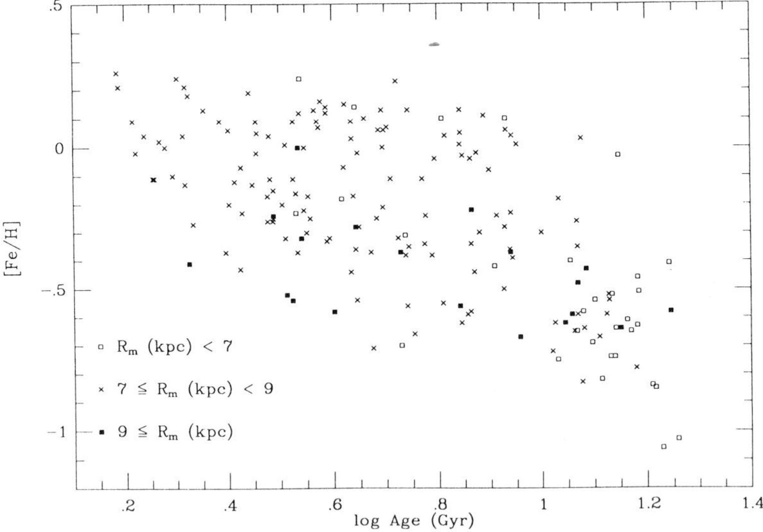

Fig. 3. The iron abundances plotted relative to ages for solar-type stars in orbits at different mean distances from the Galactic Centre (from Edvardsson et al. 1992). The large spread in [Fe/H] at a given age is certainly real.

Centre where the star was once formed, assuming the initial orbit to be rather circular. It turns out that the scatter in [Fe/H] for each age-R_m group is still significantly greater (about 0.2 dex) than the small observational error in [Fe/H] (about 0.05 dex). On the other hand, the scatter in abundances *relative to iron*, e.g. in [α/Fe] which denotes the mean of the logarithmic abundances of the so-called alpha-elements Mg, Si, Ca and Ti relative to the solar values, is very small ($\approx$0.05 dex) within each age-R_m group. The conclusion from this is that the ejecta from SNe II and SNe Ia must be very well mixed in the Galaxy at every given time and radius from the Centre before new stars form. E.g., it does not seem to happen very often that star formation is triggered by a supernova in such a way that the individual SN remnant is able to significantly affect the chemical composition of the stars. Stars that are possible exceptions from this will be discussed below. Astonishingly, this mixing is much more efficient than the mixing with hydrogen - a fact that is hard to understand as long as the formation of alpha elements and iron, respectively, is ascribed to phenomena with a time delay as long as billions of years. Obviously, this result has strong implications as regards the dynamical conditions in the Interstellar Medium.

A closer look at the spread around the mean relations of Fig. 2 discloses interesting structure. E.g., one finds that for small R_m , i.e. for stars probably formed inside the solar galactic orbit, there is an excess in [α/Fe] at a given [Fe/H], as compared with stars with greater R_m. This indicates that star formation occurred more vigorously in the inner Galaxy, which led to a rapid build up of alpha elements. This interpretaion is supported by the results by Burkert, Truran and Hensler (1992) for their chemodynamical model of the evolution of the Galactic Disk. The time scale for the formation of iron and nickel, however, is set by the evolution time for intermediate mass stars that form SNe type Ia, a time scale which is similar at different galactocentric distances.

I have attempted to determine empirically what the yields are from supernovae of different types for the different elements observed. Essentially this can be done from the slopes in Fig. 2, by using a suitable model of the chemical evolution of the Disk. Such a model, with infall of gas and allowance for a time delay between the promt production of α elements by Type II SNe and the slower production of iron etc. from Type Ia SNe, has been described by Clayton (1988) and modified by Pagel (1989). This one-zone model has the virtues of being simple and analytical but still reproduces essential properties of the Disk population such as the metallicity distribution for G dwarfs.

Using this model relations between the abundance ratio, [X/II], of any element X relative to an element II, assumed to be only produced in SNe II, and [II/H] can be calculated, cf. Fig. 4. These relations are dependent on the true yields of X from SN Ia and II, respectively. We denote the true yield of SN II, relative to the sum of type I and II yields, by R_{II}. By plotting the observed abundances in Fig. 4 we can thus estimate R_{II} for different elements.

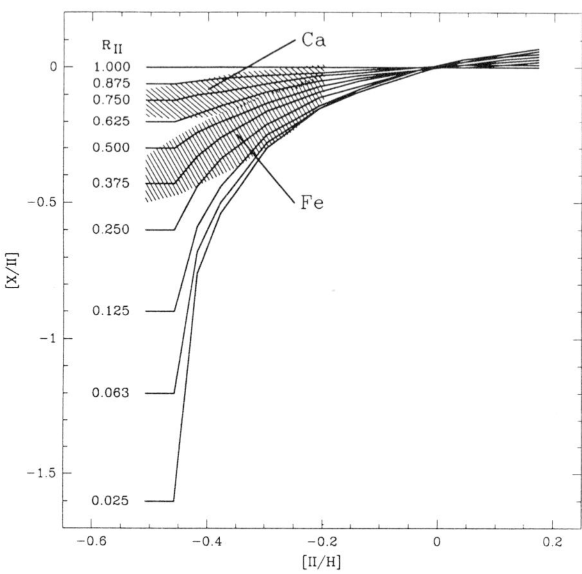

Fig. 4. The abundance [X/II] of an element X, relative to that of another element (II), as a function of [II/H]. Element II is supposed to be formed entirely in SNe Type II. The curves correspond to calculations, made using the model for the Galactic Disk by Pagel (1989), for different values of the fraction of the true yield of X from SN Type II, relative to the sum of the yields from SNe type Ia and II. Observed abundances in disk stars for X=Ca and Fe, respectively, and II=O and Mg are shown by the hatched areas.

In doing this in practice I have adopted the parameters of the model of Pagel (1989). In particular this means a normalised time delay parameter $\omega\Delta$ of 0.5, which with a value of ω of 0.3 Gyr^{-1} (Clayton 1988) corresponds to a delay between Type II and Type Ia SNe of 1.7 Gyr. The particular choice of model parameters is, however, not very important for the results below. As Type II element I have chosen oxygen *and* magnesium (i.e. performed the anaysis with both alternatives and then averaged the result). Both elements are, according to the SN models by Thielemann et al. (1990) almost entirely formed in SNe Type II. This choice was due the fact that oxygen observations only exist for half the number of stars and moreover the oxygen abundances were scaled to results for the oxygen line judged to be most reliable, namely the forbidden 630 nm line, which was observed for even fewer stars. The Mg abundances also suffer from uncertainties, in particular due to the small number of lines observed.

The resulting empirical R$_{II}$ values are given in Table 1 and compared with theoretical values, estimated from supernovae models by Thielemann et al. (1990).

Table 1. The fraction of the true yields of SN II relative to SN I + SN II, R_{II}, estimated
from the abundance results and compared to estimates from supernova models.

Element	R_{II}(abundances)	R_{II}(SN models)
O	1.00	0.98
Na	0.49	1.00[1]
Mg	0.87	0.96
Al	0.62	0.97
Si	0.65	0.69
Ca	0.74	0.56
Ti	0.76	0.71
Fe	0.29	0.25
Ni	0.33	0.19

1) Not given for SN Ia by Thielemann et al. (1990).

The agreement for most elements between yields estimated from the observed
stellar abundances and those derived from SN models is quite satisfactory, in view of the
fact that the theoretical SN yields have been derived from just two models at particular
masses. Thus, no integrations accross the mass spectrum has been performed, neither has
the gradual increase of the initial abundance of heavy elements been considered. The two
elements that depart seriously, Na and Al, are both quite uncertain in the nucleosynthesis
calculations for SNe. They are assumed to be formed in explosive carbon and neon burning,
respectively, and their yields are dependent on the neutron excess and thus on the Ne
abundance which, through helium burning, is dependent on the initial oxygen abundance of
the pre-SN star. The   calculations of the synthesis of Na and Al by Woosley and Weaver
(1982) in SN II models, giving results for two different overall metal abundances that
differ by a factor of 100,   suggest that the discrepancy for Al may well be due to this
dependence on initial abundances while the calculated difference in Na production between
initially metal rich and metal poor Type II SNe seems far too small to account for the low
R_{II} value observed. Other known sources of Na, such as the NeNa cycle probably operating
on the main sequence for massive stars and leading to detectable enrichment of Na in
supergiant atmospheres (Boyarchuk et al. 1988), should not contribute enough Na to
significantly affect galactic abundances (cf. Denisenkov and Ivanov 1987). The Na
production, possibly occurring in the dying deflagration wave of carbon burning in SN Ia,
is "practically unknown" (Nomoto 1992). Nomoto et al (1992) derive a solar Na/O
abundance ratio from SN II models, when integrating over the proper mass distribution.
This indicates that Type II SNe are the main contributors of Na. In contrast to this our
results in Fig. 2 and Table 1 suggest a considerable   contribution from "delayed sources",
such as Type Ia SNe. In fact, there is independent evidence for that. We thus find a tendency

for the Mg abundance to be greater relative to Na in the inner part of the Galaxy, as compared with the outer part, at a given [Mg/H]. This is shown in Fig. 5. The corresponding tendency for the alpha elements relative to Fe was interpreted above as a result of differences in star formation rate and evolution time for SNe of different types.

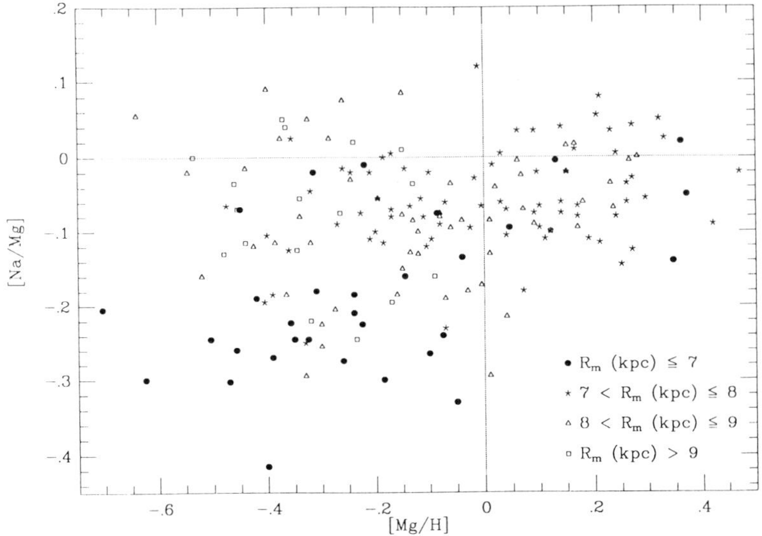

Fig. 5. The abundance ratios [Na/Mg] relative to [Mg/H] plotted for disk stars in orbits at different mean distances from the Galactic Centre (from Edvardsson et al. 1992). A tendency for the stars with R_m < 7 kpc to have systematically low Na abundances relative to Mg at a given (low) Mg abundance is clearly seen.

Another interesting phenomenon may have implications for nucleosynthesis. David Lambert has traced a new type of peculiar stars in our sample that are enriched in Na, Mg and Al relative to Fe by about 0.10 dex, normal in Si and Ca and poor in Ba. The few stars of this category for which oxygen has been measured do not show any oxygen enrichment. They are metal rich relative to the Sun and constitute about 5% of the sample. These stars contribute to the scatter and the "upward bending" of the relations for Na, Mg and Al in Fig. 1 at [Fe/H] > 0.0. It is interesting to see that, according to the SN nucleosynthesis calculations, this group of stars might be formed from a gas mix that is relatively more enriched by SN II than usual. However, this explanation does not seem viable, if not the metallicity dependence of the Na production is much stronger than predicted by Woosley and Weaver (1982).

Incidently, it is tempting to speculate that the occurance of NaMgAl stars may be related to the existence of Na and Al rich stars in globular clusters, a phenomenon which is thought to indicate abundance inhomogeneities in the gas from which the stars were once formed (cf., e.g., Drake, Smith and Suntzeff 1992, and V.V. Smith´s contribution to the present proceedings).

Four the four elements Si, Ca, Fe and Ni the stellar spectroscopy results should be most reliable. We therefore ascribe some significance to the empirical findings that, contrary to what the SN model calculations suggest, there may be somewhat more of Type II origin for Ca than for Si, while Fe and Ni are similar in this respect.

Non-trivial effects can also be traced for the s-elements Y and Ba. The abundances of these elements, relative to iron, are found to increase with time in the Galaxy in an interesting way, dependent on the iron abundance itself as well as on R_m. Much of the scatter in the s-elements, e.g. for Ba in Fig. 2, is in fact due to this systematic variation. Obviously another, even longer time scale is operating here, probably corresponding to the evolution time for less massive stars forming red giants where most of these elements may be formed. We note in passing that our data suggest delay times Δ of at least about 2 Gyrs, which indicate initial stellar masses of less than 2 solar masses for the red giants contributing the s-elements to the Galactic Disk. This mass is considerably less than suggested by Pagel (1989); however, it is natural to suppose that it was greater earlier in the history of the Galaxy.

3. The light elements in the oldest stars

The possibility that observable amounts of some of the elements heavier than Li, like Be and B, were formed in the early Universe as a result of inhomogeneities has received considerable attention in recent years (see contributions by Malaney and Bürger et al. in the present proceedings for further references). Most recently, much of this attention was stimulated by the discovery of Be in the prototype metal-poor dwarf HD 140283 ([Fe/H] = -2.8) by Gilmore, Edvardsson and Nissen (1991). The Be abundance observed, of about 1 part in 10^{13} as compared with hydrogen, was about one order of magnitude higher than expected from models of Be production by spallation processes in the interstellar medium of the Early Galaxy. The question was whether this abundance in fact revealed the existence of a plateau of a constant abundance of Be, independently of [Fe/H] provided that the latter is sufficiently small, as was discovered ten years ago for Li (Spite and Spite 1982).

In order to find the answer to this question we attempted identification of Be in several metal-poor stars with different metal abundances (Gilmore et al. 1992). The only spectral lines that are available for this purpose are the two Be II lines at 313 nm. They are situated in a difficult spectral range, distorted by lines originating in the terrestrial atmosphere and close to the atmospheric ultraviolet cut-off. Moreover, the spectral regions are crowded, so that the weak Be lines get rather blended, and the programme stars are not so bright in the ultraviolet. We used the Anglo-Australian 3.9 m telescope to obtain spectra

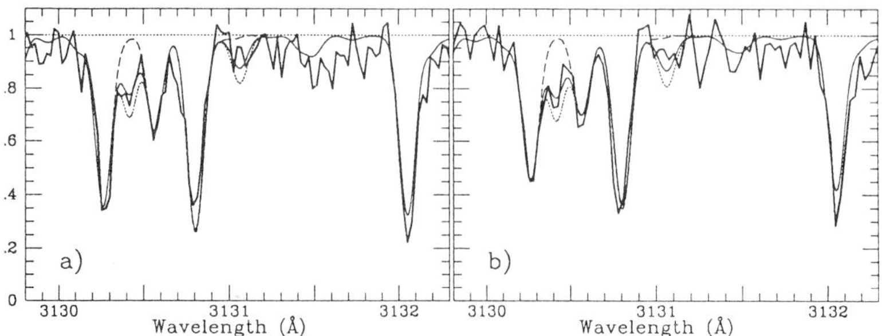

Fig. 6. Observed and synthetic spectra for HD 116064 ([Fe/H]=-2.2) and HD 213657 ([Fe/H]=-2.3). The observed spectra are shown by thick solid lines, while the thinner lines show synthetic spectra with different Be abundances: dashed lines = no Be, thin solid lines = adopted abundance, dotted lines = 1.6 x adopted (from Gilmore et al. 1992). The problems of obtaining useful spectra for Pop II dwarfs in the ultraviolet spectral region, even with large telescopes, should be clear, e.g. after comparison with Fig. 1.

for 11 stars, 2 of which are displayed in Fig. 6. The observed spectra were analysed by comparison with synthetic ones from model atmospheres. The data and parameters behind these model spectra have been determined on independent grounds - there are no free fitting parameters in the comparison shown in Fig. 6 except for the Be abundance. In each model spectrum displayed in Fig. 6 about 250 spectral lines from different elements were included. In spite of the rather noisy spectra we consider the identification of the beryllium lines to be safe and estimate that their strength (scaling roughly linearly with the abundance) can be determined with relative errors on the order of 25%.

The Be abundances ε(Be) resulting from this investigation were found to be proportional to the oxygen abundances ε(O) (determined from neighbouring lines of the OH molecule) as is shown in Fig.7. A linear relation was also traced for boron from observations with the Hubble Space Telescope by Duncan, Lambert and Lemke (1992). These interesting results do not support any primordial origin of the Be or B, however, nor did they agree with current models for the build-up of light elements in the Galaxy through spallation. Those models predicted a quadratic variation of ε(Be) and ε(B) with ε(O), basically reflecting the fact the the increase $d\varepsilon$(Be,B)/dt is proportional to the cosmic-ray flux, assumed to be proportional to the SN rate, times ε(O) in the interstellar medium which is proportional to the integrated SN rate.

We proposed that the linear dependence may be explained if spallation of the fresh heavy nuclei takes place close to the SN remnants before the nuclei are mixed into the general interstellar medium. This possibilitiy was further investigated by Feltzing and Gustafsson (1992) who found that enclosure times of the cosmic rays in the oxygen enriched SN remnants of at least 10^3 years are needed. An alternative explanation, based on the fact that the cosmic-ray particles of greatest significance for production of Be and B in the metal-poor interstellar gas in the Early Galaxy should be heavy nuclei, colliding with the interstellar protons and helium nuclei, was proposed by Duncan, Lambert and Lemke (1992). A third scenario, involving a halo and a disk phase, with cosmic rays efficiently confined in the halo phase and with a flat spectrum, has been described by Prantzos, Cassé and Vangioni-Flam (1992, see also Prantzos's contribution in the present proceedings). There are some interesting possibilities to distinguish between these hypotheses, as well as other alternatives (see Malaney's contribution in the present proceedings) by observing more metal-poor stars and isotopic ratios for Li and B.

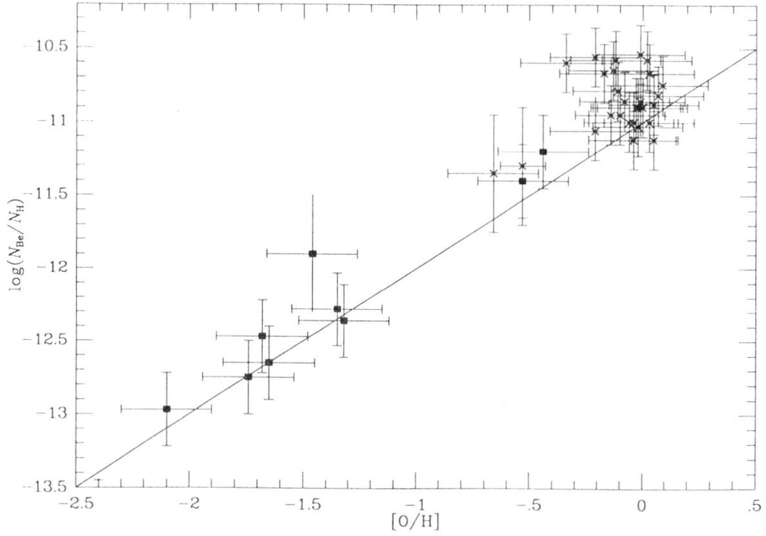

Fig. 7. Beryllium abundance vs. logarithmic oxygen abundance relative to the Solar value, [O/H], for Pop. II dwarfs and disk dwarfs (from Gilmore et al. 1992). The line is ambigously drawn with slope unity.

As is seen in this example, abundance studies based on a very small number of spectral lines for single chemical elements and a number of different stars may give interesting results, even if the S/N and the spectral resolution are not as high as one would wish. It is also seen, however, that one does not always learn about what one was primarily interested in. Instead of findings about the conditions in the Universe some microseconds from the origin of time and the character of the phase transition from a quark-gluon plasma to a hadron plasma, we find new bounds on cosmic rays and the dynamical conditions in the young Galactic Halo. True enough, this general unpredictability concerning the outcome of scientific study is not astonishing to anybody involved in empirical studies of complex systems. Our third example will, however, illustrate this further.

4. Molecular abundances in shells around carbon stars

In an investigation of the chemical composition of (N-type) carbon stars, based on high resolution high S/N infrared spectra from the stellar photospheres, we found that the reason why the spectra of these stars are entirely dominated by carbon molecules of different types is their (in most cases moderately) increased atmospheric abundance of carbon, which produced by helium burning and mixed to the surface in the AGB phase (Lambert et al. 1986). (The alternative, that the oxygen abundance could be considerably reduced in these stars so that the stable CO molecule would not be able to bind almost all carbon available, was excluded.) In at least one respect the result of our investigation was astonishing: we did not find any tendency for nitrogen to be enhanced above its solar value, not even for ^{13}C rich stars, as one would expect due to dredge up of CNO processed material in earlier red-giant stages. Although the nitrogen abundances were based on lines from several molecules (CN, NH, HCN) we considered this result to be preliminary in view of a number of uncertainties in molecular and stellar data.

Later, we discovered that the vast majority of the bright N-type stars show CO emission from circumstellar shells (Olofsson et al. 1992a). Moreover, for many of these shells we could also observe millimeter wavelength lines of HCN, CN and CS as well as of their counterparts with ^{13}C, using the (sub)millimeter telescopes at Onsala Space Observatory, and ESO (SEST). It was then of interest to see whether these lines could be used for estimating the elemental abundances in the shells, e.g. for checking the photospheric N abundances or, more importantly, to deduce carbon abundances for shells that are so dense and dusty that photospheric spectra can not be obtained. The outcome of such a study may be of significance, e.g. for deciding about the origin of carbon in the Galaxy.

In undertaking this investigation we used the elemental abundances derived from the photospheric spectra by Lambert et al. (1986) to predict the abundances of CO, HCN and CS in the shells (Olofsson et al. 1992b). The latter abundances, relative to CO, are thought to be the result of processes in equilibrium in the upper photosphere and are then relatively easy to predict. An important uncertain parameter is, however, the mass loss rate, which is essentially estimated with a simplified shell model from the CO observations (cf. Olofsson et al. 1982).

A main result from this is that the predicted HCN and CS abundances in the shells are almost one order of magnitude smaller than those observed. We do not think that this is a result of the fact that for some stars the HCN lines show some maser characteristics. Also the lines of CN molecules, presumably formed by photodissociation of HCN in the shells, tend to be much stronger than expected in that all available photospheric nitrogen or more would have to be consumed in order to explain the observed CN molecular abundance.

Instead of suggesting systematic errors in the photospheric abundances - which might be tempting in view of the revised nitrogen abundances that this would imply - we claim that these results show that our understanding of the structure and line formation in the circumstellar shells is not satisfactory (see also Omont´s contribution to the present proceedings). The most probable failure, also suggested by other circumstances (cf. Olofsson et al. 1992a), is that the mass loss rate has been strongly underestimated by almost one order of magnitude. Thus, we find that our efforts to study the build-up of chemical elements in carbon stars and its galactic significance instead give clues to a better understaning of mass loss from AGB stars. Admittedly, this may also be very significant for nucleosynthesis but in a more indirect way.

5. Concluding remarks

Hopefully, our examples have demonstrated the need for and possibilities to carry out systematic empirical studies of galactic nucleosynthesis, and the often unexpected outcome of such studies. Like other historical sciences this study in general seems to develop from beautiful simplicity to some frustration. As in archeology and history the focus then changes from philosophical interest and belief in general basic principles to a systematic empirical mapping of some regularities but more irregularities, gradually opening up a rich complexity which is certainly interesting but difficult to easily understand or even grasp intuitively. Fortunately, Nature, like History, still generously presents well defined problems to us - Why was this golden coin buried at this remote place? or Why is that beryllium here? Why did this empire fall apart so profoundly and rapidly? or Why was this mixing of supernovae products so effective? But, in order to reach deeper and more satisfactory understanding of the complexity, we now see that far more extensive and systematic studies are probably needed. The means *are* there or possible to develop, at least for stellar spectroscopy and, as many beautiful results to be presented at this conference seem to indicate, for nuclear physics. What is needed is scientists willing to devote themselves to this work which is hard and extensive and therefore does not always seem so attractive, not even for funding agencies.

New theory is needed too, not only more realistic numerical modelling of nuclei, gas and stars in galactic interplay but also analytical models and other means, so that we can understand what we observe and what we calculate. Otherwise, we will develop the Midas syndrome of modern astrophysics further - i.e., turn everything we touch upon into moving shadows across a computer screen - into, as somebody phrased it, "just weather". Nucleosynthesis is more than weather, more important as a fundamental boundary condition for the existence of all of us, and probably even more interesting.

Acknowledgements. Hardly anything of the work described above in which I have had the great pleasure to participate would have been thought of or, which more important, carried out, without a group of excellent collaborators: J. Andersen, U. Carlström, B. Edvardsson, K. Eriksson, S. Feltzing, G. Gilmore, K. Hinkle, D. Lambert, P. E. Nissen, H. Olofsson and J. Tomkin. I would like to use this opportunity to thank them all. K. Nomoto and S. Woosley are thanked for helpful discussions and G. Hammarbäck for assistance with Fig. 4.

References

Boyarchuk A A, Gubeney I, Kubat I, Lyubimkov L S, Sakhibullin N A 1988 *Astrophysics* **28** 202

Burkert A, Truran, J W, Hensler G 1992 *Astrophys. J.* **391** 651

Clayton D D 1988 *Monthly Notices Roy. Astron. Soc.* **234** 1

Denisenkov P A, Ivanov V V 1987 *Pis´ma Astron. Zh.* **13** 52

Drake J J, Smith, V V, Suntzeff N B 1992 *Astrophys. J. Letters*, submitted

Duncan D, Lambert D L, Lemke M 1992 *Astrophys. J.*, submitted

Edvardsson B, Andersen J, Gustafsson B, Lambert D L, Nissen, P E, Tomkin J 1992 *Astron. & Astrophys.* submitted

Gilmore G, Edvardsson B, Nissen P E 1991 *Astrophys. J.* **378** 17

Gilmore G, Gustafsson B, Edvardsson B, Nissen P E 1992 *Nature* **357** 379

Grenon M 1987 *J. Astron. Astrophys.* **8** 123

Feltzing S, Gustafsson B 1992 *Astrophys. J.*, submitted

Lambert D L, Gustafsson B, Eriksson K, Hinkle K H 1986 *Astrophys. J. Suppl.* **62** 373

Nomoto K 1992 private communication

Nomoto K et al. 1992 this conference

Olofsson H, Johansson L E B, Hjalmarson Å, Nguyen-Q-Rieu 1982 *Astron. & Astrophys.* **107** 128

Olofsson H, Eriksson K, Gustafsson B, Carlström U 1992a *Astrophys. J. Suppl.*, submitted

Olofsson H, Eriksson K, Gustafsson B, Carlström U 1992b *Astrophys. J. Suppl.*, submitted

Pagel B E J 1989 *Rev. Mex. Astr. Astrofis.* **18** 161

Prantzos N, Cassé M, Vangioni-Flam E 1992 *Astrophys. J.*, submitted

Spite F, Spite M 1982 *Astron. & Astrophys.* **115** 357

Thielemann F-K, Nomoto K, Hashimoto M 1991 in *Supernovae, Les Houches, Session LIV 1990*, eds. J Audouze, S Bludman, R Mochkovitch and J Zinn-Justin, Elsevier Science Publ.

Woosley S, Weaver T A 1982 in *Supernovae: A survey of Current Research*, eds. M Rees and R J Stoneham, D Reidel Publ. Co., p 79

The surface composition of chemically peculiar stars

Verne V. Smith
Department of Astronomy
and McDonald Observatory
The University of Texas
Austin, TX 78712 USA

ABSTRACT: The surface chemical composition of a star can be altered by different mechanisms, such as internal nucleosynthesis and the mixing of this material to the star's surface, mass transfer in a binary system, or compositional differences in the gas from which stars form. In this review we choose an example of each of the above processes and discuss how understanding the origins of chemical peculiarities wrought by the above mentioned events increases our understanding of stellar and chemical evolution and nucleosynthesis.

I. INTRODUCTION

The title assigned to me as a speaker concerns the abundances in "chemically peculiar" stars, which is certainly a very broad topic! Our first task is to pare this discussion down to a more manageable size. In its most general form, chemically peculiar could be taken to be any object that does not show the same abundances as some "standard abundance distribution": this distribution is usually taken to be the solar-system abundances, either solar photospheric or meteoritic. We note here that for a few elements, most notably lithium, the solar and meteoritic abundances are quite different, with the photospheric abundance of Li being some 100 times less than the meteoritic value (Anders and Grevesse 1989). As it is the Sun that has destroyed Li below the base of the outer convection zone, one could argue that, in terms of its Li abundance, the Sun is peculiar relative to a standard distribution based upon the meteoritic abundances. This rather silly example is meant to illustrate how the broad topic of chemical peculiarities can be applied perhaps to almost any star, or population of stars. A famous astronomer once responded to this speaker's desire to pursue abundance studies as a first-year graduate student with the statement that "Abundances, bah...you can slice them as thin as you'd like, just like baloney!". While my choice of topics to review may be slicing some abundances thin, in the sense that many elements and isotopes may seem a bit obscure, I hope to demonstrate how some of these "chemical peculiarities" tell us a great deal about stellar and chemical evolution and nucleosynthesis.

We note that there are a number of ways that can lead to a star having an abundance pattern that could be labelled "peculiar" in some way. For example, the following general processes can be identified:

    1) Diffusion
    2) Fractionation
    3) Internal
    4) External
    5) Primordial

The first two examples do not involve nuclear processes, *per se*, and will not be discussed in a conference devoted, in large part, to nuclear astrophysics. In the last three cases, we will pick examples to illustrate how these three processes, or really three sites of nucleosynthesis, can lead to peculiar abundances on the surface of a star. As an "Internal" process we will discuss the element lithium in AGB stars, where the abundance has been altered by nuclear processes inside a star and this altered abundance has then been mixed to the star's surface. For an "External" process we will discuss heavy-element abundances in Barium and CH stars: classes of stars whose peculiar abundances are due to mass transfer from a companion star. Finally, as an example of peculiar abundances which probably existed in the gas from which certain stars formed (hence, "Primordial"), we will discuss sodium, aluminum, and oxygen abundance variations in globular cluster stars.

II. LITHIUM PRODUCTION AND HOT-BOTTOMS IN ASYMPTOTIC GIANT BRANCH STARS

As established by the Li surveys of G, K, and M giants (Lambert, Dominy, and Sivertsen 1980; Luck and Lambert 1982; Brown *et al.* 1989), the vast majority of Galactic red giants have a rather weak Li I 6707 Å doublet feature and low Li abundances due to convective dilution of Li on the red giant branch (RGB); this result holds also for most of the thermally-pulsing asymptotic giant branch (AGB) stars of spectral types S and C (Torres-Peimbert and Wallerstein 1966; Boesgaard 1970; Catchpole and Feast 1976; Denn, Luck, and Lambert 1991; Abia *et al.* 1991). Because both ^{6}Li and ^{7}Li are destroyed by (p,α) reactions at low temperatures (T $\gtrsim$ 2 x 10^6 K), Li is destroyed throughout most of the interior mass of a main sequence star; during subsequent RGB evolution, the deepening convective envelope will mix large amounts of material devoid of Li to the stellar surface and the observable Li abundance will decline by a large factor (Iben 1967). The above mentioned surveys have confirmed, in a very general way, the convective dilution of Li in red giants, although there are fascinating subtleties in Li destruction on the main sequence and in giants (for a thorough review see Michaud and Charbonneau 1990) that we will not disucss here.
One striking feature of a small fraction of S and C stars, that has been known for decades (McKellar 1940), is the presence of strong Li I lines, in contrast to the norm for most red giants of weak or absent Li I. Cameron (1955) proposed the beryllium transport mechanism as a potential path for ^{7}Li production: ^{3}He(α,γ)^{7}Be(e$^-$,ν)^{7}Li. The trick with this mechanism is that ^{7}Be production must take place near T $\gtrsim$ 10^7 K, yet be rapidly convected to lower temperatures (T $\lesssim$ 2 x 10^6 K) where the ^{7}Li created from electron captures can survive. Cameron and Fowler (1971) suggested that the thermally-pulsing AGB stage of stellar evolution might be a potential site where the ^{7}Be transport mechanism might operate. This asociation of AGB evolution with ^{7}Li synthesis has been explored by a number of investigators (Ulrich and Scalo 1972; Scalo and Ulrich 1973; Iben 1973; Sackmann, Smith, and Despain 1974; Scalo, Despain, and Ulrich 1975) using stellar models. In general, it is found that the deep and massive convective envelopes of luminous AGB stars develop very high base temperatures (T ~ 40-80 x 10^6 K) in between thermal pulses: Scalo *et al.* (1975) labelled these convective regions as "hot-bottom" envelopes. Light-element nucleosynthesis was followed by Scalo *et al.* (1975), and more recently by Sackmann and Boothroyd (1992), using time-dependent convective diffusion, and both studies find large enhancements of surface ^{7}Li due to operation of the Be transport mechanism. Lithium production occurs only at high luminosities (M_{bol} ~ -6) in these models and one would then presume that the Li-rich S and C stars in the Galaxy are the higher luminosity AGB stars; however, determining the luminosities of field red giants is a difficult and uncertain task, so it is not possible to accurately compare the luminosity at which hot-bottom burning occurs in real Galactic AGB stars with the models.
Determining luminosities of red giants in the Magellanic Clouds is not as problematic as in the Galaxy because the distances to the Magellanic clouds are known and recent spectroscopy of red giants in the Clouds has revealed the luminosities at which envelope burning commences in AGB stars using Li as a signature. One persistent puzzle about AGB

stars in the Magellanic Clouds was the absence of predicted luminous C stars, which are the intermediate-mass (4-8 $M_\odot$) thermally-pulsing AGB stars (Blanco, McCarthy, and Blanco 1980; Mould and Aaronson 1982,1983,1986; Reid, Tinney, and Mould 1990). The lack of luminous C stars ($M_{bol} \lesssim -6$) led to some speculation that severe mass loss terminated AGB evolution towards higher luminosities. Contrary to this speculation, Wood, Bessell, and Fox (1983) discovered luminous ($-7 \lesssim M_{bol} \lesssim -6$) Mira variables with strong ZrO bands; the ZrO tags them as S stars, thus they are high-luminosity thermally-pulsing AGB stars. Instead of being carbon stars, however, these stars are oxygen rich. Wood *et al.* (1983) suggested, as one alternative, that hot-bottom burning might be responsible for the conversion of ^{12}C, produced in ^{4}He-burning thermal pulses and mixed into the envelope, to ^{14}N via the CN cycle at the base of the deep convective envelope: this would keep the ratio of C/O < 1 and prevent the formation of a C star.

Strong evidence for hot-bottom burning is provided by the spectroscopy of Smith and Lambert (1989,1990) and Smith, Lubowich, and Lambert (1992) of some of the Wood *et al.* stars in the Magellanic Clouds, which they find to have strong Li I lines. The pattern of the appearance of the Li I doublet is that it is strong in the luminous, oxygen-rich stars in the Clouds ($-7 \lesssim M_{bol} \lesssim -6$), while being absent in the lower-luminosity AGB stars and the very luminous and massive core-burning supergiants ($M_{bol} \lesssim -7$). These studies interpret the presence of a substantial Li I line as being due to the nucleosynthesis of ^{7}Li in the hot bottoms of the most luminous AGB stars.

Using improved abundance analysis techniques (improved molecular opacities and spherical model atmospheres) Plez, Smith, and Lambert (1992) are carrying out detailed abundance analyses (including Li, the s-process, ^{12}C, ^{13}C, and N abundances) of a number of Small Magellanic Cloud (SMC) AGB stars. In Figure 1, we show a short region of the

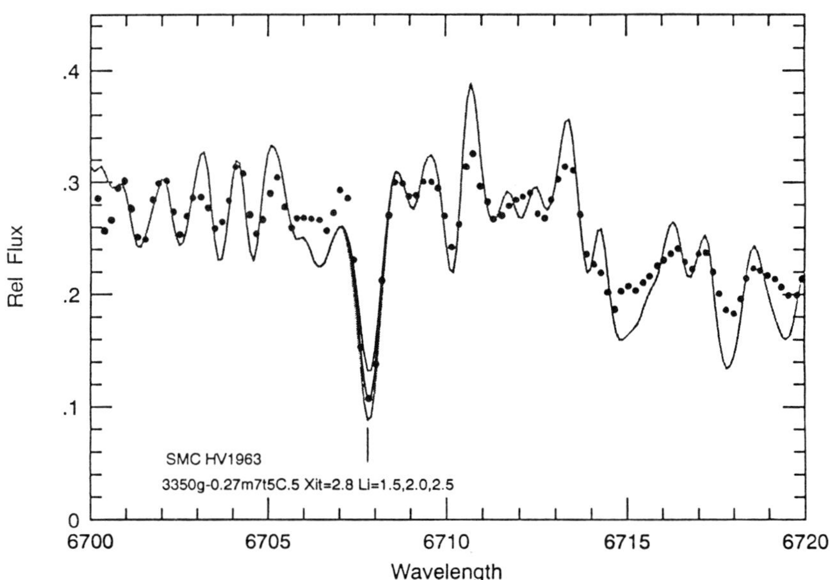

Figure 1. Sample spectrum of the Li I 6707.8 Å doublet in the SMC luminous AGB star HV 1973 (filled circles) Solid curves are three synthetic spectra with differing Li abundances.

spectrum near the Li I 6707.8 Å line in the luminous SMC AGB star HV 1963: the observed spectrum consists of the filled circles while the continuous curves are synthetic spectra with three different Li abundances of log ε(Li) = 1.5, 2.0, and 2.5 defined on a scale where log ε(H) = 12.0. The Li I doublet is clearly visible with log ε(Li) = 2.0 providing a reasonable fit.

In their recent work on modelling the formation of Li-rich red-giants, Sackmann and Boothroyd (1992) predict the Li abundance from their models as functions of luminosity for different stellar masses and metallicities. In Figure 2 we compare the predictions of Li abundance versus luminosity from their solar metallicity 4.0 $M_\odot$ and 6.0 $M_\odot$ models with Li abundances in 5 SMC Li-rich AGB stars from Plez *et al.* (1992). It is clear from Figure 2 that both the observations and models are in general agreement on both the general enrichment of Li at similar luminosities. It would seem that hot-bottom burning is responsible for Li production in luminous AGB stars and probably also the fact that these stars remain S and not C stars.

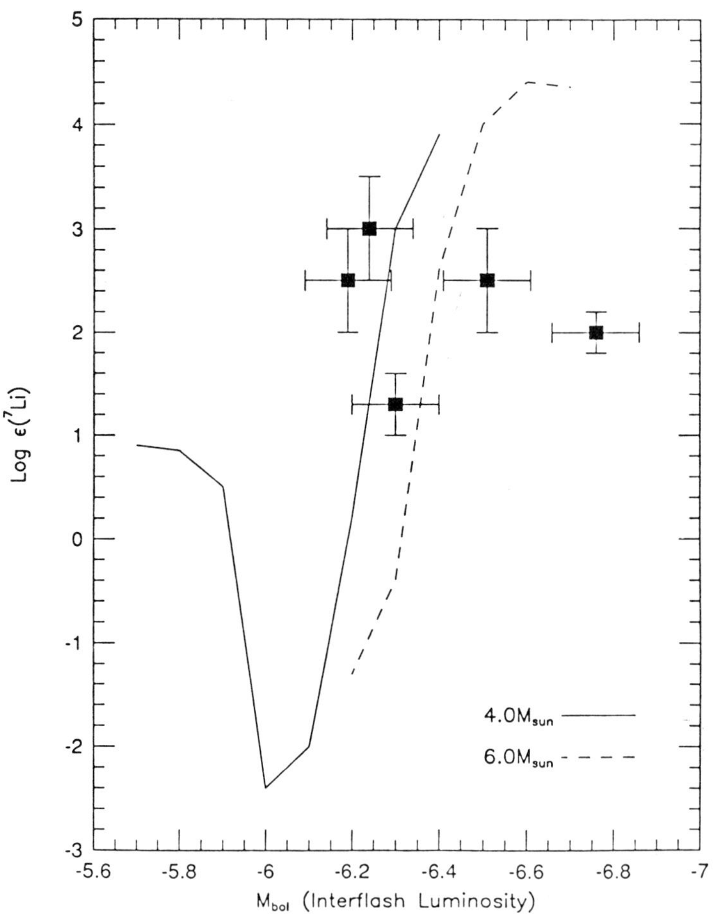

Figure 2. Li abundance as a function of luminosity for solar-metallicity AGB models from Sackmann and Boothroyd (1992) compared to Li abundances derived by Plez *et al.* (1992) for five SMC luminous oxygen-rich AGB stars.

III. BARIUM AND CH STARS AS PROBES OF THE s-PROCESS NEUTRON SOURCE

A complete understanding of the origins of the s-process elements requires, as initial data sets, accurate abundances of a wide range of heavy elements, as well as certain isotopic species. The connection of s-process production to the ^{4}He-burning thermal pulses in AGB stars has been established observationally by the spectral sequence of M-MS-S-SC-C stars: this sequence is one of increasing C/O ratio as well as increasing s-process abundances. The relation between increasing ^{12}C, from the ^{4}He-burning shells, and the increase in the total s-process abundance, has been put on quantitative grounds largely through abundance analyses of these AGB stars (*e.g.*, Lambert *et al.* 1986; Dominy, Wallerstein, and Suntzeff 1986; Smith and Lambert 1990). Although sufficient for deriving very general heavy-element trends, these cool AGB stars have extremely complicated spectra, with much of the spectrum obliterated by TiO and ZrO molecular bands (for the O-rich AGB stars), or CN and C_2 bands (for the C stars); these difficulties decrease both the accuracy of the abundance analyses, as well as the number and variety of heavy elements which can be observed. If one is interested in deriving very accurate abundances for a large number of heavy s-process nuclei, the S and C stars are difficult, if not impossible, to do at this time.

Fortunately, there are classes of stars which are hotter and of lower luminosity than the true AGB stars, and thus have less complicated spectra, yet exhibit the same abundance peculiarities as the S and C stars: these are the Barium and CH stars. For many years these lower luminosity peculiar stars (a mixture of giants, subgiants, and main-sequence stars of largely spectral types F, G, and K) presented a serious challenge to the theory of stellar evolution, as they were of much too low a luminosity to be, or have ever been, AGB stars, yet they carried the abundance signatures expected, and indeed observed, in the luminous AGB stars. It has now been established observationally that the Barium and CH stars are all binaries with (probably) white-dwarf companions (McClure, Fletcher, and Nemec 1980; McClure 1983; McClure and Woodsworth 1990) and the currently observed abundances on what is now the system primary are the result of mass transfer when the current secondary star in the system was an AGB star. Reviews of various aspects of the Barium and CH stars can be found in Lambert (1985, 1988), McClure (1989), and Smith (1992). These systems contain much potential information on the evolution of binary stars, the dynamics of mass transfer, and the interaction of main-sequence or red-giant stars with a hot, degenerate companion; however, from an abundance point-of-view, the Barium and CH stars provide an opportunity to study the chemical compositions of AGB stars in unprecedented detail, due to the relative simplicity of an F, G, or K spectrum when compared to an S or C star. Of particular interest is the ability to characterize the heavy s-process abundance distribution accurately using a large number of heavy elements; this abundance distribution is a sensitive function of the exposure of the material to the flux of neutrons and provides information on the neutron source, as well as conditions within the site of the s-processing.

The parameter which most controls the overall distribution of heavy s-process elements (such as Ba, La, Ce, Nd, or Sm) to light s-process elements (such as Sr, Y, Zr, Nb, or Mo) is the neutron exposure, or neutron fluence, which is defined as

$$\tau = \int_0^{t_i} N_n(t) v_T(t) dt \tag{1}$$

where v_T is the neutron thermal velocity, t_i is the duration of neutron irradiation, and N_n is the neutron density. If the stellar structure where s-processing occurs is independent of metallicity, as might be expected if the s-process occurs in the intershell region between the H- and ^{4}He-burning shells in an AGB star, then the duration and temperatures in a thermal pulse may not vary much. In this situation, an increase in N_n could enhance the neutron exposure which will characterize the s-process abundance distribution. As pointed out by Clayton

(1988), $N_n(t)$ may be a function of metallicity depending upon whether the neutron source is $^{13}C(\alpha,n)^{16}O$ or $^{22}Ne(\alpha,n)^{25}Mg$. Clayton notes that the product $N_n\text{-}v_T$, which is the neutron flux, Φ, can be written as

$$\Phi = \frac{S_n}{\sum\limits_A \sigma_A N_A} \tag{2}$$

where S_n is the neutron-source rate, σ_A is the neutron-capture cross-section of neutron absorbing isotope, N_A, and the product $\sigma_A N_A$ is summed over all neutron absorbers. Clearly, the total number of neutron absorbers is proportional to overall metallicity, $\sum \sigma_A N_A \propto Z$, while Clayton argues that the ^{13}C and ^{22}Ne lead to very different neutron-source rates as functions of Z. Concerning ^{22}Ne, it is noted that the abundance of ^{22}Ne in the intershell region of an AGB star is proportional to the initial total CNO abundance, which is proportional to Z, so $S_n(^{22}Ne) \propto Z$ and $\Phi(^{22}Ne)$ will not be a function of metallicity. For ^{13}C, the situation is different due to the fact that the ^{13}C supply available for the neutron source arises from the mixing of protons into the intershell region via $^{12}C(p,\gamma)^{13}N(\beta^+,\nu)^{13}C$: this mixing mechanism is still an active research topic and we simply refer here to the recent papers by Hollowell and Iben (1988,1989). Because the ^{12}C is synthesized within the AGB star itself, the ^{13}C neutron-source, to some approximation, can be taken to be independent of Z, so $\Phi(^{13}C) \propto 1/Z$ as suggested by Clayton (1988). Thus, it is predicted in this simple picture that τ may increase with decreasing metallicity.

This possible difference in τ as a function of Z can be tested with published abundance analyses of the Barium and CH stars. Recent work by Vanture (1992) on the metal-poor CH stars has made possible a comparison of the neutron exposure in these metal-poor s-process enriched stars with higher metallicity Barium and CH stars from previous studies. A convenient and simple way to parameterize τ is to compute the ratio of the average overabundances of the heavy s-process elements (Ba through Sm) to the light s-process elements (Sr, Y, Zr), [hs/ls]: we use spectroscopic bracket notation where [A/B] = log $(N_A/N_B)_{Program\ Star}$-log $(N_A/N_B)_{Standard}$. This ratio of [hs/ls] was used in a recent survey study of CH and Barium stars by Luck and Bond (1991). As Vanture (1992) shows, [hs/ls] is a sensitive function of τ and increases with increasing τ. To compare to Vanture's (1992) study of the metal-poor CH stars, we choose seven other studies of Barium and CH stars that have been conducted with high-S/N spectra obtained with solid-state detectors (Reticons or CCD's) and using model atmospheres in the abundance analysis. We show [hs/ls] versus [Fe/H] from these studies (identified in the figure caption) in Figure 3. Vanture's (1992) results at the low-Z end ([Fe/H] $\lesssim$ -1) show clearly an increasing trend in [hs/ls] as [Fe/H] decreases. Any compilation of results from a large number of studies may have some systematic effects in the comparison due to variations in the quantity and quality of both the data and the analysis, however, the increase in [hs/ls] in Vanture's results is too large to be dismissed as systematics and suggests strongly an increase in τ as Z decreases: the same type of behavior as predicted by Clayton's (1988) discussion of the ^{13}C neutron-source ($\tau \propto 1/Z$).

The simple interpretation of the results presented in Figure 3 is that the trend shown here is evidence of the $^{13}C(\alpha,n)^{16}O$ neutron source operating in AGB stars. We note, however, that Luck and Bond (1991) have identified 4 "metal-deficient Barium stars" which have [Fe/H] ~ -1.2 to -1.8 with [hs/ls] ~ -0.3 to +0.2; this is very different from the behavior of the metal-poor CH stars. Are the metal-deficient Barium stars examples of an s-process distribution from the ^{22}Ne neutron-source? These four stars deserve much closer scrutiny to resolve this possible difference between the metal-poor Barium and CH stars!

Taken together, the numerous analyses of the MgH line profiles in the S and Barium stars, along with Vanture's (1992) results of the s-process abundance distribution in the metal-poor CH stars, suggests that the s-process in the majority of the AGB stars is driven by $^{13}C(\alpha,n)^{16}O$, which itself requires the mixing of protons down into the intershell region, *e.g.*, Hollowell and Iben (1988). Remaining hopes for ^{22}Ne perhaps rest with closer

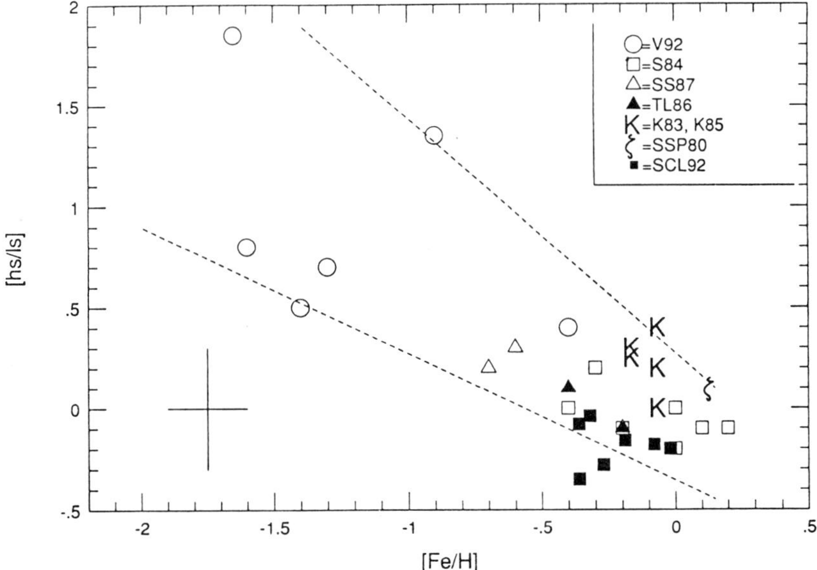

Figure 3. The heavy s-process to light s-process abundance ratios ([hs/*l*s) vesus metallicity ([Fe/H]) for a smple of Barium and CH stars; [hs/*l*s] is proportional to the neutron exposure. A typical estimated uncertainty in the abundances is shown and te dashed lines merely illustrate the range of the observed points. Key: V92 = Vanture (1992); S84 = Smith (1984); S587 = Smith and Suntzeff (1987); TL86 = Tomkin and Lambert (1986); K83 and K85 = Kovacs (1983,1985); SSP80 = Smith, Sneden, and Pilachowski (1980); SCL92 = Smith, Coleman, and Lambert (1992).

analyses of Luck and Bond's (1991) metal-deficient Barium stars or the Li-rich S stars (*i.e.*, massive AGB stars).

IV. THE Na, Al, AND O MYSTERY IN GLOBULAR CLUSTERS: A PRIMORDIAL PROCESS?

For many years now, it has been demonstrated that globular clusters are not chemically homogeneous groups of stars. The stars within globular clusters can show a wide range of abundance variations: from large chemical inhomogeneities in virtually all elements (such as C, N, O, Na, Al, Ca, Fe, Sr, or Ba) in ω Cen to only variations in C and N, which may be common to all globular cluster red giants and may only reflect the operation of the first dredge-up. Recent reviews of these abundance variations can be found in Smith (1987,1989), Suntzeff (1989), and Pilachowski (1989). There are interesting patterns in these "chemical peculiarities", such as the apparent well-defined bimodal distributions of CN strengths found in many intermediate metallicity clusters (-0.7 $\gtrsim$ [Fe/H] $\gtrsim$ -1.6) which suggest two distinct populations of red giants: "CN-srong" and "CN-weak" stars. The origins of these CN distributions within individual clusters has remained a controversial topic: do these differences in CN band-strengths arise from purely internal stellar evolution within the individual giants (*i.e.*, C and N abundance variations due to convective mixing of CN cycle material on the red giant branch), or do they reflect primordial differencs in the C and N abundances in some of the protocluster gas from which a certain fraction of the cluster stars formed? Recent observations of CN and CH variations in main-sequence stars ($M_v \simeq +6$) in the globular clusters 47 Tuc (Briley, Hesser, and Bell 1991) and NGC 6752 and 47 Tuc (Suntzeff and Smith 1991) lend some weight to the idea of primordial variations, as it is difficult to mix CN-cycle material to the surface of a low-luminosity main sequence star.

An additional pattern associated with the CN variations is the correlation of the strength of Na I and Al I lines with CN band-strength in giants of similar color and magnitude (*e.g.*, Peterson 1980; Cottrell and Da Costa 1981; Norris *et al.* 1981). Due to the fact that many of these earlier studies of Na I and Al I variations relied upon strong resonance lines, the interpretation of these observations has remained somewhat controversial. Two potential obstacles to an interpretation of these variations as simple abundance variations have been invoked: 1) non-LTE effects, and 2) differences in the atmospheric structure in the CN-strong and CN-weak giants due to the importance of C-, N-, and O-containing molecules in molecular blanketing. Beginning with the second potential obstacle, recent and detailed model-atmosphere calculations by Drake, Plez, and Smith (1992) employing opacity sampling and spherical geometry, demonstrate that the Na I and Al I line variations cannot be the result of atmospheric temperature structure. Non-LTE effects have been studied by Drake, Smith, and Suntzeff (1992) in an abundance analyis of four giants in the intermediate-metallicity

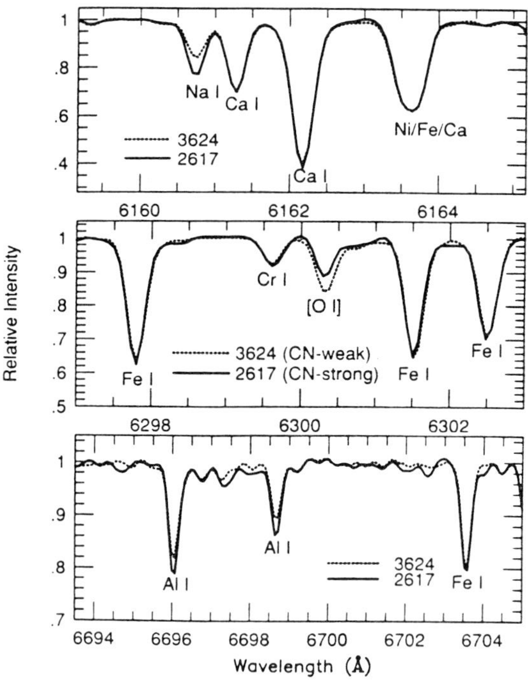

Figure 4. The spectral regions illustrating the Na I, [O I], and Al I lines in two M4 giants. The top panel shows λ6162Å in the CN-strong 2617 and CN-weak star 3624 illustrating the great similarity in the Ca I lines, but large differences in the Na I line. In the middle panel is shown the weakened [O I] line in the CN-strong star 2617 relative to the CN-weak star 3624. The bottom panel demonstrates differences in the Al I doublet near 6698 Å: note the similarities in the Fe I line, while the differences in the weak features are due to CN.

cluster M4 ([Fe/H] = -1.1). These four giants were of similar temperature and luminosity, with two being CN-strong and two CN-weak; abundances were presented for Fe, Ca, Na, Al, and O, with non-LTE calculations carried out for Fe, Ca, and Na. They ruled out any significant non-LTE effects, describing the results of detailed non-LTE calculations for Na which suggested that relaxing the LTE assumption altered the Na abundance by 0.02 dex.

Drake *et al.* (1992) based their analysis upon high-S/N high-resolution (S/N $\approx$ 200, $\lambda/\Delta\lambda$ = 35000) spectra and line variations in weak lines (those most sensitive to chemical abundance) between stars of similar T_{eff} and log g are readily apparent in such spectra. We illustrate, in Figure 4, differences in Na I, Al I, and [O I] lines between a CN-strong and CN-weak giant with similar T_{eff} and log g in M4: the CN-strong exhibits stronger Na I and Al I lines, relative to the CN-weak giant, but a weaker [O I] line. In these four M4 giants the Na and Al abundances are increased by about 0.3 dex in the CN-strong giants relative to the CN-weak giants, while O is decreased by about 0.3 dex in the CN-strong giants: the Na/Al abundances appear anti-correlated with O. In a much larger survey, using similar quality spectra, Kraft *et al.* (1992) present Na and O abundances in the globular clusters M3 and M13 plus a sample of field halo giants; we illustrate their Na and O abundances in Figure 5, along with the Drake *et al.* (1992) M4 results. The "bracket notation" of [O/Fe] and [Na/Fe] is defined as the logarithmic difference in the abundance ratios of O/Fe and Na/Fe in the program stars relative to the solar ratio. There is a clear, strong anti-correlation of [Na/Fe] to [O/Fe] which is an abundance anti-correlation of Na with O: both groups find remarkably homogeneous Fe abundances within each of the three globular clusters, with the standard deviations of the Fe abundances being 0.03 dex for both M3 and M4 and 0.06 dex for M13! The giants with increased Na abundances are also the CN-strong stars, thus, Na and Al almost certainly correlate with the N abundance.

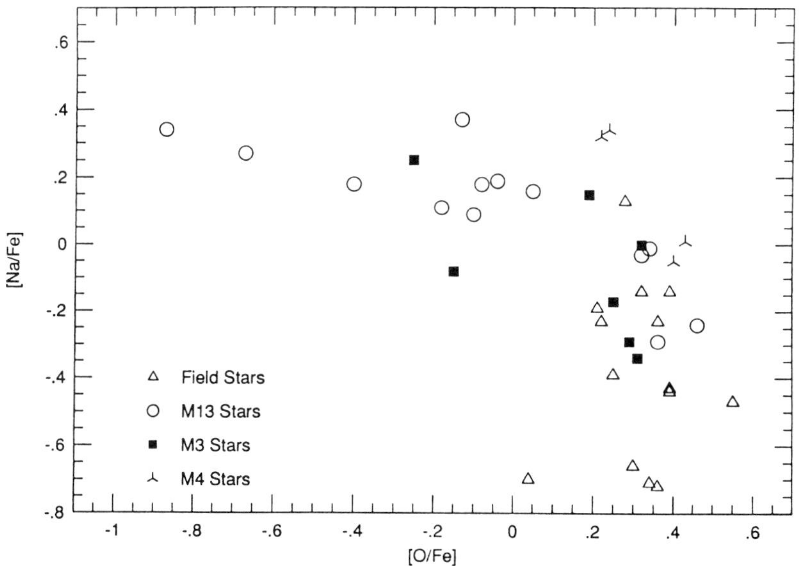

Figure 5. The Na-O anti-correlation observed in three globular clusters and a sample of halo field giants: from Kraft *et al.* (1992) and Drake *et al.* (1992). It is clear that as [O/Fe] decreases, [Na/Fe] increases.

Having established these abundance trends to accuracies approaching 0.05 dex, the challenge is now to identify the event, or events, responsible. The presence of an oxygen underabundance may lead one to suspect internal origins via the dredge-up of ON-cycle material; Denisenkov and Denisenkova (1990) have suggested that Na enrichment will accompany the dredge-up of O-poor ON-cycle material. They ascribe Na production to proton capture onto ^{22}Ne: ^{22}Ne(p,γ)^{23}Na. This can occur at layers in the star where the ON cycle runs. In addition to Na, if temperatures are high enough, Al can be synthesized from ^{25}Mg via ^{25}Mg(p,γ)^{26}Al(p,γ)^{27}Si(β^+,ν)^{27}Al. Thus, the rise in Na and Al abundances with decreasing O abundances may be ascribed to "Internal" processes: why do we use this as an example of a "Primordial" process? In standard, non-rotating stellar models, mixing during the first dredge-up is confined to material exposed only to the CN cycle on the main sequence and no ON-cycle material appears at the surface. Of course, stellar models are not real stars! However, in the case of the globular cluster CN-strong and -weak giants, which seem to correlate with the Na/Al/O abundance variations (*i.e.*, CN-strong giants have Na and Al enhanced and O depleted), analogues exist on the main sequence in 47 Tuc and NGC 6752 (Briley *et al.* 1991; Suntzeff and Smith 1991). It is virtually certain that the CN and CH variations in these low-luminosity stars are not internal. Although Na and Al variations have yet to be observed in these globular cluster main sequences (the stars are faint), we suggest that determined observations will reveal such variations.

If, as we suspect, the Na/Al/O abundance variations are primordial, the site of this nucleosynthesis remains to be determined. Perhaps the mechanism for Na and Al synthesis during the ON cycle, as proposed by Denisenkov and Denisenkova, operates in massive, or intermediate mass, stars during the early history of a globular cluster and this early generation of stars evolves quickly enough to pollute remaining cluster gas (enriched in N, Na, and Al, but depleted in O) while low-mass stars are still forming. Cottrell and Da Costa (1981) proposed that Na and Al nucleosynthesis may occur as the result of the s-process operating in intermediate mass (M ~ 5-8 $M_\odot$) AGB stars during ^{4}He-burning with ^{22}Ne(α,n)^{25}Mg as the neutron source. Mass loss from such stars may be expected to also contain material processed through the CNO cycles, with enhanced N and depleted O. The identification of the initial site of the Na, Al, O, and N abundance variations may rest upon careful analyses of other elements, such as Ca, Mg, Sc, or the heavier s-process and r-process isotopes.

V. CONCLUDING REMARKS

We hope that the examples chosen here illustrate how chemical peculiarities due to Internal, External (here, mass transfer) and Primoridal effects can be used to probe both stellar and chemical evolution, as well as nucleosynthesis. Due to space and time limitations, we restrict ourselves to three examples, but the study of "chemical peculiarities" in stars encompasses an enormous range of topics as evidenced by the large number of reviews referenced here.

I gratefully acknowledge useful discussions with Grant Bazan, Mike Briley, Bob Kraft, David Lambert, Bertrand Plez, Chris Sneden, and Nick Suntzeff which increase greatly my meager understanding of the topics discussed here. This research is supported in part by the U. S. National Science Foundation (AST 91-15090) and the Robert A. Welch Foundation of Houston, Texas.

REFERENCES

Abia, C., Boffin, H. M. J., Isern, J., and Rebolo, R. 1991, *A&A*, **245**, L1.
Anders, E., and Grevesse, N. 1989, *Geochim. Cosmochim. ACTA*, **53**, 197.
Blanco, V. M., McCarthy, M. F., and Blanco, B. M. 1980, *ApJ*, **242**, 938.
Boesgaard, A. M. 1970, *ApJ*, **161**, 163.
Briley, M. M., Hesser, J. E., and Bell, R. A. 1991, *ApJ*, **376**, 51.
Brown, J. A., Sneden, C., Lambert, D. L., and Dutchover, E., Jr. 1989, *ApJS*, **71**, 293.
Cameron, A. G. W. 1955, *ApJ*, **212**, 144.
Cameron, A. G. W., and Fowler, W. A. 1971, *ApJ*, **164**, 111.
Catchpole, R. M., and Feast, M. W. 1976, *MNRAS*, **175**, 501.
Clayton, D. D. 1988, *MNRAS*, **234**, 1.
Cottrell, P. L., and Da Costa, G. S. 1981, *ApJ*, **245**, L69.
Denisenkov, P., and Denisenkova, S. 1990, *SvAL*, **16**, 275.
Denn, G. R., Luck, R. E., and Lambert, D. L. 1991, *ApJ*, **377**, 657.
Dominy, J. F., Wallerstein, G., and Suntzeff, N. B. 1986, *ApJ*, **300**, 325.
Drake, J. J., Smith, V. V., and Suntzeff, N. B. 1992, *ApJ*, **395**, L95.
Drake, J. J., Plez, B., and Smith, V. V. 1992, in preparation.
Hollowell, D., and Iben, I., Jr. 1988, *ApJ*, **333**, L25.
______. 1989, *ApJ*, **340**, 966.
Iben, I., Jr. 1967, *ApJ*, **147**, 624.
______. 1973, *ApJ*, **185**, 209.
Kraft R. P., Sneden, C., Langer, G. E., and Prosser, C. F. 1992, *AJ*, **104**, 645.
Kovacs, N. 1983, *A&A*,
______. 1985, *A&S*, **150**, 232.
Lambert, D. L. 1985, in *Cool Stars with Excesses of Heavy Elements*, ed. M. Jashek and P.
    C. Keenan (Dordrecht: Reidel), p. 191.
Lambert, D. L. 1988, in *The Impact of Very High S/N Spectroscopy on Stellar Physics* , ed.
    G. Cayrel de Strobel and M. Spite (Dordrecht: Kluwer), p. 563.
Lambert, D. L., Dominy, J. F., and Sivertsen, S. 1980, *ApJ*, **235**, 114.
Lambert, D. L., Gustafsson, B., Eriksson, K., and Hinkle, K. H. 1986, *ApJS*, **62**, 373.
Luck, R. E., and Lambert, D. L. 1982, *ApJ*, **256**, 189.
Luck, R. E., and Bond, H. E. 1991, *ApJS*, **77**, 515.
McClure, R. D. 1983, *ApJ*, **268**, 264.
______. 1989, in *Evolution of Peculiar Red Giant Stars*, ed. H. R. Johnson and B.
    Zuckerman (Cambridge: Cambridge University Press), p. 101.
McClure, R. D., Fletcher, J. M., and Nemec, J. M. 1980, *ApJ*, **238**, L35.
McClure, R. D., and Woodsworth, A. W. 1990, *ApJ*, **352**, 709.
McKellar, A. 1940, *PASP*, **52**, 407.
Michaud, G., and Charbonneau, P. 1992, *Space Sci. Rev.*, in press.
Mould, J. R., and Aaronson, M. 1982, *ApJ*, **263**, 629.
______. 1983, *ApJ*, **273**, 530.
______. 1986, *ApJ*, **303**, 10.
Norris, J., Cottrell, P. L., Freeman, K. C., and Da Costa, G. S. 1981, *ApJ*, **244**, 205.
Peterson, R. 1980, *ApJ*, **237**, L87.
Pilachowski, C. A. 1989, in *The Abundance Spread within Globular Clusters: Spectroscopy
    of Individual Stars*, ed. G. Cayrel de Strobel, M. Spite, and T. Lloyd Evans (Paris:
    Observatoire de Paris), p. 1.
Plez, B., Smith, V. V., and Lambert, D. L. 1992, in preparation.
Reid, N., Tinney, C., and Mould, J. 1990, *ApJ*, **348**, 198.
Sackmann, I.-J., Smith, R. L., and Despain, K. H. 1974, *ApJ*, **187**, 555.
Sackmann, I.-J., and Bothroyd, A. I. 1992, *ApJ*, **392**, L71.
Scalo, J. M., and Ulrich, R. K. 1973, *ApJ*, **183**, 151.

Scalo, J. M., Despain, K. H., and Ulrich, R. K. 1975, *ApJ*, **196**, 805.
Smith, G. H. 1987, *PASP*, **99**, 67.
Smith, G. H. 1989, in *The Abundance Spread within Globular Clusters: Spectroscopy of Individual Stars*, ed. G. Cayrel de Strobel, M. Spite, and T. Lloyd Evans (Paris: Observatoire de Paris), p. 63.
Smith, V. V. 1984, *A&A*, **132**, 326.
Smith, V. V. 1992, in *Evolutionary Processin in Interacting Binary Stars*, ed. Y. Kondo, R. F. Sistero, and R. S. Polidan (Dordrecht: Kluwer), p. 103.
Smith, V. V., and Lambert, D. L. 1989, *ApJ*, **345**, L75.
______. 1990a, *ApJS*, **72**, 387.
______. 1990b, *ApJ*, **361**, L69.
Smith, V. V., and Suntzeff, N. B. 1987, *AJ*, **93**, 359.
Smith, V. V., Sneden, C., and Pilachowski, C. A. 1980, *PASP*, **92**, 809.
Smith, V. V., Coleman, H., and Lambert, D. L. 1992, in preparation.
Smith, V. V., Lubowich, D., and Lambert, D. L. 1992, in preparation.
Suntzeff, N. B. 1989, in *The Abundance Spread within Globular Clusters: Spectroscopy of Individual Stars*, ed. G. Cayrel de Strobel, M. Spite, and T. Lloyd Evans (Paris: Observatoire de Paris), p. 71.
Suntzeff, N. B., and Smith, V. V. 1991, *ApJ*, **381**, 160.
Tomkin, J., and Lambert, D. L. 1986, *ApJ*, **311**, 819.
Torres-Peimbert, S., and Wallerstein, G. 1966, *ApJ*, **146**, 724.
Ulrich, R. K., and Scalo, J. M. 1972, *ApJ*, **176**, L37.
Vanture, A. D. 1992, *AJ*, in press.
Wood, P. R., Bessell, M. S., and Fox, M. W. 1983, *ApJ*, **272**, 99.

Orbital elements of a new sample of barium stars

Alain Jorissen†[1] and Michel Mayor‡

†European Southern Observatory, Garching bei München, Germany

‡Observatoire de Genève, Switzerland

1. Introduction

Barium stars are a class of peculiar red giants of spectral types G and K whose envelopes show overabundances of s-process elements (e.g. Lambert 1988). The origin of these abundance peculiarities remained obscure, until McClure et al. (1980) and McClure (1983) announced the discovery of a binary frequency of at least 85% in a sample of 20 barium stars, compared to about 30% in a reference sample containing normal K giants (Harris & McClure 1983). This finding led to the suggestion that binarity plays a key role in the formation of a barium star. Accretion by the barium star of the heavy element-rich wind from an asymptotic giant branch (AGB) star has been proposed as a possible explanation of the barium syndrome (Boffin & Jorissen 1988).

We present here the results of a radial velocity monitoring of a complete sample of barium stars with strong anomalies, confirming the suspicion that binarity is a necessary condition for producing a barium star (see Jorissen & Boffin 1992 for a discussion as to whether binarity can also be considered as a *sufficient* condition). Interesting constraints on the mass-transfer mechanism involved can be derived from the comparison of the orbital elements of these chemically peculiar red giants with those of normal giants.

2. CORAVEL monitoring of a sample of strong barium stars

All 27 southern barium stars with strong anomalies (i.e. classified as Ba4 or Ba5 on the scale of Warner 1965) in the list of Lü et al. (1983) have been monitored since 1984 with the CORAVEL radial velocity spectrometer (Baranne et al. 1979) installed on the Danish 1.54 m telescope at ESO. Preliminary results obtained after 3 years monitoring were presented in Jorissen & Mayor (1988), and yielded a binary frequency of 24/27. After 8 years monitoring, all 27 stars exhibit significant radial velocity variations with a standard deviation ranging from 0.46 km s^{-1} to 11 km s^{-1} (while the uncertainty on one measurement is of the order of 0.25 km s^{-1}). HD 19014 is the only strong barium star which does not exhibit any clear evidence of binary motion, although a significant jitter is present in the measurements. This jitter is reminiscent of the one observed in CH stars (McClure 1984). Orbital elements have been derived for most of these systems and will be published elsewhere (Jorissen & Mayor, in preparation).

[1] E-mail: ajorisse@eso.org

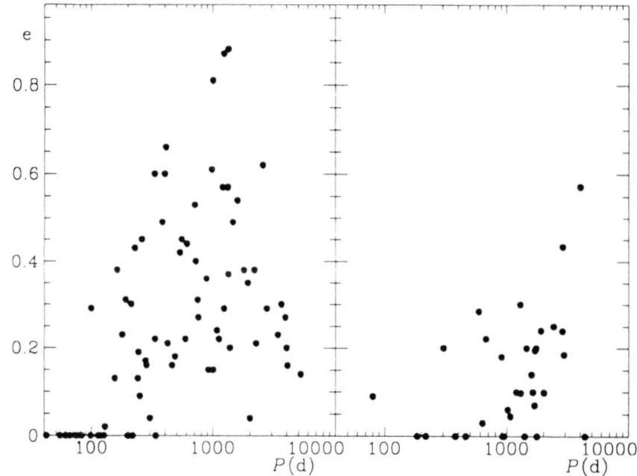

Figure 1. The (period, eccentricity) diagram for barium stars (right panel, from McClure & Woodsworth 1990 and Jorissen & Mayor, in preparation) and for cluster giants (left panel, from Mermilliod & Mayor 1992).

3. Period – eccentricity diagram

Figure 1 presents the (period, eccentricity) diagrams for the sample of barium stars (right panel) and for a reference sample of late-type giants in stellar clusters (left panel, from Mermilliod & Mayor 1992). It is immediately apparent that the average eccentricity (for a given orbital period) is smaller for the barium stars than for the normal giants. In addition, the average orbital period for systems in the range 0 to 3000 d (where the detection rate is nearly complete) is significantly larger for barium stars ($< P >\sim$ 1000 d) than for normal giants ($< P >\sim$ 400 d). Note in particular how circular orbits found among normal giants in the range 0 to 300 d redistribute among barium stars in the range 200 to 4000 d. Finally, the gap in the cluster distribution at $P > 300$ d and $e < 0.15$ (also present in the sample of G dwarfs from the solar neighbourhood studied by Duquennoy & Mayor 1991) has been (at least partially) filled in by the barium stars.

These trends suggest that the orbits of barium stars suffered a period increase and, possibly, an eccentricity decrease.

4. Orbital period vs. intensity of the chemical peculiarities

Figure 2 presents the relation between the $\Delta(38-41)$ color index and the orbital period. The color index $\Delta(38-41) = C(38-41) - 0.89\,C(42-45) + 1.32$ (where $C(38-41)$ and $C(42-45)$ are colour indices in the DDO system; McClure & van den Bergh 1968, Lü 1991) is a good indicator of heavy element overabundances, since Jorissen & Boffin (1992)

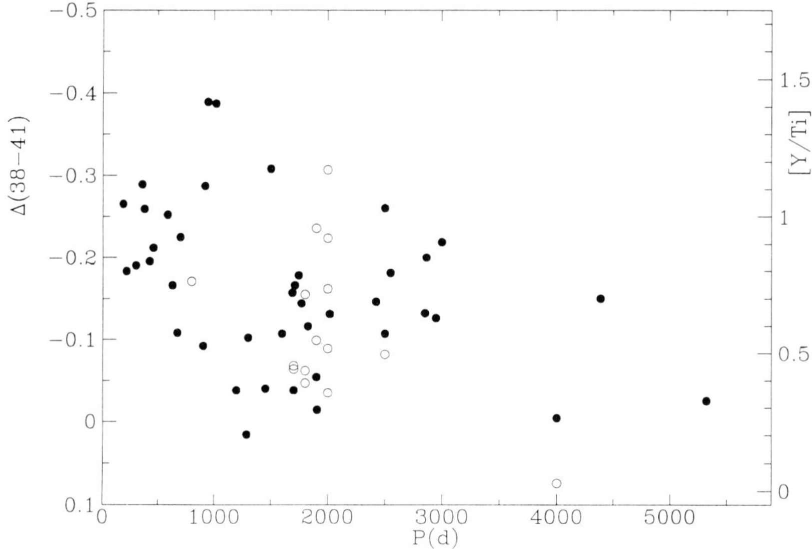

Figure 2. Orbital period vs. $\Delta(38-41)$ (see text) for the combined sample of barium stars from McClure & Woodsworth (1990) and Jorissen & Mayor (1988 and in preparation). The barium star with the longest orbital period (16 Ser = HD 139195; $P = 5324$ d) is from Griffin (1991). Open circles correspond to systems with only a lower limit available for the period. The right-hand axis is labelled in terms of [Y/Ti], according to the relation between $\Delta(38 - 41)$ and [Y/Ti] derived by Jorissen & Boffin (1992).

showed that it is correlated to the logarithmic Y overabundance through the relation [Y/Ti]$= -3\,\Delta(38 - 41) + 0.25$. Strong barium stars (Ba4 or Ba5 on Warner's scale) typically have $\Delta(38-41) \leq -0.1$, whereas mild barium stars have $-0.1 \leq \Delta(38-41) \leq 0.0$.

A rather loose correlation exists between the orbital period and the intensity of the chemical peculiarities, as shown by Fig. 2. As discussed in Sect. 5, such a correlation may be expected for a wind accretion scenario, where the efficiency of the accretion decreases with increasing orbital separation, i.e. with increasing orbital period (for a given system mass).

5. Merits of the wind accretion scenario

We show in this section that the wind accretion scenario is able to account for the trends observed in the orbital elements of barium stars and described in Sects. 3 and 4. To follow the temporal evolution of the orbital elements during the wind accretion process, we use Eqs. (31) to (33) of Huang (1956). They describe the process in terms of two basic parameters, namely the *accretion rate* $\alpha = -\mathrm{d}m_2/\mathrm{d}m_1$ (where $\mathrm{d}m_2 > 0$ is the mass gained by the accreting star and $\mathrm{d}m_1 < 0$ is the mass lost in a spherically symmetric way by the former asymptotic giant branch primary) and the reduced *transverse momentum*

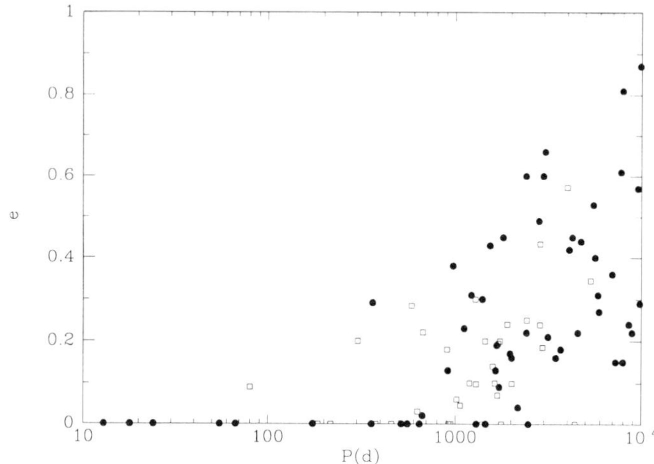

Figure 3. Comparison of the (e, P) diagram of barium stars (squares) with that of cluster giants (black dots) transformed according to the recipes for wind accretion described in the text.

λ imparted to the accreting star by the (accreted and passing) wind. More precisely, $\lambda = -(P/2\pi A)\mathrm{d}(m_2 r\dot\theta)/\mathrm{d}m_2$ (where $\mathrm{d}(m_2 r\dot\theta)$ is the transverse momentum change of the accreting star in the reference frame centered on the primary star, due to the interaction with the wind, A is the semi-major axis of the orbit and P is the orbital period) is positive when the accreting star is decelerated. Wind accretion always leads to $\lambda > 0$ (Theuns et al., in preparation). The equations describing the evolution of P, A, e and m_2 are integrated with m_1 as the independent variable till it reaches its final value of 0.6 M$_\odot$ (in the white dwarf stage), starting from 4.5 M$_\odot$. The parameter λ is kept constant during the process, while the mass accretion rate varies as $\alpha = \min(0.5, \alpha_0 (m_2/m_{2,0})^2/(A/A_0)^2)$. This dependency is derived from Bondi & Hoyle's accretion rate (Bondi & Hoyle 1944, Boffin & Jorissen 1988). It is then possible to use the (e, P) diagram of cluster giants as initial conditions and see how the resulting diagram, computed according to the previous recipes, matches that of the barium stars.

A good match is obtained for $m_{1,0} = 4.5$ M$_\odot$, $m_{2,0} = 1.5$ M$_\odot$, $\lambda = 0.01$, $\alpha_0 = 0.05$ and $A_0 = 1$ AU (Fig. 3). With these parameter values, which are suggested by a detailed numerical simulation of wind accretion (Theuns et al., in preparation), the orbital evolution corresponds to a period increase (for most of the systems; see below) at nearly constant eccentricity, and the results are not very sensitive to λ as long as α remains small. Note also how the barium systems with zero eccentricities and large periods can be accounted for by the period increase of circularized giants formerly in the range $P < 300$ d. Those among these giants which had $A_0 < 1$ AU (corresponding to the adopted value for that parameter entering the definition of α) actually see their period *decrease*. This results from the large α value experienced by these systems, leading to a large amount of mass accreted by the companion, and to the period decrease implied in such conditions by Eq. (32) of Huang (1956). It is likely that these short period systems will at some point disappear from the sample due to cataclysmic Roche lobe overflow.

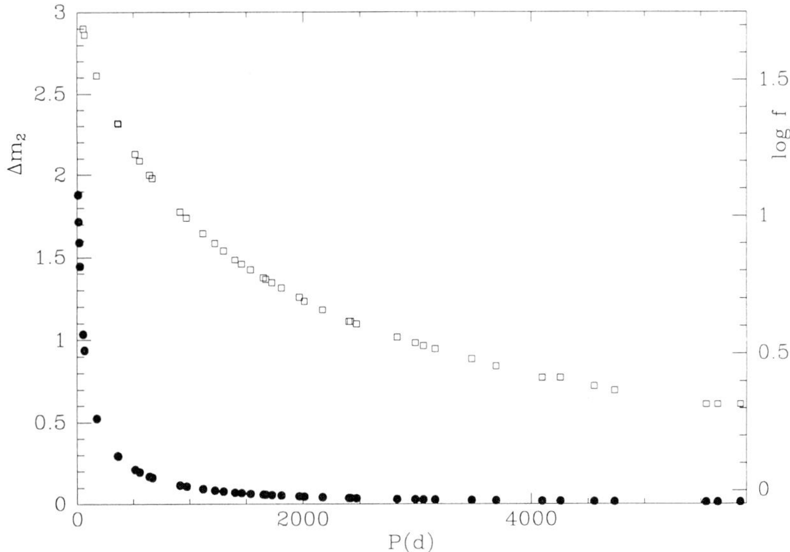

Figure 4. The amount Δm_2 of mass accreted as a function of orbital period (black dots), for the same conditions as those of Fig. 3. The open squares refer to the right-hand scale, and give the overabundance observed in the barium star for an element that is overabundant by a factor 100 in the accreted wind.

We note, however, that such a moderate period decrease could be useful in order to explain the peculiar location of HD 77247, at $P = 80$ d and $e = 0.1$.

Another interesting prediction of the same model is presented in Fig. 4 which shows the amount of mass accreted Δm_2 as a function of the orbital period. This quantity relates directly to the level of pollution of the barium-star envelope by the heavy-element-rich wind:

$$f = \frac{g\Delta m_2 + m_{2,0} - m_{2,\mathrm{core}}}{\Delta m_2 + m_{2,0} - m_{2,\mathrm{core}}},$$

where g is the overabundance factor of a given element in the wind accreted from the AGB companion, while f is the overabundance resulting from the mixing of the accreted matter in the envelope of mass $(m_{2,0} - m_{2,\mathrm{core}})$. To fix the ideas, we adopted $g = 100$ and $m_{2,\mathrm{core}} = 0.4$ $M_\odot$. The logarithmic overabundance factor $\log f$ presented in Fig. 4 should then be compared with the observed trend of [Y/Ti] with orbital period presented in Fig. 2. The agreement is seen to be satisfactory, except maybe for the shortest periods $(P < 500$ d). The larger scatter observed in Fig. 2 is due in part to the fact that the real systems cover some range in mass, whereas the calculations were performed for given masses (see Jorissen & Boffin 1992 for a discussion of other factors responsible for the scatter observed in Fig. 2).

In conclusion, the orbital elements of barium stars as well as their overabundance levels are in good agreement with the predictions derived from the dynamics of the wind accretion scenario, when one adopts reasonable values for all parameters involved. This conclusion lends support to the wind accretion scenario as an explanation for the chemical peculiarities exhibited by barium stars. A definite proof requires, however, that other questions be examined carefully, like the detailed comparison of the abundances of barium and AGB stars, or the ability for the system to remain detached while the primary is evolving on the AGB, or the existence of dwarf barium or carbon stars in the Galaxy with the frequency predicted by the wind accretion scenario.

References

Baranne A, Mayor M and Poncet J L, 1979, Vistas in Astronomy 23, 279

Boffin H M J and Jorissen A, 1988, A&A 205, 155

Bondi H and Hoyle F, 1944, MNRAS 104, 273

Duquennoy A and Mayor M, 1991, A&A 248, 485

Griffin R F, 1991, The Observatory 111, 29

Harris H C and McClure R D, 1983, ApJ 265, L77

Huang S S, 1956, AJ 61, 4

Jorissen A and Mayor M, 1988, A&A 198, 187

Jorissen A and Boffin H M J, 1992, in: Binaries as tracers of stellar formation, eds A Duquennoy & M Mayor, Cambridge Univ Press, in press

Lambert D L, 1988, in: The Impact of High S/N Spectroscopy on Stellar Physics (IAU Symp 132), eds G Cayrel de Strobel & M Spite, Dordrecht: Kluwer, p 563

Lü P K, 1991, AJ 101, 2229

Lü P K, Dawson D W, Upgren A R and Weis E W, 1983, ApJS 52, 169

McClure R D, 1983, ApJ 268, 264

McClure R D, 1984, ApJ 280, L31

McClure R D and van den Bergh S, 1968, AJ 73, 313

McClure R D, Fletcher J M and Nemec J M, 1980, ApJ 238, L35

McClure R D and Woodsworth A W, 1990, ApJ 352, 709

Mermilliod J C and Mayor M, 1992, in: Binaries as tracers of stellar formation, eds A Duquennoy & M Mayor, Cambridge Univ Press, in press

Warner B, 1965, MNRAS 129, 263

CNO in yellow supergiants

B. Barbuy[1], J.R. Medeiros[2], A. Maeder[3] *

1 IAGUSP, Av. Miguel Stefano 4200, 04301-904, São Paulo SP, Brazil
2 Universidade Federal Rio Gde do Norte, Natal 59000, Brazil
3 Observatoire de Génève, Sauverny, Switzerland

Abstract. Carbon, nitrogen and oxygen abundances were measured in a sample of 14 yellow supergiants selected according to their rotation velocities: 9 of them show v sin i < 20 km s^{-1} and the remainder 5 show v sin i > 30 km s^{-1}.

1. Introduction

For long, only the location and the star number in the H-R diagram has been used to study the evolutionary status and internal properties of yellow supergiants in relation with their rotational properties. However, recent evolutionary models show that surface C/N, O/N and ^{12}C/^{13}C abundance ratios constitute most constraining tests of internal evolution. As a result of mass loss, convective dredge-up and possibly rotational mixing, these abundance ratios are quite different from the cosmic ones. Dredge-up mixing is predicted to occur first as stars become red (super)giants following the hydrogen exhaustion in the core, reducing the surface abundance of C, and increasing that of N, but providing no important change in the O abundances. As a general rule, the larger the initial masses, the more significant are the deviations from standard deviations in a given evolutionary stage. Up to now, what is not clear is the exact importance of rotation in the evolution of supergiant stars and, in particular, in what extent CNO abundances are connected with the rotational velocity of supergiant stars. These studies have remained unclear, admitedly by the paucity of rotation velocity data for the generally slowly rotating F, G and K supergiants. This deficiency has been remedied in part by de Medeiros & Mayor (1990, 1992). By using the CORAVEL spectrometer, these authors have measured projected rotational velocities for more than 230 supergiants of luminosity class Ib and Ib-II. These observations have enabled de Medeiros & Mayor (1990) to show that yellow supergiants can be separated in two well distinct groups of rotational velocities: one with very low rotation (v sin i < 20

---

* Observations collected at ESO, La Silla, Chile

km s^{-1}) and the other group with high and moderate rotations (v sin i > 30 km s^{-1}). The discontinuity corresponds to spectral type F9Ib.

In the present paper we try to determine the CNO abundances in these two well defined groups, in order to see to what extent anomalous C/N and O/N ratios, namely the signature of the CNO cycle, are connected with high or low supergiant rotation. The combined information from the observed abundances and from the H-R diagram, in connection with evolutionary models are used to try to understand the evolutionary status and to assess the exact importance of rotation in the evolution of supergiant stars.

The present results have also direct implications regarding the internal mixing and chemical yields of those stars which are known as the main agent of nucleosynthesis in galaxies.

2. Observations

High resolution CCD spectra were obtained at the 1.4m CAT telescope of the European Southern Observatory (ESO) at La Silla, Chile.

The CES (Coudé Echelle Spectrometer) was used with the short camera and the high resolution (1024x512) RCA CCD ESO # 9.

In Table 1 are indicated the wavelength regions observed, their respective resolution, and dates of observations. All stars were observed in the indicated walevelength region at one same night. The reductions were carried out using the IHAP system at ESO, and the EVE package at the VAX 8530 of the Astronomy Department of the University of São Paulo. The flatfield used was the measurement of a B star, presenting no lines in the regions considered. This procedure not only corrects better than the lamp flatfield, but also this constitutes an efficient way to eliminate telluric lines.

In the 870 nm region, pronounced CCD fringes were difficult to be corrected for, and only one star was observed in this region.

The program supergiants Ib stars were selected from the Coravel vsini survey by de Medeiros & Mayor (1990, 1992). For the vsini measurements these authors have used the calibration of Benz & Mayor (1981) with an extension of this one for the supergiant stars (de Medeiros, 1990), taking into account the increase of the atmospheric turbulence with luminosity. The Coravel vsini precision for this class of luminosity is typically of 2,0 km s^{-1} (de Medeiros, 1990).

The observed program stars are indicated in Table 2, together with their coordinates, vsini measured by de Medeiros & Mayor, and (B-V) colours.

3. Stellar parameters and CNO abundances

Detailed analysis was only available for HD 75276 by Russell & Bessell (1989). For all other stars, temperatures were estimated based on their spectral types. Curves of growth and synthetic spectra were then built to derive their gravities and metallicities.

The code for spectral synthesis by Spite (1967) and Barbuy (1982) was used to derive CNO abundances. Model atmospheres by Kurucz (1979) were employed. The synthetic spectra were convolved with FWHM corresponding to the given vsini values, and then compared to the observed spectra.

4. Results

4.1 Low and high rotation velocity supergiants

The initial intent of this work was to derive CNO abundances for low and high rotation yellow supergiants. The derivation for high rotation ones, however, is highly masked since the lines are too broadened. For the present, no definitive results are given here. The effect of rotation can be seen in Figure 1.

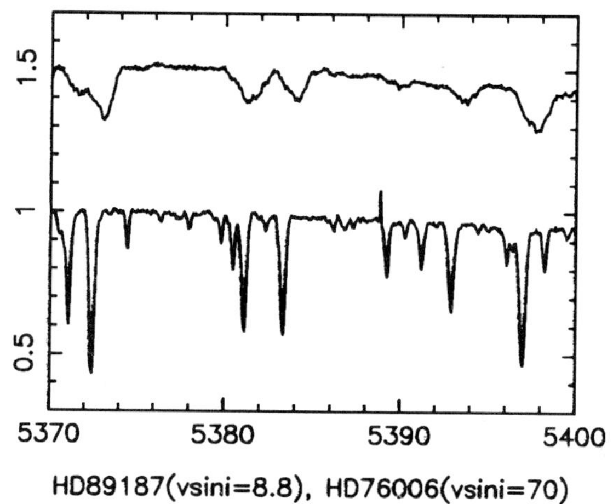

Figure 1 - Spectra of one low- and one high-rotation star

4.2 CNO abundances

Preliminary results on the C, N and O abundances are given in Table 3. Carbon, nitrogen and oxygen abundances were derived using the following lines: $CI\lambda538.032$ nm, $NI\lambda821.096$ nm, $NI\lambda821.630$ nm and $[OI]\lambda630.031$ nm.

Nitrogen-to-iron abundances obtained are solar, a result different from the overabundance found by Luck & Lambert (1985, hereafter LL85). If our result is correct, this would imply that no mixing with CNO processed material has occurred in these stars.

Carbon abundances are also approximately solar, again disagreeing with results by LL85, where carbon is deficient relative to iron.

We suggest that our supergiant stars are more massive and less evolved than those of LL85, so that, in our sample, no convective mixing has occurred yet, whereas it did occur in the sample stars by LL85.

The oxygen abundances show a deficiency of $[O/Fe] \approx -0.2$, in agreement with results by LL85. A We point out that an oxygen deficiency of $[O/Fe] = -0.2$ to -0.3 dex is also found for galactic HII regions. This puzzling difference between the oxygen abundance in the Sun and in young objects is still unexplained.

References

Barbuy, B.: 1982, PhD thesis, Univ. Paris VII

Benz, W., Mayor, M.: 1981, A&A, 93, 235

de Medeiros, J.R.: 1990, PhD thesis, Geneva Observatory

de Medeiros, J.R., Mayor, M.: 1990, ASP Conference Series 9, 404 ed. G. Wallerstein

de Medeiros, J.R., Mayor, M.: 1992, *A Catalogue of Rotational Velocities for Evolved Stars*, in preparation

Kurucz, R.: 1979, ApJS 40, 1

Luck, R.E., Lambert, D.L.: 1985, ApJ, 298, 782

Russell, S.C., Bessell, M.S.: 1989, ApJS 70, 865

Spite, M.: 1967, Annales Astrophys. 30, 211

Spite, M., Barbuy, B., Spite, F.: 1989, A&A, 222, 35

Table 1 - Log-book of the observations: wavelength region, resolution ($\lambda/\Delta\lambda$) and wavelength resolution ($\Delta\lambda$), date and main features present. All program stars were observed in these 6 wavelength regions (the 870 region was only observed for 1 star).

wavelength region (nm)	R $\Delta\lambda$ (Å)	date	feature
536 − 540	60.000 0.085	11.02.91	CIλ538
628.5 − 633.3	50.000 0.126	12.02.91	[OI]λ630
550.3 − 555	37.000 0.15	20.02.91	OI
613.2 − 618	41.000 0.15	16.02.91	CIλ538
641 − 646.4	43.000 0.15	17.02.91	FeII, OIλ645.4
819 − 826.3	55.000 0.15	11.02.91	NI
867.5 − 873.5	43.000 0.2	16.02.91	NI

Table 2 - Magnitudes, coordinates, rotation (v sin i) for the program stars

HD n°	V	α_{2000}	δ_{2000}	l, b	v sin i	ST	(B − V)
75276	5.72	$08^H47^M18.84^S$	−46°09′19.8″	265.61, −1.73	9.7	F0Ib	0.56
89187	8.4	$10^H15^M30.49^S$	−60°44′15.9″	284.99, −3.42	8.8	F3Ib	0.57
89874	7.7	$10^H20^M32.03^S$	−59°38′11.1″	284.90, −2.16	8.0	F8Ib/II	0.50
90452	7.8	$10^H25^M10.31^S$	−57°03′49.6″	284.03, +0.34	12.8	F3Ib	0.26
92662	8.1	$10^H40^M34.54^S$	−59°32′36.8″	287.03, −0.77	7.1	F2Ib	0.64
93001	8.6	$10^H43^M06.05^S$	−58°08′49.6″	286.65, +0.61	15.6	F7Ib	0.78
100278	7.69	$11^H31^M50.42^S$	−61°46′23.4″	293.67, −0.34	7.2	F8Ib/II	0.74
103807	7.6	$11^H57^M06.78^S$	−63°40′35.7″	296.95, −1.44	19.7	F7Ib	0.65
114489	6.74	$13^H11^M55.49^S$	−61°07′02.1″	305.40, +1.66	8.7	F2Ib/II	0.46
76006	7.33	$08^H51^M32.30^S$	−49°33′27.8″	268.70, −3.33	70.0	F5Ib	0.66
82346	7.3	$09^H29^M37.44^S$	−50°36′13.5″	273.63, +0.42	68.3	F3Ib	0.67
95933	7.52	$11^H03^M04.65^S$	−60°49′29.3″	290.10, −0.68	32.8	F5Ib/II	0.58
122036	8.4	$14^H02^M04.32^S$	−69°02′25.8″	309.21, −7.03	65.7	F2Ib/II	0.23
148586	8.0	$16^H32^M08.25^S$	−57°14′07.0″	329.48, −6.22	69.5	F8Ib/II	0.21

Table 3- CNO abundances

HD n°	[O/Fe]	[N/Fe]	[C/Fe]
75276	− − −	0.00	−0.10
89187	−0.30	− − −	−0.10
89874	−0.15	− − −	−0.15
90452	−0.05	− − −	0.00
92662	−0.15	− − −	0.00
93001	−0.30	− − −	−0.10
100278	−0.25	0.00	0.00
103807	− − −	−0.30	−0.10
114489	−0.30	0.00	0.00

Neutron-rich species in the halo cool dwarf G14–32

B Barbuy *

IAGUSP, Av. Miguel Stefano 4200, 04301-904, São Paulo SP, Brazil

Abstract. The abundances of neutron rich species ^{23}Na, ^{27}Al, ^{25,26}Mg are obtained for the metal-poor cool dwarf G14-32. G14-32 shows a metallicity of [Fe/H] = -1.0 which although not as extreme as given by Carbon et al (1987) ([Fe/H] = -2.4), appears to be a halo star.

1. Introduction

The neutron enrichment in the Galaxy can be studied through the enhancement of the odd-even effect as a function of metallicity. The neutron-rich species ^{23}Na, ^{25,26}Mg, ^{27}Al, which are essentially products of the hydrostatic carbon burning, are suitable to indicate the history of neutron enrichment in the Galaxy, through abundance ratios relative to ^{24}Mg, the relevant primary species.

Determinations of Al/Fe, Na/Fe and ^{25,26}Mg/^{24}Mg ratios as a function of metallicity (or age of the Galaxy) were presented by François (1986) and Spite & Spite (1992) for Na and Al and by Barbuy, Spite & Spite (1987, hereafter BSS87) for ^{25,26}Mg/^{24}Mg isotopic ratios.

The star G14-32 was selected from the survey by Carbon et al. (1987).

2. Observations: selection of G14-32

The star G14-32 of V = 8.35^{mag} and at coordinates $\alpha_{1950} = 13^h5.8^m$, $\delta_{1950} = $ -7°03' was selected from the list by Carbon et al. (1987). In BSS87 the most metal-poor dwarfs studied in terms of magnesium isotopes were presented: HD 25329 with [M/H] = -1.56 and G23-1 with [M/H] = -1.35, where the observations were carried out at the 3.6m CFH Telescope at Hawaii. We have observed MgH lines in a sample of metal-poor dwarf candidates to show MgH lines (which only appear in the cooler ones). G14-32 was selected for the reasons that (a) it appeared to be a metal-poor cool dwarf, and (b) it showed MgH lines.

*** Observations collected at the European Southern Observatory, ESO, La Silla, Chile**

The wavelength regions observed are $\lambda\lambda$ 5120-5160, 5503-5550, 6130-6195, 6471-6541, 6681-6733, 7356-7412 Å. Sandage & Fouts (1987) give for this star $\delta_{.6}$ = +0.22 and radial velocity v_r = +67.3 km s^{-1}, in agreement with our radial velocity, measured from the spectra, of v_r = +53 km s^{-1}.

3. Stellar Parameters and Results

The stellar parameters given by Carbon et al. (1987) are T_{eff} = 4700 K, log g = 4.25, [M/H] = -2.41, and their deduced carbon and nitrogen abundances were [C/Fe] = +0.47, [N/Fe] = -0.3. The equivalent widths of 90 lines were measured and curves of growth were built using the code by Spite (1967). Figure 1 shows the curve of growth for Fe1. The resulting metallicity of [Fe/H] = -1.0 is considerably higher than that reported by Carbon et al. The final stellar parameters adopted are T_{eff} = 4700, log g = 4.3, [M/H] = -1.0. Curves of growth and spectral synthesis of the observed regions show a good agreement with these parameters. The abundance pattern of G14-32 is typical of a halo star, except for the calcium abundance: [Al/Fe] = +0.5, although not expected for a metal-poor star, has been found for halo stars of this metallicity, if the red lines are used, as shown by Spite & Spite (1992). [Mg/Fe] = +0.45 and [Na/Fe] = 0.0 at [Fe/H] = -1.0 are in agreement wit the data by François (1986a,b). [Si/Fe] = +0.4 and [Sc/Fe] = 0.0 as typical of the halo, are in agreement with Gratton & Sneden (1991). [Y/Fe] = 0.0 and [Ba/Fe] = 0.0 for [M/H] = -1.0 were also found in Barbuy, Spite & Spite (1985). The carbon abundance derived from C_2 lines is [C/Fe] = +0.4, in agreement with the same ratio given by Carbon et al, despite the difference in metallicity.

Solar isotopic ^{24}Mg:^{25}Mg:^{26}Mg = 79:10:11 are obtained, a result in agreement with Barbuy et al. (1987) for these metallicities.

References

Barbuy, B., Spite, F., Spite, M.: 1985, A&A 144, 343

Barbuy, B., Spite, F., Spite, M.: 1987, A&A 178, 199

Carbon, D.F., Barbuy, B., Kraft, R.P., Friel, E.D., Suntzeff, N.B.: 1987, PASP 99, 335

François, P.: 1986a, A&A 160, 264

François, P.: 1986b, A&A 165, 183

Gratton, R., Sneden, C.: 1991, A&A 241, 501

Sandage, A., Fouts, G.: 1987, AJ 93, 592

Spite, M.: 1967, Ann. Astrophys. 30, 211

Spite, M., Spite, F.: 1992, in *The Stellar Populations of Galaxies*, IAU Symp. 149, eds. B. Barbuy & A. Renzini, Kluwer Academic Publishers, 123

The Li-rich giant CPD-55^{0}395: Li enrichment in the Galaxy

B. Barbuy, J. Gregorio-Hetem, J.A. de Freitas Pacheco *

IAGUSP, Depto de Astronomia, CP 9638, São Paulo 01065, Brazil

Abstract. In a survey of stars selected from IRAS colours using criteria to find new T Tauri stars by Gregorio-Hetem et al. (1992) a few Li-rich giants were found. CPD-55°395 is a moderately Li-rich giant with $\log \epsilon(\mathrm{Li}) \approx 1.3$. We obtain its Li, C, N, O and ^{12}C:^{13}C abundances. We further try to interpret the Li enrichment in the Galaxy.

1. Introduction

Gregorio-Hetem et al. (1992) used the IRAS Point Source Catalog to select candidates to be T Tauri stars, using a colour criterion $0.95 < \mathrm{S}_{25}/\mathrm{S}_{12} < 3.40$ and $0.50 < \mathrm{S}_{60}/\mathrm{S}_{25} < 3.30$, where S_x are the non colour corrected IRAS fluxes. The selected candidates were then systematically observed spectroscopically in the Hα plus Li λ 670.7 nm region ($\lambda\lambda$ 655-675 nm) in order to check their nature.

An interesting sub-product of this study was the discovery of a few Li-rich giants. The stars HD19745, IRAS13539-4153, IRAS17554-3822, IRAS19285+0517 listed in Gregorio-Hetem et al. (1992) are being analyzed by da Silva & de la Reza (1992). We have observed CPD-55°395 trying to know if this star which shows a strong Li line, and Hα and [OI] lines in absorption, could be a post T Tauri star (e.g. Pallavicini, Randich & Giampapa, 1992) or a Li-rich giant.

Since the discovery by Wallerstein & Sneden (1982) of a very Li-rich giant, a few other ones were detected, amounting now to about 20 of them with very strong Li - for a more complete description of the available data see Brown et al. (1989 and references therein).

*** Observations collected at the European Southern Observatory, La Silla, Chile**

2. Observations

The observations were carried out at the 1.4 m Coudé Auxiliary Telescope (CAT) of the European Southern Observatory (ESO) at La Silla, Chile. The Coudé Échelle Spectrometer (CES) was used, in conjunction with the short camera and a RCA CCD (1024x640 pixels). High resolution spectra were obtained for CPD-55°395 of V = 9.4 and spectral type K6, in the following wavelength regions: $\lambda\lambda$ 561.0 - 565.5 nm, $\lambda\lambda$ 607.5 - 612.5 nm, $\lambda\lambda$ 613.6 - 618.4 nm, $\lambda\lambda$ 628.5 - 633.0 nm, $\lambda\lambda$ 635.4 - 641.0 nm, $\lambda\lambda$ 653.7 - 659.0 nm, $\lambda\lambda$ 667.2 - 672.5 nm, $\lambda\lambda$ 668.0 - 673.0 nm, $\lambda\lambda$ 796.5 - 803.0 nm. The observations were carried out from 25 to 30 october 1991. The reductions were done using the IHAP reduction system at the HP 1000 computers of ESO.

3. Analysis

In order to carry out a detailed analysis, the equivalent widths of about 100 lines were measured in the wavelength regions listed in Sect. 2.

The effective temperature T_{eff} = 4300 K as derived from the spectral type K6 was initially used, and further determined through excitation equilibrium, using curves of growth and intensity of selected FeI lines. The curves of growth however showed a large scatter, due to unidentified features all along the spectra, affecting equivalent widths. The [OI]λ630.3 nm line was used to verify its giant nature, resulting in a final gravity log g = 0.75. The metallicity obtained is [Fe/H] = -1.0.

A microturbulent velocity of v_t = 2.0 km s^{-1}, typical for a giant, was adopted.

The abundances of C, N, O and Li, and isotopic ratios ^{12}C:^{13}C, are obtained through a fitting of synthetic spectra to the oberved spectra. The resulting abundances corresponding to the stellar parameters T_{eff} = 4300 K, log g = 0.75, [Fe/H]= -1.0 are: [C/Fe] < 0.0, [N/Fe] > 0.0 [O/Fe] = +0.2, ^{12}C:^{13}C $\approx$ 10, log ϵ(Li) = 1.3.

4. Discussion

In a sample of 644 giants, Brown et al. (1989) found only 10 giants showing Li abundances above Iben (1967a,b)'s predictions of Li dilution. For a solar metallicity star, supposing that its initial Li abundance is log ϵ(Li) = 3.1 (value in the interstellar medium, ISM), in the giant stage, a star of 1 M$_\odot$ should have log ϵ(Li) = 1.7, and a star of 3 M$_\odot$ would show log ϵ(Li) = 1.3, with depletions of factors of 28 and 60 respectively. The ^{12}C:^{13}C $\approx$ 30 in a first ascent giant branch star. Our star show ^{12}C:^{13}C $\approx$ 10, which may indicate that it is a AGB star: in that case Li should be almost completely destroyed.

D'Antona & Matteucci (1991) model reproducing log ϵ(Li) vs [Fe/H] shows that log ϵ(Li) = 2.3 at [Fe/H] = -1.0. The Li abundance in CPD-55°395 would then be destroyed by an amount 30-60, if the star is in the first ascent of the giant branch and if calculations for solar metallicity stars are applicable, resulting in log ϵ(Li) = 0.3-0.8; if it is an AGB star, Li should be all destroyed.

In Brown et al's Fig. 18, the value log ϵ(Li) = 1.3 found for our star was only detected in a few stars of their sample.

Therefore, either there was an "abnormal preservation of the star's initial Li", one of Brown et al's hypotheses, or there was enrichment possibly via the reaction ^{3}He$(\alpha,\gamma)^7$Be$(e^-\nu)^7$Li (Cameron & Fowler 1971), wherefrom ^{7}Li is brought to the surface.

Supposing that a Li enrichment occurs along the red giant branch (RGB), and supposing further that at some stage it reaches values of log ϵ(Li) = 2.0 to 3.0, this Li being lost to the ISM by stellar winds, we now investigate on the possibility of Li enrichment in the Galaxy by red giants.

5. Model of Li enrichment

Primordial ^{7}Li is seen in old-halo stars (non-evolved) with an abundance log ϵ(Li) = 2.3 (Spite & Spite 1982). The ^{7}Li abundance, without any other source, would decrease in stars of higher metallicity (which have not yet destroyed ^{7}Li) as a consequence of the 'astration' mechanism. However, the present ^{7}Li abundance is about log ϵ(Li) = 3.2 in the ISM, requiring that other sources produce lithium along the Galaxy lifetime. Cosmic rays (CR), by spallation interactions with CNO nuclei in the ISM, are a source of ^{7}Li. In a simple model, considering a constant CR intensity through the age of the disk, the ^{7}Li abundance a(^{7}Li) varies according to the equation

$$\frac{\mathrm{d}\ \mathrm{a}(^7\mathrm{Li})}{\mathrm{dt}} = -\frac{\mathrm{x}}{\tau_\mathrm{g}}\ \mathrm{a}(^7\mathrm{Li})\ +\ \mathrm{Q_{CR}} \tag{1}$$

where the first term on the right side represents destruction due to astration and the second, CR production. x $\approx$ 0.3 is the gas fraction returned to the ISM by the stars; τ_g is the time scale for gas conversion into stars (2.3 Gyr for the galactic disk) is the spallation production rate per H atom (the above rate includes the ^{7}Be production which decays into ^{7}Li through ^{7}Be$(e^-,\nu)^7$Li). From the above equation we estimate that the present ^{7}Li abundances as log ϵ(Li) $\approx$ 2.04 (about 25% is due to the primordial ^{7}Li). Clearly this about one order of magnitude less than the ISM value, requires an additional source of ^{7}Li.

Novae have been suggested as a possible source of ^{7}Li (e.g., Clayton, 1981). However, the ^{7}Li yield from novae is quite uncertain since it depends on the unknown ^{3}He concentration in the prenova envelope. In any case, we cannot disregard novae as potential sources of ^{7}Li although more detailed models are still required to compute the ^{7}Be produced in the ejecta.

The analysis of the previous section indicates that F-K giants may have ^{7}Li enriched atmospheres. These objects have typical mass loss rates of the order of 10^{-7} M$_\odot$ yr^{-1} and if they reach abundances of the order of log ϵ(Li) = 2 - 3, the ^{7}Li injection rate per star is about 10^{33} nuclei/sec.

The density of F-K giants in the absolute magnitude interval -2 to +2 is about 5.4×10^{-4} pc^{-3}. However, from the data of Brown et al. (1989), only 2% of their sample show $\log \epsilon(\mathrm{Li}) > 2.0$. Since these giants have typical mass loss rates of the order of 10^{-7} $M_\odot$ yr^{-1}, the ^{7}Li injection rate is about $Q_{F-K} \approx 10^{-27}$ cm^{-3} s^{-1}. Including this source term in eq. (1), the resulting abundance is $\log \epsilon(\mathrm{Li}) \approx 2.5$ (9% primordial; 26% due to cosmic rays; 65% due to F-K giants) comparable within uncertainties of our crude estimates, to the ISM value.

References

D'Antona, F., Matteucci, F.: 1991, A&A 248, 62

Brown, J.A., Sneden, C., Lambert, D.L., Dutchover Jr., E.: 1989, ApJS 71, 293

Cameron, A.G.W., Fowler, W.A.: 1971, ApJ 164, 111

Clayton, D.D.: 1981, ApJ 244, L97

Gregorio-Hetem, J., Lépine, J.R.D., Quast, G.R., Torres, C.A.O., de la Reza, R.: 1992, AJ 103, 549

Pallavicini, R., Randich, S., Giampapa, M.S.: 1992, A&A 253, 185

da Silva, L., de la Reza, R.: 1992, in preparation

Spite, F., Spite, M.: 1982, A&A 115, 357

Wallerstein, G., Sneden, C.: 1982, ApJ 255, 577

Interstellar lithium and the (^{7}Li/^{6}Li) ratio toward ρ Oph

M. Lemoine, R. Ferlet, A. Vidal-Madjar, C. Emerich, P. Bertin
Institut d'Astrophysique, CNRS, 98 bis bvd Arago, 75014 Paris, France

Abstract. We report here on new observations of the $\lambda6708\text{Å}$ doublet of interstellar Li I toward ρ Ophiuchi, which have led to the first direct detection of ^{6}Li outside the solar system. We derive abundances for both lithium isotopes (^{7}Li/H)$= 3.4 \times 10^{-9}$, (^{6}Li/H)$= 2.7 \times 10^{-10}$, yielding the isotopic ratio (^{7}Li/^{6}Li)$=12.5$ (9.3-15.3 at a 95% confidence level) for the ρ Oph diffuse cloud. This value is in strikingly good agreement with the ratio measured in the chondritic meteorites, representative of the early solar system value, but differs largely from the only previous estimation of Ferlet and Dennefeld (1984) (^{7}Li/^{6}Li)≥25 toward ζ Oph. Recent models of lithium evolution in our Galaxy cannot as yet explain this new value toward ρ Oph.

1. Introduction

Even if "it is slightly unsatisfactory that anything as scarce and unromantic as lithium should be so important" (Trimble, 1991), ^{7}Li is now considered as a crucial nucleus in the cosmological nucleosynthesis, either canonical (Walker et al., 1991 and references therein) or inhomogeneous, inspired from the quark-hadron phase transition (Reeves et al., 1990 and references therein). These various models lead to a wide range of primordial lithium mass fraction, from $X_7 \simeq 6 \times 10^{-10}$ in the standard one, to $X_7 \simeq 10^{-7}$ in the most extreme inhomogeneous case (fraction of the closure density in baryons $\Omega_b = 1$, and density contrast between the high and low baryonic density phases $R = 100$).
The determination of the primordial abundance of lithium is therefore a key-milestone in observational cosmology. If the real primordial abundance is the flat level found in 5600–6250 K extremely metal deficient, unevolved Population II stars, $(Li/H) = 1-2 \times 10^{-10}$ (Spite and Spite, 1982; Rebolo, Molaro and Beckman, 1988 and references therein), then, inhomogeneous nucleosynthesis is ruled out. However, this is open to debate since the surface abundance in warm Pop II stars could be lowered by lithium destruction through some form of internal mixing (Vauclair, 1988).
On the other hand, the (Li/H) value seen in unevolved, unmixed Population I stars is around 10^{-9}, also close to the value found in Type I carbonaceous chondritic meteorites likely representative of the abundance at the time of the solar system formation. The determination of the present abundance of lithium is therefore another key-milestone in chemical evolution of galaxies. Especially, measurements of the interstellar $(^7Li/^6Li)$ ratio are crucial in evaluating the relative weight of potential production and destruction mechanisms and thus in reconstructing the lithium history.

Up to now, the ($^7Li/^6Li$) isotopic ratio had however never been measured precisely outside the solar system, as ^{6}Li has never been clearly detected at the surface of stars or in the interstellar medium.
We report here on new observations toward ρ *Oph*, which have led to the first direct detection of interstellar ^{6}Li and thus to the first precise measurement of the lithium isotopic ratio outside the solar system (see also Lemoine et al., 1992).

2. Observations

The only resonance line of lithium accessible to ground-based telescopes is the doublet of Li I ($3^2P_{3/2,1/2}-2^1S_{1/2}$) at 6707.761 Å and 6707.912 Å for ^{7}Li, ^{6}Li having the same structure but shifted by 0.160 Å toward longer wavelengths. These lines are extremely weak and require observations at high spectral resolution, with a very clean instrumental profile, and a very high signal-to-noise ratio (> 1000). We chose to observe ρ Oph at the ESO-La Silla (Chile) 3.6m Telescope, linked via fiber optics to the Coudé Echelle Spectrometer, designed to provide these characteristics with a resolving power $\lambda/\Delta\lambda \simeq 10^5$. ρ Oph combines the following criteria: brightness (m_v=5.0), featureless in the Li region (spectral type B2IV), relatively simple interstellar lines, and the highest possible HI column density ($> 10^{21}$cm^{-2}) together with a high reddening (E(B-V)=0.47).
Raw images were reduced individually using the ESO Image Handling and Processing Software (IHAP). 13 individual spectra of one hour integration time each were then added together, taking into account all the relevant wavelength shifts. Figure 1 shows the final plot of the ρ Oph spectrum in the vicinity of λ6708Å. The signal-to-noise ratio is 2700 per pixel; the spectral resolution is $\simeq$ 3 km/s, implying a limiting detectable equivalent width of 30 μÅ (3σ). The total equivalent width of the Li I doublet is 2.2 mÅ.
Because the ^{6}Li and ^{7}Li lines are partially superimposed, and because the equivalent width provides only an integrated column density which could be systematically in error when more than one absorbing region is present on the line of sight, the data were analyzed with a profile analysis method, which allows the extraction of all the data contained in the line structure. This method is based on an iterative process, which computes a theoretical Voigt profile for possibly several clouds on the line of sight, convoloves it with an instrumental profile, and minimizes the sum of the squared departures between the theoretical points and the observed data points. The instrumental profile was determined by fitting sevral thorium lines extracted from several comparison spectra. Figure 2 shows the final best fit of the Li I lines, where two interstellar components have been introduced in ^{7}Li, and only one in ^{6}Li, the extra blue component being negligible in ^{6}Li. This fit of ^{7}Li gives also an excellent fit of the potassium λ7699Å line toward ρ Oph. We checked on the potassium lines that the ^{6}Li absorption was not due to a high radial velocity ^{7}Li absorbing cloud.
As a final check, we subtracted from the observed Li I profile the best solution found for the ^{7}Li absorption alone. This procedure should reveal the actual detection of the ^{6}Li doublet toward ρ *Oph*. The result, shown in Fig.3 (Top), is extremely convincing :

at the correct position, with the correct separation and the correct oscillator strengths ratio, appears directly the ^{6}Li doublet well out of the noise. And again, this observed ^{6}Li profile in Fig.3 is well fitted by the ^{6}Li solution found above (Fig.3, Middle); after subtracting this ^{6}Li absorption from the observed profile in Fig.3, just the noise alone remains (Fig.3, Bottom) !

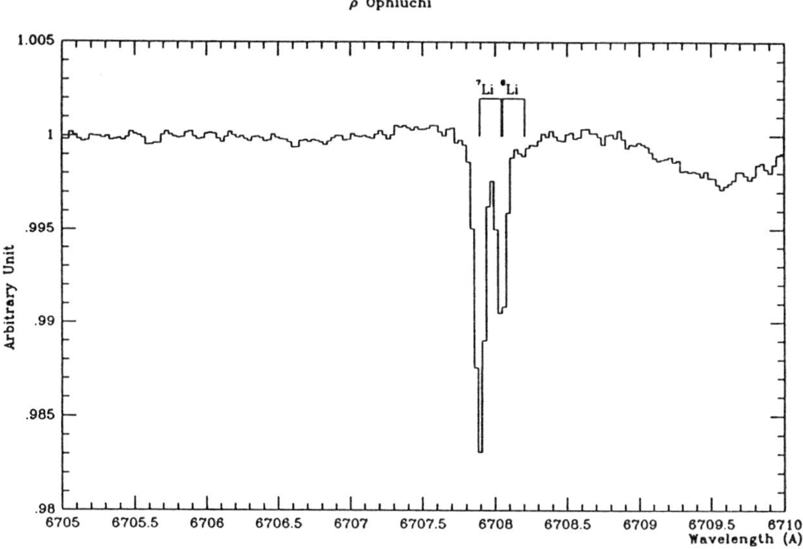

Fig.1: Resulting spectrum of ρ Oph at $\lambda6708$; the total equivalent width of the Li I lines is 2.2 m$\AA$, the signal-to-noise ratio is 2700 per pixel, at $\Delta\lambda/\lambda=3$ km/s, and the limiting detectable equivalent width is 30 $\mu\AA$ (3 σ), for a total integration time of 13h toward ρ Oph (m$_v$=5.0).

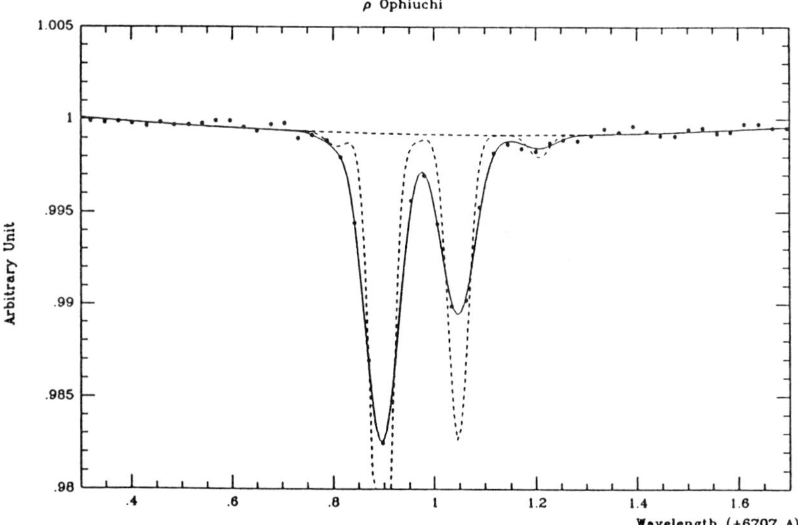

Fig.2: Final best fit for ^{7}Li and ^{6}Li. The data points represents the histogram pixels of Fig.1, the dashed lines represent the theoretical Voigt profiles, and the solid lines the Voigt profiles convolved with the instrumental profile. Two interstellar components are needed to fit the ^{7}Li lines; the second blue component is negligible in ^{6}Li.

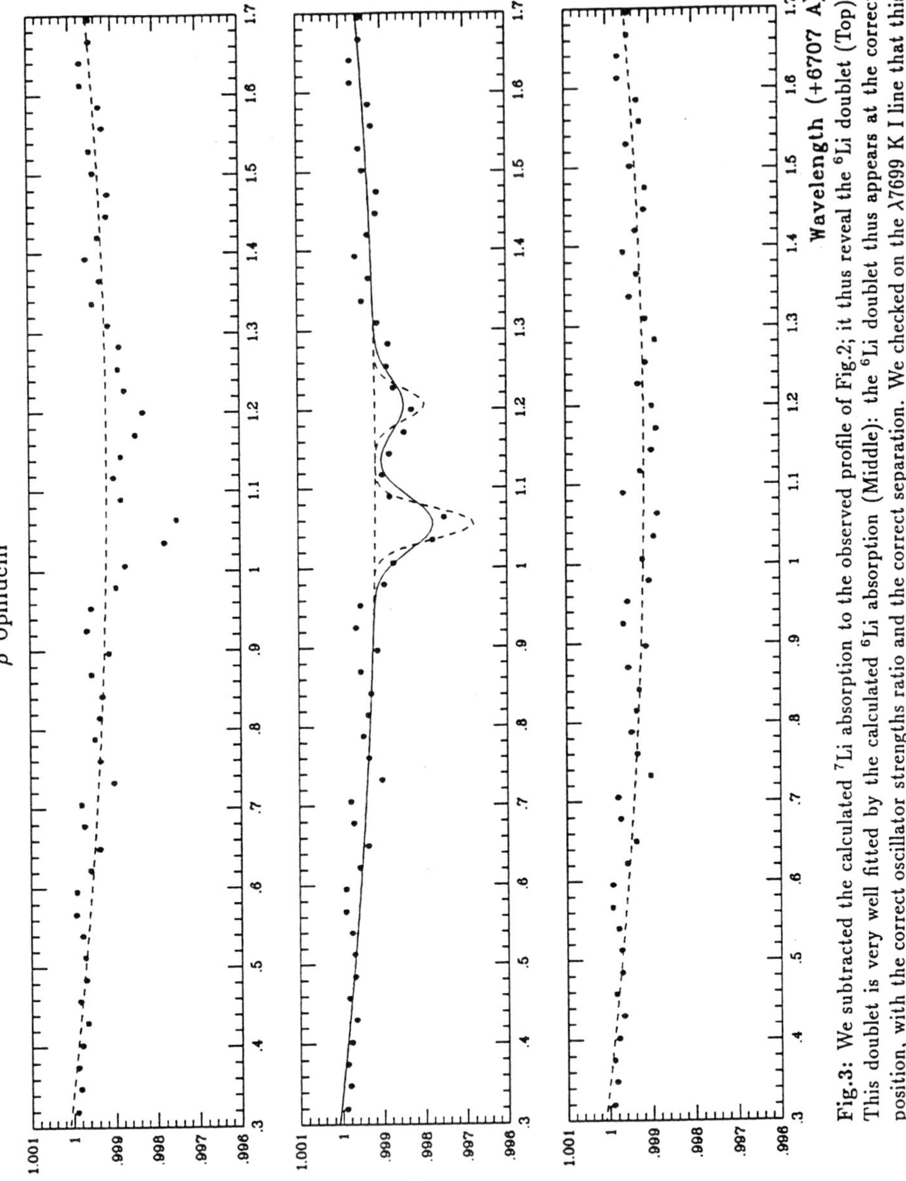

Fig.3: We subtracted the calculated ^{7}Li absorption to the observed profile of Fig.2; it thus reveal the ^{6}Li doublet (Top). This doublet is very well fitted by the calculated ^{6}Li absorption (Middle): the ^{6}Li doublet thus appears at the correct position, with the correct oscillator strengths ratio and the correct separation. We checked on the λ7699 K I line that this doublet was not due to a high radial velocity ^{7}Li interstellar component. When the ^{6}Li absorption is removed, only the noise remains (Bottom)...

3. Discussion

We derive the error bar on $N(^7Li)$ according to the limiting detectable equivalent width, i.e. 20 $\mu\overset{\circ}{A}$ with 95% confidence. We derive the error bar on $N(^6Li)$ by fitting the ^6Li doublet after subtraction of the ^7Li doublet for both upper and lower limit of the ^7Li error bar, and by taking into account in each case the limiting detectable equivalent width. We thus derive the following lithium isotopic ratio toward ρ Oph:

$(^7Li/^6Li) = 12.5^{+2.8}_{-3.2}$ with 95% confidence.

After correcting for the important ionisation of lithium in the interstellar medium, using $N(CaII)$, $N(CaI)$, and for its depletion onto interstellar grains, using $\delta_{Li} = \delta_K = -0.93$ dex, we derive the resulting effective abundances:

$^7Li/H \simeq 3.4 \times 10^{-9} \ (\geq 2.4 \times 10^{-9})$
$^6Li/H \simeq 2.7 \times 10^{-10} \ (\geq 1.6 \times 10^{-10})$

Our value of the isotopic ratio is in strikingly good agreement with the value measured in the meteorites CC I, representative of the solar system formation epoch: $^7Li/^6Li=12.5\pm0.2$, and in disagreement with the only previous estimation of Ferlet&Dennefeld (1984): $^7Li/^6Li\simeq 38 \ (\geq 25)$ toward ζ Oph, though ^6Li was not detected in that case, but only constrained through a profile analysis. Our present measurement is therefore more precise, as we were able to clearly detect the ^6Li absorption, and this high value of the interstellar lithium isotopic ratio toward ζ Oph needs to be confirmed. As well, this value of the isotopic ratio toward ρ Oph is not explained by recent models of galactic lithium evolution (Abia&Canal, 1988, D'Antona&Matteucci, 1991) which rather imply an increase of this ratio in the last 4.5 Gyr, perhaps up to a very high value. When related to the "demodulated" energy integrated flux of galactic cosmic rays, our present measurement confirms the existence of a stellar source of ^7Li, necessary to maintain the interstellar lithium isotopic ratio to a relatively constant value in the last 4.5 Gyr, and permits to constrain the yield of this stellar source (Reeves, 1992). Such a stellar source would then render unlikely scenarios with a high ^7Li primordial abundance, according to Reeves (1992), and would instead confirm the choice of the ^7Li primordial abundance close to the Pop II value for comparison with the Big-Bang nucleosynthesis calculations. Therefore, new measurements of the $(^7Li/^6Li)$ ratio on different lines of sight are crucial, and now underway, to check for the homogeneity of its value in the interstellar medium. Better constrained, these chemical evolutionary models should shed some light on the ^7Li primordial abundance, a fundamental input in the Big-Bang nucleosynthesis models.

References
Abia, C., Canal, R. 1988, A&A **189**, 55
D'Antona, F., Matteucci, F. 1991, A&A **248**, 62
Ferlet, R., Dennefeld, M. 1984, A&A **138**, 303
Lemoine, M., Ferlet, R., Vidal-Madjar, A., Emerich, C., Bertin, P. 1992, submitted to A&A
Rebolo, R., Molaro, P., Beckman, J.E. 1988, A&A **192**, 192
Reeves, H., Richer, J., Sato, K., Terasawa, N. 1990, ApJ **355**, 18

Reeves, H. 1992, submitted to A&A
Spite, F., Spite, M. 1982, A&A **115**, 357
Trimble, V. 1991, A&AR, vol.3
Vauclair, S. 1988, in *Dark Matter, ed J. Audouze and J. Tran Thanh Van*, p. 269
Walker, T. P., Steigman, G., Schramm, D. N., Olive, K. A., Kang, H. 1991 ApJ **376**, 51

The ESO key programme on supernovae: a systematic approach to supernova studies

Paolo A. Mazzali[1], L.B. Lucy[2] and I.J. Danziger[1]

[1]European Southern Observatory, Garching, FRG

[2]ST-ECF, Garching, FRG

Abstract. The main achievements of the ESO Key Programme on Supernovae are presented. On the observational side, the Programme has allowed the accumulation of an impressive wealth of spectroscopic and photometric data on recent supernovae of all types, and the building of an archive of older SN data. Modelling work has been very active within the Programme, and has been devoted mostly to spectral modelling. Early and late time spectra of both SNe II (SN 1987A) and Ia (SNe 1990N, 1986G) have been successfully modelled, with considerable gain of insight into the properties of these objects. The light curve of SN 1987A has also been successfully modelled assuming a ^{57}Co/^{56}Co ratio of $1.5\times$ solar, recently confirmed by γ-ray observations.

1. Introduction

The ESO Key Programme on supernovae gives a group of european astronomers first priority for the use of the ESO telescopes in order to observe, photometrically, spectroscopically and through direct imaging whenever possible, supernovae of all types appearing in the southern sky. Collaboration with CTIO observers allows access to their southern sky data as well, while the Asiago Observatory is involved for the northern objects. This programme is providing excellent coverage for many recent supernovae.

The presence of theoretical expertise in the collaboration allows first hand analysis of the data, in a major effort to understand the nature of supernovae through the analysis of their spectra. Theoretical modelling of the spectra is based on a Monte Carlo spectrum synthesis code which can be applied, with the appropriate changes in the density structure and composition, to the various kinds of supernovae.

For type II supernovae, the wealth of data on SN 1987A has spurred a thorough investigation into the abundances of the s-process elements in this object. The study required the introduction for the first time of the effect of line-blocking in the computation of the radiation field within the SN envelope. An accurate knowledge of the radiation field is necessary to calculate realistic NLTE occupation numbers. We showed that the s-process elements are present in the supernova envelope in an amount consistent with nucleosynthesis calculations in exploding massive stars.

For type Ia supernovae, the analysis of a series of early-time spectra of the 'standard' object SN 1990N has been undertaken. Excellent spectral fits have been obtained, which have allowed an almost complete identification of the line spectrum. This success supports the validity of the standard SN Ia explosion model of a white dwarf accreting matter beyond the Chandrasekhar limit. The fits have also produced light curves which can be compared to those obtained from the explosion models, and thus allow a critical discussion of the standard candle properties of classical SNe Ia. The application of this work to 'peculiar' SNe Ia is expected to yield knowledge on the possible lack of uniformity among this type of SNe.

In addition, the analysis of the late time nebular spectrum of the type II SN 1987A, has produced abundance determinations of key elements, ranging from oxygen to cobalt, that are reasonably consistent with the predictions of the most sophisticated explosion models. In the case of type Ia supernovae, reliable independent determinations of the iron mass have been obtained using similar methods.

Finally, the light curve of SN 1987A has been successfully modelled, yielding predictions on the abundances of radioactive elements.

2. The database

During the past two years, a wealth of supernova data has been gathered under the auspices of the Key Programme (Cappellaro *et al* 1991). This work has two basic branches: on the one hand, the telescope time allotted at La Silla, and the availability of the CTIO and Asiago telescopes, are used to observe and monitor new and recent SNe, both spectroscopically and photometrically, with the aim of covering their evolution as completely as possible. The availability of large telescopes allows us to monitor SNe also when they are faint, before and after maximum. In particular, spectroscopic observations of SNe Ia before maximum have been possible, which are opening up knowledge on a previously scarcely known epoch for SNe Ia, and have made it possible to identify examples of lack of uniformity among SNe of this type.

The best example of such work is the collection of spectra of SN 1991T, obtained at La Silla from about 2 weeks before maximum (Fig.1), showing that this object was peculiar at this early phase in showing only few spectral features, interpreted as due to Fe III (Ruiz-Lapuente *et al* 1992), although at later epochs its spectrum became that of a typical SN Ia (Filippenko *et al* 1992). This peculiar behaviour is probably correlated to the supernova's high luminosity at maximum (Phillips *et al* 1992). If the SN 1991T event was indeed peculiar, this will affect the way SNe Ia are viewed when they are considered as standard candles. Only a complete coverage of the spectral evolution can lead to the identification of the deviant objects and to their understanding and hopefully placing in the context of the general SN Ia event. Other SNe Ia for which we have obtained early and late time spectra are SNe 1990N, 1991bg and 1992A.

Alongside with the accumulation of new data is the retrieving, reducing and archiving of older data. First and foremost here is the collection of spectra of the type II SN 1987A obtained at ESO. Concerning SNe Ia, on the other hand, we have gathered

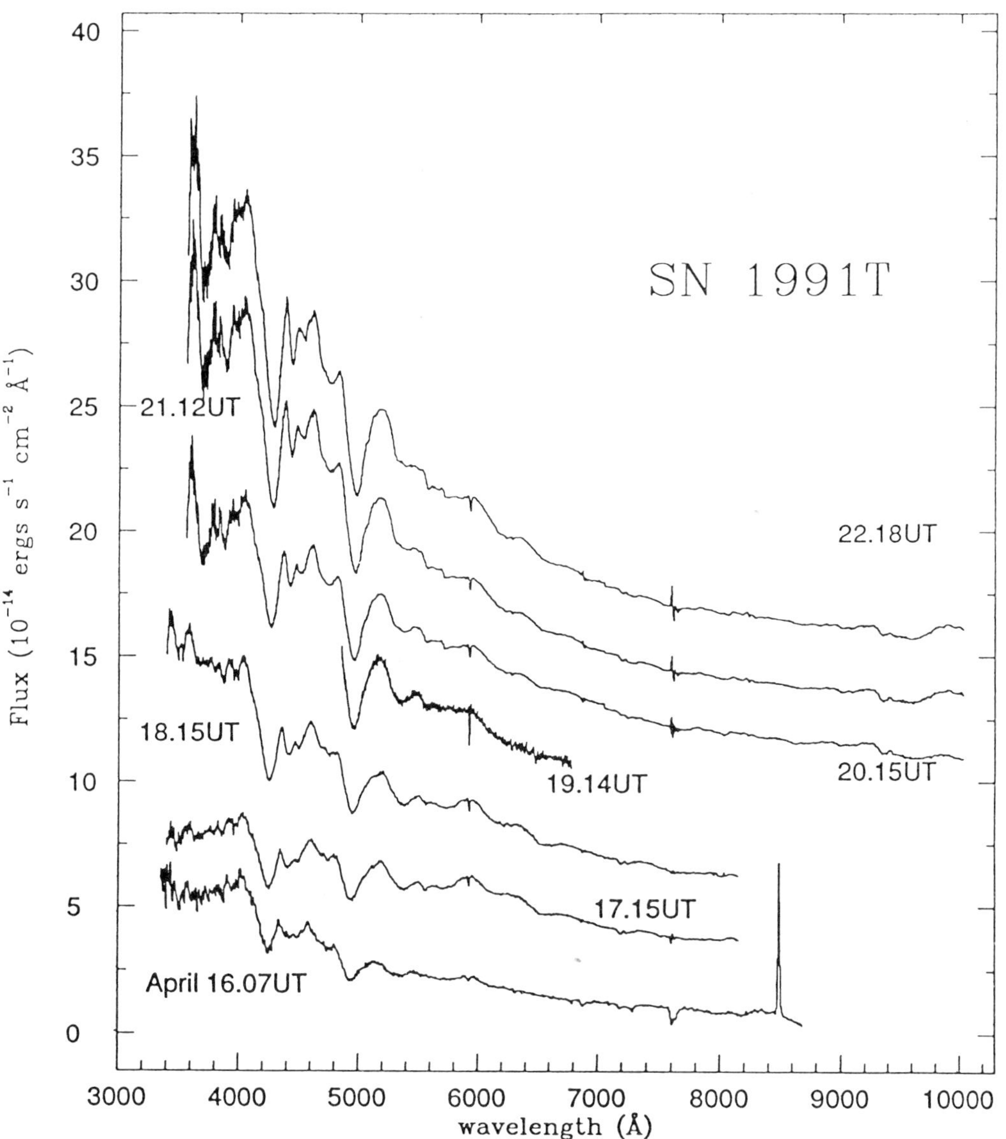

Figure 1. ESO optical spectra of SN 1991T during the first week after discovery (*B* maximum was reached on April 27). Each spectrum is shifted upwards by 2.5×10^{-14} erg cm^{-2} s^{-1} Å^{-1} with respect to the one immediately below.

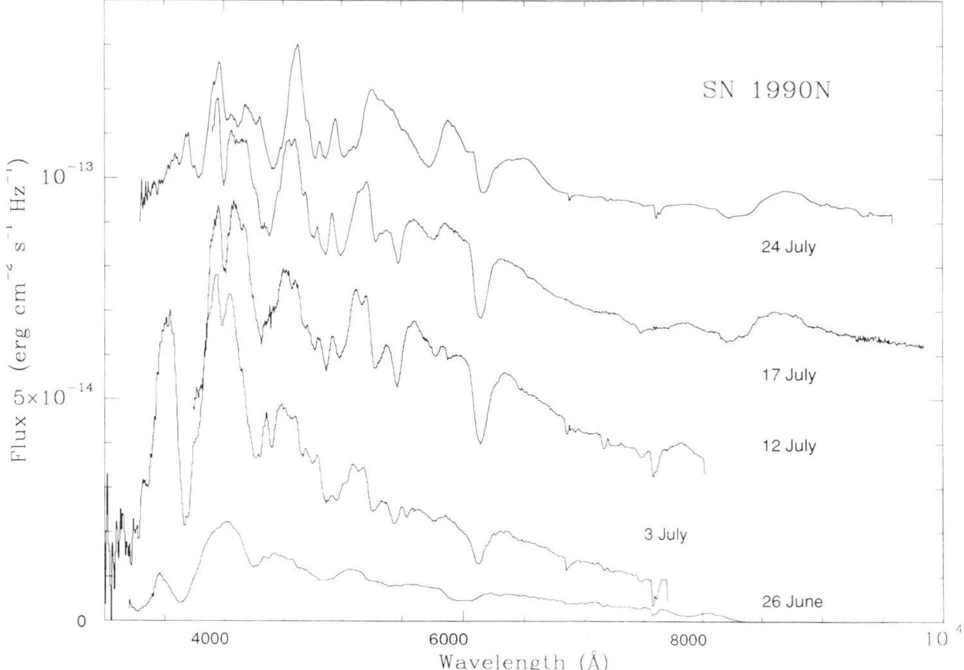

Figure 2. Optical spectra of SN 1990N in the photospheric epoch. B maximum occurred on July 10. The June 26, July 3 and 17 spectra were published by Leibundgut *et al* (1991), the July 24 one by Phillips *et al* (1992), while the spectrum of July 12 was obtained at Asiago Observatory. Each of the upper three spectra is shifted upwards by 3×10^{-14} erg cm^{-2} s^{-1} Å^{-1} with respect to the one lying immediately below. The two lower spectra are not shifted.

good spectral sequences for SN 1990N, which can be regarded as a 'standard' SN Ia, using data from various observatories (Fig.2), and for SN 1989B, which was also a fairly standard object (Barbon *et al* 1990).

Such a vast collection of data calls for an effort in understanding quantitatively the nature of SNe through the analysis of their spectra, which is the task of the 'theoretical group' within the Key Programme collaboration, so that the interpretation of the spectra through theoretical modelling is an important part of the work done within the framework of the Key Programme.

3. Early-time models

In the first several weeks after explosion, the stellar matter set in free expansion by the SN explosion can be divided in two regions: the outer part, where the largest velocities (up to several times 10^4 km/s) are reached, is optically thin in the continuum, and thick at selected line wavelengths only. Further within the ejecta, the density and temperature become sufficiently high that the continuum opacity becomes larger than one, so that an effective photosphere is formed, which emits a black body-type radiation at a characteristic temperature. No further information can be obtained from points below the photosphere. Spectral lines form in the overlying optically thin matter, the envelope. These will appear superposed on the photospheric continuum. Since the envelope is expanding, the lines take the shape of P-Cygni profiles. This situation is usually referred to as the photospheric (or early-time) epoch.

Both type II supernovae, which show strong H-Balmer lines, and type I SNe, which do not show the H lines, evolve similarly through a photospheric epoch and a later nebular phase. The main differences between the two SN types are in the light curves and the spectra (Oke & Searle 1974), which reflect differences in the type of progenitor stars. Among type I SNe, type Ia SNe show a strong Si II line near 6100 Å.

Models of early-time supernova spectra are computed by means of a Monte Carlo code, described in Lucy & Mazzali (1992, in preparation). The code originates from an algorithm built to treat stellar winds (Abbott & Lucy 1985), which in many respects resemble an expanding supernova envelope in the photospheric epoch.

The code assumes the envelope to be of the Schuster-Schwarzschild type, i.e. that a (black body-type) continuum is emitted at a sharply defined photosphere, above which no further energy deposition occurs. The envelope is assumed to be undergoing homologous spherical expansion (i.e. $v \propto r$), which becomes a good approximation very soon after the explosion. A typical SN density structure must be incorporated in the model. This is derived from SN explosion medels. For type II SNe, which are thought to originate from the explosion of massive stars (i.e. SN 1987A), the density stratification computed by Arnett (1988) has been used, while for type Ia SNe, thought to be the result of the explosion of mass-accreting C-O white dwarfs, the standard W7 model (Nomoto *et al* 1984) is normally adopted. These density structures are usually given for a representative epoch, and they must be rescaled to the required epoch after explosion, which is an input to the code. This is done using the homologous expansion property of SN ejecta. Element abundances are also taken from the aforementioned models.

The remaining physical parameters the code needs as input are the luminosity and the photospheric radius. These define the photospheric velocity, density and temperature. A temperature structure in the envelope is then calculated with an approximation to radiative equilibrium in the Monte Carlo code, assuming that the radiation field in the envelope takes the form of a diluted black body (i.e. $J_\nu = W B_\nu$).

The Monte Carlo experiment, which is also used for the computation of the emergent spectrum, consists of the study of the propagation in the envelope of a number of energy packets, of different frequency, emitted at the photosphere. The envelope is assumed to be of a purely scattering type, i.e. continuum opacity is neglected. Both line and electron scattering can occur, and these processes contribute to building up the opacity a packet encounters in flight. This is compared to randomly generated opacity values to establish which process interrupts a free flight episode, and at which location in the envelope the event takes place. Line and electron scattering opacities are calculated with an approximate NLTE treatment of excitation and ionisation (Lucy & Mazzali 1992, in preparation), which takes into account the major ionisation edges as sources of continuum opacity. Line formation by coherent scattering in the fluid frame is treated in the narrow line limit (Sobolev approximation), which is amply justified given the large velocities in a SN envelope. The master line list used in the synthesis calculations is derived from the works of Abbott (1982) for typical hot star wind ions, and Kurucz & Petreymann (1975) for species of lower ionisation. Missing ions and lines have been added, and errors in the original line list have been corrected for as we became aware of them. A packet of rest frequency ν is continuously redshifted, and encounters lines of progressively lower frequency as it scatters in the differentially expanding envelope. Thus the radiative transfer problem is of a multi-line type, especially so given the large envelope velocities. A Monte Carlo code is ideally suited to tackle this kind of problem, since it treats lines separately, unlike the formal integral approach (Puls 1987).

As in hot stars, the spectrum of a SN at early times is composed of a number of P-Cygni type lines, which are often the result of the blending of many line transitions. The emission peaks seen in the SN spectrum are just regions of low line opacity, where photons scattered by many transitions further to the blue can escape. Emergent model spectra can be compared to observed ones if the distance and reddening to the object under study are known. Since the emergent spectrum is computed by studying the propagation of photons belonging to a black body spectrum emitted at the photosphere with a certain luminosity and temperature, a fit of the flux and the features is then also a fit to the distance and luminosity of the supernova, if the intervening reddening is known. In the following, we discuss two examples of the application of this modelling work.

4. Barium and other s-process elements in SN 1987A

The good fits to the optical (Lucy 1987) and UV (Mazzali & Lucy 1990) spectrum of SN 1987A in the Large Magellanic Cloud showed that the Monte Carlo code can be used successfully to model the formation of a SN II spectrum at early times. This powerful diagnostic tool was then used to tackle one of the most interesting features of the optical spectrum of SN 1987A: the presence of strong lines identified as due to s-process elements (Ba II, Sr II, Sc II). Williams (1987) identified several features as due to these elements and, on the basis of LTE models, suggested that thay should have overabundances of factors 2 to 8 for Ba, 6 to 12 for Sc, 8 to 15 for Sr, compared to the LMC metal abundance, for the lines to form with a strength comparable to that observed. NLTE models by Höflich (1988) required for barium an even higher overabundance, of factors 10–20. These values were larger than those given by nucleosynthesis models (Prantzos *et al* 1988), which predict s-process elements to form in the He-burning core of massive stars. The discrepancy was especially large for Ba, for which Prantzos *et al* (1988) gave overabundances of factors 3 to 5 with respect to iron.

We used the Monte Carlo code to compute spectra of SN 1987A and reproduce the s-process elements' lines (Mazzali *et al* 1992a). It was clear that the standard model could not possibly reproduce the strength of the Ba II lines. Before concluding that the Ba abundance had to be increased, we pursued the possibility that the ionisation balance was altered in favour of Ba II with respect to our approximate treatment, which predicted Ba III to dominate. We thus built a model barium atom and solved the statistical equilibrium equations for the Ba II and Ba III levels in NLTE, and obtained photoionisation rates by sampling the photon distribution from the Monte Carlo model. It is in fact quite easy to compute with the code moments of the radiation field at various

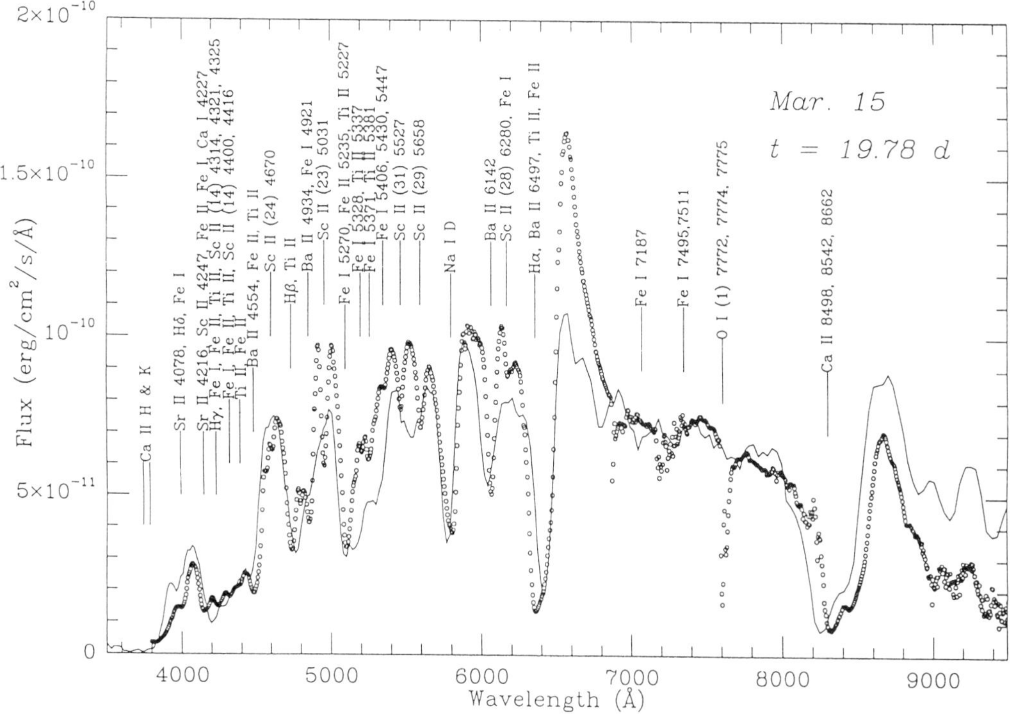

Figure 3. March 15, 1987 ESO optical spectrum of SN 1987A (circles) and Monte Carlo model with NLTE barium with s-process elements' overabundances: Ba/Fe=5, Sr/Fe=2, Sc/Fe=1 and confinement of these elements to $v \leq 6500$km/s, i.e. 89% in mass (continuous line). This model correctly reproduces both the strength and the blueshift of the observed lines.

depths in the model envelope. This method takes into account the effect of line-blocking on the photoionisation rates, i.e. the fact that the UV continuum is optically thick because of the large line opacity in the envelope, although no strong source of continuum opacity is present in that wavelength region, so that not many UV photons can propagate far from the photosphere. Since blocking is more effective at short wavelengths, the photoionisation rates from the low lying Ba II levels are reduced. New photoionisation cross sections have also been computed using algorithms from the Opacity Project, and they turned out to be larger than what simple estimates yield. The combination of the weak radiation field and the larger cross sections favours recombination to the low lying levels of Ba II, and leads to a larger Ba II/Ba III ratio, although Ba III remains the dominant ion. Since the resulting model spectra still do not fit the observations, a Ba overabundance must actually be invoked, but it is only of a factor ~ 3.

Although a correct line strength was obtained with this overabundance, the wavelengths of the model s-process elements' lines were bluer than the observed ones. The analysis of a series of spectra also shows that the wavelengths of the Ba II lines remained fairly constant in time, instead of becoming redder as the line-forming part of the envelope, which lies just above the photosphere, drops to lower velocities with the photosphere itself, as observed for most other lines. This was taken to indicate that the s-process elements are not uniformly distributed in the envelope of SN 1987A, but rather that they are confined to the inner part, so that they reach smaller velocities than the other lines. Numerical experiments showed that a good fit to the line position could be obtained if the s-process elements were confined to the inner 89% of the ejecta's mass. Obviously, the overabundance factors had to be increased somewhat to preserve the fit in line strength: the final values we derived are Ba/Fe = 5, Sr/Fe = 2, Sc/Fe = 1, where the ratios represent overabundance with respect to the average LMC metal abundance. The values are in good agreement with the nucleosynthesis prediction, and so is the confinement level: Prantzos *et al* (1988), in fact, locate the He-burning shell at 78% enclosed mass.

Fits were obtained for the spectrum of March 8, 1987 (12 days after outburst), where the s-process lines were still weak, and that of Mar 15, 1987, where the lines had reached maximum strength (Fig.3). The consistency between the two fits supports our conclusions concerning the abundance distribution of these elements, which is an important result towards the understanding of the physics of the interior of massive stars.

5. Models for the SN Ia 1990N

As we mentioned above, SN 1990N appears to have been a 'standard' SN Ia. For this SN, we have a series of 5 optical spectra covering a 4 week period, from 2 weeks before to 2 weeks after maximum, at intervals of roughly one week, obtained with various telescopes (Fig. 2). For all spectra, we have performed a flux calibration in B and V. Unfortunately, the B and V calibration factors are not always unique, although the difference is never very large. Thus, we plot the spectra twice, correcting the spectrum once for each of the two factors, and compare the B and V regions of our model spectra with the appropriately corrected observed spectrum.

As the time elapsed from explosion increases, the decreasing density due to the expansion makes the apparent photosphere recede further within the ejecta, to regions of lower velocity, so that the photosphere forms at a roughly constant density, but the lines form at increasingly low velocities. The combination of the position of the photosphere and the luminosity determines the photospheric effective temperature, and thus, through the temperature's control on the ionisation balance, the type of spectral lines which form.

Following the evolution of a standard SN Ia with a series of models is a sound method to determine the evolution of the SN parameters, in particular the luminosity and the photospheric velocity and temperature. The direct use of these quantities in the calculation of the model spectra is one of the most attractive features of our spectrum synthesis code.

A series of fits have indeed been obtained, all of which are of rather good quality. One example of such fits is shown in Fig.4 for a spectrum near maximum light. All fits are discussed in Mazzali *et al* (1992b), where for the first time model and observed spectra have been compared on a linear scale, such that any lack of accuracy in the fits is not hidden by the plotting method. The success of our model in following the evolution of the observed spectra shows that the model describes rather well the physics of SNe Ia

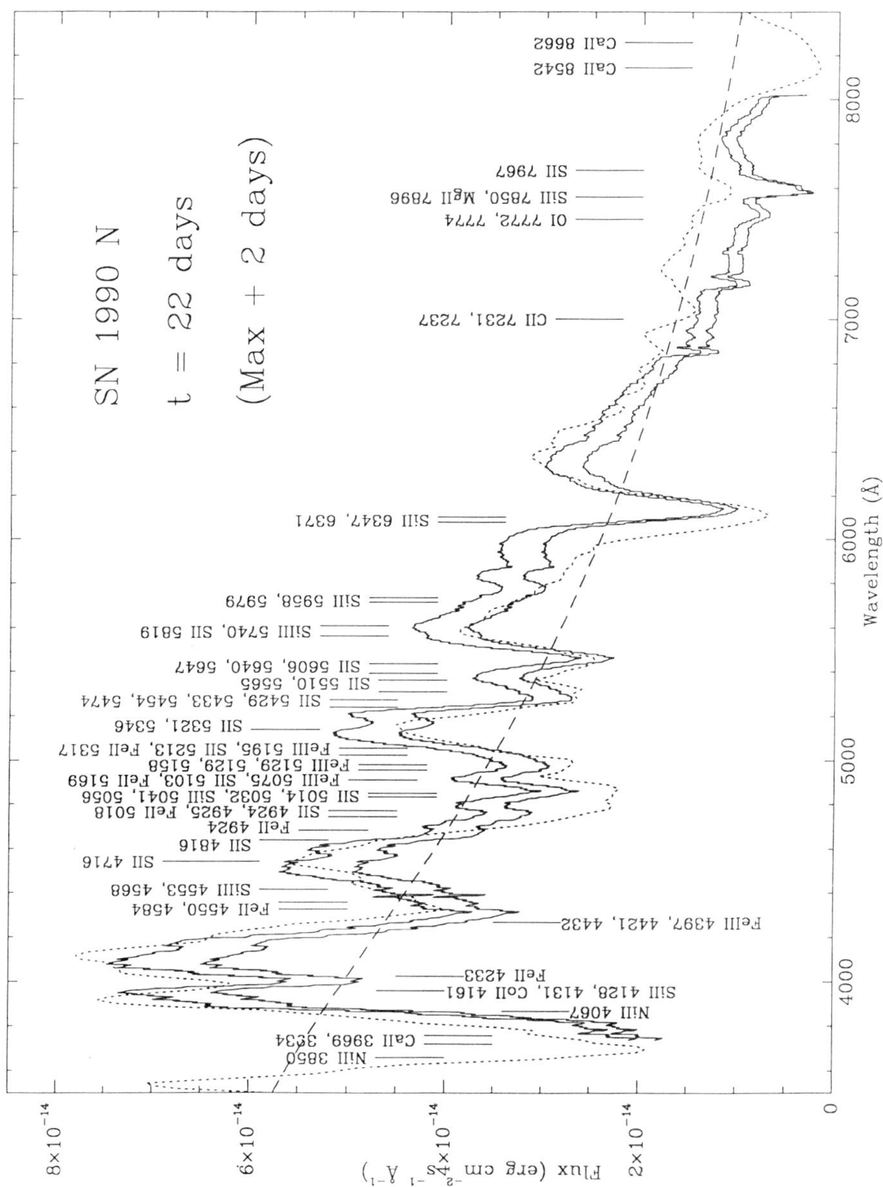

Figure 4. The optical spectrum of SN 1990 obtained at Asiago Observatory on July 12, 1990, corresponding to an epoch 22 days after outburst, is displayed with the two different photometric corrections applied: top continuous line – B photometry, bottom continuous line – V photometry. This is compared to our model spectrum (short dashes) and the underlying photospheric spectrum (long dashes). Main identifications are marked.

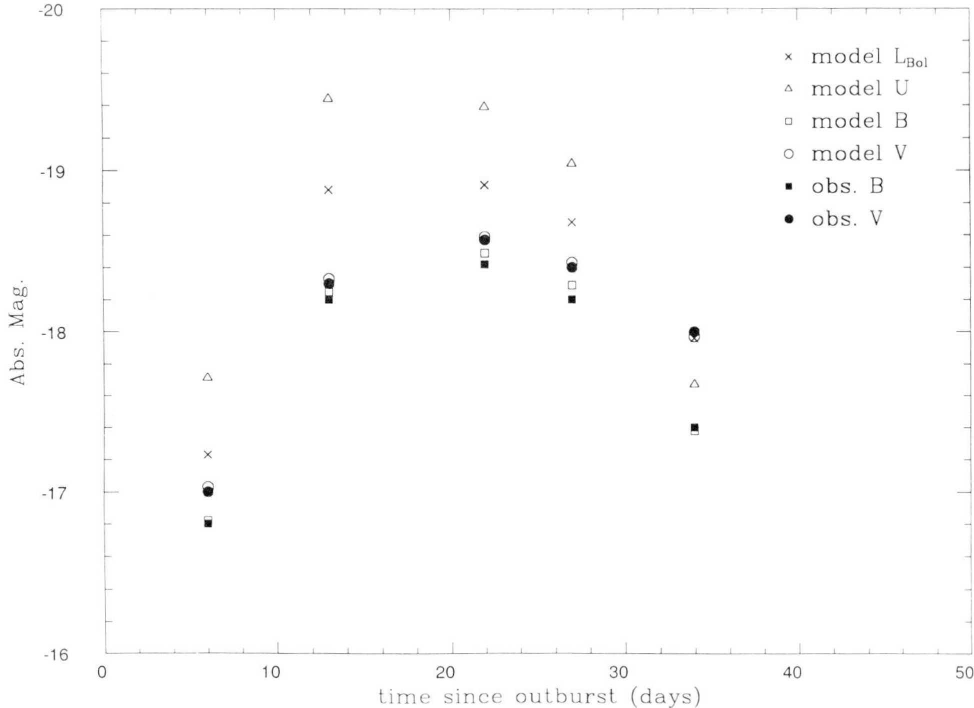

Figure 5. The colour bands and bolometric light curves of SN 1990N. The observed values are from Leibundgut *et al* (1991), while the model values are those given by our 'best fit' models.

envelopes. The fits yield an almost complete line identification for all the spectra, and allow several interesting problems related to line formation to be isolated.

Our model fits were obtained adopting for the host galaxy of SN 1990N, NGC 4639 in the Virgo cluster, a distance modulus of 31.2 mags, as an average among the 'short' distance values in the literature (e.g. Pierce & Tully 1988). The fits resulted in the light curves shown in Fig.5, where $B(max) \approx -18.5$ mag, in agreement with the most recent calibrations (Fukugita & Hogan 1991). Although B and V maxima are reached about 20 days after outburst, bolometric maximum appears to have occured earlier, somewhere between days 13 and 20, since at the earlier epochs most of the flux is emitted in the near UV, as shown by the U curve in Fig.5, which is higher at day 13 than at day 20. Also, the bolometric light curve is at least 0.4 mag higher than the B one near maximum. This shows that the commonly made assumption that the observed B curve follows closely the bolometric one obtained from the explosion models is not correct. Thus, distance moduli obtained by rescaling the observed B curve to the predicted L curve are overestimated by at least 0.4 mag.

Since the luminosity is an input to the code, our fits indicate that the combination we adopted of distance and luminosity is compatible with the observations. The adoption of a larger distance would require a higher luminosity. This in turn would lead to higher envelope temperatures and thus to the formation of lines different from those observed. The temperature could be lowered if a larger photospheric radius were also selected, but this would lead to larger line velocities and to a smaller overlying column density, and thus to a poorer fit. Testing the effect of various distance–luminosity–density combinations on the emergent spectrum is potentially a powerful method for using the full information given by the SN spectra to determine the distance to these objects. We intend to explore such calculations in the future, and we also intend to apply our code to non-standard SNe Ia and to other SN types.

6. Models for late time spectra

A few months after outburst, the density in the SN ejecta becomes so low that the apparent photosphere disappears, and SNe spectra become of a nebular type. Basic information regarding the SNe, such as abundances, distance and reddening, can be obtained from these spectra with the aid of simple modelling. A spectrum synthesis code for SNe in the nebular phase has been developed that includes the basic physics needed to reproduce the spectra. Level populations are computed in LTE, but the possibility of solving NLTE matrices for relevant species exists.

For SN 1987A, fits to the nebular spectra (Danziger *et al* 1991) have produced abundance determinations of elements from oxygen to cobalt that are in reasonable agreement with the predictions of the most sophisticated explosion models (Thielemann *et al* 1990).

Also, models of the strength of the [Co II] 10.52μ line (Danziger *et al* 1991), as well as models of the light curve of SN 1987A (Bouchet *et al* 1991) have led to the prediction that the $^{57}Co/^{56}Co$ ratio is $1.5\times$ solar (Danziger *et al* 1992), which has been recently confirmed by OSSE observations on GRO (Kurfess *et al* 1992; Wamsteker, private communication).

The nebular spectrum of SNe Ia is basically composed of emission lines of Fe II and Fe III. Since iron is a decay product of the nickel generated by the core collapse, fitting the nebular spectrum can cast light on both the iron and the original nickel mass. A fit depends on three basic paramenters, the determination of which bears important consequences for the understanding of the physics of SNe: the Ni mass produced, the distance and reddening to the SN. These three paramenters have correlated effects on the spectrum, and thus least squares methods can be employed to determine the set which yields the best fit. Reliable determinations of the Fe and Ni mass in SN 1986G have been obtained using this method (Ruiz-Lapuente & Lucy 1992), which is promising for its potential use in order to determine distance to SNe. For SN 1986G in Cen A, the best fit results were $M(^{56}Ni) = 0.38 \pm 0.03 M_\odot$, $d = 3.3 \pm 0.3 \mathrm{Mpc}$, $E(B - V) = 1.09 \pm 0.02$.

References

Abbott D C 1982 *Astrophys. J.* **259** 282

Abbott D C and Lucy L B 1985 *Astrophys. J.* **288** 679

Arnett W D 1988 *Astrophys. J.* **331** 377

Barbon R, Benetti S, Cappellaro E, Rosino L and Turatto M 1990 *Astron. Astrophys.* **237** 79

Bouchet P, Danziger I J and Lucy L B 1991 *Astron. J.* **102** 1135

Cappellaro E, Bouchet P, Della Valle M, Danziger I J, Fransson C, Gouiffes C, Lucy L, Mazzali P, Phillips M and Turatto M 1991 in: *SN 1987A and other Supernovae*, ed. I J Danziger and K Kjär, ESO Conf. Proc. 37. (ESO: Garching) p 623

Danziger I J, Lucy L B, Gouiffes C and Bouchet P 1991 in: *Supernovae*, ed. S Woosley (Springer-Verlag: New York) p 69

Danziger I J, Bouchet P, Gouiffes C and Lucy L B 1992 in: *SN 1987A and other Supernovae*, ed. I J Danziger and K Kjär, ESO Conf. Proc. 37. (ESO: Garching) p 217

Filippenko A V, Richmond M W, Matheson T *et al* 1992 *Astrophys. J.* **384** L15

Fukugita M and Hogan C J 1991 *Astrophys. J.* **368** L11

Höflich P 1988 *Proc. ASA* **7** 434

Kurfess J D, Johnson W N, Linzer R L *et al* 1992 *Bull. Am. Ast. Soc.* **24** 750

Kurucz R L and Petreymann E 1975 *Smithsonian Ap. Obs. Special Report* 362

Leibundgut, B., Kirshner, R.P., Filippenko, A.V., et al., 1991, *Astrophys. J.* **371**, L23

Lucy L B 1987 in: *ESO Workshop on SN 1987A*, ed. I J Danziger (ESO: Garching) p 417

Mazzali P A and Lucy L B 1990 in: *Evolution in Astrophysics*, ESA SP-310 p 483

Mazzali P A, Lucy L B and Butler K 1992a, *Astron. Astrophys.* **258** 399

Mazzali P A, Lucy L B, Danziger I J, Gouiffes C, Cappellaro E and Turatto M 1992b *Astron. Astrophys.* , submitted

Nomoto K, Thielemann F-K and Yokoi K 1984 *Astrophys. J.* , **286** 644

Oke J B and Searle L 1974 *Ann. Rev. Astron. Astrophys.* **12** 315

Pierce M J and Tully R B 1988 *Astrophys. J.* **330** 579

Phillips M M, Wells L A, Suntzeff N B *et al* 1992 *Astron. J.* , **103** 1632

Prantzos N, Arnould M and Casse M 1988 *Astrophys. J.* **331** L15

Puls J 1987 *Astron. Astrophys.* **184** 227

Ruiz-Lapuente P, Cappellaro E, Turatto M, Gouiffes C, Danziger I J, Della Valle M and Lucy L B 1992 *Astrophys. J.* **387**, L33

Ruiz-Lapuente P and Lucy L B 1992 *Astrophys. J.* in press

Thielemann F-K, Hashimoto M and Nomoto K 1990 *Astrophys. J.* **349** 222

Williams R E 1987b in: *Proc. IAU Colloq. No 108: Atmospheric Diagnostics of Stellar Evolution*, ed K Nomoto (Springer-Verlag: New York) p 118

Observations in circumstellar envelopes

A. Omont

Institut d'Astrophysique de Paris, CNRS

98bis, Boulvard Arago, F–75014 Paris, France

Abstract. During the very late stages of the red giant phase (late "Asymptotic Giant Branch" AGB), the mass–loss rate considerably increases and generates massive and thick circumstellar envelopes. They are very luminous infrared sources and emit many millimeter molecular lines. Their studies have developed relatively recently with the advent of millimeter and infrared astronomy, in particular with the infrared satellite IRAS and the IRAM radiotelescopes.

They are quite important from many points of view : to understand the final evolution and nucleosynthesis of stars of low and intermediate mass; as one the major contributors of return to the interstellar medium of many nuclei; as the main known site of formation of cosmic dust; as a particularly interesting case of cosmic chemistry, especially if the C/O ratio$>$1.

This review will briefly address these different points in relation to infrared and millimeter observations. A special emphasis will be put i) on millimeter observations of molecular lines,   ii) on features relevant for the abundances of nuclei in the Cosmos : circumstellar abundances and isotopic ratios reflecting nucleosynthesis and dredge–up in the late AGB phases, mass return to the interstellar medium, dust formation and properties (SiC, carbon, etc) of possible relevance for meteorite micrograins.

1. General features

Massive circumstellar envelopes (CSE's) ejected by red giants at the very end of their evolution are particularly interesting objects to understand many aspects of nucleosynthesis related to red giants : element and isotope abundances there directly reflect nucleosynthesis and dredge–up in the very late stages of the asymptotic giant branch (AGB); such abundances are especially relevant for galactic chemical evolution, because massive CSE's are believed to be one of the main contributors returning mass to the interstellar medium (ISM) and to dominate the return of many elements such as ^{12}C, ^{14}N, ^{13}C and s–isotopes; in addition, mass–loss dominates and, indeed, terminates, stellar evolution in these last AGB phases, so that the history of the mass–loss rate, $\dot{M}(r)$, visible in the envelope structure, is the best indicator of the evolution of the whole star; finally, they are the best identified site of cosmic dust, and a fraction of the micrograins formed in CSE's are believed to survive to be incorporated in meteorites where they directly inform on the composition of the AGB gas.

Most studies of CSE's have been developed relatively recently, with the advent of millimeter and infrared astronomy. They are in particular one of the most numerous class of sources detected by IRAS. This review will address in turn these various aspects of CSE's of interest for nucleosynthesis. These are infrared observations by IRAS giving information on dust with a brief discussion of the conditions of formation of dust in CSE's, in Section 2, estimates of mass–loss rates of CSE's and of the total amount of mass that they return to the ISM in Section 4 and determination of the abundances of elements and of isotope ratios by millimeter observations, in Sections 5 and 6, respectively. Detailed descriptions of the properties of CSE's will be found in books on the subject (e.g. Morris and Zuckerman 1985, Johnson and Zuckerman 1989, Menessier and Omont 1990, Schwartz 1992) or reviews (e.g. Omont 1991). Their evolution is also discussed in these references, and more specifically in Iben and Renzini (1983), Iben (1986), Weideman and Schönberner (1990) and references therein.

For the clarity of the present review, it is worth to recall the main characteristics shared by the various classes of CSE's : they represent a very brief, but crucial, phase, which probably terminates the evolution of every low and intermediate mass star at the transition between red giants and planetary nebulae. The star of most of these molecular CSE's is on the late asymptotic giant branch (AGB) of red giants (although some of them are post–AGB stars or red supergiants). Its core has then properties very close to those of a white dwarf. The high temperature induces a powerful burning of hydrogen in a shell around the core with convective transport of energy.

CSE's have thus very high luminosities ($\sim 10^4$ $L_\odot$ for AGB CSE's) and most of them are long period variables of different kinds (irregular, semi–regular, Miras, infrared) with typical periods of a few hundred days. At time intervals $\sim 10^4$–10^5 years, the temperature of the helium shell around the core becomes high enough to briefly ignite the reaction $3\,{}^4\mathrm{He} \longrightarrow {}^{12}\mathrm{C}$ ("thermal pulses"). A powerful convection (dredge–up) brings to the surface of the star different products of the nucleosynthesis which are ${}^{12}\mathrm{C}$, s–isotopes and products of CNO H–burning (${}^{14}\mathrm{N}$, ${}^{13}\mathrm{C}$, etc). Thus, the photosphere of the star ($\mathrm{T}_{eff} \leq 3000$ K) and the CSE itself present a variety of chemical compositions dominated by the value of the C/O ratio : O–rich (M) stars for C/O < 1, carbon stars for C/O > 1 and S stars for C/O ~ 1.

Many details of this brief and crucial phase of evolution, which is dominated by the large mass–loss rate often called a superwind, remain obscure. In particular, the conditions of the transition from O– to C–richness are still debated (de Jong 1989, 1990, Zuckerman and Madalena 1989, Barnbaum et al. 1991a). The red giant phase ends when the mass–loss has almost exhausted the gas of the star. The effective temperature of the star then increases until its radiation becomes sufficient to ionize the CSE into a planetary nebula, before the final evolution of the latter into a white dwarf. The transition objects between the AGB and planetary nebulae, "pre–planetary nebulae" (PPN), are especially interesting, both for the understanding of stellar evolution and for their particular structure and physical and chemical conditions.

AGB CSE's themselves include optically thin envelopes around visible stars (O–

and C–rich Miras, semi–regular and irregular variables), as well as infrared stars (OH/IR stars, cold carbon stars, etc.). As concerns the abundances of elements and isotopes, optical (or near infrared) methods, when possible, are generally more sensitive than millimeter studies. However, the latter are the only ones possible for cold CSE's. This review

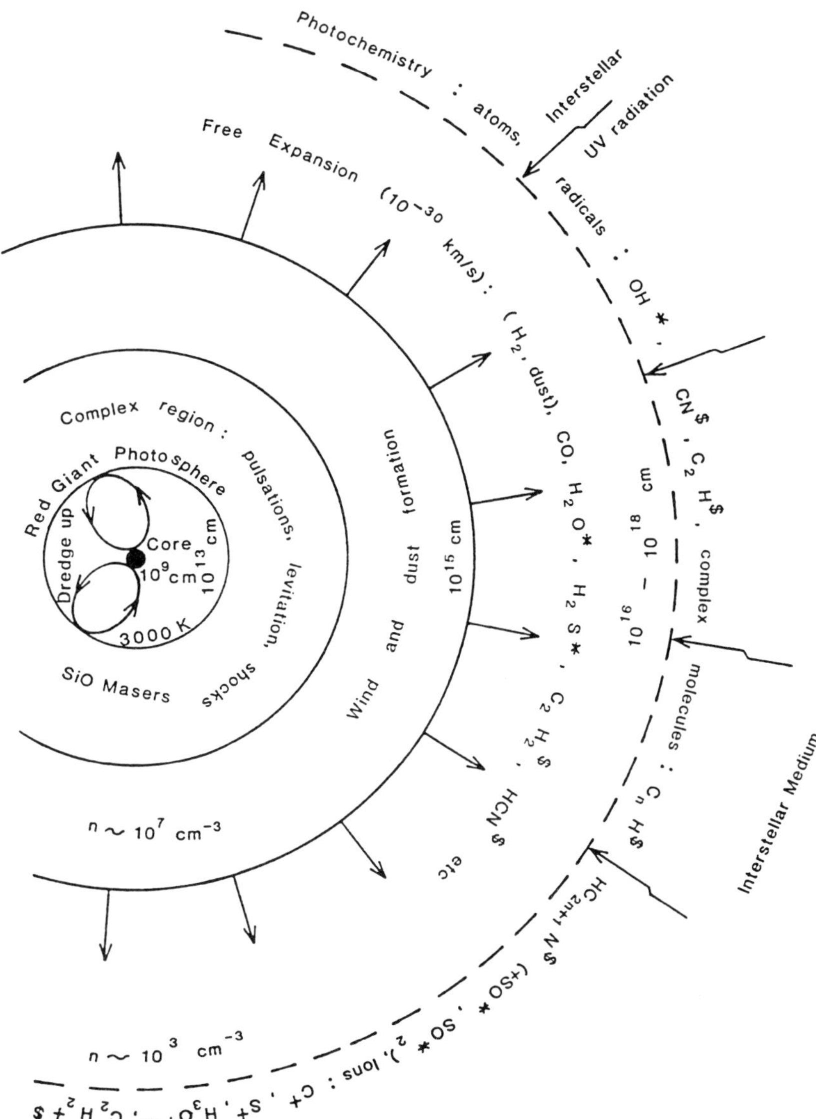

Figure 1. Sketch of an AGB circumstellar envelope. Important species of O–rich and C–rich envelopes are denoted by $\star$ and $\$$, respectively

is focussed on such cold objects and their millimeter studies. They trace a nucleosynthesis stage which can generate abundances different from visible AGB stars, and which is particularly important since it dominates the return of mass to the ISM (Section 4). The different shells of such objects with their physical and chemical conditions are sketched in Figure 1.

2. Infrared observations. Properties and formation of dust .

2.1. **IRAS** *colour–colour diagram*

Colour–colour diagrams are a good way to visualise the properties of infrared sources and to distinguish between broad classes. One of the most useful ones relates the 25/12 μm and 60/25 μm IRAS colours. Figure 2 reproduces the diagram proposed by van der Veen and Habing (1988). It is divided in regions tailored to distinguish the various classes of CSE's. The location in Figure 2 of the CSE's where millimeter lines of CO have been detected, visualises their distributions showing relatively well defined sequences : O–rich CSE's with increasing mass–loss rates from Miras in region II to very cold OH/IR stars in region IV and a concentration of C–rich CSE's in region VII with extensions to regions VIa, IIIa and IIIb. In both cases, it has been suggested that such sequences correspond to evolutionary tracks (Olnon et al. 1984, Willems and de Jong 1986 , 1988). However, many points of both evolutions remain debated.

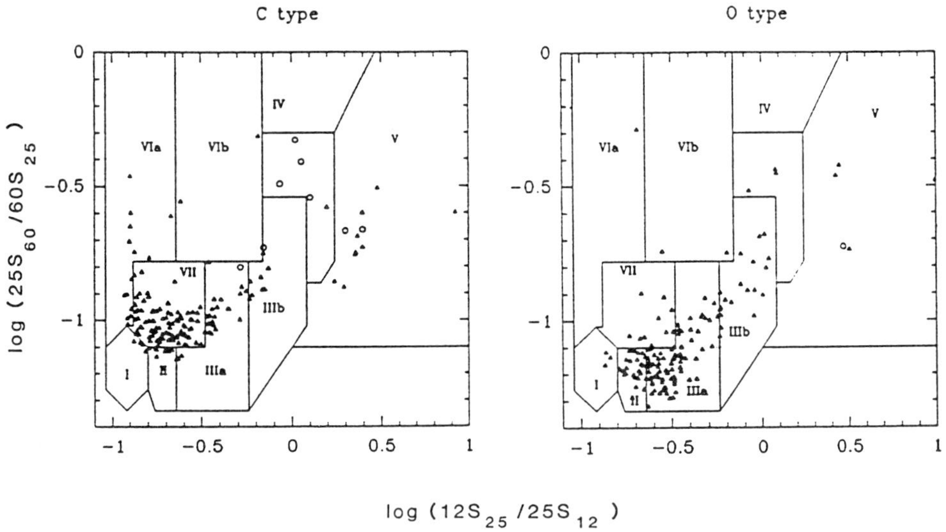

Figure 2. IRAS colour–colour diagram 12/25 μm versus 25/60 μm, with regions defined by van der Veen and Habing (1988), and location of the envelopes where millimeter lines of CO have been detected (from Loup et al. 1992).

2.2. **IRAS LRS** *spectra and dust characterization*

Another extremely important source of information on CSE's and their dust composition
are the IRAS Low Resolution Spectra (LRS) (Fig. 3). Their published atlas (IRAS Team
1986) provides 7.7–22.6 μm spectra of 5425 sources, $\sim$ 80% of which are CSE's. The
complete LRS data base contains useful information on many more sources. The 10
classes and 72 subclasses of the LRS atlas allowed a first classification of CSE's, which
will certainly be refined in the future. The most prominent absorption or emission
features are attributed to : i) silicates (9–12 and 15–21 μm), whose strong absorption in
extreme CSE's was recently modeled (Schutte and Tielens 1989) in order to infer dust
properties and the mass–loss rate. ii) Broad SiC emission at 11.4 μm in C–rich CSE's.

This 11.4 μm feature attributed to SiC is particularly prominent in the spectrum
of most C–rich CSE's. It is generally agreed that dust of C–rich CSE's is mostly solid
carbon which is probably amorphous rather than graphite. However, solid carbon has

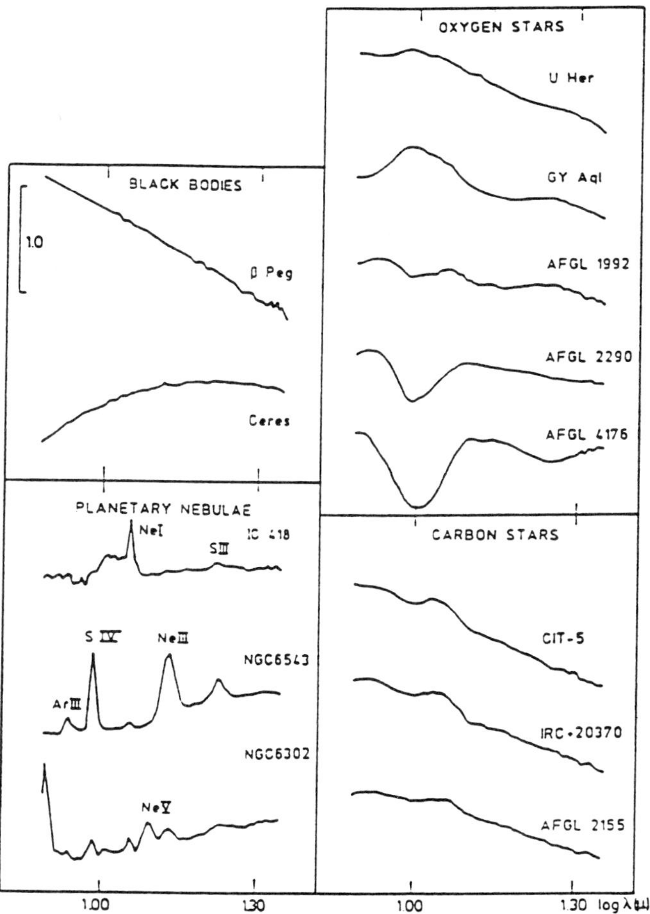

Figure 3. A sample of low resolution spectra from the IRAS Atlas (1986). From
Circumstellar Matter, Reidel–Dordrecht 1987 p. 204.

practically no infrared spectral feature. In any case, SiC is probably fairly abundant in AGB C–rich dust, maybe $\sim$2–10% per mass, as one can infer from the modelling of Baron et al. (1987) and Chan and Kwok (1990). This is consistent with the observed depletion of silicon in the gas phase of C–rich CSE's such as IRC+10216 (Morris 1975). Even in the coldest C–rich CSE's which have featureless LRS spectra (Omont et al. 1992), the absence of the 11.4 μm feature of SiC is probably mainly due to radiative transfer effects (Chan and Kwok 1990).

2.3. Dust Formation

Despite the progress brought by IRAS, our information on AGB dust remains quite meagre. For instance, the state of metals in C–rich dust is unknown. It is hoped that IR spectroscopy with the future ESA satellite ISO will somewhat improve the situation, but in a limited way. Unfortunately, there is little hope to realistically model detailed properties of circumstellar dust from the physics of its formation.

Modelling the formation of dust in the cooling outflows of CSE's is an extremely difficult problem (see e.g. Omont 1990). Even the exact treatment of the ideal situation of thermodynamical equilibrium would be hard because of the complexity of the gas and solid compositions. But thermodynamical equilibrium is certainly never achieved in this range because of supersaturation effects and the rapidity of the envelope evolution. However, thermodynamical calculations, with simplified compositions, are a useful first step and a guide to foresee possible compounds and the temperature hierarchy of the condensation.

Different attempts have been made to describe dust formation in CSE's with the use of nucleation theory (see e.g. Lefèvre 1988, Seldmayr 1990 and references therein). Condensation of gas into solid particles must go through all the intermediate steps of clusters with N particles. The latter are less stable than the bulk solid since the average binding energy of a particle is reduced because of surface effects. Only homogeneous nucleation theory is relatively simple and well developed ; it accounts for the case where a single chemical compound is present. Then, the supersaturation factor $(p/p_{sat}(T) \gg 1)$, required to overcome the cluster instability for N$\leq$10, can be computed if one can estimate the cluster binding energies. But the required inhomogeneous nucleation theory is much more difficult because many chemical species are present, and one must incorporate the detailed complicated reactions between gas and clusters. Several additional complications arise; for instance because of the importance of radiation there is no unique temperature $(T_{kin} \neq T_{dust}$, etc). The inner temperature of a cluster or of a very small grain fluctuates when it absorbes a photon or accretes an atom or a molecule. There are great uncertainties in the gas parameters and their evolution : pulsations, shocks and clumpiness of the transition layers, instabilities related to the dust itself on the dynamics, temperature and chemistry of the gas. There is a complex time evolution of the size distribution of grains and clusters, with important coagulation processes. On the other hand, the hierarchy of condensation of the different elements depending on whether they are more or less refractory, simplifies the condensation of the most volatile products which can just condense on preexisting more refractory grains. It is thus possible that the nucleation problem is completely irrelevant for the most abundant compounds. Laboratory simulations are also very difficult because of the time constants involved in CSE's. Accordingly it is very likely that most of progress in our knowledge of dust formation in CSE's will continue to come from observations. It is thus fundamental to obtain detailed information on the spatial distribution of dust, and of its temperature and composition, in the different types of CSE's.

3. General features of millimeter observation

Millimeter radioastronomy has, since more than 20 years, developed studies of rotational lines of interstellar and circumstellar molecules. Compared with interstellar clouds, CSE's are generally more compact. Circumstellar lines are thus often weaker, but the advantage of large antennas, such as IRAM–30m, is increased. Because of the well defined spherical expanding structure, circumstellar mm lines have characteristic profiles (Figure 4). With a total width 2 V_{exp}. They are also characterised by a peculiar rotational excitation through near IR vibrational lines. Many circumstellar programs were carried out in the last years at the IRAM–30m and SEST–15m.

Table 1 gives an idea of the molecules detected and of the importance of the various programs (see also Olofsson 1992, Lucas 1992). The main ones concern either relatively common, abundant molecules in large samples of sources, or exotic molecules in IRC+10216.

CO is by far the most studied molecule because of its stability and large abundance (see Section 4). HCN and SiO come next in C– and O–rich CSE's, respectively (note that both molecules display maser emission, in vibrationally excited states, close to the photosphere). Other commonly studied molecules are CS, CN, C_2H, HC_3N, HNC, SiS, SiC_2 and SiO in C–rich CSEs, H_2S, SO, SO_2 and HCN in O–rich ones.

IRC+10216 (CW Leo) is exceptionally close to the Sun ($\sim$150 pc) for a massive CSE. This fact, together with the richness of the carbon chemistry, has made extremely fruitful the search for new cosmic molecules such as polyynes HC_nN, radicals C_nH, C_nS, cycles (C_3H_2, SiC_2), metals (ClNa, ClK, ClAl, FAl). Indeed many of these exotic unstable molecules have been discovered there, generally even before their laboratory studies (see e.g. Omont 1992 for a list and abundances). Many recent discoveries of new molecules

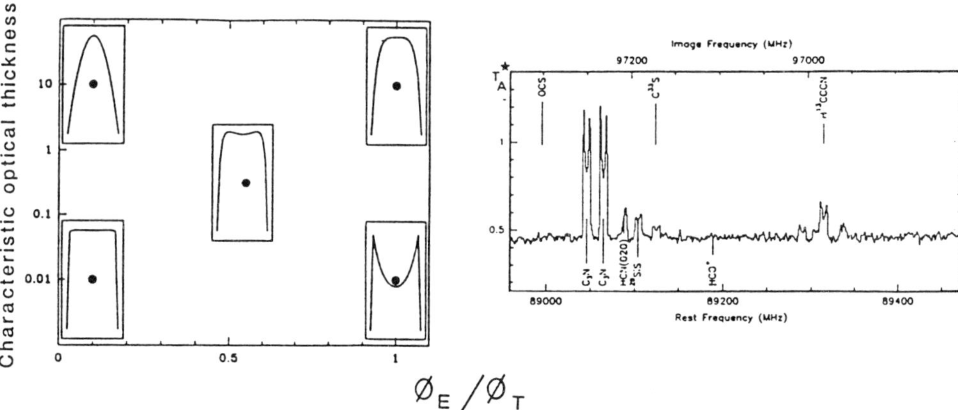

Figure 4. Circumstellar line profiles. **Left:** Theoretical profiles (from Kahane *et al.* 1992) for different values of the optical thickness and of the ratios ϕ/ϕ of the angular diameter of the envelope (ϕ) and the angular beam of the telescope ϕ. **Right:** Example of observed profiles in IRC+10216 (from Lucas et al. 1986).

were made with the IRAM 30m–telescope, where a rather systematic frequency survey of this source was carried on.

The high angular resolution of large single dishes or millimeter interferometers enable one to derive the spatial distribution of these molecules in the envelopes. It is thus found that most stable molecules have rather constant ("frozen") abundances in the whole outer molecular envelope, from the region of dust formation to the photodissociation regions (Figure 1). On the other hand, radicals (such as C_2H, CN, C_4H, SiC) and complex unstable molecules (such as HC_5N, HNC) are distributed in hollow shells characteristic of outer synthesis by photochemistry. This wealth of results has prompted detailed modelling of circumstellar chemistry, which accounts reasonably well for the observations (see e.g. reviews by Glassgold and Huggins 1988, Millar 1988, Omont 1990, 1992).

4. Millimeter CO observation. Mass–loss Rates and Return of Matter to the ISM

CO has now been detected in more than 400 CSE's (see catalog by Loup et al. 1992). CO is generally the easy molecule to detect. It often provides the first determination of the expansion and LSR velocities of the sources. The intensity of CO lines also provides one of the best determinations of the mass–loss rate $\dot{M}$.

The relation between I_{CO} and $\dot{M}$ was discussed in detail by Knapp and Morris (1985). This theory was then refined by various authors. The best discussion is probably now by Kastner (1992). For a broad class of sources, there is an approximate proportionality between $\dot{M}$ and $I_{CO} \times D^2$, where D is the distance of the source. However, I_{CO} is very weak for stars with very large $\dot{M}$, such as very cold OH/IR stars (Heske et al. 1990); and it is then almost impossible to derive $\dot{M}$ from I_{CO}. Another method uses the far IR fluxes measured by IRAS to estimate $\dot{M}$. Within reasonable approximations, $\dot{M}$ is roughly proportional to $S_{60\mu m} \times D^2$. In both methods the distance is often a major source of uncertainty.

As the CO surveys are quite complete for the solar neighbourhood (D≤1 kpc), it is easy to deduce from the values of $\dot{M}$ the total rate of mass returned to the ISM by CSE's in the solar neighbourhood. The various estimates (Knapp and Morris 1985, Jura and Kleinmann 1989, Kastner 1990) range from 3.10^{-4} to 7.10^{-4} $M_\odot/kpc^2/yr$.

5. Abundances of molecules and elements

For optically thin lines, their intensity is directly proportional to the total number of molecules in the upper level of the molecular transition in the telescope beam. If the rotational excitation can be reasonably estimated, molecular abundances can be directly deduced from millimeter line intensities (the case of CO is peculiar, because its lines are generally optically thick). However, some care has to be exercised because the rotational excitation generally strongly depends on the position. It is generally dominated by absorption/emission in near IR vibrational bands which requires elaborate modelling (Morris 1975). The derivation of molecular abundances from millimeter intensities also needs the knowledge of the total density directly related to the mass–loss rate, and the line strength involving the permanent dipole momentum. As a result such observational

Table 1. Molecules Detected at Radio Wavelengths in CSE's

Molecule	Approx. # of Sources		Molecule	Approx. # of Sources	
	C–rich	O–rich		C–rich	O–rich
CO	200	200	C_2H	22	
CN	15		C_3H	2	
CS	20	7	C_3H_2	3	
SiO ("Thermal")	8	40	C_4H_2	1	
SiS	10	4	C_4H	7	
SiC_2	10		C_5H	1	
SO		15	C_6H	1	
SO_2		18	C_3N	5	
H_2S	1	15	C_2S	1	
OCS		1	C_3S	1	
HCO^+	3	2	$HSiC_2$ (HSC_2) (?)	1	
H_2CO	1	1	$NaC\ell$	1	
HCN	100	20	$A\ell C\ell$	1	
CH_3CN	3		$KC\ell$	1	
HNC	10	3	$A\ell F$	1	
NH_3	2	1	SiC	2	
HC_2N	1		SiN	1	
HC_3N	10		C_4Si	1	
HC_5N	7		CP	1	
HC_7N	2		OH (maser)		1500
HC_9N	1		H_2O (maser)	4	200
$HC_{11}N$	1		SiO (maser)		140

abundances remain quite uncertain, possibly a factor ~ 3 in the best cases. Relative abundances of similar molecules can be slightly more accurate.

Inferring atomic abundances from observed molecular abundances implies the knowledge of the abundances of all major species containing the element considered. The main uncertainties generally are dust and the molecules without millimeter lines (N_2, H_2O, C_2H_2). The dominant species for the main elements, inside the molecular envelope are the following:

	C–rich	O–rich
Oxygen	CO	H_2O, CO + silicates
Carbon	CO + dust ($+C_2H_2$)	CO
Nitrogen	N_2 + HCN	N_2
Sulfur	CS, SiS + dust	H_2S, SO, SO_2 + dust ?
Silicon	SiS, SiO + dust	silicates (+ SiO)

6. C/O ratio

The value of the C/O ratio is a major issue both for circumstellar chemistry and as an indicator of the amount of matter processed in 3α combustion, brought to the surface of the star by dredge–up. Its value is always uncertain because it involves dust and the unobserved H_2O or C_2H_2 molecules. There are nevertheless good criteria for the fundamental discrimination between C/O>1 and C/O<1 (see e.g. Omont et al. 1992 and references therein). In most cases, dust signatures (in IRAS LRS spectra) are quite clear : silicates for C/O<1, SiC for C/O>1. The presence of OH, H_2O or SiO masers is also a proof that C/O<1; but the absence of masers does not imply that C/O>1. The value of the ratio of the intensities of the millimeter lines of CO and HCN is also generally a good criterion for distinguishing oxygen–rich and carbon–rich envelopes. It seems (Bujarrabal et al. 1993) that the ratio I(HCN)/I(SiO) is still a better indicator, since it is apparently always one order of magnitude larger in C–rich envelopes than in O–rich ones.

S stars have C/O~1. They are difficult to identify for massive envelopes, whose opacity makes the observation of the visible spectrum hard. Indeed, practically no S star is yet identified among infrared stars, although the proportion of S stars could be 30% of carbon stars. Values of I(HCN)/I(SiO) intermediate between those of C–rich and O–rich envelopes could be a good indicator (Bujarrabal et al. 1993).

It is possible that some C–rich envelopes have values of C/O much larger than IRC+10216. As the main reservoir of carbon outside of CO is believed to be dust, it should result in smaller gas to dust ratios. However, the determination of the latter is still too uncertain to give definite evidence. An issue which could be related, but not necessarily so, is the abundance of carbon outside CO in the gas phase. The main reservoir of carbon there is probably C_2H_2, which has no millimeter lines. However, its abundance can be inferred from that of C_2H, its direct photodissociation product. A recent survey that we have made at IRAM of the C_2H lines (Tejero et al. 1993), shows that a few stars have ratios I(C_2H)/I(CO) much larger than IRC+10216. Such stars generally also have large values of I(HCN)/I(CO) and large values of the expansion velocity. It is possible that they also have large initial masses (Barnbaum et al. 1991). However, further studies are needed to obtain a full understanding of all these properties.

7. Isotope Ratios

7.1. Introduction

Contrary to the abundance ratios of different elements, the values of the ratio of the abundances of two different isotopic varieties of the same molecule are believed to be approximately free of chemical effects and to be equal to the ratio of the elements of the corresponding elements (e.g. $^{12}C^{32}S/^{13}C^{32}S = {}^{12}C/^{13}C$). Indeed, some differences in the reaction rates could lead to some chemical fractionation effects as in the interstellar medium; but, such effects are believed to be negligible in circumstellar envelopes, except possibly for some CO isotopes. However, such measurements of ratios of molecular isotope varieties are not free from problems. Obviously, the abundance of a rare isotope must be large enough, so that at least one molecule carrying this isotope is detectable. In case of large isotope ratios, there is then a good chance that the corresponding millimeter line of the abundant variety is optically thick and saturates. The determination of its abundance is then quite uncertain, and often only a lower limit can be determined. However, the use of "double" isotope ratios can overcome this difficulty, by complying with the rule that optically thin lines allow much better abundance determinations. For instance, while the lines of $^{12}C^{32}S$ are often optically thick, those of $^{13}C^{32}S$ and $^{12}C^{34}S$ are generally optically thin, so that one can consider with a good accuracy that

$$\frac{^{12}C\ ^{34}S}{^{13}C\ ^{32}S} = \frac{^{12}C}{^{13}C}\ \frac{^{34}S}{^{32}S}$$

If $^{32}S/^{34}S$ is determined from another molecule, (e.g. SiS), the measurement of the lines of $^{12}C^{34}S$ and $^{13}C^{32}S$ provides an accurate value for $^{12}C/^{13}C$ (see e.g. Kahane et al. 1988).

Because of the relative abundances of the molecules and the isotopes, such determinations are presently limited to ratios involving the following "rare" isotopes : ^{13}C, ^{17}O, ^{18}O, ^{33}S, ^{34}S, ^{29}Si, ^{30}Si, and marginally ^{15}N and ^{37}Cl. They are quite complete only in IRC+10216 (section 7.2). In other stars, except for ^{13}CO which has been detected in tens of stars (7.3), the only quantitative studies concern ^{17}O and ^{18}O, which have been measured in half a dozen of stars.

7.2. Isotope Ratios in IRC+10216

Because millimeter molecular lines are typically stronger in IRC+10216 than in any other star, by at least an order of magnitude, it allows much more detailed study. While several studies were carried out since the beginning of millimeter astronomy (see e.g. Wannier and Linke 1978 and references therein and in Kahane et al. 1988), the most comprehensive one is by far that of Kahane et al. (1988) with the IRAM 30m telescope.

The molecules used and the main results are summarized in Table 2. For each molecule, several rotational lines of each of the available isotopic varieties were generally measured.

Table 2. Isotope Ratios in IRC+10216

Element	Molecules	Main Results
Silicon[1]	SiS SiC$_2$	$^{28}Si/^{29}Si$, $^{28}Si/^{30}Si$=solar
Sulfur[1]	SiS	$^{32}S/^{33}S$, $^{32}S/^{34}S$=solar
Carbon[1]	$^{13}C^{32}S/^{12}C\ ^{34}S$	$^{12}C/^{13}C$=44±3
Nitrogen[1]	$H^{13}C^{14}N/H^{12}C^{15}N$	$^{14}N/^{15}N$>4400
Oxygen[2]	CO (see section 7.4)	$^{17}O/^{18}O$=1.47±0.30
Chlorine[3]	ClNa, ClAl	$^{35}Cl/^{37}Cl$=solar

[1] Kahane et al.(1988) [2] Kahane et al.(1992) [3] Cernicharo and Guélin(1987)

For silicon and sulfur the measured ratios are essentially the same as in the Sun, within the experimental uncertainties ($\sim$10–20%). This is in agreement with most nucleosynthesis models. It is not surprising that the extreme conditions, required to affect these elements and bring the products to the surface of the stars (see Brown and Clayton 1992), are not met in an ordinary AGB star such as IRC+10216. It is also normal that the chlorine isotopes measured by Cernicharo and Guélin also display the solar value.

The ratio $^{12}C/^{13}C$ measured by Kahane et al. (1988) is more accurate than, and in reasonable agreement with previous determinations. Its value which is smaller than the solar (89) and local interstellar values (see e.g. Wilson and Matteucci 1992), is probably representative of average values of $^{12}C/^{13}C$ in AGB C–rich envelopes (section 7.3), and in visible AGB carbon stars (Lambert et al. 1986).

Although the J=1–0 line of $N^{12}C^{15}N$ has probably been detected, one can only infer a lower limit for $^{14}N/^{15}N$, because not only the $H^{12}C^{14}N$ line, but also the $H^{13}C^{14}N$ line, is optically thick. Such a value (>4400), which is much larger than the terrestrial (277) and interstellar ones, proves that most of the gas of the star has circulated through regions hot enough to destroy ^{15}N.

I postpone to section 7.4 the discussion of the oxygen isotopes.

7.3. ^{13}C in other objects

^{13}CO and $H^{13}CN$ are easy to detect with large radiotelescopes. ^{13}CO has been detected in tens of objects (many results are still unpublished), and could be detected in hundreds. However, very few accurate values of $^{12}C/^{13}C$ have been determined from millimeter lines because, as a rule, the lines of ^{12}CO and of $H^{12}CN$ are optically thick. The "double ratio" method has not yet been applied to any other star than IRC+10216.

From the observed values of the intensities of ^{12}CO and ^{13}CO, it appears that many stars are similar to IRC+10216. However, there is also direct proof that the range of variations of $^{12}C/^{13}C$ in circumstellar envelopes is comparable to that observed in visible carbon stars (Lambert et al. 1986) and in meteorite inclusions (see e.g. Stone et al. 1991, Zinner et al. 1989).

It is interesting to note that the largest values are found in two planetary nebulae, which are exceptionally young and strong CO emitters, where one can see the full effect of 3α burning. Indeed, because the lines of ^{12}CO are optically thick, only lower limits are derived for $^{12}C/^{13}C$: 65 in NGC 7027 (Kahane et al. 1992), 100 in IRAS 21282+5050 (Likkel et al. 1988).

On the other hand, millimeter observations have confirmed the very small value derived for $^{12}C/^{13}C$ by Lambert et al. (1986) from optical lines, for 3 carbon stars of J–type (Y CVn, RY Dra and WZ Cas). Both $^{13}CO/^{12}CO$ observations (Jura, Kahane, Omont 1988) and $^{13}CS/^{12}CS$ observations (Kahane 1989) show that $^{12}C/^{13}C$ is close to 3, and even possibly smaller.

7.4. Oxygen Isotopes

The J=1–0 and J=2–1 lines of $^{13}C^{16}O$, $^{12}C^{17}O$ and ^{12}C ^{18}O have been detected with a good signal–to–noise ratio by Kahane et al. (1992) in five massive C–rich envelopes (IRC+10216 (CW Leo), CIT6 (RW LMi), CRL 2688, CRL 618 and NGC 7027). They thus derived quite accurate values of $^{12}C^{17}O/^1C^{18}O$, and hence probably of $^{17}O/^{18}O$ (Table 3), since there is no reason to suspect different fractionation for $^{12}C^{17}O$ and $^{12}C^{18}O$.

Because of the saturation of $^{12}C^{16}O$ and possibly of $^{13}C^{16}O$, the values of $^{16}O/^{17}O$ (300–800) and $^{16}O/^{18}O$ (300–1600) are much more uncertain (to be compared with 2630 and 470, respectively, in the solar system).

The $^{17}O/^{18}O$ ratios measured by Kahane et al. (1992) (Table 3) are much larger than the solar and interstellar values but similar to those obtained from infrared observations in less opaque C–rich and O–rich envelopes by Harris, Lambert et al. (see references in Kahane et al. 1992). Since it is believed that such massive C–rich envelopes make an important contribution to the total mass–loss of AGB red giants, it is concluded by Kahane et al.(1992) that AGB red giants contribute little to the enrichment of the interstellar medium in ^{17}O.

Table 3. Isotopic ratio $^{17}O/^{18}O$ (Kahane et al. 1992)

NGC 7027	CRL 2688	CRL 618	IRC+10216	CIT6	Solar System
$1.58^{+0.36}_{-0.26}$	$1.12^{+0.15}_{-0.13}$	$1.66^{+0.83}_{-9.46}$	$1.47^{+0.33}_{-0.25}$	>1.1	0.18

Acknowledgments. I thank Mrs M.C. Pantalacci for typing and editing the manuscript and Mrs A. Placenti for drawing the figures.

References

Barnbaum C, Kastner J H and Zuckerman B 1991 *Astron. J.***102** 289

Barnbaum C, Morris M, Likkel L and Kastner J H 1991a *Astron. Astrophys.* **251** 79

Baron Y, de Muizon M, Papoular R and Pegourié B 1987 *Astron. Astrophys.* **186** 271

Brown L E and Clayton D D 1992 (these proceedings).

Bujarrabal V, Fuente A, Omont A, Alcolea J 1993 in preparation

Cernicharo J and Guélin M 1987 *Astron. Astrophys.* **183** L10

Chan S J and Kwok S 1990 *Astron. Astrophys.* **237** 354

de Jong T 1989 *Astron. Astrophys. Lett.* **223** L23

de Jong T 1990 *From Miras to Planetary Nebulae : Which Path for Stellar Evolution* M O Menessier and A Omont eds (editions Frontières Gif sur Yvette 1990) p 289

Glassgold A E and Huggins P J 1988 *The M–type Stars* H R Johnson H R and Querci F eds (NASA–CNRS 1988) p 291

Heske A, Forveille T, Omont A, van der Veen W E C J, Habing H J 1990 *Astron. Astrophys.* **239** 173

Iben I and Renzini A 1983 *Ann. Rev. Astron. Astrophys.* **21** 271

Iben I 1986 Quat. *J. roy. Astr. Soc.* **26** 1

IRAS Team, LRS Atlas 1986 *Low Resolution Spectra Atlas* Olnon F M and Raymond E *Astron. Astrophys. Supp.* **65** 607

Johnson H R and Zuckerman B eds 1989 *Evolution of Peculiar Red Giants* (Cambridge University Press)

Jura M, Kahane C, Omont A 1988 *Astron. Astrophys.* **201** 80

Jura M, Kleinmann S G 1989 *Ap. J.* **341** 359

Kahane C, Gomez–Gonzales J, Cernicharo J and Guélin M 1988 *Astron. Astrophys.* **190** 167

Kahane C 1989 Thesis Université Joseph Fourier Grenoble

Kahane C, Cernicharo J, Gomez–Gonzalez J and Guélin M 1992 *Astron. Astrophys.* **256** 235

Kastner J H 1990 Ph. D. Thesis UCLA

Kastner J H 1992 *Ap. J.* in press

Knapp G R, Morris M 1985 *Ap. J.* **292** 640

Lambert D L, Gustafsson B, Eriksson K and Hinkle K H 1986 *Ap. J. Supp.* **62** 273

Lefèvre J 1990 *The M–Type stars* Johnson H R and Querci F eds (NASA–CNRS 1988) p 225

Likkel L, Forveille T, Omont A and Morris M 1988 *Astron. Astrophys.* **198** L1

Loup C, Forveille T, Paul J F and Omont A 1992 *Astron. Astrophys. Supp.* in press

Lucas R, Omont A, Guilloteaus S and Nguyen-Q-Rien *Astron. Astrophys.* **154** L12

Lucas R *Astrochemistry of Cosmic Phenomena* 1992 IAU symposium n°150, Singh P D ed. (Kluwer) p 389

Menessier M O and Omont A eds *From Miras to Planetary Nebulae : Which Path for Stellar Evolution* (Editions Frontières Gif sur Yvette 1990)

Millar T J 1988 *Rate Coefficients in Astrochemistry* Millar T J and Williams D A eds (Kluwer Academic Pub. 1988) p 49

Morris M and Zuckerman B eds *Mass Loss in Red Giants* (Reidel, Dordrecht 1985)

Morris M 1975 *Ap. J.* **197** 603

Olnon F M, Baud B, Habing H J, de Jong T, Harris S, Pottasch S R 1984 *Ap. J.* **278** L41

Olofsson H. 1992 La Serena *Mass–Loss on the AGB and Beyond* Schwarz H E ed (ESO Workshops Pub.)

Omont A 1990 *Chemistry in Space* Greenberg M and Pironello V eds (Kluwer, Dordrecht)

Omont A 1991 *Late Stage of Stellar Evolution* de Loore C B ed (Springer–Verlag, Heidelberg)

Omont A, Loup C, Forveille T, te Lintel Hekkert P, Habing H., Sivagnanam P 1992 *Astron. Astrophys.* in press

Omont A 1992 Faraday symposium N°28 *Chemistry in the Interstellar Medium* Smith I ed

Schutte W and Tielens A G G M 1989 *Ap. J.* **343** 369

Schwarz H E ed 1992 La Serena *Mass–Loss on the AGB and Beyond* (ESO Workshops Pub.)

Seldmayr E 1990 *From Miras to Planetary Nebulae : Which Path for Stellar Evolution* Menessier M O and Omont A eds (Editions Frontières Gif sur Yvette 1990) p 179

Stone J, Hutcheon I D, Epstein S and Wasserburg G J 1991 *Earth Planet. Sci. Lett.* **107** 570

Tejero J, Cernicharo J, Omont A, Maillard J P 1993 in preparation

van der Veen W E C J and Habing H J 1988 *Astron. Astrophys.* **194** 125

Wannier P G , Linke R A 1978 *Ap. J.* **225** 130

Weideman V and Schönberner D 1990 *From Miras to Planetary Nebulae : Which Path for Stellar Evolution* Menessier M O and Omont A eds (Editions Frontières Gif sur Yvette)

Willems F J, de Jong T 1986 *Astron. Astrophys.* **309** L39

Willems F J, de Jong T 1988 *Astron. Astrophys.* **196** 173

Wilson T L and Matteucci F 1992 *Astron. Astrophys. Rev.* **4** 1

Zinner E, Tang Ming and Anders E 1989 *Geochimica et Cosmochimica Acta* **53** 3273

Zuckerman B and Madalena R J 1989 *Astron. Astrophys. Lett.* **223** L20

Chemical and isotopic composition of cosmic rays

P. Ferrando

Service d'Astrophysique, C.E. Saclay,
91191 Gif-sur-Yvette, Cedex,   France

Abstract. Important progress has been made these last years in deriving the composition of the Galactic cosmic rays at their source. This is largely due to a systematic effort made in measuring the fragmentation cross sections necessary to correct the observed cosmic ray abundances for propagation effects between their source and the Earth. This paper presents up to date chemical and isotopic abundances at the source of cosmic rays, taking advantage of the improved cross sections and of the latest cosmic ray measurements. Some of the models proposed for the cosmic ray sources are also discussed in relation with these observations.

1. Introduction

The origin of cosmic rays is still an enigma despite the wealth of data collected since their discovery [1] and that of heavy nuclei in this radiation [2, 3]. Cosmic rays loose the direction information about their sources when following the complicated structure of the Galactic magnetic field, so that their sources can be identified on composition features only. Cosmic rays suffer nuclear interactions in collisions with the atoms of the interstellar medium (ISM) when travelling from their source to the Earth, which results in profound modifications of their composition. Although measured with a good accuracy, the observed abundances are by no means the Galactic Cosmic Ray Source (GCRS) abundances. In order to derive these last ones, the "secondary" flux due to fragmentation of cosmic rays in the ISM must be substracted from the observation. One thus needs a reliable model to describe the cosmic ray propagation in the Galaxy, and an accurate set of the fragmentation cross sections involved at GeV/nucleon (GeV/n) energies. These last years have seen drastic improvements in the secondary flux computation mainly due to the measurements of numerous key cross sections, both in H [4] and He [5]. This in turn has resulted in more accurate GCRS abundances for isotopes (and elements) with a large secondary part.

This paper presents up to date chemical and isotopic source abundances taking into account the progress made in the propagation parameters and using the latest cosmic ray data available. Section 2 gives the usual frame of CR interpretation and presents the standard propagation model. Sections 3 and 4 review the source abundances in comparison with Solar System abundances. Finally, some of the models proposed for the GCRS are discussed in Section 5.

2. Frame of cosmic ray interpretation and propagation model

Although the cosmic ray energy spectrum extends up $\sim 10^{20}$ eV (e.g. [6]), restricted composition data are available up to $\sim 10^{15}$ eV [7, 8], and acccurate measurements up to ~ 50 GeV/n only [9, 10], an energy region where cosmic rays are of Galactic origin. From the composition data, the following five stages describing the cosmic rays life were inferred :

i) synthesis in the sources of the primary cosmic rays,

ii) extraction from their source to bring them into a state allowing acceleration,

iii) acceleration presumably by supernovae shocks (e.g.[11]),

iv) propagation in the Galaxy, giving rise to secondary cosmic rays,

v) modulation of the spectra by interaction of cosmic rays with the magnetized solar wind when they propagate in the heliosphere.

In this last step, cosmic rays are strongly decelerated in the expanding solar wind, loosing from ~ 200 to ~ 500 MeV/n in the heliosphere, depending on the solar activity. This implies that a cosmic ray measured near Earth with an energy of ~ 100 MeV/n had a much higher energy in interstellar space (e.g. [12]). Practically, no cosmic ray below ~ 500 MeV/n in the interstellar medium can reach our detectors, and the interstellar cosmic ray spectra and composition are completely unknown at these low energies.

This is the frame considered in this paper, including the so called "Leaky-Box" propagation model described below. However, the possibility that cosmic ray gain a *small* fraction of their energy during the propagation phase has been recently emphasized, and models including such a "reacceleration" are hotly debated (e.g. [13]). They are discussed by M. Giler in this volume. Also, the validity of the standard Leaky-Box model is still questioned at low energy [14, 15, 16]. Fortunately, this problem and reacceleration are not expected to have a significant effect on the determination of GCRS abundances as long as : i) pure secondaries are well accounted for by the propagation models, and ii) reacceleration is small, as believed.

The "Leaky-Box" model describing the propagation of cosmic rays in the Galaxy assumes that they fill a volume surrounding the Galaxy, with a small but finite probability of escaping this volume per unit time [17]. This implies an exponential distribution for the amount of matter encountered by cosmic rays, referred as the pathlength distribution and characterized by its average value, the "escape length" Λ_e. For stable nuclei, this model is equivalent to diffusion models, provided the storage volume is larger than the Galactic disk in which are supposed to be located the sources [e.g. 18]. In this simple (but successful) Leaky-Box model the flux J_i of a stable nuclei i is given by :

$$J_i \ (1/\Lambda_e + 1/\Lambda_d) + \text{energy loss} + \text{decay loss} = Q_i + \Sigma_j (J_j/\Lambda_{ji}) + \text{decay (j->i)}$$

LHS terms describe the losses of cosmic rays through escape (Λ_e), nuclear fragmentation (Λ_d), energy losses and decay (for unstable isotopes). RHS terms are the production of cosmic rays i, by the source (Q_i) and by fragmentation (Λ_{ji}) or decay of other cosmic rays. The terms Λ_d and Λ_{ji} are proportional to the fragmentation cross sections in the ISM. Applying this equation to pure secondary elements (Q_i=0) sets the Λ_e value which is then used to derive source abundance for other elements.

The energy dependence of Λ_e is determined from the B/C ratio, B being a pure secondary and C its main progenitor [14, 15, 16]. Its value depends on the composition of the ISM considered, H and He being the only components of importance [5]. When assuming an ISM composition of 90 % of H and 10 % of He (by number), and when considering the ionized fraction of H (important for energy losses at low energy), the B/C data require that [15] : $\Lambda_e = 35.7 \ \beta \ R^{-0.62}$ g/cm^2 for R > 4.35 GV, and $\Lambda_e = 14.3 \ \beta$ g/cm^2 for R < 4.35 GV, R being the rigidity of cosmic rays, and β their speed relative to the speed

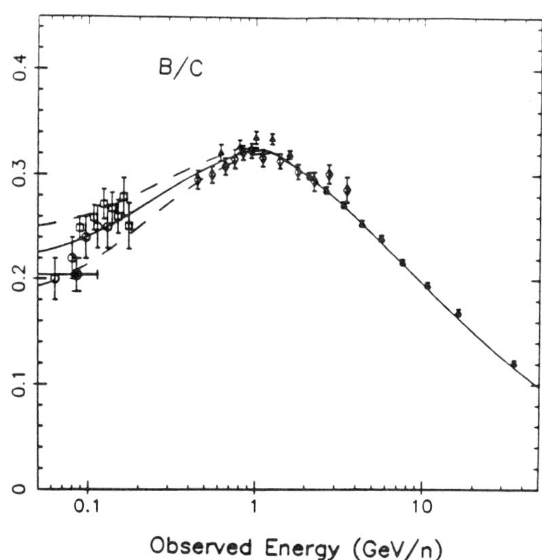

Fig.1: Selected observations of the B/C ratio as a function of energy (adapted from [16]. The curves are the predictions of the standard Leaky-Box model with the parameters given in the text. The scatter of measurements at low energy is due to different level of solar modulation at the time the measurements were made. The effect of an increasing level solar modulation is shown by the different curves at low energy, calculated from bottom to top for modulation parameters of 300, 450, and 600 MV.

of light. As shown in Fig.1, this fits the B/C data on the full energy range when the difference in solar modulation is taken into account [16].

The accuracy on Q_i depends on the accuracy of the computed secondary production, i.e. on the accuracy on the cross sections. Recently, an important systematic effort has been made in measuring the most crucial cross sections in H and He for cosmic ray propagation, and new parametric formulae were derived for predicting unmeasured cross sections [4, 5]. For some elements, the uncertainty due to unmeasured cross sections is reduced to a very small fraction [19], leading to a much better determination of the GCRS abundances.

Finally, the study of stable nuclei only allows to derive a "grammage" (g/cm^2), i.e. and integrated ISM density. The cosmic ray life-time is determined from ^{10}Be ($t_{1/2}$ = 1.6 Myr) and ^{26}Al ($t_{1/2}$ = 0.9 Myr) measurements, and is found to be ~ 10^7-10^8 years, depending on the propagation model adopted (Leaky-Box or diffusion models [18, 20, 21]). In any case the cosmic ray material is much younger than the Solar System material.

3. Chemical abundances at the source of cosmic rays

First, the observed cosmic ray spectra for elements heavier than He can be fitted with the assumption of a single source spectral shape, a power law in momentum with an index of 2.23 ± 0.05 [9], although small deviations may be present for some elements like Ne [19].This unique source spectral shape allows the determination of a GCRS composition independent of energy. In table 1 are given the GCRS abundances for elements from C to Ni derived in [9] from the accurate HEAO3-C2 data covering 0.6 to 35 GeV/n, and with the use of the propagation model and cross sections described above.

In table I are also indicated the chemical abundances for N and Ca derived from isotopic measurements [22, 23, 24]. For these elements, it is believed that only one isotope has a significant GCRS abundance (^{14}N and ^{40}Ca). Krombel and Wiedenbeck [23] estimate that 60 % of Ca is secondary, while the secondary part of ^{40}Ca alone is 4 % only, so that the GCRS abundance derived from the isotopic measurement is in principle more accurate than

<u>Table 1</u> : Galactic Cosmic Ray Source and Solar System abundances

Element	GCRS	SS	GCRS/SS
C	424.9 ± 12.1	1023 ± 99	0.415 ± 0.42
N	25.4 ± 8.8	316 ± 31	0.080 ± 0.029
^{14}N	20.0 ± 5.3	315 ± 31	0.063 ± 0.018
O	526.3 ± 5.6	2399 ± 201	0.219 ± 0.019
Ne	58.0 ± 3.4	347 ± 90	0.167 ± 0.044
^{20}Ne	40.0 ± 2.3	323 ± 84	0.124 ± 0.033
Na	3.23 ± 1.26	5.75 ± 0.41	0.56 ± 0.22
Mg	103.8 ± 2.6	107 ± 5	0.970 ± 0.051
Al	7.78 ± 1.46	8.5 ± 0.4	0.915 ± 0.177
Si	100.0 ± 1.30	100 ± 5	1.000 ± 0.052
P	0.77 ± 0.33	1.05 ± 0.10	0.73 ± 0.32
S	13.1 ± 0.9	52.5 ± 6.4	0.250 ± 0.035
Ar	2.23 ± 0.63	10.2 ± 0.6	0.219 ± 0.083
K	0.40 ± 0.34	0.38 ± 0.03	1.05 ± 0.90
Ca	6.01 ± 0.93	6.17 ± 0.44	0.97 ± 0.17
^{40}Ca	7.06 ± 1.7	6.1 ± 0.3	1.14 ± 0.29
Fe	100.8 ± 1.9	91.2 ± 2.1	1.105 ± 0.033
Co	0.19 ± 0.06	0.23 ± 0.02	0.83 ± 0.27
Ni	5.68 ± 0.22	5.0 ± 0.24	1.13 ± 0.07
Ge	$90\ (+12,-8)\ 10^{-4}$	$120 \pm 12\ 10^{-4}$	$0.75\ (+0.13,-0.10)$
Pb	$1.6 \pm 0.4\ 10^{-4}$	$3.16 \pm 0.23\ 10^{-4}$	0.51 ± 0.13

chemical measurement (for equal statistics !). These isotopic results are in agreement with the chemical result of Engelmann *et al* [9].

Gupta and Webber [24] found that $(N/O)_{GCRS} = 0.038 \pm 0.010$ fits both isotopic and chemical results. This apparent discrepancy between their determination and that of Engelmann *et al* [9] is due to their use of non final results for HEAO3-C2 (see discussion in [9]). Although apparently higher by ~ 1 %, the $(^{14}N/O)_{GCRS}$ derived from the elements (4.8 ± 1.7 %) is consistent within errors with the "isotopic" value (3.8 ± 1 %).

It has been usual to compare GCRS abundance with our best reference, the Solar System (SS) abundances. The photospheric and meteoritic abundances have been compiled by Anders and Grevesse [25]. While for most of the elements both determination agree, some important discrepancies were noted. Fe is one example with a photospheric abundance 1.45 times larger than the meteoritic one. However, the Fe photospheric abundance was recently revised [26, 27] and found in excellent agreement with the meteoritic value. Important controversy also exist for Ge and Pb (less abundant by factors 1.66 and 1.58 in the solar photosphere than in the meteorites). In view of the Fe example, we postulate that the SS abundances to compare with the GCRS are the meteoritic ones. They are given in Table 1, with the computed GCRS/SS ratios normalised to Si.

As suggested originally by Cassé and Goret [28] the GCRS abundances show a marked correlation with the SS abundances when ordered according to the First Ionisation Potential (FIP) of the elements, as in Fig.2. The elements with a low FIP (below ~ 9 eV) have relative abundances close to SS, while elements with a high FIP are depleted. This can

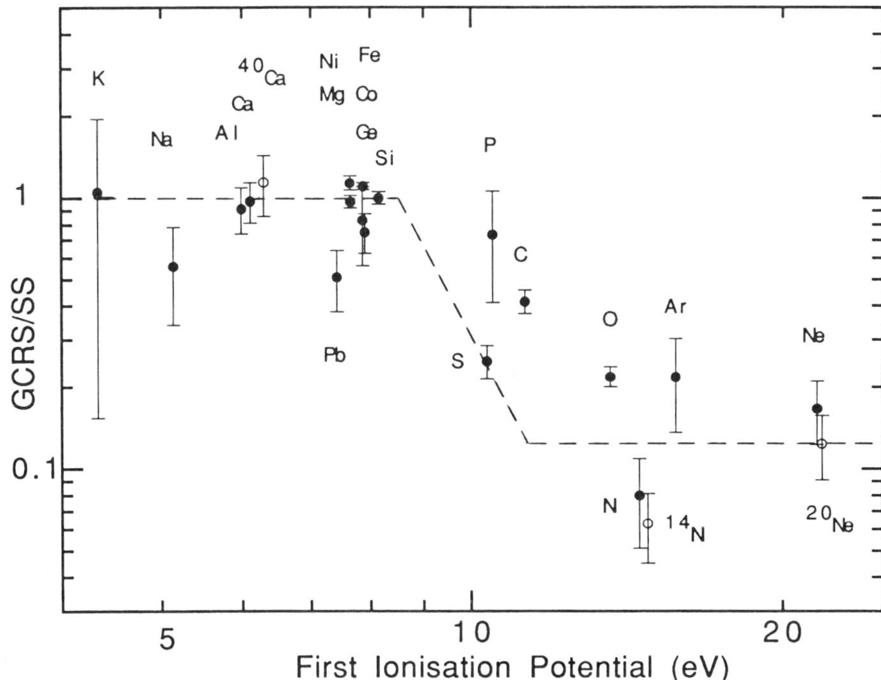

Fig.2 : GCRS/SS ratio. Data are from Table 1. The FIP fractionation dashed line is adjusted on ^{20}Ne.

be interpreted as a fractionation mechanism operating in a thermal plasma of about 10^4 K [28, 29]. The same pattern is found for Ultra Heavy nuclei [30] with the remarkable exceptions of Ge and Pb, also given in Table 1, and of the Pt group enhanced by a factor of about 2.

4. Isotopic measurements

Isotopes, rather than elements, are the genuine tracers of cosmic ray history, since almost all processes suffered by cosmic rays are nuclear (nucleosynthesis, decay, fragmentation during propagation). Moreover, the chemical fractionation noted above is not expected to play any role in the isotopic fractions of the same element. Unfortunately, isotopic measurements by a direct method have been limited to ~ 100-500 MeV/n. Results at higher energy exist for the mean mass of the elements [31, 32] but it has not yet been possible to derive any significant isotope spectrum. Mewaldt [33] gave a comprehensive review of the isotopic ratios in the Galactic cosmic rays, which I will not repeat here. I rather concentrate on new results relevant to cosmic ray source composition, obtained thanks to the improvement in cross sections and new measurements. The summary of the GCRS isotopic ratios is given in Fig.7 updated from [33] and [34]. The updates are discussed in the following.

4.1 Isotopic ratios at the source of cosmic rays

4.1.1 C isotopes. The large secondary part in the observed ^{13}C ($\sim$ 90 %) has hampered the determination of its source abundance as long as the propagation uncertainty was estimated to be $\sim$ 30 % [35]. At low energy, where ^{13}C/^{12}C is measured, the propagation uncertainty arises from cross sections, Λ_e and modulation. Webber and Soutoul [35] noted that although the uncertainties on Λ_e and modulation can be large, the uncertainty on their combination is of the same order than the uncertainty on the B/C ratio itself. Taking advantage of this, and using the new set of cross sections, they showed that the secondary fraction of ^{13}C is determined within $\sim$ 8 % and they derived a (^{13}C/^{12}C)$_{GCRS}$ ratio of 0.3 ± 0.5 % (adopted in Fig.7), 1.4 sigma lower than the SS value of 1.1 %. Since their work, additional data [36, 37] also suggest a low (^{13}C/^{12}C)$_{GCRS}$ ratio.

4.1.2 O isotopes. The reduced uncertainty on cross sections has permitted the determination of a significant ^{18}O GCRS abundance. From an average of experiments, Gibner *et al* [36] reported a (^{18}O/^{16}O)$_{GCRS}$ ratio enhanced by a factor 3.75 ± 1.2 compared to SS, the error including the estimated propagation uncertainty. This factor is adopted in Fig.7. An enhancement of ^{18}O was also found by Lukasiak *et al* [37] measurement at 22 AU with a somewhat smaller factor. Fig.3 displays resolution selected measurements with the propagation expectation for a (^{18}O/^{16}O)$_{GCRS}$ ratio of 0.2 % (solar) and 0.75 %. Gibner *et al* [36] point out that $\sim$ 50 % of the production cross sections of ^{18}O are measured, and that a systematic increase by 34 % of all secondary production cross sections is required in order to account for the data with a solar GCRS isotopic ratio. This seems very large compared to the expected accuracy of the parametric cross section formulae. The ^{18}O secondary fraction is however very important as can be seen in Fig.3, so that confirming some of calculated cross sections by a measurement would strengthen the isotopic anomaly of ^{18}O.

4.1.3 Ne, Mg, and Si isotopes. Webber *et al* [19] reexamined the Ne, Mg, and Si isotopes with the new set of cross sections. Compared to the review of Mewaldt [33], important changes were found in the GCRS isotopic ratios. The largest one is for Si for which the data are consistent with a solar isotopic composition. The secondary production of neutron rich Si isotopes was previously largely underestimated.

They confirm a large enhancement of ^{22}Ne in the GCRS, with a value of 0.46 ± 0.06 for the (^{22}Ne/^{20}Ne)$_{GCRS}$ ratio. The SS value is not well defined, an depends on the standard adopted [25]. The lowest value (0.0730 ± 0.0015) is found in the present Solar Wind, the largest ($0.132 + 0.032 - 0.025$) in high-energy Solar Energetic Particles [38]. These two extremes are considered in Fig.7.

For Mg, Webber *et al* found enhancements of the ^{25}Mg/^{24}Mg and ^{26}Mg/^{24}Mg ratios in the GCRS compared to the SS, but with factors smaller than estimated before (their value is adopted in Fig.7). The recent *Voyager* data [37] do not confirm this. Indeed, the scatter of the measurements is large even for experiments claiming a good isotopic resolution as selected in Fig.4. The question of the Mg isotopic anomaly might still be opened.

4.1.4 Fe group isotopes Since the review of Mewaldt new measurements have been presented on Fe, Co, and Ni isotopes [39, 40, 41]. The best combination of statistic and mass resolution is achieved by Leske *et al* [39], whose results are discussed below.

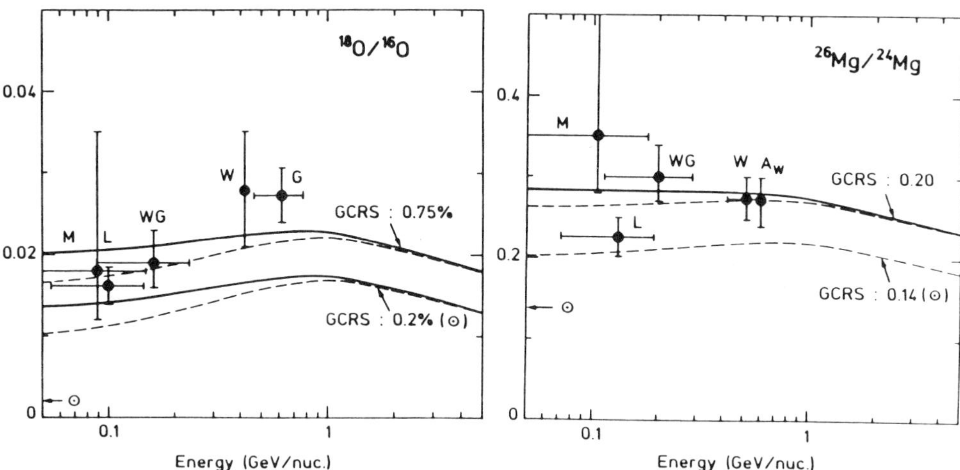

Fig.3 and 4 : Isotopic measurements of $^{18}O/^{16}O$ and $^{26}Mg/^{24}Mg$. For O, the data are from [63] (M), [37] (L), [60] (WG), [61] (W), [36] (G). For Mg, the data are from [59] (M), [37] (L), [60] (WG), and [61] (W), other results being included in the average A_W from [62]. The solar ratio is shown by the arrow on the y axis. The lines are propagation calculations assuming GCRS ratios as indicated. Dashed lines are for 300 MV of modulation, full lines for 600 MV.

Their $(^{54}Fe/^{56}Fe)_{GCRS}$ ratio is 0.73 ± 0.32 times the SS ratio, and represents the most accurate determination so far. Their error bar includes the uncertainty on ^{54}Mn decay (with an unknown period). The weigthed average of this measurement with the previously adopted average $(1.15 \pm 0.5, [33])$ results in a $(^{54}Fe/^{56}Fe)_{GCRS/SS} = 0.85 \pm 0.27$, and is adopted in Fig.7. Other isotopic ratios of Fe have still only upper limits.

For Co, they found a significant ^{59}Co abundance in the cosmic rays, implying that ^{59}Ni had time to decay by electron capture before acceleration of cosmic rays [42]. This is in agreement with the finite source abundance for the Co element [9]. The time elapsed between ^{59}Ni synthesis and acceleration is estimated to be larger than $\sim 10^5$ years [39].

For Ni, they report the first definitive enhancement of ^{60}Ni in the GCRS, with the $^{60}Ni/^{58}Ni$ ratio being 2.8 ± 1.0 times the SS value. Combined with the previously adopted average [33], $(^{60}Ni/^{58}Ni)_{GCRS}$ is found enhanced by a factor 2.5 ± 0.8 when compared to the SS value. This last factor is adopted in Fig.7.

4.2 A new determination of the GCRS isotopic abundances. The Nitrogen problem.

The isotopic ratios are measured in an energy range where the solar modulation has an important effect (e.g. Fig.1, 3, and 4), making uneasy the comparison of different data, and adding a non negligible modulation uncertainty [16, 37]. Since the modulation is depending on the A/Z ratio and on the interstellar spectral shape, Soutoul and Webber [43] remark that the abundance ratio of isotopes with the same A/Z value and largely secondary (to have the same spectrum) will not depend on the modulation level. Moreover, it will also be largely independent of the exact value of Λ_e since the secondary flux of each isotope is proportional to Λ_e. The Λ_e and modulation uncertainties are thus greatly minimized by their approach.

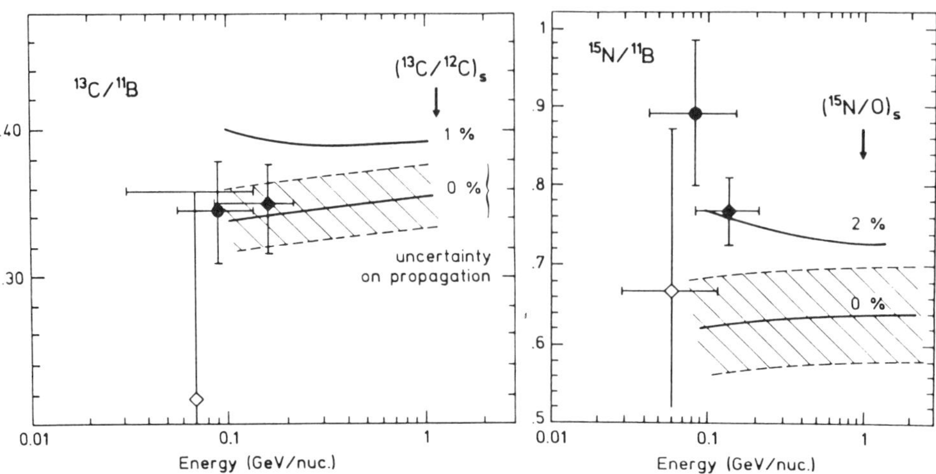

Fig.5 and 6 : Figures from [43] showing observations of $^{13}C/^{11}B$ and $^{15}N/^{11}B$ by a single experiment. Data points are from [22], [63], [22] combined with [64], [16] combined with [37]. The curves show propagation predictions for the indicated GCRS values.

They studied the $^{13}C/^{11}B$, $^{14}N/^{10}B$, $^{15}N/^{11}B$, and $^{18}O/^{11}B$ ratios, selecting published data from a single experiment. This presumably minimize systematic biases that could occur when mixing results from different experiments, and ensures that the modulation level is indeed the same for both isotopes. We show in Fig.5 and 6 two examples of what they obtained. On these figures are indicated the GCRS abundance corresponding to the propagation lines. The hatched area is an estimated uncertainty of 10 % (due to cross sections only) on the calculated ratio. The low $^{13}C/^{12}C$ ratio noted in § 4.1.1 is recovered in Fig.5.

Their result for $^{15}N/^{16}O$ (Fig.6) is quite interesting since it shows for the first time that a $(^{15}N/^{16}O)_{GCRS}$ ratio of 2 % ($\pm$ 1% roughly) is required to account for the data. This value was already suggested in previous works [22, 24], but a zero source abundance of ^{15}N was within the uncertainties which included the modulation and Λ_e uncertainties. This ^{15}N abundance in the GCRS would also bring in closer agreement the low and high energy determination of the $(N/O)_{GCRS}$ (respectively based on $^{14}N/O$ and N/O measurements). However this implies a $(^{15}N/^{14}N)_{GCRS}$ ratio 30 to 250 times larger than in the SS, and other hypothesis have to be investigated before claiming such an anomaly. If cross sections were in error, this should be such that the $^{15}N/^{11}B$ ratio is augmented by ~ 20 %. The fact that the $^{11}B/B$ and B/C ratios are correctly predicted by the Leaky-Box [22] rules out a large error on ^{11}B production cross sections. Thus, ^{15}N production cross sections should be in systematic error by ~ 20 %, which is much larger than believed [24]. The only possibility left is to incriminate the propagation model, but no viable alternative solving the ^{15}N problem has yet been found. In particular, reacceleration of cosmic rays during propagation [44] has been ruled out [45]. The situation is thus very much puzzling, and may be the sign that the understanding of propagation is not yet correct at low energy.

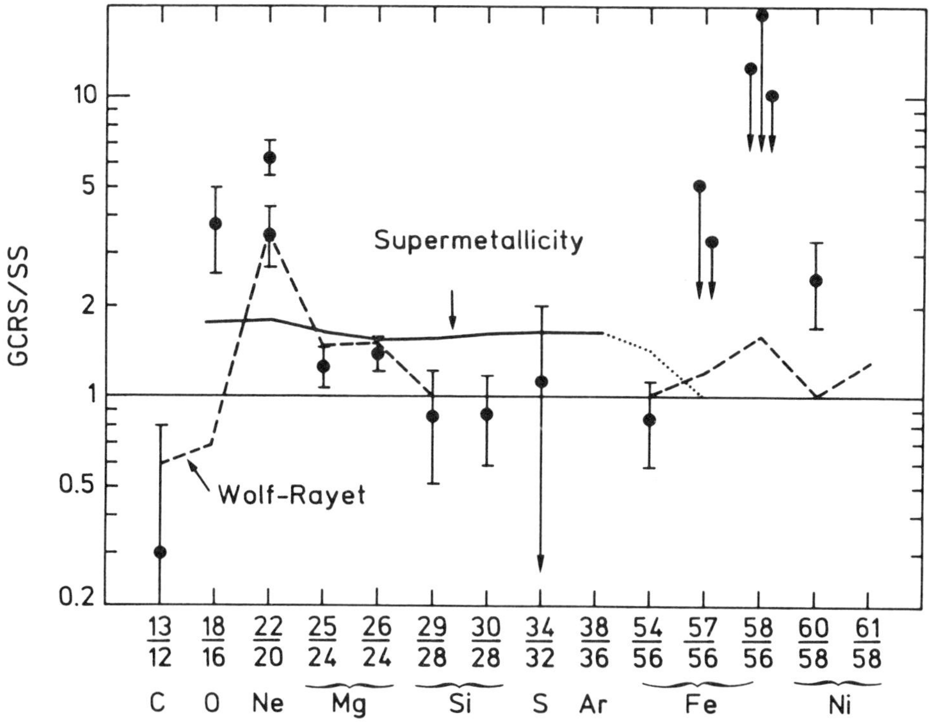

Fig.7 : Summary of the GCRS isotopic ratios. Both Solar Wind and Solar Energetic Particles data are considered for SS Neon. The lines are from the models discussed in Sect. 5.

5. Cosmic ray sources

The composition data place severe constraints on the nature of the GCRS. In the following we review some of the models suggested for the GCRS, and their problems.

The first pattern of the GCRS abundances is the strong similarity with the SS abundances for elements and isotopes. This excludes any specific cycle of nucleosynthesis at the originic of the bulk of cosmic rays (e.g. [46]). As far as the elements are concerned, Meyer [29, 46] stressed that, as in the GCRS, a depletion of high FIP elements is found in Solar Energetics Particles (SEP). He thus strongly suggested that Galactic cosmic rays are extracted from stars with chromospheric temperatures similar to that of the sun, and then accelerated with a mechanism with no charge bias. This scenario cannot explain the isotopic anomalies as well as the excess of C and O when referred to similar FIP elements. They require the addition of a specific source greatly enriched in ^{22}Ne, C and O so that a mixing of a small amount of matter from this source with SS like material keeps the overall similarity of GCRS with SS abundances. The best candidates found for this specific source are the Wolf-Rayet stars [47], and detailed calculations [48] showed that mixing ~ 98 % of SS material with ~ 2 % of the material expelled by these peculiar stars would quantitatively account for the isotopic anomalies of ^{22}Ne and ^{25,26}Mg as well as for the overabundances of C and O. The isotopic anomalies expected in this model are shown by the dashed line in Fig.7, and indeed the recent determinations of C and Si isotopic ratios are in striking agreement with the predictions. However, the new GCRS chemical abundances presented in Fig.2 have cast some doubt on this model, the major problem being the low GCRS abundances of ^{14}N, Na, Ge and Pb. If an excess, such as ^{22}Ne, can be explained by a small contribution of a very peculiar source, there is no room in this model for any underabundance with respect to the "FIP" reference level set by ^{20}Ne (dashed line in Fig.2). H and He, which have a low abundance in the GCRS, are also not explained by this model and are considered to have a different story than heavier cosmic rays. In order to cure these problems, Silberberg and Tsao [49] proposed to add a third ingredient, a rigidity dependent depletion of light nuclei occuring at the boundary of the stellospheres. They account rather well the GCRS data, including H and He, but the physical process responsible for the depletion is still to find.

Another possibility which was suggested is that cosmic ray material originate from fresh supernovae (SN) ejecta (e.g. [50]). The overabundance of r process material in the GCRS and the slight excess of Fe and Ni seems in favor of this. Isotopic anomalies of O, Ne, Mg, and Si isotopes can be obtained if cosmic rays originate from inner regions of the Galaxy where the metallicity is enhanced [51]. But similar isotopic enhancements are predicted for O, Ne, Mg, and Si (full line in Fig.7) contrary to observation. If metallicity is considered to be solar, the strong isotopic anomaly of ^{22}Ne has to be explained by an extra source made of He-burning zones material. This is proposed by Prantzos [52] who suggests a mixing of 80 % "SN" with 20 % "Wolf-Rayet" material, the SN material being "FIP" depleted, a major assumption of his model. However, the SN hypothesis has fundamental difficulties. First, as stressed several times [e.g. 29, 53], the overall similarity of SS and GCRS abundances requires a mixing of type I and II SN's in *exactly* the same proportions than the one which gave birth to the SS material, a rather strong constraint. Second, and most important, the delay of more than 10^5 years inferred from ^{59}Co between synthesis and acceleration completely rules out the acceleration of fresh SN material only, which has to be mixed with interstellar matter.

In the above models, the high FIP depletion in the GCRS is considered as a key feature to explain. For most of the elements the FIP is correlated with the volatility, low FIP elements being generally refractory, and high FIP ones volatile [54]. Only a few elements do not obey this rule, Na, Ge, and Pb being the clearest ones [54]. These three elements are underabundant (Fig.2) compared to other low FIP elements. Although the significance of the underabundance for each of them can be argued in terms of propagation uncertainty, the fact that all of them (and only them) are underabundant is rather striking. This seems to indicate that volatility rather than FIP is the organising parameter of the GCRS abundances. Since refractories, locked in interstellar grains, are enhanced in the GCRS compared to volatile elements, cosmic rays could be interstellar grain destruction products as suggested by Cesarsky and Bibring [55]. This would explain the depletion of volatile elements and the low abundance of H and He in the GCRS. As noted before (e.g. [46]), this attracting scenario cannot account for the "normal" noble gases abundances (in particular Ne/O). Interstellar grains cannot be the unique source of cosmic rays.

Another suggestion is that cosmic rays are accelerated directly out of the hot and diffuse phase of the ISM by SN shocks. In that case, particles with a high A/Z* ratio are favoured by the acceleration process, Z* being the effective charge depending of the temperature of the plasma (e.g. [56]). This explains the underabundance of H relative to heavy ions and the trend for the other elements. This is also favored by the high ^{59}Co GCRS abundance. Moreover, detailed computation of spectra and abundances of solar wind particles accelerated by the Earth bow shock are in excellent agreement with the observations [57], giving a strong observational support to this model. But it predicts that the GCRS abundances are a smooth function of A/Z^* contrary to observations (e.g. [58]). Clearly, to be viable, this hypothesis must be accompanied by the adjunction of some other material. It remains to see if mixing of accelerated ISM material with material from another reasonable source can account for the GCRS data.

6. Conclusion

None of the cosmic ray source models proposed up to now is satisfactory. Each of them has a serious problem either in its ability of fitting all the GCRS composition data, or in the more or less ad-hoc physical assumptions made (rigidity or FIP depletion for the models based on SN ejecta). Since the development of these models, new constraints have appeared. The GCRS isotopic ratios have been strongly revised for C, Mg, and Si, and new isotopic anomalies have been found for O and Ni. An excess of the Pt group was also reported, and the ^{59}Co GCRS abundance has set a long time delay between synthesis and acceleration. The previously admitted FIP fractionation is questioned, volatility seeming to better organise the chemical data.

In all the models discussed above, the ^{22}Ne excess is explained by the contribution of Wolf-Rayet stars to the source of cosmic rays. This ingredient, identified by ^{22}Ne only, is likely to stay in the future models. This examplifies the strong potential of the isotopic compared to chemical measurements. New accurate isotopic data from satellite experiments should be available in the coming years [34], greatly improving the knowledge of the GCRS abundances. Cosmic ray sources may then be identified, and it may well turn out that they are not of a unique type, but are rather a complicated mixture of astrophysical objects.

Acknowledgements : It is a pleasure to thank Dr. Franz Käppeler whose kind hospitality made this work possible. I am greatly indebted to Aimé Soutoul for essential discussions and for providing results and figures before publication. I also thank C. Cesarsky, N. Prantzos, and M. Cassé for fruitful discussions.

References

[1] Hess V S 1912 Physik. Z. **13** 1084
[2] Freier P *et al* 1948 Phys. Rev. **74** 1818
[3] Bradt H and Peters B 1948 Phys. Rev. **74**, 1828
[4] Webber W R *et al* 1990 Phys. Rev. C **41** 520, 533, 547, 566
[5] Ferrando P *et al* 1988 Phys. Rev. C **37** 1490
[6] Lloyd-Evans J 1991 Proc. 22nd Int. Cosmic Ray Conf. (Dublin) **5** 215
[7] Burnett T H *et al* 1990 Astrophys. J. **349** L25
[8] Müller D *et al* 1991 Astrophys. J. **374** 356
[9] Engelmann J J *et al* 1990 Astron. Astrophys. **233** 96
[10] Binns W R *et al* 1988 Astrophys. J. **324** 1106
[11] Blandford R D and Ostriker J P 1980 Astrophys. J. **237** 793
[12] Goldstein M L *et al* 1970 Phys. Rev. Letters **25** 832
[13] Cesarsky C J 1987 Proc. 20th Int. Cosmic Ray Conf. (Moscow) **8** 87
[14] Garcia-Munoz M *et al* 1987 Astrophys. J. Sup. **64** 269
[15] Soutoul A *et al* 1990 Proc. 21st Int. Cosmic Ray Conf. (Adelaide) **3** 337
[16] Ferrando P et al 1991 Astron. Astrophys. **247** 163
[17] Cowsik R *et al* 1967 Phys. Rev. **158**, 1238
[18] Berezinsky *et al* 1990 "Astrophysics of cosmic rays" (North Holland) p 72

[19] Webber W R *et al* 1990 Astrophys J **348** 611
[20] Simpson J A and Garcia-Munoz M 1988 Space Sci. Rev. **46** 205
[21] Ferrando P et al 1991 Proc. 22nd Int. Cosmic Ray Conf. (Dublin) **1** 588
[22] Krombel K E and Wiedenbeck M E 1988 Astrophys. J. **328** 940
[23] Krombel K E and Wiedenbeck M E 1985 Proc. 19th Int. Cosmic Ray Conf. (La Jolla) **2** 92
[24] Gupta M and Webber W R 1989 Astrophys. J. **340** 1124
[25] Anders E and Grevesse N 1989 Geochimica and Cosmochimica Acta **53** 197
[26] Holweger H *et al* 1991 Astron. Astrophys. **249** 545
[27] Biémont E *et al* 1991 Astron. Astrophys. **249** 539
[28] Cassé M and Goret P 1978 Astrophys. J. **221** 703
[29] Meyer J P 1985 Astrophys. J. Sup. **57** 173
[30] Binns W R *et al* Astrophys. J. **346** 997
[31] Dwyer R and Meyer P 1979 Proc. 16th Int. Cosmic Ray Conf. (Kyoto) **12** 97
[32] Ferrando P *et al* 1988 Astron. Astrophys. **193** 69
[33] Mewaldt R A 1989 "Cosmic Abundances of Matter", AIP Conf. Proc. N°183 (New York: AIP) p 124
[34] Wiedenbeck M E 1990 Proc. 21st Int. Cosmic Ray Conf. (Adelaide) **11** 57
[35] Webber W R and Soutoul A 1989 Astron. Astrophys. **215** 128
[36] Gibner P S *et al* 1992 Astrophys. J. **391** L89
[37] Lukasiak A *et al* 1991 Proc. 22nd Int. Cosmic Ray Conf. (Dublin) **1** 584
[38] Mewaldt R A and Stone E C Proc. 20th Int. Cosmic Ray Conf. (Moscow) **3** 255
[39] Leske R A *et al* 1992 Astrophys. J. **390** L99
[40] Grove J E *et al* 1990 Proc. 21st Int. Cosmic Ray Conf. (Adelaide) **3** 53
[41] Hesse A *et al* 1991 Proc. 22nd Int. Cosmic Ray Conf. (Dublin) **1** 596
[42] Soutoul A *et al* 1978 Astrophys. J. **219** 753
[43] Soutoul A and Webber W R 1992 in preparation
[44] Letaw J R 1987 Astrophys. J. **317** L69
[45] Guzik T G *et al* 1987 Proc. 20th Int. Cosmic Ray Conf. (Moscow) **9** 226
[46] Meyer J P 1986, "Advances in nuclear astrophysics" (Editions Frontières) p 393
[47] Cassé M and Paul J 1982 Astrophys. J. **258** 860
[48] Prantzos N *et al* 1985 Proc. 19th Int. Cosmic Ray Conf. (La Jolla) **3** 167
[49] Silberberg R and Tsao C H 1990 Astrophys. J. **352** L49
[50] Soutoul A *et al* 1991 Proc. 22nd Int. Cosmic Ray Conf. (Dublin) **2** 408
[51] Woosley S E and Weaver T A 1981 Astrophys. J. **243** 651
[52] Prantzos N 1992, this symposium
[53] Cesarsky C J and Soutoul A 1992 "Origin and Evolution of the elements" (Cambridge) in preparation
[54] Meyer J P 1981 Proc. 17th Int. Cosmic Ray Conf. (Paris) **2** 281
[55] Cesarsky C J and Bibring J P 1980 IAU Symp. 94 "Origin of Cosmic Rays" (Dordrecht: Reidel) p 361
[56] Ellison D C *et al* 1981 J. Geophys. **50** 110
[57] Ellison D C *et al* 1990 Astrophys. J. **352** 376
[58] Cassé M 1983 "Composition and origin of cosmic rays" (Reidel) p 193
[59] Mewaldt R A *et al* 1980 Astrophys. J. **235** L95
[60] Wiedenbeck M E and Greiner D E 1981 Phys. Rev. Letters **46** 682
[61] Webber W R *et al* 1985 Proc. 19th Int. Cosmic Ray Conf. (La Jolla) **2** 88
[62] Wiedenbeck M E 1984 Adv. Space Res. **4** 15
[63] Mewaldt R A *et al* 1981 Astrophys. J. **251** L27
[64] Wiedenbeck M E and Greiner D E 1981 Astrophys. J. **247** L119

Elemental and isotopic composition of the cosmic radiation measured with the balloon experiment ALICE

A. Hesse, B.S. Acharya*, U. Heinbach, W. Heinrich,
M. Henkel, B. Luzietti, C. Koch, M. Simon

University of Siegen, Physics Department,
Adolf Reichwein Str., 5900 Siegen, F.R.Germany,
*Tata Institute, Bombay, India

V.K. Balasubrahmanyan, L.M. Barbier, E.R. Christian,
J.A. Esposito** , J.F. Ormes, R.E. Streitmatter

Goddard Space Flight Center, NASA, Greenbelt, MD 20771, USA
**Bartol Research Institute, DE 19716, USA

Abstract. The ALICE balloon experiment utilizes the Cherenkov-range technique to determine the isotopic composition of the cosmic radiation in the elemental range magnesium through iron. The instrument operates in the energy range 350 MeV/nucleon to 800 MeV/nucleon. Particle charge and energy are determined from scintillation and Cherenkov radiation emissions within light diffusion chambers. Trajectories of incoming particles are determined by means of large area gas drift chambers. The stopping range of the particles is measured in a cellulose nitrate range stack. This paper briefly reports the experimental technique and the elemental and isotopic composition results of the balloon flight .

1. Introduction

Studies of the elemental and isotopic abundances of any kind of matter in the universe are of fundamental astrophysical importance to understand the nucleo-synthesis processes and the evolution of elements in our galaxy particularly. Cosmic rays are a unique probe of freshly synthesized matter (age 10^7 a). They have gone through acceleration processes and encountered a variety of different interactions during propagation through the interstellar medium. Additionally the studies of cosmic ray species provide reliable information about the structure and strength of the galactic magnetic fields, the density of the interstellar gas and acceleration conditions.

ALICE (A Large Isotopic Composition Experiment) was launched 1987 from Prince Albert, Canada and collected about 800 000 nuclei during 14 h at an altitude of 38500 m.

2. Instrument description

2.1. Experimental concept

The configuration of the Alice instrument is depicted in Fig. 1. It has a geometry factor of 0.88 m^{2}sr and a total mass of roughly 2000 kg. The Alice instrument measures the charge, the velocity, the range and the trajectory of the incoming particle by means of: two scintillators S1 and S2 , one Cherenkov counter C1, a cellulose nitrate range stack and a total of 12 drift chambers separated in two packages D1 and D2 (Esposito et al., 1985). The Cherenkov range technique is used for isotope analysis. It is an experimental goal to correlate the stopping end of a particle track in the passive range stack with the response of the scintillation and Cherenkov counters. This is achieved by the use of drift chambers which measure the trajectory of incident particles. Below the range stack another scintillation counter detects high energy particles penetrating the instrument. Additionally this detector provides a veto to identify events in which single charged particles with a high penetration power were produced inside the experimental setup by nuclear collision.

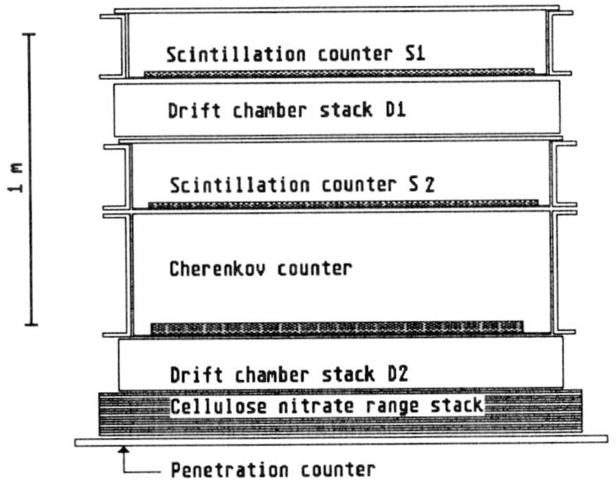

Fig. 1. Configuration of the ALICE detector setup

2.2 Scintillation counters

The two scintillation counters (1 cm of organic scintillation material P 10) were designed as diffusive light boxes with 16 Photomultiplier tubes attached to the sides (Esposito et al., 1985). Four additional fast photomultipliers in each chamber provide a fast trigger for measurement timing. Each phototube has an individual readout. For charge analysis the average signal is evaluated after a number of corrections: 1. time variation in gain, 2. saturation effects, 3. mapping of light collection, 4. angle correction and 5. fragmentation effects. The drift chamber information is used for angle and light collection efficiency correction. Fig. 2 shows the cosmic ray charge histogram measured with the scintillation counters. the charge resolution is $\sigma = 0.11$ Z for oxygen and $\sigma = 0.16$ Z for iron respectively.

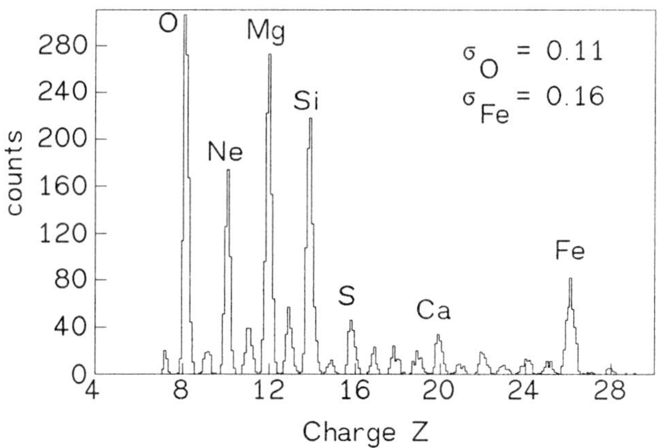

Fig. 2 . Cosmic ray charge histogram

2.3 Drift chambers

The Alice instrument utilizes two drift chamber stacks which consist of 6 individual drift chamber layers each. The individual drift chambers have a size of 135 cm by 135 cm with a gap of 2cm. The maximum drift path is 10 cm (Simon et al., 1984). The chambers are filled with an argon/ methan (90/10) mixture. The large area drift chambers provided well defined trajectories through the instrument. The measured spatial resolution was typically 250 microns for the lighter elements (O, Ne, Mg, Si). It decreased slowly with charge to 600 microns for iron.

2.4. Range stack

The range of incoming cosmic ray particles is measured in the cellulose nitrate (CN) stack, wich consists of two stacks of 425 foils, each 250 microns thick, giving a combined sensitive area of 1.44 m^2. This results in a column density of 14.85 g/cm^2 which defines the energy range for the instrument. The energy intervals analysed are 350 - 570 MeV/n for silicon and 420 -800 MeV/n for iron at the top of the Cherenkov radiator. When the energy loss due to ionisation is greater than 1 $GeV/g/cm^2$, the molecular structure of the CN is locally changed by short-range knock-on electrons.

After the flight and the disassembly of the instrument these foils are etched in a 40 degree temperature controlled NaOH solution for 18 h, generating etch channels through the foils along the particle track. The hole pattern is then transferred to an aluminum covered sheet of paper by a high-voltage spark copying technique. The resulting dots are then scanned by a computer controlled video camera (Heinrich et al., 1981). Track fitting software reconstructed about 10000 particle tracks in the range stack with etchable track lengths between 2 mm (silicon) and 20 mm (iron). The precise drift chamber trajectory enabled the tracks in the stack to be matched with the active electronic data with high effiency.

2.5. Cherenkov counter

The Cherenkov counter is designed as a square flat box of 1.2 m by 1.2 m. The radiator in the bottom is 5 cm thick. Designed as a diffuse light collection box the inside of the detector is covered with a paint of high reflective $BaSO_4$. 24 photomultipliers, 6 on each side above the radiator (pilot425) , overview the radiator and collect the Cherenkov light.

Because of the required high velocity resolution (better than one percent for all charges) for isotope analysis a number of corrections were applied to the raw Cherenkov data: 1) Nonlinearities in the ADCs used to digitize the PMT signals were measured and and corrected out of the data. ; 2) Thermal effects caused a drift in the high voltage of the PMTs during the flight, with subsequent gain changes. These time profiles were monitored and, combined with the calibrated gain/voltage characteristics, were used to correct the time dependant responses; 3) The ADC pedestals (zero signal digital response) also drifted during the flight, but were also monitored in regular intervalls by applying random gates to the ADCs. 4) a complete response map of all 24 PMTs in the detector was calibrated by exposing the detector to a 14GeV/n silicon beam at Brookhaven National Laboratory. The fully corrected Cherenkov signals are uniform over the whole detector area and throughout the flight to better than 0,5 percent.

2.6. Penetration counter

A scintillation radiator with four photomultipliers mounted in the bottom of the experiment detects high energy particles which did not stop in the range stack. Because of the high sensitivity for detection of single charged particles this detector provide a veto for fragmentation of nuclei within the experiment.

3. Isotope analysis

3.1. Cherenkov - range technique

Particles of same velocity and therefore same energy per nucleon scale in total energy proportional to their mass. Since deacceleration in the instrument occurs by ionisation losses which are related to the charge of the particle, isotopes of the same element and the same incident velocity separate by different total ranges in the range stack. The total range is the sum of traversed matter in the instrument from the top of the Cherenkov radiator down to the last etch cone in the range stack which represents the particle stop.

3.2 Analytic instrument description

A detailed computer simulation was set up to describe the instrument function, including energy losses und resulting ranges and Cherenkov light production with background light estimates. The parameters used were derived from the flight data. This simulation predicts an instrument performance in agreement with final results and is used to define a mass scale for all charges and isotopes.

3.3 *Silcon and Iron mass analysis*

500 (220) tracks of stopping silicon (iron)particles were matched by help of the driftchamber trajectory. After geometry cuts like restricted incident angle of less than 40 degrees and a Cherenkov penetration position within the mapped area 283 (134) events are the data base for isotope analysis. Because the nuclei generate more Cherenkov light with increasing incident angle, all signals were normalized to vertical incidence by empirically measuring and interpolating the light increase in separated energy intervalls. A plot of silicon events in Cherenkov - range space compared to the computer simulation is shown in figure 3. In this model supported approach, the mass scale attached to the data points is differential and insensitive to minor errors in the matching of the simulation and measured datapoints.

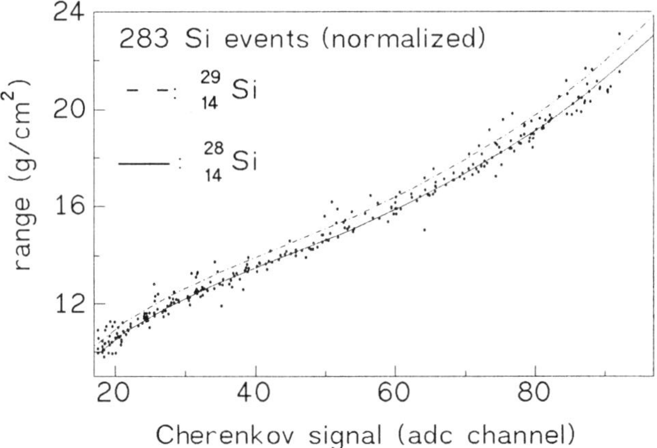

Fig. 3 Corrected Cherenkov response vs range for silicon events. The solid and dashed line are the curves obtained by the simulation for ^{28}Si and ^{29}Si respectively

The resulting mass histograms are documented in figure 4. The distribution is assumed to be a sum of Gaussian distributions over single isotopes. A Maximum Likelihood free parameter fit determines the individual abundances and the common width of the Gaussians.

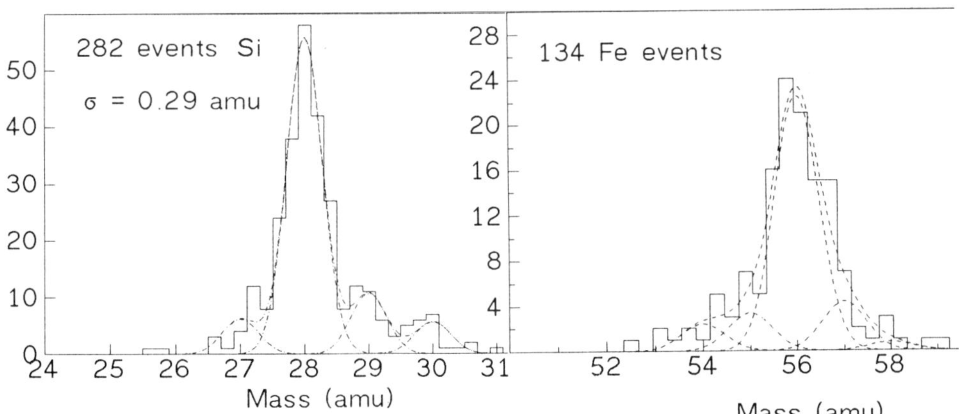

Fig. 4 Silicon (Iron) Mass histogram of 282 (134) events. The width of the fitted Gaussians is 0.29 (0.43) amu

4. Propagation and Sorce composition

While travelling through the interstellar medium the chemical and isotopic composition of Galactic Cosmic Rays are significantly changed by nuclear interactions. Furthermore the energy spectra for element species vary individually due to energy losses by ionisation. From the top of the atmosphere down to the final stop in the range detector the particle penetrate slabs of residual atmosphere and the instrument materials which additionally have to be considered. The Computer algorithm to derive GCRS compositions is based on an iterative method starting with the best known Composition and then propagating this sample through pass length distributions and energy losses as predicted by energy loss mechanisms. Charge- and mass-changing interactions are implemented using energy - dependant total and partial crossections as measured in accellerator runs ore predicted by theoretical models.

This work has not been completed yet for the ALICE measurement and final results will be published soon.

5. Conclusions

The elemental and isotopic distributions determined from ALICE flight have resolutions within the stated goals of the instrument (Esposito 1988). We are confident that the analysis techniques developed can be extended to the less abundant elements in this charge range. Current cross section data are now being incorporated into the propagation model and conclusions pertaining to the dominant source(s) and mode of injection of galactic cosmic rays will then be addressed.

Acknowledgements

The authors thank the Deutsche Forschungsgemeinschaft (DFG) for supporting this work and also thanks for a travel support by a NATO grant 0325/84

References

Esposito, J.A., B.S. Acharya, V.K. Balasubrahmanyan, B.G. Manger, J.G. Ormes, R.E. Streitmatter, W. Heinrich, M. Henkel, M. Simon, H.O. Tittel (1985) 19th Int. Cosmic Ray Conf., La Jolla, Vol. 3, 278

Heinrich, W., M. Simon, H.O. Tittel, J.F. Ormes, V.K. Balasubrahmanyan, R.E. Streitmatter (1982) Solid State Nuclear Track Detectors, Proceedings of the 11th Int. Conf., Bristol, Editor P.H. Fowler and V.M. Clapham, Pergamon Press, 863

Simon, M., M. Henkel, R. Hundt, K.D. Mathis, G. Schieweck, Th. Suck (1984) Nucl. Instr. Meth. 221 466 - 471.

An overview about initial results from the Compton observatory GRO

V. Schönfelder

Max-Planck-Institut für extraterrestrische Physik, D-8046 Garching bei München, FRG

Abstract. Gamma-ray astronomy provides a diagnostic tool to study high energy processes in the Universe. At present, NASA's Compton Observatory is performing the first complete sky survey in gamma-ray astronomy. The observatory was launched on April 5, 1991 by the Space Shuttle Atlantis into a near-earth orbit. An overview about the first highlight results from this mission is given. Special topics of presentation will be: pulsars (especially the Crab and Vela pulsars), the diffuse galactic gamma-ray emission from interstellar space, the nuclei of active galaxies (especially quasars), the 511keV and 1.8 MeV gamma-ray line emissions from the central region of the galaxy, the puzzle of cosmic gamma-ray bursts and finally, the Sun during solar flare acitvities.

1.   Introduction:

Gamma-ray astronomy is at present making a major step forward. This progress is due to NASA's Gamma Ray Observatory GRO, now called Compton Observatory, which was brought into a near-Earth orbit by the Space Shuttle 'Atlantis' on April 5, 1991. Gamma-ray astronomy is now about 25 year old: The first detection of cosmic gamma radiation was made by OSO-III in 1967.

Since gamma radiation represents the most energetic part of the electromagnetic spectrum it provides information about the most energetic processes and phenomena in the Universe. From previous gamma-ray missions we know that the gamma-ray fluxes from cosmic objects are extremely small. This is easily understandable, since a single 100 MeV gamma-ray photon has the same energy as 10^{10} infrared photons together.

In order to make these low fluxes detectable, the telescopes must have large collecting areas, high detection efficiencies, and the exposure times must be long (weeks or even sometimes months). In so far it is not surprising that the Compton Observatory is the heaviest scientific payload ever brought into space by a shuttle.

2.   The Gamma Ray Observatory GRO:

The Gamma Ray Observatory GRO is the first satellite mission that covers the entire space astronomy gamma-ray range from about 100 keV to 30 GeV, more than five orders of magnitude in photon energy. Therefore, simultaneous observations over the full dynamic range are possible. The coverage of such a broad spectral range cannot be achieved by one single instrument. Instead, GRO carries four different instruments with complementary properties.

A schematic view of GRO is shown in Fig. 1. The platform carries the three major instruments OSSE, COMPTEL and EGRET (from left to right) and the fourth instrument BATSE, which actually consists of 8 detectors - two at each corner of the spacecraft.

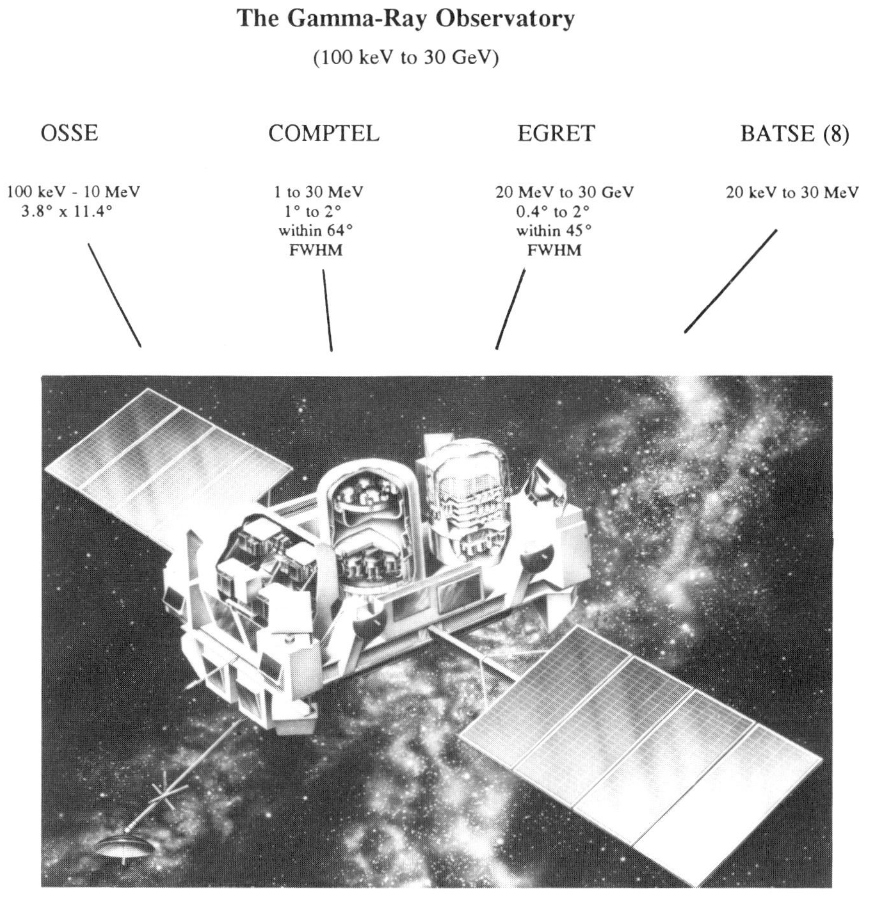

The Gamma-Ray Observatory

(100 keV to 30 GeV)

OSSE	COMPTEL	EGRET	BATSE (8)
100 keV - 10 MeV 3.8° x 11.4°	1 to 30 MeV 1° to 2° within 64° FWHM	20 MeV to 30 GeV 0.4° to 2° within 45° FWHM	20 keV to 30 MeV

Fig. 1.   Schematic View of COMPTEL

OSSE is a collimated scintillation spectroscopy experiment in the transition region between hard X-ray and low energy gamma-ray astronomy, namely between 100 keV and 10 MeV. The collimated field-of-view is 4 x 11 degrees (Kurfess et al., 1983). COMPTEL is a Compton telescope in the energy range 1 to 30 MeV. It has a wide fiel-of-view of about 1 steradian and an angular resolution within this field of 2.5 to 5 degrees FWHM (Schönfelder et al, 1984, 1992). EGRET is a sparkchamber experiment in the high energy range from 20 MeV to 30 GeV. Its field-of-view is about half a steradian and its angular resolution of the order of 1 degree (Fichtel et al., 1983). BATSE is an all sky monitor for burst and transient source events in the energy range 20 keV to 30 MeV (Fishman et al., 1985). Table 1 lists the institutes under whose responsibilities the four telescopes were built.

TABLE 1.   The 4 GRO Telescopes

OSSE (Oriented Scintillation Spectroscopy Experiment)	Naval Research Laboratory (USA) North Western University (USA)
COMPTEL (Compton Telescope)	Max-Planck-Institut f. extraterrestrische Physik (FRG) Laboratory for Space Research (The Netherlands) University of New Hampshire (USA) Space Science Department of ESA (The Netherlands)
EGRET (Energy Gamma Ray Telescope Experiment)	Goddard Space Flight Center (USA) Max-Planck-Institut f. extraterrestrische Physik (FRG) Stanford University (USA)
BATSE (Burst & Transient Source Experiment)	Marshall Space Flight Center (USA) Goddard Space Flight Center (USA) University of California, San Diego (USA)

The mission concept of GRO is illustrated in Fig. 2. The Shuttle Atlantis carried the Observatory into a 450 km orbit (28.5 day inclination). During the second day of the Shuttle mission the Astronauts onboard released the Observatory. It will stay in orbit for at least 2 years. GRO has a self-contained propulsion system to maintain the 450 km orbit. In addition, the system will allow the 15 tons heavy spacecraft to undergo a controlled re-entry at the end of the mission.

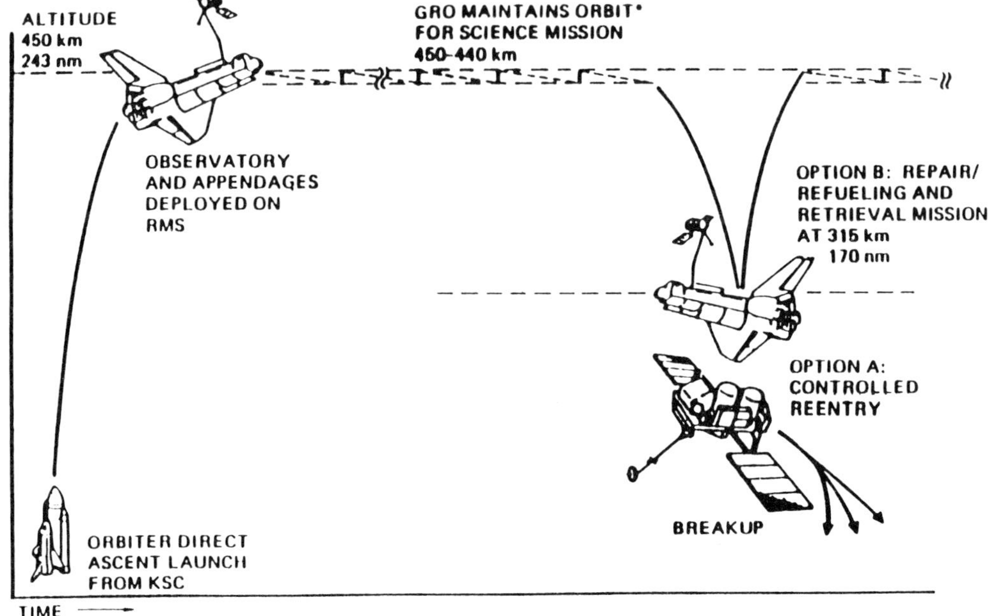

Fig. 2.   GRO Mission Concept

The GRO mission started with an all-sky survey (the first one in gamma-ray astronomy at all). The survey began on May 16, 1991 and will last until November 15, 1992. After the survey selected celestial objects will be studied in more detail. The gamma-ray observations are complemented by ground based observations in other spectral ranges.

3.   Production Processes of Cosmic Gamma Radiation

As mentioned in the introduction, gamma-ray astronomy allows to study the most energetic processes in astronomy. Fig. 3 illustrates, how gamma radiation is produced in the Universe.

The first process considered is the matter-antimatter annihilation. If electrons and positrons annihilate directly, they generate two photons at 511 keV and eventually a three-photon continuum via the formation of positronium. If protons and antiprotons annihilate at rest, they produce a number of pions (≥ 3). The π°-mesons practically immediately decay into two gamma-rays. The resulting gamma-ray spectrum peaks at $m_\pi c^2/2 \approx 68$ MeV and ranges from about 5 to 900 MeV.

Gamma-ray production via π°-decay, also occurs, if energetic protons of the cosmic radiation interact with matter, mainly hydrogen. Again, the gamma-ray spectrum peaks at 68 MeV, however, it extends to higher and lower energies (because the pions are not limited to energies below 923 MeV as in case of p-p-annihilation at rest). If *low* energy cosmic ray particles (below the pion production threshold) interact with matter, then gamma-ray lines may be emitted from excited nuclei. Finally, unstable radioactive nuclei may also emit gamma-ray lines.

Energetic electrons also produce gamma-radiation. The processes are either bremsstrahlung with matter, inverse Compton scattering with photons, or synchrotron losses. In case of the inverse Compton effect an energetic electron hits a low energy photon (e.g. a star light photon), and transfers part of its energy to this photon, which then moves on as an energetic X- or gamma-ray photon. Gamma-ray production by the synchrotoron process becomes important only in very strong magnetic fields like those near the surface of neutron stars. In all these processes the gamma-ray spectrum has the shape of a power-law, if the electron spectrum has such a shape; this very often is the case.

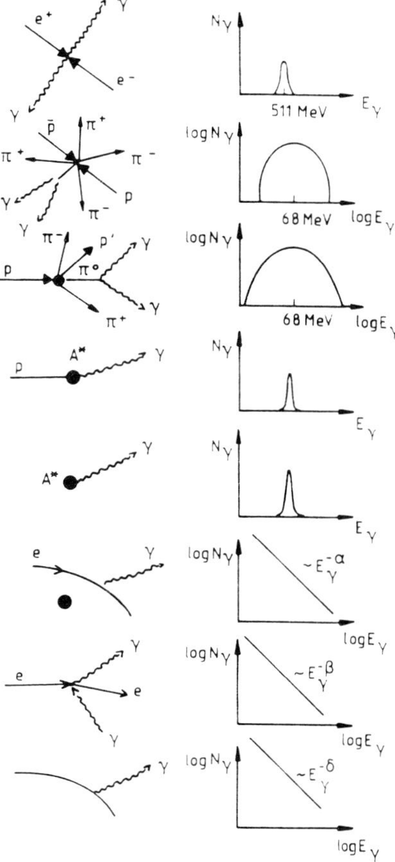

Fig. 3. Various gamma-ray production processes and the resulting gamma-ray energy spectra (from Ramana Murthy and Wolfendale (1986)).

All these processes (with the exception of the baryonic matter-antimatter annihilation) have already been observed in the various celestial objects. The most important and interesting targets of interest to gamma-ray astronomy are: compact objects in the galaxy (mainly radio pulsars and X-ray binaries), the diffuse galactic gamma-ray emission from interstellar space, external galaxies (especially the nuclei of active galaxies), the diffuse cosmic gamma-ray background and as transient gamma-ray emitters, the cosmic gamma-ray bursters and the Sun during solar flares.

Many of these objects do not only show a continuum, but also gamma-ray line emission. In the next chapter, a description of some of the most imprortant results of the Compton observatory is given.

4.   Initial Scientific Results from the Compton Observatory

4.1.   Gamma Ray Pulsars

Prior to the launch of GRO radio pulsars were the only class of compact objects within the Milky Way that was known to contain gamma-ray sources. Pulsed gamma-ray radiation from the Crab and Vela pulsars had been seen by the previous gamma-ray astronomy projects SAS-2 and COS-B and a few balloon borne instruments. The question was, whether GRO would be able to detect pulsed emission from more than these two pulsars. The answer is 'yes'. Up to now (July 1992) GRO has detected pulsed emission from three further pulsars. These are:

$$\begin{array}{ll}
\text{PSR 1509-58} & (P = 151 \text{ msec}) \\
\text{PSR 1706-44} & (P = 102 \text{ msec}) \\
\text{Geminga} & (P = 237 \text{ msec}).
\end{array}$$

PSR 1509-58 was discovered at gamma-ray energies by BATSE below $\approx$ 1 MeV. OSSE has confirmed this detection. COMPTEL and EGRET were so far unable to observe any pulsed emission from this object above 1 MeV.

PSR 1706-44 is a radio pulsar, which was discovered only 2 years ago during a radio pulsar survey. Its pulsed gamma-ray emission was discovered by EGRET. This object coincides with one of the so far unidentified COS-B sources.

The Geminga source is one of the strongest high-energy gamma-ray sources in the sky (at energies above several 100 MeV it is stronger than the Crab). It was discovered in 1973 by SAS-2 and later studied a great deal by COS-B. The gamma-ray source had been tentatively identified with the Einstein X-ray source 1E 0630+178 (Bignami et al., 1983), and this object was recently found to be a 237 sec X-ray pulsar (Halpern and Holt, 1992). With the knowledge of the X-ray pulsar period the EGRET scientists (Bertsch et al., 1992) were then immediately able to confirm the pulsed emission also at gamma-ray energies above 100 MeV. In the meantime, the pulsed gamma-ray emission has also been found in the archival SAS-2 and COS-B data.

Apart from these new discoveries the Compton observatory has made observations of the previously known two pulsars Crab and Vela with unprecedented accuracy. As an example, Fig. 4 shows the lightcurves of the Crab pulsar as measured by BATSE, OSSE, COMPTEL and EGRET, and Fig. 5 the energy spectrum of the pulsed Crab emission as measured by OSSE, COMPTEL and EGRET.

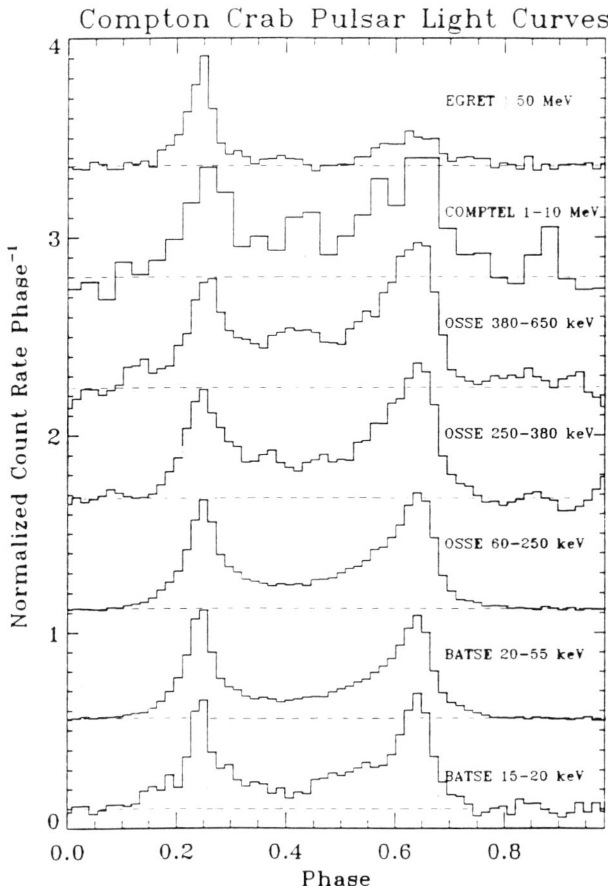

Fig. 4. Crab pulsar lightcurves as measured by GRO

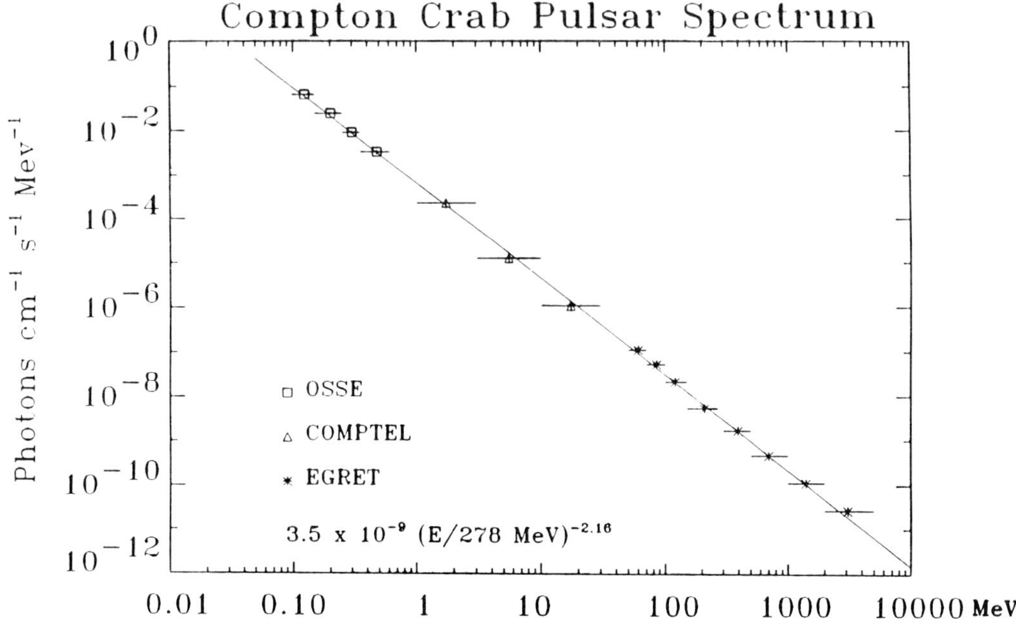

Fig. 5.   Energy spectrum of pulsed Crab-emission are measured by OSSE, COMPTEL
and EGRET.

Radio pulsars are known to be isolated rotating neutron stars. The gamma radiation is supposed to be produced by relativistic electons, interacting with strong magnetic fields or other photon fields, either near the magnetic polar caps or in more outer regions of the magnetosphere of the neutron star. Crucial for an understanding of the pulsar radiation mechanism is the fact that at least for the two previously known gamma-ray pulsars Crab and Vela their maximum of luminosity is located at gamma-ray energies. The gamma-ray luminosity of these two pulsars is 4 to 5 orders of magnitude higher than the radio-luminosity. It is important to note that also the other three newly discovered gamma-ray pulsars do have their peak luminosity at gamma-ray energies.

4.2.   Diffuse Galactic Gamma-Ray Emission

The diffuse galactic gamma radiation is produced in interstellar space by interactions of cosmic rays (primarily cosmic ray protons and electrons) with interstellar matter (mainly interstellar hydrogen). Questions to be addressed by the study of the diffuse gamma-ray emission are: *'How are cosmic rays distributed throughout the Galaxy?'* and *'What can we learn from the distribution of cosmic rays about their sources?'* These questions had already extensively been discussed after the SAS-2 and COS-B data became available; but still many questions remain open. EGRET is now repeating the galactic survey of COS-B with about ten-times higher sensitivity and somewhat improved angular resolution. COMPTEL at present is surveying the galactic plane at low gamma-ray energies between 1 and 30 MeV. This MeV-survey will provide unique information about low-energy (< 100 MeV) cosmic ray electrons, which cannot be studied through direct particle measurements (because of solar modulation) or non-thermal radio observation (because of synchrotron self-absorption). The results from EGRET and COMPTEL on the diffuse galactic gamma-ray emission are not yet available for publication. Preliminary results from the galactic center region are of excellent quality and great promise.

4.3.   Gamma-Ray Line Spectroscopy

The field of gamma-ray line spectroscopy is closely related to the question of nucleosynthesis in the Universe. During the formation of the chemical elements, not only stable, but also radioactive elements were formed. Some of these are gamma-ray emitters; these can be studied by gamma-ray spectrometers. Prior to the launch of GRO, the following lines had been detected: the 511 keV-annihilation line from the galactic center region, the 1.8 MeV line from radioactive ^{26}Al (also from the galactic center region), and a series of lines from SN 1987a resulting from the ^{56}Ni -> ^{56}Co -> ^{56}Fe decay chain. First - still unpublished - results on gamma-ray line observations by OSSE and COMPTEL have been presented at various conferences.

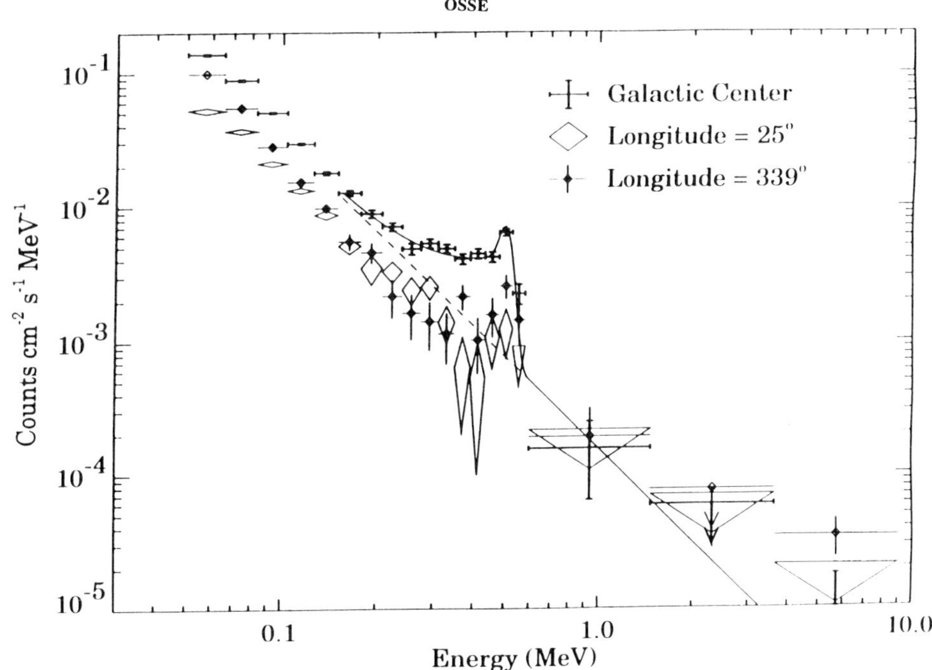

Fig. 6. OSSE measurement of the energy spectrum from the galactic center region (from Johnson et al., 1992).

Fig. 6 shows the OSSE-spectrum of the galactic center region. The 511 keV annihilation line is clearly visible together with the three-photon continuum. At l = O° the line is much stronger than at l= 25° and l = 339°. A longitude distribution of the line flux over the entire galactic plane is not yet available. This distribution will be crucial to answer the question about the origin of the annihilation line.

Initial COMPTEL results on the 1.8 MeV-line from radioactive ^{26}Al were presented by Diehl et al. (1992) at the Toulouse Conference in March 1992. Fig. 7 shows the background-subtracted energy spectrum from a 6°-wide longitude interval at the galactic center: the 1.8 MeV stands out clearly; the measured line width corresponds to the instrument energy resolution.

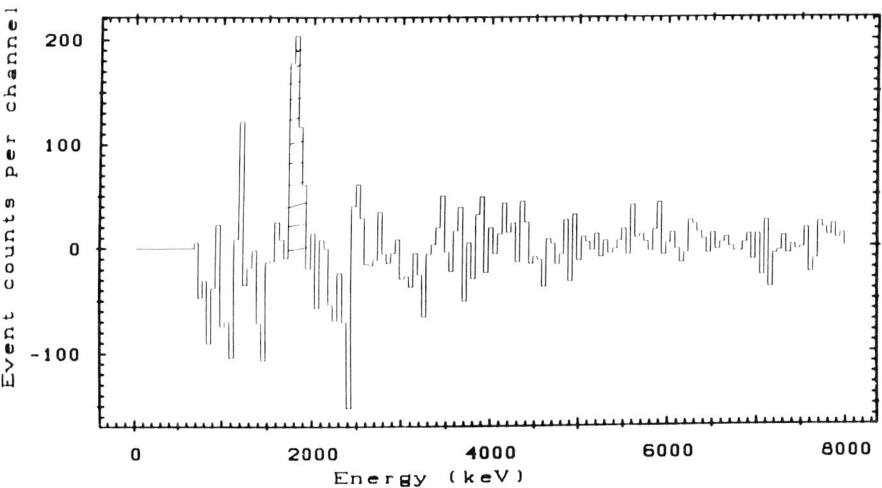

Fig. 7. Background subtracted energy spectrum from a 6°-wide longitude range around the galactic center region as measured by COMPTEL. The 1.8 MeV lines stands out clearly.

Similar measurements exist from other regions of the galactic plane as well. The resulting - still preliminary - longitudes distribution of the line-flux shows a peak at or close to the Galactic Center and a broad ridge of about ± 20° around the galactic center. But also in other parts of the galactic plane the line-flux does not drop to zero. It is expected that the longitude and latitude distributions of the line-flux will finally allow to identify the objects in which the ^{26}Al is produced. The most promising candidates are supernovae, novae or special stars like Wolf-Rayet stars. The ^{26}Al, which is produced within these objects, is ejected into interstellar space and then remains detectable for its mean decay time of about one million years.

4.4. <u>Active Galactic Nuclei</u>

Active galactic nuclei are the most powerful compact objects in the Universe. It is supposed that a supermassive object - probably a black hole - is located in its center and that it is powered by accretion of matter from its surrounding. Due to the angular momentum of the accreted matter, an accretion disk developes, and often a jet of plasma stream is visible perpendicular to the plane. For the formation of the relativistic jet, strong magnetic fields play a fundamental role.

One of the great surprises of GRO is the discovery of many quasars and BL Lac objects by EGRET to show high-energy gamma-ray emission. Table 2 lists 11 of these objects. All of them are radio-loud emitters; about half of them show superluminal motions. Many of them are time-variable in the gamma-ray emission. The redshifts of these objects range from nearly zero to more than 2. Obviously we are not only seeing the nearest quasars. If the gamma-ray emission is assumed to be emitted isotropically, the luminosities are as high as 10^{48} erg/sec (see last column of Table 2).

TABLE 2.

EGRET DETECTIONS OF ACTIVE GALACTIC NUCLEI (Apr. 91 - Feb. 92)						
OBJECT	l	b	z	Obs.	$I(> 100 \text{ MeV})$ $\times 10^{-6}\ cm^{-2}\ s^{-1}$	$L(> 100 \text{ MeV})$ $\times 10^{48}\ erg\ s^{-1}$
3C279	305	57	0.538	Jun 91 Oct 91	0.6 to 4.9	0.3 to 2
3C273	290	64	0.158	Jun 91 Oct 91	0.3 ± 0.1	0.004
0208-512	276	-62	1.03	Sep 91	0.3 ± 0.1	2
4C38.41	61	42	1.81	Sep 91	1.0 ± 0.1	6
0528+134	191	-11	0.691	Apr 91	0.8 ± 0.1	
0836+710	144	34	2.17	Jan 92	0.15 ± 0.04	1.1
MKN 421*	180	65	0.031	Jun 91	0.11 ± 0.03	0.0001
0537-441*	250	-31	0.894	Jul 91	0.34 ± 0.09	0.2
0716+714*	144	28		Jan 92	0.20 ± 0.04	
3C454.3	86	-38	0.859	Jan 92		
CTA 102	77	-38	1.037	Jan 92		

* BL Lac Object

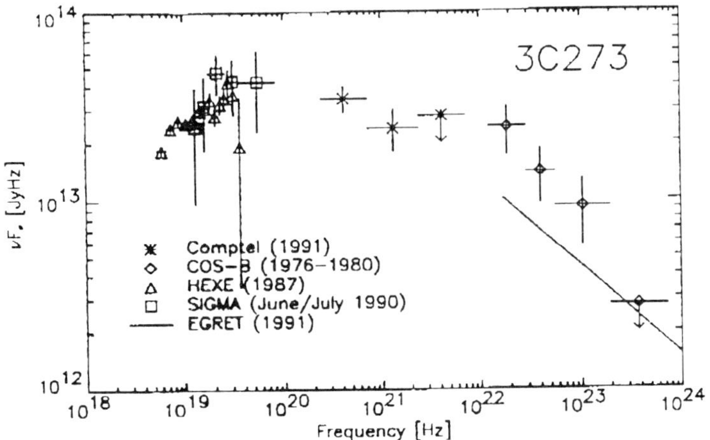

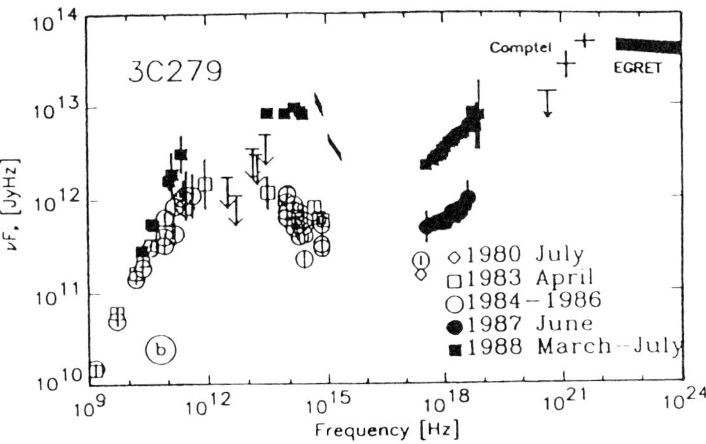

Fig. 8. Power spectra of 3C273 (top) and 3C279 (bottom).

Two of the objects (3C273 and 3C279) were not only observed by EGRET, but also by OSSE and COMPTEL. Fig. 8 shows the energy spectra of both quasars (top: 3C273, bottom: 3C279). Both quasars do have their peak luminosities at gamma-ray energies, and both quasars show spectral breaks in the COMPTEL energy range. These breaks may provide a key for the understanding of the gamma-ray emission.

In case of a spherical isotropic and central emission region of a few light days/weeks diameter photon-photon interaction would be an unavoidable consequence of the compactness; they would cause a break in the power spectra at that energy, at which the optical depth becomes '1'. At least, in case of 3C279 the observed break at about 10 MeV cannot be the result of such photon-photon interactions, otherwise the absorbed part of the gamma-ray spectrum would have to be re-emitted in other parts of the electromagnetic spectrum with the same luminosity. This is not the case. Therefore, it has been suggested that the gamma radiation must be produced in more outer, optically thin regions.

One possible 'outer' origins are the knots in the relativistic jets of these objects (as suggested by Dermer et al., 1992, Camenzind & Dreisigacker, 1992, and Mannheim & Biermann, 1992). Relativistic electrons or protons, which are accelerated within these knots are able to explain the entire emission from radio to gamma-ray energies.

Due to the strong beaming of the jet emission, the luminosity is reduced by the Doppler-Lorentz factor $D^{(3+\alpha)}$ (where α is the spectral power index and $D = \sqrt{\frac{1-\beta^2}{1-\beta\cos\Theta}}$), which turns out to be a factor of 10^4 for 3C279. Such jet-models can be tested by contemporary observations from radio to gamma-ray energies.

4.5. Gamma-Ray Bursts

The gamma-ray burst sources are the brightest gamma-ray emitters in the sky during the time of their short outburst. Prior to the launch of GRO most people had thought that in some ways neutron stars might be the sources of these bursts, though the trigger mechanism was not known. After the first results on bursts from BATSE became available, this hypothesis got in serious difficulties, and the question of the burst origin is entirely open again.

BATSE is at present measuring the global distribution of bursts in the sky. First results were published by Meegan et al., 1991. Fig. 9 shows the celestial distribution of 153 bursts, measured by BATSE. It does not show any statistically significant deviation from isotropy. Fig. 9 (bottom) shows the log N - log S diagram of these bursts. Instead of the fluence S (erg/cm^2 the relative countrate C is plotted on the abscissa. Since the number of bursts $N(>S)$ scales like $\approx R^3$, for a spherical distribution and since the fluence $S \approx 1/R^2$, the log N - log S curve should have an -3/2 slope for a spherical distribution. This is not found in the measurements, the curve bends towards weaker bursts. The surprising point was that also these weaker bursts are distributed isotropically.

Obviously, these weaker bursts are not from a disk-distribution (as would be expected from neutron stars). So, the question is open again: what is the origin of gamma-ray bursts? If they are produced in the solar system (e.g. in the Oort's cloud), the amount of energy required to explain their luminosity, would be modest ($\approx 10^{28}$ erg sec). If the bursts are galactic, the sources would have to fill a halo at least up to 50 kpc, and an energy of $\approx 10^{43}$ erg sec would be required per burst . A new population of galactic objects would be needed. This hypothesis is not supported by any excesses in the number of bursts from the Magellanic Clouds and the Andromeda nebula (though the statistical significance at this time is not yet conclusive).

BATSE   Distribution of Bursts

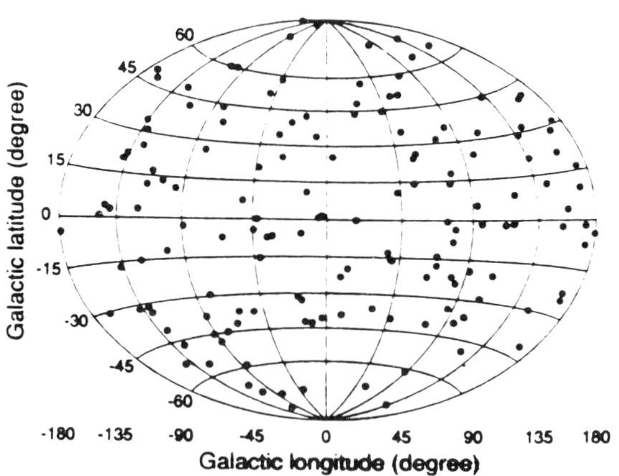

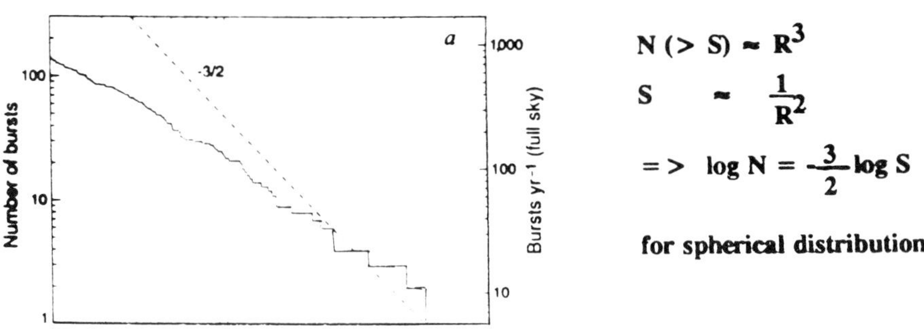

$$N \,(> S) \approx R^3$$
$$S \approx \frac{1}{R^2}$$
$$=> \ \log N = -\frac{3}{2}\log S$$

for spherical distribution

Fig. 9. BATSE results on gamma-ray bursts. Top: distribution of bursts in the sky. Bottom: log N - log S diagram.

GRB 910503

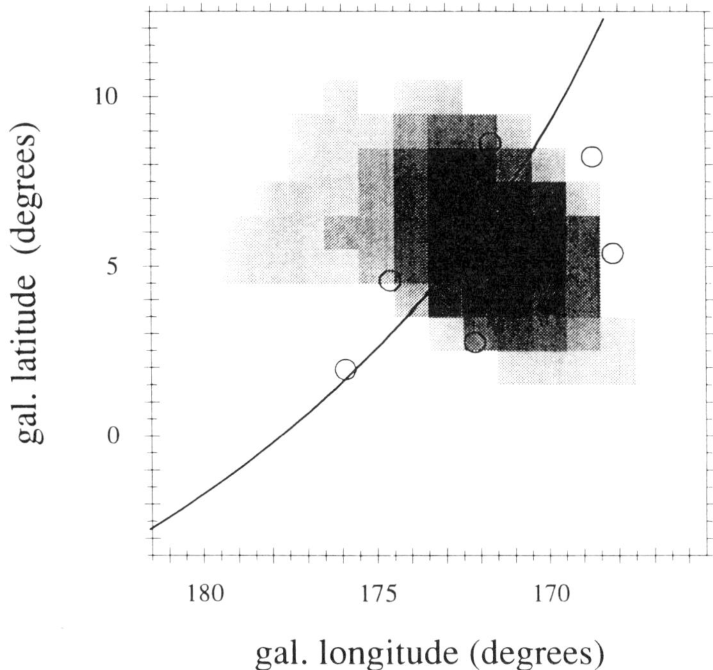

Fig. 10. The location of GRB 910503 from the COMPTEL image and the triangulation circle.

The third possibility is an extragalactic or cosmological origin. The energy release would be tremendeous: 10^{51} to 10^{53} erg/sec. The bending of the log N - log S curve would have to be explained cosmologically. Neutron star - neutron star or black hole - neutron star merging are presently discussed as possible sources, though there are serious difficulties with such models as well (e.g. time structure of bursts, how to explain cyclotron lines, how to explain repeaters).

Perhaps the most attractive models to explain the bursts are multi-component models, like e.g. the two-component model of Lingenfelter and Higdon (1992), in which two luminosity classes of galactic sources are assumed to exist (one releasing 10^{39} erg/sec, the other 10^{44} erg/sec). The first class would be visible up to 300 pc, the second one up to 100 kpc. Lingenfelter and Higdon propose that both these classes are represented by neutron stars, but that the bursts in both classes are produced in different ways (thermonuclear run away in the first case and starquakes in the second case).

A major step forward in solving the question of the origin of gamma-ray bursts would be the identification of a burst source in other spectral ranges. GRO was able to locate a few burst sources very accurately and, therefore, a counterpart search after the burst seemed promising. For illustration, Fig. 10 shows an image of the burst GRB 910503. The image was produced by COMPTEL. The source location accuracy from this image is ± 1°. Also shown are six high-energy events from this burst as measured by EGRET (the six open circles) and a one-dimensional location of the burst by triangulation, using the burst arrival times aboard GRO and Ulysses. The combination of the COMPTEL location uncertainty with the triangulation uncertainty defines an error box of about 3 arc min x 2 degrees. Attempts to find a counterpart in the ROSAT data so far remained unsuccessful (J. Greiner, private communication). But further attempts of counterpart searches for other bursts are continuing.

5.   Conclusion

Exciting results have been obtained from the early phase of the GRO-mission, which have stimulated many discussions and new theories. Whereas gamma-ray astronomy for many years was a field restricted to a small group of specialists, it has now found broad attention within the whole community of astronomers and astrophysicists.

References

Bertsch, D.L. et al., 1992, Nature 357, 306

Bignami, G.F. et al., 1983, Ap.J. 272, L9

Camenzind, M. and Dreisigacker, O., 1992, submitted to A&A

Camenzind, M. and Krockenberger, M., 1992, A&A 255, 59

Dermer, C.D., Schlickeiser, R. and Mastichiadis, A., 1992, A&A 256, L27

Diehl, R. et al., 1992, submitted to A&A Suppl.

Fichtel, C. et al., 1983, Proc. of 18th Int. Cosmic Ray Conf., Vol **8**, 19

Fishman, G.J. et al., 1985, Proc. of 19th Int. Cosmic Ray Conf., Vol **3**, 343

Halpern, J.P. and Holt, S.S., 1992, Nature 357, 222

Hermsen, W. et al., 1992, submitted to A&A Suppl.

Johnson, W.N., 1992, submitted to A&A Suppl.

Kurfess, J.D. et al., 1983, Space Research, Vol. **3**, No. 4, 109

Mannheimer, K. and Biermann, P.L., 1992, A&A 253, L21

Meegan, C.A., et al., 1991, Nature 355, 143

Lingenfelter, R.E. and Higdon, J.C., 1992, Nature 356, 132

Ramana Murthy, P.v. and Wolfendale, A.W., in 'Gamma-Ray Astronomy'
        (Cambridge Univ. Press, 1986), p. 2

Schönfelder, V. et al., 1984, IEEE-Transactions on Nucl. Science, NS-31, No. 1, 766

Isotopic astronomy from anomalies in meteorites: Recent advances and new frontiers

Charles L Harper Jr

Department of Earth and Planetary Sciences, Harvard University, 20 Oxford St., Cambridge, MA 02138 USA

Abstract. The study of isotopically anomalous materials in meteorites is a rapidly developing subdiscipline of astronomy providing primary data on stellar and supernova nucleosynthesis. Several types of presolar circumstellar grains have been analyzed and exhibit wide variations in isotopic signature, probably indicating many stellar sources. Refractory Ca,-Al-rich inclusions (CAIs) are thought to have formed within the solar system but surprisingly preserve a variety of anomaly signatures, one of which points towards a component originating in neutron-rich nuclear statistical equilibrium (NNSE) deep inside one or more supernovae. The uncertain cosmochemical mode of origin of the CAI anomalies, however, remains an important unsolved problem. Barium r-process anomalies provide critical insights because cosmo-chemically volatile holdups at ^{137}Cs ($T_{1/2}$ = 30 a) and ^{135}Cs ($T_{1/2}$ = 2.3 Ma) constrain, respectively, the time of Ba condensation relative to nucleosynthesis and the age of any 'exotic' Ba component at the time of CAI formation ~4566 Ma ago. An important frontier for future innovation is the development of mass analysis techniques for ultra-small trace element samples from individual presolar grains and micro-CAIs.

1. Introduction

An introductory overview is given of the rapidly developing field of isotopic astronomy from meteorite analysis, touching upon issues of astrophysical interest and some highlights of recent research. Most meterorites come from remnants of 'planetesimals' formed in the very early solar system when planet-forming processes leading towards isotopic homogenization were incomplete. Because meteorites preserve components which either wholly or partially escaped isotopic homogenization, it is possible to isotopically identify presolar sources which contributed material to the protosolar reservoir. Some high temperature processes in the solar nebula also preserved the isotopic signatures of presolar components in refractory objects.

1.1. Isotopic Anomalies

Isotopic anomalies are non-mass-dependent isotopic shifts relative to normal (terrestrial or standard) compositions and are usually quantified as relative differences in isotope ratios. If, for an element Q, the ratio $^{i}Q/^{j}Q$ denotes the ratio of a nuclide of interest i, to an index mass j, for the sample (s) and normal (n), then the conventional representation of an anomaly ^{ij}A is:

$$^{ij}A_N = [[(^{i}Q/^{j}Q)_s \ / \ (^{i}Q/^{j}Q)_n] - 1]_N \times F$$

where F is a magnification factor and N indicates that a normalization has been applied to remove natural and machine-derived mass-dependent fractionation. In the standard notation, anomalies are parameterized by 'δ-units' ($F = 10^3$) and 'ε-units' ($F = 10^4$). The terms 'ppm-deviation' and '%-deviation' are also used for $F = 10^6$ and 10^2. For C and N, the large deviations measured in presolar grains are presented as direct ratios ($^{12}C/^{13}C$ and $^{14}N/^{15}N$). Some examples of ε- and ppm-level anomalies in an Allende CAI are shown in Figure 1a,b,c.

1.2. 'Isotopic Astronomy'

The laboratory study of stardust in meteorites is a new kind of astronomy analogous to the study of isotopic abundances from stellar spectra. Circumstellar, nova and supernova condensates---each tiny grain carries a wealth of information hidden away in its 'isotopic library.' Grains do not carry direct information on stellar mass, temperature, luminosity or distance, but major element isotopes can be determined to much higher precisions than spectral analysis allows. A wide range of trace metal isotopes can be investigated in grains, in contrast to the few that (occasionally) can be obtained from spectral analysis (Mg, Ti, Rb, Zr).

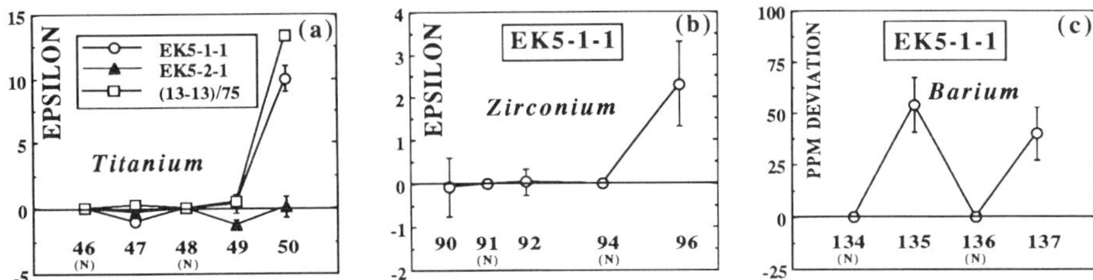

Figure 1 (a, b & c). Isotopic anomalies in Ti, Zr and Ba. EK5-1-1 is an Allende CAI exhibiting the NNSE signature in ^{50}Ti. Excess ^{96}Zr and ^{135,137}Ba are also present. EK5-2-1 is an Allende CAI without excess ^{50}Ti. It has no excess ^{96}Zr within error and no excess ^{135,137}Ba relative to the bulk composition of Allende. '13-13' is a Murray hibonite with a 10% ^{50}Ti anomaly (Ireland 1990). Its anomaly pattern is similar to EK5-1-1 depite being 100x larger.

Studies of isotopic anomalies in CAIs in primitive meteorites are also a subdiscipline of isotopic astronomy. However, as the weight of evidence on these objects suggests that they formed within the solar system, it is important to assess the epistemological limitations intrinsic in the study of CAI anomalies. Extensive studies of the minerologies, morphologies, major and trace element compositions, mass-dependent isotopic fractionations in O, Mg, Si, Ca and Ti and apparent initial $^{26}Al/^{27}Al$ ratios in these objects have not settled the question of the mode (or modes) of origin of their anomaly patterns, but do indicate complicated nebular processing histories involving multiple cycles of evaporation, melting and condensation (*cf.*, Niederer and Papanastassiou 1984; Niederer *et al.* 1985; Jungck *et al.* 1984; Niemeyer and Lugmair 1984; Fahey *et al.* 1987a,b,c; Hinton *et al.* 1988; Davis *et al.* 1990; Ireland 1988, 1990; Ireland *et al.* 1991a). These observations cast doubt on the validity of identifying anomalous components with single astrophysical sites. One commonly observed anomaly pattern present in both cm-sized objects and 10 µm grains does correspond qualitatively with predictions for nucleosynthesis in a near-the-mass-cut zone in type II supernovae. This correspondence (*cf.*, section 3.2) encourages a suspicion that CAI anomalies may record site-specific information. However, until the basic uncertainty over their cosmochemical origin is settled, the anomalies should not be treated as anything more than 'nucleosynthetic hints.'

Studies of extinct radionuclides (*e.g.*, Wasserburg 1985) are also an integral part of isotopic astronomy as they provide constraints on the astrophysical site of the origin of the solar system, but will be reviewed elsewhere. Discussion of the 'radiogenic anomalies' will be limited to topics directly relating to the interpretation of nonradiogenic anomalies.

1.3. Isotopically anomalous materials in primitive meteorites

Isotopic anomalies in elements with $Z > 18$ have been observed at the >10 ppm level in Ti, Cr and Ba in bulk samples of carbonaceous chondrites (Niemeyer 1988; Harper and Wiesmann, 1992) and range up to orders of magnitude in individual interstellar grains. In general, but with some noteworthy exceptions, the smaller the isotopically anomalous class of object, the larger the range of anomalies observed. In most millimeter to centimeter-sized CAIs in carbonaceous meteorites (Grossman, 1980; Mason and Taylor 1982), anomalies in Ti, Ca, Cr, Ni, Zn, Zr and Ba are observed at the 0.2-25 ε level (Lee *et al.* 1978, 1979; Niederer *et al.* 1980, 1981; Niemeyer and Lugmair 1981; Birck and Allegre 1984; Jungck *et al.* 1984; Niederer and Papanastassiou 1984; Niemeyer and Lugmair 1984; Niederer *et al.* 1985; Papanastassiou 1986; Birck and Lugmair 1988; Harper *et al.* 1990; Loss and Lugmair 1990; Harper *et al.* 1992). Larger anomalies in these and additional elements (Fe, Sr, Nd, Sm, Dy) have been observed in members of a rare group referred to as 'FUN' (= Fractionated, Unknown Nuclear) inclusions (Papanastassiou and Wasserburg 1978; McCulloch and Wasserburg 1978a,b; Lugmair 1978; Niederer and Papanastassiou 1984; Niederer *et al.* 1985; Papanastassiou 1986; Papanastassiou and Brigham 1989; Loss *et al.* 1990; Völkening and Papanastassiou 1989; Völkening and Papanastassiou 1990). In small (~100 μm) micro-CAIs: hibonite ($CaAl_{12}O_{19}$) grains, hibonite-bearing micro-inclusions (MacDougall 1979; MacPherson *et al.* 1983) and micro-spherules (Ireland *et al.* 1991a), isotopic anomalies in Ti and Ca are much larger (Hutcheon *et al.* 1983; Ireland *et al.* 1985; Fahey *et al.* 1985, 1986, 1987a,b; Zinner *et al.* 1986; Hinton *et al.* 1987, 1988; Podosek and Brannon 1988; Ireland 1988, 1990), ranging up to 10% and 27% in the most neutron rich isotopes of Ca and Ti, respectively, in the most anomalous hibonite sample (Ireland 1990. *Cf.*, Figure 1a).

Primitive meteorites also contain interstellar organic materials (Zinner 1988), 'presolar' or 'interstellar' grains of graphite (Amari *et al.* 1990; Hoppe *et al.* 1992a,b), silicon carbide (SiC) (Zinner and Epstein, 1987; Bernatowitz *et al.* 1987; Zinner *et al.* 1989; Lewis *et al.* 1990) and corundum (Al_2O_3) (Huss *et al.* 1992), in addition to the nanometer-sized interstellar microdiamond which carries anomalous Xe ('HL' *cf.*, Lewis *et al.* 1987, 1989; Blake *et al.* 1988; Huss 1990). As expected, isotopic compositions in the carbon-bearing grains are totally different from the normal compositions. Variations in $^{12}C/^{13}C$ span almost four orders of magnitude (Tang *et al.* 1989; Zinner *et al.* 1989; Alexander *et al.* 1990; Bernatowitz *et al.* 1991; Stone *et al.* 1991; Virag *et al.* 1992; Amari *et al.* 1992a,b,c; Hoppe *et al.* 1992a,b)! Presolar corundum is a tentative discovery based upon a ^{26}Mg anomaly in one grain indicating an initial $^{26}Al/^{27}Al$ ratio of 9 x 10-4 (Huss *et al.* 1992), 18x higher than the 'canonical' value observed in CAIs and refractory micro-inclusions (*cf.*, Podosek *et al.* 1991).

2. Recent advances in the study of presolar grains: Some highlights

The isolation, identification and isotopic analysis of presolar grains has been a *tour de force* of meteoritic science in recent years. Three highlights are the following:

2.1. SiC with an s-process signature

The existence of s-process Xe in dust from red giants was predicted by Clayton and Ward in a manuscript rejected by *Geochimica et Cosmochimica Acta* in 1975 as being too speculative and unrelated to meteoritics (Clayton and Ward 1978; Clayton 1992). Indeed, no compelling evidence existed in 1975 for the presence of 'stardust' in meteorites. Nor was it expected because most cosmochemists then thought the solar nebula originated as a hot gas. However, once s-process Xe was observed in a meteorite residue (Srivinasan and Anders 1978), a quest began to isolate the carrier. The completion of this search took 10 years of effort before SiC (along with microdiamond and graphite microspherules) was isolated (Yang and Epstein 1983, 1984; Zinner and Epstein 1987; Lewis *et al.* 1987, 1989; Tang and Anders 1988a,b,c; Tang *et al.* 1988; Huss 1990). Subsequent studies have observed the s-process signature in additional elements: Kr, Rb, Sr, Ba, Nd, Sm (Ott *et al.* 1988; Lewis *et al.* 1990; Ott and Begemann, 1990; Ott 1990; Zinner *et al.* 1991; Richter *et al.* 1992, this volume; Prombo *et al.* 1992). Thus far, however, data for elements above iron have been obtained only on grain aggregates.   Nonetheless, these exhibit well-defined s-process features and variations attributable to neutron density (at ^{80}Kr, ^{86}Kr, ^{134}Xe and ^{148}Nd via the ^{79}Se, ^{85}Kr, ^{133}Xe ^{147}Nd and ^{147}Pm branch points) and exposure (at ^{138}Ba, which has a closed n-shell and very small n,γ cross-section). Interestingly, Kr correlates with grain size: the ^{80}Kr/^{82}Kr and ^{86}Kr/^{82}Kr neutron densiometers indicate an apparent increase in neutron density with grain size (Lewis *et al.* 1990), while the ^{138}Ba/^{136}Ba neutron exposure monitor indicates a decrease in exposure with an increase in grain size (Zinner *et al.* 1991b,c; Prombo *et al.* 1992). Kr, however, may be dominated by a minor population of gas-rich grains (Nichols *et al.* 1992).

A good case for an association of most meteoritic SiC with low mass thermally pulsing asymptotic giant branch (AGB) carbon (C) stars rests upon: (i) the direct observational identification of low mass AGB stars with 'main component' s-process abundances (Käppeler *et al.* 1989) and unstable s-nuclides (^{93}Zr, Tc) in their atmospheres (*cf.*, Lambert 1991); (ii) the presence of the 11.2 μm SiC emission feature in S, SC and C stars (Cohen 1984; Little-Marenin 1986; Martin and Rogers 1987; Little-Marenin and Little 1988; Blanco *et al.* 1988); (iii) the consistency between the high trace element abundances in SiC (Amari *et al.* 1992) and the large heavy element overabundances relative to Si in AGB atmospheres; (iv) the correspondence between apparent initial ^{26}Al/^{27}Al ratios in SiC (Zinner *et al.* 1991a) and ^{26}Al/^{27}Al predicted in low mass AGB models (Forestini *et al.* 1991); (v) the consistency of the ^{12}C/^{13}C range for most SiC grains (20-100, Amari *et al.* 1992c) and that observed in s-element enhanced MS, S giants and N-type C stars (Dominy 1985; Smith 1989; Smith and Lambert 1990); (vi) the clear s-process signature observed in Xe, Ba, Sm and Nd isotopes in bulk SiC fractions (Ott *et al.* 1988; Lewis *et al.* 1990; Ott and Begemann, 1990; Zinner *et al.* 1991b,c; Richter *et al.* 1992, these proceedings; Prombo *et al.* 1992), and specifically; (vii) the high neutron densities (~10^8 cm^{-3}) indicated by the ^{79}Se, ^{85}Kr, ^{133}Xe and ^{147}Nd branch crossings inferred, respectively, from the ^{80}Kr/^{82}Kr (low), ^{86}Kr/^{82}Kr (high), ^{87}Rb/^{85}Rb (high), ^{134}Xe/^{130}Xe (>0) and ^{148}Nd/^{146}Nd (>0) ratios of the inferred s-process component in bulk SiC (Ott *et al.* 1988; Ott 1990; Lewis *et al.* 1990; Gallino *et al.* 1990; Käppeler *et al.* 1990; Richter *et al.* 1992). Models of the core He-burning s-process in massive stars (including WC stars, which might conceivably condense circumstellar SiC in the presence of s-elements) suggest average neutron densities too low (<10^6 cm^{-3}) to cross the ^{79}Se and ^{85}Kr branch points (*e.g.*, Raiteri *et al.* 1991a), whereas abundance determinations of Rb/(Zr, Sr) in AGB stars with s-element overabundances are consistent with higher neutron densities of 10^7-10^8 cm^{-3} (Tomkin and Lambert 1983; Smith and Lambert 1984; Malaney and Lambert 1988; Beer and Macklin 1989; Wallerstein 1992) which bypass ^{85}Rb (which is otherwise strongly produced---due to its ~10x lesser n-capture cross section than ^{87}Rb---when the s-process flows through it via ^{85}Kr decay). Subsequent C-burning in massive stars may produce sufficiently high neutron densities (Raiteri *et al.* 1991b, these proceedings; Beer *et al.* 1992), but bulk yield estimates of s-Kr in massive stars (Raiteri *et al.* 1992) suggest they are not the source of SiC Kr. WC stars also should have high ^{12}C/^{13}C (>10^2) in their outflows (Prantzos *et al.* 1986). Group-4 graphite microspherules (Hoppe *et al.* 1992a) have ^{12}C/^{13}C

>10^2, but most analyzed SiC grains have $^{12}C/^{13}C$ <10^2 (Amari *et al.* 1992c). Graphite emission features are observed in WC outflows, but SiC features are not (Cohen *et al.* 1989).

One outstanding problem is the origin of the $\delta^{29}Si$-$\delta^{30}Si$ anomaly correlation (slope ~1.4) for most single SiC grains (Zinner *et al.* 1989; Ireland *et al.* 1991b; Stone *et al.* 1991; Virag *et al.* 1992), which is not consistent with a neutron capture trend as expected in AGB stars (slope ~0.5, Brown and Clayton 1992). The observed trend can be modelled by adding in $^{26}Mg(\alpha,n)^{29}Si$ reactions at the higher temperatures expected in more massive AGB stars (Brown and Clayton 1992). Associated predictions include a total burnout of ^{22}Ne and much higher $^{26}Al/^{27}Al$ than inferred from Mg isotopes and Al/Mg in the grains (Zinner *et al.* 1991; Virag *et al.* 1992). The higher-T scenario requires the Ne-E component observed in SiC to be due to ^{22}Na decay rather than ^{22}Ne implantation, whereas single grain measurements (Nichols *et al.* 1992) appear to contain the non-radiogenic Ne-E('H') component predicted in AGB models (Gallino *et al.* 1990) due the expected abundance complement of implanted ^4He. Ne isotopic measurements also indicate that implanted Ne-E(H) is present in bulk SiC fractions (Lewis *et al.* 1990). In the meantime a new class of rare SiC grains, 'Y' (Amari *et al.* 1992a) have been discovered which do show the expected slope-0.5 neutron capture trend in Si.

Much better constraints will be possible when trace element isotopes are measurable in single grains. Sr, Rb, Zr and Ba should be particularly interesting because they carry diagnostic information via their neutron density and temperature dependent branchings at ^{79}Se, ^{85}Kr, ^{86}Rb, ^{95}Zr and $^{134,135}Cs$ (Klay and Kappeler 1988; Beer and Macklin 1989; Toukan and Käppeler 1990; Käppeler *et al.* 1990; Beer 1991) and the closed-shell neutron exposure monitors ^{88}Sr (N = 50) and ^{138}Ba (N = 82). Zr is particularly interesting as it is one of the most abundant trace elements in large SiC grains (Virag *et al.* 1992; Amari *et al.* 1992b) and the only s-element for which isotopic abundances in AGB stars are known. Sr, Xe, Ba, Sm are also interesting because their p- and r-only masses monitor neutron exposure.

2.2. SiC from a supernova?

Recent ion probe investigations (Zinner *et al.* 1991; Ireland *et al.* 1991b; Amari *et al.* 1992a,c) have discovered 5 large (>1 µm), isotopically distinct SiC grains ('X') carrying the signature of explosive nucleosynthesis: heavy N, light C and light Si relative to solar. These grains also have very high apparent initial $^{26}Al/^{27}Al$ ratios ranging from 0.1 to 0.6. One grain, X2, has a

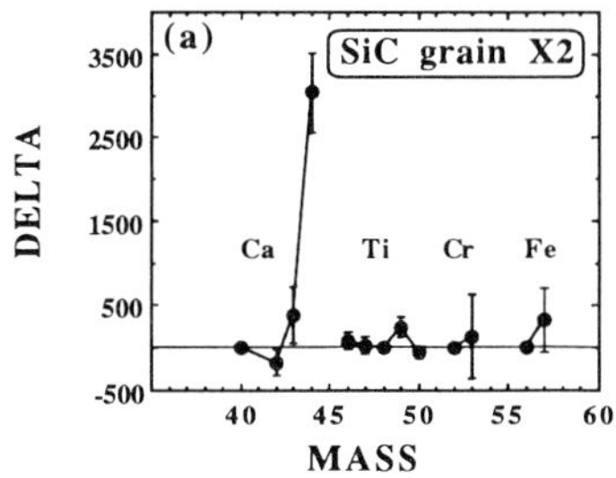

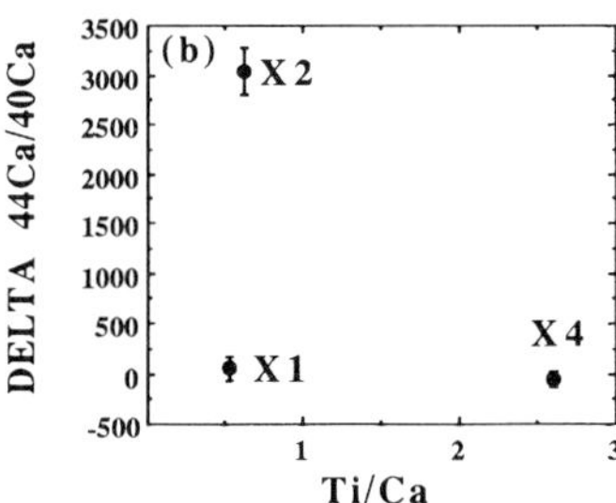

Figure 2a & b. a) A large excess of ^{44}Ca in grain X2 appears to be due to *in situ* decay of ^{44}Ti. Most other measured iron peak masses are normal within error. A supernova is the most likely source of freshly synthesized ^{44}Ti. b) However, there is no correlation between excess ^{44}Ca and Ti/Ca (Zinner *et al.* 1991b; Amari *et al.* 1992c), as expected if grains X1, X2 and X4 condensed contemporaneously with common $^{44}Ti/^{48}Ti$.

large ^{44}Ca anomaly probably due to *in situ* decay of ^{44}Ti ($T_{1/2}$ = 52 a, Fig. 2a). Others with measured Ca (X1, X4 and X5), however, do not show the anomaly, despite having similar

Ti/Ca ratios (Fig. 2b). Ti is also anomalous, with the largest effect at mass-49, which could be from *in situ* ^{49}V ($T_{1/2} = 337$ d) decay. The origin of these interesting grains remains a mystery because ^{44}Ti is thought to be an exclusive indicator of supernova nucleosynthesis, but the apparent initial ^{26}Al/^{27}Al is more than 10x too high for the bulk yields expected in supernovae (Arnould *et al.* 1980). ^{26}Al/^{27}Al ratios closer to those observed have been predicted for massive star outflows (Walter and Maeder 1989). A summary of the published isotopic data for Ca, Ti, Cr and Fe obtained for X2 is shown in Figure 2a. As a distinct ^{44}Ca anomaly is observed in submicron SiC aggregates (Amari *et al.* 1991), grains like X2 may be common in the smaller size fractions (which are too small for single-grain ion-probe analysis).

2.3. Interstellar graphitic spherules from red giants, novae and Wolf-Rayet stars?

The discovery of graphitic microspherules was the result of a two decade search for the carrier phase of Ne-E(L), an apparently pure ^{22}Ne component thought to be the product of *in situ* ^{22}Na ($T_{1/2} = 2.6$ a) decay (Black 1972; Amari *et al.* 1990). Ion probe studies at St. Louis and Canberra subsequently have revealed large anomalies in C, N, O, Mg and Ti, but normal Si (Amari *et al.* 1990; Bernatowitz *et al.* 1991; Zinner *et al.* 1991a; Hoppe *et al.* 1992a,b; Ireland *et al.* 1992). Mg isotopes indicate inferred non-zero initial ^{26}Al/^{27}Al ratios up to ~0.1 (Zinner *et al.* 1991a; Hoppe *et al.* 1992a,b). Single grain microanalysis with a gas source mass spectrometer at St. Louis has confirmed the ^{22}Na identification: Ne-E with ^{20}Ne/^{22}Ne < 0.01 and no accompanying ^{4}He (Nichols *et al.* 1992) in fourteen grains. Graphitic microsperules probably originate from a variety of different astrophysical source types, including red giants, Wolf-Rayet stars and possibly novae. A number of isotopically well-characterized grains match the isotopic expectations in CNO for Wolf-Rayet outflows at the early WC stage when He-burning products appear in the photosphere: ^{26}Al/^{27}Al ~0.1, ^{12}C/^{13}C trending from 40-4000, and ^{16}O/^{18}O < solar (Hoppe *et al.* 1992b).

3. Studies of refractory grains and Ca, Al-rich inclusions

3.1. Stubborn interpretive uncertainties

High precision U-Pb dating and initial ^{26}Al/^{27}Al determinations of CAIs and little-metamorphosed ordinary chondrites indicates ages consistent with formation of CAIs within the early solar system (Chen and Wasserburg 1981; Manhes *et al.* 1988; Ireland 1988, 1990; Ireland *et al.* 1990; Anders *et al.* 1991; Göpel *et al.* 1991; Podosek *et al.* 1991; Virag *et al.* 1991; Zinner and Göpel 1992). The cosmochemical mode of origin of CAI anomalies is an important unsolved problem because their nucleosynthetic interpretation will be unclear until the issue is settled. Models can be classified for heuristic purposes as either macroscopic or microscopic. Macroscopic models involve large isotopically distinct regions in the nebula and the identification of anomalies with isotopically anomalous ISM zones (R N Clayton and Mayeda 1977; Lee *et al.* 1978). One specific scenario of this type is inhomogeneous admixture of late-stage supernova ejecta in the context of induced star formation. Another interpretation (Hinton *et al.* 1988) involves large-scale chemical and isotopic inhomogeneities preserved in subnebular reservoirs in the context of the turbulent protoplanetary accretion disk model of Morfill (1983). Contrasting microscopic scenarios are based upon the idea of cosmochemical sorting of isotopically distinct presolar grain populations during the formation of refractory residues and condensates (R N Clayton *et al.* 1973, 1977; D D Clayton 1978; Consolmango and Cameron 1980; Lee *et al.* 1980; Jungck *et al.* 1984; Niemeyer and Lugmair

1984; Zinner *at al.* 1986; Fahey *et al.* 1987a,b; Niemeyer 1988; Ireland 1990). 'Microscenarios' have been discussed by Don Clayton and co-workers as a 'cosmic chemical memory' (CCM) paradigm which interprets the isotopic anomalies as a 'natural consequence of the aggregation and chemical transformations of presolar dust that routinely carries chemical and isotopic properties reflecting its own formation and interstellar evolution' (D D Clayton 1992). Neither the 'macro' or 'micro' scenarios are mutually exclusive.

3.2. The n-rich e-process anomaly suite and correlations with r-process anomalies

The most comprehensive studies of the NNSE anomaly suite have been made on Allende CAIs. By far the largest known effects in ^{50}Ti and ^{48}Ca are present in Murchison and Murray hibonites (*e.g.* Fahey *et al.* 1987a; Ireland 1990), but the small sizes of these grains and limitations on mass resolution in SIMS have prevented ion probe studies of other elements. Thermal ionization studies of the Allende FUN inclusion EK1-4-1 have identified a suite of positive anomalies in the iron peak elements Ca (Niederer and Papanastassiou 1984), Ti (Niederer *et al.* 1985), Cr (Papanastassiou 1986), Fe (Völkening and Papanastassiou 1989) and Zn (Völkening and Papanastassiou 1990) at all of the masses expected from NNSE calculations (Hartmann *et al.* 1985): ^{48}Ca (+134±2ε), ^{50}Ti (+38.0±1.6ε), ^{54}Cr (48.5±0.8ε), ^{58}Fe (+292±16ε) and ^{66}Zn (+16.7±3.7ε). Very high precision mass analyses of Ca, Ti, Cr, Ni and Zn in a suite of less anomalous ('ordinary') Allende CAIs have also observed a similar anomaly suite at lower levels (Birck and Lugmair 1988; Loss and Lugmair 1990). Both anomaly suites exhibit relative anomaly magnitudes within the range of that predicted in the model of Hartmann *et al.* (1985), with the exception of Fe and Zn, which in EK1-4-1 are higher and lower than model predictions, respectively. Zinc is a volatile element and as such may have been subject to isotopic equilibration.

Additional studies of Sr (Papanastassiou and Wasserburg 1978), Ba (McCulloch and Wasserburg 1978a), Nd (McCulloch and Wasserburg 1978a; Lugmair *et al.* 1978) Sm (McCulloch and Wasserburg 1978b; Lugmair *et al.* 1978) and Dy (Lugmair 1978) in EK1-4-1 have found anomalies in all cases. Ba and Sm exhibit positive anomalies at their unshielded masses relative to shielded s-only normalizations. The anomalies in Nd also fit with this scenario with a judicious choice of the fractionation correction. (Only one Nd mass is shielded, but the favored r-anomaly interpretation fits very well with the s/r decomposition based upon SiC Nd measurements. *Cf.,* Begemann 1992) The pattern of anomalies is consistent with an excess of r-process matter (Lee 1988). Curiously the stable p-process isotope of Sm (mass-144) shows a positive anomaly, whereas the p-isotopes of Ba (130, 132) are normal within uncertainty. This difference does not allow a distinction between the r-excess and s-deficit interpretations. In general, the EK1-4-1 data indicate an association between NNSE and the r-process. High precision work on Zr and Ba isotopes in a suite of 'ordinary'-type Allende CAIs confirm this association (Harper *et al.* 1990, 1991, 1992). Three inclusions with positive ^{96}Zr anomalies all had positive ^{50}Ti anomalies, whereas one inclusion without excess ^{96}Zr also had no excess ^{50}Ti. Ba anomalies similarly correlate with ^{50}Ti. Their relative magnitudes at mass-135 and mass-137 are in agreement with the near equal-atom (^{135*}Ba/^{137*}Ba ~1.0) r-component inferred for normal solar system Ba using an s/r decomposition based on the s-Ba composition inferred from the SiC measurements of Ott and Begemann (1990). They are not in agreement with the EK1-4-1 Ba anomaly pattern, which suggests an exotic r-component with ^{135*}Ba/^{137*}Ba ~0.6. Decomposition of the normal Zr abundances into s, r and p components also indicates that the apparent r-process excess at mass-96 is not accompanied by a p-process excess where it would be expected at mass-90 (Harper *et al.* 1991). The isotopic polarization appears to be r/(s+p), as in EK1-4-1

Ba, and suggests an r-excess rather than an s-deficit. However, the level of the Ba anomalies is too low to resolve associated effects in the very low abundance p-nuclides.

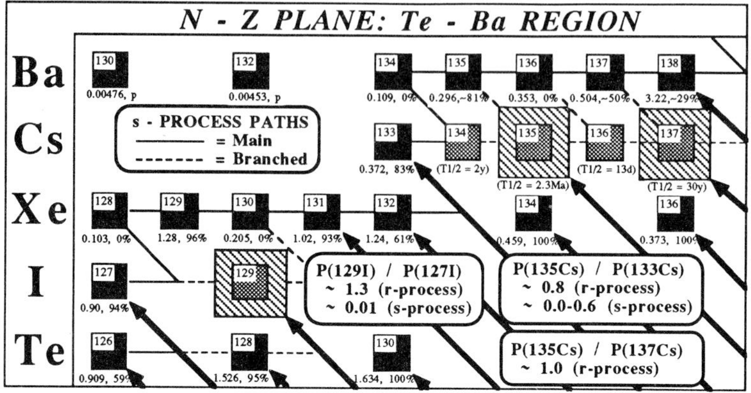

Figure 3. Nucleosynthesis in the Ba region. Heavy arrows represent decay to stability after freezeout of the r-process. The solar abundance distribution is given under each stable nuclide (normalized to Si = 10[6]) on the left. Estimated r-process fractional yields (in %) are given on the right. The r-yields are residuals after subtraction of s-process estimates: for Ba from SiC data (Ott and Begemann 1990, case 2), and from a main component best fit model of Gallino (pers. comm.) for Te, I, Xe and Cs.

I have described the correlation between NNSE and r-process anomalies as a 'cosmo-chemical association' (Harper *et al.* 1990, 1991). It is a hint that the astrophysical sites of these processes are similar or identical and supports arguments for locating the r-process site in the high entropy near-the-mass-cut 'hot bubble' predicted in core collapse supernovae driven by the delayed mechanism (Woosley and Hoffman 1992; Meyer *et al.* in press).

3.3. A late-input origin of the n-rich e-process anomaly suite? --radionuclide constraints

The recent discovery of the ^{60}Fe decay signature in meteorites (Shukolyukov *et al.* 1992) suggests that the NNSE component may have been synthesized near in time to origin of the solar system. Iron-60 is predominantly an NNSE product (Thielemann, pers. comm.) and is therefore a potential chronometer of a putative 'late input' event. Its inferred solar system *ab initio* abundance (^{60}Fe/^{56}Fe~2 x 10^{-6}) is an order of magnitude greater than that predicted for a standard galactic evolution model with no free-decay interval:

$$^{60}\text{Fe}/^{56}\text{Fe} \sim [P(^{60}\text{Fe})/P(^{56}\text{Fe})][(\kappa+1)\tau/T_g] \sim 2 \times 10^{-7}$$

where τ is the ^{60}Fe mean life (2.2 Ma), $\kappa = 1.5$ is Clayton's preferred galactic evolution infall parameter (D D Clayton 1984), $T_g \sim 10^4$ Ma is the presolar galactic age, and $P(^{60}\text{Fe})/P(^{56}\text{Fe})$ is obtained as 2 times the bulk solar system ^{50}Ti/^{56}Fe ratio (1.6 x 10^{-4}) [Assuming $P(^{60}\text{Fe}) \sim 2P(^{50}\text{Ti})$, from Hartmann *et al.* (1985), and total NNSE production of ^{50}Ti and ^{60}Fe]. This result indicates a large ^{60}Fe overabundance. As r-process anomalies correlate with NNSE anomalies, overabundances in the shorter-lived r-process chronometers are also expected.

The most diagnostic elements for investigating the age of an r-process component are U, Pu and Ba. U and Pu are useful diagnostics of provenance age because freshly-synthesized r-process matter will have ^{235}U/^{238}U ~1.2 and ^{244}Pu/^{238}U ~ 0.4 (Cowan *et al.* 1991). These ratios are much higher than in the bulk solar system 4.57 Ga ago (^{235}U/^{238}U

~0.03 and $^{244}Pu/^{238}U$ ~0.007. Hagee *et al.* 1990). Initial $^{244}Pu/^{238}U$ and $^{244}Pu/^{232}Th$ ratios estimated in hibonites from their fission track records and U, Th concentrations (Rajan and Tamhane 1982; Fahey *et al.* 1987b) indicate $^{244}Pu/^{238}U$ ratios <0.1 and (more importantly) $^{244}Pu/^{232}Th$ <10^{-3}, in grains having up to a 10% excess of ^{50}Ti (δ_{50Ti} ~103). These results do not support late input macroscenarios for the NNSE-r association in CAIs. However, the ^{244}Pu constraint may not be strict because r-process events may not all produce actinides.

Ba is a particularly interesting element because it carries information on the initial presence or absence of the unshielded 'hold-up' chronometers ^{135}Cs ($T_{1/2}$ = 2.3 Ma) and ^{137}Cs ($T_{1/2}$ = 30 a). Cs is a volatile element and its presence at hold-ups in the β-decay chains leading to ^{135}Ba and ^{137}Ba (Figure 3) means that any r-process Ba condensed into high-temperature refractory condensates (*viz.*, formed during the adiabatic cooling phase: ~1-5 yr), will be a bi-isotopic mixture of ^{137}Ba and ^{138}Ba with $^{137}Ba/^{138}Ba = [P(^{137}Ba)/P(^{138}Ba)][e^{-0.023t}]$, where t is time of grain condensation in years and 0.023 a^{-1} is the decay constant of ^{137}Cs. Furthermore, $^{135}Ba/^{137}Ba$ will be close to zero in these grains if the r-process yield dominates the presupernova 'initial' Ba in the unprocessed envelope and produced by s-process synthesis during presupernova evolution. (A Ba overabundance presumably from the presupernova s-process was observed in SN 1987A. *Cf.*, Mazzali *et al.* 1992.) In contrast to this prediction, the trend of the CAI anomaly data is consistent with the r-process production ratio $P(^{135}Ba)/P(^{135}Ba)$ ~0.96±0.12 estimated from the decomposition of bulk solar Ba using Ott and Begemann's (1990) SiC s-process Ba composition. This appears to rule out origin of anomalies models involving 'shrapnel'-like injection of supernova condensate dust grains into the protosolar nebula (Lattimer *et al.* 1978; Schramm 1978; Margolis and Falk 1979).

The critical question is whether or not r-process Ba is expected to condense into high temperature supernova condensates. The most plausible r-process site presently appears to be at the base of the ejected mass in massive star core collapse supernovae (Woosley and Hoffman 1992; Bitouzet *et al.* in press; Meyer *et al.* in press). Clayton (1987) argues that r-process nuclides in the basal ejecta are isolated from oxygen and cannot condense into refractory oxides. He also expects the basal ejecta to be heated by the pulsar and predicts that the deep zones do not condense during the adiabatic cooling phase. These arguments assume a chemically stratified 'onion'-type models of ejecta transport, which now appear improbable for Type II and Ib supernovae. Several lines of evidence from supernova light curve analysis, especially SN 1987A, clearly indicate macro-mixing within the ejecta (*cf.*, Benz and Thielemann 1990). Hydrodynamic models (Herant and Benz 1992) suggest that mixing is strongest in the innermost zones where the r-process is expected to occur. Very little is known about the long-term fate of refractory elements in supernova ejecta. It is clear that refractory condensates will form only at high gas temperatures during the early adiabatic cooling of the ejecta (t ~1-3 a). Dust formed from ~450 days onwards in SN 1987A, in agreement with predictions in condensation models based upon homogeneous nucleation theory (Kozasa *et al.* 1989, 1991). Dust also is observed in metal-rich filaments of the crab nebula (SN 1054), 938 years after explosion, but the critical factor---the condensation efficiency for refractory elements---is not well constrained (Fesen and Blair 1990), and may be quite small (Davidson and Fesen 1985). The ~330 year-old Cas A remnant contains similar metal-rich filamentary 'knots' from the inner zones of its type II SN progenitor (Chevalier and Kirschner 1979). Dwek and Warner (1981) predicted that these would contain condensate dust, but observations (Dinerstein *et al.* 1987; Greidanus and Strom 1991) do not provide strong constraints on condensation efficiencies. SN 1987A observations, however, are consistent with condensation efficiency factors >~0.3 (Kozasa *et al.* 1991; Lucy *et al.* 1991; Spyromilio and Graham 1992), indicating a substantial partitioning of refractory metals into dust. The degree of grain destruction during transit into the ISM is not known. The state of r-process elements entering the ISM remains an important but essentially unanswered question.

If r-process Cs and Ba enter the ISM in the gas phase, and if the gas phase residence time scale is long enough for ^{135}Cs decay, then r-Ba accretes into grain mantles in the ISM with ^{135}Ba/^{137}Ba = P(135)/P(137) ~1. Mixing polarizations with respect to other dust populations (*e.g.*, SiC from red giants) would then lead to a correlation between ^{135}Ba/^{137}Ba = P(135)/P(137)-type anomalies and ^{48}Ca-^{50}Ti. This prediction matches the observed anomalies and is consistent with the CCM interpretation of D D Clayton (1987, 1992).

Next we consider an alternate macroscenario: large-scale nebular inhomogeneities fed by compositional gradients in the parental molecular cloud from admixed supernova ejecta (R N Clayton and Mayeda 1977; Cameron and Truran 1977; Schramm and R N Clayton 1978; Lattimer *et al.* 1978; Wasserburg and Papanastassiou 1982; Fowler 1984; Wasserburg 1985; Arcoragi 1989). Ejecta knot mixing into collapsing molecular cloud cores may be a realistic scenario for star formation in compressed shells around supernova remnants and especially in Bok globules. Protostars have been observed in a distribution around a 'molecular shell'-type supernova remnant in the dense region compressed outwards by the wind of its progenitor (Junkes *et al.* 1992). Protostars are also observed in Bok Globules formed by the exposure of molecular cloud cores by *uv* erosion from massive stars (Reipurth *et al.* 1992; Scarotti *et al.* 1992; Sahu and Sahu 1992; Block 1990; Neckel and Staude 1990; Yun and Clemens 1990; Duvert *et al.* 1990; Petterson 1987; Reipurth and Gee 1986; Turner 1986; Reipurth and Bouchet 1984; Reipurth 1983; Brand *et al.* 1983; Schneps *et al.* 1980). The wind from a massive star is a plausible source for ^{26}Al in the early solar system (Prantzos 1991). When the star explodes, ejecta knots traverse the low density bubble and mix into the shell region where a small fraction conceivably could be sucked into collapsing protostellar cores along with the stellar wind material (Reeves 1978, 1979; Arcoragi 1989). Ba is a diagnostic element for these scenarios because: (i) most CAIs (including those shown in Figure 1) are depleted in Cs and enriched in Ba: Cs/Ba ~0; and (ii) the r-process Ba composition in any putative exotic component is a function of its age set by an appropriate timescale from ^{135}Cs-decay:

$$[^{135}Ba*/^{137}Ba*] = [P(135)/P(137)][1-e^{-\lambda_{135}\Delta t}]$$

For a free decay interval of one half life (2.3 Ma), the exotic Ba component will have ^{135}Ba*/^{137}Ba* ~0.5. This is close to that inferred from the relative magnitudes of the large Ba anomalies in EK1-4-1 (McCulloch and Wasserburg 1978a). But it is not consistent with the pattern of the smaller 'ordinary' CAI anomalies shown in Fig. 1c, which indicate an approximately equal-atom exotic Ba component. The ordinary CAI anomalies imply a free decay interval >~5 Ma and do not support injection (macro-)models. Further work with Ba isotopes in meteorites will be required to determine the *ab initio* ^{135}Cs/^{133}Cs ratio of the solar system and investigate a possible relation to the source(s) of ^{60}Fe and the NNSE component.

4. New frontiers in 'isotopic astronomy'

The discoveries of presolar diamond, SiC, graphite and corundum and the NNSE anomalies demonstrate that meteorites are a unique source of astrophysical information. Experience demonstrates a correlation between progress, availability of novel samples and innovative analytical equipment. New developments will require further innovation in both directions.

4.1. New samples

The isotopically anomalous inclusions and presolar phases discovered thus far initially attracted attention either by being particularly conspicuous (*white* Allende CAIs and *blue* Murchison hibonites) or impervious to acid dissolution treatments (diamond, graphite, SiC, Al_2O_3, and Si_3N_4---Lee *et al.* 1992). Primitive meteorites probably contain many varieties of undiscovered presolar grains and anomalous objects. Predicted circumstellar and supernova condensates not yet found in meteorites include carbides of Al, Ti, Mg, Ca and Fe, metallic Fe and Ni, nitrides of Al and Ti, oxides of Ti, Cr, Fe, Zr, spinel, perovskite and various silicates (Gilman 1969; Lattimer *et al.* 1978; Tarafdar 1987; Sharp and Huebner 1990).

4.2. New instrumentation: 'Cosmochemical observatories'

As presolar grains are mostly submicron objects, precise mass analysis of even their major elements presents a formidable technical challenge. Optimal instrumentation for presolar grain studies will make direct sample mass analyses with 'total information extraction' across the entire mass range of the nuclides. The development of new innovative mass spectrometers with improved performance in this direction will make the study of isotopic anomalies in 'cosmochemical observatories' an exciting and challenging field for many years to come.

References

Alexander C M O'D, Swan P and Walker R M 1990 *Nature* **348** 715
Amari S, Hoppe P, Zinner E and Lewis R S 1992a *Lunar Planet. Sci.* **XXIII** 27
Amari S, Hoppe P, Zinner E and Lewis R S 1992b *Meteoritics* **27** 197
Amari S, Hoppe P, Zinner E and Lewis R S 1992c *Ap. J.* **394** L43
Amari S, Zinner E and Lewis R S 1991 *Meteoritics* **26** 14
Amari S, Anders E, Virag A and Zinner E 1990 *Nature* **345** 238
Anders E, Virag A. Zinner E and Lewis R S 1991 *Ap. J.* **373** L77
Arcoragi J-P 1989 In: *Astrophysical Ages and Dating Methods* (eds. E Vangioni-Flam, M Casse J Audouze
     and J Tran Thanh Van. Paris: Editions Frontieres) 457
Arnould M, Norgaard H, Thielemann F-K and Hillebrandt W 1980 *Ap. J.* **237** 931
Beer H 1991 *Ap. J.* **375** 823
Beer H, Walter G and Käppeler 1992 *Ap. J.* **389** 784
Beer H and Macklin R L 1989 *Ap. J.* **339** 962
Begemann F 1992 preprint (Reeves Symposium: *Origin and Evolution of the Elements*)
Benz W and Thielemann F-K 1990 *Ap. J.* **348** L17
Bernatowitz T, Amari S, Zinner E and Lewis R S 1991 *Ap. J.* **373** L73
Bernatowitz T, Fraundorf G, Tang M, Anders E, Wopenka B, Zinner E and Frauendorf P 1987 *Nature* **330** 728
Birk J-L and Lugmair G W 1988 *Earth Planet. Sci. Lett.* **90** 131
Birk J-L and Allegre C J 1984 *Geophys. Res. Lett.* **11** 943
Bitouzet J-P, Thielemann F-K, Kratz K-L, Möller P and Pfeiffer B 1992 preprint
Black D C 1972 *Geochim. Cosmochim. Acta* **36** 377
Blake D F, Freund F, Krishnan K F M, Echer C J, Shipp R, Bunch T E, Tielens A G, Lipari R J,
     Hetherington C J D and Chang S 1988 *Nature* **332** 611
Blanco A, Borghesi A, Bussoletti E, Colangeli L, Fonti S and Orfino V 1988 In: *Experiments on Cosmic
     Dust Analogues* (eds. E. Bussoletti *et al.* Amsterdam: Kluwer) 167
Block D L 1990 *Nature* **347** 452
Brand P W J L, Hawarden T G, Longmore A J, Williams P M and Caldwell J A R 1983 *Mon. Not. R. astr.
     Soc.* **203** 215
Brown L E and Clayton D D 1992 *Ap. J.* **392** L79
Cameron A G W and Truran J W 1977 *Icarus* **30** 447
Chen J H and Wasserburg G J 1981 *Earth Planet. Sci. Lett.* **52** 1
Chevalier R A and Kirshner 1979 *Ap. J.* **233** 154

Clayton D D 1992 *Meteoritics* **27** 5
Clayton D D 1984 *Ap. J.* **285** 411
Clayton D D 1987 *Lunar Planet. Sci.* **XVIII** 183
Clayton D D and Ward R A 1978 *Ap. J.* **224** 1000
Clayton R N and Mayeda T K 1977 *Geophys. Res. Letts.* **4** 295
Clayton R N, Onuma N, Grossman L and Mayeda T K 1977 *Earth Planet. Sci. Letts.* **34** 209
Clayton R N, Grossman L and Mayeda T K 1973 *Science* **182** 485
Cohen M 1984 *Mon. Not. R. Astron. Soc.* **206** 137
Cohen M, Tielens A G G M and Bregman J D 1989 *Ap. J.* **344** L13
Consolmagno G J and Cameron A G W 1980 *Moon and Planets* **23** 3
Cowan J J, Thielemann F-K and Truran J W 1991 *Phys. Rep.* **208** 267
Davidson K and Fesen 1985 *Ann. Rev. Astron. Astrophys.* **23** 119
Davis A M, Hashimoto A, Clayton R N and Mayeda T K 1990 *Nature* **347** 655
Dinerstein H L, Lester D F, Rank D M, Werner M W and Wooden D H 1987 *Ap. J.* **312** 314
Dominy J F 1985 *Publ. Ast. Soc. Pac.* **97** 1104
Duvert G, Cernicharo J, Bachiller R and Gomez-Gonzalez J 1990 *Astron. Astrophys.* **233** 190
Dwek E and Werner M W 1981 *Ap. J.* **248** 138
Fahey A J, Goswami J N, McKeegan K D and Zinner E 1987a *Ap. J.* **323** L91
Fahey A, Goswami J N, McKeegan K D and Zinner E 1987b *Geochim. Cosmochim. Acta* **51** 329
Fahey A, Zinner E K, Crozaz and Kornacki A S 1987c *Geochim. Cosmochim. Acta* **51** 3215
Fahey A, Zinner E and Kurat G 1986 *Meteoritics* **21** 359
Fahey, Goswami J N, McKeegan K D and Zinner E 1985 *Ap. J.* **296** L17
Fesen R A and Blair W P 1990 *Ap. J.* **351** L45
Forestini M, Paulus G and Arnould M 1991 *Astron. Astrophys.* **252** 597
Fowler W A 1984 *Science* **226** 922
Gallino R, Busso, M. Piccio G and Raiteri C M 1990 *Nature* **348** 298
Gilman R C 1969 *Ap. J.* **155** L185
Göpel C, Manhes G and Allegre C J 1991 *Meteoritics* **26** 338
Greidanus H and Strom R G 1991 *Astron. Astrophys.* **249** 521
Grossman L 1980 *Ann. Rev. Earth Planet. Sci.* **8** 559
Hagee B, Bernatowitz T J, Podosek F A, Johnson M L, Burnett D S and Tatsumoto M 1990 *Geochim. Cosmochim. Acta* **54** 2847
Harper C L and Wiesmann H 1992 *Lunar Planet. Sci.* **XXIII** 489
Harper C L, Nyquist L E, Shih C-Y and Wiesmann H 1990 In: *Proc. Int. Symp. Nuclear Astrophys. 'Nuclei in the Cosmos'* (eds. H Oberhummer and W Hillebrandt. Garching: MPI Phys. Astrophys.)P4 138
Harper C L, Wiesmann H, Hartmann D, Meyer B, Howard W M, Gallino R, Raiteri C M 1991 *Lunar Planet. Sci.* **XXII** 517
Harper C L, Wiesmann H and Nyquist L E 1992 *Meteoritics* **27** 230
Hartmann D, Woosley S E and El Eid M F 1985 *Ap. J.* **297** 837
Herant M and Benz W 1992 *Ap. J.* **387** 294
Hinton R W, Davis A M, Scatena-Wachel D E, Grossman L and Draus R J *et al.* 1988 *Geochim. Cosmochim. Acta* **52** 2573
Hinton R W, Davis A M and Scatena-Wachel D E 1987 *Ap. J.* **313** 420
Hoppe P, Amari S, Zinner, E and Lewis R S 1992a *Lunar Planet. Sci.* **XXIII** 553
Hoppe P, Amari S, Zinner, E and Lewis R S 1992b *Meteoritics* **27** 235
Huss G R 1990 *Nature* **347** 159
Huss G R, Hutcheon I D, Wasserburg G J and Stone J 1992 *Lunar Planet. Sci.* **XXIII** 563
Hutcheon I D, Steele I M, Wachel D E S, MacDougall J D and Phinney D 1983 *Lunar Planet. Sci.* **XIV** 339
Ireland T R 1990 *Geochim. Cosmochim. Acta* **54** 3219
Ireland T R 1988 *Geochim. Cosmochim. Acta* **52** 2827
Ireland T R, Amari S, Hoppe P, Zinner E, and Lewis R S 1992 *Meteoritics* **27** 237
Ireland T R, Fahey A J and Zinner E K 1991a *Geochim. Cosmochim. Acta* **55** 367
Ireland T R, Zinner E and Amari S 1991b *Ap. J.* **376** L53
Ireland T R, Compston W, Williams I S and Wendt I 1990 *Earth Planet. Sci. Lett.* **101** 379
Ireland T R, Compston W and Heydegger H R 1985 *Geochim. Cosmochim. Acta* **49** 1989
Jungck M H A, Shimamura T and Lugmair G W 1984 *Geochim. Cosmochim. Acta* **48** 2651
Junkes N, Fürst E and Reich W 1992 *Astron. Astrophys.* **261** 289
Käppeler F, Gallino R, Busso M, Piccio G, and Raiteri C M 1990 *Ap. J.* **354** 630

Käppeler F, Beer H and Wisshak K 1989 *Rep. Prog. Phys.* **52** 945

Kozasa T, Hasegawa H and Nomoto K 1991 *Astron. Astrophys.* **249** 474

Kozasa T, Hasegawa H and Nomoto K 1989 *Ap. J.* **344** 325

Klay N and Käppeler F 1988 *Phys. Rev.* **C38** 295

Lambert D L 1991 In: *Evolution of Stars: The Photospheric Abundance Connection* (eds. G Michaud and A. Tutukov, IAU) 299

Lattimer J M, Schramm D N and Grossman L 1978 *Ap. J.* **219** 230

Lee M R, Russell S S, Arden J W and Pillinger C T 1992 *Meteoritics* **27** 248

Lee T 1988 In: *Meteorites and the Early Solar System* (eds. J F Kerridge nd M S Matthews, Tucson: Univ. Arizona Press) 1063

Lee T, Mayeda T K and Clayton R N 1980 *Geophys. Res. Lett.* **7** 493

Lee T, Russell W A and Wasserburg G J 1979 *Ap. J.* **228** L93

Lee T, Papanastassiou D A and Wasserburg G J 1978 *Ap. J.* **220** L21

Lewis R S Amari S and Anders E 1990 *Nature* **348** 293

Lewis R S Anders E and Draine B T 1989 *Nature* **339** 117

Lewis R S, Tang M, Wacker J F, Anders E and Steel E 1987 *Nature* **326** 160

Little-Marenin I R 1986 *Ap. J.* **307** L15

Little-Marenin I R and Little S J 1988 *Ap. J.* **333** 305

Loss R D and Lugmair G W 1990 *Ap. J.* **360** L59

Loss R D, Lumair G W, MacPherson G J and Davis A M 1990 *Lunar Planet. Sci.* **XXI** 718

Lucy L B, Danziger I J, Gouiffes C and Bouchet P 1991 In: *Supernovae* (ed. S E Woosley. New York: Springer) 82

Lugmair G W 1978 *USGS Open File Rpt.* **78-701** 262

Lugmair G W, Marti K and Scheinin N B 1978 *Lunar Planet. Sci.* **IX** 672

MacDougall J D 1979 *Earth Planet. Sci. Lett.* **42** 1

MacPherson G J, Bar-Matthews M, Tanaka T, Olsen E and Grossman L 1983 *Geochim. Cosmochim. Acta* **47** 823

Malaney R A and Lambert D L 1988 *Mon. Not. R. astr. Soc.* **235** 695

Manhes G, Göpel C and Allegre C J 1988 *Comptes Rendus de l'ATP Planetologie* 323

Margolis S H and Falk S W Jr 1979 *Astrophys. Space Sci.* **65** 119

Martin P G and Rogers C 1987 *Ap. J.* **322** 364

Mason B and Taylor S R 1982 *Smithson. Contrib. Earth Sci.* #25

Mazalli P A, Lucy L B and Butler K 1992 *Astron. Astrophys.* **258** 399

McCulloch M T and Wasserburg G J 1978a *Ap. J.* **220** L15

McCulloch M T and Wasserburg G J 1978b *Geophys. Res. Lett.* **5** 599

Meyer B S, Howard W M, Mathews G J, Hoffman R and Woosley S E 1992 preprint

Morfill G E 1983 *Icarus* **53** 41

Neckel T and Staude H J 1990 *Astron. Astrophys.* **231** 165

Nichols R H Jr, Hohenberg C M, Hoppe P, Amari S and Lewis R S 1992 *Lunar Planet. Sci.* **XXIII** 989

Niederer F R and Papanastassiou D A 1984 *Geochim. Cosmochim. Acta* **48** 1279

Niederer F R, Papanastassiou D A and Wasserburg G J 1985 *Geochim Cosmochim Acta* **49** 835

Niederer F R, Papanastassiou D A and Wasserburg G J 1981 *Geochim Cosmochim Acta* **45** 1017

Niederer F R, Papanastassiou D A and Wasserburg G J 1980 *Ap. J.* **240** L73

Niemeyer S 1988 *Geochim. Cosmochim. Acta* **52** 2941

Niemeyer S and Lugmair G W 1984 *Geochim. Cosmochim. Acta* **48** 1401

Niemeyer S and Lugmair G W 1981 *Earth Planet. Sci. Lett.* **53** 211

Ott U 1990 In: *Proc. Int. Symp. Nuclear Astrophys. 'Nuclei in the Cosmos'* (eds. H Oberhummer and W Hillebrandt. Garching: MPI Phys. und Astrophys.) **P4** 151

Ott U and Begemann F 1990 *Ap. J.* **353** L57

Ott U, Begemann, Yang J and Epstein S 1988 *Nature* **332** 700

Papanastassiou 1986 *Ap. J.* **308** L27

Papanastassiou D A and Brigham C A 1989 *Ap. J.* **338** L37

Papanastassiou D A and Wasserburg G J 1978 *Geophys. Res. Lett.* **5** 595

Petterson B 1987 *Astron. Astrophys.* **171** 101

Podosek F A and Brannon J C 1988 *Lunar Planet. Sci.* **XIX** 941

Podosek F A, Zinner E K, MaCPherson G J, Lundberg L L, Brannon J C and Fahey A J 1991 *Geochim. Cosmochim. Acta* **55** 1083

Prantzos N 1991 In: *Gamma-Ray Line Astrophysics* (eds. P Duroucheux and N Prantzos. New York: AIP) 129

Prantzos N, Doom C, Arnould M and De Loore C 1986 *Ap. J.* **304** 695
Prombo C A, Podosek F A, Amari S and Lewis R S 1992 *Lunar Planet. Sci.* **XXIII** 1111
Raiteri C M, Gallino, R and Busso M 1992 *Ap. J.* **387** 263
Raiteri C M, Busso M, Gallino R, Piccio G and Pulone L 1991a *Ap. J.* **367** 228
Raiteri C M, Busso M, Gallino R and Piccio G 1991b *Ap. J.* **371** 665
Rajan R S and Tamhane A S 1982 *Earth Planet. Sci. Lett.* **58** 129
Reeves H 1979 *Ap. J.* **231** 229
Reeves H 1978 In: *Protostars and Planets* (ed. T Gehrels. Tucson: Univ. of Arizona Press) 399
Reipurth B 1983 *Astron. Astrophys.* **117** 183
Reipurth B, Heathcote S and Vrba F 1992 *Astron. Astrophys.* **256** 225
Reipurth B and Gee G 1986 *Astron. Astrophys.* **166** 148
Reipurth B and Bouchet P 1984 *Astron. Astrophys.* **137** L1
Richter S, Ott U and Begemann F 1992 *Lunar Planet. Sci.* **XXIII** 1147
Sahu M and Sahu K C 1992 *Astron. Astrophys.* **259** 265
Scarrott S M, Draper P W, Rolph C D and Wolstencroft R D 1992 *Mon. Not. R astr. Soc.* **257** 485
Schneps M H, Ho P T P and Barrett A H 1980 *Ap. J.* **240** 84
Schramm D N 1978 In: *Protostars and Planets* (ed. T Gehrels. Tucson: Univ. Arizona Press) 384
Schramm D N and Clayton R N 1978 *Sci. Am.* **239/4** 124
Sharp C M and Huebner W F 1990 *Ap. J. Suppl.* **72** 417
Shukolyukov A and Lugmair G W 1992 *Lunar Planet. Sci.* **XXIII** 1295
Smith V V 1989 In: *Cosmic Abundances of Matter* (ed. C J Waddington. New York: Am. Inst. Phys.) 200
Smith V V and Lambert D L 1990 *Ap. J. Suppl.* **72** 387
Spyromilio J and Graham J R 1992 *Mon. Not. R. astr. Soc.* **255** 671
Srinivasan B and Anders E 1978 *Science* **201** 51
Stone J, Hutcheon I D, Epstein S and Wasserburg G J 1991 Earth Planet. Sci. Lett. **107** 570
Tang M and Anders E 1988a *Ap. J.* **335** L31
Tang M and Anders E 1988b *Geochim. Cosmochim. Acta* **52** 1235
Tang M and Anders E 1988c *Geochim. Cosmochim. Acta* **52** 1245
Tang M, Anders E, Hoppe P and Zinner E 1989. *Nature* **339** 351
Tang M, Lewis R S, Anders E, Grady M M, Wright I P and Pillinger C T 1988 *Geochim. Cosmochim. Acta* **52** 1221
Tarafadar S P 1987 In: *Astrochemistry* (eds. S P Tarafadar and M S Vardya. Dordrecht: Reidel) 559
Tomkin J and Lambert D L 1983 *Ap. J.* **273** 722
Toukan K A and Käppeler F 1990 *Ap. J.* **348** 357
Turner D G 1986 *Astron. Astrophys.* **167** 157
Virag A, Wopenka B, Amari, Zinner E, Anders E and Lewis R S 1992 *Geochim. Cosmochim. Acta* **56** 1715
Virag A, Zinner E, Amari S and Anders E 1991 *Geochim. Cosmochim. Acta* **55** 2045
Völkening J and Papanastassiou D A 1990 *Ap. J.* **358** L29
Völkening J and Papanastassiou D A 1989 *Ap. J.* **347** L43
Wallerstein G 1992 *Publ. Astr. Soc. Pacific* **104** 511
Walter R and Maeder A 1989 *Astron. Astrophys.* **218** 123
Wasserburg G J 1985 In: *Protostars and Planets II* (eds. D C Black and M S Mathews. Tucson: Univ. Arizona Press) 703
Wasserburg G J and Papanastassiou 1982 In: *Essays in Nuclear Astrophysics* (eds. C A Barnes, D D Clayton and D N Schramm, Cambridge: Cambridge Univ. Press) p 77
Woosley S E and Hoffman 1992 *Ap. J.* **395** 202
Yang J and Epstein S 1984 *Nature* **311** 544
Yang J and Epstein S 1983 *Geochim. Cosmochim. Acta* **47** 2199
Yun J L and Clemens D P 1990 *Ap. J.* **365** L73
Zinner E 1988 In: *Meteorites and the Early Solar System* (eds. J F Kerridge nd M S Matthews, Tucson: Univ. Arizona Press) 956
Zinner E and Göpel C 1992 *Meteoritics* **27** 311
Zinner E and Epstein S 1987 *Earth Planet. Sci. Lett.* **84** 359
Zinner E, Amari S, Anders E and Lewis R S 1991a *Nature* **349** 51
Zinner E, Amari S and Lewis R S 1991b *Meteoritics* **26** 413
Zinner E, Amari S and Lewis R S 1991c *Ap. J.* **382** L47
Zinner E Tang M and Anders E 1989 *Geochim. Cosmochim. Acta* **53** 3273
Zinner E, Fahey A J, Goswami J N, Ireland T R and McKeegan K D 1986 *Ap. J.* **311** L103

s-Process isotope abundance anomalies in meteoric silicon carbide: new data

S Richter, U Ott, and F Begemann

Max-Planck-Institut für Chemie (Otto-Hahn-Institut), Saarstraße 23, D-6500 Mainz, F.R.G.

Abstract. Silicon carbide contained in primitive meteorites with an abundance of a few parts per million contains trace elements that carry the signature of the s-process. We report new and improved data for s-process neodymium and samarium in SiC. The composition of s-process samarium indicates little bypassing of ^{147}Sm due to neutron capture by ^{147}Pm and little decay of ^{151}Sm into ^{151}Eu. Furthermore, the results suggest that there are serious problems with the neutron capture cross sections for Nd.

1. Introduction

During the past few years researchers in the field of meteoritics have succeeded in isolating from primitive meteorites rare (abundances down to a few ppm) graphite, diamond, and silicon carbide and identifying these phases as presolar in origin (cf. e.g. Ott 1992). Of these, presolar silicon carbide carries trace elements that clearly show the signature of the s-process of nucleosynthesis. Judging from the isotopic composition of these trace elements, the process responsible for their production resembles the so-called main component thought to occur during thermal pulses in AGB stars (Gallino et al., 1990). Elements for which results were obtained first were the noble gases xenon and krypton (Srinivasan et al. 1978; Ott et al. 1988), followed by barium and strontium (Ott et al. 1990a,b). The results for krypton and barium in particular have allowed to put constraints on the parameters of the s-process - neutron density and temperature (based on the ^{79}Se and ^{85}Kr branchings that affect the abundances of ^{80}Kr and ^{86}Kr - Ott et al. 1988; Gallino et al. 1990) and the neutron exposure (based on the abundance of ^{138}Ba with its extremely small neutron capture cross section - Ott and Begemann 1990a; Gallino et al. 1992). Here we report new data obtained for Nd and Sm that add to and improve previously reported results.

2. Neodymium

Nd and Sm data for interstellar SiC were first obtained by ion microprobe (Zinner et al. 1991). A major obstacle for drawing conclusions about the composition of s-process Nd and s-process Sm from these results was the fact that isotopes with atomic masses 148 and 150 exist for both elements (r-only for Nd, s-only for Sm), and that both elements contributed about equally to the signal measured at these masses. We have shown that, using thermal ionization of Sm and Nd from directly-loaded SiC, the difference in ionization yields for the two elements as a function of temperature allows to measure SiC-Nd virtually without compromising Sm interferences (Richter et al. 1992). Nd isotopic anomalies reported by Richter et al. (1992) approached, but did not quite reach, those from the ion probe work. In the meantime we have performed an additional purification of our SiC sample. The new data (labelled '5') are shown together with our old data for isotopes 143 to 150 in Fig. 1. The size of the isotopic anomaly is virtually identical now with the one obtained in the ion probe work suggesting that the s-process elements introduced into the SiC were contaminated by r-process nuclides from a previous astration. A similar result was previously obtained for

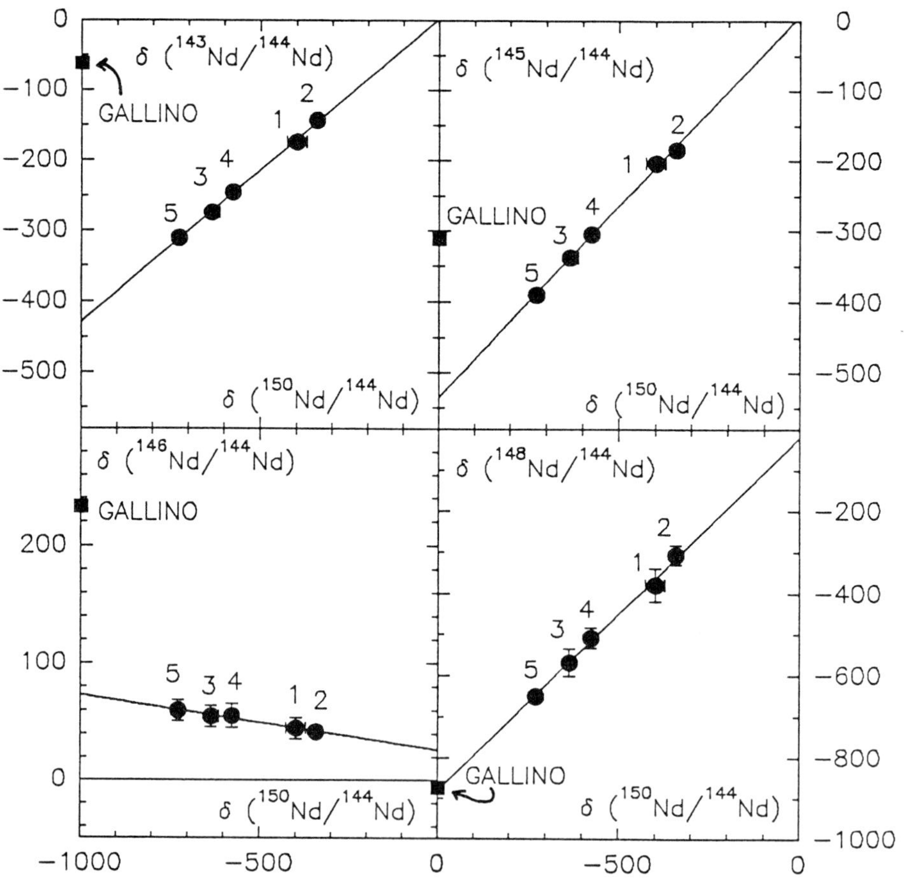

Fig. 1: Nd isotopic composition (deviation from normal in o/oo) measured in SiC from the Murchison meteorite

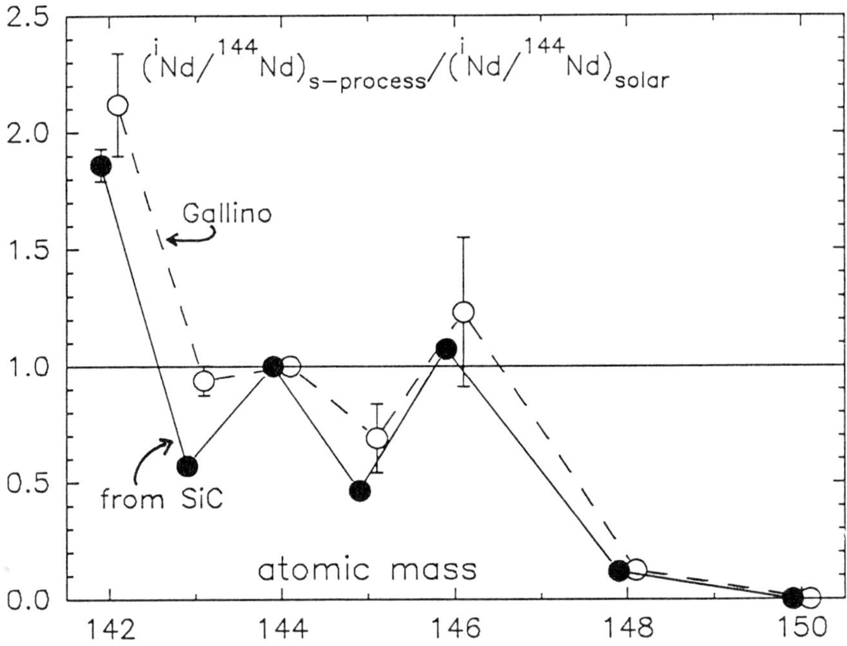

Fig. 2: Composition of s−process neodymium from meteoritic SiC normalized to isotope 144 and normal composition. For comparison the predictions by Gallino et al. (1992) are shown.

s-process Xe and Kr (Ott et al. 1988; Lewis et al. 1990). The composition of s-process Nd derived by extrapolating to r-only $^{150}Nd \equiv 0$ (i.e. $\delta(^{150}Nd/^{144}Nd) \equiv -1000$ o/oo) is $^{142}Nd/^{143}Nd/^{144}Nd/^{145}Nd/^{146}Nd/^{148}Nd/^{150}Nd = 2.13 \pm 0.08/0.293 \pm 0.006/\equiv 1/0.161 \pm 0.005/0.775 \pm 0.009/0.0281 \pm 0.0058/\equiv 0$ (Fig. 2), virtually identical with the composition we have reported previously (Richter et al. 1992).

The uncertainty in the abundance of ^{142}Nd does not include a systematic error ($<6\%$) that may arise from the correction of the ^{142}Nd signal for an isobaric interference from ^{142}Ce. The above value $^{142}Nd/^{144}Nd$ is based on the assumption that for ^{140}Ce the mixing ratio 'normal' Ce/s-Ce is the same as it is measured for ^{144}Nd.

As mentioned above, Zinner et al. (1991) had to correct their ion probe Nd data for interferences from both Ce *and* Sm. They did so by assuming the mixing ratios 'normal' Ce/s-Ce = 'normal' Sm/s-Sm = 'normal' Ba/s-Ba where the latter could be measured in the course of their experiment. The composition of pure s-process Nd was then calculated by assuming the same mixing ratio also for the corrected measured Nd. We find it remarkable that the composition they derived in this way is virtually identical, within their larger errors, with ours. It is unclear at present whether this is fortuitous or whether this might be telling us something about the mixing processes in the envelopes or surroundings of AGB stars.

An interesting observation from our Nd spectrum is that there is a finite contribution from the s-process to ^{148}Nd, which is considered r-only in the classical picture. This observation is in agreement with the results from recent calculations that try to reproduce in a realistic manner the conditions in a thermally pulsing AGB star (Gallino et al. 1992; note the similar case of ^{134}Xe) and which are shown in Figs. 1,2 for comparison. In contrast to ^{148}Nd/^{144}Nd, differences appear in the other isotopic ratios, especially ^{143}Nd/^{144}Nd (Figs. 1,2). Since these should not depend sensitively on details of the s-process, it appears that for Nd there are problems with the accuracy of the reported neutron capture cross sections, which have large nominal errors already.

3. Samarium

Samarium is more interesting than Nd from the point of view of s-process nucleosynthesis since its composition, like that of Kr, is sensitive to branchings: at ^{147}Pm (where ^{147}Sm can be bypassed) and ^{151}Sm (determining the degree to which ^{152}Sm is produced). The respective halflives under laboratory conditions are ~2.6a and ~93a , however both are shorter under stellar conditions (Takahashi and Yokoi 1987).

Measuring Sm by thermal ionization of directly-loaded (chemically not further treated) SiC without Nd interference is more difficult than vice versa. In order to minimize Nd interferences, we were forced to collect data with very low ion currents so that statistical errors are larger than for Nd. In addition, interferences at masses 144 (still due to Nd) and 154 (unidentified) prevented us from obtaining reliable data for these isotopes. Data for the other isotopes are shown in Fig. 3. Unlike the case of Nd, where data points are shown for total samples loaded, data shown here are for single blocks (averages of 10 ratios) during individual analyses.

The interferences mentioned above prevent us from extrapolating to p-only ^{144}Sm$=0$ or r-only ^{154}Sm$=0$. We have, therefore, calculated an isotopic composition for s-process Sm with the assumption that one of the ratios not affected by branching, ^{150}Sm/^{149}Sm is equal to the Gallino et al. (1992) prediction (3.04 for [Fe/H] = -0.12, but not sensitive to the latter choice), which differs from the value obtained via the local approximation (3.25) only slightly due to the effect of the finite neutron exposure

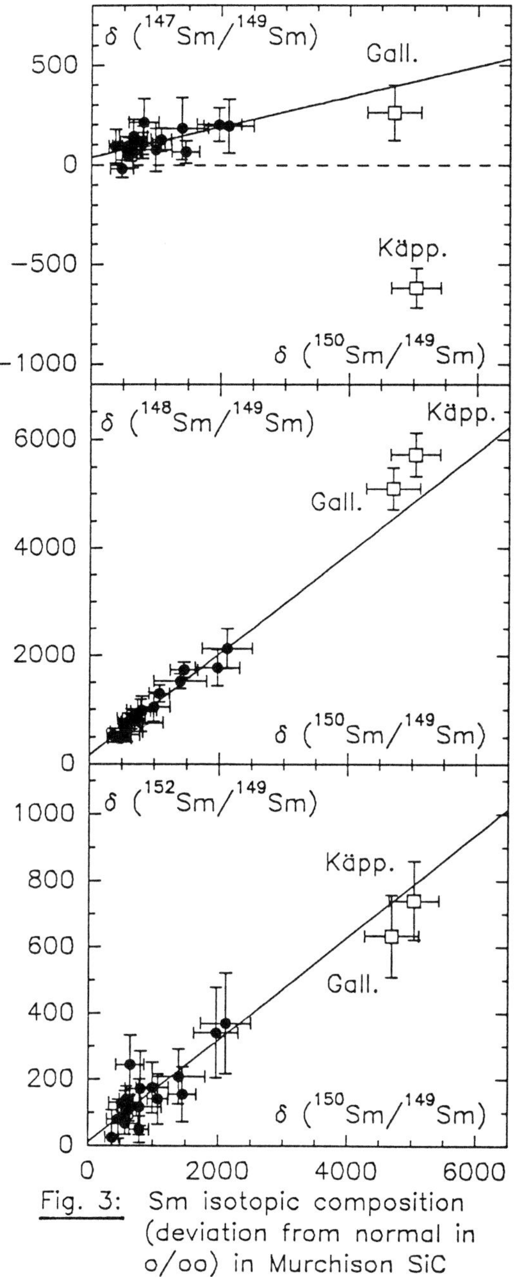

Fig. 3: Sm isotopic composition (deviation from normal in o/oo) in Murchison SiC

in their model. The result ^{147}Sm/^{148}Sm/^{149}Sm/^{150}Sm/^{152}Sm = 1.48 ± 0.12/4.53 ± 0.21/≡ 1/≡ 3.04/ 3.31 ± 0.25) is shown graphically in Fig. 4, where it is compared with the calculations of Gallino et al. (1992). (Note that we have assigned errors to the calculated values that reflect the uncertainties in the individual cross sections.) Figs. 3 and 4 show good agreement with the s-process isotopic ratios calculated by Gallino et al. (1992). Both our and their results closely resemble what would be expected from the 'local approximation' (abundance inversely proportional to neutron capture cross section) - implying little bypassing of ^{147}Sm (i.e. essentially complete decay of ^{147}Pm) and little effect on the ^{152}Sm abundance due to decay of ^{151}Sm into ^{151}Eu (i.e. almost no decay of ^{151}Sm). However, relative to the results of the classical model as listed in Käppeler et al. (1989) where a large effect on the abundance of ^{147}Sm due to the ^{147}Pm branching is expected, there exists a significant difference.

4. Summary

Improved data for s-process Nd and s-process Sm contained as traces in SiC from the Murchison meteorite have been obtained. The results indicate little bypassing of ^{147}Pm due to n-capture by ^{147}Pm and little decay of ^{151}Sm into ^{151}Eu. They suggest that there are problems with the neutron capture cross sections for the Nd isotopes.

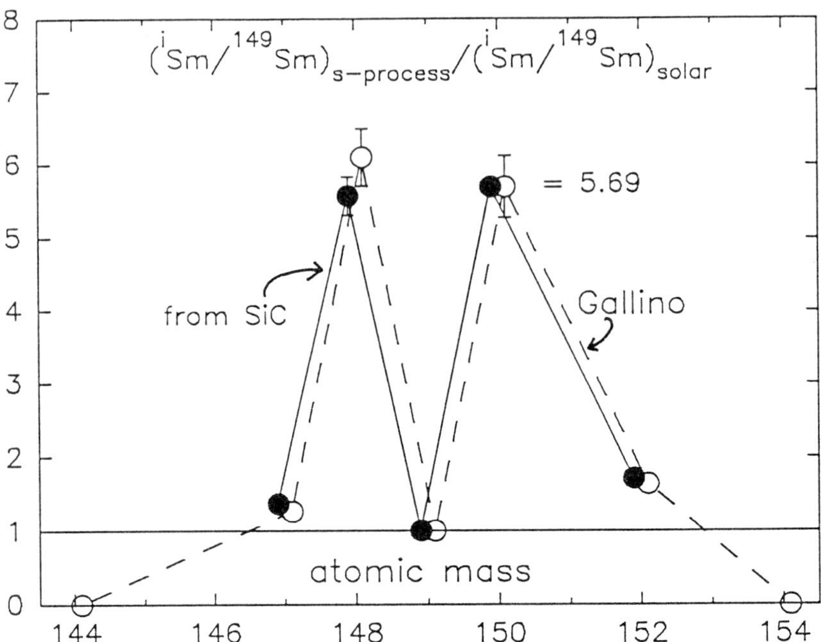

Fig. 4: Composition of s–process samarium in meteoritic SiC normalized to isotope 149 and normal composition (for details see text). Predictions by Gallino et al. (1992) are shown for comparison.

References

Gallino R, Busso M, Picchio G, and Raiteri C M 1990 *Nature* **348** 298.

Gallino R, Raiteri C M, and Busso M 1992 *Ap. J.* subm.

Käppeler F, Beer H, and Wisshak K 1989 *Rep. Prog. Phys.* **52** 945

Lewis R S, Amari S, and Anders E 1990 *Nature* **348** 293

Ott U 1992 In: *Protostars and Planets III* (eds. M S Matthews, E H Levy and J I Lunine, Tucson: University of Arizona Press), in press

Ott U and Begemann F 1990a *Ap. J. (Letters)* **353** L57

Ott U and Begemann F 1990b *Lunar Planet. Sci.* **XXI** 920

Ott U, Begemann F, Yang J, and Epstein S 1988 *Nature* **332** 700

Richter S, Ott U, and Begemann F 1992 *Lunar Planet. Sci.* **XXIII** 1147

Srinivasan B and Anders E 1978 *Science* **201** 51

Takahashi K and Yokoi K 1987 *Atomic Data Nucl. Data Tables* **36** 375

Zinner E, Amari S and Lewis R S 1991 *Ap. J. (Letters)* **382** L47

Present experimental evidence for 'last minute' nucleosynthesis in solar matter

K. Marti
Univ. of Calif., San Diego, La Jolla, CA 92093-0317

ABSTRACT: The abundances of transuranium nuclides in solar system matter are reviewed and the evidence for "last minute" nucleosynthesis is assessed. The upper limit $^{248}Cm/^{235}U \leq 5$ x 10-5 which constrains the mixing ratio of freshly synthesized r-process actinides does not conflict with the observed levels of extinct short-lived activities ^{26}Al ^{53}Mn, ^{60}Fe and ^{107}Pd. Observed actinide abundances are consistent with a continuous synthesis model, followed by a decay interval of $\sim 10^8$ years. The production ratio $^{146}Sm/^{144}Sm$ and inferred extinct ^{146}Sm abundance ratios are also consistent with a continuous p-process synthesis and do not require a late spike addition.

1. Introduction

The history of nucleosynthesis of the elements involves timescales beyond the 4.56Ga age of the solar system and long-lived radionuclides and extinct radionuclides in solar system matter provide important information. Inferred activities of extinct nuclides were sometimes used as evidence for a nucleosynthetic spike shortly before solar system matter was isolated from stellar activity in the galaxy. R-process chronological studies are based on the observed abundances of actinides (^{232}Th, ^{235}U, ^{238}U, ^{244}Pu) in the early solar system and inferred upper limits of Cm isotopes, while p-process studies are based on the inferred abundance of the p-nuclide ^{146}Sm in solar system matter.

The classical r-process mechanism is based on a flow concept similar to that of the s-process, but much higher neutron densities are requird as the β-decay lifetimes of neutron-rich isotopes are short (Burbidge et al, 1957). The neutron binding energy decreases and the rapid addition of neutron stops at the neutron drip line, at a temperature where the rate of photodisintegration (γ, n) is equal to the (n, γ) caputure process. The r-process bypasses nuclei with natural α-radioactivity and stops beyond the actinide elements. There is considerable uncertainty as to what the largest r-process mass should be, as a result of uncertainties in fission barrier predictions and β-delayed fission rates. Systematic calculations of the r-process yields require a detailed knowledge of basic nuclear properties of r-process nuclei. These properties are not currently known experimentally, since these nuclei are far from the valley of stability. However, information on the abundance distribution and on the even-odd nuclide mass systematics can be extracted from solar system nuclide abundances (e.g. Marti and Suess, 1988). Semiempirical extrapolations of nuclear properties are being refined in order to provide fits to these abundance systematics. Supernovae have long been assumed to provide sites for r-process synthesis, but the specifics of such an environment have remained unclear.

Recently, a new site, a "hot" bubble that is calculated to form during a supernova explosion, was suggested (Bethe and Wilson, 1985).

The rare neutron-deficient nuclides in solar system matter which can not be produced by neutron capture reactions are classically referred to as the p-nuclei. At the high temperatures which are required for the synthesis of the p-nuclei, photodisintegrations are important in determining the abundance pattern of p-nuclides. As a plausible site for the p-process the oxygen/neon layers in supernovae have been suggested (Rayet and Arnould, 1991). In their model the s-process synthesis that develops during core He burning could provide the seed nuclei for the p-process, but they suggest that a more solar-like distribution of seed nuclei might better reproduce p-isotopes of Mo and Ru, which are most abundant among p-nuclides in solar system matter.

Since specific supernova environments are implicated for the synthesis of p- and r-nuclides and solar system abundances probably represent a blend of products from multiple SN events it is of considerable interest to compare the respective chronologies of r- and p-process synthesis. The chronometers generally used in r-process chronologies are the transbismuth nuclides and extinct ^{129}I. The only chronometer for p-process nucleosynthesis is the ^{146}Sm nuclide, first suggested by Audouze and Schramm (1972) and studied by Lugmair et al (1975).

2. R-Process Synthesis: Actinide Chronometers

The chronology of r-process synthesis is based on the observed abundances of transbismuth nuclides in meteorites and estimated r-process production ratios . The actinide production ratios are obtained as the sums of contributions from a number of progenitors and uncertainties include possible effects of delayed fission in r-process production. The discoveries of the extinct short-lived radionuclides ^{107}Pd ($\tau = 9.4$ Ma) (Kelly and Wasserburg, 1978) and of ^{60}Fe ($\tau = 2.2$ Ma) (Shukolyukov and Lugmair, 1992) has redefined the time scale for events immediately before formation of the solar system. However, it is important to note that all these nuclides are not only made in the classical r-process, and that the chronology of r-process nuclides should be restricted to actinides and possibly ^{129}I.

The existence of extinct ^{244}Pu in the early solar system has been established in numerous investigations of fission Xe components in meteoritic minerals which match the mass yields of ^{244}Pu derived fission xenon (Alexander et al, 1971) and the identification of fossil fission tracks attributable to ^{244}Pu in meteorites (Drozd et al, 1977). The initial abundance of ^{244}Pu relative to U or the rare earth elements is observed to be slightly variable in meteorites and indicates some chemical fractionation in meteoritic phases. Lavielle et al (1992) adopt an upper limit ^{244}Pu/^{238}U ≤ 1.0 x 10^{-2}, 4.56 Ga ago.

Chronometric information based on the nuclide ^{247}Cm ($\tau = 22.5$ Ma) may be obtained from the ^{235}U/^{238}U ratio if the Cm/U ratio was fractionated prior to or during formation of meteoritic minerals and phases (Blake and Schramm, 1973). The isotopic composition of U in Allende inclusions and in meteoritic phosphates was studied by Chen and Wasserburg (1981) and they obtained an upper limit ^{247}Cm/^{235}U ≤ 4 x 10^{-3} at the time of meteorite formation. This limit is consistent with calculated ratios based on a time interval of $\sim 10^8$ a between the last r-process nucleosynthetic event and the formation of the solar system. Lavielle et al (1992) studied the Forest Vale chondrite to search for evidence of previously claimed products of ^{248}Cm ($\tau = 0.5$ Ma) in the early solar system and calculated and upper limit for ^{248}Cm by partitioning the measured fission Xe spectrum into ^{244}Pu- and ^{248}Cm-derived components. They obtained an upper limit ^{248}Cm/^{244}Pu ≤ 1.5 x 10^{-3} which again helps to constrain the timing of r-process actinide nucleosynthesis. Using the ^{248}Cm/^{244}Pu limit coupled with solar system ratios ^{235}U/^{238}U $= 0.3176$ and ^{244}Pu/^{238}U ≤ 1.0 x 10^{-2} (4.56 Ga ago), a limit ^{248}Cm/^{235}U ≤ 5 x 10^{-5} is calculated. This limit is nearly two orders-of-magnitude lower than the ^{247}Cm/^{235}U limit further constraining the evidence for any late addition of freshly synthesized actinide elements just prior to solar system-formation. This actinide record reveals that little general mixing and local production did occur in the 10^8 years prior to solar system formation. Supporting documentation has been secured from the ^{129}I chronometer in solar system matter. Although the ^{129}I ($\tau = 23$ Ma) nuclide may possibly have experienced additional late synthesis, the "classical" r-process contribution which is recorded in the low observed initial solar system ratios ^{129}I/^{127}I $\approx 10^{-4}$ is not consistent with a continuous r-process synthesis and requires a 10^8 year decay interval. Furthermore, recent results which indicate lower ^{129}I/^{127}I ratios in primitive type 3 chondritic materials compared to those observed typically in equilibrated chondrites (Swindle and Burkland, 1992), imply that either the solar system ^{129}I/^{127}I was not uniform at the time of formation, or a late minor addition of

^{129}I could have taken place. A comparison of the ^{129}I and ^{244}Pu chronologies in the same meteoritic minerals is important and may be feasible (Nichols et al, 1992).

3. P-Process Synthesis: The ^{146}Sm Chronometer

Neutron-poor isotopes of many heavy elements are produced by p-processes (Burbidge et al, 1957) or γ--processes (Woosley and Howard, 1978). Some of these nuclides are radioactive with halflives long enough to have survived in the early solar system. The ^{146}Sm nuclide was proposed as a possible p-process chronometer (Audouze and Schramm, 1972) and Lugmair et al (1975) set an upper limit for the ratio ^{146}Sm/^{144}Sm ≤ 0.013 at the time of solidification of the Juvinas meteorite 4.56 Ga ago. The first positive evidence for live ^{146}Sm in the solar system was observed in the Angra dos Reis meteorite (Lugmair and Marti, 1977), in the form of a systematic shift by 6 parts in 10^5 in measured ^{142}Nd/^{144}Nd ratios in pyroxene separates relative to those in phosphates. More recent measurements (Prinzhofer et al., 1989; 1992; Nyquist et al, 1991; Lugmair and Galer, 1992) clearly established the presence of ^{146}Sm in the early solar system. In order to obtain an initial ^{146}Sm/^{144}Sm abundance ratio we have to normalize the measured isotope ratios to a fixed solar system time. Lugmair and Galer (1992) calculate an average ratio ^{146}Sm/^{144}Sm $= 0.0070 \pm 0.0010$ 4.56 Ga ago.

TABLE 1: PRODUCTION RATIOS P_{146}/P_{144}

P_{146}/P_{144}	Reference
0.35 - 0.60	Audouze and Schramm (1972)
0.57	Lugmair et al. (1975)
~1	Audouze and Truran (1975)
0.024	Woosley and Howard (1978)
0.01 - 0.4	Woosley and Howard (1990)
0.4 - 0.7	Rayet and Arnould (1991)

In order to obtain a ^{146}Sm-based chronology of p-process nucleosynthesis we require a knowledge of the relative production rate of ^{146}Sm and of its time-dependence. Table 1 lists published estimates for the production ratio ^{146}Sm/^{144}Sm and reveals that the various estimates are not exactly consistent. Estimates by Audouze and Schramm (1972) and Lugmair et al (1975) were based on semi-empirical interpolations using elemental and isotopic systematics of p-process abundances in solar system matter. Audouze and Truran (1975) calculated the production ratio from an explosive synthesis model using solar system seed abundances. Woosley and Howard (1978) used γ-process calculations and calculated abundances expected by photodisintegration reactions. The systematics of photodisintegration reactions was reexamined by these authors (Woosley and Howard, 1990) and they infer substantially larger relative ^{146}Sm yields due to the extra binding energy in the magic ^{146}Gd nucleus, which decays to ^{146}Sm. Rayet and Arnould (1991) considered the extra binding energy of ^{146}Gd and calculate the ratio for three "realistic" SNII models and obtain a range of ratios which are essentially identical to the semi-empirical estimates obtained 20 years ago.

4. Conclusions

It is clear from the above discussion that the actinide abundances which can only be produced in the classical r-process must be considered separately from products which may be synthesized under different conditions. The shorter-lived nuclide ^{129}I whose presence in the early solar system is well established could provide critical information on possible minor admixtures of "last minute" products if ^{129}I/^{127}I and ^{244}Pu/^{238}U chronologies can be successfully coupled. The nuclides ^{26}Al, ^{53}Mn ^{60}Fe and ^{107}Pd can all be produced in different environments.

The r-process chronology proper is consistent with a nearly continuous synthesis model, provided a decay interval of about 10^8 years is added before solar system formation.

Interesting conclusions can now be based on the ^{146}Sm chronometer. Lugmair et al (1975) concluded that the then available upper limit for ^{146}Sm/^{144}Sm < 0.013 is consistent with a uniform p-process synthesis as well as a decay interval of ~10^8 years. The 10^8 year decay interval changes the ^{146}Sm activity only by a factor of 2, and since the relative production of this nuclide is not well constrained at present, no firmer conclusions are possible at this time than in 1975. The similarities in the time-dependencies $p(\tau)$ for r- and p-process products is interesting since rather different environments are implicated.

Acknowledgements

I thank Günter Lugmair and Paul Pellas for numerous stimulating discourses on this subject over the years. Partial support was provided by NASA.

5. References

Alexander E, Lewis R, Reynolds J and Michel M 1971 *Science* **172** 837
Audouze J and Schramm D N 1972 *Nature* **237** 447
Audouze J and Truran J W 1975 *Ap. J.* **202** 204
Bethe H A and Wilson J R 1985 *Astrophys J* **14**
Blake J B and Schramm D N 1973 *Nature Phys Sci* **243** 138
Burbidge E M, Burbidge G R, Flowler W A and Hoyle F 1957 *Rev. Mod. Phys.* **29** 15
Drozd R J, Morgan C J, Podosek F A, Poupeau G, Shirck J R and Tayulor G.J 1977 *Ap J* **212** 567
Kelly W R and Wasserburg G J 1978 *Geophys Res Lett* **5** 1079
Lavielle B, Marti K, Pellas P and Perron C 1992 *Meteoritics* ***in press***
Lugmair G W, Scheinin N B and Marti K 1975 *Earth Planet. Sci. Letters* **27** 79
Lugmair G W and Marti K.1977 *Earth Planet. Sci. Letters,* **35** 273
Lugmair G W and Galer S 1992 *Geochim. Cosmochim. Acta* **56** 1673
Marti K and Suess H E 1988 *Astrophys and Space Sci* 507
Nichols R H Jr, Hohenberg C M and Marti K 1992 *Meteoritics* abstract 55th Ann Met Soc meeting *in press*
Nyquist L E, Wiesmann H, Bansal B, Shih C-Y and Harper C L 1991 *Lunar Planet. Sci.* **XXII**, 989-990
Prinzhofer A, Papanastassiou D A and Wasserburg GJ 1989 *Astrophys. J.* **344** L-81-L84
Prinzhofer A, Papanastassiou D A and Wasserburg G J 1992 *Geochim. Cosmochim. Acta* **56** 797-815
Rayet M and Arnould M 1991 *Radioactive Nuclear Beams* **347**
Shukolyukov A and Lugmair G W 1992 *Lun. Plan. Sci. Conf.* **XXIII** 1295-96 and private communication
Swindle T D and Burkland M K 1992 Meteoritics Abstract 55th Ann Met Soc meeting, *in press*
Woosley S E and Howard W M 1978 *Ap. J. Suppl.* **36** 285
Woosley S E and Howard W M 1990 *Ap. J.* **354** L21

Section 2: Nuclear Physics

Charged-particle thermonuclear reactions

H.-P. Trautvetter

Institut für Experimentalphysik III, Ruhr Universität Bochum, Germany

1. Introduction

In this presentation no details on specific problems will be given, they can be found in recent review articles [1, 2] or text books, e. g. [3, 4] and references therein. Here, an overview will be given to summarize the activities in the field as it is represented in the various contributions, where new results are discussed and to which will be referred to by the name of the first author.

It is well known that the energy dependence of charged–particle reaction cross sections for low energies are governed by the Coulomb penetration factor and the De Broglie wave length. Therefore it is common to transform the cross section into the so called astrophysical S–factor:

$$S(E) = \sigma(E) \cdot E \cdot \exp(2\pi\eta), \tag{1}$$

where η is the Sommerfeld parameter ($\eta = Z_1 Z_2 e^2/\hbar v$). A typical S–factor curve is shown schematically in figure 1. The low energy tail of broad resonances or non–resonant processes can be studied experimentally down to a limiting energy E_l. However, the most interesting region for astrophysics is the vicinity of the Gamow peak, which arises at the overlap region of the Maxwell–Boltzmann distribution of the stellar plasma and the exponentially decreasing cross section [1, 4]. It is this region where the reaction rate is needed, but it is also the region which is hardly accessible by experiments. Therefore, the non–resonant contributions have to be extrapolated (dotted line in figure 1) and the properties of the subthreshold resonances as well as other resonances located near the Gamow energy have to be investigated by other means.

In chapter 2 possible techniques to approach reliable reaction rates will be sketched together with references to contributions within these proceedings, where more details can be found. In chapter 3 the most critical reactions for the solar ν–problem will be revisited and possible improvements, which could be achieved e. g. in the Gran Sasso underground laboratory, will be discussed.

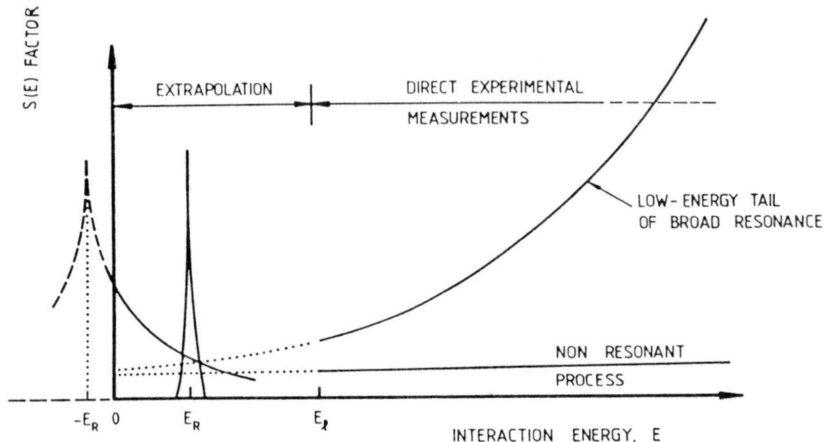

Figure 1. The steep drop of $\sigma(E)$ due to the Coulomb barrier results in a lower limit in energy, E_l, below which direct measurements are not feasible. The $S(E)$ data are extrapolated to zero energy with the guidance of theory and any other available knowledge.

2. Approaches to thermonuclear reaction rates

2.1. Direct measurements

So far, there is only one measurement which reaches the Gamow-peak energy region: the ^{3}He+^{3}He–reaction was measured down to $E = 24$ keV by [5] (Fig. 2) and this reaction will be discussed in more detail in chapter 3. The major difficulties are to cope with the extremely low cross sections. In principle, a solution could be achieved in some cases with i) high current low energy accelerators (see e. g. Greife [6]), ii) high efficiency in the detection devices, iii) 4π–geometry and iv) appropriate background reduction. The last point has basically two sources: a) beam induced background which originates mostly from impurities in the target. This problem can be overcome by implantation techniques, where the substrata have to be prepared extremely carefully. Such techniques are presently being developed in Bochum. b) cosmic ray background, which could be overcome either by active shielding or by moving the entire apparatus underground, e. g. to the Gran Sasso laboratory (chapter 3 and Greife [6]). Contributions using the direct approach can be found by Angulo [6], Greife [6], Dababneh [6], Hammer [6], Safi [6], Arkis [6] and Bucka [6]. screening on the nuclear cross sections cannot be neglected at very low energies. Enhancement of the cross section can occur already at beam energies, which are about a factor of 100 higher than the electron binding energies. Cross sections near such energies can now be measured. This process is discussed in more detail by Angulo [6] and in [7]. The screening in a stellar plasma is quite different to laboratory situations. It is therefore of great importance to understand this effect in full detail in order to extract reliably the behavior of the S-factor curve at the very low energies, which become now accessible to experiment. pushed to the energy region of astrophys-

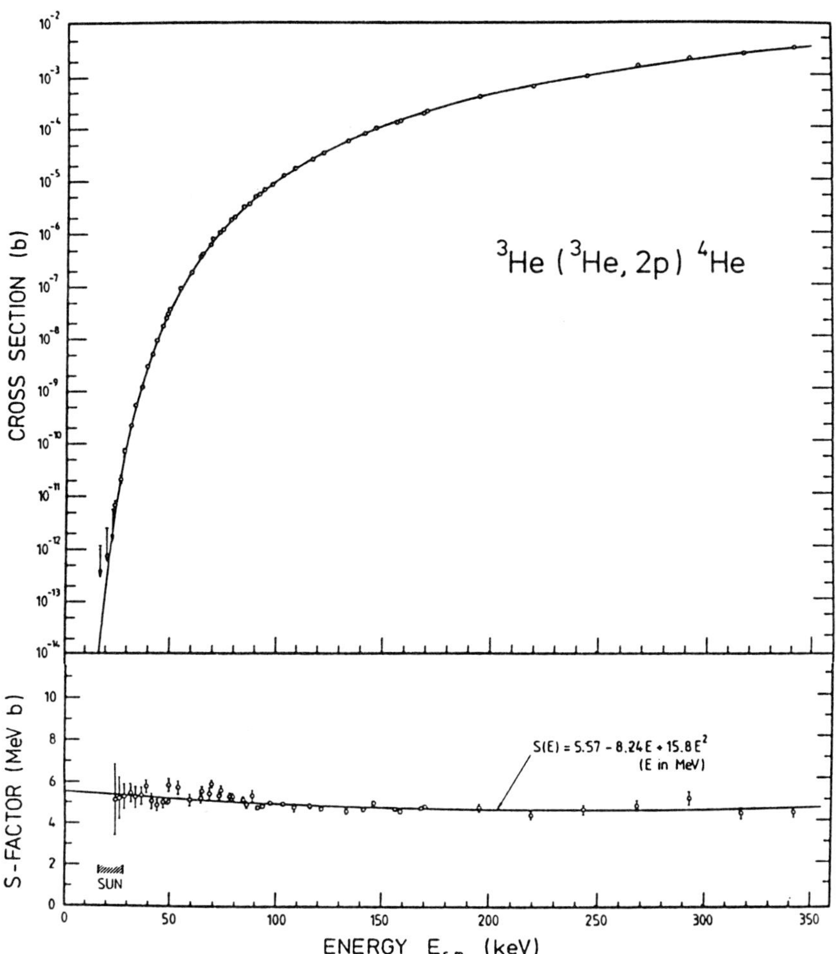

Figure 2. The cross section and S–factor as a function of energy E for the ³He+³He –reaction. Note that the measurement extends down to the gamow–peak region of the sun (shaded bar labeled sun)

ical interest, as has been pointed out in the introduction. It is therefore necessary to understand theoretically all the processes, which might contribute to the cross section near the Gamow peak region, in order to extrapolate the existing data reliably. Recently, good successes have been achieved in some cases with microscopic model calculations, Descouvemont [6], as well as with potential models and folding procedures in calculating S-factors, Krauss [6]. Although excitation energies of nuclear levels still cannot be calculated with the necessary precision, such information could be obtained using indirect methods (section 2.5). In many cases, however, theoretical considerations are the only source to obtain the level properties, which are needed e. g. to determine resonance strengths not accessible to experiment (section 2.5). Finally, theory is of course the only means to obtain information for nuclei far from stability, see e. g. Sorlin [6] and Kratz [6]. Other theoretical work can be found by Doll [6], Sahrling [6], Kruglanski [6] and Typel [6]. radioactive nuclei can be studied by preparing radioactive targets, if the life time is long enough. One example is the ^{7}Be+p–reaction, which plays an important role in the solar ν–problem (see chapter 3). However, for short life times, such as for reactions relevant for the rp-process, it is necessary to produce radioactive beams. First experiments have now been performed, and results are reported by Leleux [6], Garret [6] and Aguer [6].

2.1.1. Coulomb dissociation. known process in nuclear physics. If the center-of-mass energy is high enough to excite the nucleus of interest (usually the projectile) above the particle threshold, then this nucleus can break up and the fragments can be measured yielding a cross section, which in turn can be converted to a fusion cross section via detailed balance. The method has the advantage of being very efficient, if E1 radiation is the dominant γ–decay mode for the excited level of investigation. Disadvantages are that only ground state transitions can be measured with this method and also that results are not completely model independent. Promising first results will be described by Shotter [6], Rebel [6] and Aguer [6], while the latter uses also radioactive beams. ^{12}C$(\alpha,\gamma)^{16}$O–reaction is of great interest in astrophysics. The extrapolated $S(0.3\ \mathrm{MeV})$–value ranges between 0.02 and 0.48 MeV b [2]. An attempt will now be made to measure the shape of the α–particle energy distribution from the decay of the ^{16}O excited state ($J^{\pi} = 1^{-}$, $E_x = 9.63$ MeV). This state is populated by the β–decay of ^{16}N, which in turn has been produced by some other nuclear reaction. This experiment is described by Buchmann [6]. to obtain in some cases information on the resonant thermonuclear reaction rate even if a resonance is located in the low energy region which is not accessible to direct experimental investigations (figure 1). Such a rate has the form $\langle \sigma v \rangle \simeq \omega\gamma \cdot \exp\left(E_r/kT\right)$, where $\omega\gamma \simeq \Gamma_{\mathrm{in}} \cdot \Gamma_{\mathrm{out}}/\Gamma_{\mathrm{tot}}$ and ω is a statistical factor. The energy of excited states can often be determined precisely by γ–spectroscopy using Ge–detectors after populating the state of interest via various reaction mechanisms and hence, E_r can be deduced with the known Q–value.

The partial width of the incoming channel Γ_{in} is usually very small for states of astrophysical interest compared to the partial width of the outgoing channel Γ_{out} such that $\Gamma_{\mathrm{tot}} \simeq \Gamma_{\mathrm{out}}$ and hence $\omega\gamma \simeq \Gamma_{\mathrm{in}}$. The partial width is given by $\Gamma \simeq$ penetration-factor $\cdot \Theta^2$, where Θ^2 is the reduced particle width. The latter quantity can be extracted from transfer reactions such as (d,p), (d,n), (^{3}He,t), (^{3}He,d), (^{6}Li,d) or (^{7}Li,t) and thus the resonance strength $\omega\gamma$ can be composed. Such experiments are described by Bucka [6], Wiescher [6], Decrock [6] and Iliadis [6]. Also direct capture reactions into such levels of interest measure the reduced particle width [1, 4]. Considerable theoretical analysis is involved (section 2.3) in these methods.

3. Solar neutrino–problem solved?

3.1. Standard solar model and experimental results

Extensive theoretical studies have been carried out [8, 9] to predict the ν–flux or ν–capture rate for the experiments presently in operation. In [8] the capture rate for the ^{37}Cl and ^{71}Ga detectors are quoted to be 8.0 ± 3.0 SNU and 131.5^{+21}_{-17} SNU, respectively, and a flux of the ^{8}B–neutrinos $\Phi(^8\mathrm{B}) = 5.7(1\pm0.43)\cdot10^6$ cm^{-2} s^{-1}, whereas the prediction of [9] are 5.8 ± 1.3 SNU and 125 ± 5 SNU and $\Phi(^8\mathrm{B}) = 3.8(1 \pm 0.29) \cdot 10^6$ cm^{-2} s^{-1}, respectively. These predictions face now 4 experiments: the ^{37}Cl–capture rate of 2.1 ± 0.3 SNU [10], the ^{71}Ga–capture rate of 20^{+15}_{-20} statistical ±32 systematical of [11] as well as the very recent ^{71}Ga–capture rate of 83 ± 19 statistical ±8 systematical of Hampel [6]. The flux of the ^{8}B–neutrinos was measured via e-ν scattering [12] to result in $\Phi(^8\mathrm{B}) = 2.7(1 \pm 0.29$ stat. ± 0.34 sys.$) \cdot 10^6$ cm^{-2} s^{-1}.

Comparison of these theoretical predictions with experimental results reveal clearly that the ^{37}Cl ν–capture prediction is much higher than observation and this is also true for the direct flux measurement. The new ^{71}Ga–experiment by Hampel [6] is about a factor of 3 higher than [11] but has a much higher precision. The statistical error of [11] is of similar size as the value itself and the quoted systematical error is even larger. The result of Hampel [6] is low compared with [8], but in agreement within the quoted errors, showing directly for the first time the operation of the pp and pep processes in the sun.

In [13] a prediction of the ^{71}Ga experiment was made by using only the luminosity constraint for the sun and otherwise varying the experimental values of the ^{37}Cl [10] and e-ν–scattering [12] experiment within the quoted errors resulting in a distribution for the expected ^{71}Ga capture rate between 80 and 120 SNU, in excellent agreement with Hampel [6]. This means that all experimental observations are in good agreement, if è. g. the ^{8}B–(^{7}Be–) and CNO–neutrino flux is reduced by a factor 0.3 (0.35) and the pp and pep neutrino flux increased by only 1.05.

In the above discussion no assumptions about the properties of neutrinos were made [13]. It has been suggested that matter enhanced neutrino oscillations MSW [14] are the cause for the still remaining problem of a much too small ^{8}B–neutrino capture rate, compared with the standard solar model (SSM). However, looking at the predictions of the two SSM's [8] and [9], a substantial difference can be observed for the ^{37}Cl–capture and the ^{8}B–neutrino flux prediction. Inspecting the two references [8, 9] the major difference can be found in the nuclear physics input parameters. For this reason, we review the two most critical reactions leading to the discrepancy between the two SSM's as well as to the discrepancy with the observation of the solar neutrinos.

3.2. Nuclear physics input

Looking at the 3 branches of the pp–chain in Figure 3 together with the presently quoted accuracies [3], two reactions are of special interest: $^7\mathrm{Be}(\mathrm{p},\gamma)^8\mathrm{B}$ and $^3\mathrm{He}(^3\mathrm{He},2\mathrm{p})^4\mathrm{He}$.

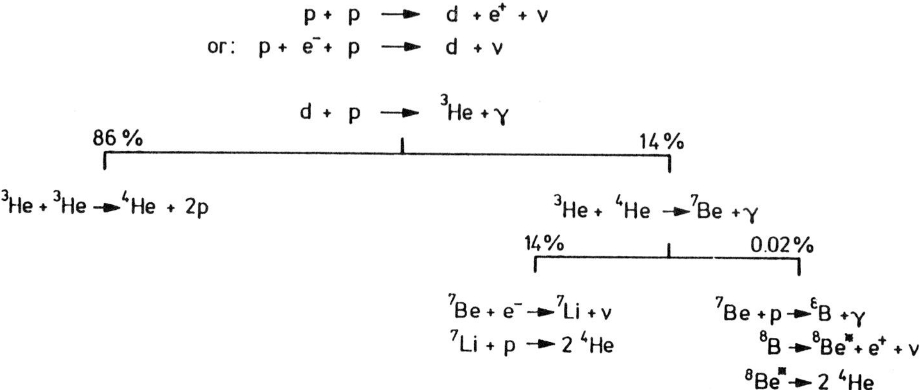

Figure 3. Hydrogen burning in the sun: the pp–chain

3.2.1. The $^7Be(p,\gamma)^8B$–reaction. The ^{7}Be(p,γ)^{8}B–reaction leads directly to ^{8}B and thus to the high energy ^{8}B–neutrinos to which the ^{37}Cl detector and the ν–e scattering are mainly sensitive. The accuracy of the S–factor at zero energy, S_{17}, is quoted [3] to be $\pm15\%$. This error will not explain the discrepancies mentioned above (sect. 3.1). However, a controversial discussion [15] on the analysis of this reaction is referred to in [9]. The basic arguments are the following: all measurements determine the area density of the ^{7}Be–target via the ^{7}Li(d,p)^{8}Li reaction. The cross section determinations for the latter reaction are all linked together by using the same source for the stopping cross sections of H in Li, which is difficult to measure. In [15] three measurements of the ^{7}Li(d,p)^{8}Li reactions, which are independent of the stopping cross section determination mentioned above, are averaged. Also, a new theoretical analysis of the energy behavior for the non–resonant cross section of the ^{7}Be(p,γ)^{8}B–reaction was performed in [15] considering a hitherto not included d–wave contribution to the cross section. Such an inclusion changes primarily the energy dependence above 1.5 MeV and results in a greatly improved fit to the data between 1.5 and 4 MeV.

In summary, the $S_{17} = 0.017$ keV b value quoted in [15] is substantially lower than that used in [8], $S_{17} = 0.024$ keV b, and even lower than the one used in [9] with $S_{17} = 0.021$ keV b. It seems that an independent check on this critical reaction rate is urgently needed.

3.2.2. The $^3He(^3He,2p)^4He$ reaction. It has been suggested by [16] that a resonance for the ^{3}He(^{3}He,2p)^{4}He reaction might be located near the particle threshold of ^{6}Be corresponding to a state at $E_x = 11.6$ MeV. Such a resonance would be located close to the Gamow peak and thus would enhance the reaction rate of ^{3}He(^{3}He,2p)^{4}He, thereby explaining the missing neutrino flux from ^{8}B. However, searches for this state, using a variety of nuclear reactions, were not successful [3]. A direct experimental search would help to clarify the situation, since all those investigations are of the type of indirect methods as described above (sect. 2.5). This has been done recently [5] down to the energy E = 24 keV with negative result. In the same work [5] upper limits were extracted down to E = 16 keV from which a resonance could be excluded except for the lowest energy point. In this experiment the background problem was solved by using a coincidence technique, which, however, reduced the solid angle substantially.

One source of the background are the cosmic rays, which cannot be eliminated completely by active shielding. Thus, it is planned to reinvestigate the ^{3}He+^{3}He reaction using passive shielding by moving the entire experiment into the Gran Sasso underground laboratory (Greife, [6]). There it should be possible to avoid the necessity for the coincidence method to fight the cosmic ray background. With the use of single spectra, the solid angle could be largely increased, and together with a high current low energy accelerator, such as currently being developed by Greife [6], it should be possible to give direct experimental evidence for the presence or absence of a resonance suggested by [16] in near future.

4. Summary

Current activities in the field of the determination of the thermonuclear reaction rates for charged particle reactions have been sketched in chapter 2. Although much progress was achieved during recent years in direct experimental investigations of the S-factor for many light ion nuclear reactions important for astrophysics, this is not the case for reactions involving radioactive nuclei. Here, we are just at the beginning and much work has to be done in this direction.

A large step forward has been accomplished by Hampel [6] in direct observations of the pp–solar neutrinos, which also demonstrated at the same time a big success of the SSM [8, 9]. There are still open questions on the low ^{8}B–neutrino–flux. The accuracy of the nuclear physics input parameters have greatly increased over the last years [3] such that it seems unlikely to find the solution of the problem in this domain. However, as has been shown in chapter 3., there are still some open questions on the nuclear physics part, which should be looked at very carefully. A program in this direction has been started in the Gran Sasso laboratory (closely located to the GALLEX–experiment) and results are expected in the near future.

It is of course obvious that also other experimental studies would profit from the massive shielding of the underground laboratory, especially if γ–rays are involved. In such cases a large NaI–detector can serve as a 4π–γ–calorimeter. Such a detector acts as a summing device and is therefore not sensitive to specific decay modes or angular distribution effects. The capture γ–rays are summed up to an energy corresponding to the reaction Q–value plus the small bombarding energy and will be of the order of 5–12 MeV. Active shielding is not very effective for this relatively low γ–ray energy range, as first examinations in Bochum have shown, but the problem could most effectively be solved by working in the underground laboratory.

References

[1] C. Rolfs, H.P. Trautvetter and W.S. Rodney, Rep. Prog. Phys. 50 (1987) 233

[2] C. Rolfs and C.A. Barnes, Annu. Rev. Nucl. Part. Sci. 40 (1990) 45

[3] P.D. Parker and C. Rolfs 1991 in The Solar Interior and Atmosphere, ed.
     A. Cox, W.C. Livingston and M.S. Matthews (Tucson: Univ. of Arizona Press) p. 31

[4] C. Rolfs and W.S. Rodney, Cauldrons in the Cosmos – An introduction to Nuclear Astrophysics, Univ. Chicago Press (1988)

[5] A. Krauss, H.W. Becker, H.P. Trautvetter and C. Rolfs, Nucl. Phys. A467(1987)273

[6] All references quoted by name in the text can be found under this name in these proceedings.

[7] H.J. Assenbaum, K. Langanke and C. Rolfs, Z. Phys. A 327 (1987) 461

[8] J.N. Bahcall and R.K. Ulrich, Rev. Mod. Phys. 60 (1988) 297

[9] S. Turck-Chièze, S. Cahen, M. Cassé and C. Doom, Ap. J. 335 (1988) 415

[10] R. Davis Jr. et al., in Proceedings of the 21st International Sosmic Ray Conference, 12,ed.
     R.J. Protheroe (Adelaide: University of Adelaide Press), 1990, p.143

[11] A.I, Abazov et al. Nucl. Phys. B19 (1991) 84

[12] K.S. Hirata et al., Phys. Rev. Lett. 63 (1989) 16, Phys. Rev.Lett. 65 (1990) 1297, Phys. Rev. D44 (1991) 2241

[13] M. Spiro and D. Vignaud, Phys. Lett. B242 (1990) 279

[14] S.P. Mikheyev and A.Yu. Smirnow, Nuovo Cimento 9C (1986) 17
     L. Wolfenstein, Phys. Rev. D17 (1978) 2369

[15] F.C. Barker and R.H. Spear, Ap.J. 307 (1986) 847

[16] V.N. Fetisov and Yu.S. Kopysov, Nucl. Phys. A239 (1975) 511
     W.A. Fowler, Nature 238 (1972) 24

The electron screening effect on boron (p,α)-reactions; search for a low energy resonance

C. Angulo, S. Engstler, W. H. Schulte, E. Somorjai[1], U. Greife, C. Rolfs and H.-P. Trautvetter

Institut für Physik mit Ionenstrahlen, Ruhr-Universität Bochum, W-4630 Bochum, FRG

Abstract. The fusion reactions ^{10}B(p,α)^{7}Be and ^{11}B(p,α)2^{4}He have been studied over the center-of-mass energy range $E = 15 - 140$ keV. The effect of electron screening has been observed in both reactions and the screening potential has been deduced. Evidences for a low energy resonance in the reaction ^{10}B(p,α)^{7}Be have been found.

1. Introduction

1.1. General objectives

Investigations of the boron (p,α)-reactions are important in several aspects. On one hand the reaction ^{10}B(p,α)^{7}Be is a dominant destructive process for ^{10}B and may have considerable effect on the extremely low boron abundance in the universe [1]. On the other hand the importance of this reaction in the CTR problem has been discussed [2]: ^{11}B is a clean fuel via the reaction ^{11}B(p,α)2^{4}He, but natural boron contains 19.7 % of ^{10}B which produces radioactive material ^{7}Be via the reaction ^{10}B(p,α)^{7}Be. The low energy cross section of ^{10}B(p,α)^{7}Be is presumably dominated by the tail of a low energy resonance located at the center of mass energy $E = 10$ keV (corresponding to the well known state $E_\text{x} = 8.701$ MeV in the compound nucleus ^{11}C). Up to now the cross section of the reaction ^{10}B(p,α)^{7}Be has only been measured at projectile energies as low as 60 keV [1]. Thus to estimate the influence of the low energy resonance and the effect of electron screening, which will be briefly discussed below, accurate values of the cross section in the thermonuclear energy range are required. With this aim, measurements of ^{10}B(p,α)^{7}Be ($Q = 1.146$ MeV) have been performed in the energy range 18 keV $\leq E \leq$ 140 keV. The effect of electron screening has been shown to be isotopic independent [3], therefore additional measurements of the reaction ^{11}B(p,α)2^{4}He ($Q = 8.5906$ MeV) have been carried out for projectiles energies 15 keV $\leq E \leq$ 140 keV to be able to separate the influence of the 10 keV resonance.

[1] Permanent address: Institute of Nuclear Research of the Hungarian Academy of Sciencies, Debrecen, Hungary.

1.2. The electron screening effect

The cross section $\sigma(E)$ of fusion reactions drops rapidly for energies below the Coulomb barrier. To extrapolate the data to the energies corresponding to hydrostatic scenarios, it is advantageous to introduce the concept of the astrophysical S(E) factor, defined by [4]

$$S(E) = \sigma(E) \; E \; \exp(2\pi\eta) \tag{1}$$

where $2\pi\eta = 31.29 Z_1 Z_2 (\mu/E)^{1/2}$ is the Sommerfeld parameter (Z_1 and Z_2 are the charge numbers of the interacting nuclei and μ is the reduced mass in amu). In the case of reactions involving light nuclides, $S(E)$ often varies slowly with energy.

For nuclear reactions studied in the laboratory, the target in general consists of neutral atoms or molecules. The atomic (or molecular) electron clouds surrounding the target nuclei screen the Coulomb potential [5]. Therefore, an incoming projectile experiences no repulsive Coulomb force until it penetrates the electron cloud. This effect leads to a higher cross section, $\sigma_s(E_0)$, than would be expected for bare nuclei, $\sigma_b(E_0)$, at a given energy E_0. The enhancement factor can be approximated by:

$$f(E_0) = \sigma_s(E_0)/\sigma_b(E_0) \simeq \exp(\pi\eta U_e/E_0), \tag{2}$$

where U_e is the electron screening potential. For energy ratios $E/U_e \geq 1000$, shielding effects are negligible, and laboratory experiments can be considered as essentially measuring $\sigma_b(E)$. However, for $E/U_e \leq 100$, shielding effects become important for understanding low energy data. Relatively small enhancements from electron screening at energy ratios $E/U_e \simeq 100$ can cause significant errors [6] in the extrapolation of cross sections to lower energies. As the screening effect in laboratory is different to that occurring in stellar plasma, for astrophysical and other applications (fusion plasmas) the value of $\sigma_b(E)$ must be known. Recent low energy studies [7] have shown clearly such screening effects, as well as their dependence on the aggregate state of the target. Since the screening potential U_e depends only on the binding energies of the atomic electrons, there should be — in a simplified model — no isotopic effect on U_e. This assumption has been confirmed in experiments recently.

2. Experimental procedure

The measurements of the excitation functions of the reactions ^{10}B(p,α)^{7}Be and ^{11}B(p,α)2^{4}He in the energy range $E = 15 - 85$ keV were carried out at the 100 kV high current accelerator and at $E = 70 - 140$ keV at the 400 kV SAMES accelerator in the Dynamitron Tandem Laboratorium (DTL) at Bochum.

Boron films evaporated on metallic tantal backing material were used as targets. As origin material natural boron powder was used for the investigation on the reaction ^{11}B(p,α)2^{4}He while 93 % enriched ^{10}B material was used to study the ^{10}B(p,α)^{7}Be reaction. Due to the high sputtering rates at low energies, the boron targets were fabricated with a thickness larger than 80 μg/cm^2. The detailled elemental composition of the boron targets was studied using nuclear analysis techniques at the 4 MV Dynamitron Tandem accelerator (DTL). Besides boron and tantal isotopes it was expected that oxygen was also present in the targets, as both elements easily oxidize. The analysis of the oxygen distribution in the solid boron targets was performed using Rutherford backscattering spectrometry (with a He^{2+} ion beam at projectile energies in the vicinity of the 3.045 MeV resonance in the reaction $^{16}O(\alpha,\alpha)^{16}O$ and at 7.6 MeV). Due to the bombardment of the targets with a high dose of protons a certain content of hydrogen was also expected. The depth distribution of this implanted hydrogen was investigated using the reaction $^{1}H(^{19}F,\alpha\gamma)^{16}O$ at the resonance energy $E_R = 6.46$ MeV with a $^{19}F^{2+}$ ion beam. Carbon deposition during the evaporation process and during the beam irradiation and other contaminant elements possibly present in the target enviroment, i.e. fluorine and nitrogen, were investigated via deuterium induced nuclear reactions with a d^+ ion beam of 920 keV.

The thickness of the boron films was measured making use of the reaction $^{11}B(p,\gamma)^{12}C$ at the resonance energy $E_R = 163$ keV ($\Gamma_R = 5.3$ keV), at the 400 kV accelerator of the Institut für Kernphysik in Münster.

The measurements of the excitation function at low projectile energies, were performed with a target set up specially designed for low cross section experiments involving high ion currents: The ion beam passed two collimators with a diameter of 19 mm and 13 mm, respectively. This collimation ensured an ion beam spot of about 20 mm diameter at the target position. Both collimators and the target itself were directly water cooled. A liquid nitrogen cooled copper tube (65 cm length, 3 cm inner diameter) extended from the second collimator to a distance of 2 cm to the target. By means of two turbomolecular pumping units the pressure was kept lower than 5×10^{-7} mbar during the measurements. The reaction products were detected with four solid state surface barrier detectors (each of 600 mm^2 active area, 100 μm active thickness) positioned at $\theta_{\text{lab}} = 129°$ around the beam axis and close to the target, leading to a total solid angle of 1.2 sr. The detectors were electrically isolated from the cylindrical target chamber and Ni-foils (0.5 or 0.25 μm thick) were placed in front of the detectors to stop the intense flux of elastically scattered ions. The target together with the target chamber and the Ni-foils formed the Faraday cup for beam integration.

Besides the low cross section of the reaction $^{10}B(p,\alpha)^{7}Be$, the major difficulty in extending the measurements to low projectile energies arises from the low energy of the resulting α particles. Furthermore, the α particles suffer an additional energy loss when passing through the Ni-foils, thus they reach the detectors with an energy of about 400 keV. At these energies the background induced by cosmic rays and the natural background has to be considered. In addition, one reaction product, the ^{7}Be nucleus, is a radioactive isotope (decay by internal conversion, $^{7}B(e^-,\nu)^{7}Li$, half life $T_{1/2} = 53$ days). The produced gamma rays can in principle interact with the solid state detectors and contribute to an increase of the background counting rate.

In order to reduce cosmic ray induced background the target chamber was surrounded with a plastic scintilliator, which worked in anticoincidence with the particle detectors. Additionally, the natural background was reduced by a factor of 3 when shielding the target chamber with 5 cm lead bricks. In total a background reduction of a factor of 10 was achieved.

3. Results

By means of the RBS analysis of the oxygen content it could be shown that oxygen atoms contributed to more than 50 % to the composition of the boron targets. The studies utilizing (d,p)-reactions at $E_d = 920$ keV proved that no significant carbon deposition occured during beam irradiation. Furthermore, a hydrogen content in the range of 10 atomic percent could be established. A detailled discussion of the analysis methods and the corresponding results will be given in [8].

Besides the target composition also the energy resolution of the ion beam and the effects of Coulomb explosion of the molecular ions (H_2^+ and H_3^+) were taken into account in the data analysis.

The angular distributions for both reactions were assumed to be isotropic [9, 10].

For energies $E \geq 100$ keV, one expects a negligible enhancement due to electron screening. Thus, the data (from previous [9] and present works) for the reaction $^{11}B(p,\alpha)2^4He$ were fitted with the polynomial expansion:

$$S_b(E) = a + bE + cE^2 + dE^3 \tag{3}$$

at energies $E \geq 100$ keV to obtain approximately the energy dependence of $S_b(E)$ for bare nuclei. The Fig. 1a shows the astrophysical $S(E)$ factor as a function of the energy E for the reaction $^{11}B(p,\alpha)2^4He$. The solid curve was obtained from the fit at higher energies and was assumed to represent the case of bare nuclei. At low energies the effect of the electron screening on $S(E)$ factor can be seen clearly. The dashed curve represents the calculated enhancement using equation 2 with U_e as free parameter and $U_e = 620 \pm 65$ eV as the best fit.

The experimental data (from previous [1, 10] and present works) of the reaction $^{10}B(p,\alpha)^7Be$ were fitted for energies higher than 100 keV with the polynomial expansion (3) as in the case of the reaction $^{11}B(p,\alpha)2^4He$. In analysing the low energy data an initial value for the screening potential of $U_e = 620$ eV was considered taking advantage of the isotopic independence of electron screening. However, $S(E)$ exhibits an enhancement larger than expected from the screening effect and the behaviour of the $S(E)$ curve could not be explained only by this effect (Fig. 1b). This enhancement in the low energy data may be due to the effect of the $J^\pi = 5/2^+$ state at $E_x = 8.701$ MeV (natural width $\Gamma_{tot} = 16$ keV) [11]. Absolute cross sections are larger compared to the previous measurements.

4. Conclusions

For the approximations used here (in particular the assumed $S_b(E)$ form at $E \leq 100$ keV), the following conclusions may be drawn:

- The effect of electron screening on the cross sections of the fusion reactions $^{10}B(p,\alpha)^7Be$ and $^{11}B(p,\alpha)2^4He$ has been experimentally verified.

- Assuming isotopic independence of the electron screening effect, the enhancement in the low energy data for the reaction $^{10}B(p,\alpha)^7Be$ could be attributed to the effect of a low energy resonance centered around 10 keV.

Acknowledgments

The authors would like to thank H. Baumeister, K. Brand and B. Hippert for technical assistance. We are also grateful to M. Berheide, L. Borucki, M. Junker, J. Meijer, N. Piel, C. Polaczyk, G. Quathamer, G. Raimann, G. Roters, T. Schange, S. Schmidt and D. Zahnow for help during the course of the experiments.

References

[1] Szabò J, Csikai J and Varnagy M 1972 *Nucl. Phys.* **A195** 527

[2] Peterson R J, Zaidins C S, Fritts M J, Roughton N A and Hansen C J 1975 Ann. Nucl. Energy **2** 503

[3] Engstler S, Raimann G, Angulo C, Greife U, Rolfs C, Schröder U, Somorjai E, Kirch B and Langanke K 1992 *Phys. Lett.* **B279** 20

[4] Rolfs C and Rodney W S 1988 *Cauldrons in the cosmos* (Chicago: University of Chicago Press)

[5] Assenbaum H J, Langanke K and Rolfs C 1987 *Z. Phys.* **A 327** 461

[6] Langanke K and Rolfs C 1989 Mod. Phys. Lett. **A4** 2101

[7] Engstler S, Krauss A, Neldner K, Rolfs C, Schröder U and Langanke K 1988 *Phys. Lett.* **B202** 179

[8] Angulo C, *Thesis*, University of Sevilla

[9] Becker H W, Rolfs C and Trautvetter H P 1987 *Z. Phys.* **A 327** 341

[10] Youn M, Chung H T, Kim J C and Bhang H C 1991 *Nucl. Phys.* **A533** 321

[11] Wiescher M, Boyd R N, Blatt S L, Rybarcyk L J, Spiznoco J A, Azuma R E, Clifford E T H, King J D, Görres J, Rolfs C and Vlieks A 1983 *Phys. Rev.* **C28** 1431

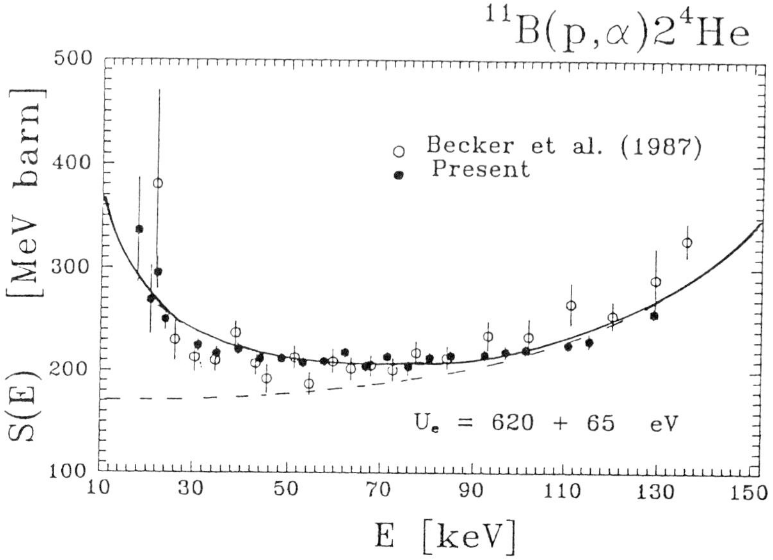

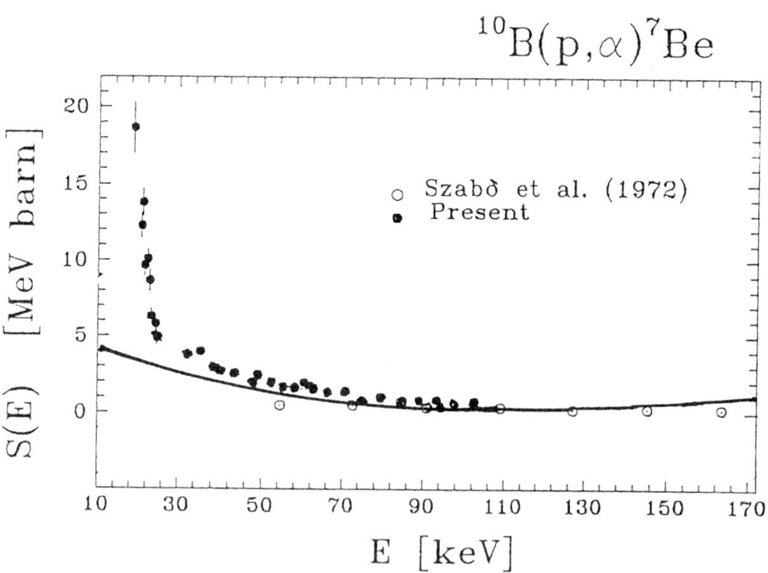

Figure 1. Astrophysical $S(E)$ factor as a function of the center-of-mass energy E for the reactions (a) $^{11}\text{B}(\text{p},\alpha)2\,^4\text{He}$ and (b) $^{10}\text{B}(\text{p},\alpha)^7\text{Be}$.

Status report on the project Laboratory for Underground Nuclear Astrophysics (LUNA)

C. Arpesella[1], C. Barnes[2], E. Bellotti[1], C. Broggini[1], P. Corvisiero[3],
N. Ferrari[1], G. Fiorentini[4], S. Fubini[5], G. Gervino[6], F. Gorris[7],
U. Greife[7], C. Gustavino[1], M. Junker[7], R. Kavanagh[2],
G. Mezzorani[8], P. Prati[3], P. Quarati[9], C. Rolfs[7], W.H. Schulte[7],
H.-P. Trautvetter[7], and D. Zahnow[7]

[1]Laboratori Nazionali del Gran Sasso, LNGS, Assergi
[2]Kellogg Radiation Laboratory, Caltech, Pasadena
[3]Dipartimento di Fisica and INFN, Genoa
[4]Dipartimento di Fisica and INFN, Ferrara
[5]ENEA, Frascati and INFN, Torino
[6]Dipartimento di Fisica and INFN, Torino
[7]Institut für Physik mit Ionenstrahlen, Ruhr-Universität Bochum
[8]Dipartimento di Scienze Fisiche and INFN, Cagliari
[9]Politecnico, Torino and INFN, Cagliari

During the last decades, experimental studies of low energy nuclear cross sections, which are relevant for nuleosynthesis in stars and the early universe, as well as for the evaluation of the solar neutrino flux, have been performed to lower and lower energies. Hereby the effects of electron screening and other effects like the Oppenheimer-Phillips Effect have also been investigated.

These measurements have been optimized by using the best experimental techniques available today. However they are basically limited by the effects of cosmic rays. This problem can be reduced significantly by carrying out experiments in an underground laboratory, such as the Gran Sasso National Laboratory (LNGS).

The underground laboratories are located alongside the 10.4 km long Gran Sasso tunnel, on the highway connection Rom to L'Aquila and Teramo. The rock overburden guarantees very efficient shielding against cosmic radiation, reducing the flux of known particles by a large factor, close to one million.

First background measurements with silicon surface barrier detectors were done in 1991, showing, that the background reduction was in the dimension of a factor 40 - 50 depending on the energy region (Fig.1).

In order to test the possibilities of an accelerator laboratory in the LNGS, a transportable 30 kV accelerator was designed which can provide a proton beam between 400 - 550 μA depending on the energy (Fig. 2).

This june 1992 the accelerator was for the first time built up in the underground laboratory for some test measurements on the reaction $^{11}B(p,\alpha)2\alpha$. First results show that after the immense reduction of the cosmic background, the next limitation will be due to intrinsic radiation of the detector or radiation out of the target and vaccuum chamber material. In spite of this limitation future experiments can reach results with higher reliability than in unshielded laboratories.

The LUNA Collaboration plans to install next year a 50 kV accelerator in order to increase the overlap with other experiments and to open the opportunity to measure the astrophysically interesting ^{3}He(^{3}He, 2p)^{4}He reaction.

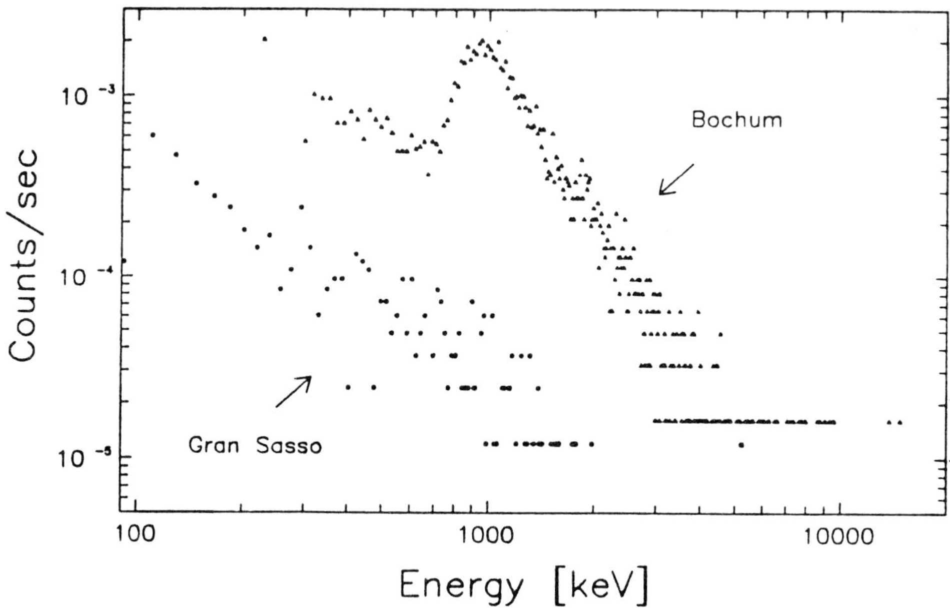

Fig.1. Spectra of a silicon surface barrier detector with a surface of 500 mm^2 and a thickness of 2000 μm, taken in the Gran Sasso Laboratory and in Bochum. The detector was shielded with 1 mm copper and 10 cm of lead.

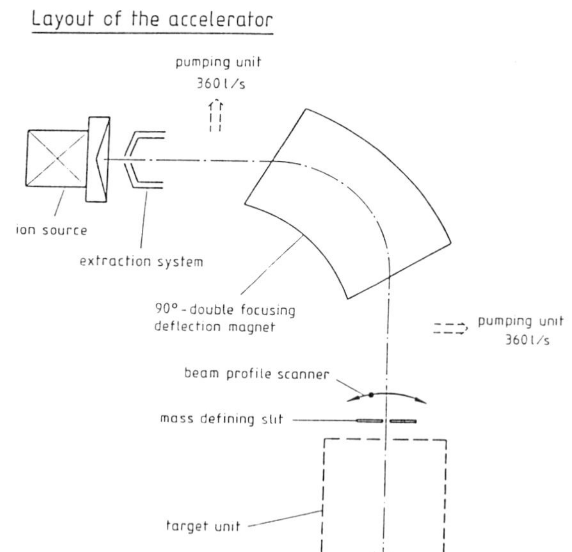

Fig.2. Layout of the accelerator setup

Studies of the nuclear reaction $^{19}F(p,\alpha\gamma)^{16}O$ in the proton energy range 0·3–3·0 MeV

S O F Dababneh, F Taffal, F M Safi, K Toukan *, H Fakhoury and I Khubeis

Department of Physics, University of Jordan, Amman, JORDAN.
* Department of Industrial Engineering, University of Jordan, Amman, JORDAN

Abstract. Different physical aspects related to the $^{19}F(p, \alpha\gamma)^{16}O$ nuclear reaction have been investigated. The excitation function in the proton energy range from 0.3 to 3.0 MeV has been measured with a high efficiency BGO scintillation detector. High resolution HPGe detector has been used for measuring gamma-ray spectra that have been carefully analyzed. The yield curves measured with thin targets (3 - 50 nm) reveal many details, especially for those resonances noticed to be having relatively large widths. Thus, additional excited states in the ^{20}Ne nucleus that have not been reported in the literature are documented and discrepancies in published data are discussed.

1. Introduction

Experimental and theoretical investigation of the $^{19}F(p, \alpha\gamma)^{16}O$ nuclear reaction in the low energy region, offers ideal opportunities for nuclear spectroscopy and for understanding reaction mechanisms and nuclear excitation levels.

The $^{19}F(p, \alpha\gamma)^{16}O$ is an example of a rearrangement reaction. It contains three of the five exit channels of the $^{19}F + p$ reaction. The five different groups of alpha particles correspond to the transition from the compound nucleus state ^{20}Ne to the ground state, pair emitting state, and three gamma-ray emitting states in ^{16}O. Of particular interest in this work are those alpha-particle groups, designated by α_1, α_2 and α_3, where alpha particle emission is followed by a γ-transition γ_1, γ_2 and γ_3 respectively, directly to the ground state in ^{16}O. Attention has been paid to discrepancies in published data, which have been analyzed and compared with our results.

2. Experiment

The experiments have been carried out using a proton beam from the Jordan University Van de Graaff Accelerator (JUVAC). The measurements have been performed at proton energies ranging from 0.3 up to 3.1 MeV with different steps according to the energy region studied. The energy calibration of the accelerator has been carried out frequently. The absolute energy (E) of a particle of mass (M) has been determined by measuring the Nuclear Magnetic Resonance NMR frequency given by

$$f = A \sqrt{ME}\left(1 + \frac{E}{2Mc^2} \right). \tag{1}$$

The constant (A) in Eq. 1 is determined using the "accurately" known resonance and threshold energies that are suitable for calibration purposes, namely the 991.90 keV resonance energy of the ^{27}Al(p, γ)^{28}Si reaction and the 1880.6 keV threshold energy of the ^{7}Li(p, n)^{7}Be reaction. Both the 340.46 and 872.11 keV resonance energies of the ^{19}F(p, $\alpha\gamma$)^{16}O reaction reported in the 1983 compilation of Ajzenberg-Selove [1] have been also used to verify the calibration procedure.

Maximum energy shift of less than 100 eV has been measured at the borders of energy bands (> 2 MeV) while this shift is only about 10 eV elsewhere (< 2 MeV). Energy spread of the beam has been estimated to be < 150 eV below 2 MeV.

Calcium, lithium and magnesium fluorides have been evaporated on carbon substrates maintained at 200° C. The evaporation rate is 5 Å/s within 3 % uniformity spread. Both HPGe and BGO have been used as detectors for the gamma photons emitted by the excited ^{16}O nuclei.

The measuring systems for both detectors have been calibrated using standard gamma-ray sources and gamma lines from other nuclear reactions. The ORTEC 7450 multi-channel analyzer has been calibrated using ^{60}Co and ^{208}Tl gamma sources as well as the ^{27}Al(p, γ)^{28}Si reaction at E_p = 2.74 MeV with E_γ = 1.78 and 9.36 MeV.

3. Results and discussion

The measurement of the excitation function for the ^{19}F(p, $\alpha\gamma$)^{16}O reaction has been done by measuring the observed γ-ray yield in the high energy portion of the γ-ray spectrum around the 6.1304, 6.9171 and 7.1169 MeV lines. The whole region containing the three main lines (γ_1, γ_2, γ_3) and their escape peaks has been considered. Contributions of other nuclear reactions namely $^{15}N(p, \alpha\gamma)^{12}C$; $^{14}N(p, \gamma)^{15}O$; $^{15}N(p, \gamma)^{16}O$; $^{12}C(p, \gamma)^{13}N$; $^{40}Ca(p, \gamma)^{41}Sc$; $^{24}Mg(p, \gamma)^{25}Al$ and $^{7}Li(p, \gamma)^{8}Be$ are insignificant due to their very low cross sections or the emitted γ-ray energies being outside the energy region of interest.

Fig. 1 shows an overall view of a thin-target excitation curve from $E_p = 300$ to 2000 keV. The counting is normalized to a LiF target containing $19\,\mu g\,^{19}F/cm^2$. This figure illustrates the energy region in which isolated and narrow resonances are found.

More accurate and detailed measurements of the excitation function have been carried out. Experimental data for the resonances observed in the $^{19}F(p, \alpha\gamma)^{16}O$

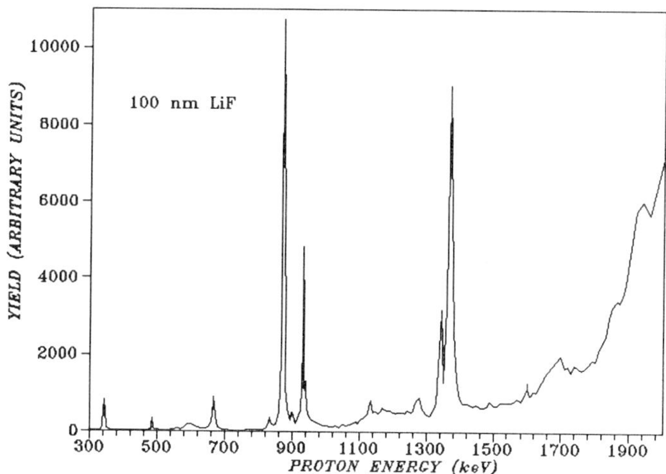

Fig. 1: Overview of the excitation curve from $E_p = 0.3$ - 2.0 MeV where isolated and narrow resonances are found

reaction are summarized in Table 1 which provides a comparison beween the data reported by Ajzenberg-Selove [1], Dieumegard et al.[2] and Ouichaoui et al.[3], along with the results obtained in this work. Since the work of Dieumegard et al. [2] is one of the main data sources for the 1983 and 1987 compilations of Ajzenberg-Selove, it seems that the resonance energies reported in the earlier compilations are noteworthy for the comparison to be more general. Also listed in Table 1 are the Γ_{obs} values measured using different targets and target thicknesses. Γ_{obs} is calculated as the 25 % to 75 % yield points in the thick target 300 nm (integrated) yield curve.

The 470 - 500 keV region has been scanned using a 3 nm CaF_2 target to investigate the existence of the 478.47 ± 0.2 keV weak resonant structure just before the 484 keV resonance. This region is illustrated in Fig. 2 where this 478.47 keV resonance with Γ_{obs} of 0.7 ± 0.3 keV is shown in the neighbourhood of the 484.22 keV resonance. This very narrow resonance has not been previously reported in the literature [1].

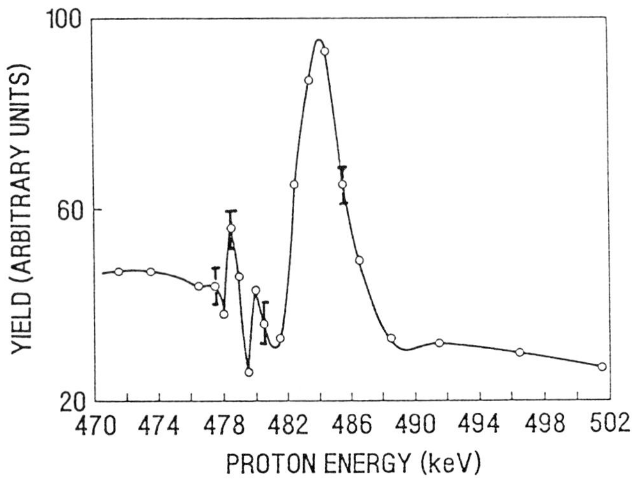

Fig. 2:   The 478.47 ± 0.20 keV new feature and its neighbouring 484.22 ± 0.07 keV resonance.

4. Energy levels of the ^{20}Ne compound nucleus

The different excitation functions and excitation energies (resonances) of the $a_{1,2,3}$ and $a_{0}\pi$ groups indicate differences in the structure of the involved compound levels. Angular momentum relations are investigated by studying the correlation between angular distributions of alpha particles and γ-radiation resulting from the reaction. Table 2 provides more detailed tabulation of these virtual states in the compound nucleus compared with those reported in the 1987 compilation of Ajzenberg-Selove [4]. This offers more opportunity for further investigation of the angular distribution of the reaction products to determine the physical parameters needed to estimate angular momentum relations.

5. Conclusion

Making use of the results obtained in this work, especially those tabulated in Table 2, provides a wide variety of choices for researchers interested in certain applications of nuclear reactions. New γ-resonances have been measured in the low E_{p}-region, e.g. the 591.8 ± 0.8, 597.6 ± 0.5 and 601.0 ± 0.8 keV instead of the

TABLE 1:   Resonances and widths measured in this work and compared with those in the literature.

| LITERATURE | | | | | This work | |
| Ajzenberg-Selove (1983) Ref. (1) | | Ref. 2 | Ref. 3 | | | |
E_p (keV)	Γ_{obs} (keV)	E_p (keV)	E_p (keV)	Γ_{obs} (keV)	E_p (keV)	Γ_{obs} (keV)
226.9 ± 3.4[c]	1.0	---	---	---	---	---
340.46 ± 0.04	2.4 ± 0.2[c]	..	---	---	340.45 ± 0.08	2.5 ± .2
---	...	---	...	---	478.47 ± 0.20	0.7 ± .3
483.8 ± 0.3[ı]	0.9 ± 0.1	483.7 ± 0.5	---	---	484.22 ± 0.07	1.3 ± .2
594 ± 3[a]	25 ± 3	594 ± 3	---	---	591.8 ± 0.8	
					597.6 ± 0.5	
					601.0 ± 0.8	
667.5 ± 2[a]	6.7 ± 0.3	667.5 ± 2	..	---	669.23 ± 0.2	3.8 ± .5
					675.05 ± 0.1	1.1 ± .2
832.1 ± 1[a]	----	832.1 ± 1	---	---	832.7 ± 0.4	7 ± 1
872.11 ± 0.2	4.7 ± 0.2[c]	872.1 ± 0.5	---	---	872.1 ± 0.23	4.3 ± .2
902.3 ± 0.9[b]	5.1 ± 1.0	898.8 ± 1	---	---	901.14 ± 0.50	5.4 ± .5
935.4 ± 1.3	8.1 ± 0.5	933.6 ± 1	---	---	935.54 ± 0.80	6.5 ± 1
1087.7 ± 1[a]	0.15 ± 0.05	1087.7 ± 1	---	---	1090.5 ± 0.5	2.5 ± .5
1135.6 ± 1[a]	---	1135.6 ± 1	---	---	1138.3 ± 0.8	3 ± 1
1280 ± 1[a]	---	1280 ± 1	---	---	1277.8 ± 2	10 ± 1
1347.7 ± 1	4.9 ± 0.7	1344.5 ± 1	---	---	1344.6 ± 2	5.5 ± .5
1371 ± 1[a]	12.4 ± 1	1370.9 ± 0.8	---	---	1370.3 ± 1	14 ± 1
1603 ± 2[a]	---	1603 ± 2	1600 ± 5	15 ± 4	1604.5 ± 2	8 ± 1
1692 ± 2[a]	35 ± 3	1690 ± 2	1692 ± 5	28 ± 7	1690.9 ± 3	27 ± 3
---	---	---	1710 ± 10	80 ± 20	---	---
---	---	---	1853 ± 10	31 ± 8	---	---
1949 ± 2.5	40 ± 10	1935 ± 3	1943 ± 5	29 ± 8	1938.5 ± 2.5	
---	---	---	---	---	1972 ± 5	
2030 ± 3	120 ± 20	2014 ± 6	2025 ± 8	27 ± 7	2030 ± 3	45 ± 5
2320	85	---	---	---	2322 ± 5	60 ± 10
2510	30	---	---	---	2520 ± 5	20 ± 5
2630	90	---	---	---	2625 ± 7	
---	---	---	---	---	2755 ± 7	
2800	60	---	---	---	2835 ± 10	50 ± 5
3020	30	---	---	---	2980 ± 7	20 ± 7

These resonances are reported in the previous compilations as:
   596.8 ± 1 (30 ± 3); 1090 ± 1 (0.7 ± 0.3), 1373 ± 1 (12.4 ± 1)
   671.6 ± 0.7 (6 ± 0.7), 1140 ± 1 (2.5), 1607 ± 1.6 (6 ± 1)
   834.8 ± 0.9 (6.5 ± 1), 1283 ± 1.4 (18.6 ± 1), 1694 ± 1.7 (35 ± 3) keV
Reported only with those in a) but not mentioned in the 1983 compilation.
Mentioned later in the 1987 compilation as [4]:
   223.99 ± 0.07 (0.99 ± 0.02), 340.46 ± 0.04 (2.34 ± 0.04)
   483.91 ± 0.10 (0.90 ± 0.03), 872.11 ± 0.20 (4.53 ± 0.16) keV

TABLE 2:   Resonance energies and the corresponding levels in the ^{20}Ne compound nucleus.

E'_R(keV)	^{20}Ne* level (MeV)		E'_R(keV)	^{20}Ne* level (MeV)	
	(This work)	(Ref. 4)		(This work)	(Ref. 4)
323.52	13.1679	13.1713	1214.2	14.0587	14.063
454.68	13.2991		1277.7	14.1222	14.128
460.14	13.3046	13.3075	1302.1	14.1466	14.150
562.30	13.4068		1524.7	14.3692	14.370
567.80	13.4123		1606.8	14.4513	14.455
571.00	13.4150		1842.1	14.6865	14.699
635.96	13.4804		1873.0	14.7184	
641.49	13.4859		1929.0	14.7735	14.776
791.30	13.6357	13.6380	2206.0	15.0510	15.050
828.70	13.6732	13.6762	2394.0	15.2391	15.230
856.33	13.7008		2494.0	15.3389	15.350
889.02	13.7334	13.736	2618.0	15.4624	
1036.2	13.8807	13.881	2694.0	15.5385	15.510
1081.7	13.9261	13.926	2831.0	15.6763	15.720

broad known one in literature [2, 4] of 594 ± 3 keV. The other resonance 667.5 ± 2 with Γ = 6.7 ± 0.3 keV [2, 4] consists of two resonances 669.23 ± 0.2 and 675.05 ± 0.1 with Γ = 3.8 ± 0.5 and 1.1 ± 0.2 keV respectively. In the higher E_p-region ($\geq$ 2 MeV), two new resonances of 1972 ± 5 and 2755 ± 7 keV have been measured. Those two resonances have not been previously reported in the literature. The other resonances of this region have been measured with higher resolution.

Acknowledgement

We thank Mr. A. Khader for his help during the measurements.

References

[1]   Ajzenberg-Selove F 1972 Nucl. Phys. **A190** 1; 1982 Nucl. Phys. **A375** 1; 1983 Nucl. Phys. **A392** 1

[2]   Dieumegard D, Maurel B and Amsel G. 1980 NIM **168** 93

[3]   Ouichaoui S et al. 1985 Il Nuovo Cimento **86A** 170

[4]   Ajzenberg-Selove F 1987 Nucl. Phys. **A475** 1

Angular distribution and yield studies of the $^{19}F(p,\alpha_0)^{16}O$ and $^7Li(p,\alpha)^4He$ nuclear reactions in the proton energy range 0·3–3·0 MeV

F Safi, F Taffal, K Toukan*, S O F Dababneh, H Fakhoury, and I Khubeis

Department of Physics, University of Jordan, Amman, Jordan
*Department of Industrial Engineering, University of Jordan, Amman, Jordan

Abstract. The angular distribution of the $^{19}F(p, \alpha_0)^{16}O$ and $^7Li(p, \alpha)^4He$ nuclear reactions have been studied for an energy range 1.1 MeV to 2.0 MeV and angular range of 35° - 160°, in addition to the excitation functions for an energy range 0.3 - 3.0 MeV at 150° and from 1.1 - 1.95 MeV at different angles. The observed angular distributions have been corrected to the center-of-mass system and analyzed in terms of Legendre polynomials. Comparison with other previous measurements is also listed, and the spins and parities of prominent resonances in the $^{19}F(p, \alpha_0)^{16}O$ and $^7Li(p, \alpha)^4He$ nuclear reactions are calculated.

1. Introduction

The theory of the angular distribution for resonance reactions has been treated in several papers [1 - 4]. Nuclear microanalysis of flourine has been carried out employing the $^{19}F(p, \alpha_0)^{16}O$ reaction [5, 6]. Some of the experimental work on the angular distribution of $^{19}F(p, \alpha_0)^{16}O$ and $^7Li(p, \alpha)^4He$ reactions has been done using thick targets [7], so that no proper information on the variation of the angular distribution with energy is given. However, results on this reaction are still incomplete. The $^7Li(p, \alpha)^4He$ reaction has been investigated up to 1.4 MeV [8]. There are still a number of questions to be settled, especially with regards to the existence of a $\cos^4\Theta$-term and its dependence on energy.

Detailed measurements of the angular distribution for both reactions have been performed at different energy intervals from 1.1 MeV to 1.95 MeV. In addition, observations on the yield curves at different angles to the incident beam direction, which are not reported previously, have been studied here at small energy intervals.

2. Experimental procedure

The experiment has been performed using the 4.75 MeV Jordan University Van de Graaff Accelerator (JUVAC). A monoenergetic proton beam with an energy spread less than 150 eV bombarded targets mainly LiF evaporated on pure carbon substrates, though MgF_2 and CaF_2 on carbon or tantalum substrates have been also used. A hundred nanometer LiF high purity deposit is evaporated on a carbon substrate. The α-peaks arising from ^7Li and ^{19}F reactions are distinctly resolved from the scattered protons.

3. Analysis of data and results

3.1.   Yield curves of the $^{19}F(p, \alpha_0)^{16}O$ reaction

When fluorine is bombarded by protons, the $^{19}F(p, \alpha)^{16}O$ reactions, elastic and inelastic proton scattering and proton capture followed by gamma ray emission, have all been observed. A study of the $^{19}F(p, \alpha_0)^{16}O$ reaction gives information both on the excited states of ^{16}O and on the excited states of ^{20}Ne above 12.9 MeV. An examination of the excitation function gives the positions and widths of the excited states of ^{20}Ne.

It is generally believed that the alpha particle coming from the $^{19}F(p, \alpha_0)^{16}O$ reaction is well separated from other alphas coming from other reactions [5, 9]. The yield curves at $\Theta = 45°, 55°, 82°, 105°, 135°$ and $142°$ in the energy range 1240 - 2000 keV are shown in Fig. (1).

Resonance locations and total widths "Γ" are determined precisely for the different angles. Summary of the results and comparison with published work [9 - 12] are listed in Table 1. It is clear from the table that some peaks are markedly shifted, which indicates strong nuclear level interference effects.

3.2.   Angular distribution of the $^{19}F(p, \alpha_0)^{16}O$ reaction

Targets of LiF of different thicknesses have been used. The yield has been normalized before each run using the 1350 keV resonance at $\Theta = 150°$. Detailed measurements of the angular distribution have been made in the present work at intervals from 20 - 100 keV, and angular range 160° - 45° in the laboratory system. Each data point has been taken after collecting 30 μC charge on the target to minimize the statistical fluctuation. The data has been collected within a solid angle $d\Omega = 0.041$ steradian, and a geometrical factor $G = 3.3 \times 10^{-3}$. The angles

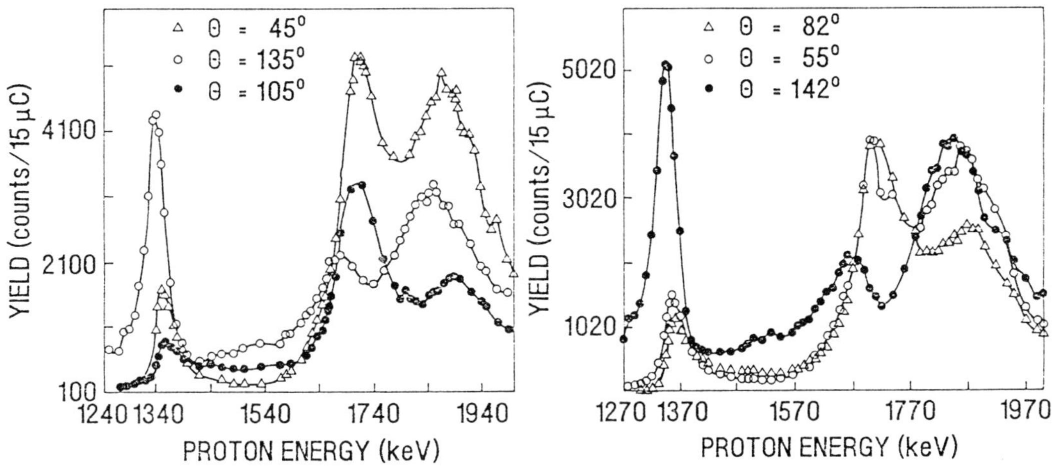

Fig. 1: The yield curves of the ^{19}F(p, α_0)^{16}O reaction as a function of bombarding energy.

and the angular distribution functions have been corrected to the C.M. system. The data are best analyzed in terms of Legendre polynomials up to fourth order. The normalized yield as a function of cosΘ is given by:

$$Y(x) = \sum_{n=0}^{4} a_n P_n(x),$$

where P_n are Legendre polynomials, x $\equiv$ cosΘ, a_n are coefficients determined by the incident channel and by selection rules for parity conservation, the total angular momentum and total magnetic quantum number, as well as by the probabilities of capture and emission of the respective particles.

The total angular momentum of ^{19}F in its ground state is 1/2, and the spin of the incoming proton is also 1/2 [13]. According to the theory of the angular distribution, the entrance channel spin is 0 or 1 and the outgoing channel spin is 0. This implies that only those states of the compound nucleus ^{20}Ne having even spin and even parity or odd spin and odd parity can decay to the ^{16}O ground state. In addition, the angular momentum of the emerging alpha particles will be equal to the spin of the compound state. Also, according to the theory of angular distribution, the asymmetric character of the angular distribution curves indicates that interference effects in the wave function of the emitted alpha-particles are being produced by adjacent or overlapping levels in the compound nucleus ^{20}Ne. Since the best fit to our observed data was up to cos$^4\Theta$, this indicates an appreciable effect from d-wave incident protons.

The angular distribution near the prominent resonance at 1350 keV has been studied as shown in Fig. 2a and determined to be:

$$Y(x) = 2.04\,P_0 - 0.502\,P_1 + 0.65\,P_2 - 1.18\,P_3 - 1.15\,P_4. \qquad [3 - 1]$$

The presence of the odd powers of $\cos\Theta$ in the equation above, indicates interference between adjacent levels in ^{20}Ne decaying to a_0 and ^{16}O in its ground state. We suggest that the relatively big term in $\cos^4\Theta$ in equation [3 - 1] is explained by comparison with the theoretical angular distribution from an assumed 2P (even parity) state in ^{19}F reacting with d-wave protons to produce a

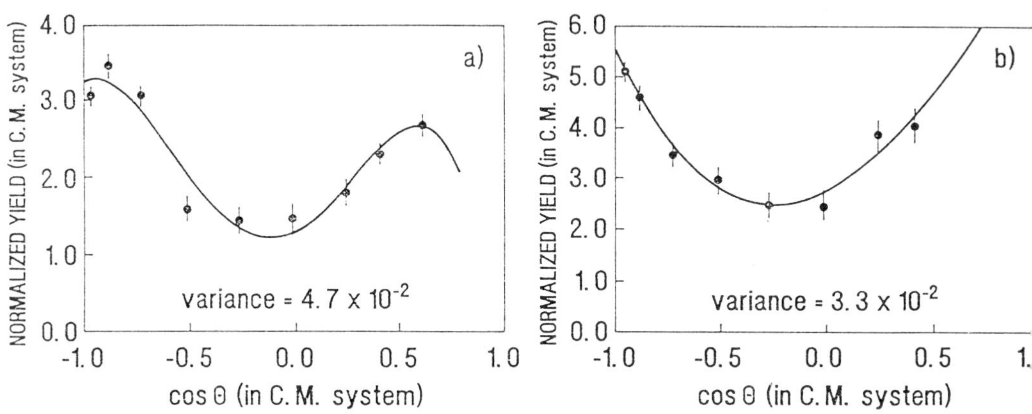

Fig. 2:   The observed angular distribution for the a_0, emerging from ^{19}F(p, a_0)^{16}O reaction. The solid curves are Legendre polynomial fits to the data.
a) E_p = 1350 keV,   b) E_p = 1660 keV

compound level of J = 2 (even parity). The odd powers of $\cos\Theta$ may be explained by interference with a J = 1, odd parity resonance at E_p = 1.1 MeV. So we give the 1350 keV resonance in ^{20}Ne an assignment of 2^+, overlapping with a 1^- resonance located at E_p = 1.1 MeV.

Concerning the 1660 keV resonance, the observed angular distribution is shown in Fig. 2b, and the corresponding angular distribution function is determined to be:

$$Y(x) = 3.97\,P_0 + 1.47\,P_1 + 2.53\,P_2 - 0.5\,P_3 - 0.05\,P_4. \qquad [3 - 2]$$

Because of the large value of "a_2" we suggest a $J^\pi = 1^-$ for this level at E^* = 14.4 MeV in ^{20}Ne. The presence of odd terms of $\cos\Theta$ in equation [3 - 2] may be explained by the interference with the 0^+ resonance at E_p = 1712 keV (see yield curves). So we suggest that the 14.4 MeV level in ^{20}Ne has a total spin of 1 (odd parity), overlapping with a spin 0 (even parity) level.

For the 1712 keV resonance, the angular distribution is represented by:

$$Y(x) = 7.08\,P_0 + 4.3\,P_1 \qquad [3 - 3]$$

Due to the relatively big coefficient of P_0, we suggest an assignment of 0^+ for this resonance.

Also the angular distribution at E_p = 1844 keV has been studied and is best represented by the equation:

$$Y(x) = 8.6\,P_0 + 6.3\,P_1 + 10.39\,P_2 + 1.96\,P_3 + 3.29\,P_4. \qquad [3 \text{-} 4]$$

The presence of relatively big odd terms of $\cos\Theta$ in equation [3 - 4] indicates the

Table 1: Summary of our results for the prominent resonances and widths of the $^{19}F(p, \alpha_0)^{16}O$ reaction as measured from the yield curves at different angles.

$\Theta°$	Resonances located at E_p (keV):	Γ(keV)	$\Theta°$	Resonances located at E_p (keV):	Γ(keV)
(a) 160°	1350 ± 4 1660 ± 2 1844 ± 4	35 ± 8 100 ± 8 135 ± 10	(b,c) 150°	1350 ± 2 1655 ± 2 1845 ± 4	35.5 ± 7 82 ± 7 117 ± 8
142°	1344 ± 5 1660 ± 6 1850 ± 6	41 ± 9 59 ± 13 117 ± 8	135°	1346 ± 6 1720 ± 15 1854 ± 4	31 ± 8 79 ± 9
120°	1353 ± 3 1710 ± 5 1897 ± 15	41 ± 8 82 ± 9 111 ± 9	105°	1360 ± 5 1720 ± 15 1890 ± 10	63 ± 8
(d) 90°	1362 ± 4 1712 ± 3 1890 ± 10	41 ± 8 88 ± 13 88 ± 13	82°	1360 ± 5 1720 ± 10 1880 ± 13	41 ± 9 100 ± 11
75°	1350 ± 3 1710 ± 5	53 ± 9	60°	1355 ± 2 1710 ± 5 1870 ± 10	53 ± 8 88 ± 11 100 ± 16
55°	1355 ± 5 1700 ± 10 1860 ± 10	41 ± 7 82 ± 11 140 ± 11	45°	1355 ± 5 1720 ± 10 1870 ± 10	41 ± 9 88 ± 11 117 ± 11

a) Ranken et al. [10] reported those lines as 1347; 1640 and 1850 keV with Γ = 48; < 115 and 140 keV respectively.
b) Dieúmegard et al. [9] reported those lines as 1347; 1652 and 1849 with Γ = 36; 90 and 122 keV respectively.
c) Ajzenberg-Selove [11] reported those lines as 1350; 1652 and 1842 with Γ = 36; 90 and 122 keV respectively.
d) Clarke et al. [12] reported those lines as 1358; 1709 and 1853 keV with Γ = 54; 148 and 132 keV respectively.

existence of adjacent levels at this energy. This resonance is given an assignment of $J^{\Pi} = 1$. The assignment of the overlapping levels is not discussed in this work, since we work up to 1950 keV proton energy.

Comparison beteween some of our angular distribution curves and those obtained by Clarke et al. [12] and Ouichaoui et al. [14] at 1660 and 1712 keV are in good agreement.

3.3. Yield curves of the $^7Li(p, \alpha)^4He$ reaction

The excitation functions for this reaction have been studied in the proton energy range 1270 - 1950 keV at the angles $\Theta = 160°, 135°, 105°, 75°, 60°$ and $45°$ to the incident beam direction. No resonances or specific features have been observed in this energy region.

3.4.   *The angular distribution of the $^7Li(p, \alpha)^4He$ reaction*

The angular distribution for the alpha particles emitted from this reaction has been studied in the proton energy range 1100 - 1950 keV. We study the angular distribution at $E_p = 1100$, 1200 keV in the range 35° - 90° and from 1270 - 1950 keV in the range 90° - 160° in the laboratory system. The angles and corresponding yields have been corrected to the C.M. system.

Our data in the given energy range, has been best fitted to the equation
$$Y(\Theta)/Y(90) = 1 + A(E)\cos^2\Theta + B(E)\cos^4\Theta.$$
A quadratic plot of $Y(\Theta)/Y(90)$ against $\cos^2\Theta$ determines the coefficients A(E) and B(E).

Since the reaction produces two alpha particles that obey Bose-statistics (even parity for the outgoing channel), this simplifies the analysis greatly. In this reaction the proton spin is 1/2 [15]. The target nucleus has an angular momentum 3/2 [15]. This results in an incident channel spin of 1 or 2. The ground state of ^7Li is assumed to be of odd parity so incoming p, f, h, ... waves can give rise to the even states of the compound nucleus ^8Be. Fig. 3 gives the angular distribution curves obtained at different energies. We can conclude from the equations representing these curves that the series ends always with a power "4". According to the theory of the angular distribution and since the intrinsic spin of α-particle is "0" this implies that any level in ^8Be which can decay into two alpha particles is obliged to have even parity and even total angular momentum. Since as mentioned above, the incident channel spin is 1 or 2, this restricts the incident orbital angular momentum to odd values of 3 & 5. This implies that $J^\pi = 2^+$.

4. Conclusions

This study of the $^{19}F(p, \alpha_0)^{16}O$ and $^7Li(p, \alpha)^4He$ nuclear reactions have focussed on two aspects, the yield curves and the angular distribution of the end products. In conclusion we point out the following:
(1)   The locations of resonanes and the measured widths of the nuclear reaction $^{19}F(p, \alpha_0)^{16}O$ are in good agreement with those reported in literature [9 - 12].
(2)   New yield curves at different angles of detection, which are not reported in the literature, are obtained, resonance locations and widths are also determined.
(3)   No resonances have been found in the yield curves of the nuclear reaction

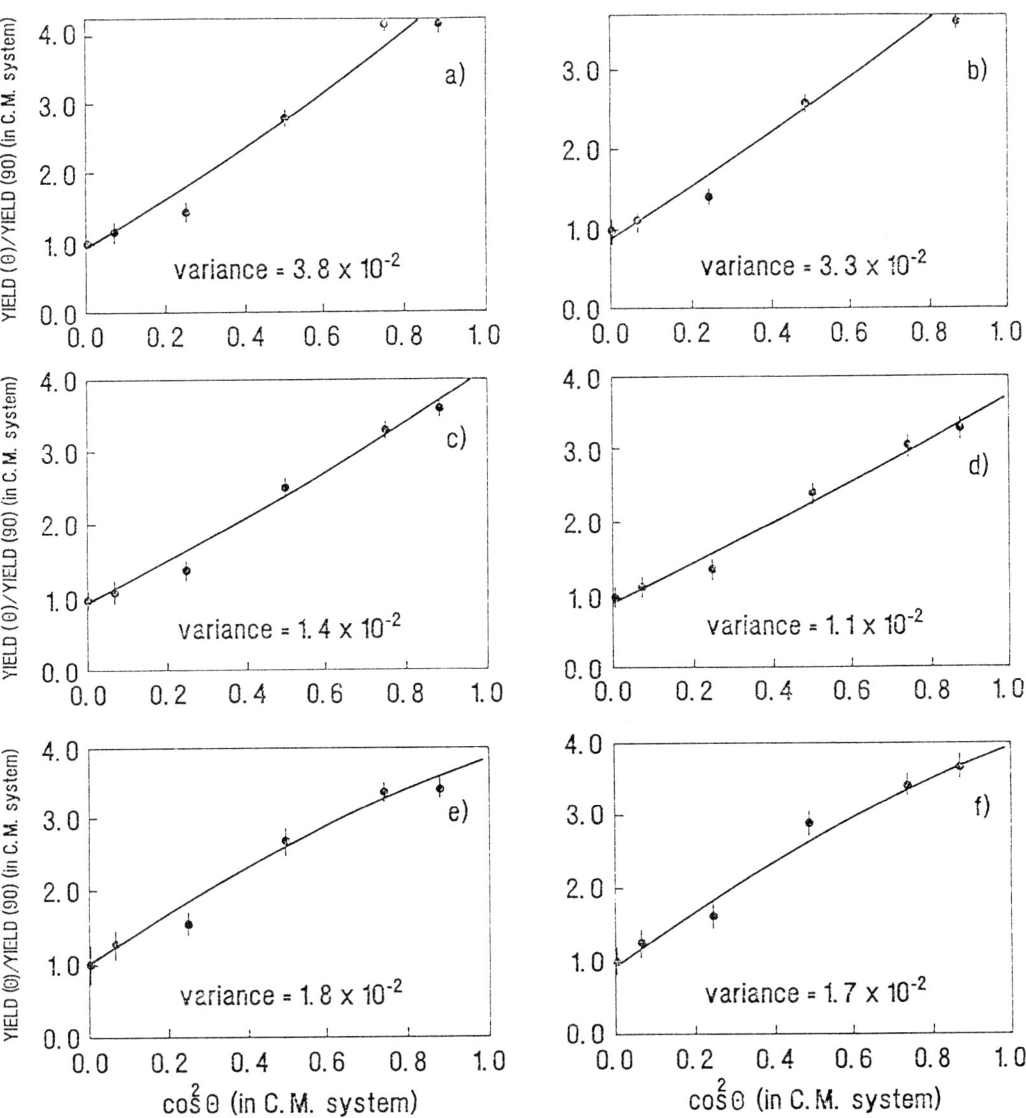

Fig. 3:   The observed angular distributions for the α emerging from the ^{7}Li(p, α)^{4}He reaction. The solid curves are the data fitted to the equation:

$$Y(\Theta)/Y(90) = 1 + A\cos^2\Theta + B\cos^4\Theta$$

a) $E_p = 1300\,\mathrm{keV}$, $Y(\Theta)/Y(90) = 0.95 + 1.30\cos^2\Theta + 1.94\cos^4\Theta$.
b) $E_p = 1370\,\mathrm{keV}$, $Y(\Theta)/Y(90) = 0.92 + 1.31\cos^2\Theta + 1.45\cos^4\Theta$.
c) $E_p = 1500\,\mathrm{keV}$, $Y(\Theta)/Y(90) = 0.91 + 1.09\cos^2\Theta + 1.41\cos^4\Theta$.
d) $E_p = 1605\,\mathrm{keV}$, $Y(\Theta)/Y(90) = 0.94 + 1.11\cos^2\Theta + 1.05\cos^4\Theta$.
e) $E_p = 1860\,\mathrm{keV}$, $Y(\Theta)/Y(90) = 0.95 + 2.01\cos^2\Theta + 0.18\cos^4\Theta$.
f) $E_p = 1950\,\mathrm{keV}$, $Y(\Theta)/Y(90) = 0.93 + 2.22\cos^2\Theta + 0.11\cos^4\Theta$.

^{7}Li(p, α)^{4}He in the proton energy range of 0.7 - 2.0 MeV and angular range of 35° to 160°.

(4)   The spins and parities of the resonances at 1.35, 1.66, 1.71 and 1.84 MeV in the ^{19}F(p, α_0)^{16}O reaction have been determined as 2$^+$, 1$^-$, 0$^+$ and 1$^-$, respectively.

(5)   The spin and parity of the excited level of ^{8}Be at 20 MeV has been determined to be 2$^+$. This is in full agreement with Kumar et al [16] who used a polarized proton beam.

(6)   In the ^{7}Li(p, α)^{4}He reaction, the data is best fitted to the equation $Y(\Theta)/Y(90) = 1 + A(E) \cos^2\Theta + B(E) \cos^4\Theta$, a result which ensures the presence of $\cos^4\Theta$ term at proton energies higher than 1.3 MeV.

Acknowledgement

The authors would like to thank Mr. A. Khader for his continuous help during the measurements.

References

[1]   Blatt J 1952 Phys. Rev. **82** 123

[2]   Wigner E P and Eisenbud J 1951 Franklin Inst. **251** 231

[3]   Wolfenstien L 1951 Phys. Rev. **82** 690

[4]   Bloch I, Hull M H, Broyles A A, Boukicius W G, Freeman B E and Breit G 1950 Phys. Rev. **80** 553

[5]   Khubeis I and Zeigler J F 1987 Nucl Inst. Meth. Phys. Res. **B24/25** 691

[6]   Jarjis R A 1978 Nucl. Inst. and Meth. **154** 383

[7]   Paul E B and Clarke R I 1952 Phys. Rev. **80** 982

[8]   Talbott F L, Busala A and Weiffenbach G C 1951 Phys. Rev. **82** 1

[9]   Dieumegard D, Murel B and Amsel G 1980 Nucl. Inst. Meth. **168** 93

[10]   Ranken W A, Bonner T W and McCrary J H 1958 **109** 1646

[11]   Ajzenberg-Selove F 1983 Nucl. Phys. **A392** 1

[12]   Clarke R L and Paul E B 1957 Can. J. Phys. **35** 155

[13]   Inglis D R 1949 Phys. Rev. **76** 1319

[14]   Ouichaoui S, Beaumevielle H, Bendjaballah N and Chami C 1985 IL Nuovo Cimento **86 A N2** 170

[15]   Rubin S 1947 Phys. Rev. **72** 1176

[16]   Kumar N and Barker F C 1971 Nucl. Phys. **A167** 434

Astrophysical S-factors for the radiative capture reaction ^{6}Li(p,γ_1)^{7}Be* at low energies

R. Bruß, O. Arkis, H. Bucka and P. Heide

Technische Universität Berlin, Institut für Strahlungs- und Kernphysik, Hardenbergstraße 36, D-1ooo Berlin 12

Abstract.   Yields for the reaction ^6Li (p,γ_1) $^7Be^*$ have been measured in the energy range 30 to 180 keV. Cross sections and S-factors are determined and compared with results of other groups. Branching ratios for the $^6Li(p,\alpha)^3He$ and $^6Li(p,\gamma)^7Be$ reactions are calculated.

1. Introduction

For nucleosynthesis models the magnitude and the ratio of the astrophysical S-factors for the reactions $^6Li(p,\alpha)^3He$ and $^6Li(p,\gamma)^7Be$ are of interest with respect to the abundance distribution of elements.

2. Experimental set-up

Using a thick ^{6}LiF-target yield measurements for the $^6Li(p,\gamma_1)^7Be^*$ reaction have been performed by means of detecting the 429 keV transition γ-rays to the ^7Be ground state with a Ge(Li) detector placed in forward direction.

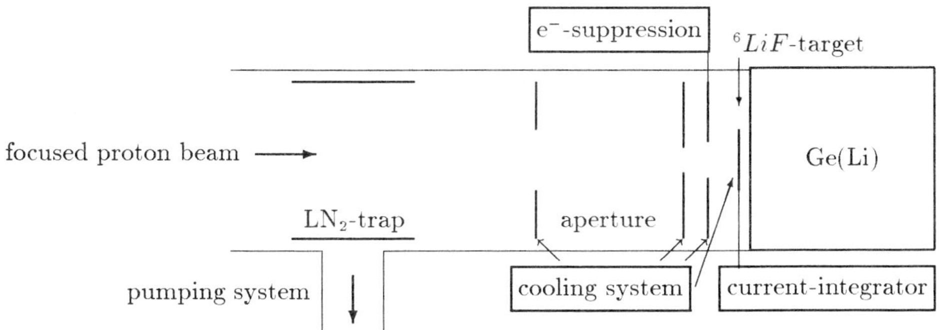

Figure 1: Schematic diagram of the experimental set-up.

The incident proton energy ranged from 30 to 180 keV, the energy uncertainty being about 1 keV. The proton beam which has been analyzed by magnetic deflection was generated by a cascade accelerator. A system of a quadrupole doublet and a vertical stearer focused the protons through an aperture on the target. The target which was produced by evaporation techniques was cooled by compressed air. A nearly 4π-shield of copper and lead protected the detector from background radiation. The detector efficiency was measured with calibrated ^{152}Eu and ^{226}Ra sources.

3. Reaction scheme

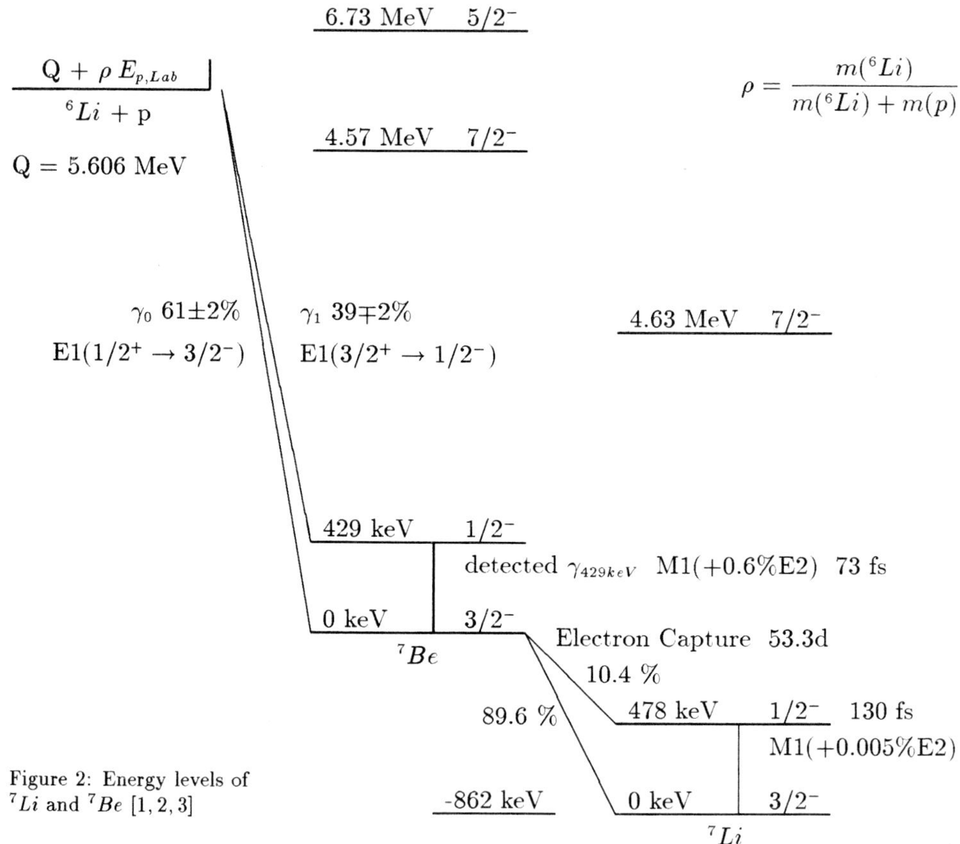

Figure 2: Energy levels of ^{7}Li and ^{7}Be [1, 2, 3]

At low proton energies no important influence of resonance effects are supposed because in the compound system ^{7}Be no energy levels with angular momentum zero are present. The angular distribution of the γ_{429keV} should be spherical.

4. Yields

$$Y_{429keV}(E_{p,Lab}) = \frac{N_{429keV}(E_{p,Lab})}{\eta_d\,\eta_t\,\eta_c\,N_p}$$

$N_{429keV}(E_{p,Lab})$: number of detected 429 keV-γ-events per mC
$\eta_d = 0.0176$: detector efficiency at 429 keV $\eta_t = 0.478$: ^6Li contents of the target
$\eta_c = 0.9977$: current efficiency (livetime to realtime)
N_p: number of protons per mC

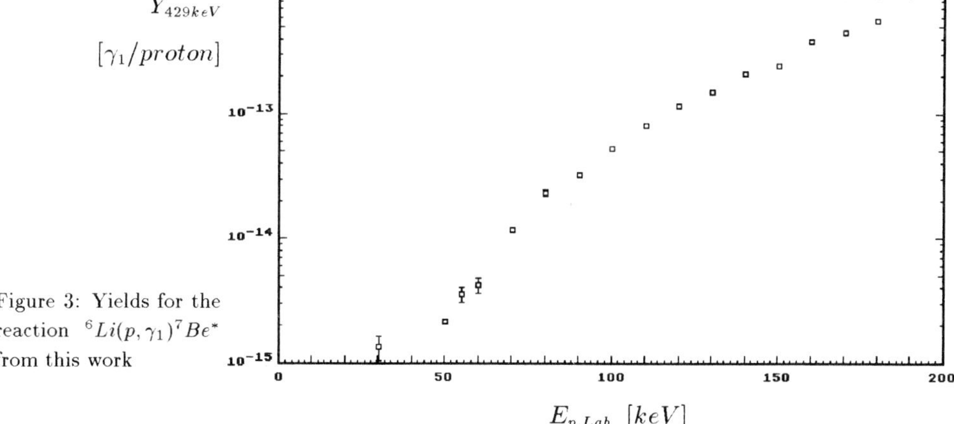

Figure 3: Yields for the reaction $^6Li(p,\gamma_1)^7Be^*$ from this work

The yield $Y_{429keV}(180\ \text{keV}) = (5.6\pm0.1)\ 10^{-13}\ \gamma_1/proton$ from this work is in good agreement with $(7.0\pm4.0)\ 10^{-13}\ \gamma_1/proton$ from [8] .

From the yields the cross section is calculated at $\tilde{E}_{p,Lab} = \dfrac{E_{p,Lab}(2) + E_{p,Lab}(1)}{2}$

$$\sigma(\tilde{E}_{p,Lab}) = \frac{Y_{429keV}(E_{p,Lab}(2)) - Y_{429keV}(E_{p,Lab}(1))}{E_{p,Lab}(2) - E_{p,Lab}(1)} B_{LiF}(\tilde{E}_{p,Lab})\ 10^{15} \quad [nb]$$

$$with\ B_{LiF} \approx \frac{B_{Li} + B_F}{2}\ being\ the\ stopping\ power\ for\ LiF\ [4]$$

5. Cross sections and astrophysical S-factors

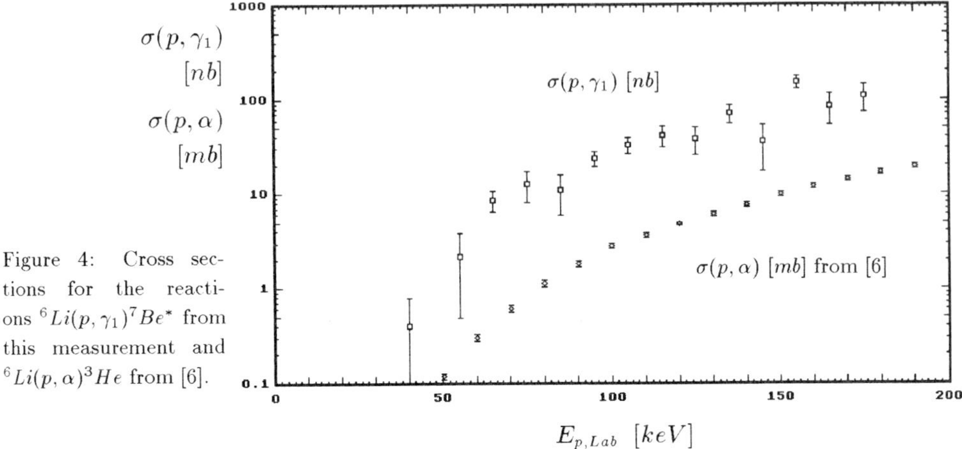

Figure 4: Cross sections for the reactions $^6Li(p,\gamma_1)^7Be^*$ from this measurement and $^6Li(p,\alpha)^3He$ from [6].

From the cross sections the astrophysical S-factor is deduced by the frequently used relation:

$$\sigma(E_{p,Lab}) = \frac{S(E_{p,Lab})}{\rho\, E_{p,Lab}} e^{-2\pi\eta(E_{p,Lab})}$$

with $\rho\, E_{p,Lab}$ being the center-of-mass energy, $\rho = \dfrac{m(^6Li)}{m(^6Li) + m(p)}$ and the Sommerfeld parameter

$$\eta(E_{p,Lab}) = 0.1575\, Z_{Li}\, Z_p\, \sqrt{\frac{m_p}{E_{p,Lab}}}, \quad E_{p,Lab}\ in\ keV, \quad m_p = 1007.3mu$$

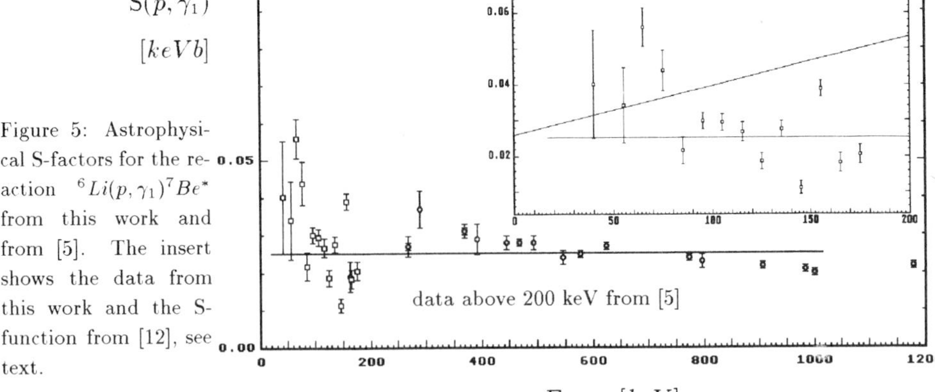

Figure 5: Astrophysical S-factors for the reaction $^6Li(p,\gamma_1)^7Be^*$ from this work and from [5]. The insert shows the data from this work and the S-function from [12], see text.

If we assume for reasons mentioned above that the S-factor for the investigated energy interval may be described in first approximation as energy independent, the measurements result as weighted average in

$$S(p, \gamma_1)=(0.0252\pm0.0025)\text{keVb}.$$

From the known branching ratio for the transitions to the ^7Be ground and first excited state $\gamma_0/\gamma_1=61\%/39\%$ [5] (fig. 2) the total radiative capture S-factor for the $p+^6Li$ reaction may be calculated

$$S(p,\gamma)=(0.0647\pm0.0065)\text{keVb}.$$

This value for S is even suited to describe the measurements of other authors [5, 7, 8, 9] approximately up to an energy of about 1 MeV within an accuracy of 20 %.

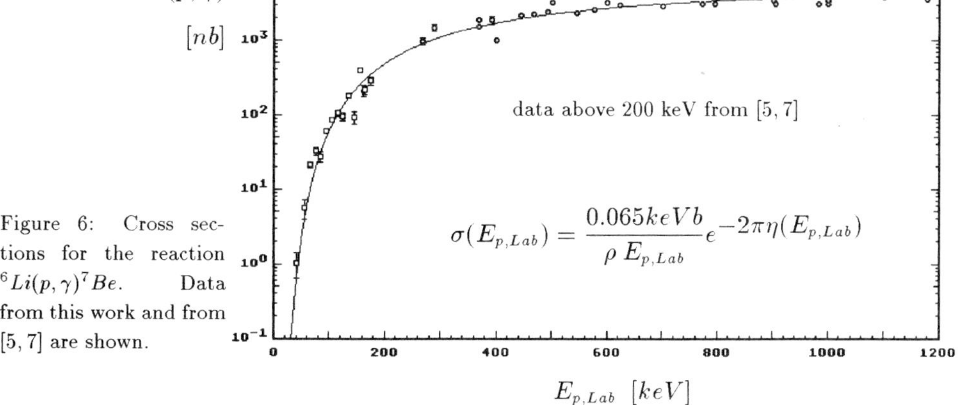

Figure 6: Cross sections for the reaction $^6Li(p,\gamma)^7Be$. Data from this work and from [5, 7] are shown.

In the approximation of a constant S-factor, $S(p,\gamma)=(0.0647\pm0.0065)$ keVb the cross section is calculated with the above relation (fig. 6) at $E_{p,Lab} = 800$ keV. The result $\sigma(p,\gamma) = 3.4\pm0.3\ \mu b$ is in good agreement with [5] $3.1\pm0.4\ \mu b$, [9] $3.9\pm0.6\ \mu b$, [8] $3.0\pm1.0\ \mu b$, and [7] $3.9\pm0.3\ \mu b$. But it disagrees with the smaller value [10] $0.35\pm0.18\ \mu b$. In a further investigation for the S-factor the result $S(p, \gamma_1)=(2.6\ 10^{-5} + 1.6\ 10^{-4}\ E)MeVb$ is published [12], E being the center-of-mass energy in MeV, with 15% uncertainty at 0.1 MeV and 23% at 0 MeV. At $E_{p,Lab}=100$ keV $S(p,\gamma_1)$ is about 1.6 times our value.

6. The branching ratio from $^6Li(p, \alpha)^3He$ and $^6Li(p, \gamma)^7Be$

With our results for the radiative capture cross section and measurements of $^6Li(p, \alpha)^3He$ from other groups [6, 11, 13, 14, 15, 16] we are able to consider the branching ratio at energies below 1 MeV. The S-factor in the energy range from 50 to 800 keV is approximately $S(p, \alpha) = (3000 \pm 160)keV\,b$ from [11]. This value for the S-factor based on measurements from [6, 11, 13]. That means for the branching ratio

$$\frac{(p, \alpha)}{(p, \gamma)} = 46000 \pm 5000.$$

References

[1] F. Ajzenberg-Selove, Nucl. Phys. A490 (1988) 1
[2] E. Browne, R. B. Firestone, Table of Radioactive Isotopes 1986
[3] C. M. Lederer, V. S. Shirley, Table of Isotopes 7th Edit 1978
[4] H. H. Andersen, J. F. Ziégler, The Stopping and Ranges in Matter, Vol. 3, (1977)
[5] Z. E. Switkowski et al., Nucl. Phys. A331 (1979) 50
[6] W. Gemeinhardt et al., Z. Phys. 197 (1966) 58
[7] R. Ostojic et al., Nuovo Cim. 76A1 (1983) 73
[8] S. Bashkin et al., Phys. Rev. 97 (1955) 1245
[9] W. E. Sweeny et al., Bull. Am. Phys. Soc. 14 (1969) 487
[10] J. B. Warren et. al., Phys. Rev. 101 (1956) 242
[11] T. Shinozuka et al., Nucl. Phys. A326 (1979) 47
[12] F. E. Cecil et al., Nucl. Phys. A539 (1992) 75
[13] H. Spinka et al., Nucl. Phys. A164 1971 1
[14] J. B. Marion et al., Phys. Rev. 104 (1956) 1402
[15] F. Bertrand et al., Report CEA-R-3428 (1968)
[16] P. H. Beaumevieille et al., Report CEA-R-2624 (1964)

The ^{10}B(p,α_0)^{7}Be reaction at thermonuclear energies

F. Knape, H. Bucka and P. Heide

Institut für Strahlungs- und Kernphysik, Technische Universität Berlin

Abstract. Cross sections were determined for the ^{10}B(p,α_0)^{7}Be reaction in the proton energy range E_{lab}=37-120 keV using direct α-detection by silicon surface barrier detectors and thick-target technique. Angular distributions between θ_{lab}=75° and 150° were measured for energies between 63 and 111 keV. The influence of the 8.701 MeV state of ^{11}C on the S-factor was clearly observed; the S-factor at the resonance energy E_R=10 keV was determined by fits to the present data. The possible enhancement of the experimental cross sections due to screening effects is discussed. The thermonuclear reaction rates deduced from the present results are compared with previously reported values. The importance of the 10 keV resonance for the reaction rates in the temperature region below T_6=500 is shown and conclusions are drawn with respect to the boron abundance.

1. Introduction

The ^{10}B(p,α_0)^{7}Be reaction in the thermonuclear energy region is important in several aspects:

a) The ^{10}B(p,α_0)^{7}Be reaction is the dominant destructive process of ^{10}B. The reaction may have considerable effect on the extremely low boron abundance in the universe and could be incorporated in the present theory of spallative generation of L-elements [1].

b) Together with recent studies of the reactions ^{9}Be(p,γ)^{10}B [2] and ^{10}B(p,γ)^{11}C [3], the study of the ^{10}B(p,α_0)^{7}Be reaction provides important input data for calculations of the production of B and C in an inhomogeneous big bang.

c) The importance of the ^{10}B(p,α_0)^{7}Be reaction for advanced fuel fusion reactors has been thorougly discussed by Peterson et al. [4]. The ^{11}B(p,3α_0) reaction is a relatively clean fusion reaction. However, natural boron contains 19.7% of ^{10}B, which produces radioactive ^{7}Be.

Experimental results of the ^{10}B(p,α_0)^{7}Be reaction in the thermonuclear energy region are quite rare. Szabo et al. [5] have measured in the laboratory energy range E_p=60-180 keV and Youn et al. [6] in the range $120 \leq E_p \leq 480$ kev. Peterson et al. [4], [7] have measured a thick target yield for E_{cm}=69 keV and from 245 keV to 3.0 MeV to provide the reaction rate for the CTR program. In the above measurements, with the exception of the measurement of Youn et al., the angular distributions were assumed to be isotropic.

176   *Nuclei in the Cosmos*

2. Experiment

The present experiment was carried out with the H^+ beam from the TU Berlin ion accelerator. The accelerator energy was calibrated by a measurement of the 163 keV resonance of the reaction $^{11}B(p,\gamma)^{12}C$ and the non-resonant capture of the $^{12}C(p,\gamma)^{13}N$ reaction in combination with a high-resolution Ge γ-ray detector. The energy uncertainty was about 1-2 keV.

Solid targets of 60 $\mu g/cm^2$ and 100 $\mu g/cm^2$ of ^{10}B (96% enriched), vacuum evaporated onto a 0.1 mm tantalum backing, were used.

Three silicon surface barrier detectors (depletion depth 100 μm, surface area 100 mm^2) masked with 0.8 μm aluminium foils were used.

The solid angles covered by the detectors were determined by using a calibrated α-source and by mechanical measurement. The detectors were mounted at $\theta_{lab}=75°$ -165°.

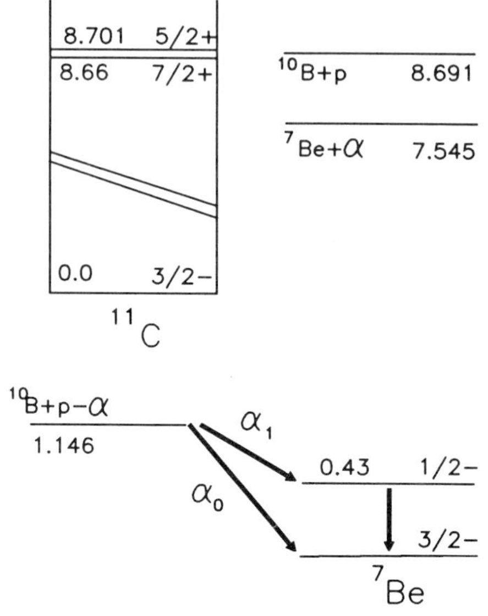

Fig. 1. Schematic diagram of the relevant level scheme and expected α-particle spectrum of the $^{10}B(p,\alpha)^7Be$ reaction.

3. Results

3.1 Angular Distribution

Angular distributions were measured in the energy range $E_p=63-111$ keV in steps of 10 keV. Fig.2 shows the angular distribution of the outgoing α-particles at $E_p=63$ keV.

The angular distributions were found to be isotropic in the entire energy range $E_p=63-111$ keV.

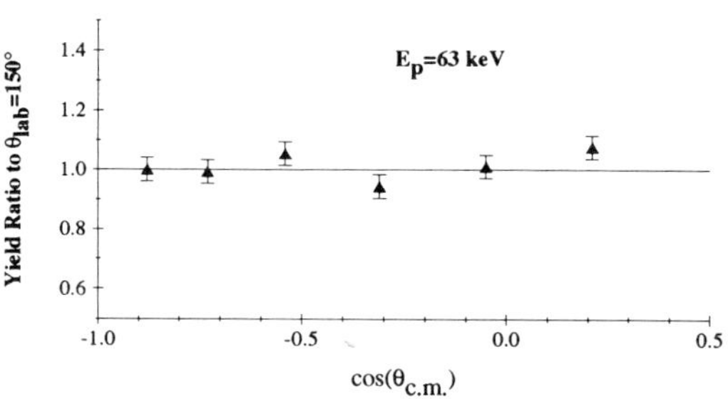

Fig. 2. Angular distribution of the outgoing α-particles at $E_p=63$ keV. The ordinate scale is normalized to the yield at $\Theta_{lab}=150°$.

3.2 Total Cross Section

The total cross sections for the $^{10}B(p,\alpha_0)^7Be$ reaction were derived from the yields at $\theta_{lab}= 150°$ on the assumption that the angular distributions were isotropic. The total cross sections obtained from the present experiment are shown in fig.3 together with previous measurements. Szabo et al. [5] have obtained two sets of data: one by activation measurement and the other by direct α-detection using solid-state detectors. Youn et al. [6] have measured in the energy region $E_p=120-480$ keV.

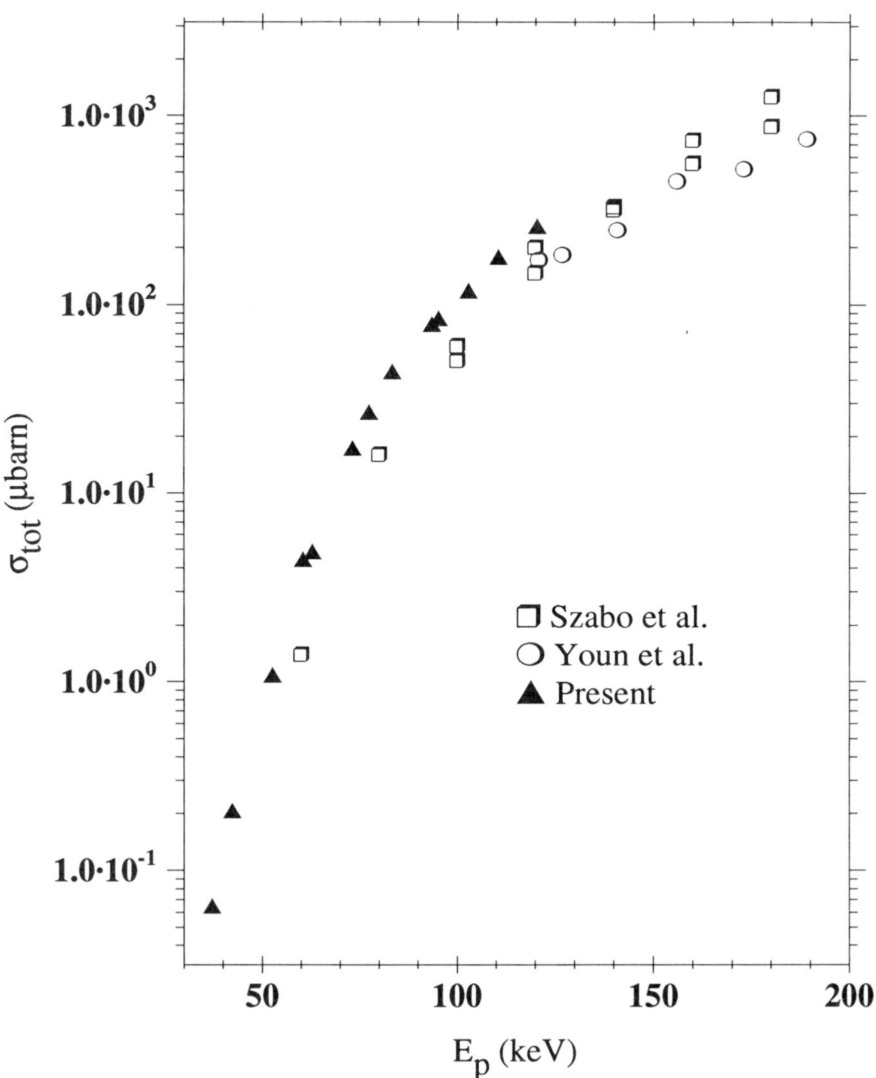

Fig. 3. Total cross sections for the reaction $^{10}B(p,\alpha_0)^7Be$ from the present measurement, together with previous results [5],[6]. Error bars are within the size of the symbols.

3.3 Astrophysical S-Factor

From the cross sections astrophysical S-factors have been calculated. As shown in Fig.4 the S-factors increase steeply with decreasing energy. This behaviour can only to a minor part be explained by screening effects, as pointed out in chapter 3.4. The S-factors of the

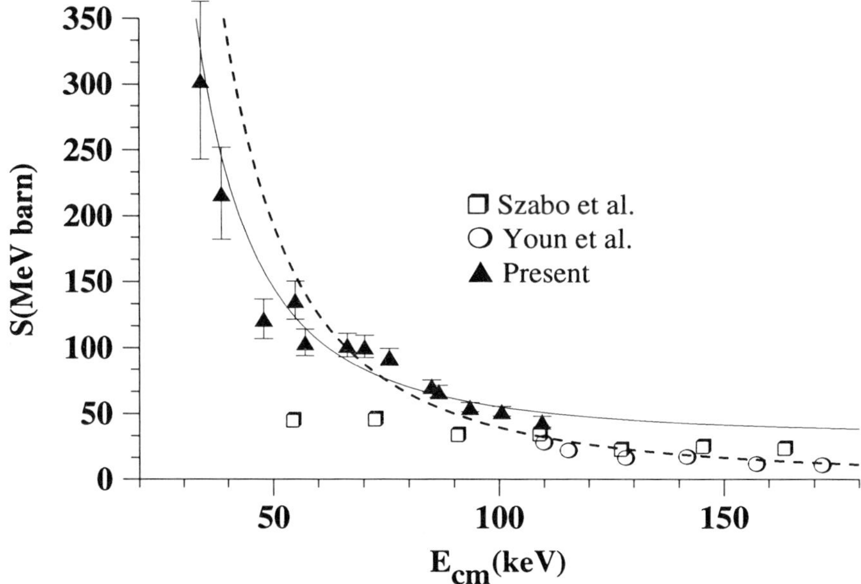

Fig. 4. Astrophysical S-factors for the ^{10}B(p,α_0)^{7}Be reaction. The error bars do include the proton energy uncertainty (1-2 keV). The fits are described in the text.

^{10}B(p,α)^{7}Be reaction in the energy range of the present measurement are rather dominated by the J_π=5/2$^+$ state at E_x= 8.701 MeV (E_r=10 keV).

We have fitted the S-factors in different ways. Since the total width $\Gamma(E_r)$=16 keV and the resonance energy E_r=10 keV are determined by previous experiments [3], only the normalization $S(E_r)$ remained to be varied.

a) Least-squares fit to the single-level Breit-Wigner shape taking into account the energy dependence of the partial widths of the entrance and exit channels and the total width as described in detail in [6]. The normalization was found to be $S(E_r)$=(3.0±0.2)·10^3 MeV·b.

b) Least-squares fit to the single-level Breit-Wigner shape using an estimation for a near-threshold resonance and neglecting the energy dependence of the exit channel partial width [8]. The fit is shown in fig.4 (dashed line). The normalization was found to be $S(E_r)$=(4.8±0.1)·10^3 MeV·b.

c) The same procedure as in (b) but considering a constant offset S_0. It was found S_0=(32±3) MeV·b and $S(E_r)$=(2.5±0.2)·10^3 MeV·b. The fit is shown in fig.4 (full line).

3.4 Screening Effect

An estimation of the energy dependence of the screening effect for the ^{10}B(p,α_0)^{7}Be reaction can be obtained by adding an effective energy increase U_e to the energy E of the projectile: E_{eff}=E+U_e. The transmission of a projectile with the energy E through a shielded Coulomb barrier is equivalent to the transmission through an unshielded barrier with the effective energy E_{eff}. The ratio of the cross sections for the energies E_{eff} and E provides a factor f, the generally called enhancement factor, which indicates the increase of the experimentally determined cross section through the screening effects:

$$f = \frac{\sigma(E_{eff})}{\sigma(E)} \approx \frac{E}{E_{eff}} \cdot \frac{\exp(-2\pi\eta(E_{eff})}{\exp(-2\pi\eta(E))} \approx \exp(\pi\eta(E) \cdot \frac{U_e}{E}) \quad . \tag{1}$$

where we have used $U_e < E$.

Based on a simple model of electron screening Bencze [9] derived the following expression for the effective energy increase:

$$U_e = 46.12 \, Z_1 Z_2 (Z_1^{2/3} + Z_2^{2/3})^{1/2} \quad . \tag{2}$$

When this formula is applied to the $^{10}B(p,\alpha)^7Be$ reaction, $U_e \approx 460$ eV can be obtained. At the proton laboratory energy of 30 keV this would correspond to an enhancement factor $f \approx 1.27$. Accordingly Assenbaum et al. [10] found $f \approx 1.17$. An estimation by Langanke [11] based on the Born-Oppenheimer method provides an increase in energy by $U_e \approx 348$ eV, at $E_p = 30$ keV the enhancement factor is therefore $f \approx 1.20$. This means that the increase of the S-factor with decreasing energy can only to a minor part be explained by screening effects.

3.5 Thermonuclear Reaction Rate

The thermonuclear reaction rates are expressed as $N_A\langle\sigma v\rangle$, where N_A is Avogadro's number, and the average of the cross section, σ, times the relative velocity, v, of the reactants is taken over the Maxwell-Boltzmann distribution as a function of the temperature T. If the equation for the S-factor is inserted, one obtains

$$N_A\langle\sigma v\rangle = (\frac{8}{\pi\mu})^{1/2} \frac{N_A}{(kT)^{3/2}} \int_0^{E_{max}} S(E) \exp(-\frac{E}{kT} - 2\pi\eta) \, dE \quad , \tag{3}$$

where μ is the reduced mass, E is the center of mass energy and η is the Sommerfeld parameter.

The thermonuclear reaction rates obtained by numerical integration of the S-factor-fit (c) from chapter 3.3 are shown in fig.5 in comparison to the numerical values of Caughlan et al. [12] and Youn et al. [6]. As is the case in our evaluation, Youn et al. have also taken the 10 keV resonance into account. The data demonstrate the strong influence of this resonance for the boron destruction at low stellar temperatures ($T_6 \leq 200$).

The nuclear reaction networks necessary for the investigation of the primordial nucleosynthesis arising in inhomogeneous cosmologies require the addition of several new reactions to the standard network. One of them is the reaction $^9Be(n,\gamma)^{10}Be(\beta)^{10}B$, which leads to our target element [13]. Together with the common reaction $^7Be(n,p)^7Li$, the reaction rates of the $^{10}B(p,\alpha)^7Be$ reaction might be useful for further calculations of the production of ^7Li which is an important indicator of the baryonic density of the universe.

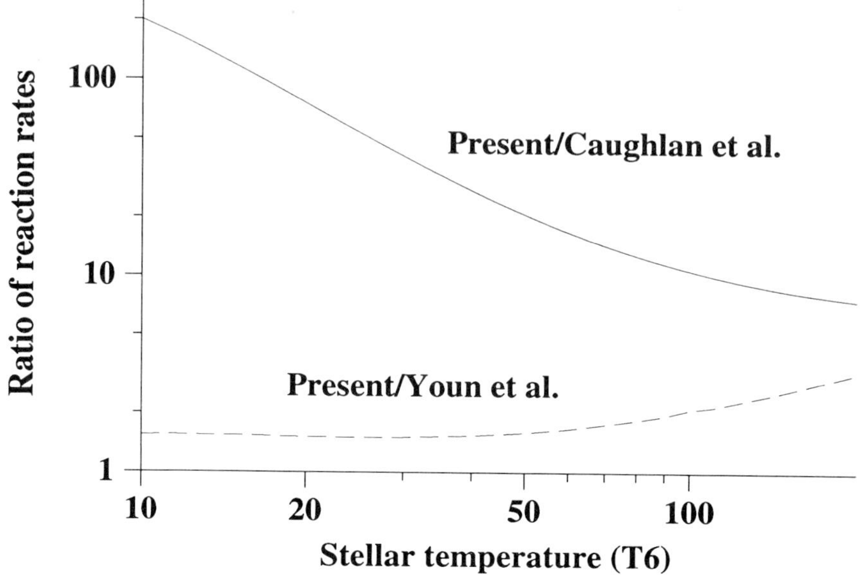

Fig. 5. Ratio of the present reaction rates for the ^{10}B(p,α_0)^{7}Be reaction to those of Caughlan et al. [12] and Youn et al. [6].

References:

[1]   H. Reeves: nuclear reactions in stellar surfaces and their relations with stellar evolution (Gordon and Breach, London, 1971)

[2]   F.E. Cecil et al. and P.D. Kunz: Nucl. Phys. A539 (1992) 75

[3]   M. Wiescher et al.: Phys. Rev. C28 (1983) 1431

[4]   R.J. Peterson et al.: Ann. Nucl. Energy Vol.2 (1975) 503

[5]   J. Szabo, J. Csikai and M. Varnagy: Nucl. Phys. A195 (1972) 527

[6]   M. Youn, H.T. Chung, J.C. Kim and H.C. Bhang: Nucl. Phys. A533 (1991) 321

[7]   N.A. Roughton et al.: Atomic Data and Nuclear Data Tables 23 (1979) 177

[8]   C.E. Rolfs und W.S. Rodney: Cauldrons in the Cosmos (Chicago-Press, 1988) 166

[9]   G. Bencze, Nucl. Phys. A492 (1989) 459

[10]   H.J. Assenbaum, K. Langanke, C. Rolfs: Z. Phys. A-Atomic Nuclei 327 (1987) 461

[11]   K. Langanke: private communication (1992)

[12]   G.R. Caughlan, W.A. Fowler: Atomic Data and Nuclear Data Tables 40 (1988) 283

[13]   R.A. Malaney: W.J. Thompson, B.W. Carney, H.J. Karsowski (eds.), Workshop on Primordial Nucleosynthesis (Singapore: World Scientific, 1990) 49

Recent results from the ^{11}B(n,α) reaction

**T. Paradellis[1], G. Doukellis[1], G. Galios[1], S. Kossionides[1],
X. Aslanoglou[2], P. Descouvemont[3], S. Schmidt[3] and C. Rolfs[4]**

[1]Institute of Nuclear Physics, NCSR "Democritos", Aghia Paraskevi, Greece.
[2]Physics Dept. University of Ioannina, Ioannina,Greece.
[3]Universite Libre de Bruxelles, Belgium
[4]Inst. fur Experimentalphysik III, Ruhr-Universität, Bochum, Germany.

Abstract.   The cross section for the $^{11}B(n, \alpha)^8Li$ reaction has been
measured with incident neutron energy resolution of 56 and 33 keV
respectively. Using the principle of detailed balance the data were
converted to the case of the $^8Li(\alpha, n_0)^{11}B$ reaction. The data revealed
five resonances in the energy region measured. Combining available data
in the literature with the results of the present analysis both lower
and upper limits for the reaction rate of $^8Li(\alpha, n)$ have been deduced. At
$T_9=1$ we obtain an average value of $N_A < \sigma v >= (3.8\pm1.5)10^5 cm^3 s^{-1} mol^{-1}$.
This reaction rate is about a factor of three larger than theoretical estimates.

The cross section for the $^{11}B(n, \alpha)$ reaction has been measured with better neutron
energy resolution and in smaller neutron energy steps than in our previous measurement[1]. Using the principle of detailed balance the data were converted to cross
sections for the $^8Li(\alpha, n_0)$ reaction. The importance of the $^8Li(\alpha, n)$ reaction to inhomogeneous Big Bang nucleosynthesis has been discussed in ref. [1].

The neutrons were produced through the D(d,n) reaction, using a gas cell[1]. Two
experimental arrangements were used in the present experiments. The first one was
the same with that used in ref. [1], that is, a $^{11}BF_3$ counter which served both as target
and detector, placed in front of the gas cell. The second one was used for cross section
measurements at low energies, where a low background detection system is required.
This detection system consisted of a 150 mm^2 silicon semiconductor detector, 50 μm
thick, placed at a distance of 4 cm from the end of the gas cell. A 323 $\mu gr/cm^2$
target, 99% enriched in ^{11}B, was placed directly in front of the detector. Four sets of
experiments were performed:

1) Neutron resolution 56 keV, Mo foil.
Here a 4.9 mgr/cm^2 Mo foil has been used as entrance foil for the deuterons to the
gas cell. The D$_2$ pressure in the gas cell was kept fixed at 250 mb, while the $^{11}BF_3$
target detector was placed 15 cm from the end of the gas cell. Under these conditions
the neutron energy resolution was calculated to be 56 keV.

2) Neutron resolution 56 keV, Ti foil.
In this experiment the Mo foil has been changed to a 2.59 mgr/cm^2 Ti foil. The presure in the gas cell was kept at p=500 mb and the detector to gas cell distance was 15 cm.

3) Neutron resolution 56 keV, Mo foil, Si detector.
The Si detector was placed at 4 cm from the gas cell. The gas cell entrance foil was the 4.9 mgr/cm^2 Mo foil and the pressure in the gas cell was kept fixed at p=500mb. Under these conditions the neutron energy resolution was 56 keV. Data were taken in the CM energy region for the $^8Li + \alpha$ system from 590 keV to 328 keV. The agreement of the measured cross sections obtained with this arrangement ,in the upper range of energy, with those obtained with the $^{11}BF_3$ detector was excellent.

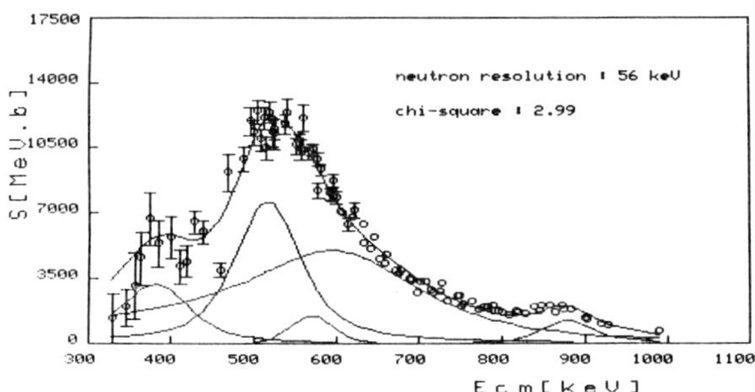

Fig. 1   *The experimental astrophysical S factor for the $^8Li(\alpha, n_0)^{11}B$ reaction as a function of energy, and with neutron energy resolution of 56 keV. The data were fitted with 5 resonances as described in the text.*

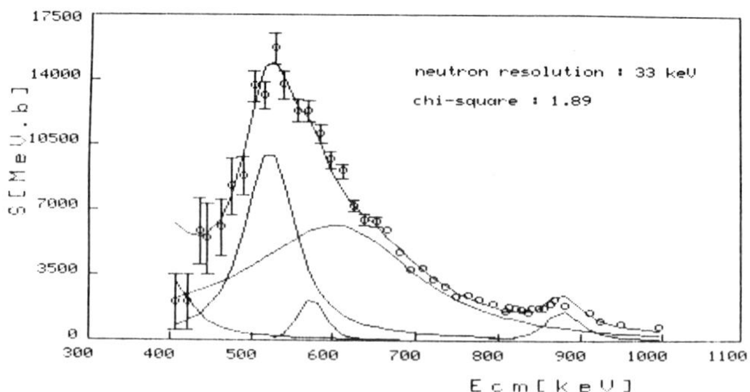

Fig. 2   *Same as in fig. 1 but with neutron energy resolution of 33 keV.*

4) Neutron resolution 33 keV, Ti foil.

Here the entrance foil was the 2.59 mgr/cm^2 Ti foil and the gas pressure was kept fixed at p=100 mb, while the $^{11}BF_3$ detector was placed at a distance of 20 cm from the gas cell.

The measured data are displayed in figs. 1 and 2. The measured cross section for the $^{11}B + n_0$ reaction is used to calculate the cross section and astrophysical S factor of the inverse reaction $^8Li + \alpha$ as described in ref. [1]. The experimentally measured S factors as a function of α energy (figs. 1 and 2) reveal several resonances, in addition to the resonance at 580 keV that was observed in ref. [1]. The line through the data points in figs. 1 and 2, is the result of a least squares fit of the parameters of five resonances. In those fits the energy spread of the incident neutrons was taken into account. The fitted parameters of the five resonaces are: a) $E_R = 395 keV, \Gamma = 60 keV, S(E_R) = 4000 MeV.b$ b) $E_R = 524 keV, \Gamma = 60 keV, S(E_R) = 10850 MeV.b$ c) $E_R = 572 keV, \Gamma < 15 keV$ d) $E_R = 620 keV, \Gamma = 244 keV, S(E_R) = 5500 MeV.b$ e) $E_R = 878 keV, \Gamma = 33 keV, S(E_R) = 2200 MeV.b$

The rate of the $^8Li(\alpha, n_0)$ reaction, at stellar temperatures around $T_9 = 1$, was obtained from the measured S factor (figs. 1 and 2). This rate is a lower limit for the $^8Li(\alpha, n)$ reaction. An upper limit was obtained using the above parameters of the five resonaces and the available data in the literature, mainly from the $^{11}B(n,n)$ reaction data of Fossan et.al.[2]. Combining these data with the present one, upper limits were set on the Γ_{n_0+n}/Γ_0 ratio which allowed the calculation of an upper limit for the reaction rate. The results of this analysis are listed in table 1. The second and third column list the limits set in the total and alpha width respectively. The fourth column lists the possible J^π values obtained and the fifth column the maximum value for the ratio Γ_{n_0+n}/Γ_0.

Table 1. *Resonance parameters obtained from the analysis discussed in the text.*

Resonance energy (keV)	Γ_t (keV)	Γ_α (keV)	J^π	$(\Gamma_{n_o+n}/\Gamma_{n_o})_{max}$
395	60	<5.6 >0.12	$1^-2^+3^-$	10.0
524	60	0.7	$3^\pm\ 4^+$	1.3
572	<1.1 >2.7	<0.91 >0.39	1^+2^+3	2.0
620	240	8.6	$1^-2^+3^-$	2.0
880	33	<33. >5.	$0^+1^\pm2^\pm3^\pm$	10.0

Fig. 3 shows the estimated lower and upper limits for the $^8Li(\alpha, n)$ reaction rate, obtained in this work, together with the theoretical estimates of Malaney and Fowler [3] and Thielmann et. al.[4]. At the astrophysically interesting region around $T_9 = 1$ the upper and lower estimated rates, in this work, are $5.4\,10^5$ and $2.3\,10^5 cm^3 s^{-1} mole^{-1}$. The average value $3.8 \pm 1.5 10^5 cm^3 s^{-1} mole^{-1}$, is two times larger than the Malaney and Fowler estimate amd about three times larger than the estimate of Thielmann et. al. Boyd et. al.[5] based on the direct measurement of the $^8Li + \alpha$ reaction at higher energies and combining their data with the measured astrophysical S factor in ref. [1], deduce a reaction rate of $5.4\,10^5 cm^3 s^{-1} mole^{-1}$, which coincides with the upper limit of the present measurement.

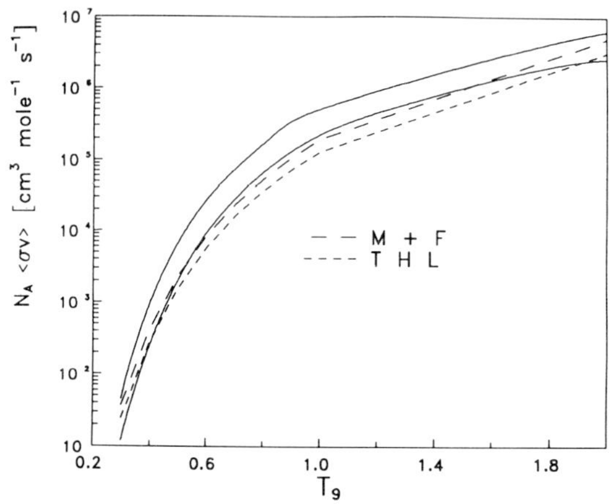

Fig. 3   *The upper and lower limits of the total rate for the $^8Li(\alpha,n)^{11}B$ reaction are shown with continous lines. The theotetical estimates of Malaney and Fowler and Thielemann are also shown.*

We believe that the use of the average value between the upper and lower limit of the reaction rates determined in the present work, reflects the true reaction rate with an uncertainty of 40% and thus hope that their use will be of help to nucleosynthesis network calculations in the framework of inhomogeneous Bing Bang models.

References

[1] Paradellis T, Kossionidis S, Doukellis G, Aslanoglou X, Assimakopoulos P, Pakou A, Rolfs C and Langanke K 1990 *Z. Phys.* **A337** 211.
[2] Fossan D B, Walter R L, Wilson W E and Barschall H H 1961 *Phys. Rev.* **123** 209.
[3] Malaney R and Fowler R W 1989 *Ap. J. Lett.* **345** 15.
[4] Thielmann F K, Arnould M and Truran J W 1987 *Advances in Nuclear Astrophysics (Gif sur Yvette: Frontiers)* p. 525.
[5] Boyd R N, Tanihata I, Inabe N, Kubo T, Nakagawa T, Suzuki T, Ynokura M, Bai X X, Kimura K, Kubono S, Shimura S, Xu H S and Hirata D 1992 *Phys. Rev. Lett.* **68** 1283.

Resonance effects in ^{6}Li(d,p)^{7}Li and ^{6}Li(d,n)^{7}Be mirror reactions at low energies

K. Czerski, H. Bucka, P. Heide and T. Makubire

Institut für Strahlungs- und Kernphysik, Technische Universität Berlin

Abstract. The influence of a broad 2^+ subthreshold resonance of ^{8}Be on the astrophysical S-factors of the ^{6}Li(d,p)^{7}Li and ^{6}Li(d,n)^{7}Be mirror reactions is discussed. A detailed DWBA analysis of the angular distributions and the excitation functions of the (d,p$_0$) and (d,p$_1$) reactions shows that this compound nucleus resonance is required to fit the experimental cross sections. The isospin impurity of this state explains the experimentally observed decrease of the (d,n$_1$) to (d,p$_1$) reaction rate ratio at low deuteron energies. A similar but stronger effect is predicted for the transitions to the ground states of the mirror nuclei.

1. Introduction

Recent investigations of the inhomogeneous primordial nucleosynthesis lead to extreme inhomogenity in the abundances of ^{7}Be and ^{7}Li nuclei. In a high-density neutron-poor region ^{7}Be dominates whereas ^{7}Li is mostly produced in a low-density neutron-rich region [1]. This anisotropy allows, in contradiction to the Standard Big Bang Model, for the formation of significantly havier (A≥12) isotopes. However, the standard nuclear networks must be supplemented by several new reactions.

In this connection the ^{6}Li(d,p)^{7}Li and ^{6}Li(d,n)^{7}Be stripping reactions can play an important role [2]. So far the nuclear rates for both reactions have been assumed to be equal in the network calculations. This assumption finds support in the isospin symmetry of nuclear forces and in the fact that ground and first excited states of ^{7}Be and ^{7}Li are isospin doublets. Therefore the cross sections of these mirror reactions should be equal for all deuteron energies. Some deviations from symmetry are implied by the different charges, binding energies and Q-values for both reactions (Q_{dp}=5.026 MeV and Q_{dn}=3.381 MeV). The influence of these effects may be computed within the distorted-wave Born approximation and gives an enhancement of the (d,n) channel of about 20% at low deuteron energies. On the other hand the relative rates of the ^{6}Li(d,n$_1$)^{7}Be and ^{6}Li(d,p$_1$)^{7}Li reactions have recently been measured [3] and it has been found that the experimental data at lower energies (E_d<500 keV) disagree with DWBA calculations. One observes a rapid decrease of the (d,n$_1$)/(d,p$_1$) cross section ratio. This effect cannot be explained with a deuteron polarization in the Coulomb field of the target nucleus, because this reduces the cross section for the (d,n) reaction only at a few percent level independent of energy [4,5].

In the present work the cross sections for the above mentioned mirror reactions to the first excited states have been measured in the deuteron energy region of special interest, this is below 200 keV. The observed enhancement of the proton channel will be explained with an isospin mixing effect by the Coulomb field in the compound nucleus ^{8}Be.

2.   Experiment

Using a thick LiF target the cross sections for the ^{6}Li(d,n$_1$)^{7}Be and ^{6}Li(d,p$_1$)^{7}Li reactions have been measured by means of the detection of the isotropic γ-decay from the first excited 1/2$^-$ states in ^{7}Li and ^{7}Be for deuteron energies from 90 to 180 keV. The γ-radiation of both reactions was detected simultaneously by one Ge(Li) detector. The experimental ratio for (d,n$_1$) to (d,p$_1$) reactions is shown in fig.1 ; it is determined solely by the ratio of the γ-counting rates corrected for the detection efficiencies. The experimental data show a decrease of the yield ratio within the investigated deuteron energy range of about 15% . A comparison with higher deuteron energy data is given in fig.4 together with a DWBA calculation which shows that the drop at very low energies for the yield ratio cannot be explained within the direct reaction theory.

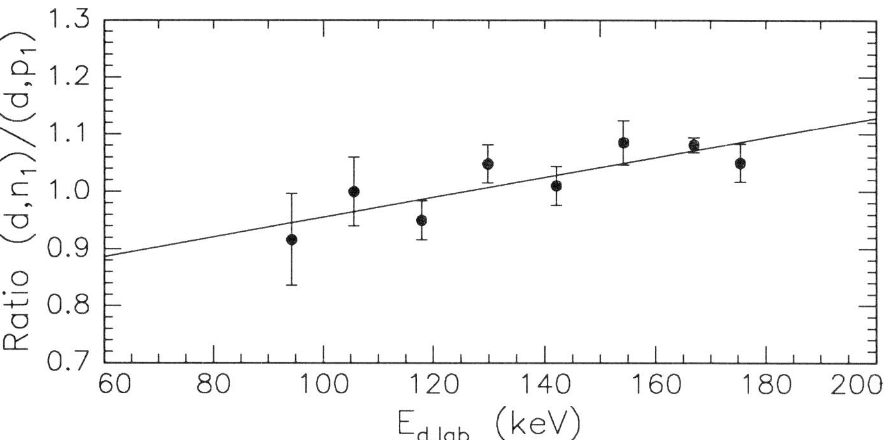

Fig.1. Experimental ratio of the ^{6}Li(d,n$_1$)^{7}Be to ^{6}Li(d,p$_1$)^{7}Li reaction yields.

3.   Resonance Contribution

This unexpected behaviour of the ratio can be explained with the influence of a subthreshold resonance state in the compound nucleus ^{8}Be. The 2$^+$ state at the excitation energy of 22.2 MeV was reported by Ajzenberg-Selove [6]; it was experimentally confirmed in various nuclear reactions. This resonance has a rather large width of 800 keV and lies 80 keV below the reaction threshold. It can be reached by s-wave deuteron capture and therefore should clearly be observed in the experimental astrophysical S-factor curve. In fig.2 and 3 the experimental data of Elwyn et al. [7] and of the present work for the ^{6}Li(d,p$_0$) and ^{6}Li(d,p$_1$) reactions are presented. Both transitions show the characteristic resonance behaviour at the low beam energies.

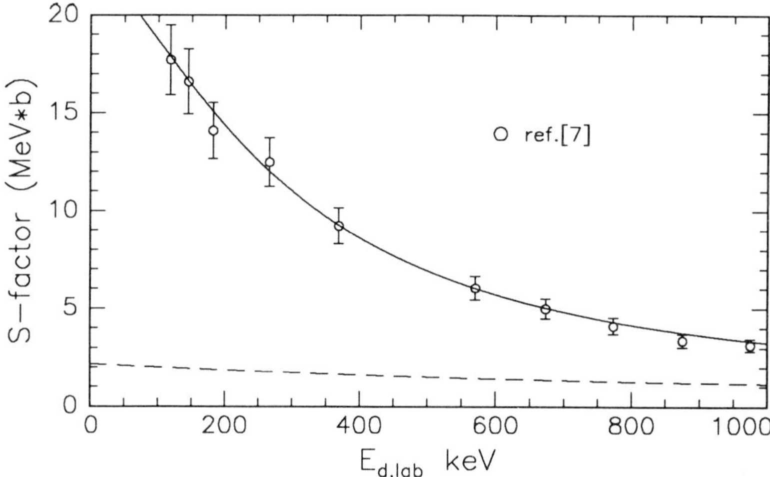

Fig.2. Astrophysical S-factors for the ^{6}Li(d,p$_0$)^{7}Li reaction. The dashed line represents the direct component (DWBA calculation) and the solid line is the sum of the direct and resonance components.

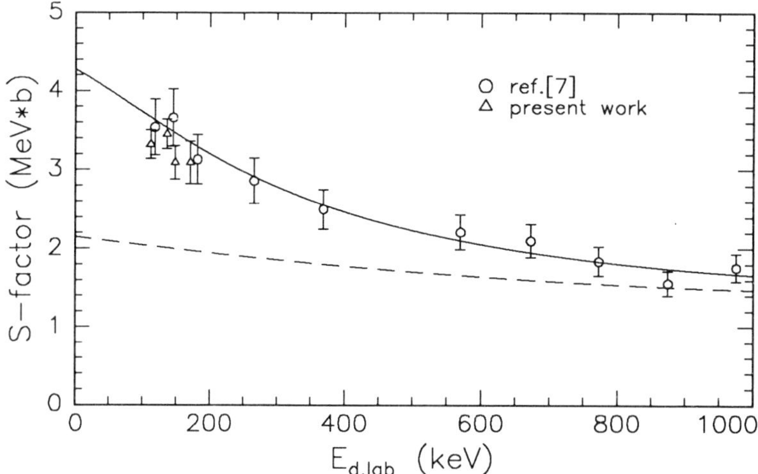

Fig.3. Astrophysical S-factors for the ^{6}Li(d,p$_1$)^{7}Li reaction. The dashed and solid lines represent the result of the calculations as in fig.2.

To determine the resonance contribution one must first calculate the direct component of the reaction cross section. This component can be calculated in the frame of DWBA. The theoretical angular distributions were compared with experimental data from [7]. The $p_{1/2}$ and $p_{3/2}$ transfers were considered for both final states, based on the theoretical spectroscopic factors [8]. The calculations were performed in the zero range approximation with conventional finite range and nonlocal parameters [9]. The optical model parameters for the proton channel were taken from a global fit for light nuclei [10] and the deuteron parameters are from [11]. With one normalization factor of 0.9 good agreement with all experimental angular distributions for energies between 145 keV and 975 keV was achieved. In the all cases, however, one had to add some isotropic contribution which simulates the resonance component. In fig.2 and 3 the dashed lines represent the calculated DWBA cross sections. The remaining part of the cross section can be fitted by a Lorentz-curve to determine the resonance contribution. The total width of the resonance was established to (650 ± 50)keV and the resonance maximum lies at the deuteron energy of (-100 ± 80)keV. The ratio of the partial widths is $\Gamma_{p0}/\Gamma_{p1}\approx 9.7\pm1.0$.

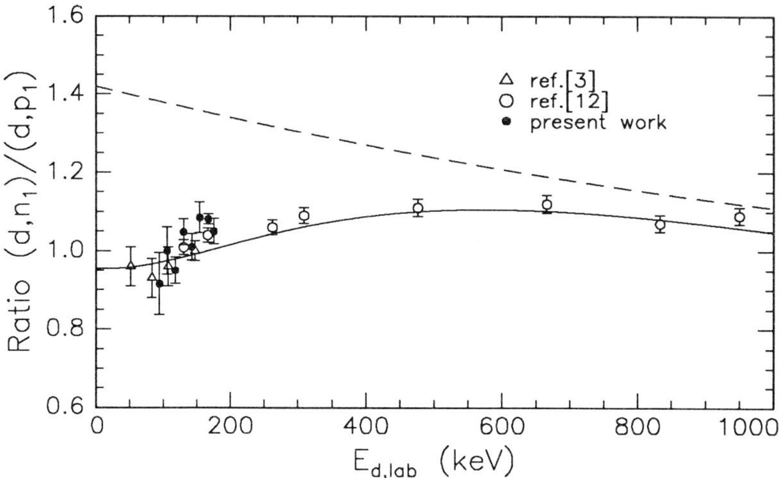

Fig.4. The comparison of the experimental and theoretical ratio of the ^{6}Li(d,n$_1$)^{7}Be to ^{6}Li(d,p$_1$)^{7}Li reaction cross sections. The dashed line represents the DWBA calculation of the direct reaction component. The solid line is the ratio predicted assuming the additional resonance contribution.

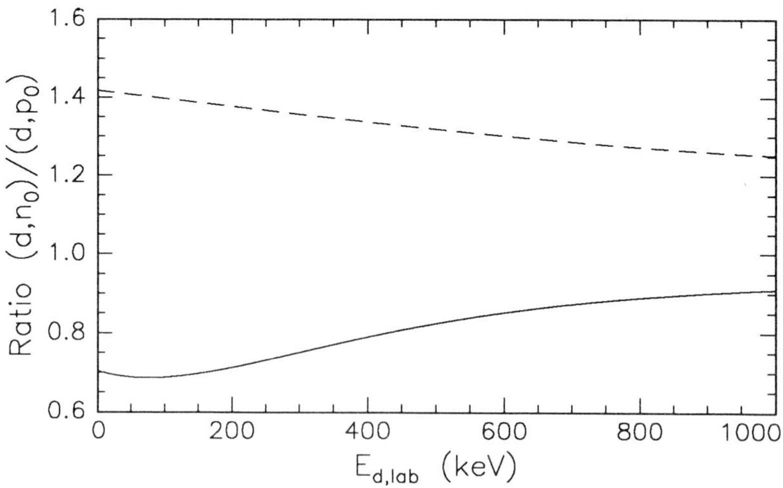

Fig.5. The theoretical ratio of the ^{6}Li(d,n$_0$)^{7}Be to ^{6}Li(d,p$_0$)^{7}Li reaction cross sections. The description of the theoretical curves is the same as in fig.4.

4. Isospin Mixing Effect

To explain the measured anisotropy in the proton and neutron channel one must assume the isospin impurity of the discussed resonance state (an admixture of isospin T=1). The two level isospin mixing model gives in the first order perturbation theory the value of the mixing parameter

$$\varepsilon = \frac{<0|\,H_{01}\,|1>}{E_1 - E_0}$$

where E_1 and E_0 are energies of the isospin states $|0>$ and $|1>$, and $<0|H_{01}|1>$ is the matrix element of a charge symmetry breaking force (Coulomb interaction). Assuming that the penetration factors for neutrons and protons are approximately equal, one gets the formula for the ratio of resonance cross sections

$$\frac{\Gamma_n}{\Gamma_p} = \frac{(\sqrt{1-\varepsilon^2}-\varepsilon)^2}{(\sqrt{1-\varepsilon^2}+\varepsilon)^2}$$

If we fit the resonance contribution to the measured (d,n_1) to (d,p_1) ratio (fig.4) we get $\Gamma_n/\Gamma_p \approx 0.1$ what corresponds to $\varepsilon = 0.46$. The mean value of the Coulomb matrix element $<0|H_{01}|1>$ for the ^{8}Be is about 100 keV [12]. This gives an estimation of the energy splitting of the isospin states of about 300 keV. The calculation with the same mixing parameter can also be performed for the transitions to the ground states. The results are shown in fig.5. In this case the enhancement of the proton channel is much stronger than for the first excited states.

5. Conclusions

It was shown that the observed rapid drop of the $(d,n_1)/(d,p_1)$ cross section ratio can be explained with the influence of the subthreshold resonance. The assumption of an isospin impurity of this state was necessary. The similar calculations for the transitions to the mirror ground states give a larger enhancement of the proton channel so that the cross section ratio is about 0.68 at 50 keV. This means that at 50 keV the ratio of the total reaction cross sections $(d,n_0+n_1)/(d,p_0+p_1)$ is about 0.72 . This value is significantly smaller than is assumed in the network calculations of the inhomogenous primordial nucleosynthesis.

References

[1] R.A. Malaney, in Nuclei in the Cosmos ed. H. Oberhummer (Springer-Verlag 1991) p.129
[2] R.A. Malaney, Workshop on Primordial Nucleosynthesis, Chapel Hill 1989 (Word Scientific Publishing Co 1990) p.49
[3] F.E. Cecil, R.J. Peterson and P.D. Kunz, Nucl. Phys. A441(1985)477
[4] S.E. Koonin and M. Mukerjee, Phys. Rev. C42(1990)1639
[5] N. Austern, Phys. Rev. C43(1991)771

[6] F. Ajzenberg-Selove, Nucl. Phys. A490(1988)1
[7] A.J. Elwyn et al., Phys. Rev. C16(1977)1744
[8] S. Cohen and D. Kurath, Nucl. Phys. A101(1967)1
[9] P.D. Kunz, code DWUCK 4, Colorado University, unpublished
[10] B.A. Watson et al., Phys. Rev. 182(1969)977
[11] D.L. Powell et al., Nucl. Phys. A147(1970)65
[12] F.C. Barker, Nucl. Phys. 45(1963)467

Low energy resonances in ^{18}O + α and ^{22}Ne + α

M Wiescher, U Giesen, J Görres, R E Azuma[1], K L Kratz[2], H-P Trautvetter[3]

University of Notre Dame, Department of Physics, Notre Dame, IN 46556, USA

Abstract. The ^{18}O(^{6}Li,d) and the ^{22}Ne(^{6}Li,d) α transfer reactions have been used to search for α-unbound levels in ^{22}Ne and in ^{26}Mg, which may be important for the resonant α capture on ^{18}O and ^{22}Ne in stellar Helium burning. To determine the resonance strengths of the observed states the ^{18}O(α,γ)^{22}Ne and the ^{22}Ne(α,n)^{25}Mg reactions were measured in the energy range below 900 keV. The resonance strengths of resonances above 600 keV were directly determined. The strengths of lower energy resonances were extracted from the results of the (^{6}Li,d) α transfer measurements. The newly observed resonance levels increase the reaction rates of ^{18}O(α,γ)^{22}Ne and ^{22}Ne(α,n)^{25}Mg significantly. The influence of the proposed reaction rates for the nucleosynthesis at temperatures $T_9 < 0.3$ is discussed.

1. Introduction

The reaction sequence ^{14}N(α,γ)^{18}F(β^-)^{18}O(α,γ)^{22}Ne(α,n)^{25}Mg is considered to be an important neutron source for the s-process in core He-burning of massive stars as well as in He-shell burning of low mass stars. Considerable efforts have been made in recent years to determine the reaction rates for the ^{18}O(α,γ)^{22}Ne α capture (Trautvetter *et al* 1978, Vogelaar *et al* 1990), and for the ^{22}Ne(α,γ)^{26}Mg and ^{22}Ne(α,n)^{25}Mg reactions (Wolke *et al* 1989, Drotleff *et al* 1991, Harms *et al* 1991). The influence of possible low energy resonances, however, has not been investigated yet because the anticipated resonance strengths (Caughlan, Fowler 1988) are too weak to be measured with current γ- and n-detector systems.

The nuclei ^{22}Ne and ^{26}Mg have a high level density above the α threshold (Endt 1990) at 9.669 MeV and 10.611 MeV, respectively. Resonant α capture reactions on the J^π=0$^+$ nuclei ^{18}O and ^{22}Ne, however, populate only natural parity states. Direct α-transfer studies via ^{18}O(^{6}Li,d),^{18}O(^{7}Li,t) and ^{22}Ne(^{6}Li,d) have been performed to

---

[1] University of Toronto, Toronto, Canada
[2] Universität Mainz, Mainz, Germany
[3] Ruhr-Universität Bochum

identify the natural parity states in ^{22}Ne and ^{26}Mg, which may contribute as low energy resonances to the reaction rates for α capture on ^{18}O and ^{22}Ne.

To search directly for the corresponding resonances in ^{18}O$(\alpha,\gamma)^{22}$Ne and ^{22}Ne$(\alpha,n)^{25}$Mg and to measure the resonance strengths both reactions were investigated in the low energy range E$_\alpha$ < 0.9 MeV.

2. Alpha-transfer reaction studies

The reactions ^{18}O$(^6$Li,d), ^{18}O$(^7$Li,t) and ^{22}Ne$(^6$Li,d) have been used to search for α-unbound natural parity states in ^{22}Ne and ^{26}Mg. The experiments were performed at the FN-Tandem accelerator at the University of Notre Dame. The (^{6}Li,d)-reactions were measured at a ^{6}Li bombarding energy of 32 MeV to assure a predominantly direct reaction mechanism. The reaction ^{18}O$(^7$Li,t) was measured only at 16 MeV beam energy.

The ^{18}O targets were prepared by evaporating 50μg/cm^2 of W^{18}O onto a 380μg/cm^2 gold foil..

For the ^{22}Ne$(^6$Li,d) experiment a gas target was used, 99.9 ^{22}Ne gas was circulated through a purification system at a constant pressure of 0.13 atm in the gas cell.

The reaction products were momentum analyzed with a 100 cm broad-range magnetic spectrograph and detected at the focal plane with a position sensitive proportional counter system.

The level structure of ^{22}Ne and ^{26}Mg was studied in the excitation range E$_x$=8.5 - 11.3 MeV and 9.3 - 12.1 MeV, respectively. Figure 1 shows a typical ^{18}O$(^7$Li,t)^{22}Ne and^{22}Ne$(^6$Li,d)^{26}Mg spectrum corresponding to the excitation region in ^{22}Ne above the α threshold. Four states in ^{22}Ne at 9.72, 9.85, 10.05 and 10.13 MeV were observed, which could contribute substantially as resonances to the ^{18}O$(\alpha,\gamma)^{22}$Ne reaction rate. Two new levels were found in ^{26}Mg at 10.69 and 10.95 MeV. The predicted level at 11.15 MeV (Berman *et al* 1969, Weigmann *et al* 1976) was not observed.

Angular distribution measurements were made over the angle range θ=7.5^o - 45°. Zero range DWBA calculations were performed to obtain information about the possible angular momentum transfer ℓ to the populated states in ^{22}Ne and ^{26}Mg. For a given ℓ-transfer the spin and parity of the populated level are determined uniquely by, J=ℓ, and π=(-1)$^\ell$.

Relative spectroscopic factors S$_\alpha$ for the populated final states have been determined from the DWBA-fit to the measured cross section

$$\frac{d\sigma}{d\Omega} = S_\alpha \cdot \frac{d\sigma_{DW}}{d\Omega} \tag{1}$$

The spectroscopic factors depend on the choice of the optical model parameters and the assumption of an (sd)^4, (sd)^3,(pf)^1 configuration of the final states.

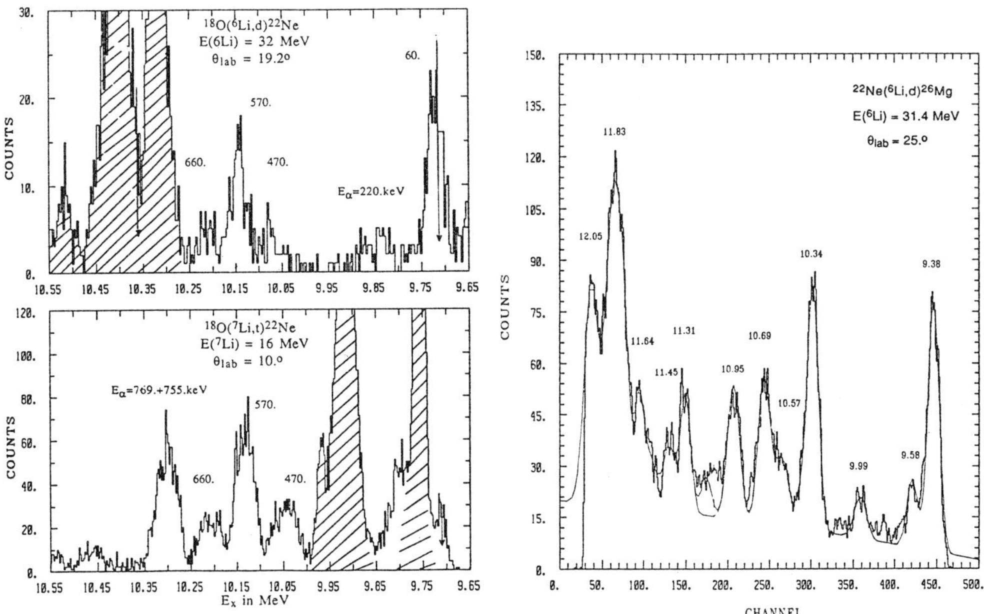

Figure 1. Typical ^{18}O(^{6}Li,d)^{22}Ne and ^{18}O(^{7}Li,t)^{22}Ne (left) and ^{22}Ne(^{6}Li,d)^{26}Mg spectra (right) populating α-unbound natural parity states in ^{22}Ne and ^{26}Mg. Shaded areas show background groups from ^{12}C and ^{16}O contaminations.

3. Alpha-capture reaction studies

The α capture reaction measurements were performed at the JN-Van de Graaff accelerator of the University of Toronto with typical beam energies between 100 and 150 μA on target.

The ^{18}O targets were prepared by anodizing 0.5 mm thick tantalum sheets in 97 The ^{22}Ne targets were prepared by implanting 210 keV ^{22}Ne ions into a 0.5 mm thick tantalum sheet with a total incident dose of 500 mC.

For the radiative capture measurements a 35% Ge-detector was positioned in close geometry at an angle of θ=55°. The efficiency of the detector was determined with calibrated γ sources and from the known resonance at 992 keV in ^{27}Al(p,γ)^{28}Si. For the ^{22}Ne(α,n) measurements the target was surrounded by a set of 31 ^{3}He-proportional counters, embedded in a polyethylen matrix to thermalize the reaction neutrons. The efficiency of the detector system has been determined with a calibrated Am-Li source to 20 $\pm$ 3%.

The ^{18}O(α,γ)^{22}Ne reaction was measured in the energy range between 350 and 800 keV to search for low energy resonances. While the previously reported resonances (Trautvetter *et al* 1978) could be verified, only upper limits were obtained for the two expected resonances at E_r = 470 and 570 keV. The reaction ^{22}Ne(α,n)^{25}Mg was measured in the energy range between 600 and 930 keV. The measurement was handicapped by severe neutron background from the ^{13}C(α,n) reaction on ^{13}C depositions on

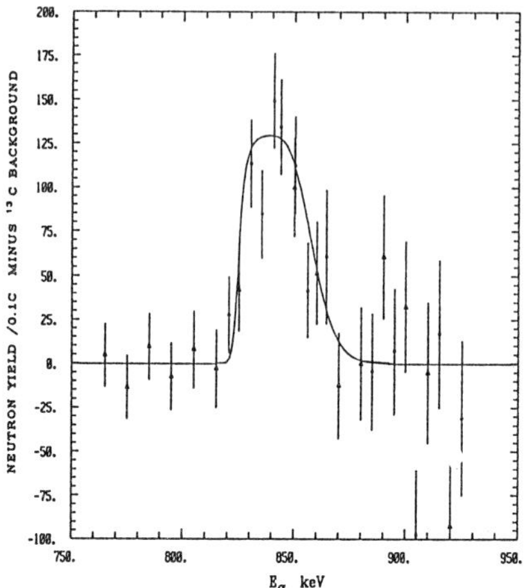

Figure 2. Resonance yield curve for the E_r = 828 keV resonance in ^{22}Ne(α,n)^{25}Mg after subtraction of beam induced neutron background. The solid line shows the expected yield calculated from the measured target thickness and stoichiometry.

target and collimators. Only the 828 keV resonance, reported previously by Drotleff *et al* (1991) was observed (see figure 2).

4. Resonance strengths

The resonance strengths $\omega\gamma$ for the observed resonances in ^{18}O(α,γ)^{22}Ne and ^{22}Ne(α,n)^{25}Mg were determined from the measured thick target yield $Y(E_\alpha)$ relative to the well known resonance strengths $\omega\gamma_{ref}$ of the E_p = 334 keV resonance in ^{18}O(p,γ) and the E_p = 640 keV resonance in ^{22}Ne(p,γ),

$$\omega\gamma = \frac{E_r \cdot Y(E_\alpha) \cdot \varepsilon(E_\alpha) \cdot N_\alpha}{E_p \cdot Y(E_p) \cdot \varepsilon(E_p) \cdot N_p} \cdot \omega\gamma_{ref}. \tag{2}$$

The resonance strengths of the unobserved resonances in ^{18}O(α,γ) and ^{22}Ne(α,γ),(α,n) were calculated from the measured spectroscopic factors S_α (see equation 1) relative to the measured resonance strength $\omega\gamma_{ref}$ of the 660 keV resonance in ^{18}O(α,γ)^{22}Ne and the 828 keV resonance in ^{22}Ne(α,n)^{25}Mg (see equation 2).

$$\omega\gamma = \frac{(2J+1) \cdot S_\alpha(E_x) \cdot P_\ell(E_\alpha)}{(2J_{ref}+1) \cdot S_\alpha(E_{x_{ref}}) \cdot P_\ell(E_{\alpha_{ref}})} \cdot \omega\gamma_{ref}, \tag{3}$$

with E_x as excitation energy, P_ℓ as α-penetrability and J as the spin of the resonances.

5. Astrophysical implications

The results of the present experiments allow a reliable calculation of the reaction rates of ^{18}O(α,γ)^{22}Ne, ^{22}Ne(α,γ)^{26}Mg and ^{22}Ne(α,n)^{25}Mg at low temperature conditions. Because of the high level density the nonresonant reaction components can be neglected.

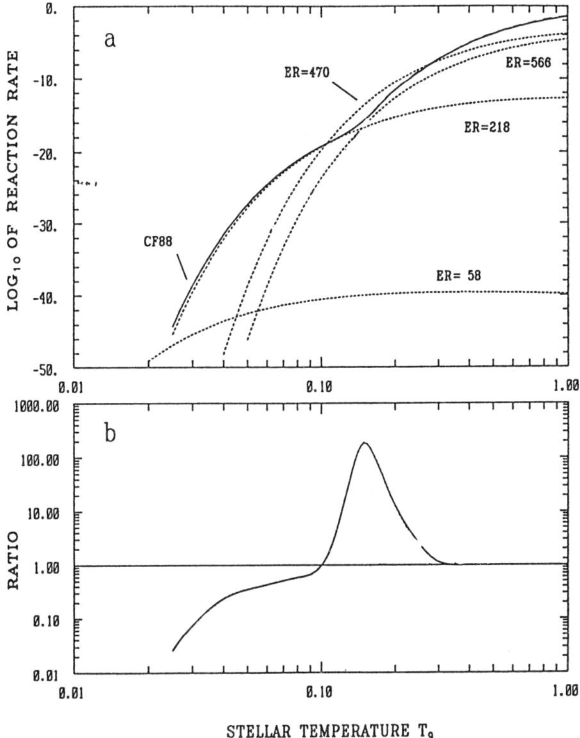

Figure 3. Contribution of single resonances to the reaction rate of $^{18}O(\alpha,\gamma)^{22}Ne$. Also shown is the tabulated rate by Caughlan and Fowler (1988). The lower part shows the comparison of the present rate with the previous one.

Figure 3 shows the contribution of the discussed low energy resonances to the reaction rate of $^{18}O(\alpha,\gamma)^{22}Ne$. The strong influence of the resonance at 470 keV increases the rate by two orders of magnitude versus the tabulated rate by Caughlan and Fowler (1988) in the temperature range of stellar helium burning, $T_9 = 0.1 - 0.3$.

The reaction rates of $^{22}Ne(\alpha,\gamma)^{26}Mg$ and $^{22}Ne(\alpha,n)^{25}Mg$ are strongly influenced by the resonance at 828 keV. The proposed resonance at 633 keV may also influence both reactions significantly. Only an upper limit could be estimated for the strength (equation 3) of the resonance. The present $^{22}Ne(\alpha,\gamma)^{26}Mg$ rate is slightly enhanced at temperature $T_9 < 0.15$ due to the two weak resonances at 98 and 401 keV. Figure 4 shows the ratio of the (α,n) and the (α,γ) rate as a function of temperature. The figure demonstrates clearly that the $^{22}Ne(\alpha,n)$ reaction dominates in the entire temperature range above $T_9 = 0.2$ despite the two low energy resonances in $^{22}Ne(\alpha,\gamma)$. Taking the 633 keV resonance into account expands the temperature range in which the $^{22}Ne(\alpha,n)$ reaction dominates down to temperatures $T_9 = 0.15$. These results indicate that in the critical temperature range $T_9 = 0.1 - 0.4$ the availability of ^{22}Ne for the neutron production is limited to a time period, which is up to two orders of magnitude shorter than thought before.

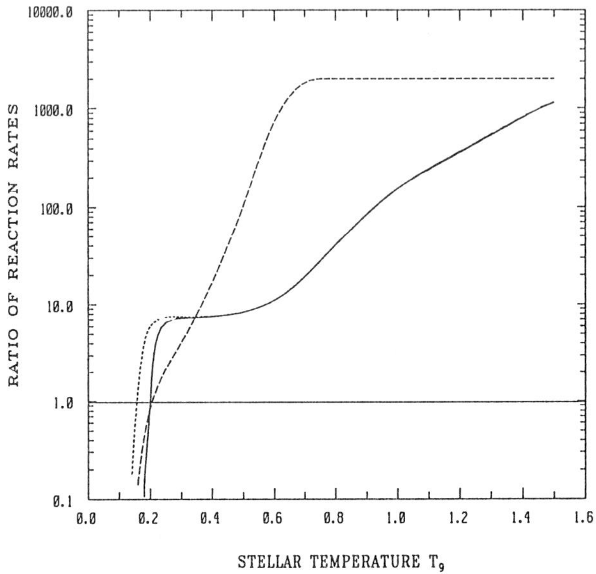

Figure 4. Reaction rate ratio of ^{22}Ne$(\alpha,\gamma)^{26}$Mg and ^{22}Ne$(\alpha,n)^{25}$Mg. The solid line shows the ratio of the present results, the dottet line shows the influence of the 633 keV resonance and the dashed line shows the ratio for the tabulated rates by Caughlan and Fowler (1988).

References

Berman B L, van Hemert R L, Bowman C D 1969 *Phys. Rev. Lett.* **23** 386

Caughlan G R and Fowler W A 1988 *At. Data Nucl. Data Tables* **40** 291

Drotleff H W, Denker A, Hammer J W, Knee H, Küchler S, Streit D, Rolfs C, Trautvetter H P 1991 *Z. Phys A* **338** 367

Endt P M 1990 *Nucl. Phys. A* **521** 1

Harms V, Kratz K L, Wiescher, M 1991 *Phys. Rev. C* **43** 2849

Trautvetter H P, Wiescher M, Kettner K U, Rolfs C, Hammer J W 1978 *Nucl.Phys. A* **297** 489

Vogelaar R B, Wang T R, Kavanagh R W 1990 *Phys. Rev. C* **42** 753

Weigmann H, Macklin R L, Harvey J A 1976 *Phys. Rev. C* **14** 1328

Wolke K, Harms V, Becker H W, Hammer J W, Kratz K L, Rolfs C, Schröder U, Trautvetter H P, Wiescher M, Wöhr A 1989 *Z. Phys. A* **334** 491

Investigation of neutron producing (α,n)-reactions relevant for the astrophysical s- and r-process

H.W. Drotleff, A. Denker, H. Knee, M. Soiné, G. Wolf and J.W. Hammer

Institut für Strahlenphysik der Universität Stuttgart, Germany

U. Greife, C. Rolfs and H.P. Trautvetter

Institut für Experimentalphysik III der Universität Bochum, Germany

Abstract : Seven (α, n)–reactions have been investigated as possible neutron source candidates for the astrophysical s– and r–process. From the viewpoint of stellar models the reactions $^{22}\mathrm{Ne}(\alpha, n)^{25}\mathrm{Mg}$ and $^{13}\mathrm{C}(\alpha, n)^{16}\mathrm{O}$ are mostly favoured. Therefore new results on these reactions are well to the fore. The excitation function of $^{22}\mathrm{Ne}(\alpha, n)^{25}\mathrm{Mg}$ has been measured from threshold ($E_{\alpha,\mathrm{lab}}= 570$ keV) up to $E_{\alpha,\mathrm{lab}}= 2\,300$ keV using improved beam–, target– and detection techniques. A sensitivity limit of 50 pikobarn could be reached in the best cases. A resonance at about $E_{\alpha,\mathrm{lab}}= 620$ keV could clearly be ascribed to the background reaction $^{11}\mathrm{B}(\alpha, n)^{14}\mathrm{N}$. On the basis of experimental data the reaction rate at helium burning temperatures was determined. Similar the excitation function of $^{13}\mathrm{C}(\alpha, n)^{16}\mathrm{O}$ has been measured in the energy range $E_{\alpha,\mathrm{lab}}= 350 - 1400$ keV over a dynamic range of 8 orders of magnitude. The S–factor of this reaction shows an increase towards lower energies due to a state in $^{13}\mathrm{C} + \alpha$ below the threshold leading to a higher reaction rate. Comparable measurements with also very low limits of sensitivity have been obtained for the other five (α, n)–reactions.

Introduction

In the most stellar evolution models the reactions $^{13}\mathrm{C}(\alpha, n)^{16}\mathrm{O}$ and $^{22}\mathrm{Ne}(\alpha, n)^{25}\mathrm{Mg}$ are considered to be the main neutron sources for the astrophysical s-process [1, 2, 3]. The reaction rate of $^{22}\mathrm{Ne}(\alpha, n)^{25}\mathrm{Mg}$ has been evaluated by Caughlan, Fowler, Harris and Zimmerman (CFHZ) [4] using experimental data above $E_\alpha= 1.9$ MeV and nuclear systematics at lower energies. Recent investigations [5] suggested a smaller role of the $^{22}\mathrm{Ne}(\alpha, n)^{25}\mathrm{Mg}$ reaction due to an enhanced (α, γ)–rate compared to the CFHZ value. Two recent papers [6, 7] suggested a resonance stucture in the (α, n)–reaction at about 620 keV. Such a low lying resonance would drastically raise the (α, n)–reaction rate, which would not be consistent with the model calculations. As discussed earlier [6] one also had to consider some possible background reactions. Therefore much effort has been made to clarify especially the energy region between the threshold (570 keV) and 850 keV.

Experiment

A ^{4}He$^+$ beam of 100 μA (at E_α= 600 keV) was provided by the 4 MV DYNAMITRON accelerator at Stuttgart [8]. The beam was guided - with about 100 % transmission efficiency - into an extended target chamber through 5 apertures of the windowless gas-target system RHINOCEROS (4 pumping stages). The system [9] was filled with 99.8 % enriched ^{22}Ne gas and the recirculating gas was purified [9]. The experimental setup and procedures are similar to those described recently [5, 6, 10]. Special solid state targets for high currents were made using enriched ^{13}C (99 %) on a copper backing. These targets were turnable and allowed accurate background measurements, when the beam hitted the bare backing material.

For ^{22}Ne$(\alpha, n)^{25}$Mg the sensitivity limit was reached at about $5 \cdot 10^{-10}$ barn by the use of an improved 4π – detector. For ^{13}C$(\alpha, n)^{16}$O the sensitivity limit was still one order of magnitude lower due to the effective differential measurement procedure. The design of the detector (two concentrical circles of 8 ^{3}He – counters each) leads to a very high absolute efficiency of $37 \pm 1\%$ and allows a rough 'spectroscopy' of the neutrons. This is done by taking into account the ratio of the count rates registrated from the inner and the outer counters [11]. Fig. 1 shows the energy dependence of this ratio which is rather sensitive in the low energy range of the neutrons ($E_n < 1$ MeV), determined as well by Monte Carlo calculations as by experimental calibrations.

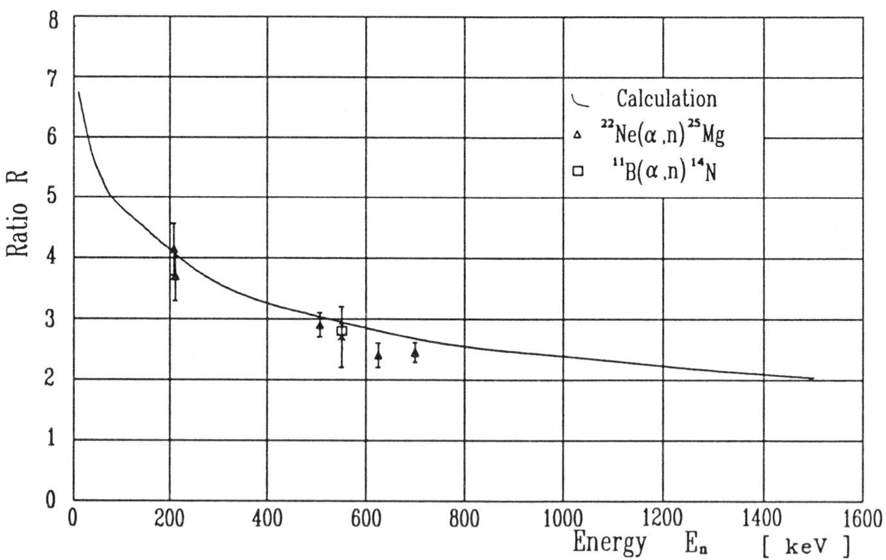

Fig. 1 :   Ratio of the count rate from the two counter circles (inner to outer counters).

An effective passive shielding reduced the background countrate of the 4π – detector to 0.08 cts/sec. The energies and shapes of the resonances have been determined during separate experimental runs.

Results and Discussion

The excitation function of $^{22}\text{Ne}(\alpha, \text{n})^{25}\text{Mg}$ in the energy range 570 keV $<$ E_α $<$ 2300 keV as a result of several different experiments is shown in fig. 2.

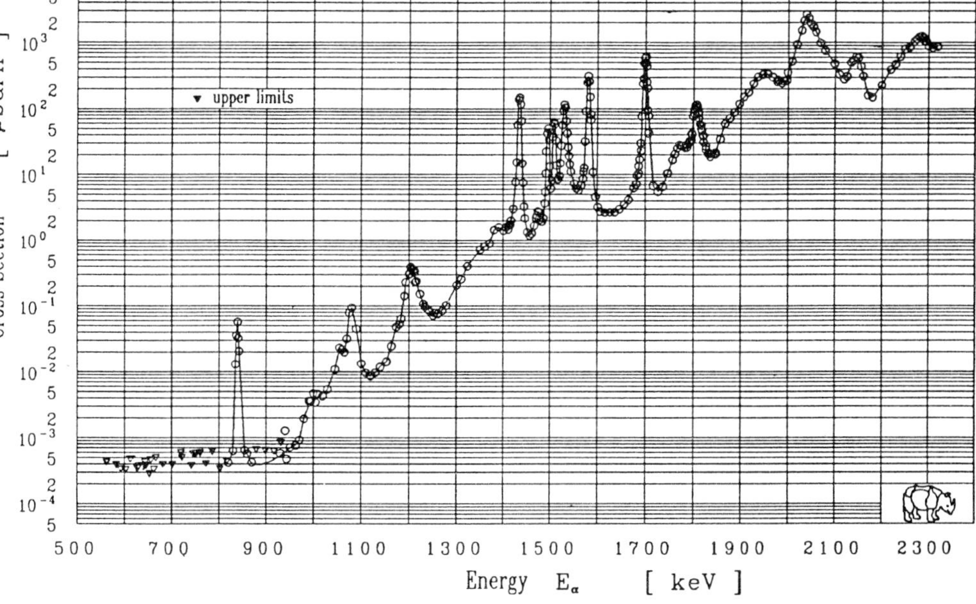

Fig. 2 :   Excitation function of $^{22}\text{Ne}(\alpha, \text{n})^{25}\text{Mg}$

One can see that the reaction $^{22}\text{Ne}(\alpha, \text{n})^{25}\text{Mg}$ proceeds mainly through resonances. For a small nonresonant contribution upper limits could be extracted from the data. The data have been corrected for energy loss in the target gas. The resonance structure observed at $E_\alpha > 1.0$ MeV is in good agreement with previous work [5], except that additional weaker resonances could be resolved in the present experiment. These resonances are listed in table 1.

The strength and width of each resonance has been evaluated by using the folding procedure ASTRA [11]. In this code the effects of a real target, energy spread and straggling of the beam are folded with a variable resonance width in order to reproduce the neutron intensity distribution in the target chamber. This distribution is then again folded with the detection efficiency as a function of the position in the target chamber, where the reaction occurs. By variing the resonance width one can simulate the observed shapes in the excitation function. The origin of the resonance at about $E_\alpha \approx 620$ keV has been investigated carefully by different means. For this purpose several experiments with other target gases (argon and neon of natural isotope abundance) were carried out. The results of these experiments proved that the origin of this resonance is indeed a background reaction. By substracting the neutron yield of these experiments the resonance structure at $E_\alpha \approx 620$ keV vanished leaving upper limits for the cross section below the resonance at $E_\alpha = 831$ keV (see fig. 2).

Energy E_α [keV]		width Γ [keV]		strength $\omega\gamma$ [meV]		Energy E_α [keV]		width Γ [keV]		strength $\omega\gamma$ [meV]	
831[a] ± 3		< 3		0.18[b] ±	0.03	1580[a] ± 3		4 ± 2		2900 ±	300
987 ± 10		30 ± 15		0.16 ±	0.1	1699[a] ± 3		3 ± 2		6035 ±	770
1076[a,c] ± 10		9 ± 5		1.90 ±	0.5	1773[a] ± 5		27 ± 5		1000 ±	240
1202[a] ± 4		19 ± 3		10.6 ±	1.5	1804[a] ± 3		15 ± 2		3010 ±	335
1384 ± 5		25 ± 10		23 ±	15	1855[c] ± 8		33 ± 5		895 ±	210
1434[a] ± 3		< 3		1105 ±	120	1950[c] ± 10		55 ± 10		31000 ±	8500
1475[a] ± 3		14 ± 5		45 ±	30	2046[c] ± 8		35 ± 5		197000 ±	33000
1495[a] ± 3		< 4		388 ±	57	2152[c] ± 10		22 ± 5		27600 ±	7000
1508[a] ± 3		4 ± 2		560 ±	60	2289[c] ± 15		50 ± 10		121000 ±	45000
1531[a] ± 3		5 ± 2		1445 ±	160						

[a]Resonance properties from the folding procedure ASTRA

[b]The strength of this resonance has changed, compared to [6], during the final evaluation of the data. The previous reported value was refered to the data of [5]

[c]from: [5]

Table 1 :   Resonance properties in ^{22}Ne$(\alpha,n)^{25}$Mg .

In additional experiments and due to the 'spectroscopy' with the 4π – detector, we could assign this resonance beyond all doubt to the background reaction ^{11}B$(\alpha,n)^{14}$N [11]. This reaction has a strong resonance at 606 keV [12]. The different energies are due to the energy loss of the scattered (straggled) projectiles. Only these can hit the wall and contribute to the background reaction.

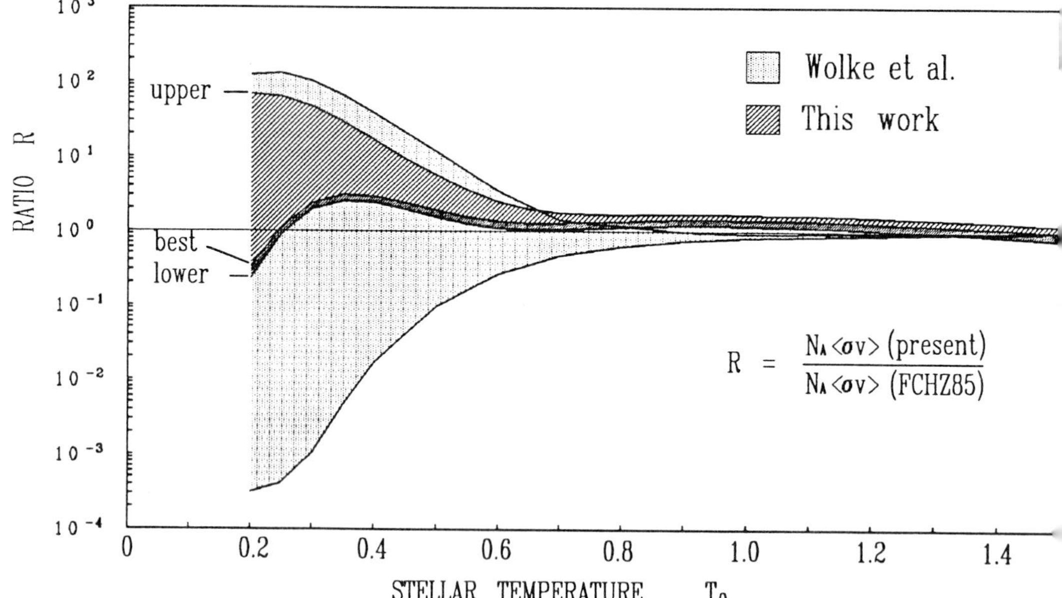

Fig 3 :   Reaction rates of ^{22}Ne$(\alpha,n)^{25}$Mg . Comparison to extrapolated data from FC 88 (FCHZ 85); lower limit in good agreement, upper limit by a factor of 70 higher.

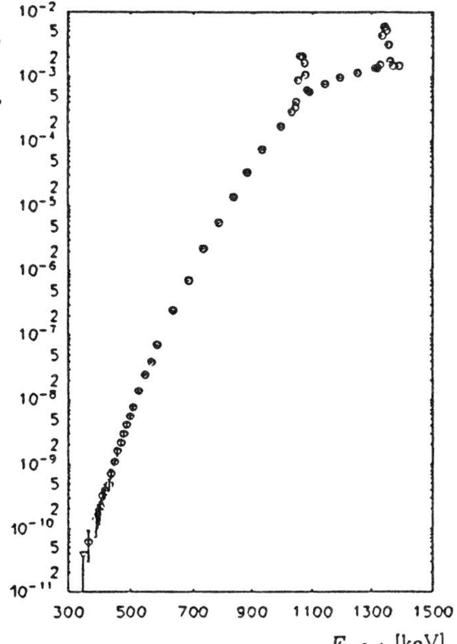

Fig. 4a: Total cross section of the $^{13}C(\alpha,n)^{16}O$–reaction as a function of $E_{\alpha,lab}$.

The resonances in the energy range around 1 MeV, and especially the resonance at 831 keV raise the reaction rate in the temperature range $0.3 < T_9 < 0.5$ slightly. For better comparison fig. 3 shows the ratio of our experimental determined reaction rate to the extrapolated values of [4]. The upper limit lies up to a factor of 70 higher than the extrapolation. This limit represents the 'worst case' because it considers all known levels in the compound system , obtained from measurements of the reaction $^{24}Mg(n,\gamma)^{26}Mg$ [13], taking maximal contribution within the limits from our measurement. The more probable case lies slightly above the lower limit, which is given by the contributions of the observed resonances. These values are in rather good agreement to the extrapolation [4], but about a factor of 3 higher at 0.35 T_9 compared with CFHZ.

For $^{13}C(\alpha,n)^{16}O$ the excitation function of fig. 4a has been obtained [15, 16]. From this the S-factor curve of fig. 4b is deduced, which shows the new experimental data points in comparison with model calculations and previous extrapolations [17, 18, 19].

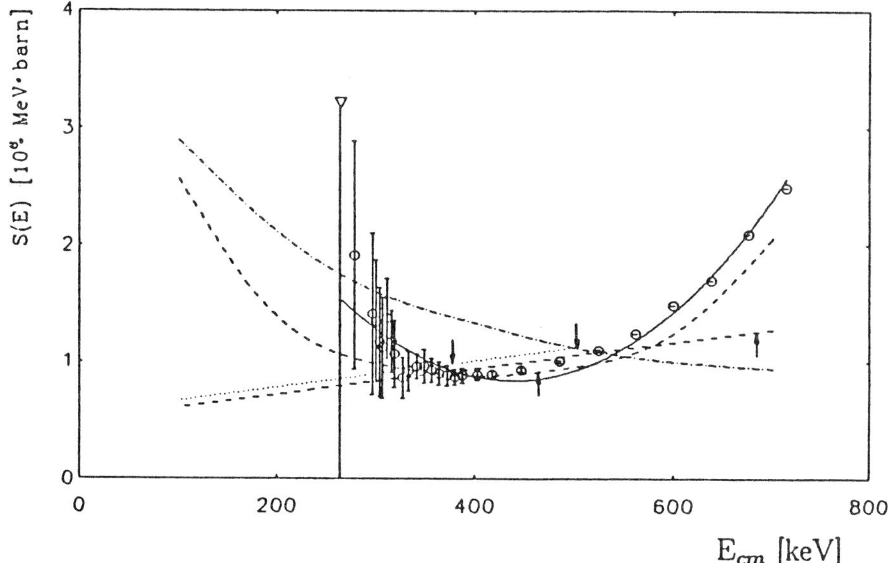

Fig. 4b: The S–factor of the nuclear reaction $^{13}C(\alpha,n)^{16}O$ as a function of $E_{\alpha\,cm}$. Experimental results of this work (circles) are compared with extrapolations from Ramström [19] (dashed line) and Davids [18] (dotted line). The arrows indicate the energy interval of the previous measurements. Results of the calculations of Descouvemont [17] based on the microscopic model (dashdotted line) and based on a Breit–Wigner–formalism (full line) are shown also. The line through the data points is to guide the eye.

Conclusion

Seven (α, n)–reactions ($^{13}C(\alpha, n)^{16}O$, $^{17}O(\alpha, n)^{20}Ne$, $^{18}O(\alpha, n)^{21}Ne$, $^{21}Ne(\alpha, n)^{24}Mg$, $^{22}Ne(\alpha, n)^{25}Mg$, $^{25}Mg(\alpha,n)^{28}Si$, $^{26}Mg(\alpha,n)^{29}Si$) have been investigated with high sensitivity methods and new excitation functions have been obtained. For $^{22}Ne(\alpha, n)^{25}Mg$ all resonances and upper limits for the yield at energies near the threshold could be determined with more sensitivity and accuracy than before. The reaction rates derived from these data are in rather good agreement with the extrapolated values. More important is, that the uncertainties in the reaction rates of this basic quantity for the stellar models could be reduced drastically, namely from several orders of magnitude to about a factor of 70 (see fig. 3) due to conclusive experimental data in the low energy region. An other important aspect for the stellar modell calculations may be that our lower limit agrees well with the extrapolation. That means the reaction rates for the reaction $^{22}Ne(\alpha, n)^{25}Mg$ are at least as high as the commonly used values.

The role of $^{22}Ne(\alpha, n)^{25}Mg$ as neutron source for the s–process is now fixed by experimental data. A significant change of this role would only be necessary if there is a resonance in the (α, γ)–reaction near or below the (α, n)–threshold which can diminish the abundance of ^{22}Ne in the stellar szenarios.

Due to a higher S–factor in the energy range 0 – 300 keV the reaction rate of $^{13}C(\alpha, n)^{16}O$ will be enhanced probably as a result of a state in ^{17}O below threshold.

References

[1]   I. Iben, Ap. J. **196**, 525 and 549 (1975)

[2]   D. Hollowell, I. Iben, Ap. J. **340**, 966 (1989)

[3]   F. Käppeler, R. Gallino, M. Busso, G. Picchio, C.M. Raiteri, Ap. J. **354**, 630 (1990)

[4]   G.R. Caughlan, W.A. Fowler, M.J. Harris and B.A. Zimmermann, At. Data Nucl. Data Tables **32**, 197 (1985)

[5]   K. Wolke, V. Harms, H.W. Becker, J.W. Hammer, K.L. Kratz, C. Rolfs, U. Schröder, H.P. Trautvetter, M. Wiescher and A. Wöhr, Z. Phys. **A 334**, 491 (1989)

[6]   H.W. Drotleff, A. Denker, J.W. Hammer, H. Knee, S. Küchler, D. Streit, C. Rolfs, and H.P. Trautvetter, Z. Phys. **A 338**, 367 (1991).

[7]   V. Harms, K.L. Kratz, M. Wiescher, Phys. Rev. **C 43**, 2849 (1991)

[8]   J.W. Hammer, B. Fischer, H. Hollick, H.P. Trautvetter, K.U. Kettner, C. Rolfs and M. Wiescher, Nucl. Instr. and Meth. **161**, 189 (1979)

[9]   J.W. Hammer et al.,Report Jül-Spez-409, ed. by W. Oelert, p. XVIII (1987) and
       J.W. Hammer et al., to be published in Nucl. Instr. and Meth.

[10]   H.W. Drotleff et al., Proc. of the Int. Conf. "Nuclei in the Cosmos" Baden bei Wien, H. Oberhummer (Ed.), Wien 1990.

[11]   H.W. Drotleff, PhD thesis, Stuttgart und Bochum 1992

[12]   T.R. Wang, R.B. Vogelaar, R.W. Kavanagh, Phys. Rev. **C 43**, 883 (1991)

[13]   H. Weigmann, R.L. Macklin and J.A. Harvey, Phys. Rev. **C 14**, 1328 (1976)

[14]   M. Arnould et al., private communication

[15]   M. Soiné, Diplomarbeit, Stuttgart 1991

[16]   A. Köhler, Diplomarbeit, Stuttgart 1988

[17]   P. Descouvemont, Phys. Rev. **C 36**, 2206 (1987)

[18]   C.N. Davids, Nucl. Phys. **A 110**, 619 (1968)

[19]   E. Ramström, T. Wiedling, Nucl. Phys. **A 272**, 259 (1976)

s-Process studies with improved cross sections

K. Wisshak, F. Voß, K. Guber, and F. Käppeler

Kernforschungszentrum Karlsruhe, Institut für Kernphysik, P.O.B. 3640
D-7500 Karlsruhe, Germany

Abstract.
New experimental techniques for the determination of neutron capture cross sections, e.g. the Karlsruhe 4π Barium Fluoride Detector, have reduced the respective uncertainties by a factor five compared to previous methods. With these improved data, detailed analyses of branchings in the s-process path allow to derive constraints on the physical parameters during the s-process. As first examples, the neutron capture cross sections of the s-only isotopes of tellurium and samarium have been determined. We discuss the results in the framework of the classical s-process approach and with a stellar model describing the s-process during helium shell burning in low mass stars. The results for tellurium confirm the prediction of the classical approach of a "local approximation" within the experimental uncertainty of 1%. The stellar model turned out to overestimate the effect of the branchings at A = 121, 122 by a factor three as a consequence of the temperature profile during the helium burning episode. The samarium cross sections provide for a considerably improved estimate of the mean neutron density.

1. Introduction

Nucleosynthesis of the heavy elements by successive neutron captures in the s-process is one of the topics in nuclear astrophysics that can be addressed in detail by laboratory experiments [1]. The most important quantities in these investigations are the isotopic abundances, N_s, and the stellar neutron capture cross sections, $<\sigma>$, averaged over a Maxwellian velocity distribution for a typical s-process temperature of $\sim 3\times10^8$K corresponding to thermal energies around $kT=30$keV. Since the s-process abundances are inversely proportional to the respective cross sections, the product $N_s<\sigma>$ is a smooth function of mass number A. The corresponding distribution, that can be calculated by means of various s-process models, must be normalized at those isotopes that are of pure s-process origin. These isotopes are shielded against the r-process beta decay chains by stable isobars. Accordingly, their s-process abundances are identical with the solar abundances, $N_\odot$. Of particular importance for s-process studies are the eight elements Kr, Sr, Te, Xe, Ba, Sm, Gd, Os with two or three s-only isotopes, since the abundance ratios

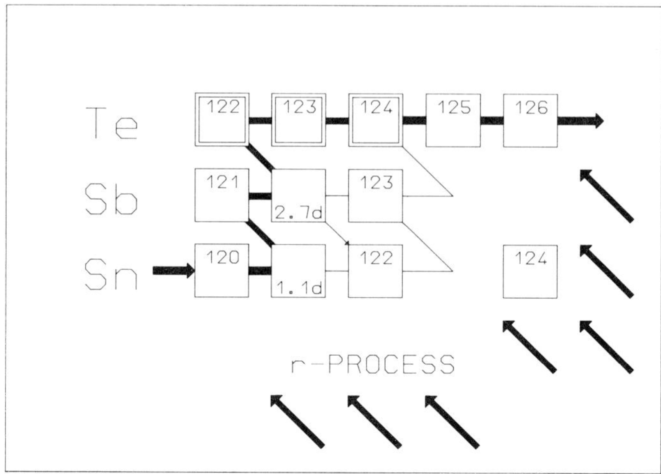

Figure 1. The s-process path in the region of the tellurium isotopes. The s-only isotopes ^{122,123,124}Te are shielded from the r-process by the stable isobars ^{122,124}Sn and ^{123}Sb. The unstable nuclei ^{121}Sn and ^{122}Sb are possible branching points.

of these nuclei are simply the isotopic ratios that are known with typical uncertainties of 0.1% [2]. These examples represent the most sensitive probes for the investigation of the s-process environment.

Information on the physical conditions during the s-process, e.g. neutron density, temperature, and electron density, can be deduced from the analysis of branchings in the s-process path [1]. The branchings result from the competition between neutron capture and beta decay whenever an unstable isotope is encountered by the s-process path, that exhibits a beta decay rate, $\lambda_\beta = \ln2/t_{1/2}$, comparable to its neutron capture rate, $\lambda_n = n_n v_T <\sigma>$. In these expressions, $t_{1/2}$ is the half-life, n_n the s-process neutron density, v_T the mean thermal neutron velocity, and $<\sigma>$ the stellar neutron capture cross section. The competition between beta decay and neutron capture defines the branching factor for the s-process flow,

$$f_n = \lambda_n/(\lambda_n + \lambda_\beta).$$

This factor depends on the neutron density, n_n, and sometimes on temperature, T, since the half-life for beta decay may be temperature-dependent under stellar conditions. Empirically, the branching factor can be determined if the involved isotopes include an s-only nucleus. The deviation of its $<\sigma>N_s$-value from the global $<\sigma>N_s$-curve then defines the branching factor. However, the best characterization of f_n can be obtained in those cases, where the branching affects one of two s-only nuclei of the same element. The 5 branchings of this type are compiled in Table 1, and are illustrated in Figure 1 by the example of the tellurium isotopes.

At ^{121}Sn and ^{122}Sb, possible branchings may cause part of the s-process flow to bypass ^{122}Te and ^{123}Te. In first approximation, the effect of these branchings is expressed by:

$$1\text{-}f_n \cong N_s<\sigma>(^{122}\text{Te})/N_s<\sigma>(^{124}\text{Te}).$$

Table 1. Branchings in the s-process path characterized by two s-only isotopes.

s-only isotopes	branching point	branching factor f_n		sensitive to
		previous	present	
^{80,82}Kr	^{79}Se	0.50±0.12		temperature
^{122,123,124}Te	^{121}Sn,^{122}Sb	0.01±0.05*	0.007±0.012	neutron density
^{128,130}Xe	^{127}Te,^{128}I	0.01±0.25*		electron density
^{134,136}Ba	^{133}Xe,^{134}Cs	0.02±0.15*		temperature, neutron density
^{148,150}Sm	^{147}Nd,^{147}Pm,^{148}Pm	0.10±0.05	0.118±0.009	neutron density
^{152,154}Gd	^{151}Sm,^{152}Eu,^{154}Eu	0.95±0.05		electron density, temperature

*Estimates from classical model much smaller than current experimental uncertainty

The uncertainty of this ratio is completely determined by the uncertainty of the cross section ratio. In a previous measurement by Macklin and Winters [3], an uncertainty of ±5% was obtained for this quantity. Since f_n is expected to range between 0.01 and 0.05 according to estimates with the classical approach and by stellar models, this is not sufficient for a quantitative analysis of the branchings at A = 121, 122.

A similar example is the branching at A = 147, 148, which is defined by the two s-only isotopes ^{148}Sm and ^{150}Sm (see Fig.2). This case is most suited for the determination of the neutron density, n_n. With a conventional technique similar to Ref.[3], Winters *et al.* [4] determined the cross section ratio for these two isotopes with an uncertainty of $\sim 4\%$. In the branching analysis, this uncertainty represents the dominant contribution to the ±40% uncertainty for the resulting neutron density. Hence, also for this branching a significant improvement can be expected from more accurate cross sections.

With this aim, the present experiment was using a new method, which allows to determine cross section ratios with an accuracy of $\sim 1\%$. This improvement will help to investigate the features of the s-process abundances in greater detail. For example, the prediction of the classical s-process model of a "local approximation", can be checked at the 1% level by comparing the $N_s{<}\sigma{>}$ values of neighboring s-only isotopes. Moreover, these data are the prerequisite for the reliable analysis of the branchings listed in Table 1, from which significantly improved information on the physical conditions during the s-process can be expected. This information will then represent relevant constraints for stellar models.

The most interesting candidates to start with are the isotopes of tellurium. Tellurium is the only element in nature with three s-only isotopes (see Fig.1). By the classical approach, only very weak branchings are predicted at A = 121 and 122. Therefore, it may provide for a unique check of the "local approximation", i.e. that the $N_s{<}\sigma{>}$-curve is flat between magic neutron numbers. The fact that one of the s-only isotopes has an odd mass number implies that it is not produced by the p-process [5,6,7]. This feature allows to exclude the possibility that a weak branching is masked by a corresponding p-process contribution, an ambiguity that has to be considered for all other branchings.

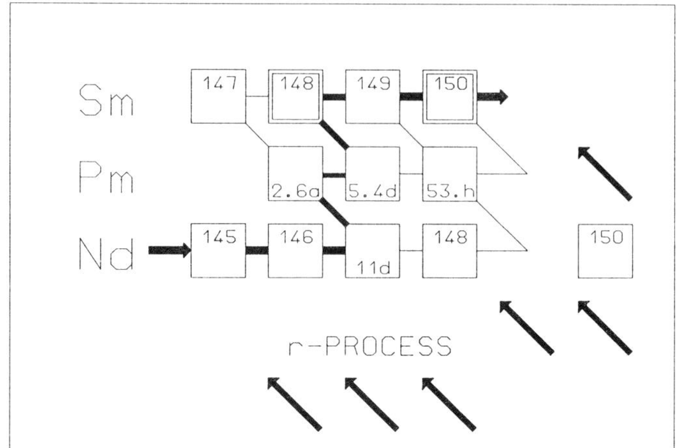

Figure 2. The s-process path in the region of the samarium isotopes. The s-only isotopes ^{148,150}Sm are shielded from the r-process by ^{148,150}Nd. Branchings at ^{147}Nd and ^{147,148,149}Pm cause part of the s-process flow to bypass ^{148}Sm.

In addition, the data are expected to yield an upper limit for the neutron density as a constraint for current stellar models, and to allow for an estimate of the p-process contribution to the even tellurium isotopes.

In a second experiment, the neutron capture cross sections of the samarium isotopes have been determined. With these data the branchings at A = 147, 148, 149 were reanalysed to derive improved estimates for the s-process neutron density (Fig.2).

2. EXPERIMENT

The experimental method with the Karlsruhe 4π BaF$_2$ detector has been published in detail in Refs.[8] and [9]. The results on tellurium are found in Refs.[10] and [11] and on samarium in Ref.[12].

The neutron capture cross sections of the tellurium and samarium isotopes were measured in the energy range from 3 to 200keV using gold as a standard. Neutrons were produced via the ^{7}Li(p,n)^{7}Be reaction by bombarding metallic Li targets with the pulsed proton beam of the Karlsruhe 3.75MV Van de Graaff accelerator. The neutron energy is determined by time of flight (TOF), the samples being located at a flight path of 78cm. Throughout the experiment the accelerator was operated with a pulse width of $\sim$1ns, a repetition rate of 250kHz, and an average beam current of 1.5μA.

The experiment was carried out in three different runs, the with proton beam energy adjusted 10, 30 and 100keV above the threshold of the ^{7}Li(p,n)^{7}Be reaction at 1.881MeV. This yields continuous neutron spectra in the energy range of interest for s-process studies, i.e.10-70keV, 3-100 keV, and 3-200keV, respectively. The use of different spectra served to optimize the signal to background ratio in different neutron energy

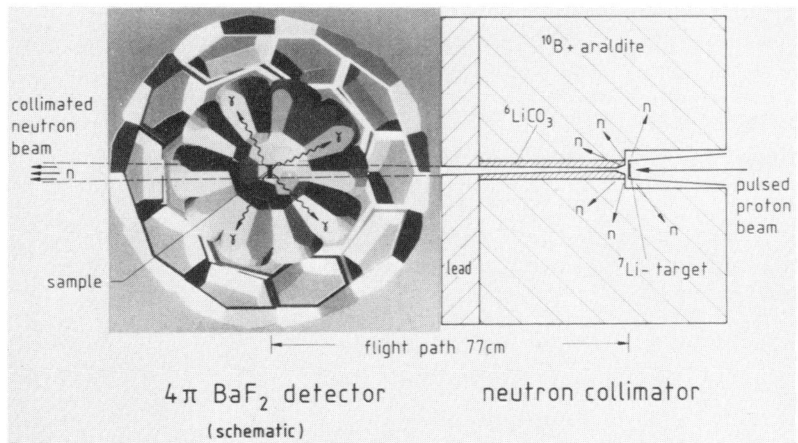

Figure 3. Experimental setup

regions . A schematic sketch of the experimental setup is shown in Fig.3. The proton beam from the accelerator falls on a metallic lithium target. A well collimated neutron beam hits the sample at the center of the detector, and capture gamma-rays are detected in the surrounding BaF$_2$ crystals.

The Karlsruhe 4π Barium Fluoride Detector was used for the registration of capture gamma-ray cascades. This detector (a comprehensive description is given in Ref.[5]) consists of 42 hexagonal and pentagonal crystals forming a spherical shell of BaF$_2$ with 10cm inner radius and 15cm thickness. It is characterized by a resolution in gamma-ray energy of 7% at 2.5MeV, a time resolution of 500ps, and a peak efficiency of 90% at 1MeV. Capture events are registered with $\sim$95% probability.

The main advantages of this detector are :

(i) The entire capture cascade is detected with good energy resolution. Thus, ambiguities in the detection efficiency due to different cascade multiplicities are avoided, and neutron capture events can be separated from gamma-ray background and background due to capture of sample scattered neutrons by selecting events with appropriate sum energy. This feature is demonstrated in Fig.4, showing the respective sum energy spectra for the various samarium isotopes. In all spectra, a narrow peak at the neutron binding energy is observed corresponding to those events where the entire cascade of capture gamma-rays was registered. In addition, the spectra exhibit more or less pronounced tails. These tails are due to events, for which at least one gamma-ray remains undetected, e.g. because it escapes through the openings for the neutron beam. Nevertheless, only a small fraction of capture events falls below the detection threshold indicated by arrows, so that the detection efficiency can still be determined with good accuracy.

(ii) The granularity of the detector allows for a further separation of capture events and background by the event multiplicity.

(iii) The short primary neutron flight path and the inner radius of the detector guarantees that the high energy part of the TOF spectra is completely undisturbed

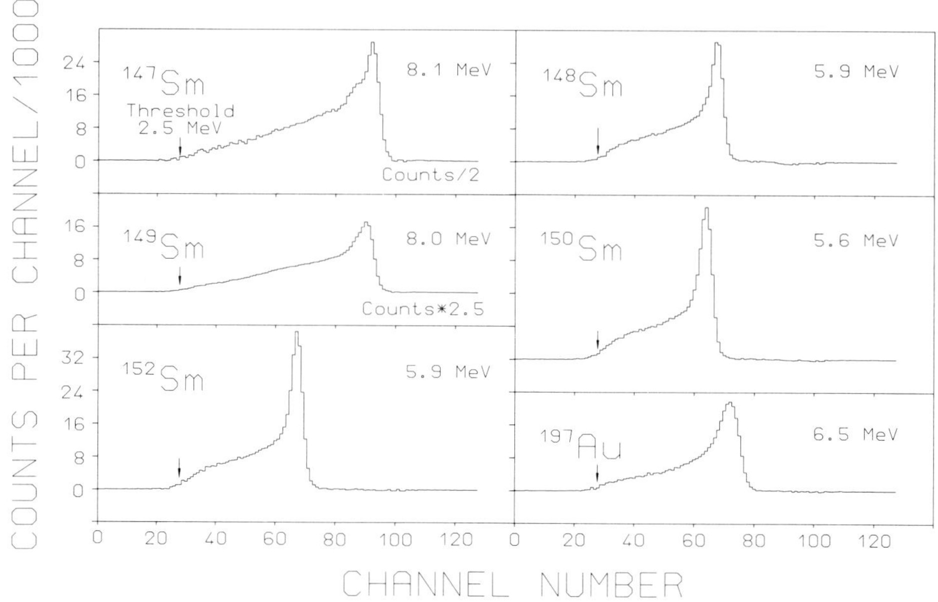

Figure 4. Sum energy spectra of the investigated samarium isotopes and of the gold standard

by background from sample scattered neutrons. This range with optimum signal to background ratio can be used to normalize the cross section.

(iv) The high detection efficiency allows the use of small samples avoiding large multiple scattering corrections.

(v) Finally, the ^{7}Li(p,n)-reaction yields neutrons exactly and exclusively in the range of interest for s-process studies.

3.   Results for the Maxwellian-averaged cross sections.

The main result of the experiment are the neutron capture cross section ratios of the tellurium and samarium isotopes versus the gold standard. These ratios were converted into absolute cross sections using the gold cross section from literature [13,14]. While the uncertainty of the cross section ratios is in general ~1%, the uncertainty of the absolute cross sections of ~2% is dominated by the 1.5% uncertainty of the standard. For a reliable determination of the Maxwellian averages at typical s-process temperatures, differential cross sections are required in the energy range from 0 to 700keV. Thus, the present experimental result had to be extended at high and low energies. This was achieved by normalization of the cross section shape from previous measurements and from model calculations [15]. The stellar cross sections were calculated for temperatures between 10 and 100keV and are reported in Refs.[10,12]. In Table 2, we present the values at kT=30keV only, including the cross section ratios that are most important for the branching analyses. Note, that the uncertainty of the gold standard cancels

Table 2. Stellar neutron capture cross sections of the s-only tellurium and samarium isotopes at kT = 30keV.

Isotope	$<\sigma>$ [mb]	
	(this work)	(Bao and Käppeler)
^{122}Te	295.4 ± 2.7*	295 ± 60
^{123}Te	831.5 ± 7.7*	822 ± 50
^{124}Te	154.6 ± 1.6*	162 ± 21
^{148}Sm	244.2 ± 2.2*	267 ± 12
^{150}Sm	422.0 ± 3.8*	447 ± 26

*The 1.5% uncertainty of the standard is not considered since it cancels out in most applications of relevance for nuclear astrophysics.

out in the cross section ratio of two tellurium or samarium isotopes. The new data are compared with the evaluation of Bao and Käppeler [6]. Apart from their better accuracy, they agree well with the previous tellurium cross sections. However, the samarium data are smaller by 7-10%, reflecting possibly a problem with the sample preparation in the previous experiment.

In addition to the pure s-isotopes listed in Table 2, measurements were also performed for ^{125}Te and ^{126}Te as well as for ^{147}Sm, ^{149}Sm, and ^{152}Sm.

4. Implications for s-process models

The new cross section data have been used to calculate the s-process abundances by means of the classical approach [14] and with the stellar model for helium shell burning in low mass stars [22,23,24]. The calculated s-process yields can then be compared with the observed abundances for testing the models, and, eventually, for interpreting the information contained in the observed abundances with respect to the physical conditions during the s-process. For this purpose a variety of input parameters are necessary:

4.1. Input data

Stellar cross sections: The neutron capture cross sections for the s-only isotopes were taken from the measurements with the Karlsruhe 4π BaF$_2$ Detector as described in Sections 2 and 3. In addition, the cross sections of ^{120}Sn, ^{121}Sb, and ^{123}Sb have recently been measured at KfK using the activation technique [8]. This work includes also an improved calculation of the cross sections for the branch point nuclei ^{121}Sn and ^{122}Sb. All other values were taken from literature [13,15,16].

Beta-decay rates: The beta decay rates of unstable isotopes may be drastically changed under stellar conditions. In the present investigation such an enhancement can be expected for the beta decay of ^{121}Sn. According to the calculations of Takahashi and Yokoi [20] the decay rate is increased by a factor 2.9 at a temperature of 3.3×10^8 K. This enhancement, however, is valid only if complete thermal equilibrium between the population of ground state and isomer is achieved. Following the procedure described by Klay *et al.* [21], the discussion in Ref.[8] showed that thermal equilibrium is reached in ^{121}Sn for temperatures T>2.4×10^8 K. At lower temperatures, the mediating gamma-transitions are strongly suppressed. This means that isomer and ground state remain decoupled for T<2.0×10^8 K. In all other cases, the assumption of complete thermal equilibrium [20] appears justified.

4.2. *s-Process Models*

Classical Approach: The classical steady s-process model [25,26,27] with the extension for the treatment of s-process branchings [28] represents a useful tool for characterizing the s-process abundances, but also the physical conditions during the s-process [1]. A detailed description of the formalism may be found in the latter reference, as well.

In this study, the parameters describing the stellar conditions are adopted from Ref.[17]: an exponential distribution of neutron exposures, $\rho(\tau) \sim \exp(-\tau/\tau_0)$ with a mean exposure $\tau_0 = 0.295 \pm 0.009$ mbarn^{-1} at a thermal energy kT $= 29 \pm 5$ keV, and a mean neutron density $n_n = (3.4 \pm 1.1) \times 10^8$ cm^{-3}.

Low Mass AGB Stars: Helium shell burning in low mass AGB stars of low metallicity [29,30,31,32] was shown to reproduce the observed s-process abundances remarkably well [22,23,24]. This scenario is described by subsequent helium shell burning episodes. In each of these thermal instabilities, two neutron bursts are produced, the first due to the ^{13}C$(\alpha,n)^{16}$O reaction at relatively low temperatures (kT $\sim$ 12keV), and the second due to the ^{22}Ne$(\alpha,n)^{25}$Mg reaction near the end of the episode, when the temperature reaches its maximum (kT $\sim$ 23keV). Though the exposure related to the second burst is comparably small, it is responsible for the final abundance patterns in the s-process branchings [17]. Near the end of each episode, freshly synthesized material is engulfed by the convective envelope. The next thermal instability follows then after 2×10^5 yr and develops in the same way.

4.3. *The Sn-Sb-Te region*

The s-process in the mass region A=120 to 124 has several important aspects: In the first place, the three s-only nuclei ^{122}Te, ^{123}Te, and ^{124}Te offer the possibilty to check the local approximation, to investigate the s-process branchings at A=121,122, and to search for possible p-process contributions in addition to the p-only isotope ^{120}Te. Secondly, tin is a proton magic element. Therefore, the (n,γ) cross sections of the tin isotopes are comparably small, and this leads to interesting consequences for the s-process flow. In particular, this region is expected to be sensitive to the more complicated neutron density and temperature profile of the stellar s-process scenario. At present, some of these aspects are still hidden, since the cross sections of most isotopes with 110<A<120 carry sizable uncertainties, but are interesing enough to motivate further experimental effort.

Classical Approach: With the parameters compiled above, the abundances N_s of the s-only isotopes ^{121,123,124}Te were calculated and the ratio $N_s/N_\odot$ to the solar abundances was determined. The results were normalized to the ^{124}Te abundance, since this isotope experiences the entire mass flow. This procedure allows to use the isotopic abundances which are given in literature [2] with an accuracy of about 0.1% (^{122}Te: 2.603(3),^{123}Te: 0.908(1), and ^{124}Te: 4.816(3) [2]). The resulting s-process abundances are compiled in Table 3. For a perfect model, all values should be unity. Obviously, the classical approach comes very close to this ideal case. For the first time, the prediction of the classical model of the "local approximation", i.e. that the product $<\sigma>N_s$ is constant for neighboring isotopes, could be checked and confirmed on the 1% level for a set of three s-only isotopes.

That the branchings at A=121,122 are so weak in the classical prescription is the consequence of the relatively high temperature rather than of the neutron density. At kT = 29 keV, the isomer in ^{121}Sn is fully thermalized and the β^+/EC branching in the decay of ^{122}Sb is suppressed by a factor three. Therefore, only a very weak branching of $\sim$ 1% would be possible in view of the $<\sigma>N_s$-values of ^{123}Te and ^{124}Te.

The comparison of the predicted s-process abundances is completely dominated by the uncertainties of the empirical ratios $<\sigma>N_s(^{122}$Te$)/<\sigma>N_s(^{124}$Te$)$ and $<\sigma>N_s(^{123}$Te$)/<\sigma>N_s(^{124}$Te$)$. The abundances being known to $\pm$0.1% implies that the abundance ratios can be determined with the $\pm$1.2 % uncertainty of the cross section ratios.

It is interesting to note that the accurate tellurium cross sections allow - for the first time - to decompose an observed abundance value into its s- and p-process components. Though the effect is similar to the respective uncertainty, the deficiency in the ^{122}Te abundance derived with the classical approach indicates a 1.6% p-process yield for this isotope. Since a comparable p-process yield is excluded for the even-odd ^{123}Te [5,6,7], the local approximation implies that also ^{124}Te should have no significant p-process component. This demonstrates that the accurate cross sections measured with the Karlsruhe 4π BaF$_2$ detector are not only needed for improved studies of the s-process branchings, but may also contribute to a more quantitative discussion of the p-process, as well.

Low Mass AGB Stars: Adopting the profiles for neutron density, temperature, and mass density from the stellar model of helium shell burning in low mass stars [23], the s-process flow through the mass region 110<A<130 was followed with the network code NETZ [33]. The results are included in Table 2 , indicating that this model predicts significant branchings at A = 121,122, which cause $\sim$ 6% of the flow to bypass ^{122}Te and ^{123}Te. The resulting discrepancy is significantly larger than the uncertainties in the data, and corresponds to four standard deviations. This is an important result, since it shows that the stellar model fails to reproduce the branchings at A=121,122 properly.

The reason for this difficulty is that the small cross sections of the heavier tin isotopes do not allow for a complete readjustment of the abundances in the second neutron burst. Therefore, the final distribution still reflects to a certain extent the situation at the end of the low temperature episode, when a relatively large ^{122}Sn abundance was built up, due to the decoupling of isomer and ground state in ^{121}Sn, and due to the stronger β^+/EC branch in the decay of ^{122}Sb. At the higher temperature during the second neutron burst, the s-process flow through ^{122}Sn is strongly reduced, but neutron captures on this seed add significantly to the ^{124}Te abundance, leading to the excess outlined above. It turns out that this problem cannot be cured by plausible modifications of the adopted stellar model parameters except by increasing the temperature also for the first neutron burst due to the ^{13}C$(\alpha,$n$)^{16}$O reaction to values above 2.5$\times10^8$ K. In the future, it will, therefore, be important to verify this result also for other temperature-dependent

Table 3. s-Process production ratios of ^{122}Te, ^{123}Te, and ^{124}Te relative to solar abundances (normalized at ^{124}Te).

Isotope	Classical model		Low mass AGB stars	
	present	previous*	present	previous*
^{122}Te	0.984 ± 0.012	1.032 ± 0.057	0.91 ± 0.01**	0.96 ± 0.06**
^{123}Te	1.003 ± 0.012	1.021 ± 0.051	0.94 ± 0.01**	0.96 ± 0.05**
^{124}Te	1	1	1	1

*Capture cross sections from Macklin and Winters[3].

**Error bars due to uncertainty of the capture cross section of tellurium isotopes. An additional uncertainty of ~0.01 comes from the cross sections of the unstable isotopes ^{121}Sn and ^{122}Sb.

branchings.

The importance of accurate cross sections in such investigations is illustrated if the above s-process studies are repeated with the respective data of Macklin and Winters[3]. In this case, the significant difference between the classical approach and the stellar model is completely masked by the larger uncertainties of $\sim 5\%$. Obviously, the new experimental technique allows for a much more detailed discussion of the information contained in the observed abundances, and, hence, represents a significant step towards a deeper understanding of the s-process and of the helium burning stages in stellar evolution.

4.4. The Branchings at A = 147 - 149

The s-only isotopes ^{148}Sm and ^{150}Sm are important for two reasons:

(i) Together, they define the branching ratio for the s-process flow at mass numbers $A = 147,148,148$. This aspect is appealing, since the beta decay rates of the branching point isotopes ^{147}Nd, ^{147}Pm, ^{148}Pm, and ^{149}Pm are practically not affected by the high temperatures during the s-process [20]. Therefore, these branchings are particularly suited for the investigation of the s-process neutron density.

(ii) By itself, ^{150}Sm represents one of the few normalization points for the global $<\sigma>N_s(A)$-curve that is not affected by a possible branching. In combination with ^{124}Te, which belongs in this category as well, the $<\sigma>N_s$-values of these nuclei allow for the determination of the mean neutron exposure, τ_0, in the classical approach.

Since these analyses are not yet finalized, only a brief summary of the preliminary results can be presented here.

Classical Approach: After the local approximation has been shown to hold within $\pm 1\%$ at the example of the ^{123}Te/^{124}Te ratio, the deviations from this behavior can now be interpreted with confidence as the result of a branching in the s-process path. In case of samarium, the cross section ratio given in Table 2 and the abundance ratio $N_s(^{148}$Sm$)/N_s(^{150}$Sm$)=1.523\pm0.001$ can be combined to yield:
$$<\sigma>N_s(^{148}\text{Sm})/<\sigma>N_s(^{150}\text{Sm}) = 0.882\pm0.009.$$

This value confirms the result of Winters *et al.* [4], who reported a ratio of 0.91 ± 0.03. Following the analysis outlined in Ref.[4], the more accurate cross section ratio determined with the Karlsruhe 4π BaF$_2$ detector allows now to deduce an improved estimate for the mean neutron density:

$$n_n = (3.3 \pm 0.3) \times 10^8 \text{ cm}^{-3} \,.$$

This is the most stringent limit for the mean neutron density that has been obtained so far. As a first step towards a refined analysis of the relevant s-process branchings, the relatively small uncertainty of this result is very encouraging. It means that the experimental efforts for a systematic inspection of the s-process abundance pattern may, indeed, reveal the physical conditions of the corresponding scenario.

Low Mass AGB Stars: First calculations for this scenario indicate that the resulting ratio $<\sigma>N_s(^{148}Sm)/<\sigma>N_s(^{150}Sm)$ appear to be systematically larger than the empirical value. However, the reason for this behavior and the possible consequences for the model need to be evaluated in more detail.

5. Summary

The results obtained with the Karlsruhe 4π BaF$_2$ detector have demonstrated that neutron capture cross sections for s-process studies can be measured with uncertainties of about 1%. This improvement by a factor five over previous techniques allows for the detailed analysis of the abundance pattern in s-process branchings, which still preserve the physical conditions of their production site, i.e. neutron density, temperature, and mass density.

First applications of the new technique have been devoted to the isotopes of tellurium and samarium, which include three and two s-only isotopes, respectively. Analysis of the Sn-Sb-Te mass region has revealed a number of features that were not accessible before, in particular the characterization of minor branchings, and the identification of p-process residuals. It was found, that the data were compatible with the classical steady flow approach, but that there are problems with the current stellar s-process model associated with helium shell burning in AGB stars of low mass and low metallicity. The recent measurement on the samarium isotopes will lead to an improved estimate for the mean neutron density in the s-process, and also to an improved determination of the mean neutron exposure.

These encouraging steps towards a more detailed study of the relevant s-process branchings will be complemented by systematic measurements on all remaining elements with two s-only isotopes. The experiment on the barium isotopes has already been completed, and the runs for Xe and Gd are in preparation. In combination with additional studies related to cross sections and decay rates of the unstable isotopes at s-process branchings, it is intended to establish a sufficiently refined data basis to arrive eventually at a quantitative picture of the s-process.

References

[1] F. Käppeler, H. Beer, and K. Wisshak, *Rep. Prog. Phys.* **52**, 945 (1989).

[2] N.E. Holden, R.L. Martin, and I.L. Barnes, *Pure Appl. Chem.* **56**,675 (1984).

[3] R.L. Macklin and R.R. Winters, *Neutron capture of* ^{122}Te, ^{123}Te, ^{124}Te, ^{125}Te *and* ^{126}Te, *report ORNL-6561 Oak Ridge National Lab.* (1989).

[4] R.R. Winters, F. Käppeler, K. Wisshak, A. Mengoni, and G. Reffo, *Ap. J.* **300**, 41 (1986).

[5] W.M. Howard, B.S. Meyer, and S.E. Woosley, *Astrophys. J.* **373**, L5 (1991).

[6] N. Prantzos, M. Hashimoto, M. Rayet, and M. Arnould, *Astron. Astrophys.* **238**, 455 (1990).

[7] M. Rayet, N. Prantzos, and M. Arnould, *Astron. Astrophys.* **227**, 271 (1990).

[8] K. Wisshak, K. Guber, F. Käppeler, J. Krisch, H. Müller, G. Rupp, and F. Voss, *Nucl. Instr. Meth.* **A292**, 595 (1990).

[9] K. Wisshak, F. Voss, F. Käppeler, and G. Reffo, *Phys. Rev.* **C42**, 1731 (1990).

[10] K. Wisshak, F. Voss, F. Käppeler, and G. Reffo, *Phys. Rev.* **C45**, 2470 (1992).

[11] F. Käppeler, W. Schanz, K. Wisshak, and G. Reffo, *Astrophys. J.*, submitted for publication.

[12] K. Wisshak, K. Guber, F. Voss, F. Käppeler, and G. Reffo, *Neutron Capture in* $^{149,150}Sm$: a *Sensitive Probe for the s-Process Neutron Density, report KfK-5067 Kernforschungszentrum Karlsruhe* 1992.

[13] R.L. Macklin, *private communication* (1982)

[14] W. Ratynski and F. Käppeler, *Phys. Rev.* **C 37**, 595 (1988).

[15] H. Beer, F. Voss, and R.R. Winters, *Astrophys. J. Suppl.* **80**, 403 (1992).

[16] Z.Y. Bao and F. Käppeler *At. Data Nucl. Data Tables* **36**, 411 (1987).

[17] F. Käppeler, R. Gallino, M. Busso, G. Picchio, and C.M. Raiteri, *Astrophys. J.* **354**, 630 (1990).

[18] H. Beer, G. Walter, and F. Käppeler, *Astron. Astrophys.* **11**, 245 (1989).

[19] J.A. Holmes, S.E. Woosley, W.A. Fowler, and B.A. Zimmerman, *At. Data Nucl. Data Tables* **18**, 305 (1976).

[20] K. Takahashi and K. Yokoi, *At. Data Nucl. Data Tables* **36**, 375 (1987).

[21] N. Klay, F. Käppeler, H. Beer, and G. Schatz, *Phys. Rev.* **C44**, 2839 (1991).

[22] R. Gallino, M. Busso, G. Picchio, C.M. Raiteri, and A. Renzini, *Astrophys. J.* **334**, L45 (1988).

[23] R. Gallino, *private communication* (1992).

[24] R. Gallino, M. Busso, and C.M. Raiteri, these proceedings.

[25] E.M. Burbidge, G.R. Burbidge, W.A. Fowler, and F. Hoyle, *Rev. Mod. Phys.* **29**, 547 (1957).

[26] P.A. Seeger, W.A. Fowler, and D.D. Clayton, *Astrophys. J. Suppl.* **11**, 121 (1965).

[27] D.D. Clayton, and R.A. Ward, *Astrophys. J.* **193**, 397 (1974).

[28] R.A. Ward, M.J. Newman, and D.D. Clayton, *Astrophys. J. Suppl.* **31**, 33 (1976).

[29] I. Iben,Jr., and A. Renzini, *Astrophys. J. Letters* **259**, L79 (1982).

[30] I. Iben,Jr., and A. Renzini, *Astrophys. J. Letters* **263**, L23 (1982).

[31] D.E. Hollowell, and I. Iben,Jr., *Astrophys. J. Letters* **333**, L25 (1988).

[32] D.E. Hollowell, and I. Iben,Jr., *Astrophys. J.* **340**, 966 (1989).

[33] S. Jaag, *Diplom thesis, University of Karlsruhe* (1990).

Neutron capture rates of light nuclei and stellar evolution

Y. Nagai, T. Shima, K. Takeda, T. Ohsaki, T. Irie and S. Seino

Department of Applied Physics, Faculty of Science, Tokyo Institute of Technology,
Oh-okayama, Meguro-ku, Tokyo 152, Japan

M. Igashira, H. Kitazawa, S. Shibata and K. Tanaka

Reserch Laboratory for Nuclear Reactors, Tokyo Institute of Technology,
Oh-okayama, Meguro-ku, Tokyo 152, Japan

T. Fukuda

Institute for Nuclear Study, University of Tokyo, Tanashi-shi, Tokyo 188, Japan

Abstract. In the nucleosynthesis of stellar evolution such as low mass asymptotic giant branch stars and of the inhomogeneous big bang models the neutron capture rates of light nuclei play critical roles. The present work aimed at measuring the rates of ^{7}Li, ^{9}Be, ^{12}C, ^{13}C and 14N by using pulsed keV neutrons and detecting prompt γ-rays from a captured state. The present results give important informations of the time evolutions of the neutron density of the AGB stars and favor the nucleosynthesis of the intermediate-mass nuclei in inhomogeneous big bang models.

1. Introduction

In the theory of stellar nucleosynthesis neutron capture reactions are considered to produce isotopes heavier than iron through both slow(s)- and rapid(r)-processes, which are assumed to proceed during He-burning phases of stellar evolution and during stellar explosions, respectively. As for the neutron sources two reactions of ^{13}C$(\alpha,$n$)^{16}$O and ^{22}Ne $(\alpha,$n$)^{25}$Mg are considered to be most important and the heavy s-process isotopes with A > 90 are claimed to be produced by capturing neutrons from the ^{13}C$(\alpha,$n$)^{16}$O reaction in low mass asymptotic giant branch stars (AGB) [1]. However, in the neutron capture rate of ^{12}C and ^{14}N would be very large, they would be neutron poisons, thus affecting the production rates of the s-process isotopes[1]. Consequently, the capture rates of the above reactions are quite important. Similarly, for the r-process like a neutrino-induced nucleosynthesis these reaction rates are also crucial, because of the possible sources of the neutron poison as discussed above[2].

On the other hand, it has been claimed that a possible first order phase transition from a quark-gluon plasma to hadrons could produce density fluctuations leading to a high density proton-rich regions and low- density neutron-rich regions[3,4]. In the latter regions a subsequent production of intermediate-mass (A > 7) nuclei was thought to be possible in the primordial nucleosynthesis in the early universe (inhomogeneous big bang models)[4,5]. These nuclei could be produced mainly through the following reaction sequence[4], H(n,γ)^{2}H(n,γ)^{3}H(d,n) ^{4}He(t,γ)^{7}Li(n,γ)^{8}Li($\alpha,$n$)^{11}$B(n,γ)^{12}B(β^-)^{12}C(n,γ)^{13}C(n,γ)^{14}C — and ^{7}Li(t,n)^{9}Be(n,γ)^{10}Be(β^-)^{10}B *etc.*
As these reaction rates have not been well known, it is quite important to measure them

to estimate quantitatively the production rates of these nuclei.

In the present experiment, the capture rates of ^{7}Li, ^{9}Be, ^{12}C, ^{13}C and ^{14}N have been measured in the neutron energies between 10 keV and 90 keV (For ^{12}C up to 250 keV).

2. Experimental procedure

The experiment was carried out by using pulsed neutrons, provided from the 3.2 MV Pelletron Accelerator of the Research Laboratory for Nuclear Reactors at the Tokyo Institute of Technology, and by detecting prompt γ-rays from a captured state. The average proton beam current was 11 μA at a repetition rate of 2 MHz.

A ^{6}Li-glass scintillation detector was used to measure the neutron energy spectrum by a time-of-flight (TOF) method. Samples of metallic Li and Be, natural C and enriched (99 %) ^{13}C and melamine were used. A Au sample was used to normalize the capture cross section, as the capture rate has been well known.

Prompt γ-rays from a capturing state were measured by an anti-Compton NaI(Tl) detector, which consists of a central NaI(Tl) detector with the size of a 7.6 cm $\phi \times$ 15.2 cm and an annular one with the size of a 25.4 cm $\phi \times$ 28.0 cm [6].

Three runs with these samples, a Au and a blank sample were carried out cyclically to smear out the possible changes of the beam energies, the beam intensities, *etc.*

3. Experimental results

The background subtracted γ-ray spectra measured at the average neutron energies of 97 keV for ^{12}C and 40 keV for ^{13}C are shown in figs. 1 and 2. The intensity of each peak was obtained by a stripping method, using the response function of the γ-ray detector. Preliminary results of the capture cross sections are obtained as 42.3(34) μb for ^{7}Li, 11.4(23) μb for ^{9}Be, 12.4(14) μb for ^{12}C and about 46 μb for ^{14}N at the average neutron energies of 30 keV and 10(4) μb for ^{13}C at the average neutron energy of 40 keV. The capture rate of the ^{12}C(n,γ)^{13}C reaction is shown in fig. 3 as a function of the neutron energy, which indicates that the neutron is captured dominantly by p-wave when the neutron energy increases.

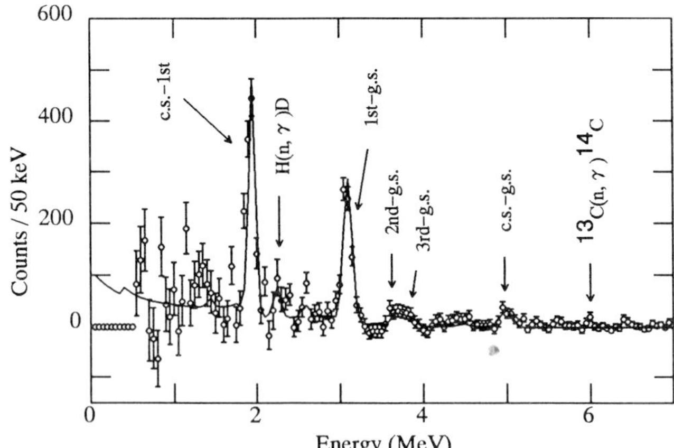

Fig.1. Background subtracted γ-ray spectrum from the ^{12}C(n,γ)^{13}C reaction measured at En = 97 keV.

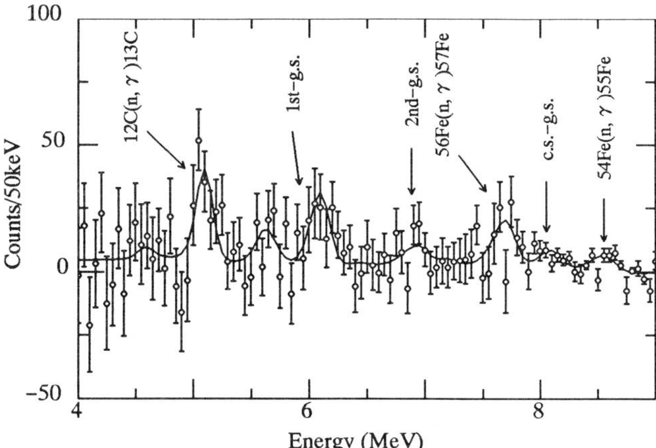

Fig.2. Background subtracted γ-ray spectrum from the ^{13}C(n, γ)^{14}C reaction measured at En = 40 keV.

4. Summary

The important roles of the neutron capture reaction of light nuclei have been discussed in the nucleosynthesis in low mass asymptotic giant branch stars and in primordial nucleosynthesis in inhomogeneous big bang models. In the present work the neutron capture rates of ^{7}Li, ^{9}Be, ^{12}C, ^{13}C and ^{14}N have been measured in the neutron energies between 10 keV and 90 keV by detecting prompt γ-rays from a captured state. (For ^{12}C the rate was measured between 10 keV and 250 keV). The results give important informations of the time evolution of the neutron density in low mass asymptotic giant branch stars. Especially, in the nucleosynthesis of the heavy s-process isotopes in the solar system the neutron recycling by ^{12}C has been considered to play an important role. Consequently, it is very interesting to see how the nucleosynthesis of the s-process isotopes is modified by the present result, which is about five times larger than the estimated value so far used. Concerning the primordial nucleosynthesis in inhomogeneous big bang models, the present results favor the production of A > 7 nuclei in inhomogeneous big bang models.

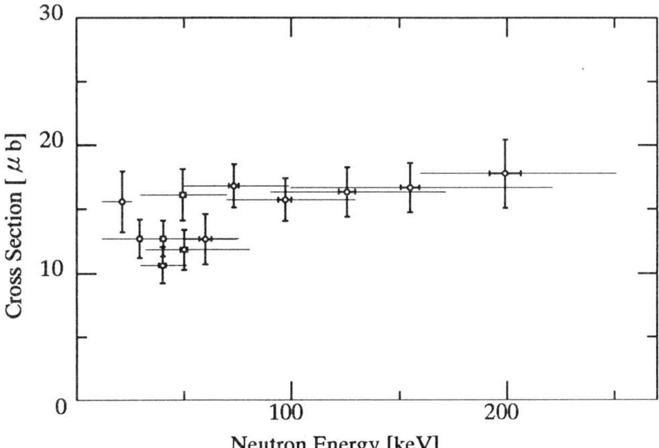

Fig.3. The energy dependence of the capture rate of the ^{12}C(n,γ) ^{13}C reaction.

References

[1] Iben I Jr and Renzini A 1982 *Ap. J. Letters* **259** L79-L83

Gallino R, Busso M, Picchio G, Raiteri C M and Renzini A 1988 *Ap. J. Letters* **33** L45-L49

Käppeler F, Gallino R, Busso M, Picchio G and Raiteri C M 1990 *Ap. J.* **354** 630-643

[2] Woosley S E, Hartmann D H, Hoffman R D and Haxton W C 1990 *Ap. J.* **356** 272-301

[3] Applegate J, Hogan V and Scherrer R J 1987 Phys. Rev. **D 35** 1151-1160

[4] Malaney R A and Fowler W A 1988 *Ap. J.* **333** 14-20

[5] Applegate J, Hogan V and Scherrer R J 1988 *Ap. J.* **329** 572-579

Kajino T, Mathews G J and Fuller G M 1990 *Ap. J.* **364** 7-14

[6] Igashira M, Kitazawa H and Yamamuro N 1986 *Nucl. Instr. Meth.* **A245** 432-437

Nucleosynthesis in the Cd–In–Sn region

Zs Németh, T Belgya, S W Yates[1], Ch Theis[2] and F Käppeler[2]

Institute of Isotopes, Budapest, Hungary

Abstract. The low-spin levels of ^{113}Cd were studied by the $(n,n'\gamma)$ reaction up to 2.5 MeV excitation energy. We identified the levels which decay towards both the ground state and the isomer. The positions of these gateway levels imply thermal equilibrium for temperatures above $\sim 10^8$ K, resulting in the destruction of the isomer under s-process conditions. This affects the branching at ^{113}Cd and, hence, the s-contributions to ^{113}In and ^{114}Sn, which were found to be 1.5% and 6%, respectively. For the r-process production of ^{113}In an upper limit of 16% was established, a value smaller than suggested earlier. Consequently, both isotopes are predominantly produced by the p-process. For ^{114}Sn, this is confirmed by recent p-process calculations, which, however, do not produce the necessary amount of ^{113}In. At A=115, 46% of ^{115}Sn, a neutron-deficient isotope, was found to be accounted for by the r-process. Since the s-process contributes less than $\sim 10\%$, a sizable fraction of ^{115}Sn must also be assigned to the p-process.

1. Introduction

The interpretation of the observed abundances in the Cd-In-Sn region in terms of the contributing nucleosynthesis mechanisms is rather complex due to a number of long-lived, low-lying β-decaying isomers in this mass range. The β-decay mode of these isomers allows the r-process to reach neutron-deficient isotopes as ^{113}In or ^{115}Sn (Clayton 1978, Ward and Beer 1981), and gives rise to branchings in the s-process path. These isomers are important for the influence of temperature on the s-process branchings at A=113 and 115. In particular, it depends on the internal level structure of the branch point isotopes, whether isomer and ground state participate independently in the s-process according to their production ratio in the (n,γ) reactions, or whether their population probabilities are readjusted in the hot stellar photon bath by thermally induced transitions. Therefore, these effects need to be considered explicitly in a quantitative discussion of the observed abundances.

The main paths of nucleosynthesis in the Cd-In-Sn region are shown in Fig. 1. There are three branching points in the s-process flow (solid lines) at ^{113}Cd, at ^{115}Cd

[1] Department of Chemistry, University of Kentucky, Lexington, USA
[2] Kernforschungszentrum Karlsruhe, Institut für Kernphysik, Karlsruhe, Germany

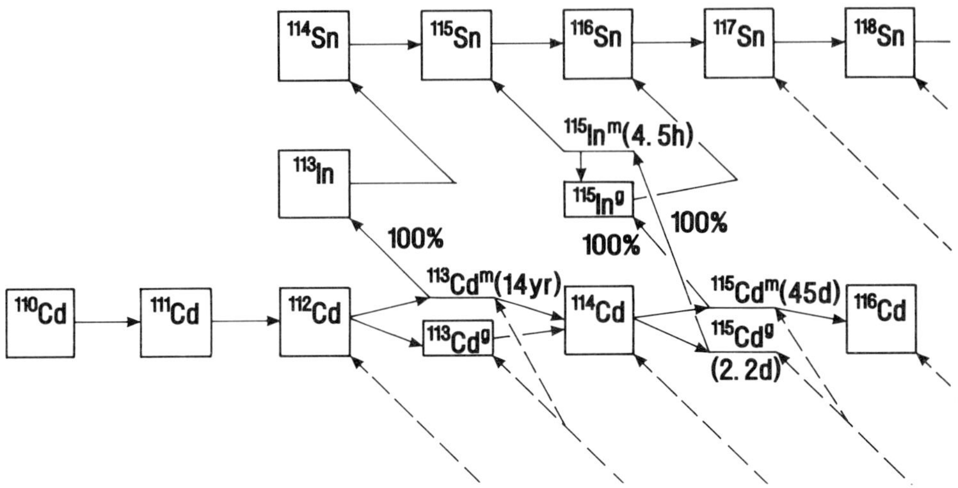

Figure 1. Possible s- and r-process paths in the Cd-In-Sn region. Only temperature-independent branchings are indicated. For a discussion of temperature effects see text. Solid lines: s-process paths, dashed lines: post-r-process β-decays.

and at ^{115}In. Of them, the first one at ^{113}Cd is the most important. The $1/2^+$ ground state of ^{113}Cd is quasistable, and there is an $11/2^-$ isomer of 14 y half-life which decays to ^{113}In by a branching ratio of 99.9%.

The effective $(^{112}\mathrm{Cd}\rightarrow^{113m}\mathrm{Cd})/(^{112}\mathrm{Cd}\rightarrow^{113g}\mathrm{Cd})$ branching ratio depends on a number of nuclear physics and astrophysics parameters. One of them is the capture cross section ratio. A direct measurement of this value is difficult because of the long half-life of the capture products. Since no such investigation has yet been carried out, values from a systematics had to be used (Ward and Beer 1981, Beer *et al* 1989).

Another important parameter is the location of the lowest-lying gateway levels which decay both towards the ground state and to the isomer. If such levels exist at a sufficiently low excitation energy, thermal equilibrium can be achieved and (γ,γ') reactions can significantly modify the branching ratio. In order to get a complete picture on the low-lying, low-spin states of ^{113}Cd we carried out an $(\mathrm{n,n'}\gamma)$ experiment. The inelastic neutron scattering is a nonselective compound-nucleus reaction at low excitation energies which can populate all levels of the target nucleus up to an angular momentum transfer of 5 to 6 units. This means, that all states of ^{113}Cd in the 1/2-11/2 spin window containing the gateway levels, can be observed. In Section 2, we describe the identification of the gateway levels, and in Sections 3 and 4 we discuss the astrophysical consequences.

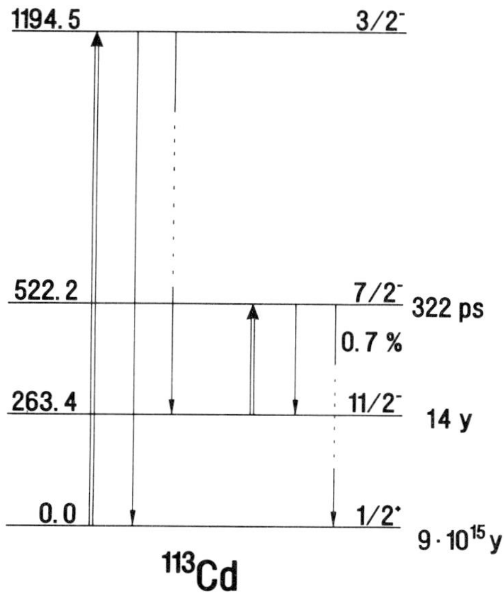

Figure 2. Simplified level scheme of ^{113}Cd indicating the first gateway level and the transitions to isomer and ground state.

2. Identification of the gateway levels in ^{113}Cd

The ^{113}Cd(n,n$'\gamma$) experiment was performed at the 7 MV Van de Graaff accelerator of the University of Kentucky in Lexington. A metallic powder target of 28.2 g with an enrichment of 94.58% was irradiated by neutrons produced via the ^{3}H(p,n) reaction. Gamma-rays were detected by a high purity Ge detector of 52% efficiency. The background was reduced by a BGO annulus. We recorded excitation functions from 1.0 to 2.5 MeV by 100 keV steps. Angular distribution measurements were also carried out at 1.7 and 2.5 MeV at 7-7 positions. From the ∼400 observed gamma-ray lines, ∼350 were assigned to transitions in ^{113}Cd.

The lowest-lying level which decays directly to the isomer and via cascades to the ground state is identified at 522.2 keV. This is a known level with I$^{\pi}$=7/2^{-} (Matumoto *et al* 1978) and 322 ps half-life (Ohya *et al* 1980). The tentative spin assignment has been verified explicitely. We also determinded the branching ratio of the weak transition bypassing the metastable state to 0.7(1)% as indicated in Fig.2. This value is somewhat smaller than the 1.22(12)% by Blachot (1990). Additional gateway levels are identified at 855.2 keV and at 1194.4 keV. These levels can be populated by multistep photoexcitations. Klay *et al* (1991) have shown that, considering the Maxwellian distribution of photons under s-process conditions, multistep excitations play an equally important role with the one-step excitations. Several gateway levels above 1.2 MeV are also found, but they are of negligible importance for nucleosynthesis. We emphasize that in our measurement *all* states of astrophysical significance are populated, thus allowing for a complete analysis of the thermal coupling between the ground state and the isomer. A full account for the ^{113}Cd(n,n$'\gamma$) results, with special emphasis on nuclear structure aspects, will be given elsewhere (Németh *et al* 1993).

3. s- and r-Process branchings at ^{113}Cd

The analysis of the decay properties of the gateway levels suggests that the 522 keV state is most important for attaining thermal equilibrium under plausible s-process conditions. For thermal energies of 12 keV and 26 keV thermal energy, which are characteristic temperatures for the s-process in low mass stars to current stellar models (Gallino *et al* 1988, Gallino 1989, Käppeler *et al* 1990), we obtain equilibration times of 67 s and 0.6 ms, respectively. Hence, thermal equilibrium is quickly achieved, leading to a strong depletion of the isomer by (γ,γ') reactions. The higher-lying gateway levels assist in the destruction of ^{113m}Cd, but their effect is orders of magnitude smaller than that of the 522 keV state.

Beer and Ward (1981) estimated the branching ratio feeding ^{113m}Cd via neutron capture on ^{112}Cd to 6.4%. Because of the thermally induced transitions the *effective* branching ratio to ^{113m}Cd is reduced to 0.2%. This value corresponds to a 1.5% and 6% s-process production of ^{113}In and ^{114}Sn, respectively. Since this s-process branching is so weak, ^{114}Sn is practically a pure p-isotope. The main branch of the s-process, therefore, continues to ^{115}Cd, the next branching point.

The 6% s-process contribution to ^{114}Sn is obtained by supposing that there is no significant branch at ^{114}In. The few experimental cross section data for ^{113}In(n,γ) (Chaubey and Sehgal (1966), Pönitz (1966)) suggest that the ^{114m}In/^{114g}In branching ratio is roughly 2:1. The 0.5% EC$+\beta^+$ branch in the decay of ^{114g}In (Andre and Depommier 1964) is suppressed in the hot stellar plasma, leaving the β^--decay to ^{114}Sn the only significant decay mode.

With ^{114m}In the situation is similar: normally, it decays predominantly via internal transitions (95.6%) with a small (EC$+\beta^+$)-branch of 4.4% (Coffman and Hamilton 1969), but both decay modes are retarded under stellar conditions. Though detailed calculations are necessary to deduce the effective branching ratio, it appears unlikely that there is considerable leakage to ^{114}Cd. We note that thermalization, which occurs at 26 keV (but not at 12 keV), would reduce the isomer and, hence, the branching ratio to ^{114}Cd.

As pointed out above, the r-process can also contribute to the nucleosynthesis of ^{113}In. The post-r-process β-decay flow is, however, complicated in this mass region, since most of the participating odd isotopes have isomers with a dominant β-decay mode. This fact, combined with the lack of experimental data makes it difficult to follow the β-decay chains quantitatively. Preliminary data of Fogelberg *et al* (1988) and a study of the decay pattern of ^{111m,g}Pd suggest that the decay of ^{113}Pd populates predominantly the isomer ^{113m}Ag. Under laboratory conditions, the decay of this isotope is dominated by internal transitions (64%) with a weaker β^- decay channel of 36% to ^{113g}Cd (Fogelberg *et al* 1990).

The further flow of β-decays is strongly dependent on how much the post-r-process conditions differ from the laboratory situation. If a couple of minutes after the r-process freeze-out the temperature is still high enough to ionize at least the L- and outer shells of the silver atoms, internal conversion, which is the dominant decay mode of ^{113m}Ag, is suppressed and the effective β-decay branching ratio will be nearly 100%. Furthermore, thermalization may also occur and can result in a modified ^{113m}Ag/^{113g}Ag production ratio.

The decay scheme of ^{113g}Ag is well-known (Matumoto *et al* 1978). Summing up the direct feeding of the ground state and the various γ-cascades, we obtain a 98.5%/1.5% branching ratio for the population of ^{113g}Cd and ^{113m}Cd, respectively. Since the γ-cascades following the β-decay of ^{113m}Ag feed exclusively ^{113g}Cd (Niizeki 1990), at most 1.0% of the post-r-process flow can bypass ^{113g}Cd. This value corresponds to an upper limit of 16% for the r-process contribution to ^{113}In. However, a value of 1 to 2% appears more plausible, since the suppression of the internal conversion decay mode of ^{113m}Ag in the (still) hot post-r-process environment decreases the effective branching ratio to ^{113m}Cd. In any case, our figure is considerably smaller than the 28% suggested by Ward and Beer (1981). The weak s- and r-process production of ^{113}In suggests that it was synthetized mostly by the p-process, an expectation that does not agree with the results of recent p-process calculations (Prantzos *et al* 1990, Rayet *et al* 1990, Howard *et al* 1991) which yield only $\sim$10% p-process contribution to the solar-system abundance of ^{113}In.

4. s- and r-Process branchings at ^{115}Cd and at ^{115}In

The s-process branching at ^{115}Cd has similarities to the one at ^{113}Cd. The effective branching ratio is again sensitive to temperature and to the nuclear physics parameters. An important difference is, however, that the ground state has a short half-life compared to the timescale of the s-process. Hence, a weak s-process branch can reach ^{116}Cd only if the isomer survives at higher temperatures.

The level scheme of ^{115}Cd is less well known than that of ^{113}Cd. The available β-decay data (Matumoto and Tamura 1978), nonetheless, allow to identify gateway levels at 719.9 and 1092.1 keV. These states are nuclear structure counterparts of the 855 keV and 1194 keV levels in ^{113}Cd, respectively. The corresponding partner of the 522 keV state is identified at 393.9 keV, but it is not clear if that acts as a gateway level. The low-energy member of a possible 33 keV-360 keV cascade, which bypasses the isomer, is well below the observational limit of Matumoto and Tamura (1978). Our calculations show that, even if such a transition does not exist, thermal equilibrium via the 719.9 keV level is quickly established at 26 keV thermal energy, resulting in a strong reduction of the isomer, similar to the case of ^{113}Cd. On the other hand, thermal equilibrium at 12 keV is achieved only if the 393.9 keV state is also a gateway level.

The above results indicate that the further s-process flow is temperature-dependent. If we use the classical approach (kT=28 keV), thermal equilibrium is reached and the isomer is strongly reduced, just as in the case of ^{113}Cd. This results in an effective branch of 0.9% to ^{115m}Cd, leaving only a marginal s-process contribution to ^{116}Cd. The decay of ^{115g}Cd feeds exlusively ^{115m}In, which is, however, thermalized above 19 keV, so that it is strongly depopulated. This means that, in the classical approach, there is no effective branching at ^{115}In. The main s-process path reaches ^{116}Sn via ^{116}In, and the contribution to ^{115}Sn through the ^{115g}Cd$\rightarrow$^{115m}In branch is less than about 1%, even smaller than through the path discussed in Section 3.

If we consider the stellar models for low mass stars, ^{115}Cd may not be thermalized during the first pulse at kT=12 keV. Hence, the laboratory branching ratio of 22% to ^{115m}Cd (Ward and Beer (1981)) remains valid. and we calculate a branching ratio of 76/24 for ^{115m}Cd(β^-)/^{115m}Cd(n,γ), which corresponds to a 5% s-process contribution to ^{116}Cd. However, if thermal equilibrium is established in ^{115}Cd, then the s-process contribution to ^{116}Cd reduces to about 2%.

The ^{115m}In, populated by the desintegration of ^{115g}Cd, decays by 95% via isomeric transitions, and by 5% via β^- to ^{115}Sn (Hansen *et al* 1974). In this case, the s-process temperatures are not high enough for thermal equilibration of ground state and isomer, but sufficient for ionizing the In atoms down to the L shell. The resulting hindrance of the internal conversion decay of ^{115m}In, which is the principal decay mode, results in an effective β-decay branching ratio of 6%. This value corresponds to a 40% s-process production of ^{115}Sn through this path. This figure is significantly different from the result obtained with the classical approach.

However, the situation changes during the second pulse (kT=26 keV). Though it is considerably shorter than the first, it is sufficient to rearrange the abundance pattern of the s-process branchings: While ^{116}Cd remains unchanged within 1%, the final ^{115}Sn abundance is reduced to 9% of the observed value at the end of the second pulse. This result is now in good agreement with the value obtained by the classical approach.

The post-r-process β-decays along the mass chain A=115 are even more complicated than in the case of the A=113 chain. The β-decay of ^{115g}Ag feeds ^{115g}Cd with 94.5% and ^{115m}Cd with 5.5% probability (Matumoto and Tamura (1978)). Supposing that the β-decay of ^{115m,g}Pd populates exclusively ^{115g}Ag (which is unlikely), this 5.5% branch is the only bypass around ^{115m}In, not contributing to the production of ^{115}Sn. Supposing that neither ^{115}Cd nor ^{115}In are thermalized, and that the laboratory β-decay branching ratio holds for ^{115m}In, we obtain a 44% r-process contribution to ^{115}Sn. The guess that ^{115m}In remains unthermalized is justified by the fact that the 2.2 d half-life of ^{115g}Cd acts as a bottleneck, allowing for sufficient cooling time.

Turning to the other limit, that the decay of ^{115m,g}Pd populates uniquely ^{115m}Ag (which seems to be a more reasonable approximation) does not mean too much of a change. Under normal conditions, ^{115m}Ag decays to 79% by IT and to 21% via the β^- branch (Fogelberg *et al* (1990)). Since the post-r-process β-decay flow reaches it already some ten second after r-process freeze-out, the environment is still hot enough for the silver atoms to be highly ionized. This results in a strong hindrance of internal conversion, so that only a weak branch of less than 5.5% can bypass ^{115m}In. Any thermalization of ^{115}Cd would reduce this branching ratio further. Even if ^{115m}In is not bypassed at all, the r-process production of ^{115}Sn is limited to 46%, only 2% more than in the previous case. This means that the r-process production of ^{115}Sn is quite insensitive to temperature. Our values agree well with those of Ward and Beer (1981) and of Beer *et al* (1989).

Compared to the total abundance, we find that the s- and r-process contributions can only account for half of the observed ^{115}Sn. This result implies a sizable p-process contribution to ^{115}Sn, in contradiction to the calculations of Prantzos *et al* (1990), Rayet *et al* (1990) and Howard *et al* (1991), where ^{115}Sn is severely underproduced.

Our results concerning the origin of the three rare, neutron-deficient isotopes ^{113}In, ^{114}Sn, and ^{115}Sn, are summarized in Table 1. Comparison with the normalized overproduction factors from recent p-process studies shows good agreement only ^{114}Sn, while the calculated yields of both odd nuclei are too low in all p-process models. It is interesting to note that the strong r-process component of ^{115}Sn: The mass chain at A=115 is the only point where the post-r-process β-decay flow crosses the s-process path and provides for a significant contribution to a neutron-deficient isotope.

Table 1. The contributions of various astrophysical processes to the nucleosynthesis of ^{113}In, ^{114}Sn and ^{115}Sn in (%) of the observed abundances. The inferred p-process fractions are compared to recent calculations.

Isotope	Our analysis[a]			p-process overproduction factors [b]		
	s	r	p	Prantzos *et al*	Rayet *et al*	Howard *et al*
	%	%	%	(1990)	(1990)	(1991)
^{113}In	1	<16	>83	0.17	0.12	0.34
^{114}Sn	6	–	94	4.0	1.6	2.4
^{115}Sn	9[c]	46	–	0.09	0.007	0.0014

[a]Typical uncertainties are 10%
[b]Normalized to mean overproduction of all p-only nuclei between A=106 and 120
[c]model for low mass stars; classical s-process yields ~5%

5. Summary

The processes contributing to the abundances of the rare isotopes ^{113}In, ^{114}Sn, and ^{115}Sn have been discussed on the basis of new information on level structure and transition rates of ^{113}Cd and with an update of the β-decay chains in this mass region. As a result, we find marginal s-process contributions for all three cases, independent of whether the classical approach or a stellar s-process model was used. In addition, a significant r-process component could be identified only for ^{115}Sn. Consequently, all three isotopes do have dominant or, at least, strong p-process components.

The fact, that the main nucleosynthesis mechanisms contribute very little to the abundances of these rare isotopes, allowed for a quantitative decomposition, resulting in reliable p-process yields. The comparison with recent p-process calculations shows that these models fail for the odd isotopes ^{113}In and ^{115}Sn, while the p-process origin is confirmed for ^{114}Sn.

This work was partly supported by the Hungarian Research Fund under Contract No. OTKA T/4182, by OMFB and by BMFT under Contract No. X238.0, and by the U. S. National Science Foundation under Grants No. PHY-9001465 and No. INT-801880.

References

Andre S and Depommiere P 1964 *J. Phys. (Paris)* **25** 673

Beer H, Walter G and Käppeler F 1989 *Astron. Astrophys.* **211** 245

Blachot J 1990 *Nucl. Data Sheets* **59** 729

Chaubey A K and Sehgal M L 1966 *Phys. Rev.* **152** 1055

Clayton D D 1978 *Astrophys. J.* **224** L93

Coffman F E and Hamilton J H 1969 *Nucl. Phys. A* **127** 34

Fogelberg B, Lund E, Zongyuan Y and Ekström B 1988 *Proc. 5th Int. Conference on Nuclei far from Stability, Lake Russeau.* Towner I S, editor (New York: American Physical Society) p 296

Fogelberg B, Zongyuan Y, Ekström B, Lund E, Aleklett K and Sihver L 1990 *Z. Phys. A* **337** 251

Gallino R, Busso M, Picchio G, Raiteri C M and Renzini A 1988 *Astrophys. J.* **334** L45

Gallino R 1989 *Proc. IAU Colloquium 106, The Evolution of Peculiar Red Giants.* Johnson H L and Zuckerman B, editors (Cambridge: Cambridge University Press) p 176

Hansen H H, de Roost E, van der Eijk W and Vaninbroukx R 1974 *Z. Phys. A* **269** 255

Howard W M, Meyer B S and Woosley S E 1991 *Astrophys. J.* **373** L5

Käppeler F, Gallino R, Busso M, Picchio G and Raiteri C 1990 *Astrophys. J* **345** 630

Klay N, Käppeler F, Beer H and Schatz G 1991 *Phys. Rev. C* **44** 2839

Matumoto Z and Tamura T 1978 *J. Phys. Soc. Jpn.* **44** 1070

Matumoto Z, Tamura T and Sakurai K 1978 *J. Phys. Soc. Jpn.* **44** 1062

Németh Zs *et al* 1993, to be published

Niizeki H 1990 *Bull. Yamagata University (Natural Science)* **12** 229

Ohya S, Mutsuro N, Matumoto Z and Tamura T 1980 *Nucl. Phys. A* **344** 382

Pönitz W 1966 *Thesis,* Kernforschungszentrum Karlsruhe

Prantzos N, Hashimoto M, Rayet M and Arnould M 1990 *Astron. Astrophys.* **238** 455

Rayet M, Prantzos N and Arnould M 1990 *Astron. Astrophys.* **227** 271

Ward R A and Beer H 1981 *Astron. Astrophys.* **103** 189

High resolution ^{138}Ba(n,γ) measurements

H. Beer*, F. Corvi, A. Mauri**, K. Athanassopulos

Commission of the European Communities, Joint Research Centre,
Central Bureau for Nuclear Measurements, B-2440 Geel, Belgium

Abstract. The Maxwellian-averaged (n,γ) cross section of ^{138}Ba, with magic neutron number 82, is crucial in determining the average neutron exposure τ_0 of the s-process. Therefore high resolution neutron capture measurements of ^{138}Ba were carried out at GELINA in the energy range 0.5-103 keV. A total of 54 neutron resonances were resolved and analysed in order to derive their capture areas $g\Gamma_n\Gamma_\gamma/\Gamma$. From these values the Maxwellian-averaged (n,γ) cross section was calculated *vs* temperature kT from 10 to 50 keV. Compared to previous data, the present ones show a steeper rise with decreasing kT : this fact indicates that for the ^{13}C(α,n) reaction an exposure which is about 30% smaller is sufficient for describing the solar abundances.

1. Introduction

The main feature of the s-process is a ledge-precipice structure in the capture cross section times abundance distribution $<\sigma_\gamma>_A N_A$ (Seeger et al. 1965). This structure is the interplay of nuclear properties at the magic neutron numbers 50, 82, 126 with the properties of the neutron exposure of the s-process. The average exposure τ_0 necessary to drive the synthesis up to Pb and Bi has to be large enough to compete with the small Maxwellian-averaged capture cross sections $<\sigma_\gamma>$ of isotopes with a closed neutron shell N = 82.

Particularly important is the case of the isotope ^{138}Ba since its cross section $<\sigma_\gamma>$ is the smallest one in the region and therefore it is crucial in determining τ_0. The following facts have to be kept in mind :

the suggested neutron sources driving the s-process, ^{13}C(α,n) and ^{22}Ne(α,n), are ignited at different temperatures, 12 keV and 25 keV, respectively. Additionally a combined burning of these sources is presently discussed (Gallino et al. 1989, Käppeler et al. 1990, Beer 1991) where the exposure of the ^{13}C(α,n)

---

* Scientific visitor from KfK Karlsruhe
** National expert from ENEA, Bologna

reaction has to carry the synthesis through the ^{138}Ba bottle-neck. The quantity $<\sigma_\gamma>$ of ^{138}Ba must, therefore, be determined *vs* temperature kT which implies the measurement of the excitation function ^{138}Ba(n,γ) from a few eV to 200 keV. The relatively large level spacing of the resonances of the compound nucleus makes an accurate prediction of $<\sigma_\gamma>$ *vs* kT only possible after a detailed determination of the resonance parameters.

Although high resolution neutron capture measurements of ^{138}Ba were already performed at ORELA (Musgrove *et al.* 1975, 1979), there are some good reasons to repeat them. First of all, the Oak Ridge measurements suffered of a very high prompt neutron background resulting in large corrections, up to 90%, for the capture areas of the strongest s-wave levels. Secondly, a systematic error should be present in the data, due to the particular detection technique used, namely the pulse height weighting method, since it was later found out that the weighting function applied at the time was incorrect (Corvi *et al.* 1988, 1991). This error could be important when normalizing the ^{138}Ba data to gold capture. Finally the neutron energy resolution can now be considerably improved since the Oak Ridge measurements were still performed with a 40 ns linac pulse width. For all these reasons we decided to repeat the ^{138}Ba(n,γ) measurements at GELINA.

2. Experimental

Time-of-flight neutron capture measurements of ^{138}Ba were performed at the pulsed electron linac GELINA. The accelerator was operated to provide 1 ns wide bursts of 100 MeV electrons with peak current 100 A and repetition rate 800 Hz, the average beam power being 7 kW. Neutrons were produced in a rotary uranium target and moderated in a 4 cm thick slab of water canned in beryllium. The flight path used was normal to the moderator face and had a length of 28.4 m. The sample consisted of 43 g of $BaCO_3$ enriched to 99.8% ^{138}Ba canned in a 8 cm diameter aluminium container with 0.3 mm thick walls. The thickness of the relevant isotope is 0.0026 at/b. The sample was viewed by two cylindrical C_6D_6 liquid scintillators to which a pulse height weighting technique was applied in order to make their efficiency proportional only to the total γ-energy emitted. Details of the detection geometry and of the weighting function finally adopted can be found in a recent paper (Corvi *et al.* 1991).

The relative neutron flux was simultaneously measured with an in-beam ionization chamber containing two back-to-back ^{10}B deposits of thickness 40 μg/cm^2. The data were normalized to gold capture in a separate run performed at 100 Hz repetition rate by applying the saturated resonance technique to the 4.9 eV resonance in ^{197}Au.

3. Analysis and Results

Two resonances were observed for the first time below 2 keV (a region not covered in former Oak Ridge measurements), namely at E_0 = 648 and 1946 eV. In the 100 Hz calibration run the capture in the 648 eV level was compared to Au(n,γ) and

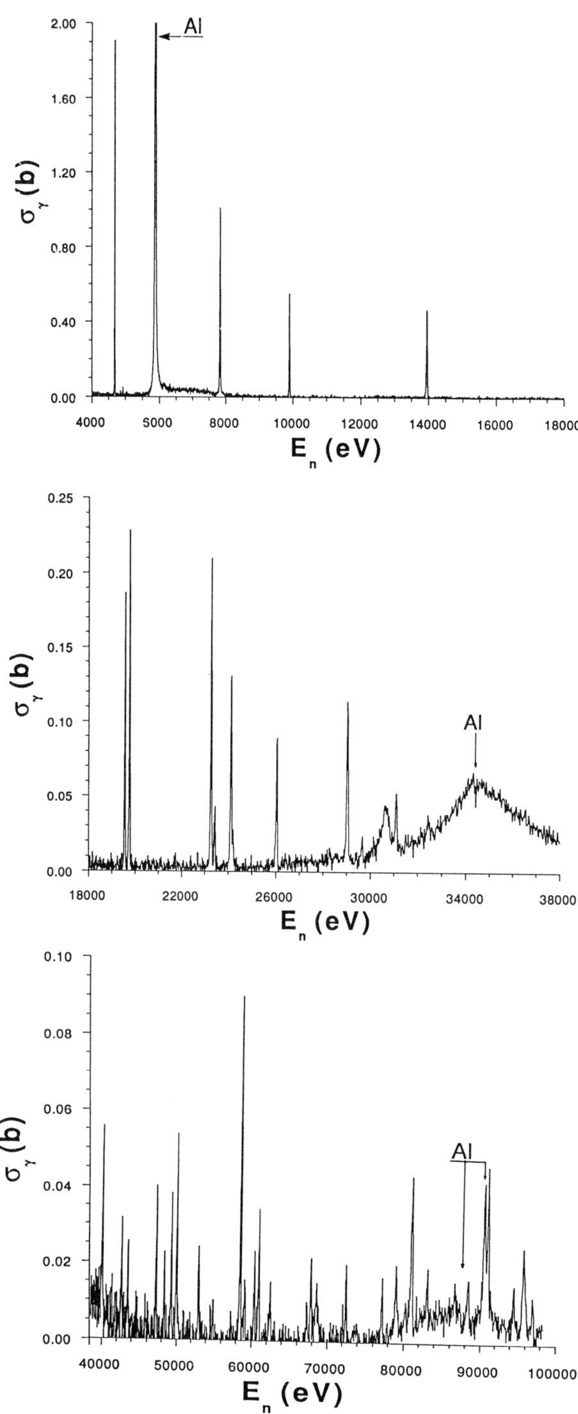

Fig. 1. The ^{138}Ba capture cross section uncorrected for Al contamination.

the value of the area $g\Gamma_n\Gamma_\gamma/\Gamma = 5.07 \pm 0.21$ meV was determined. The given error includes both statistics and systematics contributions from various sources. All the data of $^{138}Ba(n,\gamma)$ were normalized to this area value. After background subtraction, correction for the neutron flux and proper normalization, the weighting counting rate was reduced to a capture cross section. This is shown in Fig. 1 where σ_γ for ^{138}Ba is given *vs* neutron energy in the range 4-98 keV. It should be noted that the plotted curve is still Doppler and resolution broadened and is not corrected for multiple scattering, neutron self-shielding and prompt background contribution. One may also notice the large contamination due to capture in Al whose main resonances are marked with arrows.

The data were analysed with the R-matrix shape fitting code FANAC (Fröhner 1976). An iterative fit to the observed capture area was performed by varying the smaller of the input values for Γ_n or Γ_γ. Finally, the corrected capture kernel $g\Gamma_n\Gamma_\gamma/\Gamma$ was obtained. In the case of s-waves we assumed in the analysis the neutron widths values of Bilpuch *et al.* (1961). In the case of other resonances, for which no transmission data exist, we were obliged to assume arbitrary values of Γ_n, taken always much larger than Γ_γ except for the level at $E_0 = 648$ eV. This uncertainty has fortunately little influence on the value of the relevant capture area. The s-wave data were corrected for the background due to the prompt detection of scattered neutrons by using the values of the scattered neutron sensitivity measured by us as a function of energy for the same detection geometry (Corvi et al. 1983). Although not as large as in the Oak Ridge measurements this correction is sometimes considerable: in the worse case ($E_0 =$

Table 1. Capture areas of ^{138}Ba resonances in the range 0.5-103 keV

E_0 (keV)	$g\Gamma_n\Gamma_\gamma/\Gamma$ (meV)	E_0 (keV)	$g\Gamma_n\Gamma_\gamma/\Gamma$ (meV)	E_0 (keV)	$g\Gamma_n\Gamma_\gamma/\Gamma$ (meV)
0.648	5.1 ± 0.1	42.86	31 ± 5	72.53	17 ± 7
1.946	38 ± 1	43.71	27 ± 4	72.98	63 ± 9
4.698	29 ± 1	47.50	64 ± 6	77.74	66 ±10
7.850	60 ± 2	48.64	35 ± 5	79.57	149 ±19
9.924	33 ± 1	49.57	62 ± 6	80.75	36 ±10
13.99	54 ± 2	49.94	11 ± 7	81.62	233 ±17
19.63	42 ± 2	50.30	124 ±30	83.65	63 ± 9
19.84	58 ± 3	53.33	43 ± 5	87.28	38 ± 3
23.36	70 ± 3	54.93	20 ± 5	89.04	67 ±13
23.52	13 ± 2	55.25	23 ± 5	91.72	270 ±21
24.22	46 ± 3	58.89	212 ±13	95.16	85 ±13
24.29	9 ± 2	59.47	53 ±16	96.55	336 ±30
26.14	39 ± 3	60.78	56 ± 7	97.67	81 ±14
29.17	58 ± 4	61.35	73 ± 8	97.71	55 ±13
29.78	7 ± 3	62.86	65 ±22	100.59	109 ±16
30.77	73 ±35	67.74	31 ± 8	101.00	65 ±14
31.24	26 ± 3	68.26	68 ± 9	101.65	66 ±14
40.22	72 ±24	68.96	110 ±51	102.33	201 ±19

30.77 keV) it amounts to about 40% of the measured area. Capture areas are listed in Table 1 for 54 resonances, 15 of which observed for the first time. The given errors are the quadratic combination of the statistical ones, taken equal to two standard deviations, and an uncertainty on the relative neutron flux, increasing from 0 at 0.648 keV to 5% at 100 keV. For s-waves, an error equal to half the correction for prompt neutron scattering was also included.

The Maxwellian-averaged capture cross section $<\sigma_\gamma v>/v_T$ calculated from the data of Table 1 in the range 10-50 keV is listed in Table 2. The not yet measured excitation function above 100 keV was estimated using an average cross section of 2.5 mb at 100 keV derived from our data and a 1/E energy dependence. The given errors derive from those of the capture areas plus a 4.1% normalization error related to the uncertainty in the area of the $E_0 = 648$ eV reference resonance. A 50% uncertainty was assumed for the contributions to $<\sigma_\gamma v>/v_T$ from the energy region above 100 keV. The quantity $<\sigma_\gamma v>/v_T$ obtained in the present work is also plotted *vs* kT in Fig. 2 (full line) and compared to that calculated by Beer *et al.* (1992) from the Oak Ridge data. Also a Karlsruhe activation data point at 25 keV is reported (Beer and Käppeler 1980). One may notice that, while both curves agree with this last value, their behaviour vs temperature is quite different, the present data suggesting a much steeper rise with decreasing kT.

Table 2   Maxwellian-averaged (n,γ) cross section of ^{138}Ba for various kT values

kT (keV)	$<\sigma_\gamma v>v_T$ (mb)	kT (keV)	$<\sigma_\gamma v>v_T$ (mb)
10	7.64 $\pm$ 0.52	25	4.38 $\pm$ 0.34
12.5	6.53 $\pm$ 0.46	30	4.02 $\pm$ 0.30
15	5.79 $\pm$ 0.42	35	3.76 $\pm$ 0.35
17.5	5.27 $\pm$ 0.39	40	3.55 $\pm$ 0.38
20	4.90 $\pm$ 0.35	45	3.37 $\pm$ 0.40
22.5	4.61 $\pm$ 0.34	50	3.20 $\pm$ 0.43

4. Discussion

The present results must be considered under the following aspects:
the Maxwellian-averaged capture cross section of ^{138}Ba is a steeper function of kT: $<\sigma_\gamma> \sim (kT)^{-0.58}$ than the variation $<\sigma_\gamma> \sim (kT)^{-0.27}$ suggested by the ORNL data. Therefore, an s-process dominated by the burning of ^{13}C(α,n) at kT = 12 keV can drive the synthesis through the bottle-neck ^{138}Ba up to Pb and Bi more efficiently also with a reduced neutron flux ($\sim$30%) than according to the old ORNL data. A much larger capture cross section at kT = 12 keV has consequences for the ^{138}Ba s-process abundance if formed at this stellar temperature. This may concern the interpretation of the s-process nucleosynthesis in Barium stars and s-process related isotopic anomalies of Ba in meteorites.

These results should be considered preliminary: in the near future we intend to perform a more accurate data normalization, to extend the useful energy range up to 200 keV using a 60 m flight path and a larger quantity of ^{138}Ba, and to carry out transmission measurements.

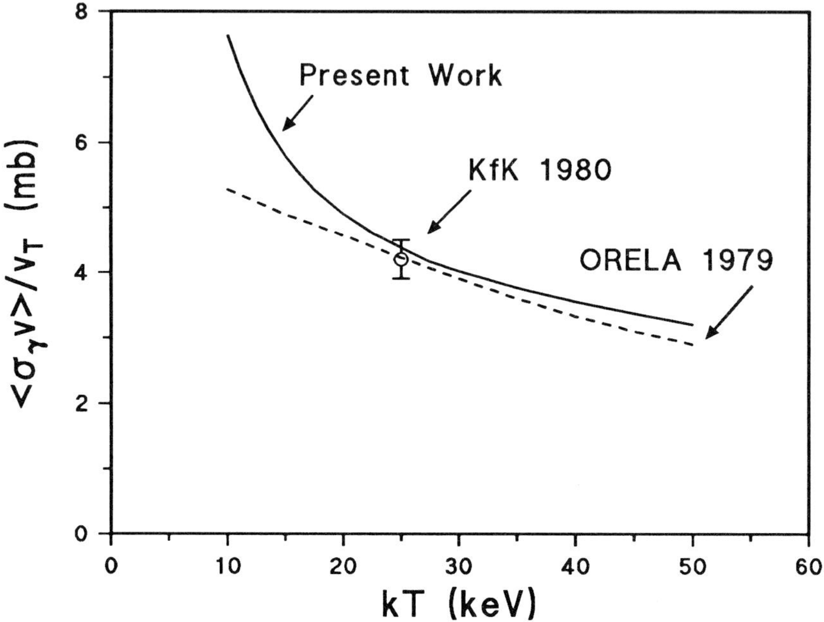

Fig. 2. The Maxwellian-averaged capture
cross section *vs* stellar temperature

References

Beer H, Käppeler F 1980 *Phys. Rev.* **C21** 534

Beer H 1991 *Ap. J.* **379** 409

Beer H, Voss F, Winters R R 1992 *Ap.J. Suppl.* **80** 403

Bilpuch E G, Seth K K, Bowman C D, Tabony R H, Smith R C and Newson H W
     1961 *Ann. Phys. (New York)* **14** 387

Corvi F, Brusegan A, Buyl R, Rohr G 1984 *Proc. Cons. Meet. on Nuclear Data for
     Structural Materials*, Vienna, Report INDC (NDS)-152L

Corvi F, Prevignano A, Liskien H, Smith P B 1988 *Nucl. Instr. Meth. Phys. Res.*
     A265 475

Corvi F, Fioni G, Gasperini F, Smith P B 1991 *Nucl. Sci. Eng.* **107** 272

Fröhner F H 1976 Report KfK-2145

Gallino R 1989 *Evolution of Peculiar Red Giant Stars* (Cambridge University
     Press) p 176

Käppeler F, Gallino R, Busso M, Picchio G, Raiteri C M 1990 *Ap. J.* **354** 630

Musgrove A R, Allen B J, Boldeman J W, Macklin R L 1975 *Nucl. Phys.* **A252**
     301

Musgrove A R, Allen B J, Macklin R L 1979, *Aust. J. Phys.* **32** 213

Seeger P A, Fowler W A, Clayton D D 1965, *Ap. J. Suppl.* **11** 121

Neutron induced reactions followed by charged particles emission

Yu.P.Popov and Yu.M.Gledenov
Frank Laboratory of Neutron Physics, JINR, Dubna,
Moscow Region, 141980 Russia

Abstract. The (n,α) and (n, p) reactions have been
investigated with resonance neutrons at the Dubna
time-of-flight facility. The description of the
(n,α) experimental data in the frame of the statistical theory is discussed. Both reactions are considered from the nucleosynthesis point of view. An
attempt was made to estimate the role of the (n, p)
reaction in the explosion of supernovae.

1. Introduction

Clear anticorrelation of average neutron capture cross sections at stellar temperatures with the isotope abundance in
nature once established has demonstrated the importance of the
s-process in nucleosyntheses and stimulated further investigations of the (n,γ) reaction with about 30 keV neutrons.

The neutron induced reactions on heavy nuclei followed
by the emission of charged particles (alpha particles, for
example) have considerably smaller cross sections due to a low
probability for α-particle penetration through the Coulomb
barrier. The emission of protons from the reaction induced by
slow neutrons is a rare event because in stable nuclei the
binding energy of the neutron and proton is, as a rule, nearly
the same and $Q(n, p) \leq 1$ MeV. As a result the proton decay of
neutron resonances in this case is a deep subbarrier process.
However, the cross sections of light and neutron-deficient
target nuclei for the reactions followed by the emission of
charged particles are higher and thus can play a notable role

in the formation of some isotopes and influence their abundance in nature.

A brief account of the results of investigations of (n,α) and (n, p) reactions being carried out for some years at our Laboratory will be given in the following sections.

2. The Investigation of the (n,α) Reaction with Resonance Neutrons

Some thirty years ago rather a large number of the experimentalists have turned their attention to a search of (n,α) reactions induced by thermal neutrons on medium and heavy nuclei. However their experiments have yielded only qualitative estimates on the compound states alpha decay characteristics. The quantitative results on α-widths of neutron resonances were obtained in Dubna for the target nuclei from Zn to Hf[1].

To stage experiments we naturally took favorable nuclei-the nuclei with large values of Q(n,α), that would result in the final nuclei with a magic or nearly magic number of neutrons (protons).

The atomic weight dependence of the ratio of experimental average α-particle widths to experimental total γ-widths is presented in Fig.1 in logarithmic scale. Actually the points indicate the upper limits of the ratios averaged over resonance cross sections of (n,α) and (n,γ) reactions in the case of s-neutron capture. The data for ^{33}S were taken from[2] and for ^{59}Ni from[3], the other data having been obtained in Dubna (see review[1]).

Fig.1 illustrates the influence of the neutron magic numbers N = 28, 50 and 82. From Fig.1 it is seen that some notable competition bet-

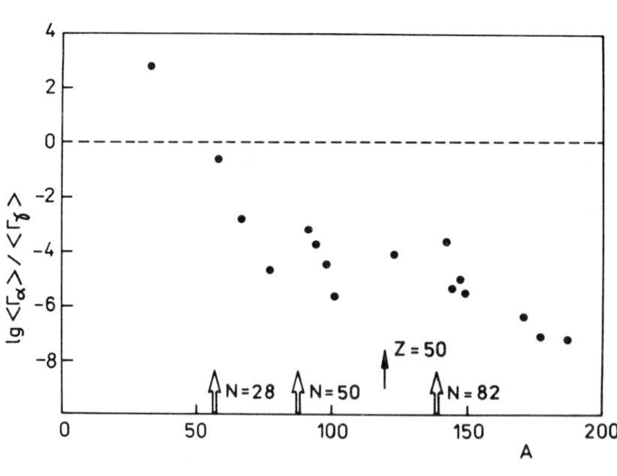

Fig.1

ween the (n, α) and (n, γ) reactions during the nucleosynthesis of stable nuclides can be expected only for the nuclei with
$A \leq 60-70$.

The analysis of the (n, α) data shows that today's experimental results can be described with the statistical model.
Partial α -particle widths obey the Porter-Thomas distribution,
that is the χ^2-distribution with a number of degrees of freedom $\nu = 1$, while for total α -particle widths one should use
$\nu_{eff} \geq 1$ to take into account the competition of different
α -decay channels.

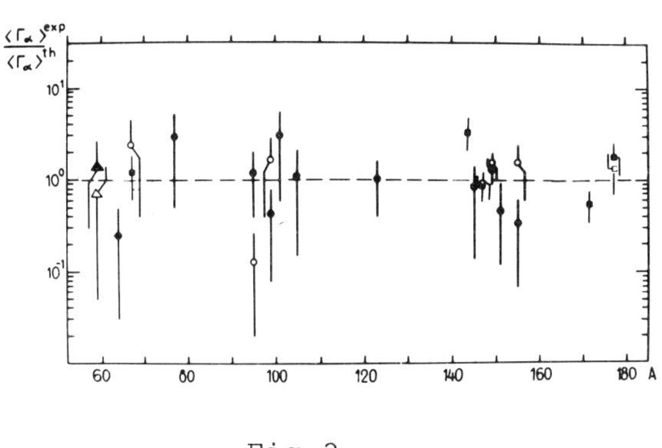

Fig.2

Fig.2 shows the ra-
tios of experimental
mean α -particle
widths (averaged, for
every nuclide, over
equal spin resonan-
ces) to calculated
ones in the cluster
model of α -decay developed by Kadmenski
and Furman[4]. The
errors at individual
points are mainly due to the small number of resonances over
which the averaging was performed.

The satisfactory agreement between the experimental and
theoretical data (see Fig.2)evidences for the validity of the
statistical description of the mean α-decay probabilities
of compound nuclear states (constancy of ratios) and for the
satisfactory choice of the parameters of the cluster model
(closeness of the ratio values to unity). To an accuracy of
a factor the ratios of α -particle widths shown in Fig.2 are
equal to the α -particle strength function of a "black" nucleus.

One can hope that the theory could make it possible to reproduce the average (n,α) reaction cross section of medium
and heavy nuclei (e.g. at neutron energies around 30 keV)
with an accuracy of a factor of 2 or 3. At the same time for
^{147}Sm there are some experimental indications of the nonsta-
tistical behaviour of the mean α-particle width averaged

over 200-500 eV bins, that is over 30-70 neutron resonances. It really demonstrates the systematic dependence of $\langle\Gamma_\alpha\rangle$ on E_n in the interval of several keV with the maximum different from minimum by about 3-5 times[5].

Measurements of α-decay spectra of low-lying resonances have unexpectedly revealed a beautiful effect of multilevel interference in the exit channel of decay. Asghar and Emsallem[6] noted that the measured partial thermal cross section for the ^{145}Nd$(n,\alpha)^{142}$Ce reaction followed by the α-particle transition to the first excited 2^+ state of the Ce nucleus, $\sigma(n,\alpha_1)$, is equal to the contribution (extrapolated by the Breit-Wigner formula) of the low-lying resonances[7] to the thermal cross section. At the same time the measured partial cross section for the α-particle transition to the ground state, $\sigma(n,\alpha_0)$ is at least 30 times smaller than the extrapolated contribution from the same low-lying resonances. This maybe interpreted as the indication of the presence of destructive interresonance interference in the (n,α) reaction. An analogous pattern was obtained later from the ^{67}Zn$(n,\alpha)^{64}$Ni reaction[8]. According to calculations[9] the suppression could not be explained by interference between only two resonances. This inter-resonance interference pattern is rather unusual, because for the ^{67}Zn and ^{145}Nd nuclei the distance between the resonances is considerably larger than the widths of the resonances ($D/\Gamma = 10$-100). Nevertheless in the case of the single channel emission of a secondary particle this effect should be taken into account while analysing the experimental data that point to the deviation of the energy dependence of the cross section from the "1/v law" in the inter-resonance region.

3. The (n, p) Reaction Induced by Resonance Neutrons

The proton emission after neutron capture is one of the most poorly investigated channels of the compound states decay. In recent years some laboratories started such experiments. Nevertheless it was the first channel in which parity violation in a neutron reaction followed by charged particles emission was discovered. The joint PINP (Gatchina) - FLNP (Dubna) experiment on the high intensity thermal neutron beam from the Gatchina reactor was staged to register the effect of P-odd correlation between neutron spin and proton emission directi-on[10].It gave: $a_{np}=-(1.5\pm0.34).10^{-4}$.The analysis of the results gives the possibility to estimate the matrix element value for the parity nonconservation interaction. Our value is w_{np} = = 0.06 eV, i.e. only ten times smaller than the single-particle estimate of the nucleon-nucleon interaction.

Investigations of the (n, p) reaction are, as a rule, performed on neutron-deficient radioactive target nuclei. They have essentially different binding energies of the neutron and proton and Q(n, p) grows up to several MeV. In the last years regular measurements of the (n,p) reaction were carried out at LANL (i.e.[12]) and our Laboratory.

The measurements of the neutron resonance proton widths have given an interesting result for some medium weight nuclei (A$\sim$ 20-40): the probability for the proton emission is comparable with that for γ -rays or neutrons or even exceeds that. This fact must be taken into account in calculation of the formation of different isotopes in nature due to the s- or r-processes of nucleosynthesis.

Popov and Rigol[13] made an attempt to compare the rates of β -decay and (n, p) reaction on ^{36}Cl and ^{22}Na in the supernovae explosion, in other words, to find the relation between:

$$\lambda \cdot n \qquad \text{and} \qquad n \int_0^\infty \sigma_{np}(E_n)\phi(E_n)dE_n,$$

where λ is the β -decay constant of the investigated isotope, n - is the isotope concentration, $\sigma_{np}(E_n)$ - the energy dependence of the (n, p) cross section, $\phi(E_n)$ - the energy distribution of the neutron flux ($n/cm^2 s$) from the supernovae explosion. If the energy distribution of neutrons obey the Maxwellian law at $10^9 K^\circ$ and the integral flux reaches $10^{24} n/cm^2 s$, then, for example, at $E_n \leq$ 10 keV (where the cross section of the ^{36}Cl(n, p)^{36}S reaction was measured) the integral flux will be up to $10^{22} n/cm^2 s$ and

$$\lambda = 7.7.10^{-14}s^{-1} \ll \int_0^{10\ keV} \sigma_{np}\phi\ dE_n = 4.10^{-3}s^{-1}.$$

So, in the supernovae explosion the nuclei ^{36}Cl "burnout" due to the (n, p) reaction but the β-decay $^{36}Cl \rightarrow ^{36}Ar$ is of no importance. Consequently, from the modern point of view it is not necessary to search in meteorites for an anomaly in the abundance of the ^{36}Ar isotope/14/. It might be necessary to take into account the $^{36}Cl(n, p)^{36}S$ reaction contribution into ^{36}S abundance to understand "the overproduction of ^{36}S" pointed to in/15/.

Note, that the "substitution" of β-decay for the (n, p) reaction may also affect the neutrino flux balance, because the latter process is a neutrino-free one.

References

/1/ Balabanov N P et al. 1990 Sov.J.Part.Nucl. 21(2) 131-152
/2/ Wagemans C, Weigmann H and Barthelmy R 1987 Nucl.Phys. A496 497-506
/3/ Harvey J A 1976 Proc.Intern.Conf. on Interaction of Neutron with Nuclei (Lowell) 143-162
/4/ Kadmenski S G and Furman W I 1975 Sov.J.Part.Nucl. 6 469
/5/ Popov Yu P 1990 Capture Gamma-Ray Spectroscopy (AIP Conf Proc 238 Ed by R Hoff) 298-307
/6/ Asghar M and Emsallem A 1982 Neutron Capture Gamma-Ray Spectroscopy and Related Topics (Ed. by T von Egidy, IOP Publishing, Bristol) 455
/7/ Popov Yu P et al. 1971 Sov.J.Nucl.Phys. 13 913
/8/ Gledenov Yu M et al. 1985 Sov.J.Nucl.Phys. 41 831
/9/ Vtjurin V A et al. 1987 Sov.J.Nucl.Phys. 45 1291
/10/ Antonov A et al. 1984 Sov.J.JETP Lett 40 209
/11/ Gledenov Yu M et al. 1985 Z.Phys.A 322 685
/12/ Koehler P E and O'Brien H A 1988 Proc.Intern.Conf.Nucl. Data for Science and Technology (Mito, Japan Ed. by S Igarasi JAERI) 1101-1105
/13/ Popov Yu P and Rigol J 1985 Preprint JINR P15-85-497 Dubna
/14/ Hudson B 1977 Meteoritic 12 258
/15/ Wagemans C, Weigmann H and Barthelemy R 1988 Capture Gamma-Ray Spectroscopy (Ed. by K Abrahams and P van Asshe, IOP, Bristol) 739

Investigation of the ^{14}N(n,p)^{14}C reaction by neutron filter technique

J Andrzejewski

Dept. of Experimental Methods of Nuclear Physics, University of Lodz, Poland

Yu M Gledenov, Yu P Popov, V I Salatski, P V Sedyshev, Li Ho Bom

Joint Institute for Nuclear Research, Dubna, Ruusia

V A Pshenichnyj

Institute for Nuclear Research, A.S. of Ukraina, Kiev

Abstract. In view of importance of the ^{14}N(n,p)^{14}C reaction for s-process nucleosynthesis, the cross section at neutron energy of 24.5 keV has been directly measured. The two-grid ionization chamber for spectrometry analysis was employed. Reported recently by Koehler et al., the cross section of this reaction is a factor of approximately 2.5 larger than the result reported by Brehm et al. at neutron energy of 25 keV. In the Institute for Nuclear Research in Kiev, to investigate the reaction, independent method was applied, with the use of interference Fe filter. Our preliminary result for the cross section of the ^{14}N(n,p)^{14}C reaction at the neutron energy of 24.5 keV is $\sigma_{np} = (2.11 \pm 0.17)$ mb.

For a long time two reactions - $^{13}C(\alpha,n)^{16}O$ and $^{22}Ne(\alpha,n)^{25}Mg$ have been considered the basic neutron sources for s-process nucleosynthesis in stars. We deal, however, with inaccuracy in the evaluation of neutron beam originated in these reactions. It results from considerable amount of ^{14}N nuclei produced in the CNO hydrogen - burning phase. This, in connection with the large $^{14}N(n,p)^{14}C$ cross section for thermal neutrons shows a strong neutron beam absorbtion by the ^{14}N nuclides. Until recently the value of this cross section for star energies has been obtained from inverse reaction [1] as well as from extrapolation of the cross section obtained for E_n = 150 keV [2] towards lower energies, with regard to thermal cross section [3]. The precise knowledge of the value is needed to define the role of ^{14}N as a neutron "poison".

The absolute cross section of the $^{14}N(n,p)^{14}C$ reaction measured by Brehm et al.[4] for neutron energy of 25 keV and 52.4 keV is 0.81±0.05 mb and 0.52±0.06 mb, respectively. These results converted to astrophysical reaction rate are a factor of approximately 3 smaller then the rate obtained in the inverse reaction [4]. On the basis of the data Brehm et al. concluded that $^{14}N(n,p)^{14}C$ reaction plays less important role in neutron capture, and that smaller amount of ^{14}C and larger amount of ^{15}N produced in the nuclear transformation cycle should be taken into account. Direct measurement of the $^{14}N(n,p)^{14}C$ reaction cross section was also made in Los Alamos by Koehler and O'Brien [5]. The time of flight method used for neutron spectroscopy made it possible to measure the reaction at energy interval from 0.61 meV to 34.6 keV. Cross section data reported by Koehler et al. are in agreement with the data obtained from measurement made in inverse reaction [1,6] with the detailed balance method. They are a factor of 2.5 higher than the data obtained by Brehm et al. for E_n = 25 keV and 52.4 keV. Because of divergence, at the Kiev Institute, the measurement of the reaction was made similarly to the measurement of the ^{7}Be(n,p)^{7}Li reaction [7].

As in Koehler's experiment, the target was adenine evaporated in vacuum - $C_5H_5N_5$. Its thickness was 96 μg/cm^2 and the area - 48 cm^2. In the measurement a double ionization chamber with two grids for spectrometry and separation of the charged particles of different types was used. The cross section was performed on the neutron beam obtained with the use of interference Fe filter. The filter lets through neutrons of the average energy E_n = 24.5 keV, at an energy spread of 2 keV. The filter consisted of 300 mm ^{56}Fe, 400 mm Al and 70 mm S. Owing to Al and S addition, passing of undesirable neutron groups through Fe layer is eliminated. Neutron beam intensity was $4.4*10^6$ n/s cm^2 [8]. Neutron distribution for Fe filter is presented in fig.1.

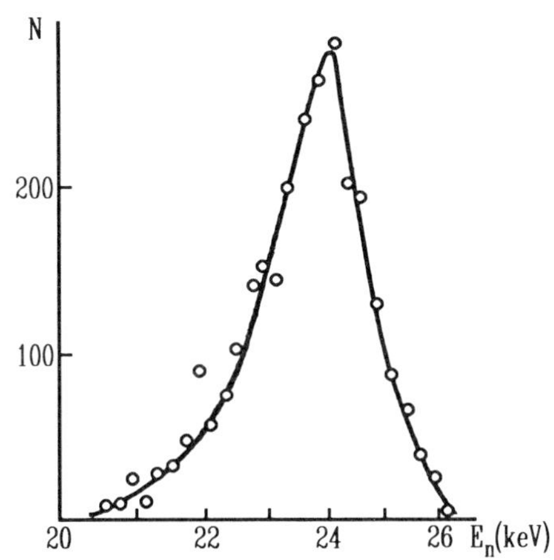

Fig.1. The transmission of neutrons through the Fe(300 mm) + Al(400 mm) + S(70 mm)

The cross section for the ^{14}N(n,p)^{14}C reaction was determined by comparing it with the cross section of the ^{6}Li(n,α)^{3}H reaction measured in the same conditions. Preliminary estimation of the cross section value of this reaction for neutron energy E_n = 24.5 keV is σ_{np}=(2.11±0.17)mb. This value is approximately a factor of 2.6

larger than Brehm's result, and in agreement with Koehler's result . Further measurements, aiming at the study of the emission angle of protons with respect to the normal of the cathode, will be made in the Kiev reactor, since it is supposed that as in ^{6}Li for $E_n = 25$ keV, this dependence is not isotropic.

References

[1] Gibbons J H and Maclin R L 1959 *Phys. Rev.* **114** 571

[2] Alley W E and Lesseler R M 1972 *Nucl. Data* **Tab.A11** 621

[3] Hanna G C, Primeau D B and Tunnicliffe P R 1961 *Can. J. Phys.* **39** 1784

[4] Brehm K, Becker H W, Rolfs C, Trautvetter H P, Käppeler F and Ratyński W 1988 *Z. Phys. A* **330** 167

[5] Koehler P E and O'Brien H A 1989 *Phys. Rev. C* **39** 1655

[6] Sanders M 1956 *Phys. Rev.* **104** 1434

[7] Andrzejewski J, Gledenov Yu M, Popov Yu P, Salatski V I and Pshenichnyj V A 1991 *Z. Phys. A* **340** 105

[8] Andrzejewski J, Vertebnyj V P, Vo Kim Thanh, Vtiurin V A, Kiriluk A L and Popov Yu P 1988 *Sov. Nucl. Phys.* **48** 20

Measurement of the ^{14}N(n$_{th}$,p)^{14}C reaction cross-section

S Druyts

CEC, JRC, Central Bureau for Nuclear Measurements, B-2440 Geel, Belgium

C Wagemans*, S Pommé

Nuclear Physics Laboratory of the University, B-9000 Gent , Belgium

P Geltenbort

Institut Laue-Langevin, F-38042 Grenoble, France

H P Trautvetter

Ruhr-University, D-4630 Bochum, Germany

Abstract.   The ^{14}N(n$_{th}$,p)^{14}C reaction cross-section has been measured at the high flux reactor of the I.L.L. (Grenoble) using various ^{14}N-samples . A preliminary value of (1.8 ± 0.1) b was obtained , confirming the generally adopted value . The astrophysical importance of this result is discussed.

1. Introduction

The ^{14}N(n,p)^{14}C-reaction is important in the astrophysical s-process for several reasons. Not only does it consume neutrons needed in the s-process, but it also produces ^{14}C and plays a role in a decrease in the production of ^{15}N. Therefore, the Maxwellian averaged cross-section value of this reaction in the energy-range corresponding to s-process temperatures (kT ≈ 20 keV) has to be determined . Until now, this has been done in three ways :

Firstly, it was determined by combining the evaluated thermal cross-section value (1.83 b [1]) with the high-energy data of the inverse reaction ^{14}C(p,n)^{14}N (Gibbons et al [2], Sanders [3]) and the ^{14}N(n,p)^{14}C-reaction above 150 keV neutron energy (Johnson et al [4]).

Secondly, the Maxwellian averaged cross-section value was measured directly, using a ^{7}Li(p,n)-neutron source with an almost Maxwellian neutron flux distribution (Brehm et al [5]). It turned out to be a factor of 2.5 lower than the value obtained using the first method.

---

*National Fund for Scientific Research, Brussels, Belgium

Thirdly, it was determined by measuring the shape of the $^{14}N(n,p)^{14}C$-cross-section between near thermal energy and ~35 keV neutron energy and normalizing it to the previously mentioned evaluated thermal cross-section value (Koehler et al [6]).The Maxwellian averaged cross-section value calculated in this way is in good agreement with the value obtained using the first method.

The thermal cross-section value might play a key role in the striking discrepancy mentioned above, since furthermore, there are no resonances in the $^{14}N(n,p)^{14}C$-reaction below ~0.5 MeV. Therefore, we performed a series of $^{14}N(n_{th},p)^{14}C$-cross-section measurements under well-defined experimental conditions .

2. Experimental method and measurement

The measurements were performed at the Institut-Laue-Langevin in Grenoble (France). We made use of the curved neutron guide H22, which provides an intense thermal neutron flux of ~5 . $10^8 n/cm^2 s$ with a negligible background of epithermal and fast neutrons and γ-rays.This enabled a very clean detection of the 0.6 MeV protons emitted in the $^{14}N(n_{th},p)^{14}C$-reaction.

A surface barrier detector (15μ, 200 mm^2) was mounted out of the neutron beam in a vacuum chamber installed at the end of the neutron guide. Its energy-calibration was done by means of the $^{10}B(n,\alpha)^7Li$, $^6Li(n,\alpha)t$ and $^{35}Cl(n,p)^{35}S$-reactions. Especially the latter reaction was very appropriate since protons are released of practically the same energy as in the $^{14}N(n,p)^{14}C$-reaction.

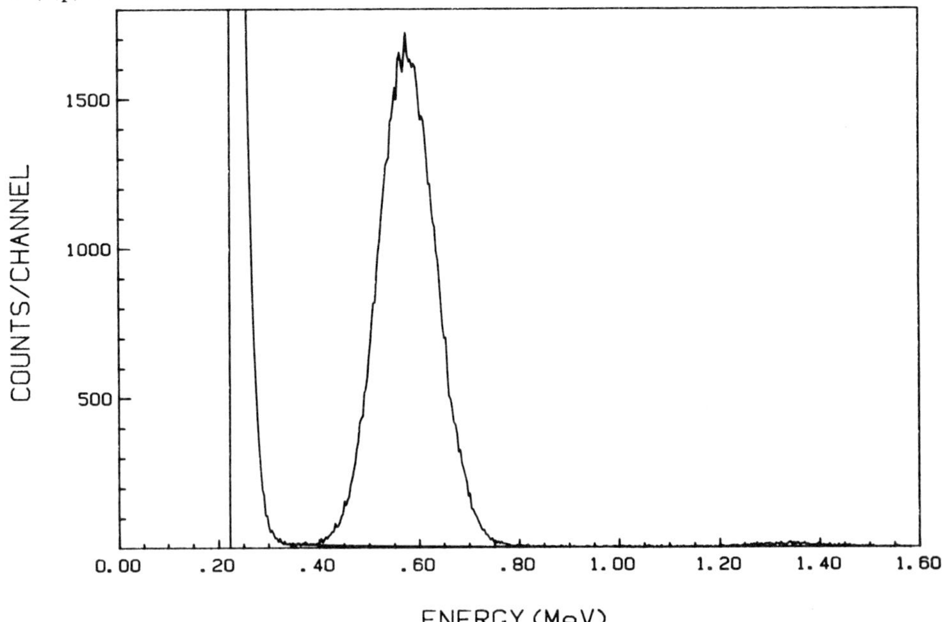

Fig 1 Spectrum of the protons , emitted in the $^{14}N(n,p)^{14}C$-reaction

The thermal neutron flux was determined using the $^{235}U(n_{th},f)$-reaction on a well-determined ^{235}U-sample, strictly maintaining the same detection geometry. Here a thermal fission cross-section value of 584.25 b was adopted, as given in the ENDF-B6 evaluated data file .

A variety of ^{14}N-samples was used (air, polyimide, adenine) .The amount of ^{14}N in the samples was (2.77 ± 0.06) µg/cm^2 for the polyimide $(C_{22}H_{10}O_5N_2)$ and (36.1 ± 1.5) µg/cm^2 for the adenine $(C_5H_5N_5)$. In fig. 1 a typical $^{14}N(n_{th},p)^{14}C$-spectrum is shown, obtained with an adenine-sample. The background-to-signal ratio is estimated to be about 2%. About one third of the background is due to the tail of the $^7Li^*$-peak of parasitic $^{10}B(n,\alpha)^7Li$-reactions. Measurements on air at several pressures (1 Torr, ~5 . 10^{-3} Torr) were used to estimate the background contribution due to the air still present in the chamber during the measurements using the solid samples. It turned out to be roughly one third of the background . This background component can be further reduced by improving the vacuum in the detection chamber.

By combining the (n_{th},p)-counting-rates (after background correction) with the corresponding $^{235}U(n_{th},f)$ counting-rates, $^{14}N(n_{th},p)^{14}C$ cross-section values of (1.75 ± 0.06) b and (1.82 ± 0.08) b respectively were measured for the polyimide-sample and the adenine-sample . Since further measurements are planned on additional samples and under improved background conditions, we recommend a preliminary cross-section value of (1.8 ± 0.1) b based on the present experiments.

Table 1:   *Survey of the $^{14}N(n_{th},p)^{14}C$-cross-section measurements*

Reference	$^{14}N(n_{th},p)^{14}C$ cross-section(b)	Detector	Normalization reaction
Hanna et al [7]	1.83 ± 0.03	ionization chamber	$^{197}Au(n,\gamma)$
Cuër et al [8]	1.91 ± 0.10	nuclear emulsion	$^{10}B(n,\alpha)$
Coon and Nobles [9]	1.90 ± 0.05	proportional counter	$^{10}B(n,\alpha)$
Batchelor and Flowers [10]	1.72 ± 0.04	ionization chamber	$^{10}B(n,\alpha)$
This work (preliminary value)	1.8 ± 0.1	surface barrier	$^{235}U(n,f)$

3. Discussion

In table 1 the $^{14}N(n_{th},p)$-reaction cross-section values of previous measurements are compared to the value obtained in this work. In the latter a better resolution and lower background are obtained than in the other measurements, since use was made of a cleaner neutron-beam and a more suitable detector. Moreover, different kinds of samples were used. By doing more measurements on additional samples and in yet improved background conditions, we hope to reduce the error on the cross-section value to 2 %. The preliminary value for the $^{14}N(n_{th},p)^{14}C$-cross-section obtained in this work is in good agreement with the generally adopted evaluated value of 1.83 b.

Since Koehler et al [6] made use of this value to normalize their cross-section data in the keV-region, this work confirms the Maxwellian-averaged $^{14}N(n,p)^{14}C$-cross-section value they obtained. This would contradict the value obtained by Brehm et al [5].

An astrophysical consequence is that the $^{14}N(n,p)^{14}C$-reaction might be more important as a neutron poison and in the production of ^{14}C. Moreover, there may be a bigger decrease in the production of ^{15}N, since there is more competition with the $^{14}N(\alpha,\gamma)^{18}F(n,\alpha)^{15}N$-reaction chain.

References

[1]   Mughabghab S F , Divadeenam M and Holden N E 1981 *Neutron resonance pameters and thermal cross-sections* (New York : Academic Press)

[2]   Gibbons H and Macklin R L 1959 *Phys. Rev.* **114** 571-80

[3]   Sanders R M 1956 *Phys. Rev.* **104** 1434-40

[4]   Johnson C H and Barschall H 1950 *Phys. Rev.* **80**   818-23

[5]   Brehm K , Becker H W , Rolfs C , Trautvetter H P , Käppeler F and Ratynski W 1988 *Z. Phys.* **A330** 167-72

[6]   Koehler P E   and O'Brien H A 1989   *Phys. Rev.* **C39** 1655-7

[7]   Hanna G C , Primeau D and Tunnicliffe P R 1961   *Can. J. Phys.* **39** 1784-806

[8]   Cuër P , Lonchamp J P   and Gorodetzky S 1951 *J. phys. radium* **12** S   6-7

[9]   Coon J H and Nobles R A 1949 *Phys. Rev.* **75**   1358-61

[10] Batchelor R   and Flowers B H 1949 *AERE -Report*   N/R 370

Determination of the ^{32}S(n_{th},α)^{29}Si and ^{33}S(n_{th},α)^{30}Si reaction cross-sections

C Wagemans[*]
Nuclear Physics Laboratory, University, B-9000 Gent, Belgium

P D'hondt
SCK/CEN, B-2400 Mol, Belgium

R Brissot[**]
Institut Laue-Langevin, F-38042 Grenoble, France

Abstract. The ^{32}S(n_{th},α)^{29}Si and ^{33}S(n_{th},α)^{30}Si reaction cross-sections have been determined at a very clean thermal neutron beam of the I.L.L.,yielding values of $\leq$ 0.5 mb and (0.11 $\pm$0.01)b, resp. Also the ^{33}S(n_{th},$\gamma\alpha$)^{30}Si reaction has been observed with a cross-section of (0.5$\pm$0.1) mb in the α-energy region from 2.45 to 2.75 MeV.

1. Introduction

The thermal neutron induced (n,α) cross-sections of ^{32}S and ^{33}S are poorly known. For ^{33}S(n_{th},α)^{30}Si e.g., the experimental cross-section values vary between 52 and 180 mb, which is unfortunate since it is known since long that this reaction is of importance for nucleosynthesis calculations in carbon stars, especially to investigate the ^{36}S production [1]. Moreover, there is a renewed interest of astrophysicists for ^{33}S(n,α)^{30}Si data for the interpretation of isotopic anomalies in meteoric SiC grains [2,3], esp. with respect to the ^{30}Si concentration.

---

* National Fund for Scientific Research
** Present address: Institut des Sciences Nucléaires, F-38026 Grenoble, France

Although mainly the keV - neutron energy region is of importance for astrophysical calculations, the thermal cross-section is a normalization value for these data [4], hence equally important. So a series of accurate ^{32}S and ^{33}S(n_{th},α) cross-section measurements was performed at the high flux reactor of the Institut Laue-Langevin at Grenoble.

2. Experimental method

2.1. Samples

The main problem for the preparation of the samples (as well as for the measurements) was the sublimation of the sulphur. Fig. 1 shows the sublimation temperature of sulphur as a function of the pressure. Obviously, sulphur sublimates at room temperature with improving vacuum.

This problem has been overcome by developing a new technique [5] in which elemental sulphur is evaporated onto a cooled formvar substrate (49 µg/cm^2) mounted on an aluminium frame. After the evaporation, the sulphur layer is covered with a thin formvar foil (17µg/cm^2) mounted on an identical frame. The foils attract each other in such a way that no air remains in between. The formvar - sulphur - formvar sandwich obtained in this way proved to be very stable under vacuum.

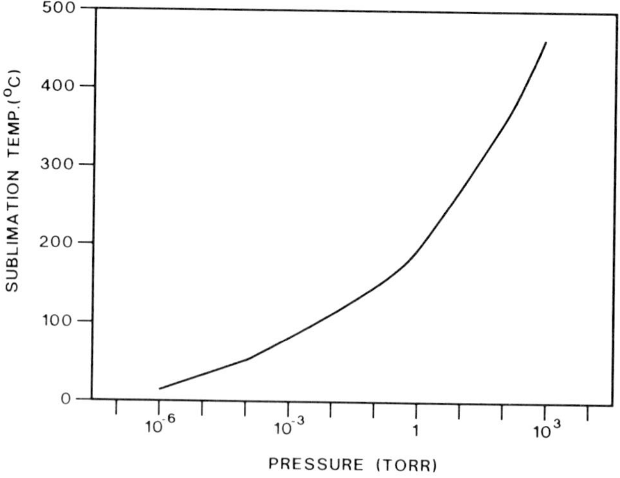

Fig. 1   Sublimation temperature of sulphur as a function of the pressure

Three natural (95.0 % ^{32}S, 0.75 % ^{33}S) and three enriched (59.15 % ^{33}S) sulphur samples were prepared with thicknesses varying between 13 and 51 µg/cm^2 (Table 1). The thicknesses were first determined by a combination of spectrophotometry and α-energy loss measurements in the sandwich [5]. Subsequently, ^{33}S(n$_{th}$,α)- counting rate comparisons were performed for all samples. After the experiments finally, a destructive chemical analysis (combustion) was done. In this way, reliable information on the number of sulphur atoms in the layer could be obtained.

2.2. Experimental procedure

The measurements were performed at the end of the neutron guide H22D installed at the high flux reactor of the Institut Laue-Langevin. At this position, a thermal neutron flux of 5x10^{8}n/cm^2.s is available with very low background, the fast neutron and γ-ray intensity of the direct beam being reduced by a factor 10^6.

A sample was mounted in the centre of a vacuum chamber installed in the neutron beam. The reaction particles were detected in a low geometry with a surface barrier detector mounted outside the beam. The energy calibration of the detector(s) was done using the well-known energies of the ^{10}B(n,α)^{7}Li and ^{6}Li(n,α)t reaction products and of α-particles emitted in the radio-active decay of ^{234}U and ^{235}U. The neutron flux calibration was done by replacing the sulphur sample by a well-known ^{235}U sample, strictly maintaining the same geometry, and adopting a value of 584.3 b for the ^{235}U(n$_{th}$,f) cross-section [6].

Table 1: Sample characteristics

Sample material	^{33}S enrichment (%)	Thickness (µg S / cm^2)
Natural S	0.75	13.1 ± 0.7 [a]
"	0.75	23.6 ± 1.2 [a]
"	0.75	24.1 ± 1.2 [a]
Enriched S	59.15	13.5 ± 0.7 [a]
"	59.15	38.4 ± 3.3 [b]
"	59.15	50.6 ± 2.6 [a]

determined (a) via combustion (b) via non-destructive analysis

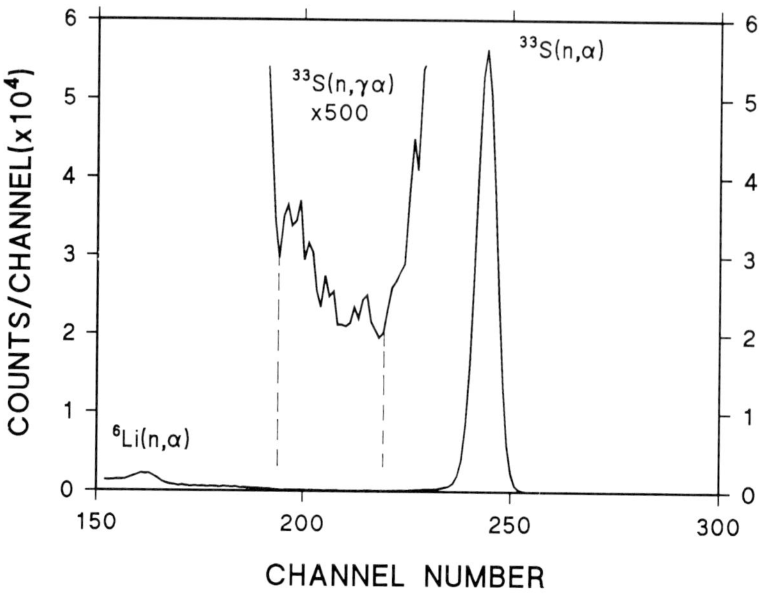

Fig. 2   Pulse-height spectrum of the $^{33}S(n_{th},\alpha)$-particles

The energy of the particles emitted in the $^{32}S(n_{th},\alpha)^{29}Si$ and $^{33}S(n_{th},\alpha)^{30}Si$ reactions is 1.34 and 3.08 MeV, resp.   Two types of measurements were performed: (i) using a 100μm thick, 450 mm² large surface barrier detector (ii) using a telescope set-up in anti-coincidence, consisting of a 30μm thick, 300 mm² large ΔE-detector and a 100μm thick, 450 mm² large E-detector. This telescope detector was used to reduce the background and to eliminate 2.73 MeV tritons produced via $^{6}Li(n,\alpha)t$ reactions in impurities in the sample. Background determinations were performed in both detector configurations.

3. Results

First the $^{32}S(n_{th},\alpha)^{29}Si$ reaction was investigated using the natural sulphur samples. The 1.34MeV $^{32}S(n_{th},\alpha)$- particles turned out to be masked by the tail of a peak of 1.48 MeV $^{10}B(n_{th},\alpha_1)$- particles due to neutron interaction with boron impurities in the sample (σ_{th} =3840 b). So for this cross-section only an upper-limit of 0.5 mb could be determined.

For the study of the $^{33}S(n_{th},\alpha)^{30}Si$ reaction all six samples mentioned in Table 1 were used. Fig. 2 shows the pulse-height spectrum obtained with the 38.4 μg/cm² enriched sulphur layer and the telescope detector. Combining all experimental data, a $^{33}S(n_{th},\alpha)$ cross-section of (0.11 ± 0.01)b is obtained.

Also the $^{33}S(n_{th},\gamma\alpha)^{30}Si$ reaction has been observed (see fig.2) in the measurements with the enriched sulphur samples using the telescope detector. The different sulphur layer thickness of the samples used results in different signal-to-background ratios in the energy region of interest for the $(n,\gamma\alpha)$-reaction. So a combination of these measurements enables a fair background-correction, resulting in a value of (0.5 ± 0.1)mb for the $^{33}S(n_{th},\gamma\alpha)$ - cross-section in the energy region from 2.45 to 2.75 MeV

Table 2 :   Experimental and evaluated values for the ^{32}S and
$^{33}S(n_{th},\alpha)$ reaction cross-sections (mb)

Reference	$^{32}S(n_{th},\alpha)^{29}Si$	$^{33}S(n_{th},\alpha)^{30}Si$
Münnich [7]	6.8 ± 1	180 ± 80
Benisz [8]	3.9 ± 0.5	151 ± 22
Harris [9]	< 6.8	52 ± 16
Asghar [10]		140 ± 30 [a]
Mughabghab [11]	7 ± 4	190 ± 80
ENDF - B6	3	
JENDL - 3	7	169
This work	≤ 0.5	110 ± 10

a) preliminary result

4. Discussion

Table 2 gives a survey of the experimental and evaluated values for the ^{32}S and $^{33}S(n_{th},\alpha)$ cross-sections.

The experimental $^{32}S(n_{th},\alpha)$ cross-section values [7,8] were obtained using the nuclear emulsion technique, which has a fairly low energy resolution. Both results are significantly higher than our value, which is likely due to the detection of $^{10}B(n,\alpha_1)$ background particles with comparable energy.

For $^{33}S(n_{th},\alpha)$, four experimental values have been reported [7-10] varying between 52 and 180 mb. The main reason for these large fluctuations is probably the low sublimation temperature of sulphur, resulting in a wrong assessment of the number of sulphur atoms irradiated. Also the evaluated values clearly need to be revised.

The ^{33}S(n$_{th}$,$\gamma\alpha$)-reaction has been observed previously by Asghar and Emsallem [10], who reported a preliminary value of 1.75mb in the α-energy region from 2.25 to 2.75 MeV. Even taking into account the larger energy interval considered, this value is at least a factor of two higher than the present result, which is probably due to the neglection of the background correction.

As already mentioned, it is especially the keV-neutron energy region which is of importance to nucleosynthesis calculations. A quantity often used in such calculations is the Maxwellian averaged cross-section at kT = 30 keV. For ^{33}S(n,α)^{30}Si, three experimental data sets are available, yielding the following Maxwellian averaged cross-sections: Auchampaugh et al. [12] (690 $\pm$ 170)mb; Wagemans et al.[4] (217 $\pm$ 20)mb and Steiniger [13] (181 $\pm$ 10)mb. The result of Auchampaugh et al.[12] was based on a thermal ^{33}S(n,α) cross-section of 140 mb, so the adoption of the value of 110 mb obtained in this work would yield a revised Maxwellian average of 540 mb, which already reduces the discrepancy.

References

[1]   Howard W, Arnett W, Clayton D and Woosley S 1972 *Astrophys. J.* **175** 201

[2]   Gallino R, Busso M, Picchio G and Raiteri M 1990 *Nature* **348** 298

[3]   Ireland T, Zinner E and Amari S , *Astrophys. J. (Lett.)*, in print

[4]   Wagemans C, Weigmann H and Barthélémy R 1987 *Nucl. Phys.* **A469** 497 (renormalized to σ_{th} = 110 mb)

[5]   Geerts K, Van Gestel J and Pauwels J 1985 *Nucl. Instr. & Meth.* **A236** 52

[6]   Peelle R and Condé H 1988 *Proc. Int. Conf. on Nuclear Data for Science and Technology (S. Igarasi, ed.), Saikon Publishing Co,* 1005

[7]   Münnich F 1958 *Z. Phys.* **153** 106

[8]   Benisz J , Jasielska A and Panek T 1965 *Acta Physica Polonica* **28** 763

[9]   Harris 1966 *Dissertation Abstracts* **B27** 2077

[10]   Asghar M and Emsallem A 1978 *Proc. Third Int. Symp. on Neutron Capture Gamma-Ray Spectroscopy, Brookhaven, N.J.* 549

[11]   Mughabghab S, Divadeenam M and Holden N 1981 *Neutron Cross-Sections, Academic Press, N.Y.*

[12]   Auchampaugh G, Halperin J, Macklin R and Howard W 1975 *Phys. Rev.* **C12** 1126

[13]   Steiniger R 1988 *Diplomarbeit, University of Karlsruhe, Germany*

$^{17}O(n,\alpha)^{14}C$—Bottle-neck for primordial nucleosynthesis?

H Schatz, F Käppeler, P E Koehler[1]**, M Wiescher**[2]**, H-P Trautvetter**[3]

Kernforschungszentrum Karlsruhe, Institut für Kernphysik III, Postfach 3640, 7500 Karlsruhe 1, Federal Republic of Germany

Abstract. The cross section of the $^{17}O(n,\alpha)^{14}C$ reaction, which is of potential importance for nucleosynthesis in inhomogeneous big bang models, was investigated in the energy range from 10 keV to 250 keV. Compared to previous measurements our results are about 100 times lower around 130 keV but show reasonable agreement in other energy regions. In terms of the astrophysical reaction rate, this discrepancy gives rise to a correction of up to a factor 2. This results imply that the $^{17}O(n,\alpha)^{14}C$ reaction plays indeed an important role in inhomogeneous big bang models and that it could hamper significantly the synthesis of heavier elements.

1. Introduction

In the standard model of primordial nucleosynthesis homogeneous and isotropic conditions during the whole synthesis-process are assumed. In such a scenario neutron densities are fairly low as determined by the weak freeze-out neutron to proton ratio and, therefore, the synthesis of heavier elements is efficiently blocked by the instability gaps at A=5 and A=8. In contrast, more recent models account for inhomogenities, possibly induced by a first order quark–hadron phase transition at about 100 MeV. Under certain conditions these models arrive at significantly different primordial isotopic abundances (Applegate *et al* 1987, Alcock *et al* 1987, Malaney and Fowler 1988, Applegate *et al* 1988 ,Terasawa and Sato 1989, 1990, Kajino *et al* 1990, Mathews *et al* 1990, Sato and Terasawa 1991, Thielemann *et al* 1991 and Kawano *et al* 1991). This is mainly due to the fact, that the resulting low density regions provide for sufficiently high neutron densities to bridge the instability gaps at A=5 and A=8 by neutron induced reactions, thus leading to the production of significant amounts of heavier elements.

[1] Los Alamos National Laboratory, Los Alamos, USA
[2] University of Notre Dame, Notre Dame, USA
[3] Ruhr-Universität Bochum

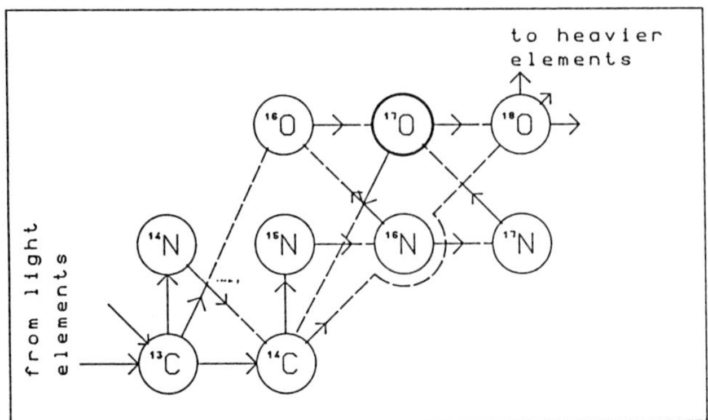

Figure 1. Mass flow in the CNO region for an inhomogeneous big bang model (simplified). The length of the arrows indicates the netto mass flow in a logarithmic scale (from Thielemann *et al* 1991).

The importance of the $^{17}O(n,\alpha)^{14}C$ reaction was pointed out first by Applegate *et al* 1988. These authors found a significant hindrance of the mass flow to heavier elements due to cycling of material between ^{14}C and ^{17}O by this reaction. Recent network calculations by Thielemann *et al* 1991 confirm indeed that the largest part of the mass flow to heavier elements passes through ^{17}O (see figure 1). Of crucial interest is the branching at ^{14}C since the $^{14}C(\alpha,\gamma)^{18}O$ reaction is the most important bypass for ^{17}O. Recent cross section measurements and theoretical investigations for the various possible reaction channels of ^{14}C (Wiescher *et al* 1990, Kawano *et al* 1991, Beer *et al* 1992, Görres *et al* 1992) show that the reactions $^{14}C(p,\gamma)^{15}N$ and/or $^{14}C(d,n)^{15}N$ are by far dominant, both leading to ^{17}O by the reaction sequence $^{15}N(n,\gamma)^{16}N(\beta^-)^{16}O(n,\gamma)^{17}O$, or alternatively by $^{15}N(n,\gamma)^{16}N(n,\gamma)^{17}N(\beta^-)^{17}O$. Another important reaction sequence bypassing ^{14}C, but leading to ^{17}O is $^{13}C(\alpha,n)^{16}O(n,\gamma)^{17}O$. Thus, the cross section of the $^{17}O(n,\alpha)^{14}C$ reaction determines the mass flow to elements heavier than oxygen, which possibly could act as an r-process seed according to Applegate *et al* 1988.

We investigated experimentally the cross section of the $^{17}O(n,\alpha)^{14}C$ reaction and calculated the astrophysical rate for this reaction to determine its influence on the mass flow in the CNO region.

2. Experiment

The $^{17}O(n,\alpha)^{14}C$ cross section was measured in the energy range between 10 keV and 250 keV relative to the thermal value. Neutrons were produced at the Karlsruhe 3.75 MV Van de Graaff accelerator via the $^{7}Li(p,n)^{7}Be$ reaction. A 10 μm thick target made of metallic Lithium was used to obtain a broad neutron spectrum with an average energy of 52 keV to cover the energy range from 10 keV to 80 keV. At higher energies neutron spectra with a FWHM of about 25 keV were produced by means of 2.5 μm thick Li targets. The resulting neutron flux was in the range of $1 \cdot 10^8$ s^{-1}cm^{-2}. The shape of the spectra was calculated according to the data of Ratynski and Käppeler 1988 and Liskien and Paulsen 1975, respectively. The energy in beam direction was determined with the time-of-flight method, operating the accelerator in pulsed mode.

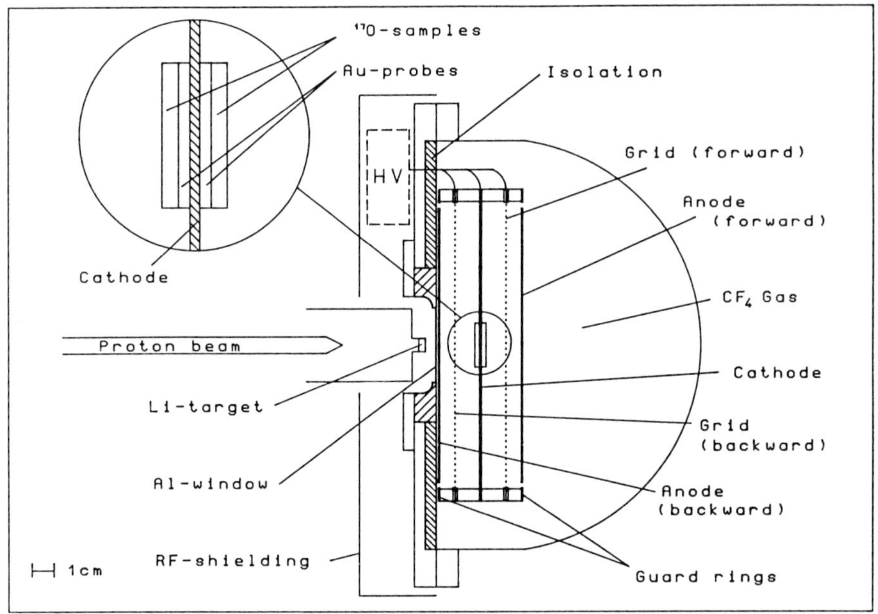

Figure 2. Experimental setup

The ^{17}O samples were made by anodic oxidation of niobium foils using water with an isotopic enrichment in ^{17}O of 37.5 %. Two samples were manufactured with a thickness of $3.7 \cdot 10^{17}$ and $2.4 \cdot 10^{17}$ ^{17}O atoms/cm^2, respectively. Both samples were irradiated simultaneously and the emitted α–particles were detected with a gridded, gas filled twin ionization chamber (see figure 2), which is described elsewhere (Schatz 1992). One sample was mounted in forward direction and the other one in backward direction with respect to the neutron beam. In this way, the reaction products were detected with nearly 4π solid angle for eliminating angular distribution effects. The ionization chamber was operated with CF$_4$, a gas pressure of 150 mbar to stop the 1.5 MeV α–particles, and at a voltage of 750 V. Figure 3 shows the α–spectrum taken with the thinner sample at an average neutron energy of 123 keV in the cross section minimum.

The neutron flux was determined via the activation technique by irradiation of two gold foils placed under each ^{17}O sample. The induced activity was counted after the measurement. The neutron flux was calculated using the gold capture cross sections of Macklin 1982 normalized to the value of Ratynski and Käppeler 1988.

The efficiency of the ionization chamber was determined with a Monte Carlo calculation to account for self absorption effects.

The sample thickness was measured with the same method as described above in a neutron beam of the Munich research reactor FRM I, using the well known thermal cross section of the ^{17}O(n,α)^{14}C reaction (Hanna *et al* 1961).

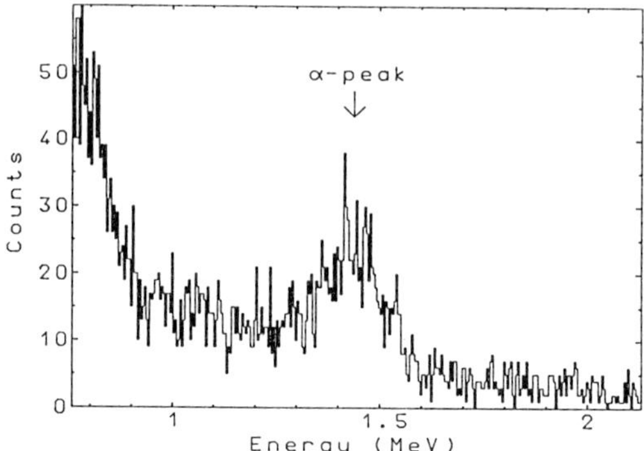

Figure 3. Spectrum for the thinner sample at an average neutron energy of 123 keV yielding the low cross section of 1.8 mb.

3. Results

There is only one other direct experimental determination of this cross section, which was recently carried out by Koehler and Graff 1991.

The comparison (figure 4) shows reasonable agreement in the energy range up to 100 keV, but large discrepancies in the energy range between 100 keV and 250 keV. Despite of the low cross section value at 123 keV measured in this work, the signal/background ratio was still good enough (see Figure 3) to exclude the resonance postulated by Koehler and Graff. Comparison with the data obtained from the cross section values of the inverse reaction (Sanders 1956) via detailed balance are in reasonable agreement with the present work in the energy range above 150 keV. The higher values of Sanders at lower energies could possibly result from background problems.

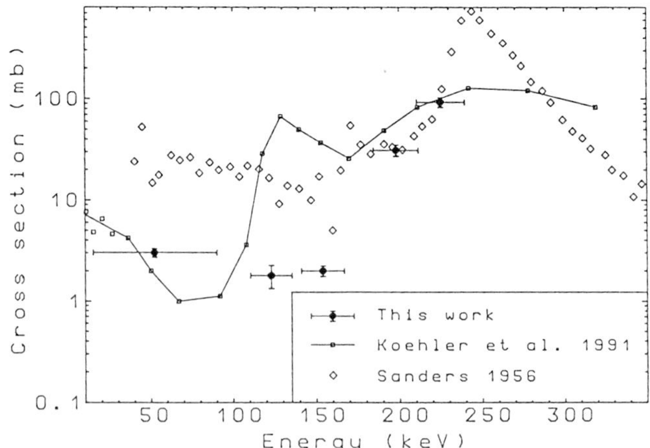

Figure 4. Comparison of cross sections with previous measurements. The energy bars in the data of this work represent the FWHM of the used neutron spectrum.

By operating the ionization chamber in twin mode, the forward/backward ratio of the emitted α–particles could be determined. The results show significant deviations from isotropy indicating strong interference between the resonances.

4. Astrophysical implications

The astrophysical reaction rate for ^{17}O(n,α)^{14}C was calculated from the new cross section data for the temperature range between $0.1 \cdot 10^9$ K and 10^9 K being relevant in inhomogeneous big bang scenarios (figure 5).

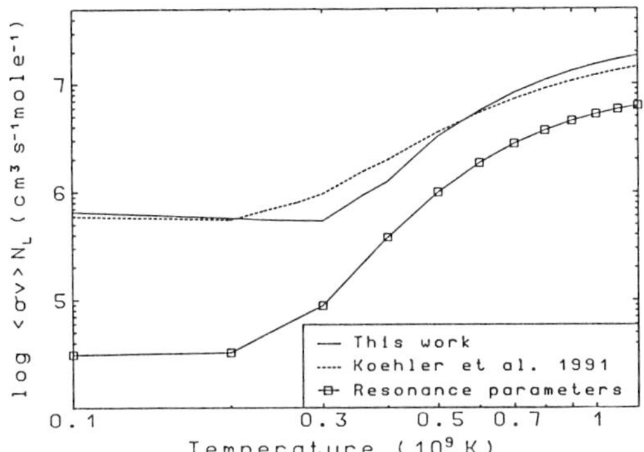

Figure 5. Astrophysical reaction rate for ^{17}O(n,α)^{14}C.

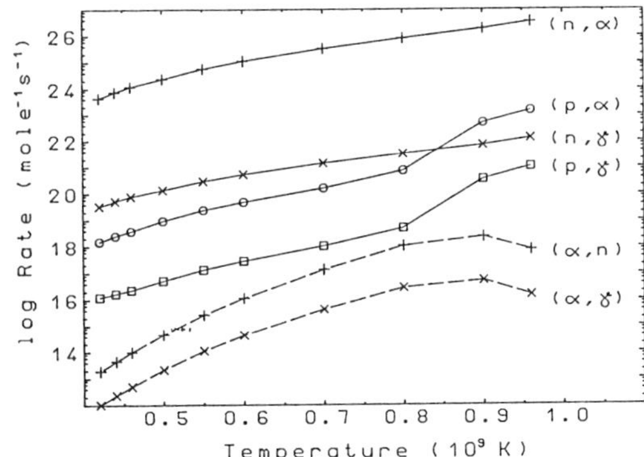

Figure 6. The rates of all important reaction channels of ^{17}O in an inhomogeneous big bang model.

For energies below 100 keV the values of Koehler and Graff 1991, normalized to our data, were used and for energies above 250 keV the data of Sanders 1956 were adopted. The maximum difference of the resulting reaction rate with respect to the data of Koehler and Graff 1991 occurs at $0.3 \cdot 10^9$ K, where the new rate is 2 times smaller. Furthermore, the new rate is about 2 to 3 times higher than predicted by calculations based on known resonance parameters (Weinman and Silverstein 1958).

Reaction rate calculations for all important reaction channels of ^{17}O, using the light particle abundances of Thielemann *et al* 1991 and the cross sections of Caughlan and Fowler 1988, show that the new rate for the ^{17}O(n,α)^{14}C reaction is about 10^4 stronger than all relevant reactions leading to heavier elements (figure 6). Therefore it is most probable that this reaction hampers the synthesis of heavier elements than oxygen, and, hence, the production of nuclei, which could act as r-process seeds.

References

Alcock C R, Fuller G M and Mathews G J 1987 *Ap. J.* **320** 439

Applegate J H, Hogan C J and Scherrer R J 1987 *Phys. Rev. D.* **35** 1151

Applegate J H, Hogan C J and Scherrer R J 1988 *Ap. J.* **329** 572

Beer H, Wiescher M, Käppeler F, Görres J and Koehler P E 1992 *Ap. J.* **387** 258

Caughlan G R and Fowler W A 1988 *Atomic Data Nucl. Data Tables* **40** 291

Görres J, Graff S, Wiescher M, Azuma R E, Barnes C A and Wang T R 1992 *Nucl. Phys.* in print

Hanna G C, Primeau D B and Tunnicliffe P R 1961 *Can. J. Phys.* **39** 1784

Kajino T, Mathews G J and Fuller G M 1990 *Ap. J.* **364** 7

Kawano L H, Fowler W A, Kavanagh R W and Malaney R A 1991 *Ap. J.* **372** 1

Koehler P E and Graff S M 1991 *Phys. Rev. C.* **44** 2788

Liskien H and Paulsen A 1975 *Atomic Data and Nucl. Data Tables* **15** 57

Macklin R L 1982 Private communication

Malaney R A and Fowler W A 1988 *Ap. J.* **333** 14

Mathews G J, Meyer B S, Alcock C R and Fuller G M 1990 *Ap. J.* **358** 36

Ratynski W and Käppeler F 1988 *Phys. Rev. C.* **37** 595

Sanders R M 1956 *Phys. Rev.* **104** 1434

Sato K and Terasawa N 1991 *Phys. Scripta* **T36** 60

Schatz H 1992 M.A. thesis Universität Karlsruhe

Terasawa N and Sato K 1989 *Phys. Rev. D.* **39** 2893

Terasawa N and Sato K 1989 *Ap. J.* **362** L47

Thielemann F K, Applegate J H, Cowan J J and Wiescher M 1991 *Nuclei in the Cosmos* ed H Oberhummer and C Rolfs (Berlin: Springer Verlag)

Weinman J A and Silverstein E A 1958 *Phys. Rev.* **111** 277

Wiescher M, Görres J and Thielemann F K 1990 *Ap. J.* **363** 340

Neutron capture cross-sections and gamma-ray strength functions

M Uhl

Institut für Radiumforschung und Kernphysik University of Vienna, Austria

J Kopecky

Netherlands Energy Research Foundation ECN Petten, The Netherlands

Abstract. Predictions of capture cross sections by Hauser-Feshbach calculations should be based on accurate gamma-ray strength functions and level densities. In the mass region $135 \le A \le 205$ gamma-ray strength function models for dipole radiation were evaluated by demanding a simultaneous reproduction of average resonance data, capture cross sections and gamma-ray production spectra. The best results for E1 are obtained with a strength derived from a "generalized Lorentzian" with a non-zero limit as the energy tends to zero and a width dependent on energy and deformation. We recommend for M1 to derive the strength from a spin-flip giant resonance of Lorentzian shape.

1. Introduction

Neutron capture cross sections are an essential input for theories of stellar nucleosynthesis [1]. Most of the cross sections required for s-process models concern stable targets and thus can be measured. However, in some case where cross sections for unstable targets and/or targets in excited states are required one has to resort to nuclear reaction theory. In the energy and mass region of interest for astrophysical applications the capture process is dominated by the compound nucleus mechanism. Though in some specific mass regions also valence capture [2] may be important most capture cross section calculations rely on the statistical model alone. We follow this practice and employ in the code MAURINA [3] the Hauser-Feshbach theory [4] in the formulation of Moldauer [5]. The mass region considered is $135 \le A \le 205$. The average cross section for the capture of a neutron (energy ε_n) with emission of one primary gamma-ray (energy ε_γ) leaving the residual nucleus in states with spin J', parity Π' around excitation energy E' is given in terms of the transmission coefficients $T_{nj\pi}(\varepsilon_n)$ and $T_{\gamma XL}(\varepsilon_\gamma)$ for neutrons and photons, respectively, and the level density $\rho_\gamma(E'J'\Pi')$ by

$$< \frac{d\sigma_{n\gamma}^{(1)}(\varepsilon_n, E'J'\Pi')}{d\varepsilon_\gamma} > = \rho_\gamma(E'J'\Pi') \frac{\pi}{k_n^2} \sum_{J\Pi} g_J \sum_{j\pi} \sum_{XL} \frac{T_{nj\pi}(\varepsilon_n) \, T_{\gamma XL}(E - E')}{\Theta_\gamma(EJ\Pi) + \Theta_n(EJ\Pi)} \, W_{j\pi, XL}^{J\Pi}, \quad (1)$$

where g_J, k_n and $W_{j\pi, XL}^{J\Pi}$ stand for the statistical factor, the wave number and the width fluctuation correction factor, respectively. The quantity $\Theta_\gamma(EJ\Pi)$ represents the sum over all open channels of the gamma-ray transmission coefficients:

$$\Theta_\gamma(EJ\Pi) = \int_0^E dE' \sum_{XL} \sum_{J'\Pi'} T_{\gamma XL}(E - E')\rho_\gamma(E'J'\Pi'), \qquad (2)$$

where $E = \varepsilon_n + B_n$, J and Π represent total energy, angular momentum and parity, respectively; an analogous expression holds for $\Theta_n(EJ\Pi)$. For the calculation of gamma-ray spectra and capture activation cross sections Eq. (1) must be supplemented by an appropriate treatment of gamma-ray cascades. At lower excitation energies the level densities in Eqs. (1) and (2) are replaced by the actual levels. The quantity $\Theta_\gamma(EJ\Pi)$ and the average resonance spacing $<D(EJ\Pi)>$ define the average total radiation width $<\Gamma_\gamma(EJ\Pi)> = (<D(EJ\Pi)>/2\pi)\,\Theta_\gamma(EJ\Pi)$.

For capture cross calculations one needs transmission coefficients and level densities. At present these auxiliary quantities cannot be determined from first principles. Rather, they require additional models and underlying parameters. The neutron transmission coefficients are calculated within the optical model. In general capture cross sections do not depend very critically on optical model parameters. Photon transmission coefficients $T_{\gamma XL}(\varepsilon_\gamma)$ for multipole type XL are related to the corresponding gamma-ray strength function $f_{XL}(\varepsilon_\gamma)$ by

$$T_{\gamma XL}(\varepsilon_\gamma) = 2\pi\varepsilon_\gamma^{2L+1}f_{XL}(\varepsilon_\gamma). \qquad (3)$$

In the order of their significance for capture calculations the most important multipoles are E1, M1 and E2. Models for gamma-ray strength functions and level densities are described in sections 2 and 3, respectively, and their impact on capture calculations is discussed in section 4.

The accuracy of the calculations is mainly restricted by the uncertainties of gamma-ray transmission coefficients and level densities. These uncertainties stem from the model parameters and, more seriously, from insufficiencies of the models themselves. In the restricted energy region of interest for nucleosynthesis an adequate representation of capture cross sections can be achieved with various combinations of models. All products $\rho_\gamma(E'J'\Pi')T_{\gamma XL}(E - E')$ yielding similar values of the $\Theta_\gamma(EJ\Pi)$ result in comparable capture cross sections (see Eq. (1)), irrespective of the absolute values and the energy dependence of the factors. In the valley of stability the data bases are sufficient to allow a meaningful interpolation of the corresponding model parameters. That means, on the other hand, that from the analysis of capture cross sections alone little can be learned on the correctness of the underlying gamma-ray strength function models. To test these models one has to consider their impact on other observables, too. In section 4 we evaluate in favourable cases different strength function models for E1 - and M1 radiation by requiring a simultaneous reproduction of average resonance capture (ARC) data, capture cross sections and gamma-ray production spectra. Models which fulfill this demand and, in addition are also supported by theory, are obviously of interest from the nuclear physics point of view. However, we hope that they are also valuable for astrophysical applications: Extrapolations away from the line of stability should be more reliable for realistic models.

2. Gamma-ray strength function models

In the following we adopt for the (downwards) strength functions the definition of Bartholomew et al. [6] and neglect any dependence on the spins.

The simplest model, the *single particle model (SP)* [7], results in an energy independent strength function. We employed this model with a strength of 1 WU/MeV for the (unimportant) contributions of M2, E3 and M3 radiation and occasionally also for M1 radiation but a with strength adjusted to ARC data (*ad-*

justed single particle model (ASP)). However, one should note that an energy independent strength is at variance with a finite energy weighted sum rule.

General reaction theory relates the downwards strength function to the photoabsorption cross section. If the latter is dominated by a giant resonance (GR) of Lorentzian shape Brink's hypothesis [8] leads to a strength function derived from a *standard Lorentzian (SLO)*

$$f_{XL}^{SLO}(\varepsilon_\gamma) = 26 \times 10^{-8} \left[\text{mb}^{-1} \text{MeV}^{-2} \right] \frac{1}{2L+1} \sigma_0 \Gamma_0 \frac{\varepsilon_\gamma^{3-2L} \Gamma_0}{(\varepsilon_\gamma^2 - E_0^2)^2 + \varepsilon_\gamma^2 \Gamma_0^2} \, . \tag{4}$$

The Lorentzian parameters (σ_0 E_0 Γ_0) respectively stand for peak cross section, energy and width of the resonance. As a consequence of Brink's hypothesis that each excited state has built on it a GR similar to that of the ground state f_{XL}^{SLO} depends on the transition energy only. The SLO model was employed for E1, M1 and E2 radiation. The giant E1 dipole resonance (GDR) parameters were taken from Ref. 9. For M1 we employed a spin-flip GR as proposed by Bohr and Mottelson [10] with simple prescriptions for E_0 and Γ_0 in terms of the mass number A: $E_0 = 41 A^{-1/3}$ (MeV) and $\Gamma_0 = 4$ (MeV). The peak cross section was adjusted to reproduce available strength function data or systematics [11]. As E2 contributions are rather small we chose the global prescriptions for the Lorentzian parameters proposed in Refs. 12 and 13.

For the dominating E1 radiation we considered some refinements on the basis of microscopic models. The theory of Fermi liquids [14] predicts an energy and temperature dependent width of the GDR : $\Gamma(\varepsilon_\gamma, T) = \beta(\varepsilon_\gamma^2 + 4\pi T^2)$, where β is a normalization factor. The first term reflects the spreading of particle-hole states into more complex states, while the second one accounts for collisions of quasi-particles. The temperature T refers to the energy of the absorbing state and can be calculated within a level density model. Kadmenskij [15] proposed to choose the constant β so that by $\Gamma(\varepsilon_\gamma \to E_0, T \to 0) = \Gamma_0$ compatibility with photoabsorption data is guaranteed. If this expression

$$\Gamma(\varepsilon_\gamma, T) = (\Gamma_0/E_0^2)(\varepsilon_\gamma^2 + 4\pi T^2) \tag{5}$$

replaces in the fraction of Eq. (4) the quantity Γ_0, the resulting strength $f_{E1}^{ELO}(\varepsilon_\gamma)$ is derived from a *Lorentzian with energy dependent width (ELO)* , as originally proposed by Dover et al. [16]. The *model of Kadmenskij et al. (KAD)* [15] includes quasi-particle fragmentation and correctly considers the low energy limit of the dipole polarization operator. The resulting E1 strength function reads

$$f_{E1}^{KAD}(\varepsilon_\gamma, T) = 8.68 \times 10^{-8} \left[\text{mb}^{-1} \text{MeV}^{-2} \right] \sigma_0 \Gamma_0 \frac{E_0}{(\varepsilon_\gamma^2 - E_0^2)^2 + \varepsilon_\gamma^2 \Gamma_0^2} \, 0.7 \Gamma(\varepsilon_\gamma, T) \, , \tag{6}$$

where again $\Gamma(\varepsilon_\gamma, T)$ is defined by Eq. (5). By adding to $f_{E1}^{ELO}(\varepsilon_\gamma, T)$ the $\varepsilon_\gamma \to 0$ of $f_{fE1}^{Kad}(\varepsilon_\gamma, T)$ limit Kopecky and Chrien [17] obtained an E1 strength function derived from a *generalized Lorentzian (GLO)*

$$f_{E1}^{GLO}(\varepsilon_\gamma, T) = 8.68 \times 10^{-8} \left[\text{mb}^{-1} \text{MeV}^{-2} \right] \sigma_0 \Gamma_0 \left\{ \frac{\varepsilon_\gamma \Gamma(\varepsilon_\gamma, T)}{(\varepsilon_\gamma^2 - E_0^2)^2 + \varepsilon_\gamma^2 \Gamma^2(\varepsilon_\gamma, T)} + 0.7 \frac{\Gamma(\varepsilon_\gamma = 0, T)}{E_0^3} \right\} \, , \tag{7}$$

which for initial energies around the neutron separation energy is very close to the result of the KAD model. It is important to realize the different low energy behaviour of these models. The E1 strength derived from the models SLO and ELO vanishes for $\varepsilon_\gamma \to 0$ while the KAD and the GLO model result in a non-zero limit. The dependence on the temperature $T = T(E - \varepsilon_\gamma)$ of the E1 strength from the models ELO, KAD and GLO implies a partial breakdown of Brink's hypothesis. Note that also the more sophisticated E1 models require the same GDR parameters

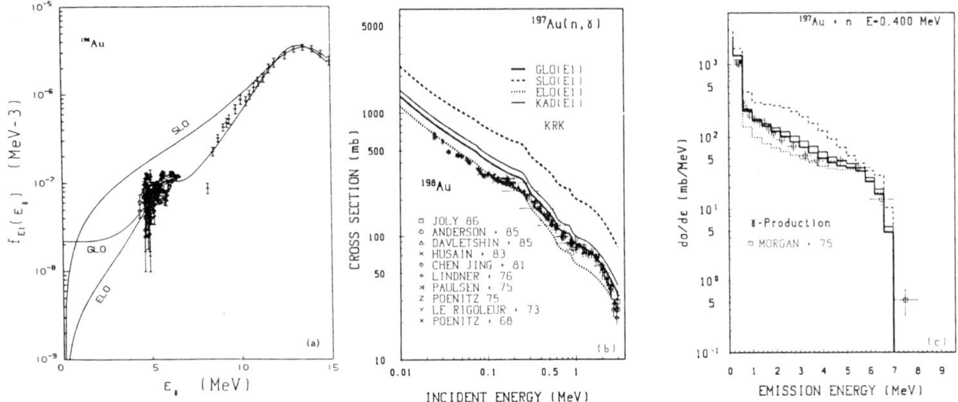

Figure 1. : (a) E1 strength functions according to models described in text are compared to ARC - and photoabsorption data for ^{198}Au . The resulting capture cross sections (b) and gamma-ray production spectra (c) are confronted with experimental data.

as the simple SLO model. For a fragmented GDR, as encountered in deformed nuclei, we used in Eqs. (4), (6) and (7) the sum of two corresponding terms.

Fig. 1a shows for ^{198}Au most of the discussed E1 strength function models compared to ARC and photoabsorption data; an E1 pygmy resonance is included in case of the ELO - and the GLO model [18]. While the SLO model is incompatible with the ARC data one cannot decide in favour of ELO or GLO on basis of these data alone. Additional information offer the results of model calculations which critically depend on $f_{E1}(\varepsilon_\gamma)$ at low energies where ARC data are lacking. Similar results for some lighter spherical nuclei are discussed in Ref. 18. However, for strongly deformed nuclei the SLO model reproduces the ARC data while the GLO model fails. This is illustrated in Fig. 3a for ^{156}Gd. For a unified description in case of deformed and spherical nuclei we proposed in Ref. 19 an empirical correction to Eq. (5)

$$\Gamma(\varepsilon_\gamma, T) = \left[k_0 + (1 - k_0)(\varepsilon_\gamma - \varepsilon_0^\gamma) / (E_0 - \varepsilon_0^\gamma)\right](\Gamma_0/E_0^2)(\varepsilon_\gamma^2 + 4\pi^2 T^2) . \tag{8}$$

The constant k_0 can be used to reproduce the ARC data around the energy ε_0^γ. In order to guarantee $\Gamma(\varepsilon_\gamma \to E_0, T \to 0) = \Gamma_0$ the new normalization constant β is energy dependent. As $k_0 = 1$ yields again Eq. (5) we can consider Eq. (8) as a generalization of Eq. (5). For $k_0 > 1$ the width is enhanced compared to the outcome of Eq. (5). Consequently, if Eq. (8) with $k_0 > 1$ is used in Eqs. (6) and (7) we refer to the resulting models as "enhanced" and change the acronyms to EELO, EKAD and EGLO. Fig. 3a displays the E1 strength from the EGLO model with $k_0 = 4$ and $\varepsilon_0^\gamma = 4.5 MeV$. The deformed and transitional nuclei considered so far required values of k_0 between 1 and 5 and energies ε_0^γ between 4 and 5 MeV.

3. Level density models

Most calculations were performed with two semi-empirical models with parameters determined by the average spacing of s-wave resonances D_0 and the number of low lying levels. They are the *backshifted Fermi gas model (BSGF)* [20] and the model by *Kataria, Ramamurthy and Kapoor (KRK)* [21]; the latter is supplemented by a conventional pairing shift and a constant temperature portion according to Gilbert and Cameron [22]. The KRK model explicitly accounts for shell effects in terms of the shell correction to the binding energy and thus its parameters can more

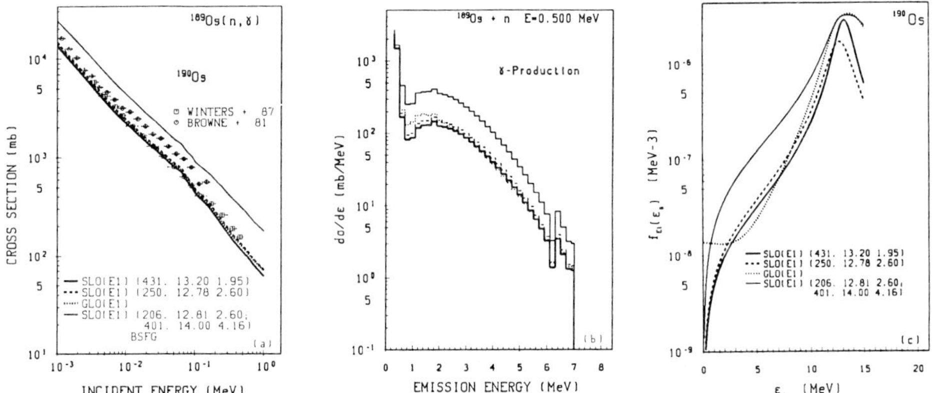

Figure 2. : The ^{189}Os(n, γ) cross section (a) and the gamma-ray production spectrum (b) calculated with the E1 strength functions displayed in (c); in (a) and (c) the Lorentzian parameters ($\sigma_0\,E_0\,\Gamma_0$) are indicated.

reliably be extrapolated to nuclei without D_0 data. This is a more difficult problem for the BSFG parameters which implicitly are affected by shell effects.

4. Results and discussion

In this section we discuss the impact of the different E1 and M1 strength function models described before. The emphasis is on E1. Unless stated otherwise, the model for M1 is SLO. The level density model employed is indicated on the plots. In general we do not attempt to improve the fit to the data by adjusting model parameters; so the agreement of experiment and theory conveys a feeling for the accuracy of model calculations.

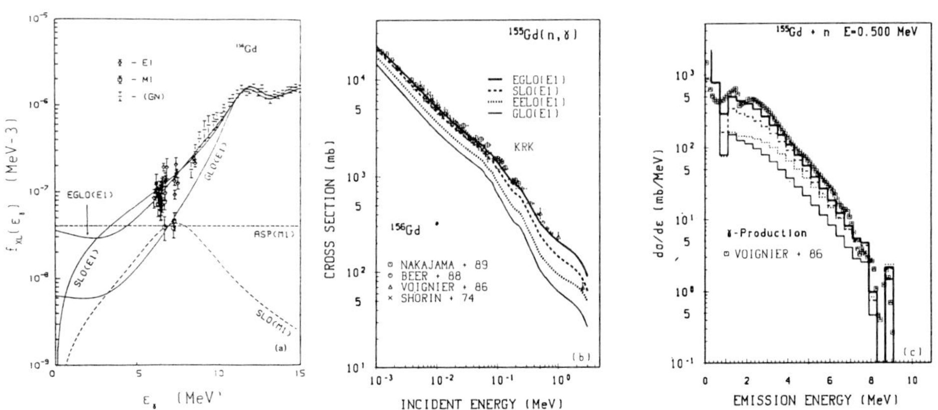

Figure 3. : (a) E1 and M1 strength functions according to models described in text are compared to ARC - and photoabsorption data (NG) for ^{156}Gd . The resulting capture cross sections (b) and gamma-ray production spectra (c) are confronted with experimental data.

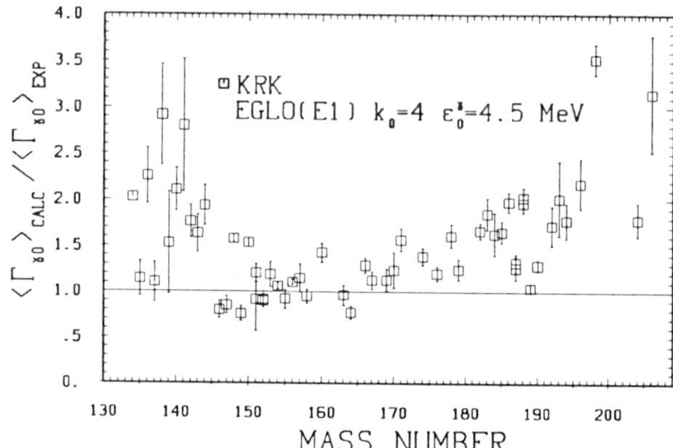

Figure 4. : Ratio of calculated and experimental average total s-wave radiation width. The calculations are based on the EGLO model for E1 with the parameters $k_0 = 4$ and $\varepsilon_0 = 4.5$ MeV.

We start with the ^{197}Au(n, γ) reaction. The gold capture cross is an important standard. Moreover there are sufficient experimental data to evaluate strength function models. The SLO model clearly overpredicts the ARC data (Fig. 1a), the capture excitation function (Fig. 1b) as well as the gamma-ray production spectrum (Fig. 1c). While the capture excitation function is compatible with ELO and GLO the gamma-ray production spectrum favours the GLO model. In spite of the quite different low energy behaviour of the E1 strength, the shapes of the resulting capture excitation functions are very similar. The slope of the spectrum is more sensitive to f_{E1} at small energies; the best reproduction is achieved with the finite $\varepsilon_\gamma \to 0$ limit characteristic for the GLO - or the KAD model. The capture cross sections with the GLO model are somewhat higher then the experimental data; also the Maxwellian-averaged cross section recently published by Ratynski and Käppeler [23] is overpredicted by about 20%. By similar arguments we found in Ref. 19 support for the GLO - or the KAD model in some other spherical nuclei: ^{94}Nb, ^{106}Pd and ^{144}Nd. According to Figs. 1b und 1c there is little difference between the results obtained with the GLO - and the KAD model. Therefore we use in the following the GLO - or the EGLO model only.

The next example, the reaction ^{189}Os(n, γ), illustrates the weak dependence of capture excitation functions on f_{E1} models. This reaction has recently been studied in the context of the Re-Os chronometer for stellar nucleosynthesis [24], [25]. The capture excitation function (Fig. 2a) is well described by two calculations using the SLO model with two different sets of Lorentzian parameters, both adjusted to fit the data. The SLO model with the Lorentzian parameters reproducing the photonuclear data [9] (actually they are for two Lorentzians) overpredicts the cross sections while the GLO model with the same parameters works quite well. The E1 strength functions, which fit the capture excitation functions, here also produce similar spectra (Fig.2b) in spite of their different energy dependence (Fig. 2c).

Fig. 3 shows that for the compound nucleus ^{156}Gd the GLO model fails to reproduce not only the ARC data but also the capture cross section and the gamma-ray spectrum as well. A good fit to all observables is achieved with the enhanced generalized Lorentzian (EGLO) with $k_0 = 4.0$ and $\varepsilon_0^\gamma = 4.5$ MeV. It is interesting to note that here the SLO model does a good job, too. Very similar results we obtained also for other strongly deformed targets as ^{151,153}Eu which are of interest for the s-process branching at A = 151 and for ^{156,157}Gd.

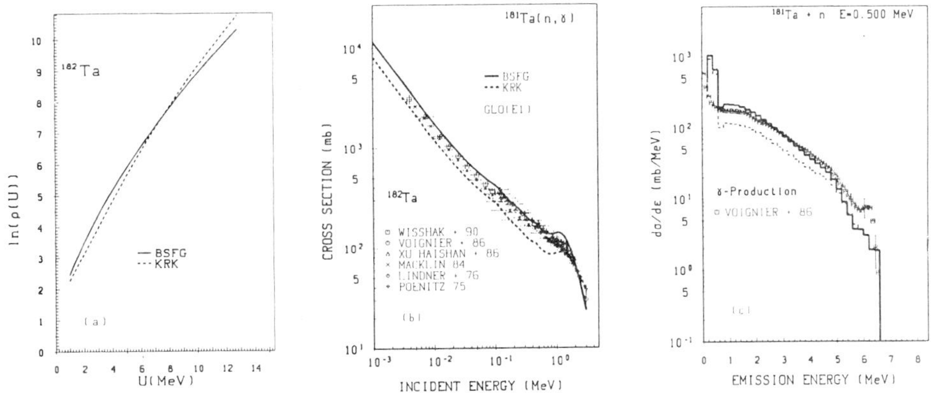

Figure 5. : The impact of different level density models (a) on capture cross sections (b) and gamma-ray production spectra (c) for ^{181}Ta(n, γ) .

The enhancement factor k_0 seems to be related to the deformation. This is illustrated in Fig. 4 by means of the average s-wave radiation width $<\Gamma_{\gamma 0}>$ in the mass region $135 \leq A \leq 205$. A good reproduction of the experimental data [26] [27] in the region of maximum deformation $150 \leq A \leq 170$ is achieved using the EGLO model with $k_0 = 4.0$ and $\varepsilon_0' = 4.5$ MeV. For lighter and heavier nuclei the calculation starts to overpredict the data. We are just trying to find a simple expression which gives the enhancement in terms of the deformation or a related quantity. However, it turns out difficult to determine a clear trend of k_0 in the region of transitional nuclei (see also Fig. 4). Here the enhancement is smaller and therefore more strongly afflicted by additional uncertainties as those caused by the level densities, the M1 contribution and the presence of other reaction mechanisms. Unfortunately nuclei with accurate gamma-ray production spectra and ARC data are scarce in this mass region.

As cross sections and spectra represent the sum of E1 and M1 contributions the conclusions on E1 strength function models depend to a certain extent on the M1 strength prescription employed. In Ref. 19 we showed for ^{197}Au(n, γ) that a the combination ELO(E1) + ASP(M1) gives nearly the same results as GLO(E1) + SLO(E1). This holds also for the deformed Gd isotopes if the enhanced models EELO and EGLO are employed for E1. The energy independent M1 strength plays a similar role as the finite $\varepsilon_\gamma \to 0$ limit of the GLO - or EGLO model (see Fig. 3a). In spite of the popularity of the ASP model for M1 we prefer the SLO model which gives a finite energy weighted sum rule. Moreover the finite $\varepsilon_\gamma \to 0$ is backed by a microscopic model [15]. The low energy behaviour of the M1 strength, in particular for deformed nuclei, may also be influenced by the collective isovector M1 ("scissors") mode [28].

The semi-empirical level density models employed represent an additional source of uncertainty. Even if the same information on low lying levels and resonance spacing is used for the parameterization different models result in appreciably different cross sections. As a characteristic example we display in Fig. 5 the results for ^{181}Ta(n, γ) . Fig. 5a shows the responsible difference in the in the total level densities occuring a few MeV below the energy E where the product $\rho_\gamma(E'J'\Pi')T_{\gamma xL}(E-E')$ in Eq. (1) is large. In this energy range the relevant part of the KRK model is usually the constant temperature portion. This unpleasant uncertainties are the price for using simple semi-empirical models. They hopefully can be reduced in future by more sophisticated level density models.

5. Conclusion

In this contribution we concentrated on capture calculations in the mass range $135 \leq A \leq 205$. Here we propose to use for the strength of M1 the SLO model and for that of the dominating E1 radiation the EGLO model with ε_0' around 4.5 MeV and k_0 around 1 for spherical and around 4 for strongly deformed nuclei, respectively. In the region of transitional nuclei the situation is complicated. More experimental data sensitive to strength function models as ARC - and gamma-ray spectra measurements would help to better establish the trend of the enhancement k_0 . We hope that strength function models which are supported by microscopic theory and various independent experimental data will offer a sound basis for extrapolations away from stability.

References

[1] Käppeler F, Beer H and Wisshak K1989 *Rep. Prog. Phys.* **52** 945-1013
[2] Allen B J and Musgrove A R de L 1978 *Adv. Nucl. Phys.*, **10** 129-95
[3] Uhl M *unpublished*
[4] Hauser W and Feshbach H 1952 *Phys. Rev.* **78** 366-73
[5] Moldauer P A 1980 *Nucl. Phys* **A344** 185-95
[6] Bartholomew G A et al. 1973 *Adv. Nucl. Phys.* **10** 229-325
[7] Blatt M and Weisskopf V F 1952 *Theoretical Nuclear Physics* (New York: Wiley)
[8] Brink D M 1955 *Ph. D. Thesis* (Oxford University)
[9] Dietrich S S and Berman B L 1988 *At. Nucl. Data Tables* **38** 199-338
[10] Bohr A G and Mottelson B R 1975 *Nuclear Structure* Vol.II (London: Benjamin)
[11] Kopecky J and Uhl M *Proc. IAEA Specialists' Meeting on the Measurement, Calculation and Evaluation of Photon Production Cross Sections, Febr. 5-7, 1990, Smolenice, CSFR, INDC(NDS)-238* 103-12
[12] Speth J and van der Woude A 1981 *Rep. Prog. Phys.* **44** 719-86
[13] Prestwitch W V, Islam M A and Kennet T J 1984 *Z. Phys.* **A315** 103-11
[14] Pines D and Noziers D 1966 *The Theory of Quantum Liquids* (New York: Benjamin)
[15] Kadmenskij S G, Markushev V P and Furman V I 1983 *Sov. J. Nucl. Phys.* **37** 165-8
[16] Dover C B, Lemmer R H and Hahne F J W 1972 *Ann. Phys. (N.Y.)* **70** 458-507
[17] Kopecky J and Chrien R E 1987 *Nucl. Phys.* **A468** 285-300
[18] Kopecky J and Uhl M 1990 *Phys. Rev.* **C41** 1941-55
[19] Kopecky J and Uhl M 1991 *AIP Conference Proceedings* **238** 607-9
[20] Dilg W, Schantl W, Vonach H and Uhl M 1973 *Nucl. Phys.* **A217** 269-298
[21] Kataria S K, Ramamurthy V S and Kapoor S S 1978 *Phys. Rev.* **C18** 549-63
[22] Gilbert A and Cameron W 1965 *Can. J. Phys.* **43**, 1446-96
[23] Ratynski W and Käppeler F 1988 *Phys. Rev.* **C37** 595-604
[24] Winters R R, Macklin R L and Hershberger R L 1987 *Astron. Astrophys.* **171**, 9-15
[25] McEllistrem M T et al. 1989 1989 *Phys. Rev.* **C40** 591-600
[26] Mughabghab S F, Divadeenam M and Holden N E 1981 *Neutron Cross Sections* Vol. 1 Part A (New York: Academic)
[27] Mughabghab S F 1984 *Neutron Cross Sections* Vol. 1 Part B (New York: Academic)
[28] Bohle D et al. 1984 *Phys. Lett.* **137B** 27-31

Radioactive nuclear beams for astrophysics

P Leleux

Université Catholique de Louvain, Institut de Physique Nucléaire
B-1348 Louvain-la-Neuve, Belgium

Abstract. Radioactive nuclei are present in all nucleosynthetic processes. New
information about nucleosynthesis is thus expected from the advent of radioactive
nuclear beams. Relative merits of radioactive beams versus radioactive targets,
requested properties of the beams, and possible detection schemes are discussed.
Results already obtained by direct and indirect methods are quoted.

1. Radioactive nuclei in astrophysics

Radioactive nuclei take part in all nucleosynthetic processes. Each time a
radioactive nucleus is produced, a bifurcation occurs in the process :
depending on the temperature (T) and density (ρ) prevailing at that moment in
the astrophysical environment, either decay of the radioactive nucleus or a
reaction with H or He will occur, with consequences for the relative nuclear
abundances and possibly for the energy generation from this process ; a given
nucleosynthetic process will thus produce different element abundances and
generate different amounts of energy depending on the existing (T,ρ)
conditions and in this sense radioactive nuclei are to a large extent responsible
for this diversity. Note that this bifurcation occurs already at very modest (T,ρ)
values, like in our sun : in the p-p chain, the ^{7}Be nuclei either decay or they
capture a proton, leading to the p-p II and III sequences, respectively, with
relative probabilities of 1 and 10^{-3}, respectively (the latter however has given
birth to the solar neutrino problem, more than twenty years ago) [1].

As decay rates of radioactive nuclei are practically independent of any
physical conditions whereas reaction rates generally increase with the energy
(or the temperature), radioactive nuclei are expected to play a much more
important role in the so-called explosive burning modes, which operate in the
higher (T,ρ) regime and at much shorter time scales than the corresponding
normal burning modes. Among the explosive modes, the hydrogen burning
has received most attention recently ; not only could the astrophysical sites be
defined with some confidence, e.g. the accretion of matter from a companion
star on a white dwarf or on a neutron star, but also specific important
reactions involving radioactive nuclei could be specified in different sub-cycles
(^{8}B(p,γ)^{9}C in the hot p-p cycle, ^{13}N(p,γ)^{14}O in the hot CNO-cycle, ^{20}Na(p,γ)^{21}Mg
in the hot Ne Na- or MgAl-cycle, ^{15}O(α,γ)^{19}Ne and ^{14}O($\alpha,$p)^{17}F in the rp- or αp-
process, this list being non limitative) [2]. The explosive

burning of elements heavier than hydrogen (He, C, O, Ne, Si) has been less
studied until now ; the specific and important reactions have still to be defined
some of those heavy explosive modes lead to the production of elements heavier
than Fe, e.g. through the r- or p-process. In the nucleosynthesis of nuclei of A
> 60, radioactive nuclei are required e.g. to determine i) rates of (n,γ) reactions
on radioactive isotopes [3] and ii) neutron binding energies at some specific
places in the chart of the nuclides [4]. Leaving the stellar nucleosynthesis for
the primordial nucleosynthesis, it is well known that in a particular scenario
of the origin of the universe, the inhomogeneous big-bang, reactions on
radioactive nuclei on the neutron-rich side of the valley of stability like ^{8}Li(α,n)
or ^{6}He(d,n) are important [5].

2. The observables

From § 1, it is clear that the rates of (p,γ) (α,γ) (α,p) ... reactions on radioactive
nuclei are very important quantities. As far as light nuclei are concerned
those reactions will proceed mostly via definite levels of the compound nucleus
(resonant states) which are located in the vicinity of the Gamow window, i.e. in
the optimal energy region for nuclear burning corresponding to the expected
temperature of the astrophysical environment, once the penetrability of the
Coulomb barrier has been taken into account. Furthermore, the energy, the
spin and the total width of those resonances have to be known as well ; these
quantities enter into the reaction rates and can be determined by the resonant
elastic scattering of radioactive nuclei on light nuclei as H, D, He, which
should always be performed before the reaction of astrophysical interest.
Finally, static properties of ground states of radioactive nuclei, like mass,
lifetime, Q_β, are needed in specific cases, e.g. to determine the path of a given
process : radioactive nuclei close to the proton drip line are particularly
important for the rp- or the αp-process.

3. Direct measurements by the two-accelerator method

3.1. Radioactive beam or radioactive target

The most direct way to obtain the reaction rate A + p → B + γ, where A is a
radioactive nucleus of lifetime τ, is i) to accelerate protons and to bombard a
target of nuclei A, or ii) to accelerate nuclei A and to bombard a proton target
(Figure 1). In both cases, the center-of-mass energy of the collision is the one
requested by the astrophysical problem. Both approaches, i.e. the radioactive
target (RT), and the radioactive beam (RB) share a common part : the
production of nuclei A by a nuclear reaction on a stable target, the extraction
the reaction products from the target, the selection of the atoms A by chemical
or cryogenic means, the ionization of the atoms and further the magnetic mass
selection of ions A. Then, in the RT approach, ions A are implanted into
backing of surface S which is simultaneously bombarded by a proton beam
intensity I ; in the RB approach on the other hand, ions A are accelerated with
a transmission factor ε and bombard a proton target of thickness t. It can be
shown that the number of reactions per unit time is larger in the RT approach

if $\tau > \dfrac{t.\varepsilon.S}{I}$, while the RB approach is more favorable if $\tau < \dfrac{t.\varepsilon.S}{I}$ (in the former case, the measuring time should be larger than $10\,\dfrac{t\,\varepsilon\,S}{I}$, which is not a very stringent condition, as will be shown below). For typical values of $t = 10^{19}$ at/cm^2, $S = 1$ cm^2, $\varepsilon = 0.05$ and $I = 10^{15}$ sec^{-1}, the border value is 5.10^2 sec. However, consideration of background in the detector render the RB approach more attractive for nuclei of τ much longer than this border value (e.g. the ^{13}N(p,γ)^{14}O reaction ($\tau = 865$ sec) was measured in the RB approach [6]). In fact, the RT approach is interesting mostly for very long τ (day or more), the target being then made off-line and subsequently bombarded.

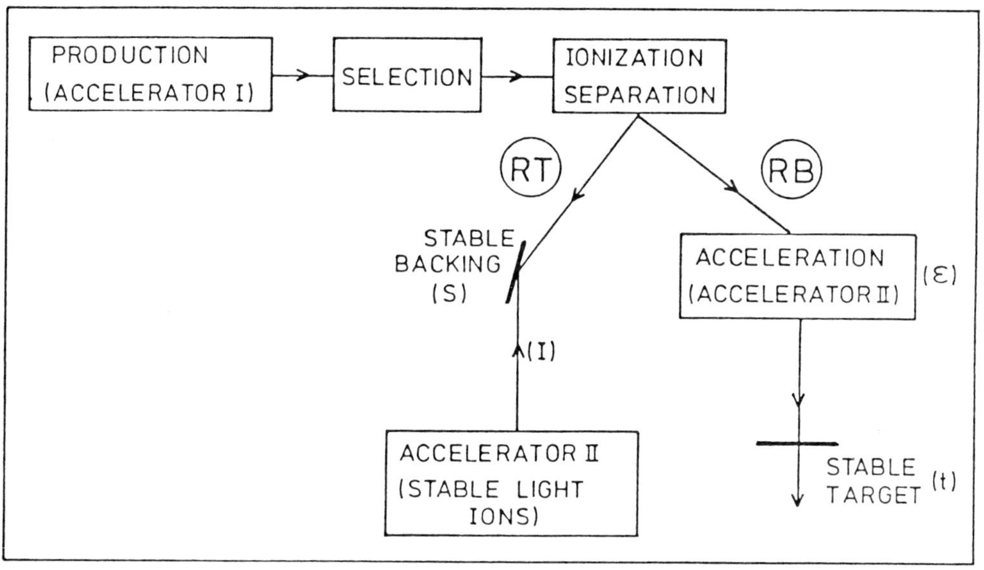

Figure 1 : Schematic set-up of a rate measurement by the two-accelerator method, using either a Radioactive Beam (RB) or a Radioactive Target (RT). Important quantities (t, ε, S, I) are defined in the text.

3.2. Quality of a RB

Having decided in favour of a RB, one could ask what qualities RB should possess. In the first generation of experiments which is presently underway, in which isolated resonant levels in light nuclei will be studied, the two most important qualities of a RB are its intensity (the goal should be 10^{10} sec^{-1}) and its purity (stable isobars of the RB being a particular problem). The energy resolution is at present less important, a precise scanning of resonances being hindered by the low beam intensity ; even an energy width equivalent to the Gamow window would be tolerable. As long as a well defined beam spot on target can be obtained, the emittance i.e. the product of the spatial and angular

dimension of the beam is not a major problem either. All those features make
a cyclotron particularly attractive as an accelerator of radioactive ions ; acting
as a powerful mass separator, a cyclotron does particularly well with respect to
the purity factor ; being a pulsed machine, it offers in addition the possibility to
select good events by a time-of-flight signature.

3.3. Detectors : a challenge for the experimentalists

Measurements with radioactive beams are not simply the extrapolation to a
"hotter" environment of measurements with stable beams ; in other words
adding more and more shielding around the detectors is not a viable solution.
Note that the $^{13}N(p,\gamma)^{14}O$ reaction cannot be considered as a representative
case : (i) the large cross section for capture ($\Gamma_\gamma = 3.8$ eV) exceeds the one of other
(p,γ) reactions of interest by at least a factor of 10 ; (ii) the high γ-ray energy (E
= 5.17 MeV) allowed to use a conventional set-up, i.e. large volume Ge detectors
[6]. On the contrary, each measurement with a RB should be the occasion of a
deep reflexion about which is the best way to obtain a clean result : examples
are (i) the direct detection of the light reaction products (γ-rays, p, ...), (ii) the
direct detection of the final nuclei and (iii) the detection of the radioactivity of
the final nuclei if any (β^+, α, ...). A gamma-ray detector with a high threshold
[7], a recoil mass spectrometer [8] and a solenoid to guide the positrons [9] are
particularly attractive tools to be used in these three methods, respectively.

Not only detectors but also detection methods could be improved : as an
example, the reverse kinematics used in the RB experiments was exploited to
develop a method of resonance scanning in one energy step, yielding the spin
energy and total width of resonant levels in the compound nucleus consisting
of the radioactive nucleus plus a proton; recoil protons from a CH_2 target were
detected ; of particular importance in this set-up was the simultaneous
measurement of the proton energy and of the proton time-of-flight with respect
to the cyclotron r.f. [10].

Figure 2 shows a recoil proton spectrum from the scattering of 11 MeV
^{19}Ne ions on a thick CH_2 target, taken at a laboratory proton angle of 20° ; this
spectrum represents a scan of the ^{20}Na excitation energy region containing the
resonance at 445 keV above the ^{19}Ne + p threshold, which is of particular
astrophysical interest. Curves are calculated spectra with as parameter the
width of this resonance : an upper limit of 1.3 keV could be already deduced
from this spectrum. This limit, which is still very far above the calculated
width, should be soon improved by the development of multistrip annular
proton detectors of large solid angle [11].

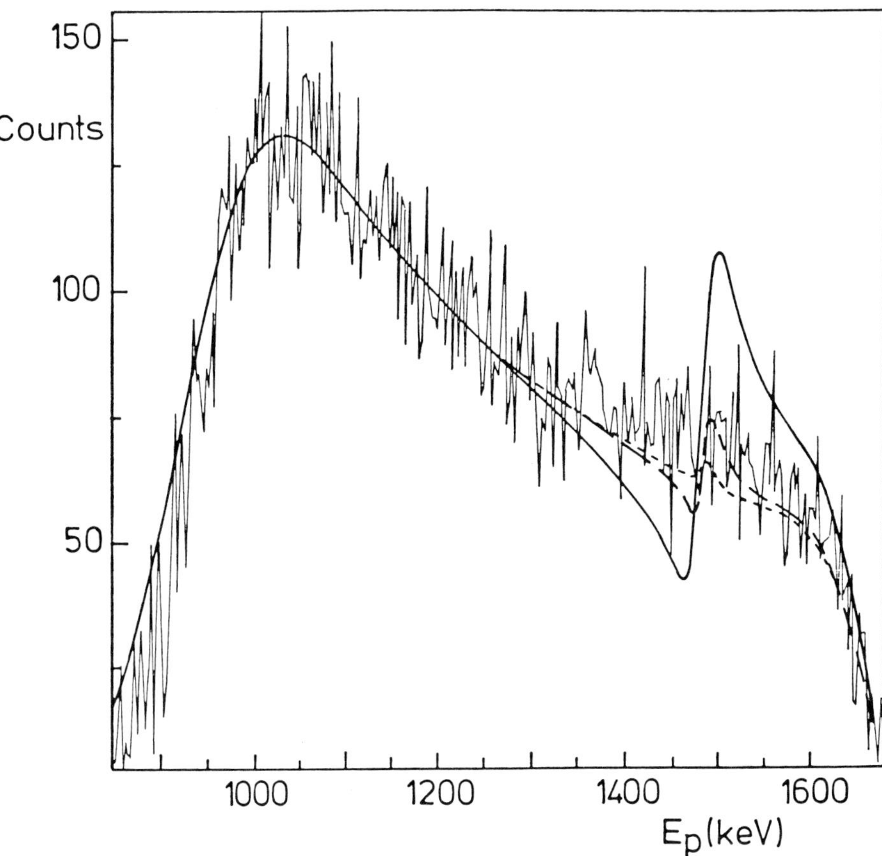

Figure 2 : Recoil proton spectrum from ^{19}Ne + p scattering. Curves represent calculations for the 445 keV resonance with a width of 10 keV (solid curve), 1 keV (long-dashed curve), 0.1 keV (short-dashed curve).

Well-established methods as well can be adapted to the radioactive beam environment : the direct component of a radiative capture reaction proceeding through a resonant state was shown [12] to be related to the spectroscopic factor of the same reaction leading to the ground state of the final nucleus ; the spectroscopic factor of the ^{13}N(d,n)^{14}O$_{g.s.}$ reaction (equivalent to the ^{13}N(p,γ)^{14}O$_{g.s.}$ reaction) has been measured by a standard activation method [13].

4. Other strategies

The acceleration of a RB up to the energy requested by astrophysics and the subsequent direct measurement of reaction rates and of properties of the levels of interest can be supplemented by other strategies through which equivalent or additional informations can be obtained. A common feature of those strategies is that a single accelerator is used, the energy of the RB being either much larger or much smaller than the requested astrophysical energy.

4.1. High-energy RB are obtained by the projectile fragmentation of a high-energy stable beam passing through a target. Radioactive nuclei of interest are selected by a recoil mass separator of high resolving power or collected by a large angle solenoid magnet[14]. Properties like the lifetime, the radioactivity, the Q_β of those radioactive nuclei can be obtained : for nuclei close to the proton-drip line, this approach is the only tractable, as the two-accelerator method could hardly cope with very short life times of the order of 0.1 sec or less.

Among recent results from this projectile fragmentation method is the β delayed proton decay of ^{20}Mg to ^{20}Na [15,16], which will improve the prediction of the (still unmeasured) reaction rate of the ^{19}Ne(p,γ)^{20}Na reaction ; also six new particle-bound isotopes from Ga to Sr have been recently discovered [17] leading to a redefinition of the path of the rp-process.

Rates of radioactive capture reactions (A + a $\rightarrow$ B + γ) can also be obtained by the so-called Coulomb dissociation method, in which a high-energy beam of nuclei B is selected as described above and is then dissociated in A + a by virtual photons in the Coulomb field of a heavy target ; A and a are detected in coincidence. This method is limited to cases where the density of particle-stable levels in A or in B is low ; the incident energy of nuclei B should also be high enough to avoid possible interpretation problems [17]. The detailed balance favours photodisintegration by a large factor as compared to capture. Recent measurements of the ^{13}N(p,γ)^{14}O reaction rate by this method [19,20] are in good agreement with the direct determination of the rate.

Note that in the rare cases where the energy region of interest is very wide and the cross section is not expected to vary rapidly with energy, the high-energy radioactive beam from a single accelerator can be decelerated or degraded to match the region of interest. The ^{8}Li(α,n)^{11}B reaction cross section was measured in that way between 1.5 and 6.5 MeV in the c.m. system, starting from the fragmentation of a GeV ^{14}N beam in a Be target [21].

4.2. It should be mentioned also that the decay properties of radioactive nuclei not too far from the valley of stability can be obtained using the common part of a RB-RT set-up, i.e. the production accelerator, the purification, ionization and separation devices. ^{16}N ions of about 20 keV have been selected, implanted and their β-delayed α emission measured, in order to get information on the E1 matrix element of the ^{12}C(α,γ)^{16}O reaction [22].

5. Conclusion

A worldwide effort to conceive and to develop projects to produce RB is presently underway. This community can note with interest that every one of the projects contains an astrophysical motivation, some of them concentrating mostly on nuclear astrophysics. New results are thus expected over the next few years and we can hope an improved knowledge of the explosive burning modes. The measurements of the $^{13}N(p,\gamma)^{14}O$ reaction rate have been already the occasion to infer interesting although not revolutionary consequences for the hot CNO cycle [23] : the boundary in the (T,ρ) plane between the cold CNO cycle, in which ^{13}N decay dominates and the hot CNO cycle, in which p capture by ^{13}N dominates, is now firmly established [23], and the $^{13}C/^{12}C$ ratio in nova ejecta are 30 to 40 % lower than previously estimated.

As mentioned above, the intensity of the RB is a decisive factor. We are presently at the beginning of such beams, and intensities are less than 10^9 sec^{-1}. To encourage those involved in this development as well as those expecting results in the near future, I would like to show what has been achieved in a neighbouring field, i.e. the polarized ion sources : intensities there were increased by more than six orders of magnitude through a continuous effort over a 25 years period (Figure 3) [24].

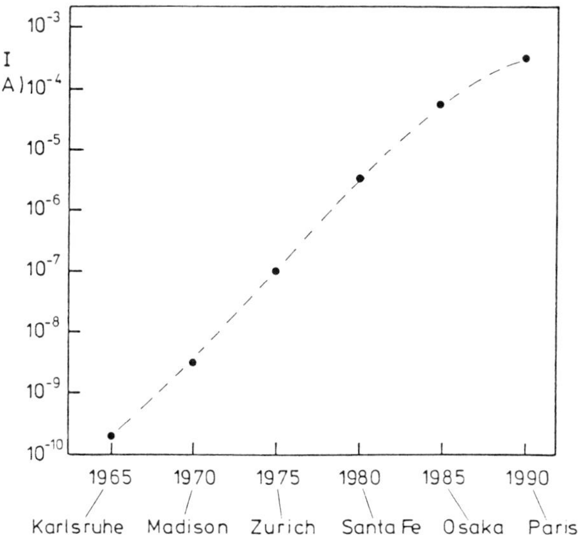

Figure 3 : Beam intensity from polarized negative hydrogen ion sources as a function of the year.

6. Acknowledgments

The author is a Maître de Recherches of the National Fund for Scientific Research, Belgium ; he wishes to thank for numerous discussions his colleagues of the Brussels, Leuven and Louvain-la-Neuve universities who have joined in an Interuniversity Attraction Pole sponsored by a grant of the national government.

References

[1] Rolfs C and Rodney W 1988 *Cauldrons in the Cosmos* (Chicago : The University Press)
[2] Arnould M and Rayet M 1990 *Ann. Phys. Fr.* **15** 183-254
[3] Käppeler F 1991 *Second International Conference on Radioactive Nuclear Beams*, Louvain-la-Neuve, Th. Delbar editor (Bristol : Adam Hilger) 299-305-310
[4] Howard WM and Meyer BS 1991 *Second International Conference on Radioactive Nuclear Beams*, Louvain-la-Neuve, Th. Delbar editor (Bristol : Hilger) 299-304
[5] Malaney RA 1992 *this conference*
[6] Decrock P et al 1990 *Phys. Rev. Lett.* **67** 808-811
[7] Seuthe S et al 1988 *Nucl. Instr. Meth.* **A272** 814-824
[8] Smith MS, Rolfs C and Barnes CA 1991 *Nucl. Instr. Meth.* **A306** 233-239
[9] Delbar Th et al 1991 *Second International Conference on Radioactive Nuclear Beams*, Louvain-la-Neuve, Th. Delbar editor (Bristol : Adam Hilger) 391-394
[10] Delbar Th et al 1992 accepted in *Nucl. Phys. A*
[11] Coszach R et al 1992 *this conference*
[12] Rolfs C 1973 *Nucl. Phys.* **A217** 29-70
[13] Decrock P et al 1992 *this conference*
[14] Sherril BM 1991 *Second International Conference on Radioactive Nuclear Beams*, Louvain-la-Neuve, Th. Delbar editor (Bristol : Adam Hilger) 3-20
[15] Wiescher M *private communication*
[16] Kubono S et al 1992 *INS-Rep-926*, to be published in *Phys. Rev. C*
[17] Mohar MF et al 1991 *Phys. Rev. Lett.* **66** 1571-1574
[18] Gazes SB et al 1992 *Phys. Rev. Lett.* **68** 150-153 ; Hesselbarth J and Knöpfle KT 1991 *Phys. Rev. Lett.* **67** 2773-2776
[19] Motobayashi T et al 1991 *Phys. Lett.* **B264** 259-263
[20] Kiener J et al 1991 *Second International Conference on Radioactive Nuclear Beams*, Louvain-la-Neuve, Th. Delbar editor (Bristol : Adam Hilger) 311-316
[21] Boyd RN et al 1992 *Phys. Rev. Lett.* **68** 1283-1286
[22] Buchmann L 1992 *this conference*
[23] Jorissen A and Arnould M 1992 *Astron. Astrophysics* **254** L9-L12
[24] Clegg TB 1990 *7th International Conference on Polarizatin Phenomena in Nuclear Physics*, A. Boudard and Y. Terrien editors (Les Ullis : Edit. de Physique) 533-540

The β-delayed α spectrum of ^{16}N and the low-energy extrapolation of the ^{12}C$(\alpha,\gamma)^{16}$O cross section

L Buchmann†, R E Azuma‡, C A Barnes§, A Chen‡, J Chen‡, J M D'Auria‖, M Dombsky‖¶, U Giesen†‡, K P Jackson†, J D King‡, R Korteling‖, P McNeely¶, J Powell‡, G Roy¶, M Trinczek‖, J Vincent†, P R Wrean§, and S S M Wong‡

† TRIUMF, Vancouver, B.C., Canada

‡ University of Toronto, Toronto, Ont., Canada

§ California Institute of Technology, Pasadena, CA, USA

‖ Simon Fraser University, Burnaby, B.C., Canada

¶ University of Alberta, Edmonton, Alta., Canada

Abstract. The low-energy part of the $\beta\alpha$-decay spectrum of ^{16}N permits the determination of the reduced α-width θ_α^2 of the 7.12 MeV subthreshold state of ^{16}O, which dominates the low-energy p-wave capture amplitude of the astrophysically important reaction ^{12}C$(\alpha, \gamma)^{16}$O. In a first run, a total of 7.1×10^3 α particles from the decay of ^{16}N nuclei, selected by the TRIUMF isotope separator TISOL, have been observed in a 10.6 μm detector in coincidence with the recoiling ^{12}C nuclei in a second detector, thus removing most sources of low-energy α background. These new data show a narrower width for the main α peak than previously determined. In addition, there is an indication of the low-energy interference anomaly predicted by several authors. The spectrum obtained can be parametrized in K- and R-matrix calculations, which also include the previously-measured ^{12}C$(\alpha, \gamma)^{16}$O cross sections and elastic scattering phase shifts. Results of a data collection run in June with much higher α yields will be presented.

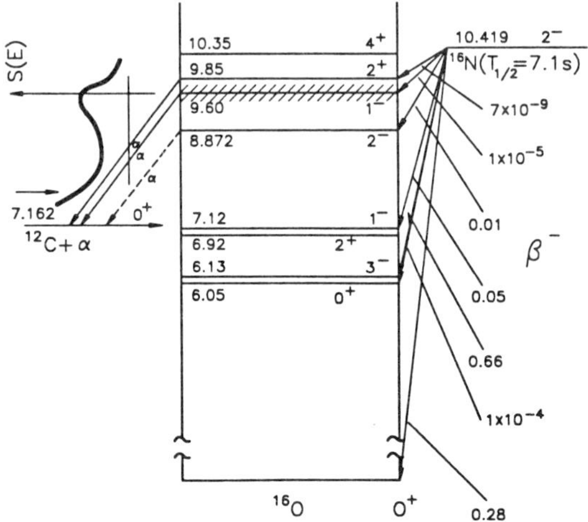

Figure 1. Level scheme of ^{16}O with respect to the ^{12}C$(\alpha,\gamma)^{16}$O reaction and the decay of ^{16}N.

1. Introduction

There has been a great deal of interest over the years in the low-energy cross section, usually expressed as the astrophysical S-factor S(E), of the radiative capture reaction ^{12}C$(\alpha,\gamma)^{16}$O [1,2,3,4,5]. The E1 and E2 components of the cross section are dominated, in the positive low α-energy region, by the tails of the $J^\pi=1^-$ and $J^\pi=2^+$ sub-threshold resonances corresponding to the bound states at 7.117 and 6.917 MeV, respectively ([6] and figure 1). However, the cross section in the experimentally accessible region above E(CM)=1 MeV is dominated by the broad resonance corresponding to the 1^- state at 9.585 MeV for p-wave capture, and by direct capture in the case of d-wave.

Although there has been an increasing number of direct measurements over the last twenty years [2,3,4], as well as theoretical efforts, there still remains a large uncertainty in the value of the cross section when extrapolated to stellar He-burning energies, E_α(CM)$\approx$300 keV. Although the cross section here is some 6 to 7 orders of magnitude smaller than that measured at the lowest experimental point, the difficulty is caused by the fact that the cross section in the energy region where measurements have been made is only weakly dependent on the properties of the subthreshold states.

The usual extrapolation of the radiative capture cross sections is performed using R- and K-matrix formulations [7,8,9]. Barker [7] pointed out that additional information from both the ^{12}C$(\alpha,\alpha)^{12}$C elastic scattering reaction and the α-particle spectrum following the β decay of ^{16}N could be incorporated into the calculations to improve the reliability of the extrapolations. Several additional studies incorporating all three reactions have subsequently appeared [10,11,12].

The elastic α-scattering data for all of these analyses were obtained from published work ([7] and references therein) whereas the data for the α-particle spectrum were a private communication (from [13]). There are at least two kinds of limitations associated with the β-delayed α-particle data. Firstly, there appears to be insufficient detail available to permit a reliable correction for possible experimental distortions of the spectrum shape. Secondly, the energy range for the data is restricted to energies $E_\alpha \geq 1.1$ MeV(Lab). The newer calculations [8,10,12] indicate that the energy range below $E_\alpha = 1.1$ MeV is especially important for the extrapolation of the radiative capture data to the region of interest. It is for these reasons that the current set of experiments was initiated. In these new experiments, the α-particle spectrum has been determined from both single counting and α-^{12}C coincidence data over the extended energy range of $600 \leq E_\alpha(\text{Lab}) \leq 2400$ keV, with particular attention being paid to the region below 1.1 MeV.

It can be seen from both the favourable branching ratio [6] and from the β-decay phase space that the ^{16}N β decay will preferentially feed the region of ^{16}O just above the α-decay threshold. This strongly enhances the influence of the reduced α-width, θ_α^2 ($\ell=1$), of the 7.117 MeV subthreshold state on the shape of the α-particle spectrum. The effect of this state [9,10,11] appears as a destructive interference anomaly in the spectrum for energies $500 \leq E_\alpha(\text{Lab}) \leq 1100$ keV. This effect may, in fact, provide a unique opportunity for the measurement of the reduced α-width of this subthreshold state, and hence of the E1 component of the α-capture reaction at astrophysically-important temperatures. In addition to selecting the p-wave part of the interaction, the ^{16}N$(\beta\alpha)^{12}$C decay may also lead to a possible f-wave contribution to the α spectrum from allowed β decay to $J^\pi=3^-$ states in ^{16}O.

In an experimental determination of the singles α spectrum, part of the region of the interference anomaly will be obscured by ^{12}C recoil events. Coincidence data on the other hand permit the accumulation of a clean α spectrum over the whole anomaly range as well as providing additional information which can be used for detailed analysis of the spectral shape.

Previous analyses, however, have excluded this particularly sensitive region of the α spectrum and were based only on the shape of the main α peak. In the analysis of the β-delayed α-particle data [7], it was found that the best fit required the inclusion of an f-wave component, attributed to the 6.130 MeV subthreshold state only. The f-wave contributions from other states were excluded on the basis of nuclear structure arguments. In subsequent analyses [10,11] an E1 "pseudo-spectrum" was used which was constructed by subtracting the f-wave component of ref. [7] from the experimental data.

It was observed that these fits were extremely sensitive to the precise shape of the main α-particle peak. In particular, the shape of the wings of the peak at the $0.1 \rightarrow 1.0$ % intensity level (especially on the low energy side) strongly influences the magnitude of both the p-wave and f-wave components and hence the reliability of the extrapolation based on these magnitudes. To help illustrate these features, the experimental data, the f-wave component [7], and the E1 contribution as calculated by Humblet et al. [9] are shown in figure 2.

In the analysis of our experimental run in January, 1992, evidence was found to show that the shape of the main α peak is indeed narrower than that reported, taking into account the corrections of reference [7]. In addition, there is clear evidence of α-particle events below $E_\alpha = 1$ MeV. In a run in June, 1992, orders of magnitude

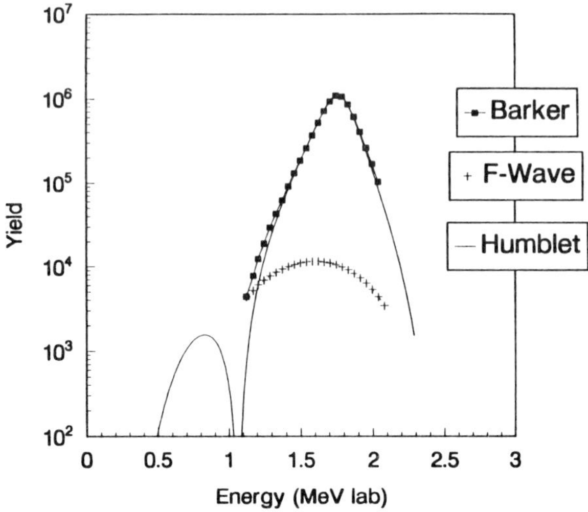

Figure 2. The experimental data of Ref. [7,13], a fit to the p-wave part of the data [9], and the f-wave used in fitting these data [7].

more α–^{12}C coincidence events were collected. A preliminary analysis of these events confirms the results of the January run.

2. Experimental set-up

In the experiments, a radioactive beam of ^{16}N ions was extracted from the isotope separator TISOL at TRIUMF [14,15,16] and implanted into a self supporting thin (20 μg·cm^{-2}–January; 10 μg·cm^{-2}–June) carbon collector foil of 1 cm diameter. After a short implantation time (7 s–January; 3 s–June) the foil was rotated into a position directly between two thin, transmission-mounted surface barrier detectors (separation $\approx$9 mm–January; $\approx$5 mm–June). Both singles and α–^{12}C coincidence spectra were recorded for a time equal to the implantation time. A front view of the detector arrangement for the June run is shown in figure 3; a cross section of the experimental set-up is shown in figure 4. During the recording time, implantation continued on another collector foil with the sequence being continuously repeated. In the June run three pairs were employed (90° rotation); in the January run only one pair was used (180° rotation).

The intensity of the radioactive beam at the collection point was typically 5×10^5 ions·s^{-1} selected at the mass A=30 (^{16}N^{14}N) position. For this, a zeolite target of 13.1 g·cm^{-2} was bombarded with a proton beam $\leq$1.5 μA at 500 MeV. At this mass setting, a small quantity of molecular ^{18}N^{12}C, $\approx 10^{-5}$ of the ^{16}N intensity, was also observed.

Thin detectors were chosen to minimize background from the β particles whose intensity is 10^5 greater than the α yield of ^{16}N. The supplier-specified thicknesses of the detectors were 10.6 μm (upstream) and 28 μm (downstream) for the January run and 10.6, 11.4, 11.8, 12.3, 12.9, 13.3, 15.8 μm for the June run. Upstream detectors were labeled with odd numbers (e.g. D1) and downstream detectors with even numbers

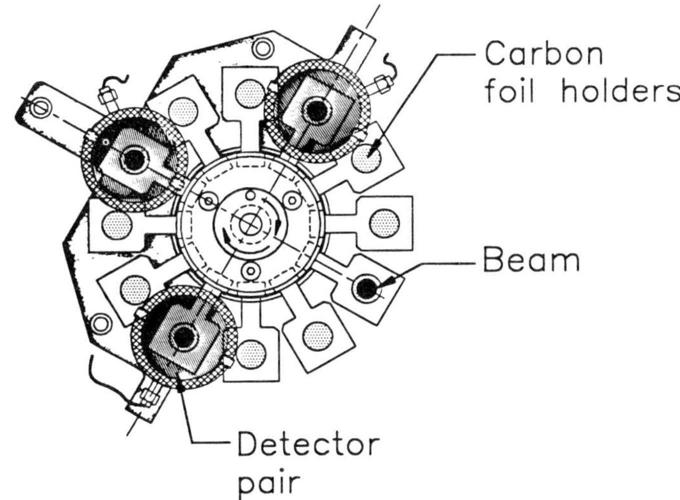

Figure 3. Front view (beam direction) of the detector station for the June, 1992 run showing the target wheel with thin carbon foils and three detector pairs.

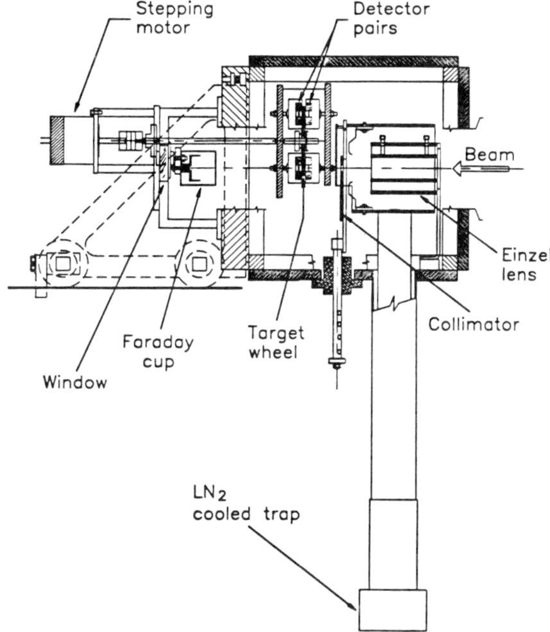

Figure 4. Cross section of the detector station for the June, 1992 run showing the extended set-up.

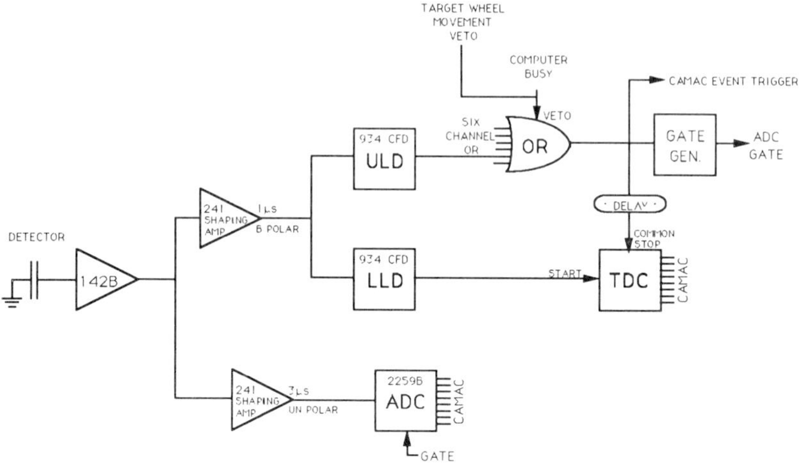

Figure 5. General electronics scheme. Note that the set-up is multiplexed for each channel used.

(e.g. D2). Extensive stopping calculations showed that, for foils with thicknesses greater than 10 μg·cm^{-2} and impinging energies of 10-12 keV, the distribution of the stopped particles in the foils is asymmetric; that is, the $\alpha/^{12}$C particles arriving at downstream detectors traverse more material than the ones hitting upstream detectors. This became important in the January run as extensive carbon build-up was observed which resulted in a shifting of peak centroids. This led to different peak shapes in the upstream and downstream detectors. With the installation of LN$_2$ cooled surfaces close to the foil in the June run, no carbon build-up or systematic shifts in the centroid of the main α peak were observed. In the June run a thick detector was mounted behind the thin detector D1 to register β activity as well as charged particles not stopped in the first detector in the case of isotopes other than ^{16}N.

Before impinging on the target, the ion beam was restricted by circular collimators of 8 or 10 mm diameter which ensured that the beam mainly hit the foil.

The general electronic scheme is shown in figure 5. The preamplifier outputs were split into fast and slow signals used for timing and energy measurement, respectively. The energy signal was further amplified in a spectroscopy amplifier with a shaping time of 3 μs and then sent to a CAMAC-based ADC. The fast signal was amplified and fed to a system of discriminators and to TDC's to obtain relative timing measurements between coincident particles. A double discriminator system was used for the fast timing. The upper level discriminator (ULD) signals were combined in a multiple OR gate the output of which acted as an event trigger and common stop for all TDC's. The lower level discriminator (LLD) provided the start signals for the individual channels of the TDC's. The advantage of this system is that the LLD threshold can be set partly in the noise to register the lowest energy ^{12}C particles without overloading the computer system. The event rate of the system was set by the trigger rate of the ULD (set well below the lowest α-particle energy of interest) and was typically a few tens per

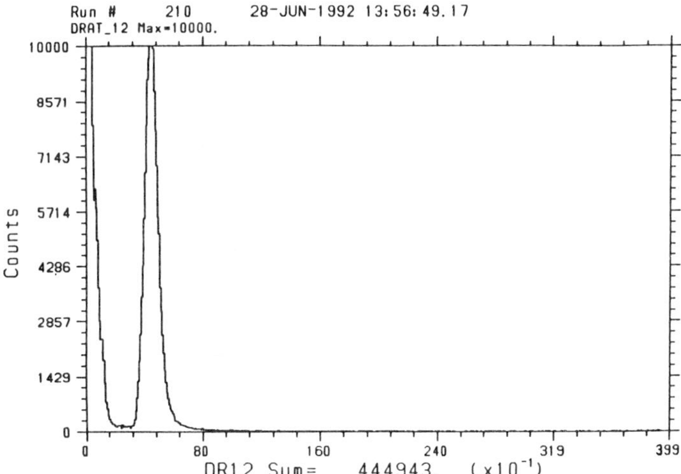

Figure 6. Pulse-height ratio spectrum for one detector pair and a subset of the data in the June, 1992 run.

second. All events were recorded on tape for further analysis. To avoid noise from the stepping motor the data acquisition was disabled while the wheel was being rotated.

3. Data analysis

The energy of the α particles was calibrated with the use of a ^{148}Gd source (E_α=3.183 MeV) and the known [17] monoenergetic decay lines of the β-delayed α decay of ^{18}N (1083 keV and 1409 keV (Lab), respectively). It was noted that the centroids of the ^{18}N decay lines were systematically shifted upwards for coincidence spectra with respect to the centroids of the singles spectra. This can be explained by the strong reduction of the low-energy tails in the coincidence spectra, as described below.

The ratio of the energies of the α- and ^{12}C-particles in the ^{16}N breakup is, neglecting minor $\beta\nu$ recoil corrections, inversely proportional to their mass ratio. The distribution of pulse height ratios of coincidence events is shown in figure 6. However, the maximum of the ratio peak is not at three but, depending on the detector, between four and five. This is attributed to pulse-height defects in the detection of the ^{12}C particles and the larger stopping power for ^{12}C particles in the entrance gold layer of the detectors (40 μm) and in the carbon foil.

Figure 7 is a two-dimensional scatter diagram of the energies of particles in coincidence for a pair of detectors. Events with the appropriate energy ratio of E1/E2$\approx$ $\frac{4.5}{1}$ or $\frac{1}{4.5}$ are seen as an intense diagonal band. The two bands extending horizontally or vertically from the main peak to low energies are attributed to coincidence events for which the α particles have been degraded in energy, whereas the ^{12}C particles have not. These energy-degraded events are normally referred to as the tail of the response function of the detector. The energy loss in such events can occur either in the detector

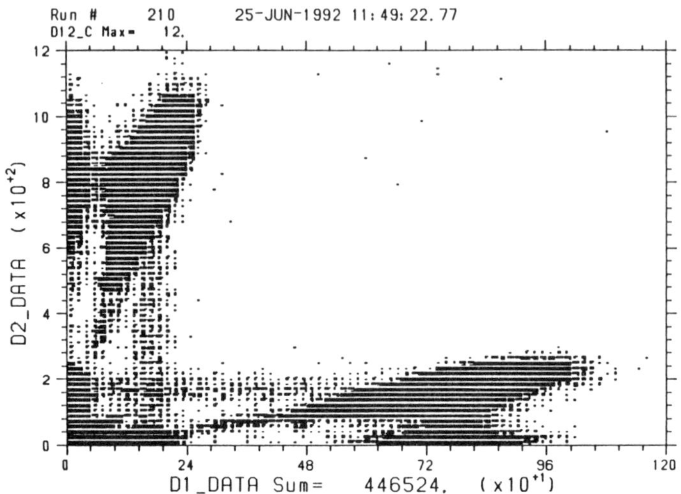

Figure 7. Coinciding event cut (pulse height) for one detector pair and a subset of the data in the June, 1992 run.

or in the collector foil. It is clear that, with careful cuts in the data analysis, the "tail" events can be excluded from the final spectra.

In addition to the above data, the time-of-arrival differences between members of coincident pairs were recorded, as shown in figure 8. Appropriate windows were set around these difference peaks to narrow the timing and therefore to lower random coincidence rates.

Occasional bursts of beam-related noise were initially a severe restriction to the experiment, especially in the low-yield, anomaly region of the α spectrum. It was observed, however, that this noise characteristically appeared in all of the detectors of the set-up. Consequently, a third detector was added to the January set-up. Since only adjacent pairs of detectors can detect real α–^{12}C coincidences, all events with "non-real" coincidences were flagged as noise pulses.

The coincidence efficiency can be determined down to an effective α energy of $\approx$1000 keV with the use of the monoenergetic peaks present in the ^{18}N $\beta\alpha$-decay spectrum. Figure 9 shows a coincidence $\beta\alpha$-decay spectrum of ^{18}N. The coincidence efficiency down to this α energy is found to be constant. Below this value the electronic efficiency can be determined with pulser signals. This component of the efficiency remains constant down to a pulse height corresponding to E_α=300 keV. However, this method does not take into account potential problems caused by pulse height defects and stopping power distributions for the ^{12}C particles. An attempt is being made to use ^{16}O nuclei and protons in coincidence from the βp decay of ^{17}Ne to determine coincidence efficiency. The proton is detected in the thin D1 detector plus the thick detector mounted behind it, while the recoiling ^{16}O nucleus is detected in the opposite, thin detector (D2). In addition, accelerator-based experiments for determining the coincidence efficiency and pulse-height defects may be necessary.

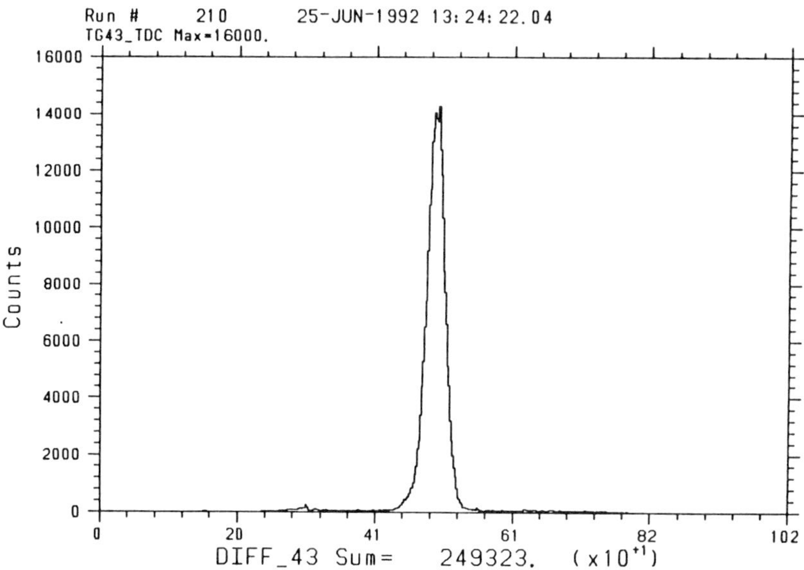

Figure 8. Time difference spectrum for one detector pair and a subset of the data in the June, 1992 run.

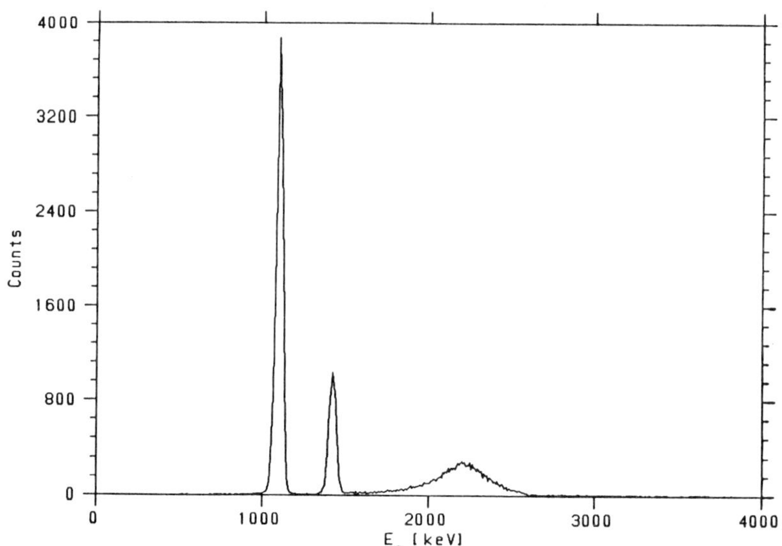

Figure 9. Energy spectrum of ^{18}N using a subset of the data in the June, 1992 run.

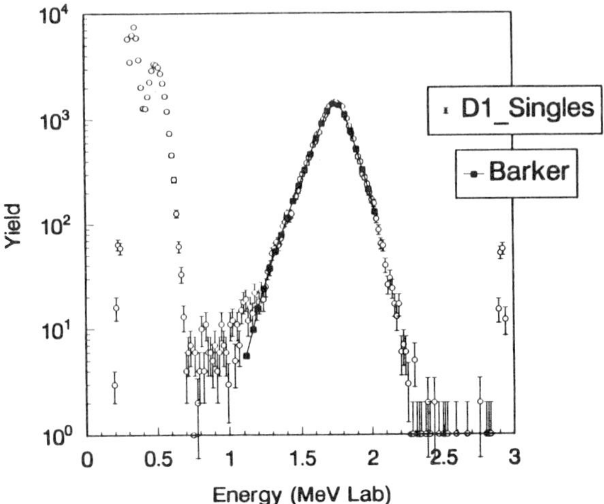

Figure 10. Singles energy spectrum for the upstream detector (D1) in the January, 1992 run.

4. Results and discussion

Figure 10 shows the total D1 (upstream) singles spectrum collected in the January, 1992 run. A clear separation of the ^{12}C peak from the β background can be observed. The singles data closely match those of ref. [7] over an intensity range of two orders of magnitude even though the present data have not been corrected for broadening due to energy resolution and straggling. Some of the additional counts in the region just above 1 MeV and at the upper energy end of the present results are due to the presence of an ^{18}N contamination. In addition to this, there nevertheless appears to be some residual tailing on the low energy side of the main α peak at the intensity level of a fraction of a percent. It is well known that particle detectors generally have a response function (see above) of this order of magnitude. It is therefore difficult to determine unequivocally from a singles spectrum that the counts in the energy region $700 \leq E_\alpha \leq 1000$ keV originate solely from the interference anomaly.

However, it can be seen from the fits of ref. [9] that the analysis of the data is crucially dependent on the precise shape of the main α peak. In addition, any tailing below the main peak at the intensity level seen here is indistinguishable from an f-wave component and would introduce an unacceptably large uncertainty into the analysis.

As noted above, the present experiment has proposed to resolve these difficulties by observing the α particles in coincidence with their recoil ^{12}C partners. The coincidence spectrum of detector D1 from the January run is shown in figure 11.

The α spectrum following the β decay of ^{16}N can be fitted [9] by the expression

$$W_{\ell\alpha}(E) = f_\beta(E)p_{\ell\alpha}^2(E)\left|\frac{\frac{B_1 g_{\alpha 1}}{E_1 - E} + \frac{B_2 g_{\alpha 2}}{E_2 - E} + \frac{B_3 g_{\alpha 3}}{E_3 - E}}{1 - ip_{\ell\alpha}^2(E)K_{\ell\alpha\alpha}}\right|^2 \tag{1}$$

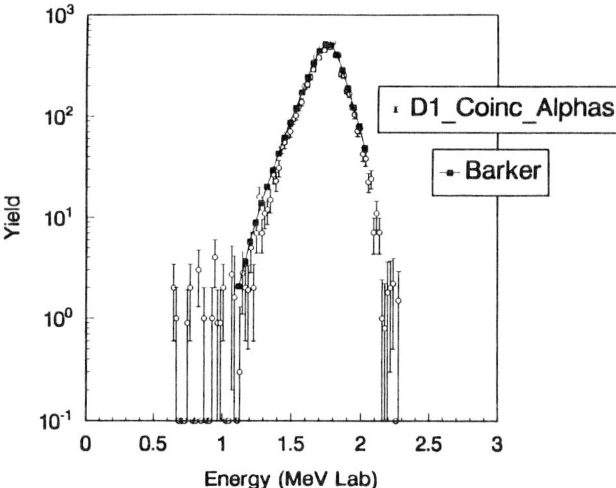

Figure 11. Coincidence energy spectrum for the upstream detector (D1) in the January, 1992 run and fit to the data.

where the subscript ℓ refers to the angular momentum of the decay (here p- or f-wave; $\ell=1$ or 3), α refers to numbers associated with the α decay of the states in question, β refers to β-decay dependent numbers, numerals in the subscript refer to particular states, $f_\beta(E)$ is the integrated (over β,ν momenta) Fermi distribution ($f_\beta(E) = f_\beta(8, W)$ with $W = (3.768 - E)/0.511$), $p^2_{\ell\alpha}(E)$ is the K-matrix penetrability [9], B_i is the β feeding factor into a particular state i, $g_{\alpha i}$ are the K-matrix partial widths of a state [9], E_i is the state energy as measured, and $K_{\ell\alpha\alpha}$ is the K-matrix of the α decay of a state and is given by

$$K_{\ell\alpha\alpha} = b_{\alpha\alpha} + \sum_{i=1}^{3} \frac{g^2_{\ell\alpha i}}{E_i - E} \tag{2}$$

with a constant background term $b_{\alpha\alpha}$.

Fig. 11 shows an example of such a fit to the January, 1992 coincidence data with a single state f-wave included. Because of the low number of counts (7100) this fit is shown only as an example of the fitting procedure.

In the June run, more than 1.25×10^6 $\alpha - {}^{12}\mathrm{C}$ coincidence events have been obtained. These data are presently under analysis.

Acknowledgements

Numerous people have contributed to this experiment. We want to thank in particular H. Biegenzein, D. Jones, P. Machule, H. Sprenger and A. Wilson for help in the mechanical set-up of the experiment, G. Sheffer and the electronics workshop for help in the requisition and setting up of the electronics components, and D. Diel and P. Green for help with the data acquisition system. The operators of the 500 MeV cyclotron are acknowledged for providing the essential proton beam. Our special thanks go to Teleglobe Canada for the donation of a 6 GHz amplifier without which the operation of the ECR source at TISOL and this experiment would have been impossible.

One of the authors (REA) wishes to express his thanks to Caltech for the hospitality received during the early stages of the experiment.

We are grateful to Triumf management for their vigorous support of the TISOL facility in general and timely assistance to this experiment in particular.

Funding for the experiment was provided by the Natural Sciences and Engineering Research Council of Canada. The Caltech participation was supported in part by the U.S. National Science Foundation.

References

[1] Bethe H A 1990 *Rev. Mod. Phys.* **62** 801

[2] Dyer P and Barnes C A 1974 *Nucl. Phys.* **A233** 495

[3] Redder A, Becker H W, Rolfs C, Trautvetter H P, Donoghue T R, Rinckel T C, Hammer J W and Langanke K 1987 *Nucl. Phys.* **A462** 291

[4] Kremer R M, Barnes C A , Chang K H, Evans H C, Fillipone B W, Hahn K H and Mitchel L W 1988 *Phys. Rev. Lett.* **60** 1475

[5] Wooseley S and Weaver T A 1986 *Ann. Rev. Astron. Astrophys.* **24** 205

[6] Ajzenberg-Selove F 1986 *Nucl. Phys.* **A460** 1; 1988 *Nucl. Phys.* **A475** 1

[7] Barker F C 1971 *Aust. J. Phys.* **24** 777

[8] Ji X, Fillipone B W, Koonin S E and Humblet J 1990 *Phys. Rev. C* **41** 1736

[9] Humblet J, Fillipone B W and Koonin S E 1991 *Phys. Rev. C* **44** 2530

[10] Barker F C 1987 *Aust. J. Phys.* **40** 25

[11] Barker F C and Kajino T 1990 *Aust. J. Phys.* (to be published)

[12] Baye D and Descouvement P 1988 *Nucl. Phys.* **A481** 445

[13] Neubeck K, Schober H and Wäffler H 1974 *Phys. Rev. C* **10** 320
Hättig H, Hünchen K and Wäffler H 1970 *Phys. Rev. Lett.* **25** 941

[14] Dombsky M, D'Auria J M, Buchmann L, Sprenger H, Vincent J, McNeely P, and Roy G 1990 *Nucl. Instrum. Methods* **A295** 291

[15] Buchmann L, Vincent J, Sprenger H, Dombsky M, D'Auria J M, McNeely P, and Roy G, 1991 *Nucl. Instrum. Methods* (to be published)

[16] McNeely P, Roy G, Soukup J, D'Auria J M, Buchmann L, McDonald M, Schmor P W, Sprenger H and Vincent J 1990 *Rev. Sci. Instrum.* **61** 273

[17] Zaoh Z, Gai M, Lund B J, Rugari S L, Mikolas D, Brown B A, Nolen J A and Samuel M 1989 *Phys. Rev. C* **39** 1985

Prospects for studies of astrophysical interest with radioactive ion beams

J. D. Garrett

Physics Division, Oak Ridge National Laboratory, Oak Ridge, TN, USA 37831-6371

Abstract. Realistic estimates of the new nuclei that will become accessible for study with the planned new generation of radioactive beam facilities are compared with the predicted paths of the various nucleosynthetic processes. These facilities should permit studies of essentially the complete set of nuclei associated with the rp- and p-processes, the CNO cycle, and big-bang nucleosynthesis, as well as the r-process below $A \cong 150$.

1. Introduction

Reaction times in exotic cosmic sites, such as the early universe, supernovae, and cataclysmic binary stellar systems, are many orders of magnitudes less than the more than four-billion year lifetime of terrestrial matter. For example, the r(apid-neutron-capture)-process nucleosynthesis that is thought to occur in the low-density, high-entropy bubble [1] of type II supernovae (SN II) is complete within a few seconds, see e.g. [2]. Furthermore, neutron or proton excesses often exist at such nucleosynthetic sites. Hence many of the nuclear reactions powering stellar processes are inaccessible to experimental study using beams and targets of stable nuclides. Such reactions can only be studied with radioactive beams and/or radioactive targets. However, the lifetimes of the nuclei participating in many of the interesting nucleosynthetic reactions are too short even for *in situ* target production [3]. Therefore, nuclear astrophysics provides a strong justification for the development of high-intensity beams of radioactive ions, see e.g. [4-6].

Indeed the era of radioactive ion beams already is here. Five talks at this symposium [7-11] discussed nuclear astrophysical measurements involving radioactive ions, and two others [12,13] were concerned with the production and use of radioactive ion beams. The present contribution will address the impact of the variety of exotic beams projected for the next generation of high-intensity radioactive beam facilities on astrophysics. In particular the "new" nuclei that can be produced by the IsoSpin Laboratory (ISL) [6] and the more modest, but funded, Oak Ridge Radioactive Ion Beam (RIB) Facility [5], and that cannot be studied with stable beams and stable targets [14], will be compared with the projected paths of the "fast" nucleosynthetic processes, i.e. the r- and the r(apid)-p(proton-capture)-processes [15,16], which occur furthest from stable nuclei.

2. Estimates of new nuclei accessible with radioactive ion beams

The distribution as a function of Z and N of nuclei that can be studied using radioactive beams of the Oak Ridge RIB Facility and the ISL, but which are not accessible with stable beams and stable targets, has been estimated. The detailed criteria for this selection are given in refs. [14,17] and are briefly described in the ensuing paragraphs.

Because of the decreasing ratio of protons to neutrons for heavy stable nuclei, the most efficient mechanisms for populating proton- and neutron-rich nuclei are different. Heavy-ion induced fusion-evaporation reactions using proton-rich projectiles and targets with nearly equal masses are best for producing proton-rich nuclei. Of course, the ratio of protons to neutrons can be increased by either proton-rich radioactive projectiles or targets or both. The criteria assumed in refs. [14,17] for determining which proton-rich isotopes are accessible with radioactive-ion beams are beam currents $> 10^8$ ions/s and production cross sections > 5 μb. These criteria are selected to allow measurements of nuclear masses, beta-decay lifetimes and end-point energies, and a few low-lying states. Because of the delicate balance between neutron and charged-particle emission, production cross sections for proton-rich nuclei are particularly sensitive to the relative number of protons and neutrons [5,6]. Often an additional neutron or one less proton will allow a rather complete nuclear structure study of low- and medium-high-spin properties of the nucleus. Both facilities, i.e. RIB and ISL, will be able to study proton-rich nuclei of interest to astrophysics, e.g. rp- and p-process nuclei. However, recent calculations [18] indicate that the previous estimates of the most proton-rich RIB beams were too optimistic. This realization is folded into the discussion in section 3.2.

The most efficient technique for producing neutron-rich nuclei is fission or spallation of neutron-rich heavy nuclei (with a large ratio of neutrons to protons) and reactions (especially neutron-pickup, proton-stripping, and isospin-increasing charge-exchange reactions) induced by such neutron-rich products. Indeed the ground-state properties of neutron-rich products and the low-lying states of their beta-decay daughters are studied at several isotope separators, see e.g. [11,19-21]. The new ingredient provided by the ISL (and other similar large ISOL facilities based on very high energy primary beams for secondary ion production) will be the use of very neutron-rich products as beams to access even more neutron-rich nuclides and to provide more complete nuclear structure information for the beam nuclei. The criteria assumed in ref. [14] for studying "new" neutron-rich nuclei that cannot be studied with stable beams and targets are 10^5 ions/s for Coulomb excitation of the beam particle and 10^8 ions/s for transfer of up to three nucleons.

3. The feasibility of studying nucleosynthetic processes with radioactive on beams

3.1. The r-process

The large densities of free neutrons in certain cosmic sites, such as SN II, will produce an r-process [15,22]. Neutron-capture rates associated with such cosmic sites are much larger than the beta-decay rates of neutron-rich nuclei near stability. Indeed how far from stability this r-process occurs is a measure of the neutron density of the environment. To obtain a quantitative estimate of this neutron density, it is necessary to know the beta-decay rates along the r-process path.

Figure 1 compares the predicted path [23] of the traditional r-process with the "new" nuclei [14] which will become available for study with the ISL [6]. The 10^5 ions/s criterion used to estimate the "new" proton-rich nuclei is a conservative estimate for the number of radioactive ions necessary for a direct measurement of the beta-decay rates critical for establishing neutron fluxes associated with the r-process. Therefore, the area of "new" nuclei shown in Figure 1 should be extended to include up to four additional neutrons for such measurements.

Radioactive beams from facilities such as the ISL not only will provide access to a much greater portion of the predicted r-process path, but they also will allow detailed studies of the N = 50 and 82 "waiting points." Since the measured and extrapolated r-process beta-decay rates apparently imply different neutron densities for the r-process paths between the different neutron closed shells [11], detailed beta-decay rates associated with the "holdup" at these closed-shell "waiting points" will provide information about the time dependence of the neutron density at the r-process site. Likewise these radioactive beams also will allow detailed studies of r-process nuclei near A = 120, where a deficiency in the current calculations relative to solar system abundances has been attributed [2] to uncertainties in nuclear structure information. Even though the radioactive beams will provide a significant improvement of r-process data over the impressive fission-product data [11], the r-process path for A > 150 will remain inaccessible to experimental study.

The experimental detection of neutrinos related to the core collapse of SN 1987A [24] and the associated development in supernova core-collapse theory [1,2,25] have led to a reassessment of the site and mechanism of r-process nucleosynthesis. Increasingly the r-process is becoming associated [2,25] with low-density, high-entropy, neutron-rich environments (the so-called "hot bubble") which are produced by neutrino-induced reactions following the core collapse of supernovae. Such neutrino-induced processes also produce significant modifications of the starting-point and the low-mass (A < 125) evolution of the r-process, see Figure 2. The modified path, less neutron rich than the traditional r-process path, is a consequence of (α,n) reactions that for light nuclei compete with neutron capture. The increased alpha-particle density is a result of (ν,α) and photonuclear reactions which disassociate a significant fraction of the presupernova heavy elements into alpha particles. Figure 2 also indicates that radioactive beams will allow experimental measurements for nearly the whole portion of the r-process where such modifications are predicted. Though at this early stage the detailed path of the modified r-process may be not correct, this prediction is indicative of the excitement and timeliness of such nuclear astrophysics studies. A supernovae explodes, neutrinos are detected fixing the neutrino flux associated with the core collapse of a SN II, stellar models are revised producing major modifications in the r-process just as a new generation of accelerators allowing detailed nuclear structure measurements on the nuclides in the modified r-process becomes technically feasible.

3.2. The rp-process

The conditions for explosive hydrogen-burning (a hot hydrogen-rich environment) have been hypothesized for a variety of stellar sites, such as thermonuclear runaway in material accreting on the surface of a white dwarf in interacting, close binary stellar systems, see e.g. [26]. This outburst is observed as a nova, or if very energetic as an x-ray burst. If the temperatures and densities are sufficiently high, such an rp-process (which also may occur

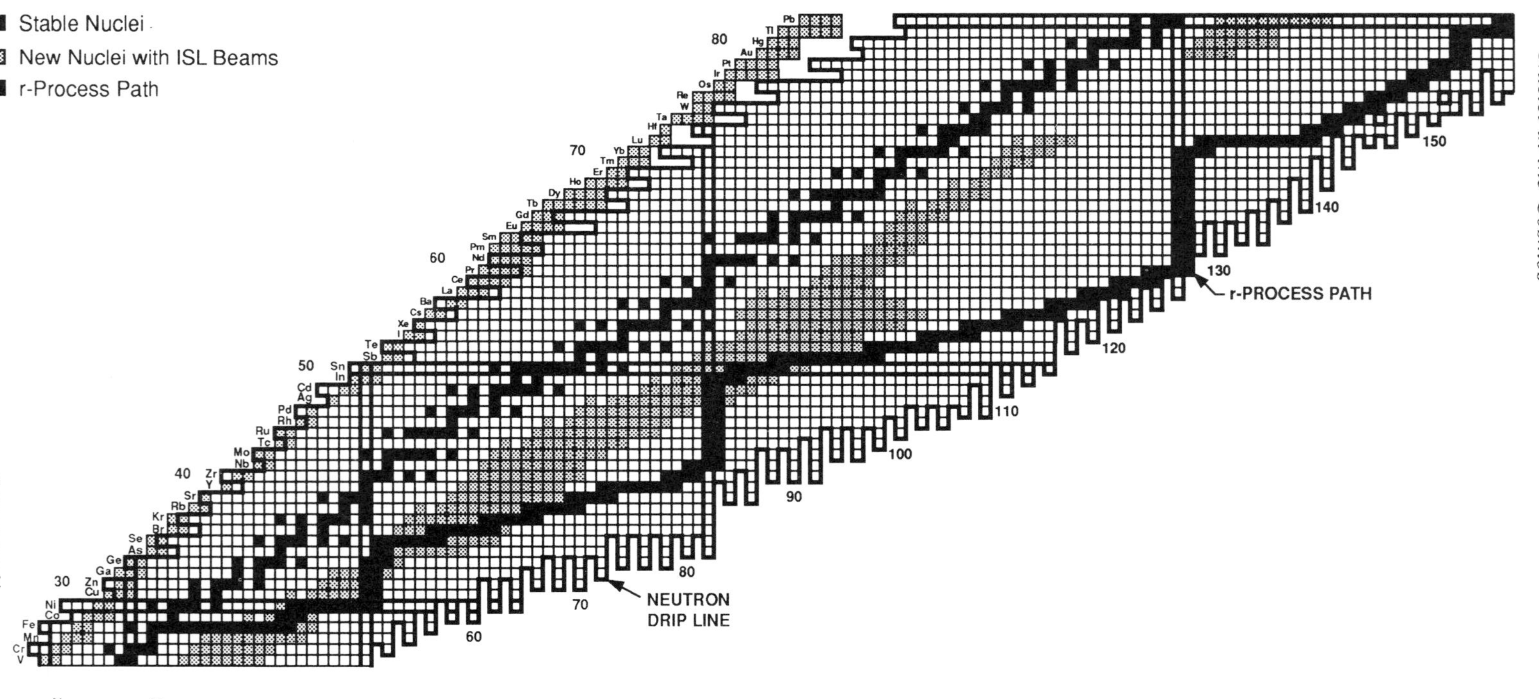

Figure 1. Comparison of the "new" nuclei that can be studied [14] using the radioactive beams of the projected IsoSpin Laboratory [6] and the traditional path of r-process nucleosynthesis [23]. The criteria for producing "new" nuclei, established for reaction studies [14], are probably too conservative for the measurement of β-decay rates; see text for discussion. For reference, the closed neutron and proton shells and the predicted [30] drip lines are shown by heavy solid lines, and stable nuclei are indicated as filled squares. The r-process path also is shown by filled squares; however, these nuclei can be distinguished since they have a large neutron excess.

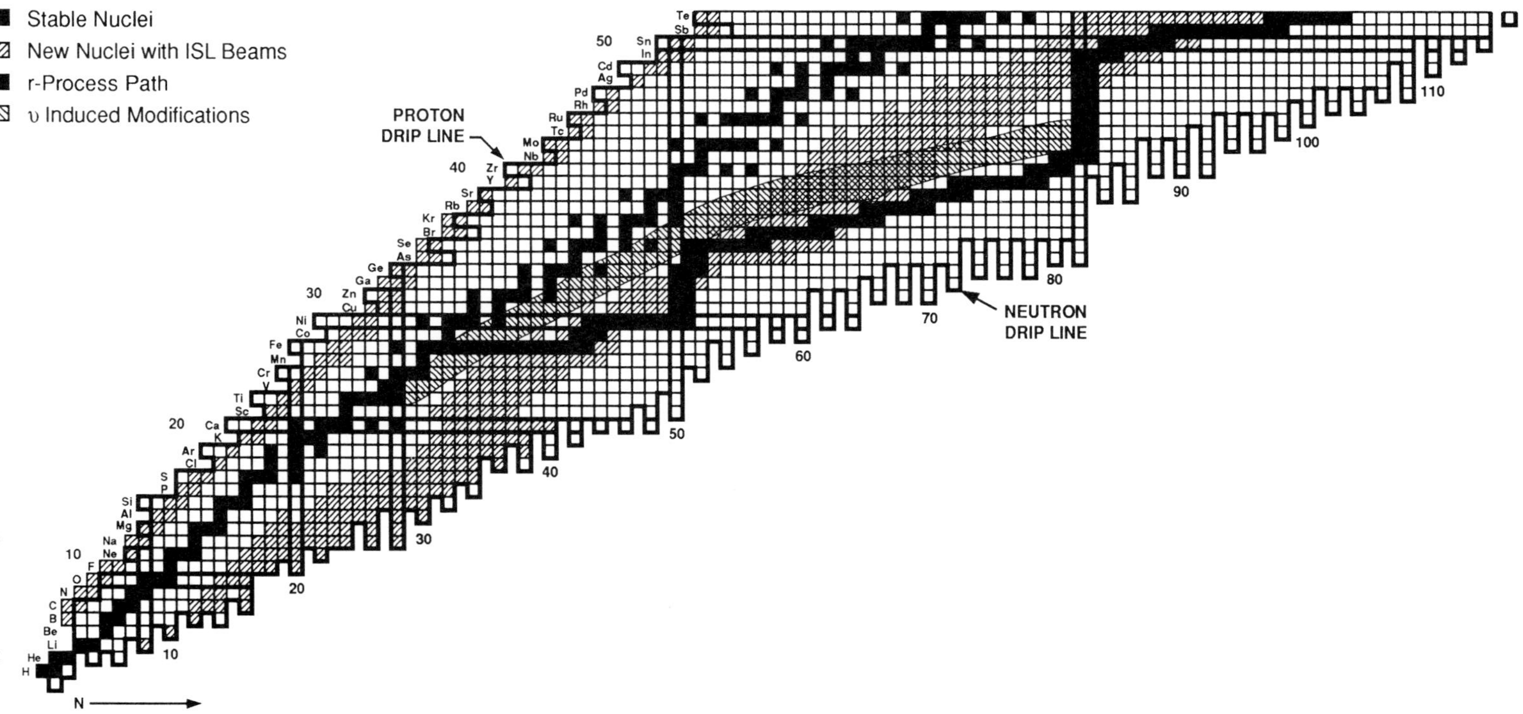

Figure 2. Comparison of recent estimates of the r-process path in light nuclei for a low-density, high-entropy, neutron-rich environment [25] (labeled υ induced modifications and diagonally shaded) with that of the traditional r-process [23] shown as the neutron-rich filled squares. Above A ≅ 125 the two paths merge. The "new" nuclei that can be studied [14] using the radioactive beams of the projected IsoSpin Laboratory [6] are shown for reference by a different kind of diagonal shading. See also the caption to Figure 1.

for accreting neutron stars and in the passage of the shock wave through the hydrogen zone of a type II supernova) may proceed all the way to mass 80 [16]. The path of this rp-process, shown in Figure 3 for rather high temperatures, 1.5 x 10^9 °K, and proton densities, 10^4 g/cm^3 [27,28], is determined by the balance between the (p,γ) reaction rates for large densities of "high-temperature" protons and β$^+$ decay. This process will proceed so long as the rate for the (p,γ) reactions is greater than the rate for β$^+$ decay. Therefore, both reaction rates and specific nuclear properties, such as masses, binding energies, quantum numbers of energy levels in the Gamow window, level densities, and isomeric states, are needed for nuclei in the rp-process path. Indeed the "new" proton-rich beams from the Oak Ridge RIB Facility can provide rates for a variety of the rp-process reactions, such as the (p,γ) reaction on ^{17,18}F, ^{21,22}Na, ^{26,27}Si, ^{29,30}P, ^{30,31}S, ^{33,34}Cl, ^{38}K, ^{58,59}Cu, and ^{65}Ge, see Figure 3, and the Isospin Laboratory should be able to provide a nearly complete set of reaction rates for this process. Among the most important reaction-rates to be measured are the ^{27}Si, ^{31}S, ^{65}Ge (p,γ) "bottle necks" and the ^{26}Si,^{30}S(pγ) "waiting points" [26,27]. Likewise nuclear structure properties of nearly all of the nuclei in the rp-process path, not accessible with stable beams and stable targets, can be studied using the radioactive beams produced with both the Oak Ridge RIB Facility, and the Isospin Laboratory.

3.3. Other nucleosynthetic processes

The emphasis of the preceding subsections has been on "fast" nucleosynthetic processes (the r- and rp-processes) which populate neutron- and proton-rich nuclei respectively that are far from stability. Radioactive ion beams also will provide (indeed already are providing) crucial information for other nucleosynthetic processes, such as the p(roton-capture)-process, the CNO cycle, helium burning and big-bang nucleosynthesis. Indeed these studies typically are easier, since nuclei nearer to stability are involved. For example, the radioactive-beam studies reported at this meeting [7-10] involve beams only one isospin unit from stability. Either new facility considered in this work, i.e. the Oak Ridge RIB Facility or the ISL, would be able to provide important new information for p-process, CNO-cycle and big-bang nucleosynthesis. Radioactive-beam CNO-cycle and big-bang studies already are in progress using first-generation (low-energy and/or low-intensity) facilities, such as that installed at Louvain-la-Neuve, Michigan State University, RIKEN, GANIL, and GSI.

4. Summary

Recent technical advances in the production and ionization of radioactive atoms allow the construction of a new generation of high-intensity, broad-mass-range radioactive beam accelerators. These facilities will not only permit nuclear-structure and reaction-rate measurements for essentially the complete set of nuclei associated with the rp- and p-processes, the CNO cycle, and big-bang nucleosynthesis, but also will allow similar measurements for the r-process up to A ≅ 150. The advent of such radioactive beam facilities will provide a sound foundation to astrophysics just at the time that observational astronomy is shedding its atmospheric restrictions. Indeed the future promises to be exciting.

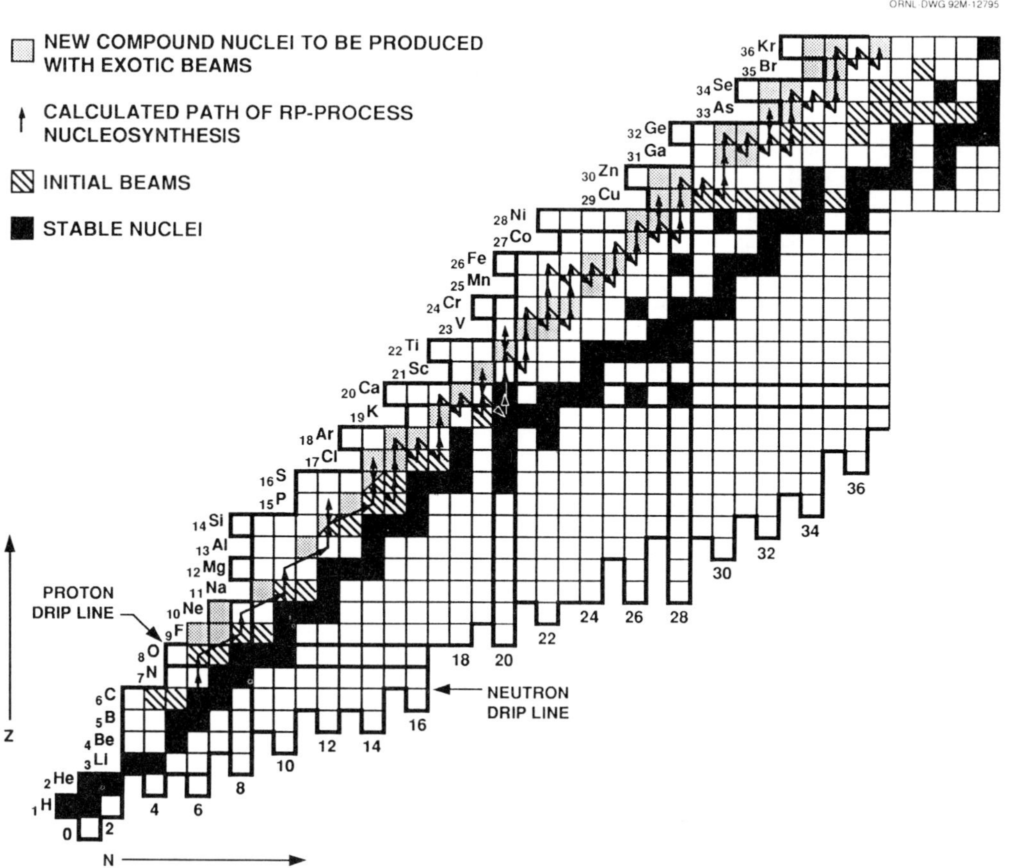

Figure 3. Comparison of the "new" proton-rich compound nuclei (shaded squares) that can be produced by the proposed Oak Ridge RIB Facility [5] and the predicted [27,28] path of a "hot" rp-process (arrows) assuming a temperature of 1.5×10^9 °K and proton densities of 10^4 g/cm^3. Projected initial radioactive beams (indicated by heavy shading) could be used with a hydrogen target for reaction-rate measurements for many of the rp-process reactions. For reference stable nuclei are shown as solid squares and proton and neutron closed shells and drip lines [29,30] are indicated by heavy lines.

Important discussions with C. Baktash, A.E. Champagne, I.Y. Lee, F.K. Thielemann, and M. Wiescher are acknowledged. Oak Ridge National Laboratory is managed by Martin Marietta Energy Systems, Inc. under contract DE-AC05-84OR21400 with the U.S. Department of Energy.

References

[1]   Mayle R W and Wilson J R 1988 *Astrophys. J.* **334** 909

[2]   Meyer B S, Howard W M, Mathews G J, Hoffman R, and Woosley S E 1992 in *Unstable Nuclei in Astrophysics* ed. Kubono S and Kajino T (World Scientific, Singapore) pg. 37

[3]   Rolfs C E and Rodney W S, *Caldrons in the Cosmos* 1988 (University of Chicago Press, Chicago)

[4]   Howard W M *et al.*1990 in *Proceedings of the Workshop on the Science of Intense Radioactive Ion Beams* ed. McClelland J B and Vieira D J, Los Alamos Report LA-11964-C, pg. 68

[5]   *A Proposal for Physics with Exotic Beams at the Holifield Heavy Ion Research Facility* February 1991 ed. Garrett J D and Olsen D K, Physics Division, Oak Ridge National Laboratory

[6]   Casten R F, D'Auria J M, Davids C N, Garrett J D, Nitschke J M, Sherrill B M, Vieira D, Wiescher M, and Zganjar E F 1991 *The IsoSpin Laboratory, Research Opportunities with Radioactive Nuclear Beams*, LALP 91-05 (Los Alamos National Laboratory, Los Alamos)

[7]   Buchmann L *et al.* in these proceedings

[8]   Coszach R *et al.* in these proceedings

[9]   Decrock P *et al.* in these proceedings

[10]   Aguer P *et al.* in these proceedings

[11]   Kratz K-L, Møller P, Pfeiffer B, Wøhr A, and the ISOLDE Collaboration, in these proceedings

[12]   Leleux P, in these proceedings

[13]   Sorlin O, Guillemaud-Mueller D, and Mueller A C, in these proceedings

[14]   Garrett J D March 1992 in *Proc. of the Workshop on the Techniques of Secondary Nuclear Beams, Dourdan, France*, (Editions Frontieres, Paris), in press

[15]   Burbidge E M, Burbidge G R, Fowler W A, and Hoyle F 1957 *Rev. Mod. Phys.* **29** 547

[16]   Wallace R K and Woosley S E 1981 *Astrophys. J. Supp.* **45** 389

[17]   Garrett, J D 1992 in *Reflections and Directions in Low Energy Heavy-Ion Physics*, ed. Hamilton J H *et al.* (World Scientific, Singapore) pg. 304

[18]   Baktash C, private communication

[19]   Kluge H J 1986 ISOLDE Users' Guide, CERN Report CERN 86-05

[20]   Buecher M, Casten R F, Gill R L, Schumann R, Winger J A, Mach H, Moszynski H, and Sistemich K 1990 *Phys. Rev. C* **41** 1115; and Gill R L, Casten R F, Warner D D, Piotrowski A, Mach H, Hill J C, Wohn F K, Winger J A, and Moreh R 1986 *Phys. Rev. Lett.* **56** 1874

[21]   Omtvedt J P, Hoff P, Hellström M, Spanier L, and Fogelberg B 1991 *Z. Phys.* **A338** 241

[22]   Cowan J J, Thielemann F-K, and Truran J W 1991 *Phys. Rep.* **208** 267

[23]   Seeger P A, Fowler W A, and Clayton D D 1965 *Ap. J. Suppl.* **11** 121

[24]   Hirata K *et al.* 1987 *Phys. Rev. Lett.* **58** 1490; Bionta R M *ibid.* 1494

[25]   Woosley S E and Hoffmann R 1991 *Astrophys. J.* in press

[26]   Champagne A E and Wiescher M 1992 *Ann. Rev. Nucl. Part. Sci.* **42** in press

[27]   Wiescher M, Gorres J, Thielemann F-K, and Ritter H 1986 *Astron. Astrophys.* **160** 56

[28]   Wiescher M, private communication

[29]   Détraz C and Vieira D J 1989 *Ann. Rev. Nucl. Part. Sci.* **39** 407

[30]   Comay E, Kelson I, and Zider A 1988 *At. and Nucl. Data Tables* **39** 235

Measurement of resonance widths in ^{20}Na using the ^{19}Ne + p elastic scattering

R Coszach*, F. Binon[x], P Decrock[+], Th Delbar*, P Duhamel[x], W Galster*, M Huyse[+], P Leleux*, E Liénard*, P Lipnik*, C Michotte*, A Ninane*, G Vancrayenest[+], P Van Duppen[+], J Vanhorenbeeck[x], J Vervier*, G Roters[o], H.P Trautvetter[o], T Davinson", R Page", P Woods" and A.C Shotter"

*   Institut de Physique Nucléaire, Université Catholique de Louvain, Louvain-la-Neuve, Belgium
+   Instituut voor Kern- en Stralingsfysika, Katholieke Universiteit Leuven, Leuven, Belgium
x   Institut d'Astronomie, d'Astrophysique et de Géophysique, Université Libre de Bruxelles, Bruxelles, Belgium
o   Institut für Experimental Physik III, Ruhr Universität Bochum, Bochum, Germany
"   Department of Physics, The University of Edinburgh, Edinburgh, Great Britain

Abstract. The excitation energy, total width and spin and parity of resonances in ^{20}Na have been measured using resonant scattering of radioactive ^{19}Ne beams on polyethylene targets. The results are analyzed in the framework of the Breit-Wigner formalism. The actual experimental limitations are discussed and future developments are pointed out.

1. Introduction

The pioneering project at Louvain-la-Neuve using the two cyclotron concept to produce intense radioactive nuclear beams (RNB) has provided ^{13}N and more recently [1] ^{19}Ne beams in the energy range $10 < E_{lab} \lesssim 80$ MeV. The beam intensity on target is of the order of 10^9 particles per second. The low-lying resonances in ^{20}Na at excitation energies of 2.6 - 3.1 MeV are of particular interest to Nuclear Astrophysics :

The radiative capture ^{19}Ne (p,γ) at $T_9 > 0.5$ leads to a breakout from the hot CNO cycles and may serve as a trigger for the rp-process [2] building up heavier elements in a rapid sequence of explosive H-burning. The capture reaction ^{19}Ne $(p,\gamma)^{20}$Na in a hot stellar environment (e.g. nova) is dominated by a single resonance at $E_{cm} \sim 450$ keV, whose spin and parity as well as its total and γ widths are unknown, the latter are expected to be of the order of a few meV [3]. A direct measurement of the γ width of this resonance is foreseen to be carried out in the near future in Louvain-la-Neuve.

2. Resonant Scattering

The resonance parameters E_{cm} (res), Γ_{tot} and spin and parity J^π can be obtained from resonance scattering using a radioactive beam on a hydrogen target. The heavy ion loses a considerable fraction of its initial laboratory energy in the thick polyethylene target ($[CH_2]_n$) and the recoiling proton scans the respective range of excitation energies. The experimental method and theoretical analyses have been described previously [4,5] and the merits of using a RNB are discussed in [6]. Preliminary results of a direct measurement searching for the lowest lying resonance above the $^{19}Ne + p$ threshold in ^{20}Na at $E_{cm} \sim 450$ keV were presented [7]. Approximate values of E_{cm} and J^π of several low-lying resonances in ^{20}Na have been given based on indirect methods [3,8,9] using the charge exchange reactions ($^3He,t$) and (p,n) on stable ^{20}Ne targets. Such measurements are valuable as they provide a first glimpse at the relevant resonance states. However, a full analysis would require the exact knowledge of the underlying reaction mechanism, the intermediate and final state interactions, etc. The direct method of resonant scattering through the nuclear state is preferable, as it allows to determine E_{cm} (res), Γ_{tot} and J^π reliably, provided the penetration through the barrier can be parametrized unambiguously ; this is usually the case for subbarrier energies, where the penetrability approaches exp $(-2\pi\eta)$ - here η is the Sommerfeld parameter.

We report on measurements using ^{19}Ne beams of $E_{lab} = 19.2$ and 20.5 MeV on 200-500 $\mu g/cm^2$ thick $[CH_2]_n$ targets. The protons recoiling into the forward hemisphere were detected in PIPS Si solid state detectors at laboratory angles of $\theta_L = 0°$, 14°, 17°, 27°, 32° and 37° ; the ^{19}Ne beam was stopped in a 3 mg/cm^2 Al foil, the β^+ of the RNB were deflected away from the detector at $\theta_L = 0°$ by a strong magnetic field. For $\theta_L > 0°$, the recoiling protons were separated from elastically scattered ^{19}Ne and ^{12}C ions by means of a time-of-flight signal, derived from the cyclotron RF and the Si detector. The energy scale was calibrated precisely using a 3 line α source and a pulser (at the low energy end) in on-line mode. In addition, resonant scattering from known resonances using an isobaric ^{19}F beam served as a reference. Two broad resonances were observed in $^1H(^{19}Ne,p)$ at $E_{cm} = 781$ and 869 keV, respectively. The resonances were analyzed in the framework of the Breit-Wigner formalism by taking into account all interference effects of the type resonance-Coulomb and resonance-resonance in the differential scattering cross section. The orbital angular momenta l = 0 and l = 2 had to be considered, to obtain a good fit for the resonance at $E_{cm} = 781$ keV. The resonance at $E_{cm} = 869$ keV required only l = 0 for a good fit. The results are summarized in table 1 and full details will be given in a forthcoming paper. The data including Breit-Wigner fits are displayed in Fig.1.

Table 1 : Resonance parameters deduced from a Breit-Wigner analysis of resonant scattering

E_{cm}(res) [keV]	Γ_{tot} [keV]	$E^*(^{20}Na)$ [MeV]	J^π
~ 450	< 1	~ 2.65	
780.5 ± 1	33.6 ± 3	2.980	1^+
869.0 ± 2	38.7 ± 6	3.068	0^+

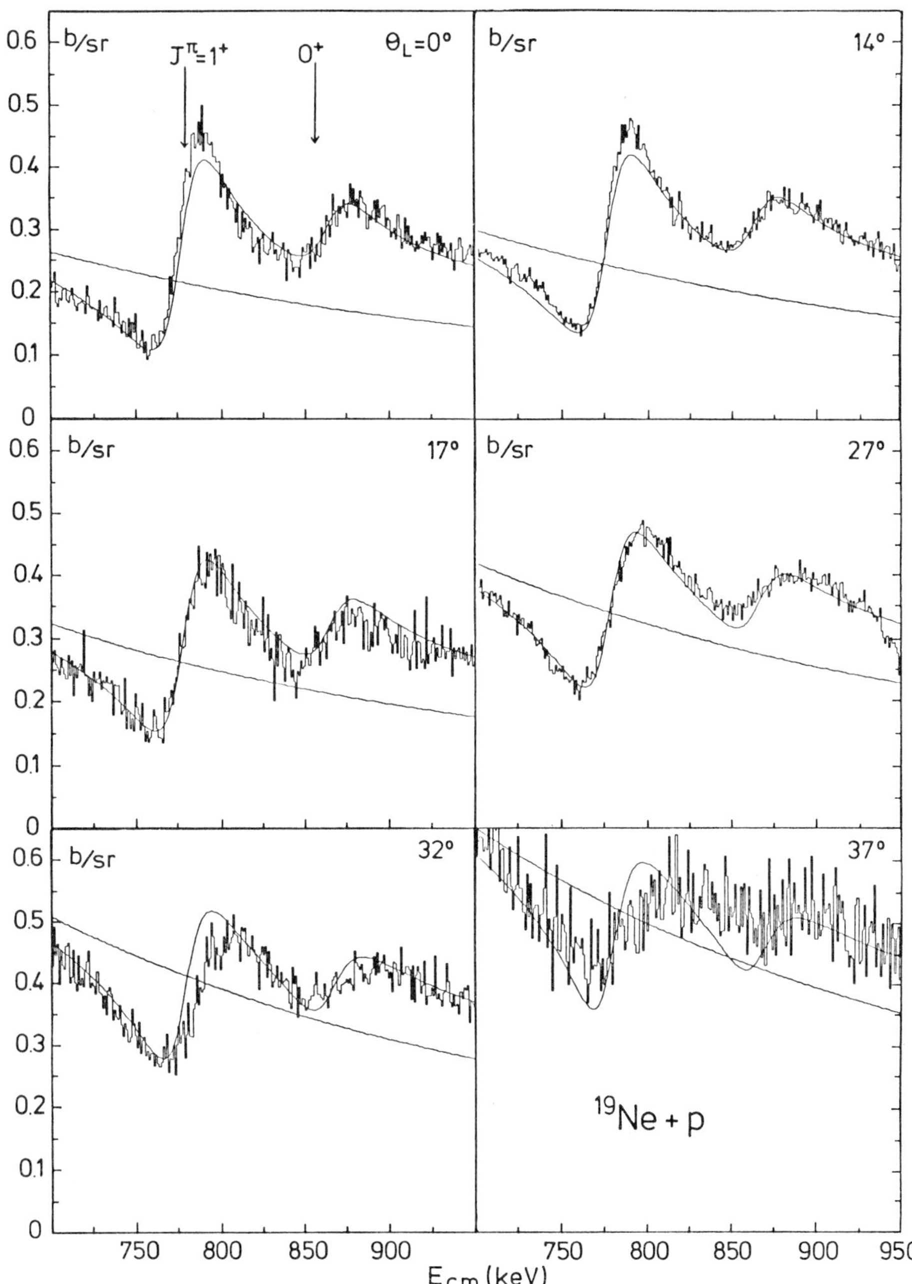

Fig.1 Excitation functions for resonant scattering ^{1}H(^{19}Ne,p) ; the full lines are Breit-Wigner fits and Coulomb scattering, respectively.

The spin assignments are in agreement with [3], however, the resonance energies given in [3] are 55 and 32 keV higher, respectively. The measured resonance energies are in rather good agreement with the ones given in [8,9], where $J^\pi = 3^-$ and $J^\pi = 1^+$ assignments were adopted and the $J^\pi = 0^+$ level was not observed. The observed discrepancies may indicate uncertainties of experimental or theoretical nature inherent in the use of charge-exchange reactions, as mentioned already.

Our final aim regarding the current investigation is to determine the resonance parameters of the lowest lying resonance at $E_{cm} \sim 450$ keV. A detailed analysis of first data [7] yields an upper limit for the total width $\Gamma_{tot} < 1$ keV [6]. This limit can be lowered to $\sim$ 0.1 keV by improving the statistical error by about a factor 10. A large microstrip detector array is under construction using integrated electronics developed jointly by the University of Edinburgh and the Rutherford-Appleton Laboratory [10]. The Louvain-Edinburgh Detector Array (LEDA) will offer an improvement in solid angle over the existing setup in excess of a factor 100 and could be operational by early 1993. This might well be sufficient to determine the energy and the width of the resonance and support the spin and parity assignment given in [11]. LEDA will be a powerful device to search for low-lying resonances using RNBs.

References

[1]   Decrock P, Delbar Th, Duhamel P, Galster W, Huyse M et al in *Radioactive Nuclear Beams* 1992 Adam Hilger Ed. Delbar Th 121
[2]   Champagne A E and Wiescher M 1992 priv. comm. and to be published *Ann Rev Nucl Part Sci*
[3]   Lamm L O, Browne C P, Görres J, Graff S M, Wiescher M et al 1990 *Nucl. Phys.* **A510** 503
[4]   Galster W, Leleux P, Licot I, Liénard E, Lipnik P et al 1991 *Phys. Rev.* **C44** 2776
[5]   Benjelloun M, Delbar Th, Duhamel P, Galster W, Leleux P et al 1992 accepted for publication *Nucl. Instr. Meth.A*
[6]   Leleux P 1992 this conference
        Rolfs C E and Rodney W S, *Cauldrons in the Cosmos* 1988 The University of Chicago Press
[7]   Galster W in *Radioactive Nuclear Beams* 1992 Adam Hilger Ed. Delbar Th 375
[8]   Kubono S, Orihara H, Kato S, Kajino T 1989 *Astrophys.* **J344** 460
[9]   Smith M S, Magnus P V, Hahn K J, Howard A J, Parker P D et al 1992 *Nucl. Phys.* **A536** 333
[10]   Thomas S L, Davinson T and Shotter A C 1989 report RAL-89-063
        Davinson T, Shotter A C, MacDonald E W, Springham S V, Jobanputra P et al 1990 *Nucl. Instr. Meth.* **A288** 245
[11]   Descouvemont P, Baye D 1990 *Nucl. Phys.* **A517** 143

Measurement of the ^{13}N(d,n)^{14}O reaction cross section

P. Decrock, M. Gaelens, M. Huyse, G. Reusen, G. Vancraeynest, P. Van Duppen and J. Wauters

Instituut voor Kern—en Stralingsfysica, K.U. Leuven, Belgium

Th. Delbar, W. Galster, P. Leleux, E. Lienard, P. Lipnik, C. Michotte and J. Vervier

Institut de Physique Nucléaire, U.C.L., Belgium

K. Gruen and H. Oberhummer

Institute of Nuclear Physics, Technical University Wien, Austria

Abstract. The cross section of the ^{13}N(d,n)^{14}O reaction has been measured by using a radioactive beam of ^{13}N with an energy of 8.2, 12.0, 16.2, and 28.5 MeV. By comparing the experimental cross sections with the results of a D.W.B.A. calculation, a value of 0.90±0.23 is obtained for the spectroscopic factor S ($[^{13}$N,p$|\,^{14}$O$_{gs}]$).

1. Introduction

Recently we have reported on a proton—capture cross section measurement using an energetic radioactive ^{13}N beam :^{13}N(p,γ)^{14}O (Decrock P. et al 1991). This reaction is dominated by a resonance at E_{cm} = 526 keV. The non—resonant direct capture contribution to the ^{14}O ground state can play a role at low temperatures ($\leq 10^8$ K) where due to the interference effect with the resonant contribution an enhancement of the total cross section by a factor of 2, with respect to the resonant contribution, might be possible (G.J. Mathews and F.S. Dietrich 1984). For direct capture reactions this non—resonant cross section (σ) can be written as the product of a spectroscopic factor (S) and a theoretical cross section (σ_{theo}) in the potential model (for example for an E1 transition : σ(E1)=S $\times$ σ_{theo}(E1)) (Oberhummer 1992).

In the case of a one particle transfer reaction with orbital angular momentum l, the cross section (σ_l) can, apart from some numerical factors, be written as the product of a spectroscopic factor (S) and a calculated cross section (σ_l(DW)) (σ_l=S $\times$ σ_l(DW)). This calculated cross section σ_l(DW) can be obtained with a Distorted Wave Born Approximation (D.W.B.A.) in the potential model (Oberhummer 1992). By comparing σ_l(DW) with σ_l a so called experimental spectroscopic factor S_{exp} is derived. It has been shown that the spectroscopic factors derived from radiative capture reactions are in good agreement with the results obtained out of one particle transfer reactions (C. Rolfs 1973, C. Rolfs and R.E. Azuma 1974, C. Iliadis et al 1992).

Transfer reactions involving short living nuclei such as ^{13}N (T1/2=9.96 m) can only be measured in reverse kinematics using radioactive nuclear beams. With the the Belgian Radioactive Ion Beam facility (Darquennes et al 1990) ^{13}N beams are available enabling these type of measurements .

2. Experimental set–up and results

A ^{13}N beam (8.2, 12.0, 16.2 and 28.5 MeV) was sent on a deuterated polyethylene target $((CD_2)_n$, 100–200 $\mu g/cm^2)$. The different beam energies correspond to 1.0, 1.50 2.0 and 3.75 MeV in the centre of mass system respectively. The reaction rate of the ^{13}N(d,n)^{14}O reaction is determined by detecting the β delayed 2.3 MeV γ ray of ^{14}O (T1/2=70.59 s). A schematical drawing of the detection set–up is shown in figure 1. The ^{13}N beam and the ^{14}O reaction products are implanted in an aluminized mylar tape which is then moved, after an implantation time of 128 s, to a decay station where the 2.3 MeV γ rays are detected by two Ge detectors (70 and 90 %). In order to reduce the 511 keV annihilation count rate from the β^+ decaying ^{13}N atoms, 3 to 4 cm of Pb is placed between the source and the detectors. Furthermore, the detectors were surrounded with 10 cm of Pb in order to reduce the ambient background.

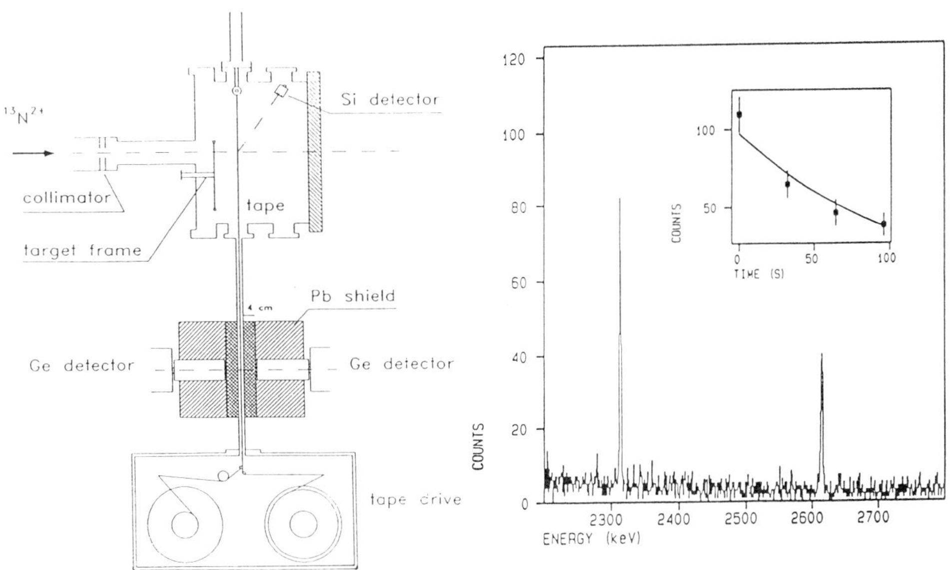

Fig. 1 *Experimental set–up*

Fig. 2 *γ–ray spectrum showing the 2.3 MeV line from the decay of ^{14}O*

     Figure 2 shows an example of a γ spectrum (energy of the ^{13}N beam = 12.0 MeV). The 2.313 MeV γ line is clearly observed and the inset shows the time behavior of its intensity which proves its assignment to the decay of ^{14}O (the curve is a fit with a fixed 70.59 s half live). In the reaction chamber, a silicon detector is mounted which records the scattered ^{13}N atoms and the recoiling deuterons. Out of these particle spectra one can obtain information concerning the target thickness, the target homogeneity and the in– and outcoming beam energies. For the three lowest beam energies the intensity of the deuteron peak can also be used for normalization of the cross section by assuming pure Rutherford scattering. The results of the total cross section measurements are shown in figure 3. The error bars on the data points contain the statistical error and the error on the detection efficiency at 2.3 MeV. An error bar of 15%, mainly due to the uncertainty in the stopping power of the nitrogen beam in the polyethylene target, should be taken into account for the absolute scaling of the cross section.

3. Distorted Wave Born Approximation

The cross section calculations are performed using the D.W.B.A. code TETRA (Bach B. et al 1990). The real part of the potential for the entrance and exit channel is calculated using a folding procedure. Also for the bound state a folding procedure is used and the depth of the potential is adjusted to reproduce the proton binding energy. For the imaginary part a potential of the Wood–Saxon type (surface) is used. The potential parameters of the imaginary part are fixed to reproduce the experimental cross sections of the reactions $^9Be(d,n)^{10}B$ and $^{14}N(d,n)^{15}O$ for which the spectroscopic factors are taken out of the literature (Park et al 1973, Miura K. et al 1992, Ritter R.C. et al 1970, Schiffer J.P. et al 1967, Phillips G.W. and Jacobs W.W. 1969).

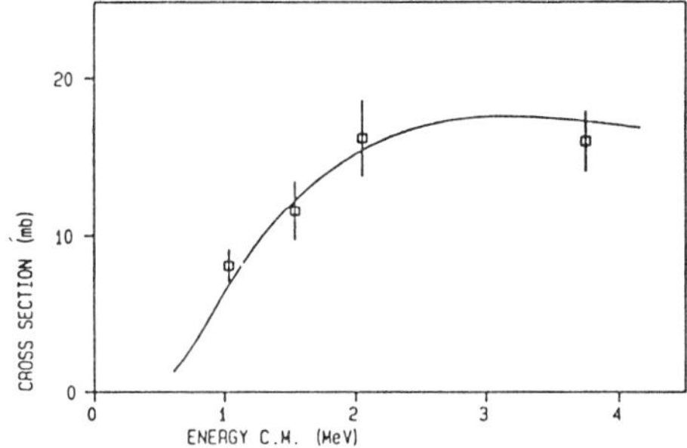

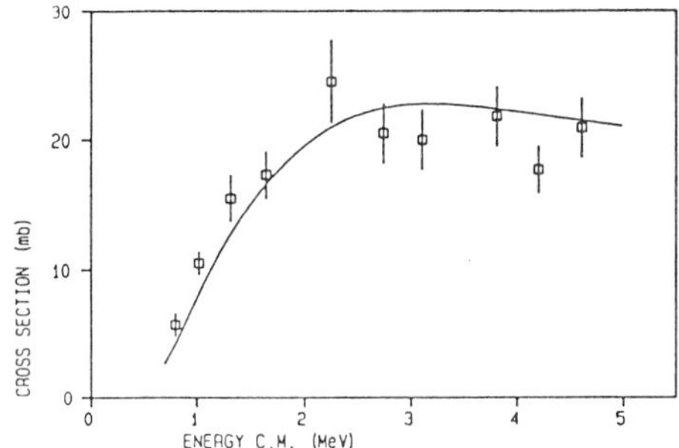

Fig. 3. $^{13}N(d,n)^{14}O$ cross section with D.W.B.A. fit

Fig 4. $^{14}N(d,n)^{15}O$ cross section with D.W.B.A. fit

Figure 4 shows as an example the experimental cross section of the $^{14}N(d,n)^{15}O$ (Retz–Schmidt and Weil 1960) reaction together with the D.W.B.A. calculation (S=0.95). The $^{13}N(d,n)^{14}O$ reaction cross section is calculated with the same potential parameters and the spectroscopic factor is adjusted in order to reproduce the experimental cross sections. The result of the D.W.B.A. calculation for the $^{13}N(d,n)^{14}O$ reaction is shown in figure 3. The calculations clearly reproduce the energy behavior of the cross sections, namely the penetration of the Coulomb barrier which lies around 2.2 MeV. As the experimental spectroscopic factors S taken from literature for $^{10}B_{gs}$ and $^{15}O_{gs}$ vary with about 20%, the same error bar should be taken into account for the obtained value for $^{14}O_{gs}$. Including 15% for the absolute scaling error of the experimental $^{13}N(d,n)^{14}O$ reaction cross section leads to an uncertainty of 25%. The derived spectroscopic factor in this way is 0.90±0.23. This result clearly deviates from the predicted theoretical value of 1.73 given by Cohen S. and Kurath D. 1967.

4. Conclusion

In this contribution, we described a cross section measurement of the $^{13}N(d,n)^{14}O$ reaction using a post–accelerated radioactive ^{13}N (T1/2=9.96 m) beam. The data were interpreted in the framework of a D.W.B.A. calculation from which a spectroscopic factor was deduced. This spectroscopic factor will be used to calculate the non–resonant proton capture cross section of the $^{13}N(p,\gamma)^{14}O$ reaction. These calculations are in progress.

References

Bach B., Gruen K., Raimann G. and Rauscher T., Technical Universty, Wien *Computer code TETRA, unpublished*
Cohen S. and Kurath D. 1967 *Nucl. Phys.* A101 1–16
Darquennes D. et al 1990 *Phys. Rev.* C42 R804
Decrock P. et al 1991 *Phys. Rev. Lett.* 67 808
Iliadis C. et al 1992 *Nucl. Phys.* A539 97–111
Mathews G.J. and Dietrich F.S. 1984 *The Astrophys. Journ.* 287 969–976
Miura K. et al 1992 *Nucl. Phys.* A539 441–450
Oberhummer H. editor 1992 *Nuclei in the Cosmos,* Springer Verlag, Heidelberg
Park Y. S. et al 1973 *Phys. Rev.* 8 1557
Phillips G.W. and Jacobs W.W. 1969 *Phys. Rev.* 184 1052
Retz–Schmidt Th. and Weil J. L. 1960 *Phys. Rev.* 3 119 1079
Ritter R.C. et al 1970 *Nucl. Phys.* A140 609–624
Rolfs C 1973 *Nucl. Phys.* A217 29–70
Rolfs C. and Azuma R.E. 1974 *Nucl. Phys.* A227 291–308
Schiffer J.P. et al 1967 *Phys. Rev.* 4 164 1274

Acknowledgements

This work was supported in part by a grant from the Belgian government (I.U.A.P. program). The authors are grateful to the cyclotron crew of Louvain–la–Neuve for delivering excellent beams.

Breakup of light ion projectiles—astrophysical interest

A C Shotter

Physics Department, The University of Edinburgh, Edinburgh EH9 3JZ, Scotland, UK

Abstract. A study of nuclear projectile breakup under the influence of the Coulomb field of a target nucleus, provides a possible indirect way to infer the strengths of some radiative capture reactions of astrophysical interest. However for such extracted information to be reliable, several stringent conditions have to be satisfied; these conditions are identified and discussed.

1. Introduction

The collision between two nuclei at an energy of about 10 MeV/A can involve a variety of complex phenomena. For central collisions (i.e. for collisions where the impact parameter is small compared to the sum of nuclei radii) most of the nucleons of both projectile and target will be involved in exchanging substantial energy and momentum between target and projectile; complete fusion can take place. For larger impact parameters only a subset of overlapping surface nucleons will initially be involved in large exchanges of energy and momentum. For even larger impact parameters, where the two nuclei surfaces just graze each other, the exchange of energy between the nuclei will be substantially reduced. Indeed inelastic collisions in this region take on the character of modified elastic collisions; hence the name for such reactions as quasielastic.

One of the the simplest quasielastic processes is inelastic excitation of either target or projectile with no transfer of mass. If the projectile is excited then it will generally move quite far away from the reaction site before decaying. For low excited states the decay will involve γ-ray emission, for higher excited states decay will occur by the emission of nucleons or cluster of nucleons.

Such inelastic excitation, followed by particle emission, is a reaction that, in recent years, has been considered as a possible indirect way to determine certain nuclear reaction strengths that have astrophysical interest. This indirect way of measuring astrophysical data forms the topic of this paper. The aim of the paper is mainly to review the work of the Edinburgh group to this field. For this reason most discussion will be concerned with nuclear collisions initiated by light ion projectiles of about 10 MeV/A.

2. Quasielastic breakup reactions

Quasielastic reactions where the projectile is first excited before subsequently decaying into several fragments are called sequential breakup reactions, and are characterised by the fact that centre of mass energy (c.m.) of the breakup fragments, ε, is fixed at a value equal to the difference between the energy of the excited state and the threshold energy for dissociation of the projectile into the fragments.

In 1981 a new type of quasielastic breakup reaction was reported for 70 MeV ^{7}Li projectiles scattering off ^{120}Sn nuclei and where the ^{7}Li broke up into an α and t fragment [1].   For this process, and in contrast to sequential breakup, the c.m. energy of the α and t, ε, was not fixed but rather extended over a range of values. This surprising result showed that it was possible for such ^{7}Li projectiles to rapidly break up into α and t fragments under the influence of the projectile-target interaction without passing through any intermediate excited states. This new process was called direct breakup. Effectively what is happening is that when the differential, or tidal force, across the projectile due to the target-projectile interaction exceeds the projectile's cohesion strength then the projectile rapidly fragments. Further work [2] to investigate this direct breakup led to the conclusion that the Coulomb field between projectile and target could largely account for the breakup process for those events where the fragments were scattered to forward angles. The main reason for this conclusion was that it was possible to account for most of the experimental features of direct breakup data of ^{7}Li by a semiclassical model of Coulomb excitation [3]. For this model it is assumed that the projectile follows an orbit simply determined by the monopole moment of the Coulomb repulsion between projectile and target, and the breakup of the projectile is determined by the other higher multipoles of the interaction.

This Coulomb theory of breakup yields a simple factored expression for the breakup cross section for fragments with c.m. energy ε:

$$\sigma_b(\theta) \propto f(\theta,\varepsilon) B(\varepsilon)$$

where $\sigma_b(\theta)$ is the differential cross section for breakup where the centre of mass of the fragments defines a trajectory at an angle θ relative to the projectile beam direction; $f(\theta,\varepsilon)$ is a function determined by the Coulomb orbit of the projectile [3], and $B(\varepsilon)$ is the reduced probability function for Coulomb excitation of the ^{7}Li projectile to the α+t continuum. It was proposed [2] that information about $B(\varepsilon)$ could be determined from the inverse radiative capture reaction i.e. fusion of an α and t of c.m. energy, ε, with the emission of a photon E_γ, specifically:-

$$B(\varepsilon) \propto \frac{\sigma_c(E_\gamma)}{E_\gamma}$$

where $\sigma_c(E_\gamma)$ is the radiative capture cross section.

For this model a simple relationship therefore exists between the cross sections for breakup and radiative capture:-

$$\sigma_b(\theta) \propto f(\theta,\varepsilon) \frac{\sigma_c(E_\gamma)}{E_\gamma} \tag{1}$$

Using available radiative α+t capture data, it was shown in [2] that the features of the experimental breakup data of ^{7}Li could be predicted in a very satisfactory way by this simple Coulomb model.

3. Astrophysical implications of breakup reactions

Following the apparent success of describing the breakup of 70 MeV ^{7}Li in terms of radiative capture [2], it is of obvious interest to ask if the relationship may be turned around, i.e. could breakup data be used to *infer* radiative capture strengths. This possibility of extracting radiative capture data has been purchased in particular by the Karlsruhe group [4].

At first consideration this possibility seems an attractive idea because of the following reasons:

I) Due to phase space considerations Coulomb breakup cross sections are considerably larger than radiative capture cross sections.

II) The breakup fragments are directed into a narrow forward cone, which makes for efficient detection.

III) Since small values of ε (the fragment c.m. energy) are related to the opening angle between the fragment paths, θ_{12}, the fragment reduced mass μ, and projectile velocity v_p, by

$$\varepsilon = 2\mu v_p^2 \sin^2\left(\frac{\theta_{12}}{2}\right),$$

information concerning radiative capture reactions near $\varepsilon \approx 0$, may be studied with breakup reactions for which $\theta_{12} \approx 0$; since the energy to mass ratios of the fragments are nearly the same as the projectile for this situation, the precision with which ε can be extracted is determined mainly by the precision of θ_{12} measurement.

This last feature, i.e. the possibility of using breakup reactions to study radiative capture cross sections at very low energy, is the reason why this method could be of particular use to study radiative capture reactions of nuclear astrophysics interest.

4. Extraction of radiative capture cross sections : problems

The relationship (eq. 1) between Coulomb breakup and radiative capture follows directly from the principle of detailed balance. The original use of this relationship was intended only to provide a rough indication of the strength of the Coulomb processes for breakup reactions [2]. However to go further and attempt to deduce accurate and reliable radiative capture cross sections from breakup cross sections, is a much more difficult task. For this to be successful it is essential that all aspects of the breakup mechanism are fully understood, and higher order Coulomb processes are fully accounted for. In general, neither of these two conditions have been fulfilled so far. In particular, for projectile -target kinetic energies greater than the Coulomb barrier, the nuclear interaction between the reacting participants plays a significant, and even dominant role. Also the relationship (eq. 1) is only valid for the simplest Coulomb assumption; higher order processes such as different final state Coulomb accelerations for the breakup fragments degrade the usefulness of eq. 1.

The following subsections discuss recent results concerning the influence of the nuclear force, and the Coulomb fragment final state interactions (f.s.i.) on the relationship between radiative capture and breakup reactions.

4.1.1 Nuclear interaction: experimental

The relationship (eq. 1) only concerns that part of the breakup process that is Coulomb in origin. If a significant breakup contribution arises from the nuclear interaction between projectile and target, then eq. 1 will fail to predict the experimental breakup cross section. To investigate the importance of this nuclear breakup, several breakup reaction situations have been investigated experimentally and cross sections have been studied in relation to eq 1.

The breakup of 50 to 70MeV ^{7}Li projectiles into the α+t channel has been studied for ^{96}Zr, ^{120}Sn and ^{208}Pb targets [5]. The experimental breakup cross sections were compared with those calculated from eq. 1 for angles smaller than the grazing angle (this is the angular range for which the Coulomb interaction is most likely to dominate). The experimental value of σ_b for ^{96}Zr and ^{120}Sn targets are in reasonable agreement with those calculated from eq. 1 using known experimental values of σ_c for α+t radiative capture. However for the ^{208}Pb target the experimental values are considerably higher than those calculated from eq. 1. For this reaction, this may indicate a considerable nuclear component to the breakup cross section.

The breakup reaction ^{9}Be$\rightarrow$^{8}Be+n is predicted by eq. 1 to have a large Coulomb cross section. This reaction has been studied at an energy of 10 MeV/A for breakup from a ^{120}Sn target [6]. For angles smaller than grazing, the form of the breakup cross section as a function of ε is reasonably reproduced by eq. 1, but the magnitude of the cross section is in poor agreement. Recent results show a similar situation applies for the breakup of 10 MeV/A ^{10}B projectiles into the ^{9}Be+p channel scattered from a ^{58}Ni target [7]. The breakup of 156 MeV ^{6}Li into the α+d channel has been studied in some depth by the Karlsruhe group [4], and they report results effectively in good agreement with eq. 1.

Further experiments concerning the breakup of ^{7}Li have been conducted by Mason et al. [8] and Utsunomiya et al. [9]. While Utsunamiya et al. do not specifically consider the nuclear component, Mason et al. conclude that the nuclear force plays an important role in the breakup process.

Studies of the breakup of polarised ^{7}Li have been undertaken in an attempt to further elucidate the reaction mechanism [10]. So far only the $^TT_{20}$ analysing powers for the breakup of 70 MeV ^{7}Li have been measured. These measurements give some support to the conclusion that the breakup is strongly influenced by the Coulomb force for angles smaller than grazing, but they cannot rule out the possibility of a significant nuclear component.

In summary, for the breakup of light projectiles where the fragments are scattered forward and inside the grazing angle, the evidence from the experimental data concerning the relative importance of the Coulomb to nuclear interaction is inconclusive. However for breakup of ^{7}Li and ^{9}Be where the fragments are scattered to wide angles there is clear and unambiguous evidence that the nuclear interaction plays the dominant role [11]. Since there is no a priori reason why this nuclear interaction should become insignificant for events where the fragments are scattered forward, it must be concluded that in general nuclear breakup will remain important even for forward scattered reactions.

4.1.2 Nuclear interaction : theoretical

An alternative way to gain an understanding of the relative importance of nuclear to Coulomb interactions for breakup reactions is to calculate the reaction transition probability from a knowledge of the structure of the projectile and fragments, and taking full account of the interaction between projectile, fragments and target. However such calculations are very difficult because of the three interacting particles in the final channel. Therefore for all such calculations approximations have to be introduced (e.g. [4], [12], [13], [17]).

Of the various methods of calculation, the continuum-discretized coupled channels (CDCC) calculation is probably the most reliable, since such calculations are very successful in predicting inelastic scattering amplitudes for those reactions which have a strong breakup component [13]. CDCC calculations [17] for the breakup of 156 MeV ^{6}Li indicate that the nuclear interaction plays a dominant role at forward angles; this is in direct contrast to the Karlsruhe group conclusions [4].

However even for the best calculations of the CDCC type, some uncertainties still exist due to the approximations used. For example, for all CDCC calculations it is necessary to limit the fragment momentum ranges that can be included in the calculation (this is necessary simply to restrict the computer calculation times to reasonable limits). This limitation means that the effects of fragment-target Coulomb and nuclear interactions are only partially accounted for. Nevertheless, even with this limitation the CDCC calculations currently provide the best guide for interpretation of breakup mechanisms.

4.2 Coulomb final state interaction effects (f.s.i.)

The relationship eq. 1 is derived from the assumption that the projectile and fragments all follow the same Rutherford scattering trajectory determined by the Coulomb repulsion between the projectile and target charge. For the case of direct breakup the dissociation of the projectile into fragments takes place near the point of closest approach to the target; at this point the kinetic problem involves not just two but three particles. Restricting the projectile and fragments to a common trajectory assumes that the solution to the three body problem is well represented by the two body solution. This will be true only if a) the charge to mass ratio for both fragments are the same, and b) the spatial separation between the fragments is zero. Condition a) is often the only one quoted when justifying the common trajectory assumption, however condition b) is just as important, and difficult to meet because, clearly, the fragments must be moving apart if breakup has taken place. If these conditions a) and b) cannot be met, then the fragments will undergo a differential relative acceleration due to their Coulomb interaction with the target. The effect of these Coulomb final state interactions will be that the final experimental measured fragment relative energy, ε_{exp}, will be different from the actual value of ε at the breakup vertex. From the experimental measurements of the two fragments it is not possible to relate ε_{exp} to ε. Further, the f.s.i. interactions also complicate how the final trajectory of the centre of mass of the fragments relates to a particular projectile impact parameter. Both the effects detailed here will reduce the accuracy with which the radiative capture cross sections can be extracted from the breakup cross sections.

Following earlier estimates of Coulomb f.s.i. effects [14], recent accurate 3 body semiclassical calculations have been undertaken for breakup fragments moving in the Coulomb field [15]. These show for example, that for the breakup of 70 MeV ^{7}Li projectiles scattered from ^{120}Sn, and for an initial breakup vertex value of ε of precisely 1 MeV, the distribution of ε_{exp} has a mean value of about 1.25 MeV and a spread of about 0.5 MeV. This difference between ε and ε_{exp} introduces a significant uncertainty in the use of eq. 1 to extract the radiative capture cross sections from the experimental breakup cross sections.

5. Future prospects and conclusions

The use of eq. 1 to extract σ_c from σ_b relies on the assumption that there cannot be any (or at least very little) contribution to σ_b from the projectile-target nuclear interaction. As discussed above, at 10 MeV/A projectile energy, most evidence indicates that σ_b contains considerable contributions from both the nuclear and Coulomb interactions.

However, this situation is likely to improve as the projectile energy increases, because the Coulomb cross section for any process becomes rapidly larger at the grazing angle with increasing energy. How fast the corresponding nuclear component increases is not certain, but it is unlikely to increase as rapidly as the Coulomb term, and is most likely to maximise at some projectile energy, and then decrease.

It should be noted that even at high energies the Coulomb f.s.i. will still cause problems as indicated in section 4.2. However, if information is to be sought only for *resonant* radiative capture, then the f.s.i. do not cause in general any serious problems, because the corresponding breakup reaction concerns breakup of the excited projectile which takes place far removed from further interaction with the target.

The original use of the relationship linking breakup processes to capture reactions was to gain some insight into breakup mechanisms. The use of this relationship to extract σ_c from σ_b is much more difficult since it requires a very detailed knowledge of the breakup mechanism; without this, there can be little confidence in the extracted values of σ_c. Can we rely on this method at all? It all depends on the reaction studied. The best chance for success would be a) to use a very high energy projectile beam, and b) confine the method to the extraction of *resonance* parameters associated with radiative capture. Recent experiments concerning the high energy breakup $^{14}O \rightarrow {}^{13}N+p$ seem to comply with these conditions [16]. However, even for this reaction the message of caution expressed in [17] should not be ignored.

Acknowledgements

I would sincerely like to thank my past and present colleagues with whom I have had the pleasure to work with on the topics discussed in this paper: D Branford, E Macdonald, V Rapp, T Davinson, I Yorkston, N Davis, M Nagarajan, P Woods, C Shepherd Themistocleous, Y El-Mohri and Q Khan.

References

[1]   Shotter A C, Bice A N, Wouters J M, Rae W D and Cerny J 1981 *Phys. Rev. Lett.* **46**,12

[2]   Shotter A C, Rapp V, Davinson T, Branford D, Sanderson N E and Nagarajan M A 1984 *Phys. Rev. Lett.* **53** 1539

[3]   Alder K, Bohr A, Huus T, Mottelson B and Winther A 1956 *Rev. Mod. Phys.* **28** 432

[4]   Kiener J, Gsottschneider G, Gils H J, Rebel H, Corcalcive V, Basu S K, Baur G and Raynal J 1991 *Z. Phys.* A **339** 489
        Shyam R, Baur G and Banerjee P 1991 *Phys. Rev.* **C44** 915

[5]   Shotter A C, Rapp V, Davinson T and D Branford 1988 *J. Phys. G: Nucl. Phys.* **14** 169

[6]   Macdonald E W, Shotter A C, Branford D, Rahighi J, Davinson T and Davis N J 1992 *Phys. Lett.* **B283** 27

[7]   El-Mohri Y 1992 *PhD Thesis, Edinburgh University*

[8]   Mason J E, Gazes S B, Roberts R B and Teichmann S G 1992 *Phys. Rev.* **45C** 2870

[9]   Utsunomiya H, Lui YW, Cooke L, Dejbakhsh H, Haenni D R, Heimberg P, Ray A, Srivastava B K and Schmitt R P 1990 *Nucl. Phys.* **A511** 379

[10]   Shepherd-Themistocleous C 1992 *PhD Thesis, Edinburgh University*

[11]   Yorkston J, Shotter A C, Davinson T, Macdonald E W and Branford D 1991 *Nucl. Phys.* **A524** 495
        Shotter A C 1989 *J. Phys. G: Nucl. Phys.* **15** L41

[12]   Thompson I J and Nagarajan M A 1983 *Phys. Lett.* **123B** 379

[13]   Sakuragi Y, Yahiro M and Kamimura M 1986 *Prog. Theor. Phys. Suppl.* **89** 136

[14]   Shotter A C and Nagarajan M A 1988 *J. Phys. G: Nucl. Phys.* **14** L109

[15]   Khan Q 1992 *Private Communication*

[16]   Kiener J, Lefebvre A, Aguer P, Bacri C O, Bimbot R, Bogaert G, Borderie B, Clapier F, Coc A, Disdier D, Fortier S, Grunberg C, Kraus, L, Linck I, Pasquier G, Rivet M F, St Laurent F, Stephan C, Tassan-Got L, Thibaud J P 1991 *Contrib. 2nd Int. Conf. Radioactive Beams, Louvain la Neuve*
        Motobayashi T, Takei T, Kox S, Perrin C, Merchez F, Rebreyend D, Ieki K, Murakami H, Ando Y, Iwasa N, Kurokawa M, Shirato S, Jian-shi Ruan (Gen), Ichihara T, Kubo T, Inabe N, Goto A, Kubono S, Shimoura S and Ishihara 1991 *Phys. Lett* **B264** 259

[17]   Hirabayashi Y and Sakuragi 1992 *Nucl. Phys. Group preprint Osaka City University OCU NP 5*

What are the conditions for the application of the Coulomb dissociation approach in nuclear astrophysics?

H. Rebel[1], G. Baur[2], H.J. Gils[1], and J. Kiener[3]

1)   Kernforschungszentrum Karlsruhe, Institut für
        Kernphysik, P.O.B. 3640, D - 7500 Karlsruhe, Germany

2)   Forschungszentrum Jülich, Institut für Kernphysik, P.O.B.
        1913, D - 5170 Jülich, Germany

3)   Centre de Spectrométrie Nucléaire et de Spetrométrie de
        Masse, F - 91405 Campus Orsay, France

Abstract : *In one-photon approximation, Coulomb dissociation experiments can be directly related to photodisintegration of the projectile, and, by time-reversal, to radiative capture. This can also be of great interest for nuclear astrophysics. The main theoretical problems in the realisation of such an attempt are the influences of nuclear excitation and higher-order electromagnetic interactions. It will be discussed, when and how these effects show up, and how they can be taken into account. Various cases of specific astrophysical interest are considered.*

1.   Introduction

Laboratory nuclear astrophysics is often a frustrating endevour. The desired cross sections are among the smallest measured in the laboratory, often requiring long data-collection times with painstaking attention to background [1]. Radiative capture reactions are among the most important reactions for the formation of some elements. Several years ago, Coulomb dissociation of fast projectiles was proposed as indirect method of several advantages to study radiative capture processes [2]. During the last years, this method has proven to be a new access to a specific class of capture processes which are relevant for nuclear astrophysics. It is the purpose of this contribution to discuss the theoretical analysis of this kind of experiments and to point out the favourable conditions for experimental investigations. The progress in the field can be judged by comparing the present contribution to a short review given about a year ago [3]. A topical review is in preparation [4]. The method and possible background effects are scrutinized with general aspects, and some illustrative examples are discussed.

2.   The Coulomb dissociation approach

The idea is to measure electromagnetic matrix-elements between bound and continuum nuclear states by means of electromagnetic excitation. These matrix-elements are directly related to those of photodisintegration, and, by time-reversal, to those of radiative capture. In order to study the process

$$b + c \rightarrow a + \gamma \tag{1}$$

one uses the reaction

$$a + Z \rightarrow b + c + Z \tag{2}$$

where the Coulomb field of a (heavy) nucleus with proton number Z provides a copious source of equivalent photons. Since one can only start with a projectile nucleus a in the ground state, the present method is restricted to radiative capture which leads to the ground state of that nucleus a. The possibility to say something about the process $a^* + \gamma \rightarrow b + c$, where a^* denotes an excited state of nucleus a, by means of multiple electromagnetic excitation in reaction (2) seems to be remote. On the other hand, radiative capture processes induced by nuclei in excited state b^* (or c^*) are in principle accessible by the Coulomb dissociation method. Such processes can be of interest for nuclear astrophysics. Another point of interest arises from the fact that bare nuclei a, b and c appear in reaction (2). The effects of screening in laboratory and astrophysical environments has been proven to be an important issue [1, 5]; the results from Coulomb dissociation can represent the limit of noscreening.

In order to reach the nuclear continuum by means of electromagnetic excitation, the equivalent photon spectrum should contain the corresponding high frequency component. The "safe bombarding condition" of (sub)-Coulomb excitation implies a severe limitation of the equivalent photon spectrum to rather low energies. Therefore it is advantageous to use projectile velocities well above the Coulomb barrier, in order to overcome the adiabaticity condition

$$\hbar\omega \le \frac{\hbar c}{b}\,(\gamma \cdot \ v/c) \tag{3}$$

where b is the impact parameter and $\gamma = (1-(v/c)^2)^{-1/2}$. Electromagnetic excitation processes occur dominantly for sufficiently large values of $b > R_1 + R_2$, the sum of the nuclear radii, i.e. at very forward angles. Nuclear excitation effects originate from grazing collisions. These grazing collisisons can in principle also lead to forward scattering angles and interfere with the electromagnetic excitation characterized by l-values much beyond grazing. Thus, the "footprints" of nuclear and electromagnetic excitation show up clearly in the angular distribution. These effects can be quantitatively studied in a DWBA approach, using well established computer codes. Simpler methods, of the eikonal type, can be very helpful to explore the physics of the interference of Coulomb and nuclear excitation [6, 7].

A severe and challenging problem arises especially for direct transitions into the continuum : *"post-acceleration"* or *"Coulomb final-state-interaction"* change the relative energy between the outgoing fragments which is determined by the asymptotic kinematics of coincident experimental observation. Since the strength of the Coulomb interaction scales with $1/v$, these effects are expected to be expecially strong for low beam energies. At sufficiently high energies, which are of actual interest, a tremendous simplification occurs due to $\tau_{coll} \ll \tau_{nucl} = \hbar/(E_n - E_0)$. Consequently we can apply the sudden approximation or Glauber theory ("frozen nucleus") [8]. These approaches have the merit that the response of the nucleus a, which is excited, is treated in a fully quantal way. The semiclassical picture of a classical relative motion of projectile and target can be retained. Interesting problems related to the time-dependent response of the projectile system to the transient Coulomb field of the target nucleus arise. It is a hope to use the time-dependent Coulomb field as a "clock" with investigating the time-dependence of nuclear excited states. The investigation of the time-dependent tunnelling problem [9] is of special interest. It is well known [10] that the radial matrix-elements of low-energy radiative capture peak at radii well beyond the range of nuclear interactions. A careful discussion of the time-dependence of the projectile wave function will be necessary to judge the relevance of the overlap of the wave-function of the break-up fragments with the target nucleus. A simple model has been recently developed [11]. However, taking into account the time-dependence of the projectile wave-function, these strong interaction effects are considerably reduced. The example of the harmonic oscillator, perturbed by an external time-dependent force, may illustrate the point (see e.q. 15 of [12]). The time-dependent probability distribution corresponds to the classical harmonic motion of the Gaussian probability distribution of the ground state.

3.   Specific features of astrophysically relevant cases

We illustrate the general theoretical implications by particular astrophysically relevant examples.

a)   *A test case with minimal post-acceleration effects :* ^{6}**Li** $\rightarrow \alpha + $**d**
A full account of the successful Karlsruhe experiment at $E_{6_{Li}} = 156$ MeV has recently been published [13], demonstrating the useful conditions of the approach.

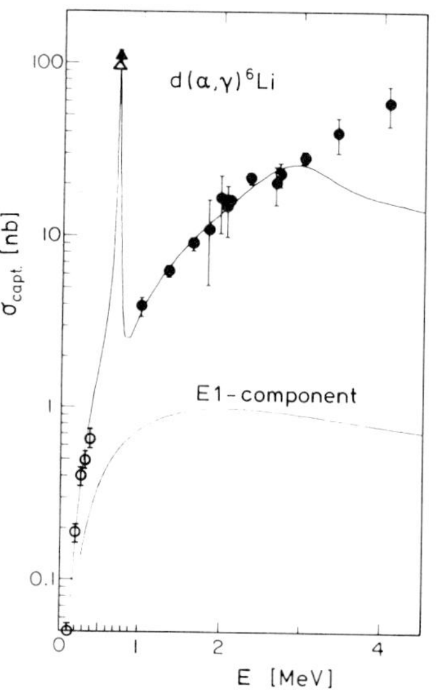

Fig. 1 :   Cross section for the d (α, γ) ^{6}Li capture reaction. The low-energy data (open circles) are deduced from Coulomb dissociation of 156 MeV ^{6}Li projectiles [13].

Though an experiment at lower energies ($E_{6_{Li}} = 60$ MeV [14]) has found integrated cross sections which are in agreement with first order Coulomb excitation theory, the angular correlations of nonresonant break-up exhibit considerable discrepancies. A more thorough theoretical analysis appears to be necessary as the experimental conditions do not guarantee the exclusion of substantial nuclear contributions for this case and higher-order effects should be included.

b)   *A case exhibiting post-acceleration effects :* ^{7}**Li** $\rightarrow \alpha + $**t**
This reaction has been investigated by Utsunomiya et al. [15]. The latest status is reported in ref. [16]. A further experiment at again

rather low beam energies, $E_{7_{Li}}$ = 54 MeV, was performed by Gazes et al. [17]. From simple classical considerations (see e.q. [18]) strong post-acceleration effects have to be expected. Qualitative agreement of the experimental results is achieved with a rather classical model, which take into account the distortions due to the target Coulomb field. However a quantal approach with convincing results is missing. At low beam energies, the sudden sudden approximation is not applicable.

c)   *A case with tractable post-accelerating effects :* $^8B \rightarrow {}^7Be + p$

A current experiment at RIKEN attempts to measure the $^8B \rightarrow {}^7Be + p$ dissociation at E_{8_B} = 50 MeV/A [19]. Since rather low equivalent photon energies are of astrophysical interest, the sudden approximation is well applicable at that beam energy. It is may be expected that a clean theoretical analysis of the potential experimental results is feasible. In view of the solar neutrino puzzle and recent GALLEX results [20], the astrophysical S-factor for the ^7Be (p, γ) reaction is of vital interest.

d)   *A well established case with an exotic beam :* $^{14}O \rightarrow {}^{13}N + p$

The results of two Coulomb dissociation experiments, at E_{14_O} = 87.5 MeV/A and E_{14_O} = 70 MeV/A [22], compare well with each other and with a recent capture experiment [23]. The thermonuclear reaction rate is essentially dominated by the Γ_γ-width of the 5.17 MeV 1^- level in ^{14}O. The experiment Γ_γ = 3.1 ± 0.6 eV [21], Γ_γ = 2.4 ± 0.9 eV [22] and Γ_γ = 3.8 ± 1.2 eV [23]. The lower error bars in the Coulomb dissociation experiments are essentially due to the better statistical accuracy thanks to the fact that the Coulomb field of a nucleus is a prolific source of equivalent photons.

e)   *Different equivalent photon spectra for different multipolarities :*
$^{12}N \rightarrow {}^{11}C + p$

The radiative capture process ^{11}C (p, γ) ^{12}N is part of the reaction network in the hot p-p chain [24]. A microscopic cluster -model study has been recently performed [25]. The ground state spin and parity of ^{12}N (τ = 11 ms) are 1^+. Above the rather low p + ^{11}C threshold at E_{thr} = 0.6 MeV there is a 2^+ level at 0.96 MeV, with an estimated γ-life-time of 260 ± 40 fs. This is an M1 transition and the corresponding resonance strength is therefore known for astrophysical purposes. Since the E2 equivalent photon numbers are larger than the M1 photon numbers at realistic experimental conditions : $(v/c)^2 \ll 1$, the Coulomb excitation $(1^+ \to 2^+)$ could proceed via E2 excitation. Angular distributions and the bombarding energy dependence may help to disentangle the two different contributions. For further potentially interesting levels, a 2^- (E = 1.19 MeV, Γ = 118 $\pm$ 14 keV) and 1^- (E = 1.8 MeV, Γ = 750 $\pm$ 250 keV) the excitation is purely by E1, and the above problem of multipole mixture is absent. The width of these two states is governed by the proton width, while the photon width is not directly known.

An experiment at GANIL is under way to determine the γ-widths of these states [26]. A ^{12}N radioactive secondary beam of about 100 MeV/A interacts with a ^{208}Pb ("Primakoff"-) target. The ^{11}C and p fragments are observed in the forward direction Due to the unequal charge to mass ratio of these fragments, noticeable post-acceleration effects might be expected. However, the decay length of the above mentioned resonances is long enough, so that the decay ^{12}N $\to$ p + ^{11}C occurs essentially outside the Coulomb field.

f)   *A really challenging case with a continuum transition of mixed multipolarities* : ^{16}O $\to$ α + ^{12}C

The importance of the ^{12}C (α, γ) ^{16}O radiative capture process for nuclear astrophysics is extensively discussed [1] with controversial results of the astrophysical S-factor at low energies. The Coulomb dissociation approach was theoretically outlined in Ref. 18. Due to the relatively high threshold (E_{thr} = -7.162 MeV) large incident ^{16}O energies are necessary to overcome the adiabaticity condition. The problem that E2 equivalent photon numbers are strongly dominating over E1 has been also emphasized (recently rediscovered [27]), and the sensitivity of angular correlations to E1-E2 interference has been pointed out [18].

4.   Outlook

The Coulomb dissociation method has been increasingly appreciated as a useful tool of photonuclear physics, using (equivalent) photons of energies below, above and in the giant resonance region (see [28]) e. g.). Applications to nuclear astrophysics are expecially rewarding. They need a precise knowledge of the relative energies between the outgoing fragments. In this respect the "magnifying glass effect" arising from the kinematical situation with three particle in the final state is very helpful, especially at the low relative energies characteristic for nuclear astrophysics. This contribution has put emphasis on the various theoretical implications of the analysis of the data. The problems require straightforward (but sometimes time consuming) procedures. There are also more subtle aspects of great intrinsic theoretical interest, e. g. the time-dependence of phenomena in a transient Coulomb field.

References

[1]   C. Rolfs and W.S. Rodney, "Cauldrons in the Cosmos", Univ. of Chicago Press (1988)

[2]   H. Rebel, "Workshop on nuclear reaction cross sections of astrophysical interest", unpublished report, Kernforschungszentrum Karlsruhe, Febr. 1985
        G. Baur, C.A. Bertulani and H. Rebel, Nucl. Phys. **A459** (1986) 188

[3]   G. Baur, "Coulomb Dissociation for Nuclear Astrophysics", Proceedings of the International Workshop on Unstable Nuclei in Astrophysics, Tokyo, Japan, June 7-8, 1991, World Scientific, eds. S. Kubono and T. Kajino, p. 223

[4]   G. Baur and H. Rebel, J. Phys. G, Topical Review, "Coulomb Dissociation as a Tool of Nuclear Astrophysics", in preparation

[5]   G. Blüge, K. Langanke, H.-G. Reusch and C. Rolfs, Z. Phys. **A333** (1989) 219

[6]   N.K. Glendenning, Classical and Quantum Mechanical Description of Heavy-Ion Interaction", Proc. of the International Conference on Reactions between Complex Nuclei", Nashville, Tennessee, June 10-14, 1984, ed. by R.L Robinson et al., North -Holland

[7]   C.A. Bertulani and G. Baur, Nucl. Phys. **A480** (1988) 615

[8]   G. Baur, C.A. Bertulani and D. Kalassa, to be published

[9]   C.G. Callan jr. and S. Coleman, Phys. Rev. **D16** (1977) 1762
        A.O. Caldeira and A.J. Legget, Ann. Phys. (N.Y.) **149** (1983) 374

[10]   C. Rolfs, Nucl. Phys. **A217** (1973) 29

[11]   E.W. MacDonald et al., Phys. Lett. **B283** (1992) 27; A.C. Shotter, J. Phys. (Nucl. Phys.) **G15** (1989) L41

[12]   G. Baur and C.A. Bertulani, Nucl. Phys. **A482** (1988) 313c

[13]   J. Kiener et al., Phys. Rev. C44 (1991) 2195

[14]   J. Hesselbarth and K.T. Knöpfle, Phys. Rev. Lett. **67** (1991) 2773

[15]   H. Utsunomiya et al., Phys. Rev. Lett. **65** (1990) 847

[16]   H. Utsunomiya, contrib. to First European Biennal Workshop on Nuclear Physics, Megève (France), March 25-29, 1991

[17]   S.B. Gazes, J.E. Mason, R.B. Roberts and S.G. Teichmann, Phys. Rev. Lett. **68** (1992) 150

[18]   G. Baur and M. Weber, Nucl. Phys. **A504** (1989) 352

[19]   T. Motobayashi and M. Gai, private communication

[20]   GALLEX collaboration (P. Anselmann et al.), Phys. Lett. B, submitted June 1, 1992

[21]   T. Motobayashi et al., Phys. Lett. **B264** (1991) 259

[22]   J. Kiener et al., submitted to Nucl. Phys. A

[23]   P. Decrock et al., Phys. Rev. Lett. **67** (1991) 808

[24]   M. Wiescher et al., Ap. J. **343** (1989) 352

[25]   P. Descouvemont and I. Baraffe., Nucl. Phys. **A514** (1990) 66

[26]   P. Aguer, J. Kiener et al., private communication

[27]   T.D. Shoppa and S.E. Koonin, Phys. Rev. C, in press

[28]   D.L. Olson et al., Phys. Rev. C44 (1991) 1862

Astrophysical radiative capture cross sections: Measurements using the breakup of radioactive beams

J. Kiener[1], P. Aguer[1], G. Bogaert[1], B. Borderie[2], A. Coc[1], D. Disdier[3], S. Fortier[2], C. Grunberg[4], L. Kraus[3], A. Lefebvre[1], I. Linck[3], M.F. Rivet[2], P. Roussel-Chomaz[4], F. St Laurent[4], C. Stephan[2], L. Tassan-Got[2], J.P. Thibaud[1]

1) C.S.N.S.M. IN2P3-CNRS, Bat 104-108, 91405 Orsay (France)
2) I.P.N. IN2P3-CNRS, 91406 Orsay (France)
3) C.R.N. IN2P3-CNRS, Univ. Louis Pasteur, B.P. 20, 67037 Strasbourg Cedex (France)
4) GANIL BP 5027 14021 Caen Cedex (France)

Abstract *The Coulomb breakup of the two short lived isotopes ^{14}O and ^{12}N has been investigated at Ganil, aiming at the determination of the radiative widths of low-lying resonances important for the nucleosynthesis in explosive astrophysical scenarios. The experimental setup including the beam production, and the results obtained are presented. As an outlook, a comparison between Coulomb breakup and capture measurements for some reactions of astrophysical importance on radioactive nuclei is given.*

1   Introduction

Radiative capture of charged particles like protons and α particles on radioactive nuclei may play an important role in explosive astrophysical scenarios like the hot CNO, NeNa, MgAl-cycles or the hot pp-chains and the rp-process. With increasing density and temperature even radiative capture on short lived nuclei ($T_{1/2} < 1h$) may be involved in the nuclear burning process, competing with the β-decay. However, experiments for determining reaction rates involving radioactive material face difficulties in dealing with a hazardous beam or target. Especially the detection of the capture γ rays in the presence of radioactive material may cause serious background problems. Instead of a direct measurement of these reactions, the time reversed Coulomb dissociation of fast projectiles on a heavy target nucleus offers the advantage of an increased cross section by the high density of virtual photons seen by the passing projectile [1]. This is especially important when

Figure 1: Experminental setup

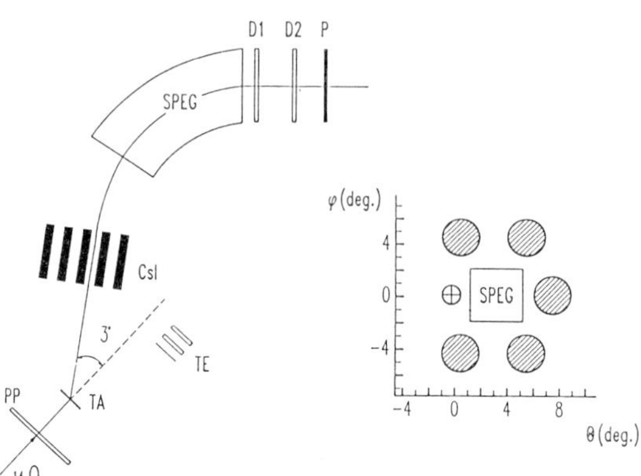

dealing with relatively low intensity radioactive beams and the application of Coulomb breakup to radioactive isotopes was recently suggested by Aguer [2].

Using the method of Coulomb breakup we investigated the two reactions $^{13}N + p \rightarrow ^{14}O + \gamma$ and $^{11}C + p \rightarrow ^{12}N + \gamma$, important in the hot CNO-cycle and hot pp-chains respectively. In both cases the thermonuclear reaction rate is believed to be dominated by resonant E1-capture via a low lying excited state in the compound nucleus. These are the 5.17 MeV 1^--state in ^{14}O at 0.55 MeV and the 1.19 MeV 2^--state in ^{12}N at 0.59 MeV c.m.-energy. The proton width of both states being already known, a determination of the radiative width leads directly to the radiative capture cross section.

2 Secondary beam and experimental setup

The desired radioactive beams have been produced by fragmentation and transfer reactions in a thick ^{12}C-target of a ^{16}O primary beam for the case of ^{14}O and a ^{14}N primary beam for ^{12}N . The secondary beam of selected magnetic rigidity was guided over a flight path of 100.4 m to a secondary target of ^{208}Pb, where the breakup reaction took place. The parameters of the targets and beams are given in table 1. The reaction products, such as ^{13}N, ^{11}C and other heavy elements were analysed in the SPEG spectrometer by a set of two drift chambers, followed by a plastic scintillator giving energy loss information and a timing signal. Protons were detected by a system of 5 CsI detectors located in the reaction chamber, 70 cm downstream the lead target, and around the aperture of the spectrometer as displayed in fig. 1.

On that drawing D1 and D2 are the drift chambers, P the scintillator and PP a parallel plate detector, used for beam normalisation. TA is the ^{208}Pb-target and TE a silicium-telescope also used for beam normalization. The SPEG spectrometer was set at a small angle covering an angular region of 4° below the grazing angle. At these angles nuclear excitation is in both cases small compared to Coulomb excitation [3]. Post-acceleration of the

Table 1: Experimental conditions for the secondary beam production

	^{12}N $T_{1/2}= 11$ ms	^{14}O $T_{1/2}= 70$ s
primary beam $E_{primary}$ $I_{primary}$ produc. target $I_{secondary}$ $E_{secondary}$ reaction target Θ_{SPEG}	^{14}N 95.6 MeV/u 1- 2 μA ^{12}C 0.8 g/cm^2 0.6 - 1.3 10^4 part./s 65.5 MeV/u ^{208}Pb 100 mg/cm^2 2.2°	^{16}O 95.6 MeV/u 1- 2 μA ^{12}C 1.0 g/cm^2 0.5 - 1. 10^6 part./s 70. MeV/u ^{208}Pb 100 mg/cm^2 3.15°

fragments after breakup, which can trouble the extraction of capture cross sections is irrelevant, since breakup occurs far away from the target nucleus in the case of the investigated E1-resonances in ^{12}N and ^{14}O .

3 Results

3.1 ^{14}O

Fig. 2 shows the relative energy spectrum of all coincidence events with a ^{13}N detected in SPEG and a proton in one of the CSI-detectors having the correct time-of-flight and sum energy corresponding to the breakup of 980 MeV ^{14}O projectiles.

We observe a large peak at about 0.55 MeV, representing the breakup via the 5.17 MeV-state and containing about 5000 events collected in 13 h. The full line is the result of a Monte-Carlo simulation of the experiment assuming pure Coulomb breakup via that level, and taking into account the beam resolution and emittance, energy and angular dispersion in the thick lead target and detector resolutions. In adjusting the height of the simulated spectrum to the experimental data, the excitation cross section for the level could be extracted and compared to Coulomb excitation computations. Our result is in agreement with two other recent measurements [4], [5]. We deduced a new thermonuclear reaction rate:

$$\mathcal{N}_A < \sigma v >= \ ((-1.727\ 10^7/T_9^{2/3})\exp[-15.168/T_9^{1/3} - (T_9/0.8324)^2]$$
$$\times(1. + 0.027\ T_9^{1/3} - 17.54\ T_9^{2/3} - 3.373\ T_9$$
$$+0.0176\ T_9^{4/3} + 0.766\ 10^{-2}T_9^{5/3})$$
$$+3.1\ 10^5\ T_9^{-3/2}\exp(-6.348/T_9))\ cm^3s^{-1}$$

Figure 2: Relative energy spectrum of the $^{14}O(\gamma,p)^{13}N$ reaction

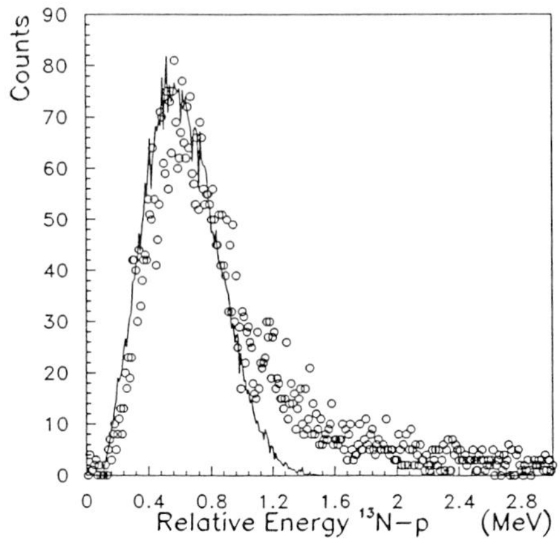

3.2 ^{12}N

The low intensity of the ^{12}N beam compared to the ^{14}O beam (table 1) looks puzzling, since in both cases the secondary beam is produced in the same way by the stripping of two neutrons. New experiments are necessary to understand if that effect has an experimental basis or if it is connected to some physical effect. Due to the low intensity ^{12}N beam much less events could be collected in this experiment. A sum energy spectrum of ^{11}C-proton coincidences from the breakup of ^{12}N is shown in Fig. 3a. The very low background below the sum energy peak is remarkable. The relative energy spectrum is displayed in fig. 3b, containing about 200 events. It can be noted, that most of the events are concentrated around 600 keV, where the breakup via the 2^--level is expected, though some contribution from neighbouring levels (2^+-level 0.36 MeV and 1^--level at 1.2 MeV relative energy) can not be excluded . In view of this, it seems premature to give a value for the radiative width of the 2^--level, however the result of a microscopic calculation of Descouvemont [6] (140 meV) could give account of the observed yield, a value given by Wiescher [7] based on γ-transition strength estimates (1.9 meV) being too low. A second experiment is planned, using more CSI-detectors to increase statistics and giving a better angular definition for ^{11}C-proton correlations to disentangle the different contributions.

4   Coulomb breakup of accelerated secondary beams

We studied other reactions of astrophysical importance involving light ($A < 30$) radioactive isotopes, which could be measured by the method of Coulomb breakup. Some of the major reactions involved in the hot pp chain or rp process and the hot cycles can be analysed using that technique. We estimated rates for *detection* of the heavy fragment in SPEG assuming a 30 MeVA

Figure 3: Sum energy spectrum and relative energy spectrum of the ^{12}N-breakup

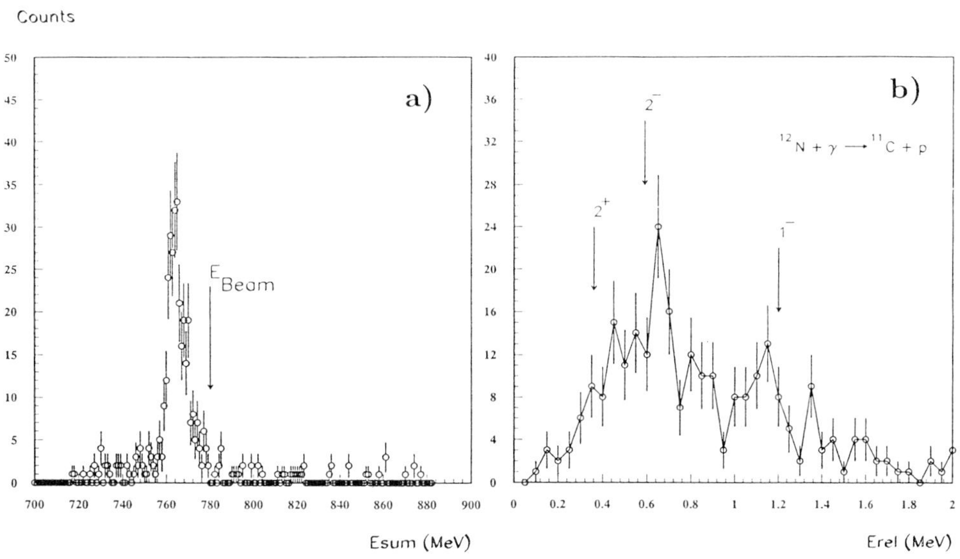

Table 2: Estimated rates for Coulomb breakup and resonant capture reactions

Reaction	E_{rel}	Breakup	Capture
$^{8}B(\gamma,p)^{7}Be$	0.64 MeV	1000/mn	3/mn
$^{9}C(\gamma,p)^{8}B$	0.92 MeV	50/mn	20/h
$^{18}Ne(\gamma,p)^{17}F$	0.59 MeV	<0.5/s	<1/s
$^{20}Na(\gamma,p)^{19}Ne$	0.45 MeV	1/mn	1/mn
$^{21}Mg(\gamma,p)^{20}Na$	0.3 MeV	30/mn	25/mn
$^{12}N(\gamma,p)^{11}C$	0.59 MeV	40/mn	5/h
$^{14}O(\gamma,p)^{13}N$	0.55 MeV	4/s	2/s

beam of the intensity 10^9/s and a 10 mg/cm^2 ^{208}Pb target. In comparison with the *reaction* rate for the direct reaction, assuming a 10^9/s beam at the correct c.m. energy on a thick target covering the measured resonance we find that already these rates are in favor of the Coulomb breakup approach. The major results of these studies are published in [8]. Some reactions are listed in Table 2.

It should be noted, that the coincidence requirement in Coulomb breakup studies results in a very low background (see figure 3a), permitting detection rates of some events per hour. In view of these advantages, Coulomb breakup of radioactive beams of relatively low intensity is very promising for the extraction of radiative capture cross sections. The method is limited however to radiative capture to the ground state and demands a careful study for each reaction if the conditions for its application are given (see e.g. [9]). Problems are connected with the possible mixture of different multipolarities like M1+E2. When dealing with nonresonant capture, the effect of post-acceleration on the outgoing fragments may spoil the relative energy information. But as a matter of fact, the nuclear burning process on short-lived radioactive nuclei is only effective where capture reactions on strong resonances can compete with their β-decay, and in such cases Coulomb breakup is certainly of great interest.

References

[1] Baur G. , Bertulani C.A., Rebel H. 1986 "N.P. **A458** 188"

[2] Aguer P. 1989 " Proceedings of The First International Conference On Radioactive Nuclear Beams, Berkeley, October 16-18, World Scientific "

[3] Kiener J. et al " submitted to Nuclear Physics"

[4] Decrock P. et al. 1991 "Phys. Rev. Lett. **67** 808"

[5] Motobayashi T. et al 1991 " Phys. Lett. **B264** 259"

[6] Descouvemont P. and Baraffe D. 1990 "preprint Université Libre de Bruxelles (1990)"

[7] Wiescher M. et al 1989 " Ap. J. **343** 352 "

[8] Aguer P. et al 1992 " Proceedings International Workshop on the Physics and Techniques of Secondary Nuclear Beams, Dourdan, March 23-25 1992 "

[9] Rebel H., Baur G., Gils H.J. and Kiener J. " this conference "

E1 and E2 contributions to the ^{7}Li $\to \alpha + t$ Coulomb dissociation cross section in semiclassical and quantum mechanical calculations

S. Typel

Institut für Theoretische Physik I, Westfälische Wilhelms-Universität, Wilhelm-Klemm-Straße 9, W-4400 Münster, Germany

Abstract. The Coulomb break-up of ^7Li has been proposed to provide information on the astrophysical relevant $^3H(^4He, \gamma)^7Li$ capture cross section. Dissociation cross sections in semiclassical and quantummechanical calculations of the reaction $^7Li + ^{208}Pb \to \alpha + t + ^{208}Pb$ are given at $\alpha - t$ relative energies below 3 MeV using the Resonating Group Method for a microscopic description of the seven nucleon system. Differences between the semiclassical and quantummechanical calculation are very small. E1 and E2 contributions are significant and have to be separated in the experimental data to obtain the astrophysical S-factor.

1. Introduction

The Coulomb dissociation of light nuclei at small relative energies E of the emitted fragments has been suggested as an alternative to determine radiative capture cross sections of astrophysical interest via the detailed balance theorem [1]. But this indirect method exhibits some difficulties in extracting astrophysical S-factors from the experimental data. The contributions of different electromagnetic multipoles in pure Coulomb break-up as well as nuclear contributions have to be disentangled. The final state interaction of the target Coulomb field may change the energy and angular distribution of the outgoing fragments, especially in the case of direct break-up into particles of different charge-to-mass ratio [2].

Because the cross section of the $^3H(^4He, \gamma)^7Li$ capture reaction which produces most of the ^7Li during the primordial nucleosynthesis is not very well known at low energies [3, 4, 5] the Coulomb dissociation of ^7Li is of special importance. Measurements were done by various groups [6, 7, 8] using different targets and projectile energies.

The contributions of different electromagnetic multipoles in the capture and break-up reactions are very different as an analysis of the $\alpha(d, \gamma)^6Li$ capture and ^6Li break-up has shown [9], hence it has to be examined in the ^7Li case. Here the capture cross section is clearly dominated by the direct capture from the $J = \frac{1}{2}^+$ s-wave into the $\frac{3}{2}^-$ ground and $\frac{1}{2}^-$ first excited state of ^7Li with only small E2 contributions at αt energies

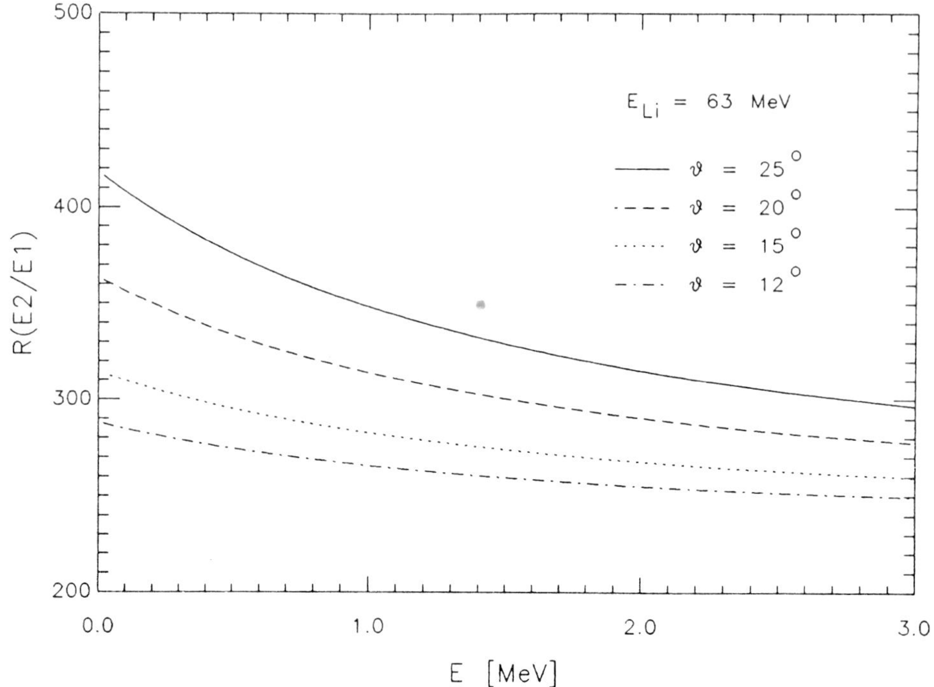

Figure 1: E2/E1 ratio of virtual photon numbers per solid angle for the breakup of ^7Li incident on ^{208}Pb

$E < 1$ MeV. In the break-up reaction from the ^7Li ground state the E2 component is very much magnified as compared to the capture reaction. This is seen from the ratio R(E2/E1) of virtual photon numbers per solid angle $\frac{dn_{E\lambda}}{d\Omega}$ seen by the ^7Li nucleus in the target Coulomb field [1]. This ratio for a projectile energy of $E_{Li} = 63$ MeV incident on ^{208}Pb at different scattering angles ϑ_{Li} as in the Utsunomiya et al. experiments [7] is given in fig. 1. As a consequence E1 and E2 multipole contributions may be observed in the breakup reaction changing the absolute value of the cross section and the angular distribution of the fragments. For an estimation of this effect a theoretical calculation is necessary because E2 contributions to the capture reaction are not known experimentally. Simultaneously the difference in the semiclassical (SC) and quantummechanical (QM) calculation of the break-up process may be examined.

2. Calculation of dissociation cross sections

The theory of Coulomb excitation of nuclei in heavy ion collisions is well known [10] and is easily adapted to break-up reactions of light nuclei [1]. In the semiclassical picture the projectile moves on a hyperbola in the target Coulomb field. The break-up cross section is calculated in first order time dependent perturbation theory. Triple differential cross sections for the angular distribution of the fragments are given by Baur and Weber [2]. The quantummechanical calculation starts from a distorted wave born approximation of the T-matrix [11]. Initial and final state wave functions are products of Coulomb wave

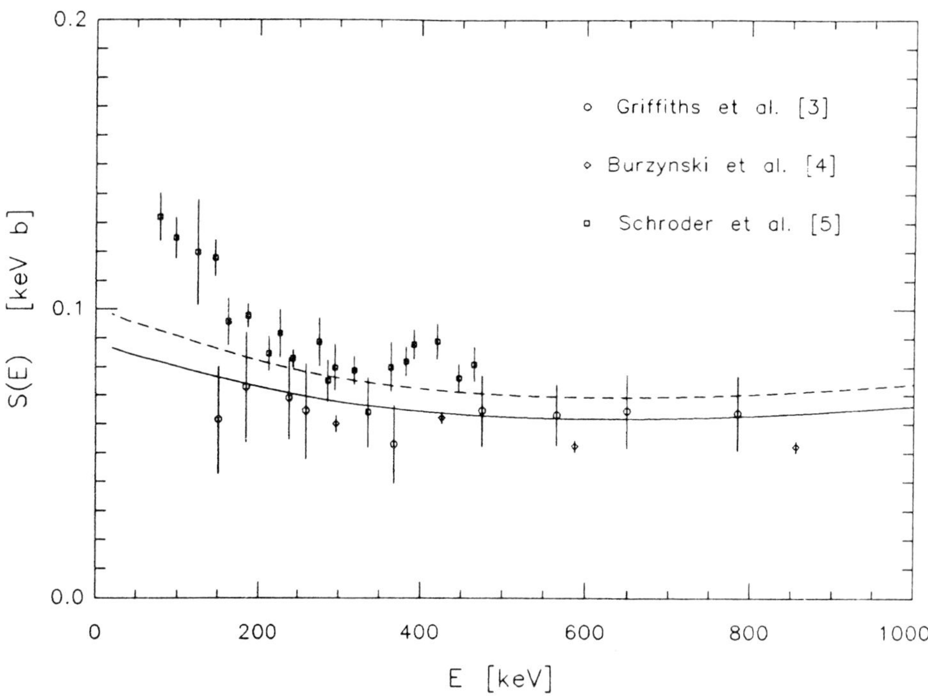

Figure 2: Comparison of the calculated astrohysical S-factor of the $t(\alpha,\gamma)^7Li$ reaction with experimental data (dashed/solid line: nuclear parameter set 1/2).

functions describing the motion of the projectile in the target Coulomb field and bound resp. scattering wave functions of ^7Li / αt. As a result the cross sections depend on the semiclassical/quantummechanical excitation functions [10] and the usual electromagnetic multipole matrix elements in the projectile system.

Bound and scattering states of the seven nucleon system are microscopically described as antisymmetrized αt-cluster wave functions using the Resonating Group Method [12] including a pseudostate to take the distorsion of the triton into account. The modified Wildermuth-Tang potential was chosen as the nucleon-nucleon interaction [13]. The two free parameters of this potential are fixed to to give the correct binding energy $E = -2.4678$ MeV of the ^7Li ground state and the energy of the $\frac{7}{2}^-$ resonance at $E = 2.16$ MeV (set 1) resp. the first excited state ($\frac{1}{2}^-$) at $E = -1.9902$ MeV (set 2) refering to the αt threshold [14].

This model gives an accurate description of the system at low energies [13, 15, 16, 17, 18]. After fixing the description of the clusters and the nuclear interaction there are no free parameters. Reduced matrix elements of E1 and E2 transition operators are calculated in the long wavelenght approximation. The relative angular momentum between α and t is taken to be $l = 0, \ldots, 3$ which is coupled with the channel spin $s = \frac{1}{2}$ to give the total angular momentum J of the bound or scattering states.

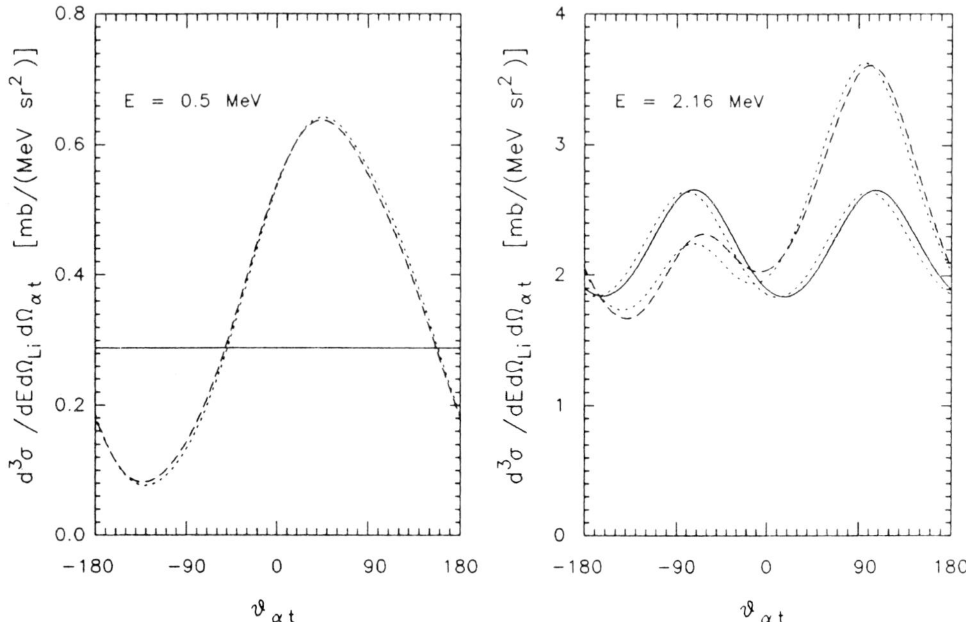

Figure 3: Angular distribution of the fragments in their rest system in the reaction plane (see text).

3. Results and conclusion

The calculated S-factor (fig. 2) fits the data rising slowly with decreasing energy similar to the earlier calculations [13, 15, 16, 17, 18] but not as substantially as the Schröder et al. data. The result with the parameter set 1 gives higher values because the capture contribution to the first excited state of ^7Li is overestimated as the binding energy is to small. But for the break-up calculation the parameter set 1 is used to give the correct energy of the $\frac{7}{2}^-$ resonance which is more important than the $\frac{1}{2}^-$ state.

Figure 3 gives the angular distribution of the fragments in the scattering plane as a function of the direction of the fragment relative momentum in their center of gravity system pointing to the triton/α-particle ($\vartheta_{\alpha t} > 0$ / $\vartheta_{\alpha t} < 0$) and orientated like the beam coordinate system. In the direct break-up region (left) a pure E1 dissociation into the s-wave results in an isotropic distribution of the fragments (solid line) but considering the break-up into all states with final relative orbital angular momentum $l_f = 0, \ldots, 3$ via E1 and E2 transitions there is a significant variation in the distribution of the outgoing particles (dashed line). In the QM calculation the resonant break-up (right) into the $\frac{7}{2}^-$ state the triple differential cross section shows the typical E2 distribution (solid line). Including all E1 and E2 transitions the cross section are distincly influenced (dashed line). The corresponding results of the SC calculation are given by the dotted line showing only small differences in the angular correlation.

When the two outgoing fragments are moving parallel to the scattered ^7Li as in the Utsunomiya et al. experiments the α-particle or the triton may have the greater velocity.

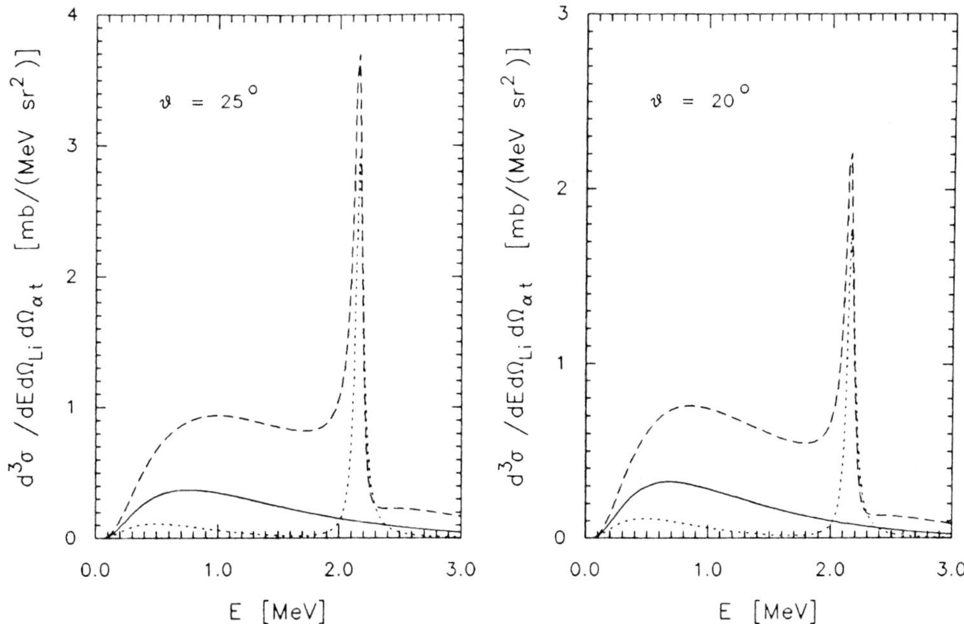

Figure 4: Triple differential cross section for the ^7Li break-up in the reaction plain. Full line: E1, $J_f = \frac{1}{2}^+$; dashed/dotted line: E1+E2, all channels ($v_\alpha < v_t$ / $v_\alpha > v_t$)

This case is examined in figure 4. Where the pure E1 break-up into the s-wave gives no asymmetry (solid line) the consideration of E1 and E2 transitions shows a marked difference (dashed/dotted line: $v_\alpha < v_t$ / $v_\alpha > v_t$). Because the experimental data are currently again evaluated there is no comparison between experiment and theory possible until updated data are available [19].

In conclusion the Coulomb dissociation reaction $^7Li + {}^{208}Pb \to \alpha + t + {}^{208}Pb$ was studied in semiclassical and quantummechanical calculations considering E1 and E2 transitions in the microscopically discribed seven nucleon system. While the difference between QM and SC results are very small the E2 contribution causes a significant change in the break-up cross section especially in the distribution of the fragments and may not be neglected in extracting S-factors for the inverse capture reaction from experimental data.

As the current calculation considers only the Coulomb dissociation process the inclusion of nuclear break-up [11] is desirable. This is easily done by an extension of the present calculation. The influence of the final state interaction which is still a theoretical problem is likewise of particular interest.

References

[1] Baur G., Bertulani C. A. and Rebel H. 1986 *Nucl. Phys.* **A458** 188

[2] Baur G. and Weber M. 1989 *Nucl. Phys.* **A504** 352

[3] Griffiths G. M., Morrow R. A., Riley P. J. and Warren J. B. 1961 *Can. J. Phys.* **39** 1397

[4] Burzynski S., Czerski K., Marcinkowski A. and Zupranski P. 1987 *Nucl. Phys.* **A473** 179

[5] Schröder U., Redder A., Rolfs C., Azuma R. E., Buchmann L., Campbell C., King J. D. and Donoghue T. R. 1987 *Phys. Lett.* **B192** 55

[6] Shotter A. C., Rapp V., Davinson T., Branford D., Sanderdon N. E. and Nagarajan M. A. 1981 *Phys. Rev. Lett.* **53** 1539

[7] Utsunomiya H., Schmitt R. P., Lui Y.-W., Haenni D. R., Dejbakhsh H., Cooke L., Heimberg P. and Ray A. 1988 *Phys. Lett.* **B211** 24
Utsunomiya H., Lui Y.-W., Cooke L., Dejbakhsh H., Haenni D. R., Heimberg P., Ray A., Srivastava B. K. and Schmitt R. P. 1990 *Nucl. Phys.* **A511** 379
Utsunomiya H., Liu Y.-W., Haenni D. R., Dejbakhsh H., Cooke L., Srivastava B. K., Turmel W., O'Kelly D., Schmitt R. P., Shapira D., Gomez del Campo J., Ray A. and Udagawa T. 1990 *Phys. Rev. Lett.* **65** 847

[8] Gazes S. B., Mason J. E., Roberts R. B. and Teichmann S. G. 1992 *Phys. Rev. Lett.* **68** 150

[9] Typel S., Blüge G. and Langanke K. 1991 *Z. Phys.* **A339** 335

[10] Alder K., Bohr A., Huus T., Mottelson B. and Winther A. 1956 *Rev. Mod. Phys.* **56** 432

[11] Bertulani C. A. and Hussein M. S. 1991 *Nucl. Phys.* **A524** 306

[12] Wildermuth K. and Tang Y. C. 1977 *A unified theory of the nucleus* (Braunschweig: Vieweg)

[13] Kanada H., Kaneko T. and Tang Y. C. 1982 *Nucl. Phys.* **A380** 87

[14] Ajzenberg-Selove F. 1988 *Nucl. Phys.* **A490** 1

[15] Mertelmeier T. and Hofmann H. M. 1986 *Nucl. Phys.* **A459** 387

[16] Kajino T. 1986 *Nucl. Phys.* **A460** 559

[17] Altmeyer T., Kolbe K., Warmann T., Langanke K. and Assenbaum H. J. 1988 *Z. Phys.* **A330** 277

[18] Kajino T., Mathews G. J. and Ikeda K. 1989 *Phys. Rev.* **C40** 525

[19] Utsunomiya H. private communication

Nuclear data for unstable isotopes

O Sorlin

Institut de Physique Nucléaire, F-91406 Orsay, France

Abstract. This paper will decribe the ways of producing unstable nuclei in laboratories. The usefulness of the production of unstable species will be emphasized through the major explosive stellar processes.

1. Introduction

Nuclear Physics and Astrophysics are both entrusted with the task of understanding nucleosynthesis and energy production in the stars. At high tempertures and densities present in explosive scenarios such as the early universe, cataclysmic binary stars (nova or accretion stars), and supernovae, the nucleosynthesis proceeds throughout unstable nuclei. In order to produce and to study the most exotic isotopes that are not accessible from stable beam - stable (or radioactive) target experiments, it is necessary to develop facilities that utilize Radioactive Nuclear Beams.
The existing methods for producing untable nuclei will be described in § 2. A review of the major explosive stellar processes (inhomogenous big bang (§ 3), rp (§ 4), r (§ 5)) will be made through some selected examples using RNB.

2. Production of unstable isotopes.

Unstable nuclei are produced in laboratories by the "interaction" of a primary beam with a production target. Protons (i), heavy-ions (ii), or neutrons (iii) can be used as primary beams leading to different induced reactions:
   (i) spallation, fission and target fragmentation.
   (ii) projectile fragmentation, multi-nucleon transfer, fusion-evaporation.
   (iii) neutron-induced fission.
The production rate of a Radioactive Nuclear Beam is characterized for all of these methods by the relation:

$$P = \sigma \ I \ N$$

σ is the production cross section in cm^2 which is reaction dependent. I is the intensity of the primary beam in pps, it should be of course as high as possible. N is the target thickness in atoms.cm^{-3}. Even if a thick target yields to a high production rate, some compromises will have to be taken into account in each method for avoiding a (i) too long extraction time of the produced nuclei or (ii) too much energy-losses in the target. The purity of the RNB produced is another important factor for the feasability of experiments. The two next subsections will describe the two first methods of producing RNB with their corresponding devices of nuclei selection. Further details concerning existing and forthcoming RNB can be found in [1, 2].

2.1 Production of RNB by proton-induced reactions.

High energy proton beam will produce spallation and fragmentation reactions. If the target nuclei are fissile, a proton or neutron beam can induce fission products. The spallation reaction can be described in two steps: (1) the target nucleus experiences a rapid ejection of nucleons due to the incoming proton and remains in an excited state, (2) it in turn dumps its excitation energy through neutron evaporation. Mainly due to this second step, this method favors the production of neutron deficient nuclei. Conversely, the neutron induced or proton induced fission is more suitable to neutron rich isotopes. The nuclei produced by one of these reactions have to be extracted from the target. Therefore, the target is heated up to a 2000 degrees and maintained to a high voltage. Then the reaction products may diffuse in a tube which is at the same voltage. During this process of drifting out of the tube, the reaction products will undergo collisions with the walls, and will thus be ionized by the mechanism of surface ionization. This very simple model of ion-source is limited to the production of isotopes with low ionization potential as Alkaline, Alkaline Earth elements, and Rare Earth. Thus, single charged ions of a well-defined velocity are produced. Due to these features, a magnetic selection will yield to a good selection in Av/Q among the species produced. This method of production is used at ISOLDE/CERN. This method gives high production rates for proton or neutron rich nuclei, and it is at the present time the only way to produce heavy nuclei very far from stability as those encountered in the r process (see §5). Nevertheless, it is important to note the different sources of loss during the ion beam formation as for instance the decay of short-lived species, and the absorption by the tubes during collisions. These losses are shown on Fig. 1.

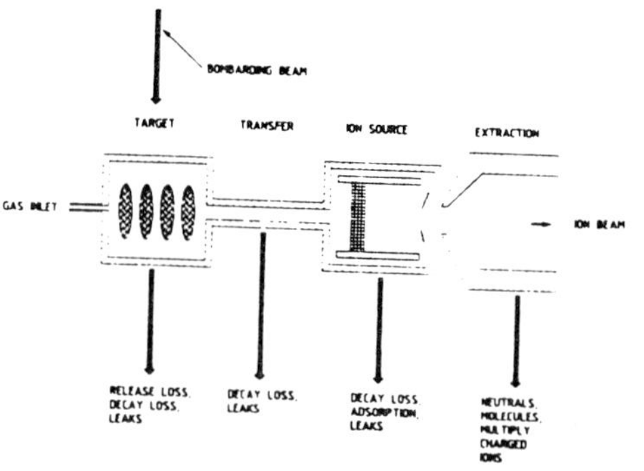

Fig. 1: The steps in the production of an ion beam, with the losses which may occur at each stage [3].

Moreover, the extraction of other species than those mentionned below requires much complicated ion-sources which have to deal with chemical properties of the desired elements. Finally, another selection would be required in order to separate the isobars transmitted by the magnet. So, even if the production rate is huge, transmission and selection limitations may hamper the production of some desired radioactive species.

2.2 Production of RNB by heavy ion-induced reactions.

Three different kind of reaction can occur depending on the energy of the incident projectile. At high energies (>200 MeV/u), a pure fragmentation of the projectile will happen. These energy domains are used for instance in Berkeley, GSI, Saturne facilities. This method will produce elements ranging from the lightest up to the beam mass. The fragmentation can be described as following by the so called Participant-Spectator Model: once the projectile collides with the target, parts of the projectile and target form a highly excited participant which travels at an intermediate energy between velocity of the target and of the beam.The remaining part of the target form the so called target spectators, whereas the remaining parts of the projectile still travel focused around the beam incident direction and with the beam velocity. At intermediate energies (between 30 and 100 MeV/u) like in GANIL, MSU, RIKEN facilities, some multi-nucleons transfers also occur (see for instance §5 which shows up the prossibility of producing ^{46,47}Cl with a ^{48}Ca beam). Due to this exchange possibility between projectile and target, a high (resp. low) N/Z ratio of the target will enhance the production rate of neutron (resp. proton) rich nuclei. At energies just above the coulomb barrier, fusion evaporation reaction lead to the production of specific species by the appropriate choice of projectile and target. In the fragmentation process, thick target are important for increasing the reaction rate. Nevertheless, this target thickness is limited because of slowing down, angular and energy stragglings of the fragments in the material. These interactions degrade the narrow velocity and angular distibutions of the fragments. This beam qualities are required for guiding the fragments into a spectrometer. This spectrometer, used for the selection of nuclei of interest, is generally composed with three parts which are: the magnetic, the degrader, and the velocity selections. This is sketched on the Fig. 2 where the F1-3 slits account for the three selections.

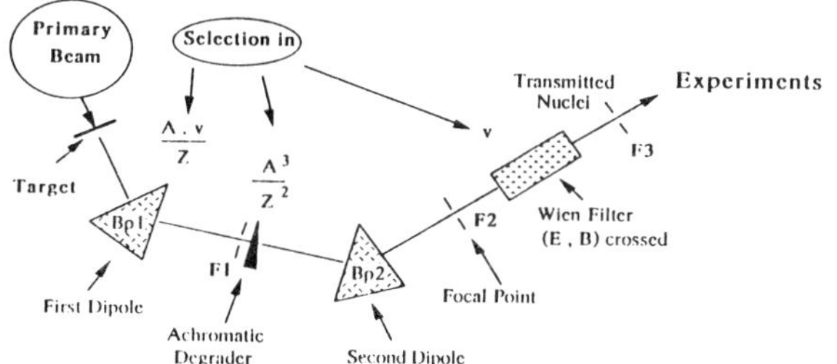

Fig. 2: Schematic view of a spectrometer composed with three magnetic selections.

The first magnet deflects the fragments according to the relation $B\rho = Av/Z$ (for fully stripped ions). The degrader is a layer of material that induces different energy-losses for each element. According to the Bethe approximation $dE/dx \propto AZ^2/E$, each nucleus of energy proportional to Z^2/E after the first magnet will loose $dE/E \propto A^3/Z^2$. The magnetic field of the second dipole has then to be matched to the output energy of the desired nucleus. It will be then preferentially transmitted. The third selection can be performed by a velocity filter, composed with electric and magnetic crossed fields. The nucleus whose velocity is equal to $(E/B)_{filter}$ will travel straightforward, others will be deflected. An example of the efficiency of these three sets of selections is shown below (Fig. 3) in the case of the fragmentation of ^{24}Mg for producing a ^{20}Mg secondary beam.

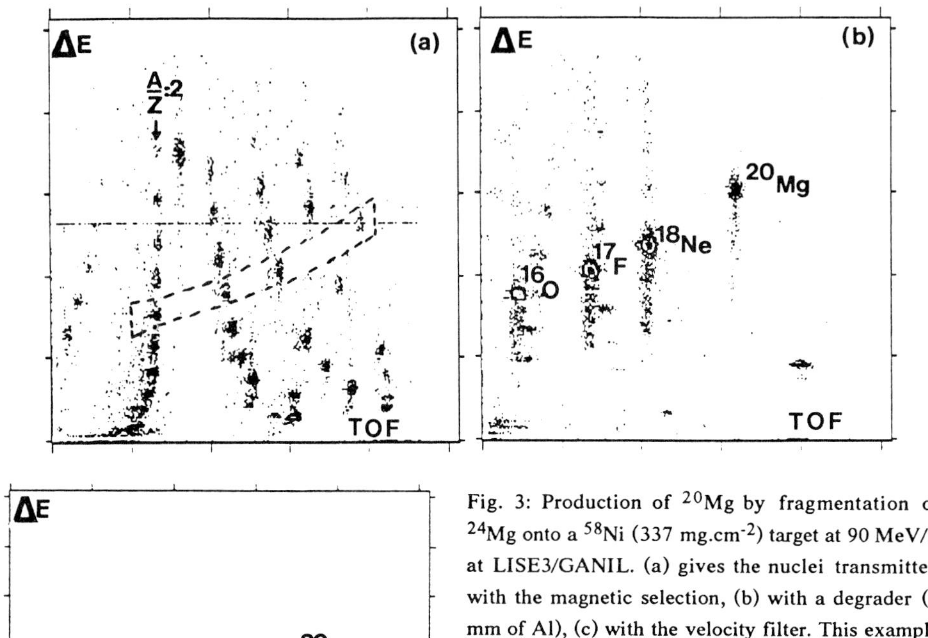

Fig. 3: Production of ^{20}Mg by fragmentation of ^{24}Mg onto a ^{58}Ni (337 mg.cm^{-2}) target at 90 MeV/u at LISE3/GANIL. (a) gives the nuclei transmitted with the magnetic selection, (b) with a degrader (1 mm of Al), (c) with the velocity filter. This example shows the possibility of producing very pure beams (here 99.6%) far off stability. The astrophysical interest of ^{20}Mg will be discussed in §4.

This fragmentation method is well suited for the production and selection of light species on the proton or neutron rich side (see §3, 4, 5). Actually, at intermediate energies, the production of unstable nuclei is limited to masses lower than ≈ 100 due to the production of charge states in the target at these energies. This effect will considerably hamper the discrimination of the fragments produced. The drawbacks of the fragmentation method are different in the case of high or intermediate energy incident beam. At the present time, high energy facilities which do not encounter charge states problems, unfortunatly have two low primary intensities to reach species very far of stability.

3. Inhomogeneous big-bang nucleosynthesis

The standard big-bang nucleosynthesis model is known to reproduce fairly well the abundance of nuclei up to (A=7, Z<4), but cannot bridge the A=5 and A=8 gaps. Therefore, it cannot account for the production of heavier nuclei. Inhomogeneous scenarios, which are characterized by a nonuniform density of the universe, may overcome this problem by successive neutron and α captures leading to higher masses. Recent measurements in the region of ^{8}Li have been performed in order to quantify the light element production. The important reactions that are thought to occur in this region, and through which all nuclides of mass A>11 must pass (in this kind of scenario), are represented on the Fig. 4. Recent experiments have been performed in Japan [4] and in the USA [5] with a ^{8}Li radioactive beam, measuring the reaction rates of ^{8}Li(α,n)^{11}B, ^{8}Li(d,n)^{9}Be and ^{8}Li(d,t)^{7}Li respectively. The first experiment shows an enhancement of this reaction rate of a factor of five as compared to inverse reaction measurement [6], due to the fact that the inverse reaction only proceeds through the ground state of ^{11}B. This factor will probably increase the production of intermediate mass element and thus strengthen the occurence of this inhomogeneous scenario.

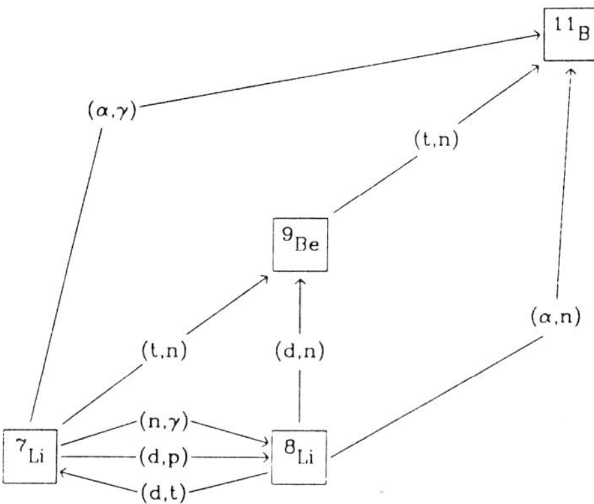

Fig. 4: Major reactions and pathways relevant for the synthesis of ^{8}Li, ^{9}Be, ^{11}B and for the building of heavier species with primordial inhomogeneous nucleosynthesis [5].

4. From the hot CNO cycle to the rp process

Although of minor importance for the synthesis of the element in the universe, processes involving proton captures are often important for explaining the nuclear energy production in stars. A comprehensive review of the explosive proton capture scenarios can be found in [7] and in the included references. In main sequence stars, the nuclear burning of hydrogen (pp chains and CNO cycle) accounts for the energy production that matches the gravitational shrink and converts H into He. At temperatures and densities much higher than those encountered in such star's interior, the hot CNO cycle (HCNO) will increase the energy generation including the ^{13}N(p,γ)^{14}O reaction. At much higher temperatures of about $T_8 = 2$ (2.10$^{8\circ}$K) several breakout reactions may happen leading to other cycles due to the bypass of the long-lived nuclei ^{14}O, ^{15}O (half-life respectively 102s and 176s). The ^{15}O(α, γ)^{19}Ne and ^{19}Ne(p, γ)^{20}Na reactions are supposed to generate a way of escape towards higher masses (about A=70). The so-called rp process is then driven by the competition of rapid proton captures and β^+ decays (Fig.5). While the β^+ decay rates are not sensitive to an increasing temperature, the proton capture rates increase exponentially, yielding to a reaction flow more or less far off stability depending of the temperature conditions ($T_8 > 2$). The reaction rates for the different capture reactions have to be known to predict the reaction path for different temperature and density conditions, to determine the resulting isotopic abundances, and to evaluate the nuclear energy production and the time scale of the process. High temperature and density conditions may occur in a large variety of astrophysical sites like cataclysmic binary systems (including novae and some X-ray bursts), and during type II supernovae explosions. The former objects are thought to be thermonuclear outbursts triggered by mass accretion onto the surface of either a white dwarf or a neutron star. The nuclear reactions are therefore critical for the understanding of the outburst phenomenon itself together with the concommitent synthesis of elements. Even if it seems, up to now, beyond the edge of feasibility, the estimated isotopic abundances deduced from an rp reaction network calculation would lead to an extremely interesting comparison with the rate of γ-ray emission of characteristic long-lived isotopes (^{22}Na, ^{26}Al, ^{44}Ti, ^{54}Mn, ^{56,57}Co, ^{65}Zn) detected by the Gamma Ray Observatory. It could also be of great interest for the understanding of some isotopic anomalies observed in primitive meteorites, as for instance the overabundance of ^{26}Mg.

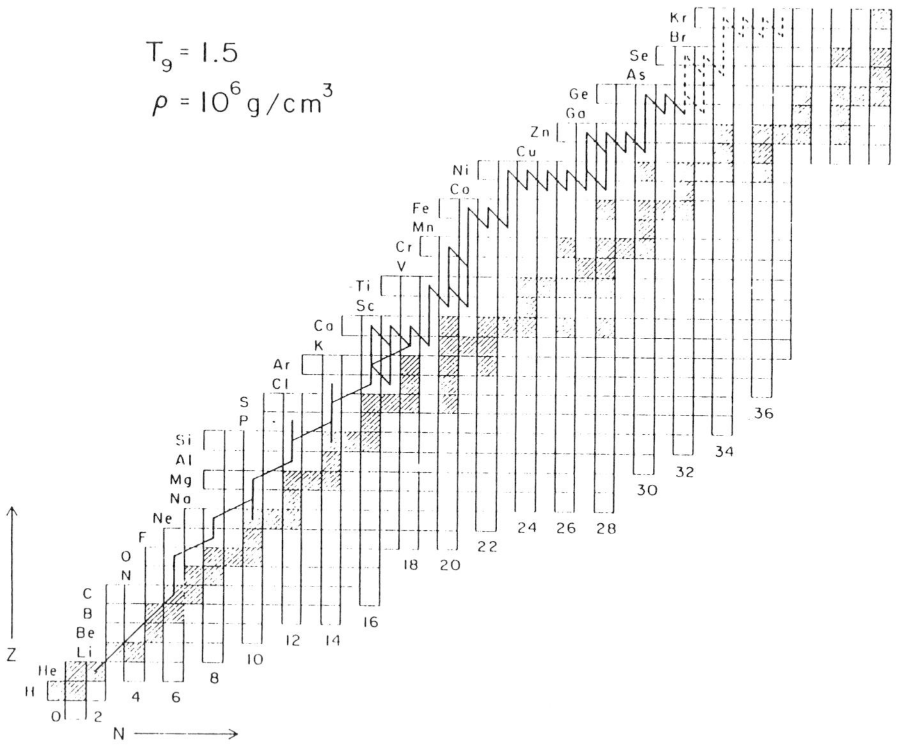

Fig. 5: Major flows in the rp process are indicated with solid arrows. The dashed arrows account for a flow of less than 1/10 of the solid arrows [7].

Because reaction cross sections at stellar energies are often extremely low, it is scarcely possible to measure the direct reaction rate in the laboratory. Nevertheless a first step has been made in Louvain-la-Neuve facility for the ^{13}N(p, γ)^{14}O reaction by means of a low energy and high intensity beam of ^{13}N. Generally, for light species, the reaction rate is dominated by the resonant term. In the case of the ^{19}Ne(p, γ) ^{20}Na reaction, it is proportionnal to:

$$\lambda p\gamma\,(^{19}\mathrm{Ne}) \propto T_9^{-3/2} \sum_r \omega\,(J(^{19}\mathrm{Ne}),\, J_p,\, J_r)\,\frac{\Gamma_p\,\Gamma_\gamma}{(\Gamma_p + \Gamma_\gamma)}\,\exp\,(-\,11.60045\,E_r(\mathrm{MeV})/T_9),$$

where Γ_p, and Γ_γ are the proton and the gamma widths respectively which characterize the formation and the decay of the compound nucleus $^{20}\mathrm{Na}^*$, and ω is the statistical factor. This reaction rate cannot, at present time, be deduced from a direct reaction. Though indirect measurements allows to determine the position of the energy resonances in the $^{20}\mathrm{Na}$. As shown in figure 6, it is of utmost importance for the knowledge of the temperature onset of the rp process. The left curve [8] shows that (taking theoritical estimates of the Γ widths and the spin parities) the determination of the resonant energy yields to a breakout temperature of about $T_8 \approx 2.2$, instead of $T_8 \approx 3.8$ deduced by Langanke [9] from the level properties of the mirror nucleus $^{20}\mathrm{F}$.

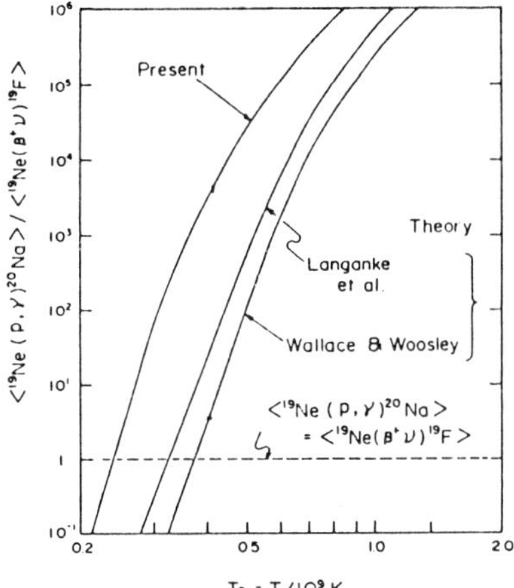
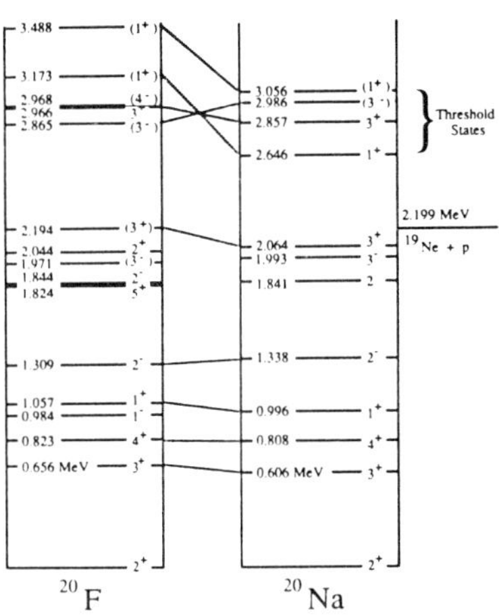

Fig. 6: The left part [8] shows the ratio of proton capture $^{19}\mathrm{Ne}(p,\gamma)^{20}\mathrm{Na}$ reaction against the decay to $^{19}\mathrm{F}$. The calculation has been made by different authors who took into account the energy levels of the mirror nucleus (Wallace & Woosley [10]), the predicted Thomas-Ehrmann shift to the excited states of $^{20}\mathrm{Na}$ (Langanke et al. [9]). The experimental determination of these energy levels (right part) [8, 11, 12, 33] by transfer reaction leads to a huge increase of the ratio (Present [8]), and thus a lower escape temperature from the HCNO cycle.

These level positions have been experimentally determined by the transfer reaction $^{20}\mathrm{Ne}(^3\mathrm{He}, t)^{20}\mathrm{Na}$ [8, 11, 12]. They are very different from the predictions of Langanke (Fig. 6). This strengthen the necessity of such measurements. The most important levels are those in the Gamow peak region for the temperature of interest. The 2.646 MeV which is thought to be a 1^{+}- or 0^- state seems to be the most promising level above the proton threshold. The observation of the delayed protons and γ rays from $^{20}\mathrm{Mg}$ (Fig. 7) and their relative ratios determine whether this first level of $^{20}\mathrm{Na}$ (above the proton reaction threshold) could enhance the reaction rate. Figure 7 shows an example of an experiment performed at GANIL looking at the β^+ decay of $^{20}\mathrm{Mg}$ through p-γ coincidences in order to correlate the protons from the $^{20}\mathrm{Na}^*$ with the excited states of $^{19}\mathrm{Ne}$. The results from this experiment are preliminary and are presently under further analysis to extract the astrophysically interesting information [13].

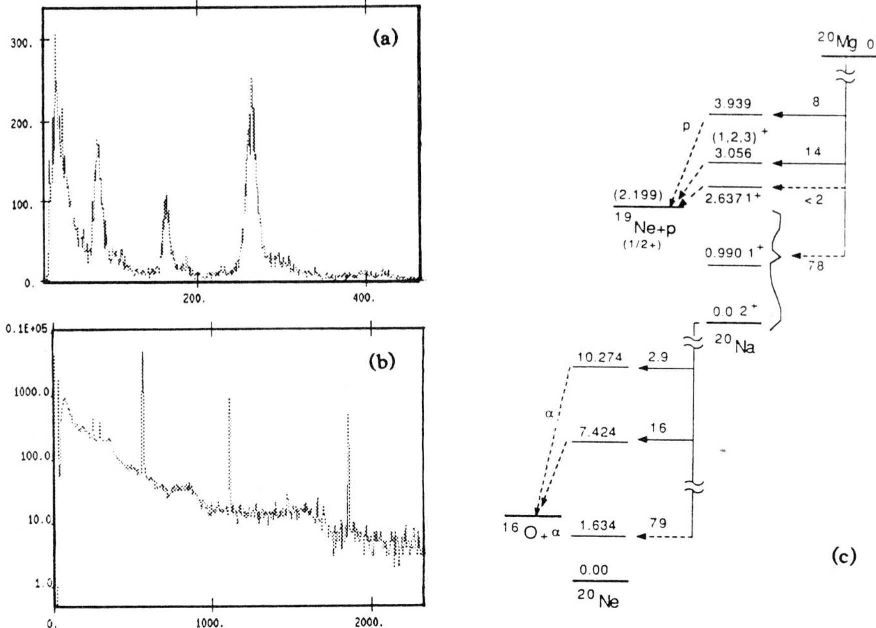

Fig. 7: The on-line proton (a) and the gamma spectra (b) of ^{20}Mg decay (see Fig. 3 for its production) have been obtained in coïncidence with a β ray in order to avoid major parts of background. The corresponding level scheme and branching ratios (c) [14], partially known before the experiment will be improved by the present much important statistics, and the possible correlation between protons from ^{20}Na* and γ rays from ^{19}Ne* may be discovered [13].

The capture reaction flow to higher masses will be hindered by photodesintegration processes for low S_p-value nuclei and then will have to wait for the β-decay of the latter. The reaction rates close to the proton drip line or near closed shells (low S_p-values) can be determined by single resonances because of the low level density above the proton threshold. Thus, in some selected cases (low S_p-values, light nuclei) the precise determination of the resonance parameters is required. In the high mass region, not only the reaction rates are unknown, but also even the nuclear properties as β$^+$ decay, masses, deformations... Many experiments are pionneering these "exotic" regions. For instance, some nuclei have been recently identified (^{43}Cr, ^{46,47}Fe), and some half-life and level schemes determined at LISE3/GANIL (see for instance the example of ^{31}Ar in Fig. 8 and [15] for the study of 23 < Z < 26 region). These level schemes experiments, devoted to the study of the β-p-γ decays of very unstable nuclei as ^{31}Ar (shown in Fig. 8) may give wealthy information on the Γ_p / Γ_γ ratios of the β-decay populated excited states above the proton threshold of the son nucleus (here ^{30}S). This kind of method could be used for the indirect study of proton capture rates on nuclei involved in the rp process path.

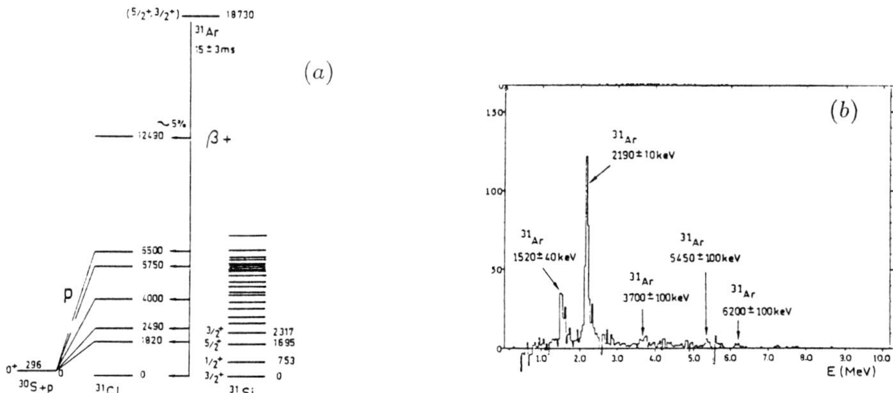

Fig. 8: Example of a part of the decay scheme of ^{31}Ar (a) taken from [16] with the corresponding proton energy spectra (b). This shows that a determination of the level schemes of very proton-rich nuclei is feasible even if it is located at the drip-line.

The investigations in the region of ^{65}As at NSCL / MSU, necessary for the knowledge of predicted bottelneck nuclei, are presented on figure 9.

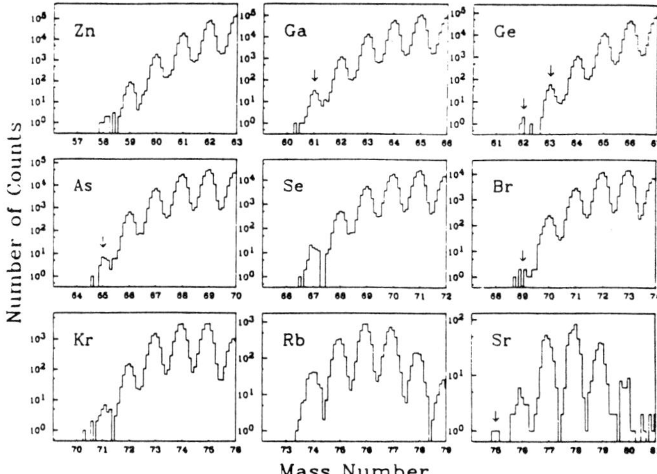

Fig. 9: Mass spectra observed for some elements observed among the ^{78}Kr fragmentation products. The newly discovered nuclei are indicated by the arrows [17]. The observation of ^{65}As and of ^{69}Br, hitherto predicted unbound, shows that the rp process can proceed through these nuclei. Conversely, the non-observation of ^{73}Rb will probably delay this process for reaching higher masses.

So, according to the production rate in the rp process path, it may be possible to study some important nuclear parameters as for instance the masses or the level scheme. These properties are necessary to refine the statistical Hauser-Feshbach calculations of the reaction rates for the nuclei which exhibit a high level density at the reaction threshold. Then, the parameters for the formation and the de-excitation of the compound nucleus are required (for $A+p \rightarrow B^* \rightarrow B+\gamma$). This includes: mass determination, level densities, mean Γ_p, the knowledge of the low energy levels in the B nucleus, and the level strengths by Giant Dipolar Resonance properties.

5. The neutron rapid captures processes.

The r process is known to synthesize about half of the elements heavier than iron. Its contribution is deduced from the predicted s process element abundance [18]. (The study of the s process lies beyond the scope of this review which deals with explosive processes). After about 30 years of efforts devoted to the knowledge of a most "plausible site for the r-process ([19])" and to explain the observed abundance curve of the elements, very recent improvements seem to show that a major step is occuring. A comprehensive review of the r process can be found in [20]. The r process is thought to develop in the hot bubble formed during a supernova (SN) explosion ([19] and references herein). The observational data suggest that low-mass type II SN may be the most likely contributors to the r elements [21]. However, less extreme conditions of density and temperature could produce rapid neutron captures. These conditions are encountered in the He layers of SN, the neutron flux being released by explosive (α, n) reactions on ^{13}C, ^{22}Ne induced by the passage of the shock wave of the explosion. The "classical" r process path drives far off stability by successive neutron captures and β^- decays (Fig. 10). The neutron capture is impeded by the photodesintegration when the path reaches neutron closed shells (because of the low S_n encountered) and have to wait until the β decay of the nucleus. These nuclei, at the different neutron closed shells $N = 50$, 82 and 126 are the classical "waiting points". An (n, γ)-(γ, n) equilibrium is reached for high temperature and density conditions. Thus, if we can determine the neutron separation energy S_n, the neutron density d_n and the stellar temperature, the abundance of these unstable progenitors is simply given by the Saha equation:

$$N(A, Z) \propto N(A-1, Z) \frac{d_n}{T^{3/2}} \exp\left(\frac{-S_n}{kT}\right)$$

The enhancements in the r element abundance curve (see insert in Fig. 10) called "r peaks" are evidences of the accumulation time at these waiting points. The understanding of the position and of the isotopic abundance ratios around these peaks is clearly related to the knowledge of the decay properties (half-life $T_{1/2}$ and neutron-delayed emission probability P_n) of nuclei around the closed shells. The measurements of these key properties by Kratz et al. allows to fix constraints upon the neutron capture time (and then on the neutron density d_n)and on the temperature necessary to match the β decay rates. The experimental work at ISOLDE / CERN concerning the waiting point nuclei $^{130}Cd_{82}$ and $^{79}Cu_{50}$ [26, 32] brings substantial informations about the conditions of temperature and neutron density required into stellar environment to exhibit an r process. An imaging picture of the correlation between the progenitors and the stable r process elements at the $N = 130$ closed shell, shown on Fig. 11.

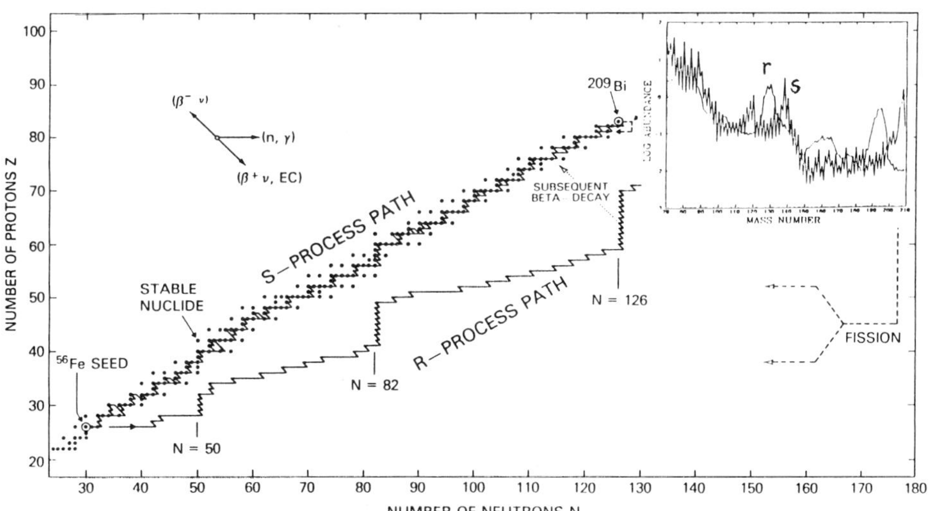

Fig. 10: Schematic view of the r process path in the (N, Z) plane [22]. This path was computed for a temperature of $T_9 = 1$ and a neutron density $d_n = 10^{24}$ n.cm^{-3}. The r process element abundance curve deduced from the estimated s process is presented in the insert.

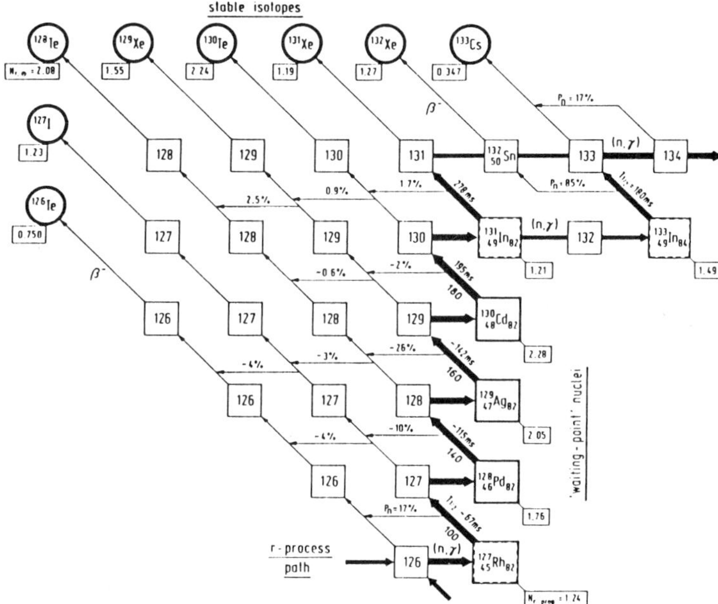

Fig. 11: Schematic view of the r-abundance features and the β-decay properties of nuclei in-or-close to the r process path in the A ≈ 130, "waiting point region" [26]. $T_{1/2}$ and P_n values given in this figure are either experimental values ($^{131,\,133}$In, ^{130}Cd) or RPA predictions.

Recent calculations including these measurements and high quality predictions for nuclear masses and β decays, show that the three peaks cannot be fitted by a global steady flow calculation, but at least three different neutron densities have to be included. Thielemann and Kratz [23, 24] point out that additionnal experimental data (masses, $T_{1/2}$, P_n, deformations, level schemes) far off stability as well as theoretical improvements are needed to obtain good agreement not only for the peaks but also for the whole abundance curve of the r elements. It seems that some discrepancies between the fitted and observed curve may reflect nuclear structure as for instance the transition from deformed to near spherical isotopes in the r process path before the neutron closed shells. Preliminary experiments have been performed in the Fe region at FRS / GSI and ILL [25]. Even if the predicted branching points (by β decay) are still out of reach, this is a first step to the study of this region.

The r process is historically related to neutron captures on an iron seed. But, as mentioned above, these neutron captures can also occur for lighter species produced by either the photodesintegration of iron in the α process model [27] or in He burning shells on "seed" nuclei. Anyway, a kind of r process has to be invoked to explain the existence and of light stable neutron-rich isotopes as ^{46,48}Ca.

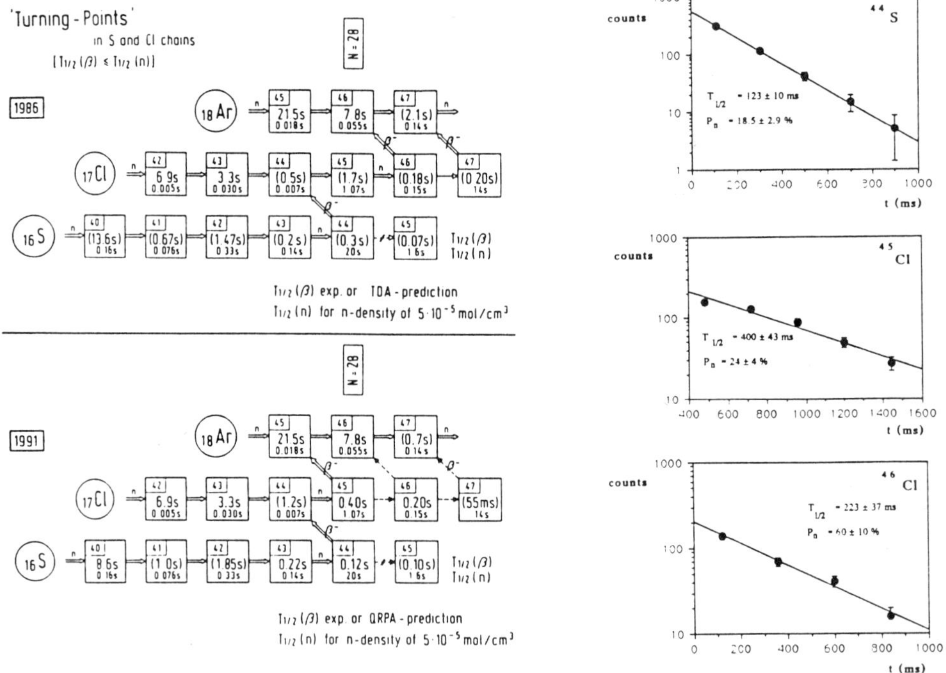

Fig. 12: Neutron capture path in the ^{16}S to ^{18}Ar chains for a stellar temperature of 8 T$_8$ and a neutron density of 5 10^{-5} mol cm^{-3}; upper part: status in 1986, lower part: present status. With the new experimental data obtained at LISE / GANIL facility [31], at both N = 28 "turning points" isotopes, ^{44}S and ^{45}Cl, β-decay starts to dominate over further neutron captures. Hence, the possible A= 46, 47 progenitors of ^{46}Ca will be produced in small amounts only. Even if the neutron capture path reaches ^{46}Cl and ^{47}Cl, their respectively higher and lower P$_n$-values largely contribute to the reduction of ^{46}Ca production. With this, large ^{48}Ca/ ^{46}Ca ratios can be obtained, as required to explain the observed abundances in EK-1-4-1 inclusion [28]. The right part shows the experimental data.

In fact, neither O and Si burning nor the s process can reach such neutron rich nuclei. In order to understand their synthesis and their abundance ratio, which has been found to be 50 in the solar system and 250 in refractory inclusions of the Allende meteorite, the decay properties of their progenitors have to be determined. Actually, the theoritical predictions cannot hold for such an underabundance of ^{46}Ca without invoking very severe stellar constraints on the neutron exposure [28, 29]. ^{46}Ca was thought to be produced mostly by the decay of ^{46}Cl (Pn = 30%) and of 100% P_n of ^{47}Cl (Fig. 12). The most striking effect given by the experimental results (shown in Fig. 12 right part) is the much shorter half-life of ^{45}Cl compared with the predictions. This nucleus then acts as a turning point and drives the path preferentially towards ^{45}Ca. The contribution of the A = 46,47 progenitors of ^{46}Ca will then be reduced (see caption of Fig. 12). It is important to note that the observed overabundances of ^{48}Ca, ^{50}Ti and ^{54}Cr seem to be correlated with those of ^{58}Fe, ^{64}Ni, ^{66}Zn [30]. Hence, astrophysical models have to show how they can produce all correlated abundances in a realistic and selfconsistent way. This emphasizes the growing complementarity between nuclear physics and isotopic anomalies observation.

Conclusion:

Nuclear Physics plays a key role in the understanding of the origin of the elements in nature and in the energy production in stars. Explosive stellar processes require the knowledge of nuclear properties of very far off stability species. Therefore, RNB facilities are necessary for their study. It has been widely outlined that experiments are successfully being performed on each side of the valley of stability. Though, improvements need to be achieved to cover a broader range of nuclei along explosive nucleosynthesis paths. RNB projects, as it is shown in [1, 2, 34] are promising and set excitements in the nuclear astrophysicists community.

Acknowledgments:

I would like to adress special thanks to D. Guillemaud Mueller, K.-L. Kratz, M. Mohar, A. C. Mueller and M. Wiescher for their fruitful comments and discussions concerning this work.

References:

[1] 1989 *Proceedings of the First International Conference on Radioactive Nuclear Beams, October 16-18, Berkeley, ed. Nitschke J M.*

[2] 1991 *Proceedings of the Second International Conference on Radioactive Nuclear Beams, Louvain-La-Neuve, August 19-21, ed. Adam Hilger.*

[3] Ravn H L and Allardyce B W 1989 *Treatise on Heavy-Ion Science, Vol. 8, ed. Bromley D A, Plenum Press New York* p. 363.

[4] Boyd R N et al. 1991 *Nucl. Phys. A* **538** 523c.

[5] Farrel et al. 1991 *Proceedings of the Second International Conference on Radioactive Nuclear Beams, Louvain-La-Neuve, August 19-21, ed. Adam Hilger* p. 287

[6] Paradellis T et al. 1990 *Z. Phys. A* **337** 211.

[7] Champagne A E, Wiescher M 1992 *Ann. Rev. of Nucl. and Part. Sci.* in print.

[8] Kubono S et al. 1989 *Ap. J* **344** 460, Kubono S 1988 *Z. Phys. A* **331** 359.

[9] Langanke K et al. 1986 *Ap. J* **301** 629.

[10] Wallace R K and Woosley S E 1981 *Ap. J Suppl.* **45** 389.

[11] Lamm L O et al. 1990 *Nucl. Phys. A* **510** 503.

[12] Smith M S et al. 1992 *Nucl. Phys. A* **536** 333.

[13] Piechaczek A et al. 1992 *Proceedings of the 6 th Conference on Nuclei far from Stability, 19-24 July, Bernkastel-Kues* to be published.

[14] Kubono S et al. 1991 *Proceedings of the Second International Conference on Radioactive Nuclear Beams, Louvain-La-Neuve, August 19-21, ed. Adam Hilger* p. 317.

[15] Borrel V 1992 *Z. Phys.* to be published.

[16] Borrel V et al. 1987 *Nucl. Phys. A* **473** 331, see also Bazin D et al. 1992 *Phys. Rev. C* **45** (1) 69.

[17] Mohar M et al. 1991 *Phys. Rev. Lett.* **66** (12) 1571.

[18] Käppeler F 1989 et al. *Rep. Prog. Phys.* **52** 945.

[19] Meyer B S et al. 1992 *Proceedings of the Origin and Evolution of the Elements, June 22-25, Paris, to be published in Cambridge Univ. Press* and submitted to *Ap. J.*

[20] Cowan J J and Thielemann F-K 1991 *Phys. Rep.* **208** 267.

[21] Mathews G J and Cowan J J 1990 *Nature* **345** 491.

[22] Rolfs C E and Rodney W S 1988 *Cauldrons in the Cosmos, ed. Univ. of Chicago Press.*

[23] Kratz et al. 1992 *Contribution to this Conference.*

[24] Thielemann F -K and Kratz K -L 1991 *Proceedings of the 22nd Mikolajki Summer School, Selected topics in Nuclear Astrophysics, ed. Szeflinska G, IOP Publishing, Bristol.*

[25] Czajkowski et al 1992 *Contribution to this Conference.*

[26] Kratz K-L 1988 *Rev. in Mod. Astron.* 1 184, Kratz et al. 1991 *Proceedings of the International Workshop on Nuclear Shapes and Nuclear Structure at low excitation energy, Cargèse, June3-7.*

[27] Woosley S E and Hoffman R 1991 *Ap. J.* in print.

[28] Kratz K-L et al. 1991 *Proceedings of the First European Biennal Workshop on Nuclear Physics, March 25-29, Megève, ed. World Scientific.*

[29] Thielemann etal. 1990 *Proceedings of the International Symposium on Nuclear Astrophysics, Nuclei in the Cosmos, June 18-22, Baden/Vienna, ed. Oberhummer H and Hillebrandt W.*

[30] Harper C L et al. *Proceedings of the International Symposium on Nuclear Astrophysics, Nuclei in the Cosmos, June 18-22, Baden/Vienna, ed. Oberhummer H and Hillebrandt W.*

[31] Sorlin O 1991 Thèse de l'Université Paris VII, IPNO -T.91.04.

[32] Kratz et al. 1992 *Ap. J.* to be published.

[33] Görres et al. *Phys. Rev. C* to be published.

[34] Garrett J 1992 *Contribution to this Conference.*

Nuclear reaction cycles in explosive hydrogen burning

C. Iliadis, J. Görres, L. Van Wormer, M. Wiescher

Dept. of Physics, University of Notre Dame

F.K. Thielemann

Harvard Smithsonian Center for Astrophysics

Abstract: Several experiments have been performed to investigate the reaction branchings $^{31}P(p,\gamma)/(p,\alpha)$ and $^{35}Cl(p,\gamma)/(p,\alpha)$ for temperatures typical for explosive hydrogen burning. The results are discussed in the framework of large scale network calculations.

. Introduction

Hydrogen burning at temperatures and densities far in excess of those attained in the interiors of ordinary main-sequence stars may occur in a variety of astrophysical sites, including Novae, type I Supernovae, and X-ray bursts. For explosive and hydrogen rich environments, the rp-process has been proposed by Wallace and Woosley (1981) as a fast nucleosynthesis reaction sequence whereby CNO material will be transferred into the Fe-Ni mass region. The resulting network of nuclear reactions involves proton and alpha capture, (p,α) reactions, and β-decay processes, as well as possible inverse reactions, and extends from the line of stability up to the proton drip line.

Several effects might impede a continous flow of nuclear material to heavier masses. For example, proton capture on isotopes at the proton drip line is inhibited because the resulting nuclei are proton unbound. In this case, the nuclear flow has to wait for a β^+-decay. Similarly, the (p,γ) reactions on even-even $T_z=-1$ nuclei (^{22}Mg, ^{26}Si, ^{30}S, ^{34}Ar, ..) with typically very low Q-values (Q<0.5 MeV) are strongly hindered by the inverse photodisintegration process. The reaction flow again has to wait for the β-decay of these nuclei. In both cases, the nuclei involved represent waiting points for the rp-process and cause a significant impedance for the continous flow towards heavier isotopes.

Nuclear reaction cycles represent another type of impedance effect. For the proton capture on even-odd $T_z=-1/2$ nuclei, the Q-values are 0.5<Q<2.5 MeV. These reactions are fairly low for low temperatures due to the corresponding low level density in the created compound

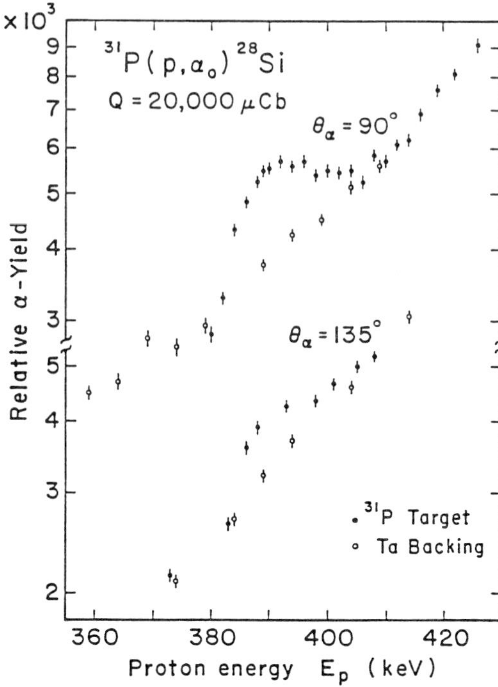

Fig. 1. Excitation function for the reaction $^{31}P(p,\alpha)^{28}Si$, measured at $\theta=90^\circ$ and 135°. The solid and open circles correspond to the yield in the expected α-energy range using an implanted ^{31}P target and a blank Ta-backing, respectively.

nucleus. Therefore, for low temperatures, $T_9<0.4$, the β-decay dominates. For the resulting odd-even $T_z=1/2$ nuclei (^{23}Na, ^{27}Al, ^{31}P, ^{35}Cl, ^{39}K, ..) both the (p,γ) and the (p,α) reaction are possible. While the (p,γ) reaction will transform material to heavier masses, the competing (p,α) reaction returns material to lighter isotopes giving rise to various reaction cycles, e.g. NeNa-, MgAl-, SiP- and SCl-cycle. The temperature dependent branching ratio $(p,\gamma)/(p,\alpha)$ determines how much material will be stored in the reaction cycle. In the extreme case $(p,\alpha)>>(p,\gamma)$ no more material will be transferred to heavier mass regions and the rp-process reaches its end point.

2. Experiments

The odd-even $T_z=1/2$ nuclei with $A<43$ are stable and the $(p,\gamma)/(p,\alpha)$ branching ratio can be studied experimentally using stable targets. In principle, all the levels above the proton threshold in the compound nucleus can contribute to the (p,γ) stellar reaction rate, whereas only levels of natural parity can decay through the α-channel. As an illustration of the experimental technique involved in the study of such reaction cycles, we discuss the investigation of the $^{31}P(p,\gamma)/(p,\alpha)$ and $^{35}Cl(p,\gamma)/(p,\alpha)$ branching ratios.

The reaction $^{31}P(p,\gamma)^{32}S$ has been studied (Iliadis et al. 1991) for proton bombarding energies E_p=280-620 keV. The target used was produced by implanting ^{31}P ions into thick tantalum backings. The emitted γ-radiation was observed by a 35% Ge detector at θ=55° in close geometry to the target. Two new (p,γ) resonances could be identified at E_R=383 and 403 keV. The outgoing α-particles from the reaction $^{31}P(p,\alpha)^{28}Si$ were observed at the same time with Si surface-barrier detectors at angles θ=90° and 135°. In order to absorb the intense yield of elastically scattered protons, 2 μm Havar foils were placed in front of the Si detectors. The total yield in the expected α-energy range measured with the implanted ^{31}P-target (solid circles) and a blank Ta-backing (open circles) is shown in Figure 1. A new (p,α) resonance is clearly observed at both angles for E_R=383 keV, corresponding to the natural parity state at E_x=9236 keV (J^π=1⁻) in ^{32}S. The derived stellar reaction rates for $^{31}P(p,\gamma)^{32}S$ and $^{31}P(p,\alpha)^{28}Si$ are experimentally established for stellar temperatures T_9>0.4. For T_9=0.4 the (p,γ) stellar rate is about 3 times as high as the (p,α) rate, suggesting weak cycling of stellar material in the SiP mass range (see

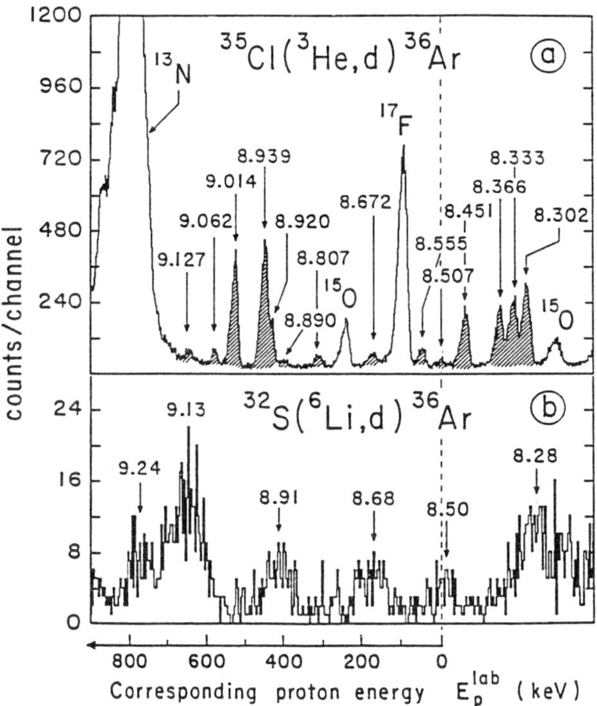

Fig. 2. Deuteron spectra from the reactions $^{35}Cl(^3He,d)^{36}Ar$ (a) and $^{32}S(^6Li,d)^{36}Ar$ (b), obtained at θ=7.5°. The peaks are labelled with our measured excitation energies in ^{36}Ar. The dotted line shows the proton threshold.

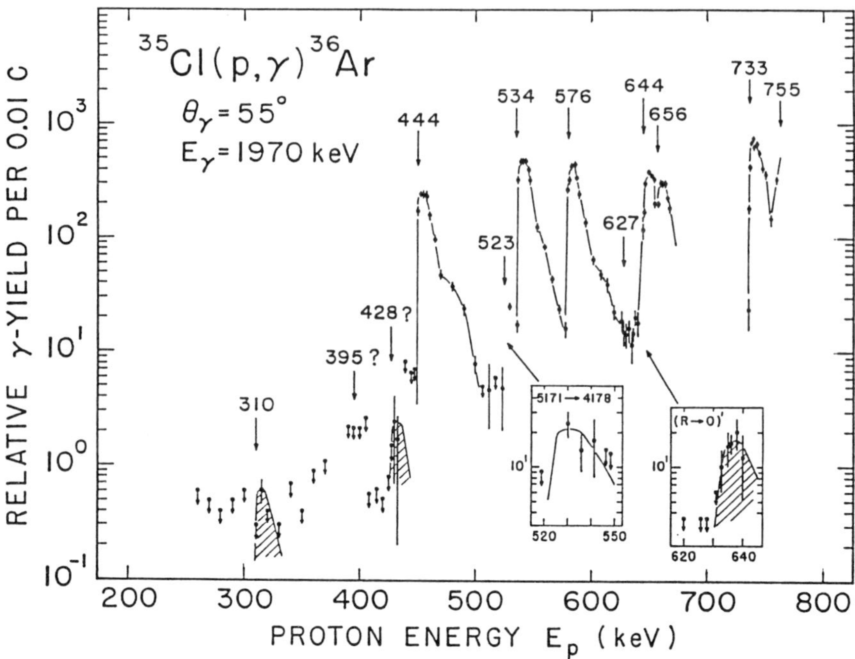

Fig. 3. Yield curve for ^{35}Cl(p,γ)^{36}Ar, measured at θ=55° for the decay of the first excited state in ^{36}Ar at E_x=1970 keV. Measured or expected resonance energies are given in keV. Hatched areas indicate new found resonances. The solid line through the data points is to guide the eye.

section 3). For lower temperatures, the stellar rates are still uncertain mainly due to the unknown contribution of the threshold state at E_x=9060 keV (J^π=(0-2)$^-$) in ^{32}S which has been observed in transfer reaction studies (Kalifa 1978). This state might increase the (p,α) reaction rate by orders of magnitude in the case of natural parity.

For the ^{35}Cl+p system only a few levels were known within 0.5 MeV above the proton threshold (Endt 1990) in the compound nucleus ^{36}Ar. In order to establish the level scheme, the reaction ^{35}Cl(^{3}He,d)^{36}Ar has been measured. The outgoing deuterons were momentum analysed with a broad-range magnetic spectrograph and detected by a position sensitive proportional counter located in the focal plane of the spectrograph. The targets were produced by implanting ^{35}Cl into thin carbon foils. A deuteron spectrum obtained at θ=7.5° is shown in Figure 2a. Two new unbound states in ^{36}Ar at E_x=8807 and 8890 keV have been observed, corresponding to (p,γ) resonances at E_R=310 and 395 keV. To search for the corresponding proton resonances the reaction ^{35}Cl(p,γ)^{36}Ar was measured in a similar set-up used for the investigation of the ^{31}P(p,γ)^{32}S reaction. The targets were produced by implanting ^{35}Cl ions into thick tantalum backings. The resulting excitation function is shown in Figure 3. At least two new (p,γ) resonances have been clearly observed at E_R=310 keV, corresponding to the new found state at E_x=8807 keV in the (^{3}He,d) reaction, and at E_R=627 keV. The existence of a new resonance at E_R=428 keV is still uncertain because no primary γ-decay could be observed and the intensity of he secondary decay of the first excited state in ^{36}Ar has poor statistics. To obtain information .bout the location of natural parity states in ^{36}Ar, the ^{32}S(^{6}Li,d)^{36}Ar reaction has been measured n the same set-up as used for the (^{3}He,d) reaction. The target was produced by evaporating InS nto thin carbon foils. A deuteron spectrum measured at θ=7.5° is shown in Figure 2b. The esults show two natural parity states above the proton threshold at E_x=8.68 and 8.91 MeV, orresponding to (p,α) resonances at E_R=170 and 420 keV. Further experiments are presently nder way to obtain the complete information needed for the calculation of the stellar reaction ates.

. Network calculations

Ve have performed numerical network calculations (Van Wormer 1991) with a fixed omposition and unique temperature density history in order to investigate the influence of npedance effects on the reaction flow in the rp-process. The network contained 946 nuclear iteractions including β^+-EC processes, (p,γ), (p,α) and (α,γ) reactions as well as possible iverse reactions between 216 stable and unstable nuclei up to the isotope Kr. For the initial otopic distribution, solar abundances were used. The stellar reaction rates were updated with

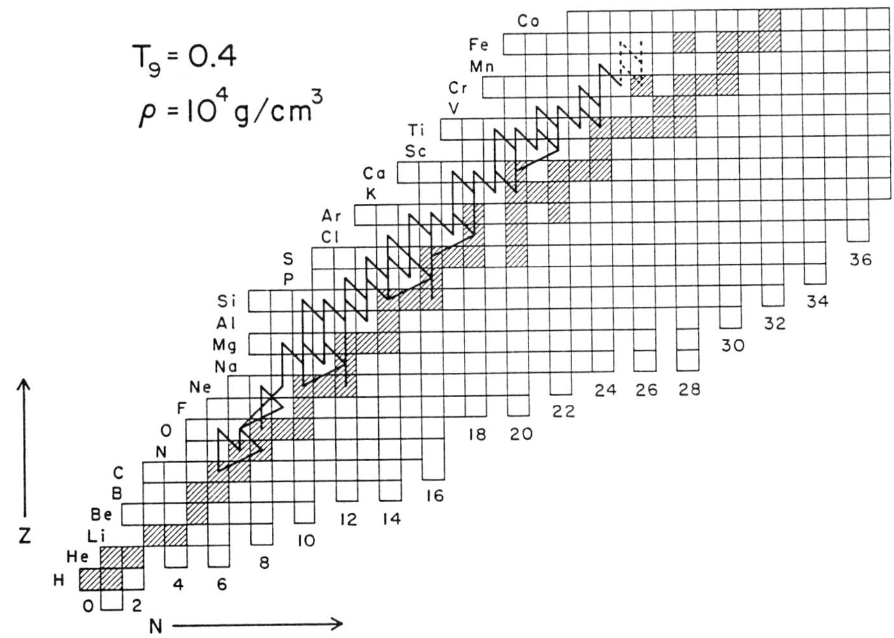

g. 4. Major nuclear flows in the rp-process integrated over a period of t=100 s. The calculation was performed for onstant temperature T9=0.4 and density ρ=10^4 g/cm^3. Solid lines indicate dominant nuclear flows. Dashed lines dicate flows roughly an order of magnitude weaker.

our experimental results. The reaction flow calculated for a constant temperature and density of $T_9=0.4$ and $\rho=10^4$ g/cm^3 corresponding to the extreme conditions for nova outbursts is shown in Figure 4. The calculations were performed for a burning period of 100 s. The nucleosynthesis for masses A<20 is dominated by the hot CNO cycle. The reaction flow to higher masses is induced by the break-out from the hot CNO-cycle via

$$^{15}O(\alpha,\gamma)^{19}Ne(p,\gamma)^{20}Na(p,\gamma)^{21}Mg$$

and by the proton capture on the initially present ^{20}Ne material. The further flow runs close to the proton drip line feeding the SiP mass range. The rp-process in the mass region A=28-36 is characterized by two reaction cycles, the SiP-cycle:

$$^{28}Si(p,\gamma)^{29}P(p,\gamma)^{30}S(\beta^+\nu)^{30}P(p,\gamma)^{31}S(\beta^+\nu)^{31}P(p,\alpha)^{28}Si$$

and the SCl-cycle:

$$^{32}S(p,\gamma)^{33}Cl(p,\gamma)^{34}Ar(\beta^+\nu)^{34}Cl(p,\gamma)^{35}Ar(\beta^+\nu)^{35}Cl(p,\alpha)^{32}S.$$

Proton capture reactions on ^{31}P, ^{31}S and ^{35}Cl, ^{35}Ar, respectively, are responsible for the leakage out of these cycles. A continous mass flow leads to the CaSc-cycle and leaks out via $^{43}Ti(p,\gamma)^{44}V$ and $^{43}Sc(p,\gamma)^{44}Ti$. The material flow finally reaches ^{52}Fe. This isotope represents the endpoint in the rp-process for the assumed burning conditions due to its long β-decay life time ($\tau=11.95$ h) and the weak depletion reaction $^{52}Fe(p,\gamma)^{53}Co$ (Van Wormer 1991). For higher temperatures the impedance due to the reaction cycles disappears because the proton capture on the even-odd $T_z=-1/2$ nuclei dominates over the competing β^+-decay, thus bypassing the material recycling (p,α) reactions on odd-even $T_z=1/2$ nuclei (see section 1). Although the reaction flow calculations indicate clearly the cycling of nuclear material in certain mass ranges, a quantitative understanding has to await further experimental results because the stellar reaction rates involved still carry large uncertainties.

References

Endt, P.M. 1990, Nucl.Phys.A521, 1
Iliadis, C. et al. 1991, Nucl.Phys. A533, 153
Kalifa, J. et al., Phys.Rev. C17, 1961
Van Wormer, L. 1991, PhD. thesis (unpublished)
Wallace, R.K. and Woosley, S.E. 1981, Ap.J.Suppl. 45, 389

Nuclear structure far from stability and isotopic r-process abundances

K-L Kratz[1], F-K Thielemann[2], P Möller[1], B Pfeiffer[1] and A Wöhr[1]

[1]Institut für Kernchemie, Universität Mainz, Germany
[2]Harvard-Smithsonian Center for Astrophysics, Cambridge, MA, USA

Abstract. Without assuming a particular astrophysical site or model, we deduce the conditions responsible for the production of r-process nuclei by making use of: (i) the solar r-process abundances $(N_{r,\odot})$, and (ii) nuclear masses, β-decay half-lives and β-delayed neutron emission probabilities for nuclei between stability and neutron drip line. We find that the $N_{r,\odot}$ distribution is the result of a superposition of several local steady-flow components with comparable time scales. Both, the successful reproduction of the r-abundance peaks as well as the remaining deficiencies are clear signatures of nuclear-structure very far from stability.

1. Introduction

The discovery of the rapid neutron-capture process goes back to the pioneering work of Burbidge et al. [1] and Cameron [2]. Since then the r-process has been related to environments with a high neutron density, where neutron captures are faster than β-decays, even for neutron-rich nuclei up to 15-30 units from stability. Magic neutron numbers are encountered for smaller mass numbers A than in the valley of stability, which shifts the r-process abundance peaks in comparison to the s-process peaks. But besides this basic understanding, the history of r-process research has been quite diverse (for a recent review, see e.g. [3]).

Several investigators (see e.g. [4,5]) noticed that the overall shape of r-abundances and the position of the peaks may be reproduced by a superposition of neutron densities. The suggested astrophysical sites were type II supernovae (SN), but until today a complete understanding of the SN mechanism is still pending. Alternatives were explosive He-burning environments, where the r-process had only to act on previously s-processed material and one neutron density was sufficient to transfer matter from each s-process peak into the next higher r-process peak (see e.g. [6]). Cameron et al. [7] performed steady-flow calculations (but without an $(n,\gamma) \rightleftharpoons (\gamma,n)$ equilibrium), independent of a particular site. Their results qualitatively reproduced the main r-peaks with decreasing abundances for increasing A. More recently, first experimental information in the $A \simeq 80$ and 130 regions (see e.g. [8-10]), together with improved accuracy in $N_{r,\odot}$ [11] made it possible to even analyze isotopic abundance patterns. The authors concluded that, indeed, an $(n,\gamma) \rightleftharpoons (\gamma,n)$-equilibrium (waiting point approximation) *and* a steady flow was required to reproduce specific features in these peaks.

The aim of the present investigation is (i) to deduce stellar conditions which reproduce the $N_{r,\odot}$ pattern, without assuming a particular astrophysical site or model, (ii) to analyze whether all of the previously mentioned scenarios are still consistent with the present knowledge of the $N_{r,\odot}$ and nuclear properties far from stability, or if more stringent constraints can be set, and (iii) whether we can learn nuclear structure from isotopic r-abundances.

2. R-process calculations

The system of differential equations for an r-process network includes terms for neutron captures, neutron-induced fission, photodisintegrations, β-decays, β-delayed neutron emission and - for nuclei with $Z>80$ - β-delayed fission. Depending upon the specific conditions, either β-decays can be faster than neutron captures and photodisintegrations like in an s-process, or vice versa when an $(n,\gamma) \rightleftharpoons (\gamma,n)$ equilibrium exists. If the β-flow (i.e. β-decay of the nuclei) from each Z-chain to $(Z+1)$ is equal to the flow from $(Z+1)$ to $(Z+2)$, then a steady-flow or β-flow equilibrium will exist.

Fig. 1 shows in the neutron number density (n_n) – stellar temperature (T_9) plane the boundary line beyond which this 'waiting-point' approximation is valid. In such a case, the fractional abundance of nucleus (Z,A) is given as a function of n_n, T_9 and the neutron binding energy $B_n(Z,A)$

$$P(Z,A) = N(Z,A)/\sum_A N(Z,A) = f(n_n,T_9,B_n(Z,A)) \tag{1}$$

In other words, the maximum abundance for a given n_n and T_9 is located at a nucleus with the same B_n in each isotopic chain. With this, B_n introduces the dependence on nuclear masses for very neutron-rich nuclei. Under the assumption of an $(n,\gamma) \rightleftharpoons (\gamma,n)$ equilibrium, we no longer need a detailed knowledge of neutron capture cross sections.

Given a steady-flow equilibrium, the total abundance in each isotopic chain is $N(Z)=\sum_A N(Z,A)$. In case of an $(n,\gamma) \rightleftharpoons (\gamma,n)$ equilibrium in addition to a steady flow, we also have

$$N(Z)\sum_A P(Z,A)/T_{1/2}^{Z,A} = N(Z)/T_{1/2}(Z) = \text{const.} \tag{2}$$

Then, the assumption of an abundance for $N(Z_{min})$ at a minimum Z-value is sufficient to predict the whole set of abundances as a function of A.

Within the waiting-point approximation, for a number of Z's at the neutron magic numbers $N=50$ and 82 only *one* isotope contains the dominant abundance, and the sum over A is reduced to one term at $N=N_{magic}$. From the known $N_{r,\odot}^A$ and known P_n-values of their neutron-rich isobars, one can then predict the abundances of the progenitor isotopes in the r-process path ($N_{r,prog}^A$), and therefore also 'predict' the $T_{1/2}^A$ for these nuclei. As was shown by Kratz et al. [8-10], these 'β-flow requests' on $T_{1/2}$ agree closely with experimental or QRPA-values, suggesting that, indeed, a steady-flow equilibrium was achieved, at least locally. The shaded area in Fig. 1 indicates those conditions for which the $N=50$ isotones $^{76}_{26}$Fe to $^{80}_{30}$Zn as well as the $N=82$ isotones $^{127}_{45}$Rh to $^{130}_{48}$Cd possess the maximum abundance in their isotopic chains. At $Z=31$, respectively $Z=49$, the r-process path finally branches off from the N_{magic}, and two even-N isotopes share the dominant abundance. In this way, the $N_{r,prog}$ ratios of ^{81,83}Ga and ^{131,133}In are related to the $N_{r,\odot}$ of the stable isotopes ^{81}Br and ^{83}Kr and ^{131,132}Xe, respectively. The known large P_n-values of ^{83}Ga and ^{133}In play an important role in these ratios. In order to reproduce the $N_{r,\odot}$ ratios, certain n_n–T_9 conditions are required, which again lie within the boundaries given in Fig. 1. All of this is accumulated evidence that the $N_{r,\odot}$ originated from freeze-out abundances of an r-process with high neutron number densities ($n_n \geq 10^{20}$ cm^{-3}) and temperatures, ($T \geq 10^9$ K) typical for an $(n,\gamma) \rightleftharpoons (\gamma,n)$ equilibrium and at least a local steady flow in the vicinity of the r-process peaks [8].

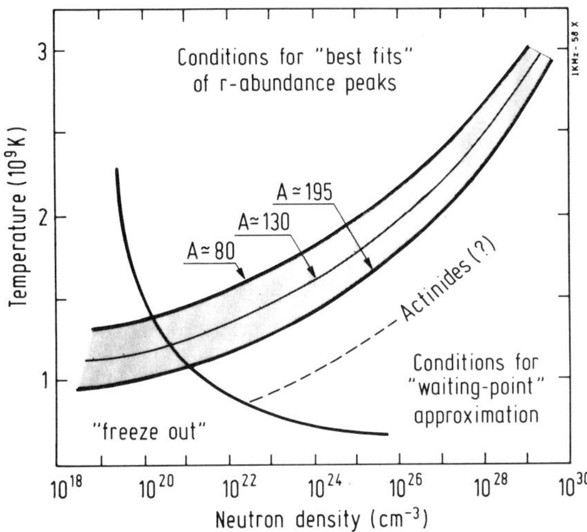

Fig. 1: *T_9-n_n conditions for the waiting-point and steady-flow approximation for different mass ranges of $N_{r,\odot}$ [11]. Since r-process nuclei beyond Bi decay via β- and α-chains, no clear features emerge for a 'best fit'; hence, the dashed line for the actinide component indicates but a first guess.*

3. Nuclear data

Since the vast majority of nuclei in or close to the r-process path is not accessible in terrestrial laboratories, a general understanding of their nuclear-structure properties may only be obtained through theoretical means. However, since a number of different quantities are needed in r-process calculations, in the past it was not possible to obtain them all from one source. Taking them from different sources, however, raises the question of consistency. Therefore, one should avoid to mix a specific mass model for predicting B_n and the r-process path, with theoretical $T_{1/2}$, P_n and P_{df}-values originating from other approaches containing different masses, shapes and/or fission barriers. In such mixed calculations, although often performed due to the lack of data (see e.g. [3,12]), occasionally nuclear-structure signatures may vanish or 'artificial' effects may occur, thus strongly limiting the predictive power far off β-stability.

Therefore, in the present paper, the *unified* macroscopic-microscopic approach of Möller et al. [13-15] has been applied, within which all properties can be studied in an internally consistent way. A folded-Yukawa single-particle model with extensions is combined with the most recent version of the finite range droplet model which includes Coulomb redistribution effects and an improved formulation of the Lipkin-Nogami pairing model. As a first step, nuclear ground-state masses and shapes are calculated. Once these quantities are known, nuclear wave functions are derived for the appropriate shapes. Matrix elements giving β-decay rates and other quantities of interest may then be determined. $T_{1/2}$ and P_n-values are deduced from theoretical Gamow-Teller (GT) strength functions calculated within the QRPA [16], using the same folded-Yukawa single-particle levels as for the shell corrections of the nuclear masses.

This *consistent* nuclear-data set is expected to yield more reliable predictions than earlier models. Nevertheless, being aware that also our new approach must have its deficiencies, we have further improved the data set by taking into account all recent experiments on Q_β, B_n, $T_{1/2}$ and P_n. Furthermore, for nearby extrapolations known nuclear-structure properties, either model-inherently not contained (e.g. p-n residual interactions) or not properly described (e.g. the onset of deformation at $A \simeq 100$) by our above - still too simplistic - approach were taken into account. For a detailed discussion, see Kratz et al. [17].

4. Fits to solar r-abundances

First, we tried to generalize our earlier fits to selected mass regions [8-10] to the whole $N_{r,\odot}$ distribution by assuming a *global* steady flow. The results of such calculations for conditions which reproduced in detail the $A \simeq 80$ peak present, however, major problems. Indeed, three r-abundance maxima are produced; but the $A \simeq 130$ and 195 peaks are much too tall and are also shifted to higher A, which means that in these regions n_n is too low. Already Kratz et al. [10] had noticed the need for different normalizations of the $A \simeq 80$ and 130 peaks. Hence, we conclude that the $N_{r,\odot}$ distribution, with declining peak heights as a function of A, *cannot* be explained by a global steady flow with an $(n,\gamma) \rightleftharpoons (\gamma,n)$-equilibrium. This result rules out the steady-flow scenario of Cameron et al. [7] as a possible solution.

In a next step we have focussed on reproducing the details of individual mass regions, and finding the break points between steady-flow areas. The results of these *static* calculations can be summarized as follows: The r-process, indeed, has reached a steady-flow equilibrium which is, however, no longer global but only local in between the r-abundance peaks. It breaks down at the top of each peak, i.e. at the $N=50$, 82 and 126 closed shells. Similar to the $N_s\sigma$-curve (see Fig. 2) for the s-process, we obtain a three- or four-step r-process $N_r/T_{1/2}-$, respectively $N_r\lambda_\beta$–curve with different neutron number densities n_n. The statistical weights of the first three $N_r/T_{1/2}$–components are 10:2.5:1. There remains, however, an uncertainty for the conditions which produce the $N_{r,\odot}$ beyond $A \simeq 200$. Since their progenitors decay via β- and α-decay chains, there is no simple shape to fit, and our study is just exploratory. Nevertheless, our results definitely rule out explosive He-burning scenarios as a possible solution and support the SN II origin of the r-process (see e.g. [18]).

Finally, the question remains whether the $N_{r,\odot}$ curve can be the result of a superposition of three (or more) local steady flows, where each component dominates one peak. This is not possible, as already indicated by our *global* steady-flow calculations. A superposition will only make sense when the abundances for a specific T_9-n_n condition, appropriate to a region $A \leq A_{peak}$, will be set to zero for $A > A_{peak}$. Such a situation can occur in time-dependent calculations, where the r-peaks containing neutron-magic nuclei with the longest $T_{1/2}$ will act as 'bottle necks', over which only small amounts of matter will pass. Fig. 3 shows the results of three time-dependent calculations for the 'best-fit' n_n-T_9 conditions of our static steady-flow fits, which reproduce the $A \simeq 80$ peak and the $85 \leq A \leq 130$ and $135 \leq A \leq 195$ regions. When starting our time-dependent calculations

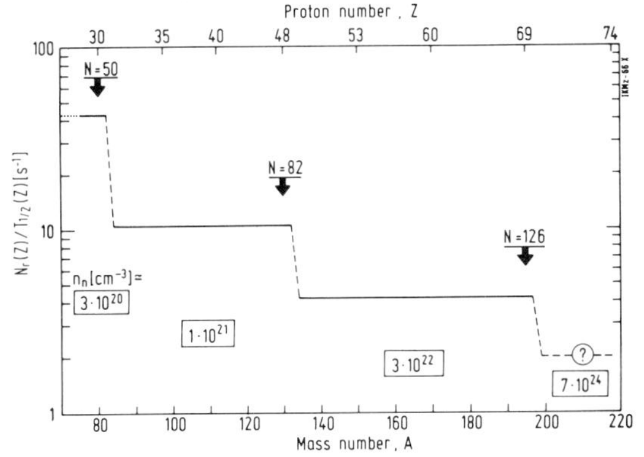

Fig. 2: *Schematic representation of a three- (or four-) step r-process $N_r/T_{1/2}$-curve with different n_n derived from static steady-flow calculations. It is seen that the β-flow equilibrium is no longer global, but only local. It breaks down at the N=50, 82 and 126 closed shells, i.e. at the top of each $N_{r,\odot}$ peak.*

with Z=26, the time scales for reaching the appropriate peaks are with 1.5 s for the first two and 2.5 s for the third component comparable, and are consistent with the expected duration of the r-process in a SN. A choice of Z_{min}=40 rather than 26 for the third component – i.e. a heavier seed nucleus, as expected from an α-rich freeze-out scenario [19] – would give the same result in 1.5 s rather than 2.5 s. When different zones with varying conditions in one SN event create these abundance patterns, a superposition (weighing the different components appropriately) can reproduce the *global* $N_{r,\odot}$ curve. This is depicted in Fig. 4, where the upper part gives the $N_{r,prog}$ distribution *before* β-decay and the lower part shows the final pattern *after* β-decay. Considerable smoothing of the initial odd-even staggering in the $N_{r,prog}$ due to β-delayed neutron emission is observed. Clearly, the main features of the $N_{r,\odot}$ distribution are well reproduced, and in certain regions even *isotopic abundances* become meaningful. However, there remain several deficiencies which – apart from $A \leq 78$ where the ($N_\odot$–N_s) residuals presumably are not of genuine r-origin – clearly indicate nuclear-structure effects very far from stability not accounted for properly in our present macroscopic-microscopic nuclear-data set.

The solution to the $A \simeq 120$ abundance trough lies mainly in the nuclear masses along the r-process path at $B_n \simeq 2$ MeV. In the Möller et al. mass model [14,15], the decrease of the B_n's occurs too slowly when approaching N=82. For our 'best-fit' n_n–T_9 conditions, this implies that there would exist *not a single* waiting-point isotope between ^{112}Zr$_{72}$ and ^{125}Tc$_{82}$. This model-deficiency presumably has its origin in an overestimation of the N=82 shell strength below $^{132}_{50}$Sn$_{82}$. It is interesting to note in this context, that the $A \simeq 180$ region is fitted quite well, indicating that the vanishing of the N=126 shell strength below $^{208}_{82}$Pb is described correctly in the Möller et al. mass model. In contrast, a deep trough below the $A \simeq 195$ peak – as in the $A \simeq 120$ region – results when using the recent ETFSI mass model [20].

In the $103 \leq A \leq 115$ and $160 \leq A \leq 180$ mass regions, sinusoidal deviations show up with 'spikes' at $A \simeq 111$ and 175 originating from the overabundant progenitor isotopes ^{112}Zr and ^{176}Sm, respectively. As is discussed in detail in [17], these deviations are due to model-deficiencies in the development of quadrupole deformation around neutron mid-shells N=66 and 104 which are clearly correlated with the other nuclear properties B_n, $T_{1/2}$ and P_n. In any case, we can show that the 'pygmy' $N_{r,\odot}$ peaks around $A \simeq 103$ and 163 have their origin in the β-decay properties of *deformed* progenitor isotopes; hence the assumption of large fission contributions seems not necessary.

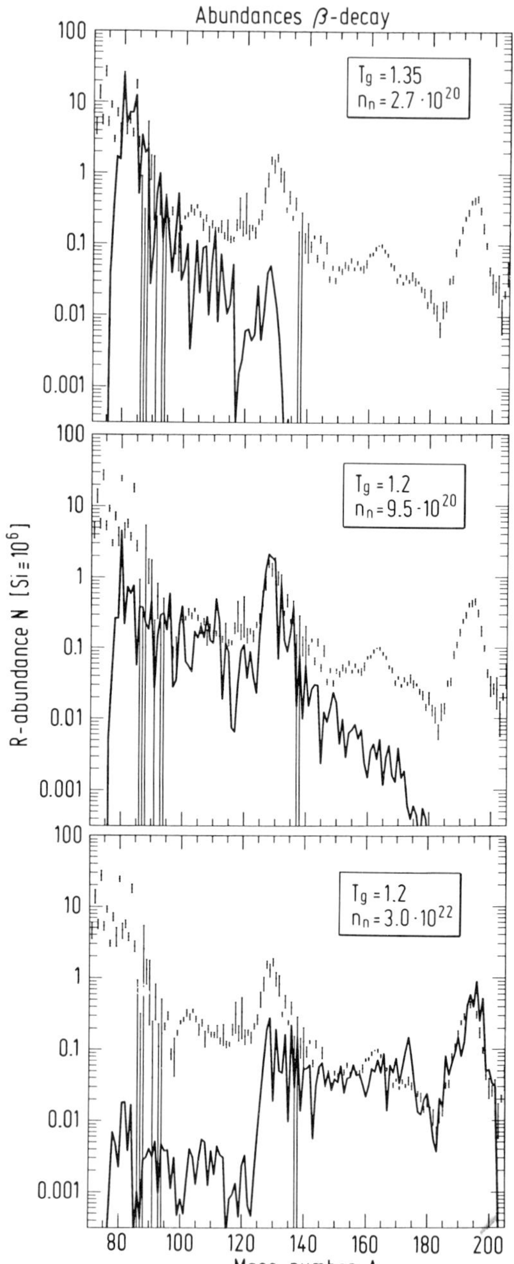

Fig. 3: *Time dependent calculations with 'best-fit' conditions from our static steady-flow calculations for a seed nucleus with Z=26. Upper part: formation of nuclei in the A≃80 peak at 1.5 s; middle part: formation of the 85≤A≤130 region at 1.5 s; lower part: formation of nuclei beyond A≃130 up to the A≃195 abundance peak after 2.5 s. If a seed nucleus with Z=40 is chosen, the same configuration occurs at 1.5 s.*

So far, we assumed that equilibrium conditions apply at one instant and neutron densities and temperatures are instantaneously frozen, thus neglecting neutron captures during freeze-out. In a realistic scenario, these neutron captures will certainly smoothen the abundance curve somewhat; however, as shown in [17], they will not distort the initial pattern completely. The strongest 'proof' for this is the odd-even staggering in the $N_{r,\odot}$ around $A\simeq80$ which is preserved almost exactly as imprinted by a steady flow and subsequent β-decay and β-delayed neutron emission. It is just in this region where most of the nuclear properties come from experiment, i.e. are most reliable. From the appendix of [3], we expect $<\sigma v>_{n,\gamma}$-values of about 10^{-19} to 10^{-20} cm^{3}s^{-1}, e.g. for ^{85}As or ^{112}Zr, corresponding to a neutron-capture time scale of $\tau_{n,\gamma}=1/(n_n<\sigma v>)\simeq0.04$ s before freeze-out. Thus, the drop in n_n at freeze-out must be short in comparison to this time scale. In consequence, the major deficiencies in our calculated N_r curve discussed above will remain. This is an additional proof that they must be due to deficiencies in nuclear properties.

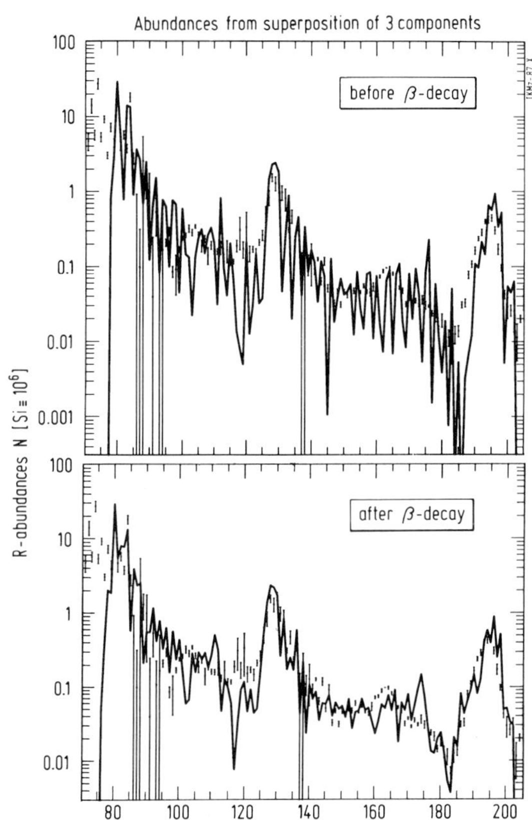

Fig. 4: *R-abundance distribution obtained from a superposition of three time-dependent calculations with the 'best-fit' T_9-n_n values for the $A\simeq80$ peak and the $90\leq A\leq130$ and $135\leq A\leq195$ mass ranges.*

We have shown here that already three n_n-T_9 components can give a decent fit to the global $N_{r,\odot}$ curve. Astrophysical reality will, however, be continuous. Recently, Meyer et al. [18] performed r-process calculations in the high-entropy bubble scenario with ten components using an exponential distribution of neutron exposures. Also in such a case one can recognize that similar deficiencies around $A\simeq120$ remain as presented here. Only when allowing besides T_9 and n_n also varying time scales for different components, one may avoid the above deficiencies (see Goriely et al. [12]). This is, however, not realistic for two reasons: Only for r-components with $B_n \geq 3.5$ MeV (corresponding to $n_n \leq 10^{18}$ cm^{-3}) the $A \simeq 120$ trough will be 'filled up'. These conditions clearly lie below freeze-out (see Fig. 1) where the assumed waiting-point concept is no longer valid. Furthermore, different neutron-exposure times will not exist in one SN event where all mass zones experience the *same* time scale. Therefore, an approach as presented at this conference by [12] is astrophysically not meaningful and prevents a nuclear-structure learning effect.

In summary, we can conclude that both the overall success in the reproduction of the r-abundance peaks as well as the local failures discussed above, indicate nuclear-structure signatures of extremely neutron-rich isotopes, the vast majority of them not accessible in terrestrial but only in stellar laboratories. This offers fascinating perspectives for experimentalists and theoreticians in the entwined fields of nuclear and astro-physics.

This work was supported by grants from DFG (Kr 806/1), BMFT (06MZ106), GSI and NSF (AST 89-13799).

References

[1] Burbidge E M *et al* 1957 *Rev. Mod. Phys.* **29** 547

[2] Cameron A G W *et al* 1983 *Ap. Space Sci.* **91** 221

[3] Cowan J J *et al* 1991 *Phys. Rep.* **208** 267

[4] Seeger P A *et al* 1965 *Ap. J. Suppl.* **97** 121

[5] Hillebrandt W 1978 *Space Sci. Rev.* **21** 639

[6] Thielemann F-K *et al* 1979 *Astron. Ap.* **74** 175

[7] Cameron A G W 1983 *Astrophys. Space Sci.* **91** 235

[8] Kratz K-L *et al* 1988 *J. Phys.* **G14** 742

[9] Kratz K-L *et al* 1991 *Z. Phys.* **A340** 419

[10] Kratz K-L *et al* 1991 *Nuclear Shapes and Nuclear Structure at Low Excitation Energies* (Plenum Press) 339

[11] Käppeler F *et al* 1989 *Rep. Prog. Phys.* **52** 945

[12] Goriely S *et al* 1992 *Contribution to this Conference*

[13] Möller P *et al* 1990 *Nuclei in the Cosmos* (MPA/P4) 226

[14] Möller P *et al* 1992 *Nucl. Phys.* **A** in press

[15] Möller P *et al* 1992 to be subm. to *At. Data Nucl. Data Tables*

[16] Möller P and Randrup J 1990 *Nucl. Phys.* **A541** 1

[17] Kratz K-L *et al* 1992 *Ap. J.* in press

[18] Meyer B S *et al* 1992 *Ap. J.* in press, and *Contribution to this Conference*

[19] Woosley S E and Hoffman R 1992 *Ap. J.* in press

[20] Pearson J M *et al* 1992 *Nucl. Phys.* **A** in press

β-Decay half-lives of neutron-rich isotopes of Fe, Co, Ni involved in the beginning of the r-process

S. Czajkowski[1], M. Bernas[1], P. Armbruster[2], H. Faust[2], J.P. Bocquet[2], R. Brissot[3], P. Dessagne[4], C. Miehe[4], C. Pujol[4], G. Audi[5], J. Lee[6], C. Kozhuharov[7], H. Geissel[7], E. Hanelt[7], G. Münzenberg[7], D. Vieira[7].

[1]IPN Orsay, [2]ILL Grenoble, [3]ISN Grenoble, [4]CRN Strasbourg, [5]CSNSM Orsay, [6]McGill Montreal, [7]GSI Darmstadt.

Abstract: The very neutron-rich Fe- to Ni-isotopes are of interest since they are located at the very beginning of the astrophysical r-process path. The β-decay half-lives of several isotopes, identified in thermal fission of ^{235}U or ^{239}Pu, have been measured at the ILL high-flux reactor using the Lohengrin spectrometer. Half-lives have been determined from time-correlations analysis between the fragment implantation and the detection of the subsequent β-particles in the same detector. With the fragment separator FRS, at GSI, the projectile fragments of ^{86}Kr have been separated. The β-decay half-life of ^{65}Fe has been measured.

1 Introduction

Studies of very neutron-rich nuclei in the Fe-Ni region provide a test for nuclear models far off the stability since they are located along the Z=28 proton shell closure before and eventually up to the N=50 neutron shell closure. These nuclei are encountered at the beginning of the astrophysical r-process path, following the very first neutron captures on the seed nucleus ^{56}Fe. β-decay half-lives of these nuclei, especially those close to the N=50 line, are usefull to determine the r-process path and, consequently, the elemental composition of the first r-peak of the abundancy distribution. These studies are of importance since the astrophysical site of the r-process is not known, and the values of $T_{1/2}$, P_n, and mass excesses enter as constraints astrophysical models to provide values of high neutron-flux densities and temperatures in stellar explosions (Supernovæ).

The half-lives of the waiting points nuclei ^{80}Zn and ^{79}Cu have been recently measured [1] using on-line mass separator techniques. However the extraction efficiency of the available ion sources is much lower for the Fe-Ni than for Zn and Cu.

In the present work, we have used in-flight separation techniques. Very asymetric thermal fission of ^{235}U and ^{239}Pu was used to produce neutron-rich Cu- to Fe- isotopes, and to measure their half-lives . An other approach was provided by projectile fragmentation of the high energy ^{86}Kr delivered by SIS at GSI: ^{65}Fe fragments were produced and separated, and ^{65}Fe β-decay half-life was determined for the first time.

2 Results obtained with thermal fission of ^{235}U and ^{239}Pu

Thermal fission has long been used as a source for neutron-rich isotopes. The ILL high-flux reactor provides a flux of thermal neutrons of $5\,10^{14}\,cm^{-2}\,s^{-1}$ at the target area. The process of very asymetric thermal fission, systematically investigated on ^{235}U [2] was shown to produce neutron-rich isotopes in the Ni region.

Kinetic energy of fission fragments ranges between 80 to 130 MeV, and their atomic states are distributed over several charge states q, from 17 to 24. Fragments are separated with the Lohengrin spectrometer [3] which consists of a magnetic dipole followed by an electrostatic condenser, with deflection planes perpendicular one to each other. Fragments are focussed on A/q lines, along which they are dispersed according to their kinetic energy, E. Thus, Lohengrin performs an A/q and E/q selection of the fission fragments. At the focal plane, the transmitted ions are analysed by a ΔE-E measurement in an ionization chamber, and therefore identified in mass A and nuclear charge Z. Eventhough the production yields become extremely low for the lightest species (10^{-9} for ^{68}Fe) the isotopic identification remains unambiguous since contamination from heavier masses which have larger fission rates occurs at higher energy and – to a smaller extent – larger energy losses (fig. 1).

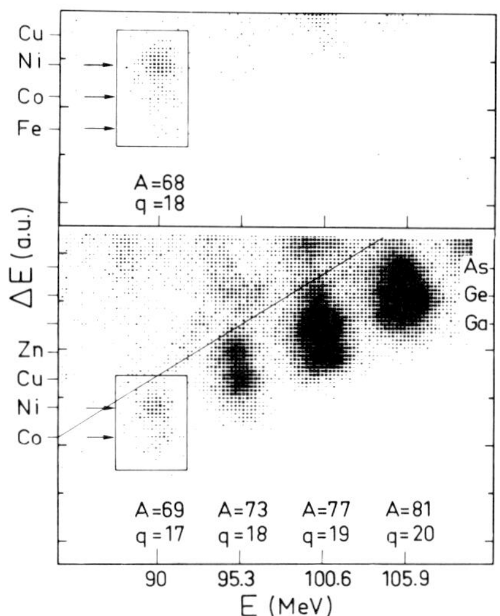

Figure 1 E-ΔE scatter plot representation of the transmitted fission fragments.

A row of 8 thin Si detectors ($0.5\,mm \times 1cm^2$) has been set inside the ionization chamber in order to implant the fragments. In the same device, emitted β-particles have been detected. The analysis of time correlations between ion implantation and the detection of subsequent β-particles provides the value of the half-life . The decay of the daughter nucleus and the background rate have been carefully taken in account. The β-decay half-lives of $^{71-74}Ni$, $^{68,69}Co$ and ^{68}Fe have been extracted using maximum likelihood procedures [4]. Since the background rate is not negligeable ($0.2\,s^{-1}$/det) and the detection efficiency is less than 0.5, only short half-lives (less than 2-3 s) can be measured. Low fragment counting rates are associated with low yields. In the worst case of ^{68}Fe, it was measured 0.15/h. However, thanks to the very fast decay, 29 fragments were enough for a first time determination of the half-life.

The main limitation of the method comes from the very low transmission of Lohengrin: The isotropy of the fission process combined with the solid angle of the collection (10^{-5}), the small energy window (1 MeV), and the selection of one charge state, lead to a transmission factor lower than 10^{-6}.

The addition of a new magnet after Lohengrin will improve the energy transmission and reduce the background. As can be seen in fig. 5, there are still numerous new isotopes to be found with this method which demonstrates that thermal fission is still a competitive way to hunt for very neutron-rich species.

3 Projectile fragmentation.

The pilot experiment was performed at GSI (Darmstadt). The ^{86}Kr beam accelerated in the heavy ion synchrotron SIS up to 500 MeV/u is impinging on a $2\,g\,cm^{-2}$ Be target. Due to reaction kinematics, projectile fragments are forward focussed, with the velocity of the projectile, ($\Delta p/p \simeq 2\%$), and fully ionised.

Fragments of interest can be selected via $B\rho$-Δp analysis [5] with the fragment separator FRS [6] : the first stage of the separator consists mainly of two dipoles. It performs a magnetic (A/Z) selection ($B\rho_1$). The transmitted fragments are decelerated according to their stopping power in an energy degrader located at the intermediate dispersive focal plane. With a second stage of dipoles, symetric to the first one ($B\rho_2$), a given isotope can be selected by tuning magnetic rigidities $B\rho_1$, $B\rho_2$ and the shape of the intermediate degrader. 2 to 4 contaminants are observed with low relative rates.

At the final focus, the transmitted ions are slowed down with an homogenous degrader of appropriate thickness in order to implanted the selected fragments in two rows of ten Si PIN diodes (0.5 mm thick), where they undergo β-decay. This implantation accuracy ($0.1\,mg\,cm^{-2}$) can be achieved although the energetic fragments (500 MeV/u, $\Delta p/p = 2\%$) have passed through $12\,g\,cm^{-2}$ of Al.

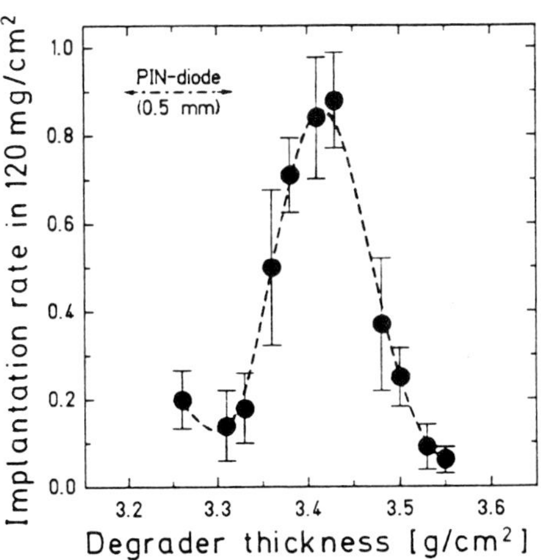

Figure 2 Implantation rate of a selected isotope (^{66}Co) inside a 0.5 mm depth of Si ($120\,mg\,cm^{-2}$ of Al eq.), as a function of the thickness of the last degrader.

This technique allows to implant only the selected isotope together with one contaminant (here the neighbour isotone ^{66}Co) in the detector array. More penetrating fragments and secondary fragments produced by nuclear reactions inside the last degrader $(3.3\,g\,cm^{-2})$ are rejected with a veto scintillator, since they have larger ranges. With the ΔE and time of flight measurement in addition to the range selection performed by the implantation, an unambiguous separation of the different species is obtained (fig. 3).

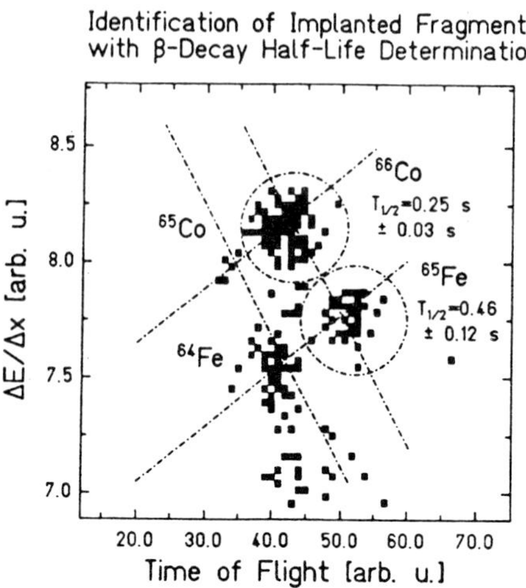

Figure 3 ΔE vs time of flight representation of the fragments detected in the PIN-diodes.

The fragment identification is confirmed by the $T_{1/2}$ determination of the main contaminant, which is found $(0.22 \pm 0.03)\,s$ in agreement with the known ^{66}Co half-life. As the daughter nucleus, ^{66}Ni, has a much longer half-life $(54.6\,h)$, one can neglect the daughter decay in the time scale of the analysis. The time spectrum of the first β detected after an implanted ^{66}Co is reported on fig. 4.

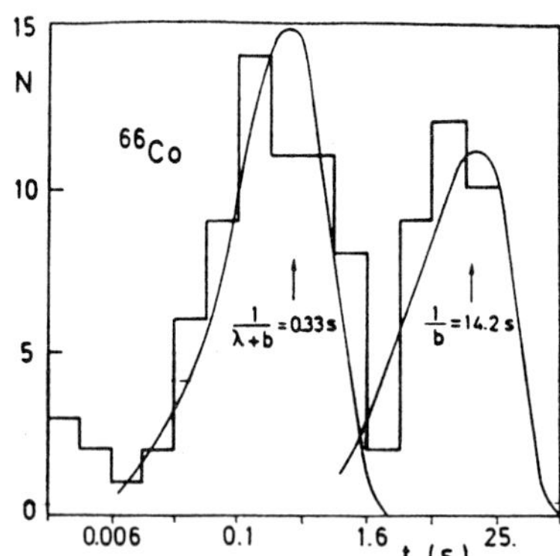

Figure 4 Time delayed coincidences between ^{66}Co fragments and first β-particles detected plotted as a function of the logarithm of the time interval. This representation [7] allows to separate the fast decay from the low frequency background $(0.05/s/det.)$

Contrary to the ^{66}Co situation, the daughter nucleus of ^{65}Fe, ^{65}Co, is short lived, and the chain of the first two decays of ^{65}Fe occurs within a time shorter than background time gaps. The β-detection efficiency evaluated from the previous ^{66}Co is 0.6. Among the pairs of β detected after the fragments, chains are selected on the basis of the shorter time criterium, which number is compatible with this efficiency. It provides half-lives of 1.12 ± 0.3 s for ^{65}Co in agreement with the known value and 0.40 ± 0.2 s for ^{65}Fe. This last value is also found with maximum likelihood analysis.

This first result was obtained after only two hours of accumulation. The success of this pilot experiment will allow to extend β-decay half-lives measurements toward more neutron-rich species.

4 Conclusion

The isotopes studied in the present work are summarised in fig. 5. If the N=50 line is still not reached for these elements, the limits of known half-lives was pushed to more than ten units off stability in the case of Ni and Co. The measured half-lives are compared with calculated ones from different models in fig. 6. The discrepancy can reach up to a factor 3 when one goes further off the known isotopes.

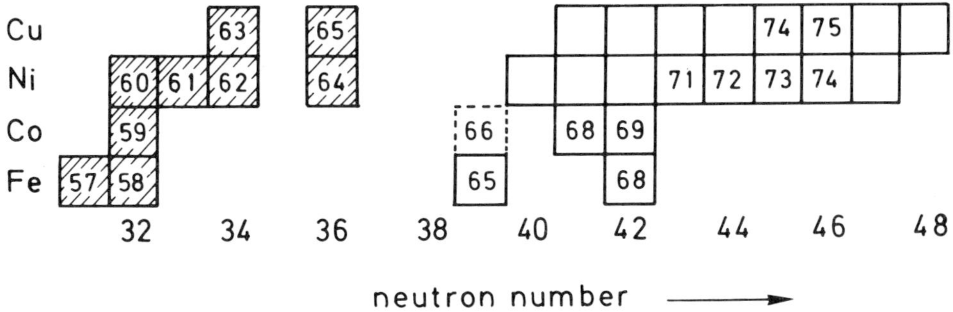

Figure 5 New isotopes observed with thermal fission. ^{65}Fe and ^{66}Co were studied with projectile fragmentation.

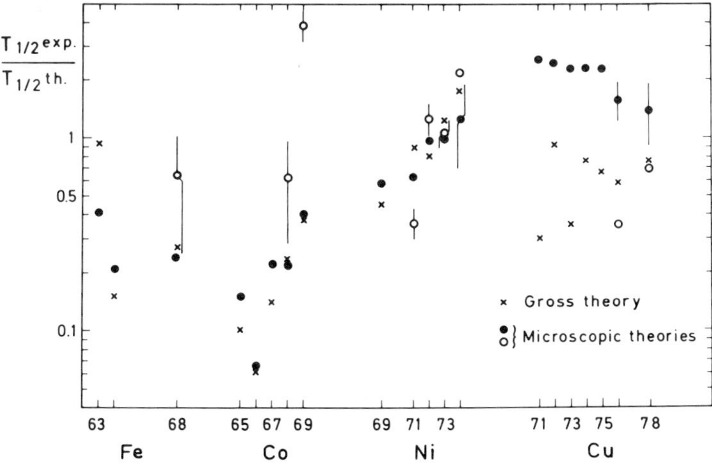

Figure 6 Ratio of measured half-lives with predicted values from a macroscopic theory [8] and from microscopic ones [9, 10] .

The r-process calculations formally need the knowledge of properties for a huge set of nuclei, the most part being up to now unknown, and are very sensitive to $T_{1/2}$, P_n and S_n values far off the stability, where the predictive power of the models decrease. Recently, K.L. Kratz et al.[11] have shown that the r-elements abundancies can be reproduced assuming a steady flow and in the waiting point approximation, through the knowledge of the half-lives of a few nuclei located close to the neutron-shell closures. β-decay half-lives measurements as close as possible to these nuclei allow more reliable extrapolations for the needed values still unknown.

On the other hand, improvement of the existing facilities, using thermal fission or projectile fragmentation as well, should allow to extend these studies further off stability, closer to the N=50 line, perhaps up to the doubly magic waiting-point nucleus, ^{78}Ni.

References

[1] Lund E., Aleklett K., Fogelberg B., and Sangariyavarish A., *Proc. of AMCO-7, THD Schrift. Wiss. und Techn.* **26**, 102, 1984
Gill R.L., Casten R.F., Warner D.D., Piotrowski A., Mach H., Hill J.C., Wohn F.K., Winger J.A. and Moreh R., *Phys. Rev. Lett.* **56**, 1874, 1986
Ekström B., Fogelberg B., Hoff P., Lund E., and Sangariyavanish A., *Phys.Scr.* **34**, 614, 1986
Kratz K.L., *Rev.Mod.Astron.* **1**, 184, 1988

[2] Sida J.L., Armbruster P., Bernas M., Bocquet J.P., Brissot R., and Faust H.R., *Nucl.Phys.A* **502**, 233c, 1989

[3] Moll E., et al. *Nucl.Instr.Meth.* **123**, 615, 1975

[4] Bernas M., Armbruster P., Bocquet J.P., Brissot R., Faust H.R., Kozhuharov C., and Sida J.L., *Z.Phys.A* **336**, 41,1990
Bernas M., Armbruster P., Czajkowski S., Faust H.R., Bocquet J.P., and Brissot R., *Phys.Rev.Lett.* **67**, 3661, 1991

[5] Schmidt K.H., Hanelt E., Geissel H., Münzenberg G., and Dufour J.P., *Nucl.Instr.Meth.A* **260**, 287, 1988 and ref. therein

[6] Geissel H., et al, *Nucl.Instr.Meth.A* **282**, 247, 1989 and ref. therein

[7] Schmidt K.H., et al., *Z.Phys.A* **316**, 19, 1984

[8] Tachibana T., Yamada M., and Nakata K., *Rep.Sci.Eng.Res.Lab., Wasada Univ.,* **88-4**, 1988 (unpbubl.)

[9] Klapdor H.V., Metzinger J., and Oda T., *At.Data Nucl.Data Tables* **31**, 81, 1984

[10] Staudt A., et al., *At.Data Nucl.Data Tables* **44**, 79, 1990

[11] Kratz K.L., et al., *Contribution to this Conference*

Stellar weak interaction rates for nuclei with A > 60

A. Ray[1], K. Kar[2], S. Sarkar[2], S. Chakravarti[3]

[1] Tata Institute of Fundamental Research, Bombay 400 005, India
[2] Saha Institute of Nuclear Physics, Calcutta 700 064, India
[3] California State Polytechnic University, Pomona, CA 91768, U.S.A.

Abstract.   Beta decay and electron capture on a number of neutron rich A> 60 nuclei at the presupernova stage have an important role in determining the core structure of massive presupernova stars and through this affect the subsequent evolution during the gravitational collapse and supernova explosion phases. They contribute to the overall changes in the electron fraction and entropy of the stellar core during its very late stage of evolution. We describe a model to calculate the beta decay rates using an average beta strength function and an electron phase space evaluated for typical presupernova matter density ($\rho = 10^7 - 10^9$ g/cc) and temperature (T = $2 - 6 \times 10^9$ ^{o}K). For the Gamow-Teller(GT) strength function we use a sum rule calculated by the spectral distribution method where the centroid of the distribution is obtained from experimental data on (p,n) reactions. In the calculation of rates we include contributions from the excited states of the mother nucleus wherever they are known experimentally. These turn out to be particularly important at high temperatures encountered in presupernova stellar cores.

1. INTRODUCTION

Beta decay (β^-) and electron (e^-) capture of neutron rich nuclei play important roles in determining presupernova core structure (Nomoto *et al* , 1991). The structure of the core is important for later gravitational collapse and supernova explosion phases. These weak interaction rates were compiled by Fuller *et al* (1982 and references therein) for a series of nuclei upto the mass number (A) of 60. Reliable rates for A > 60 are not available yet. It is important to include contributions of excited states of parent nuclei because presupernova cores are *hot and dense*. For the fp-shell, detailed shell model calculations to obtain rates are formidable for available computer resources. Bloom and Fuller (1985) tried such methods using a truncated basis space. These were for nuclei with A < 60. Among microscopic theories, best results are from complete shell model diagonalisation for A > 40 nuclei (Brown and Wildenthal 1985) and making a Quasi Random Phase Approximation for heavier nuclei (Moller and Randrup, 1990; Klapdor *et al* 1984,1990). However, extension of the QRPA calculations to non-zero temperatures relevant for Presupernova and Gravitational Collapse phases leading to Supernovae are not available yet.

2. OVERALL METHOD

We use as in Kar *et al* (1991), the Gamow-Teller Sum Rules and employ the average distribution of the quenched sum-rule strength calculated from Spectral Distribution Theory (see French and Kota 1982; Kota and Kar 1989). A continuous approximation is made for the final states. We calculate the variable phase space factors (as a function of final energy) for the reaction rates using the finite temperature Fermi-Dirac distributions. Filled Fermi-sea of electrons partially block the available phase space.

We also calculate the electron Chemical Potential (required in phase space integrals) in the *moderately* degenerate and *mildly* relativistic cases. Previous calculations in the literature have used low temperature and extremely relativistic approximations. The GT strength distribution is folded in with the variable phase space factor and integrated over the final state energy to yield the rate from a particular mother nucleus state. We then repeat the procedure to include the statistically weighted contributions from the excited states of the mother nucleus for non-zero temperatures.

The β^- decay rate from the ground state is given by:-

$$\lambda_{g.s} = \frac{\ln 2}{(g_V)^2}(6250sec)^{-1}\sum_j B(E_j)f(Q - E_j) \tag{1}$$

where the summation is over all final states with energy E_j upto the Q-value.

$$B(E_j) = g_V^2 B_F(E_j) + g_A^2 B_{GT}(E_j) \tag{2}$$

($B_F(E)$ and $B_{GT}(E)$ are the squared Fermi and GT matrix elements).

The excited states of parent nucleus are included by weighting by the factor

$$exp(-E_i/kT) \times \frac{(2J_i + 1)}{\sum_i(2J_i + 1)exp(-E_i/kT)}$$

We replace the actual microscopic strengths by the statistically averaged strength function and approximate the $\sum_j \rightarrow \int dE'$ over $Q_i = Q + E_i$ in rate equation for λ_{gs}. Therefore the decay rate from the ground state as well as the excited states is:-

$$\lambda = \frac{\ln 2}{(g_V)^2}\frac{(6250sec^{-1})}{G}\sum_i(2J_i + 1)\exp(-E_i/kT)\int_0^{Q_i}|M_i(E')|^2 f(Q_i - E')dE' \tag{3}$$

with,

$$|M_i(E')|^2 = |M_i^F(E')|^2 + |M_i^{GT}(E')|^2 = g_V^2\rho_i(E')B_i^F(E') + g_A^2\rho_i(E')B_i^{GT}(E') \tag{4}$$

Here $\rho_i(E')$ is the density of states at final energy E'.

3. GT STRENGTH SUM AND ITS ENERGY DISTRIBUTION

For the position of the IAS, the Coulomb displacement energy $\Delta_c = 1.44ZA^{-1/3}$ MeV and the width is $\sigma_c = 0.157ZA^{-1/3}$ MeV (Morita 1973). Thus it is clear that, for the fp-shell nuclei, as the very narrow Fermi resonance lies far above the energetically allowed region, Fermi operator makes vanishingly small contribution to β^- decay rates and so can be neglected for all practical purposes. However, the collective Gamow-Teller mode can be substantially split among the daughter nuclear levels as compared to the Fermi IAS. See Figure 1. We use the spectral distribution theory for evaluating Gamow-Teller strength sum and for selecting the form of the strength distribution:-

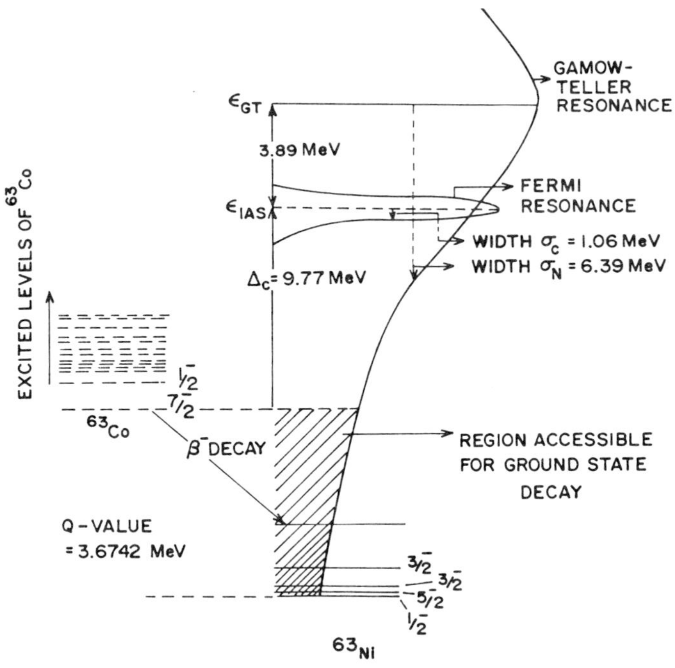

Figure 1. Beta$^-$ decay diagram for ^{63}Co with Fermi and GT resonances.

$$S_{\beta-}^{GT} = 3Z_n \sum_{nljj'} |C_{nl}^{jj'}|^2 (1- <n_{nlj}^P>) <n_{nlj'}^N> \qquad (5)$$

where $C_{nl}^{jj'} = [2(2j + 1)(2j' + 1))]^{1/2} W(l1/2j1; j'1/2)$. Here, W's are the relevant Racah coefficient encountered in the coupling of 3 angular momenta and $<n_{nlj}^P>$ and $<n_{nlj'}^N>$ are the fractional occupancies in the proton orbit (nlj) and the neutron orbit (nlj'). We then observe that the smoothed eigenvalue density distribution $\rho(E)$ of a $(1+2)$ body nuclear Hamiltonian in a large many particle space is very close to a Gaussian in energy E. This result is obtained by explicit diagonalisation of realistic Hamiltonian matrices and also proved analytically using random matrix theory and doing ensemble averaging. The statistically averaged smoothed form of the Gamow-Teller strength function is a Gaussian in energy E' because according to spectral distribution theory the strength density (i.e. the average strength times the density of states in both the initial and final spaces) has an asymptotic bivariate Gaussian form in the initial and final energies for large spaces (French *et al* 1988).

Experimentally observed total GT strength from the ground state of the target nucleus through (p, n) reactions show that the rate is consistently smaller than its shell model or other theoretical estimates. Movement of a part of the total strength upward in energy beyond the observed giant resonance region due to the tensor interaction as well as due to mixing of $N^{-1}\Delta$ (nucleon-hole Δ particle) with the low-energy $N^{-1} N$

interaction is believed to be the cause of this observed quenching. We take this into account by an empirical quenching factor Z_n in Eq(5) which, we take to be $= 0.6$.

To specify the Gaussian one needs two quantities - its centroid and its width. We note here that the physically accessible region through the Q-value of the ground state (the excited states) covers only the tail of the Gaussian. For the GT centroid we use the relation developed by Nakayama et al (1982) which gives a best-fit to the observed centroids. The expression is

$$\epsilon_{GT} = E_{IAS} + 26A^{-1/3} - 18.5(N - Z)/A \tag{6}$$

where ϵ_{GT} is the GT centroid energy and E_{IAS} is the energy of the IAS. Eq.(5) is valid for the strength distribution from the ground state. For an excited state, we extend the isobaric analog argument and add the excitation energy of the initial state to fix the centroid energy of its GT distribution. To minimise the deviation of the calculated strength distribution from the actual one, we use the width of the Gaussian as a parameter and fit the experimental halflives with predictions of the model for a number of nuclei. As the width of the GT resonance has two parts $\sigma^2 = \sigma_C^2 + \sigma_N^2$, the first part coming from Coulomb forces and the dominant second part coming from the nuclear interacations, we vary σ_N for a least square fit of the calculated halflives of a number of nuclei to their experimentally determined values. If $\tau_{1/2}^{cal}$ and $\tau_{1/2}^{exp}$ are the calculated and experimental half-lives of free decays, we vary σ_N to minimise the quantity $\sum_n |\log(\tau_{\frac{1}{2}}^{exp} / \tau_{\frac{1}{2}}^{cal})|^2$ where the summation is over all the nuclei.

4. THE PHASE SPACE INTEGRAL

The phase-space factor appearing in eq (3) is given by

$$f(T, \mu, E_0) = \int_1^{\epsilon_0} \frac{F(Z, \epsilon)\epsilon(\epsilon^2 - 1)^{1/2}(\epsilon_0 - \epsilon)^2 d\epsilon}{1 + \exp\{(\mu - \epsilon)/kT\}} \tag{7}$$

The factor in the denominator denotes the availability of electron hole in the Fermi sea to which the decay electron can go to. $F(Z, \epsilon)$ is the Coulomb correction factor and for this we use the Schenter and Vogel (1983) approximation. $F(Z, \epsilon)$ incorporates the effect of relativistic electrons in the presupernova interior conditions and also corrects for the finite size of the nucleus. The actual integrals in Eqs (3) and (7) are obtained by 24 point Gauss quadratures. The central temperature at the last nuclear burning stages approach a good fraction of an MeV. Therefore, we apply finite temperature corrections to the chemical potential at intermediate densities. The μ_e used in Eq.(7) (to the order $O(kT/m_e)^2$ and a moderate relativistic limit $p_F \geq m_e$) is obtained at a given ρ, T and Y_e (number of protons per nucleon), by solving the following equation:-

$$3\pi^2 \frac{\rho Y_e}{(kT)^3} - (\frac{\mu_e}{kT})^3 (1 + 2m_e/\mu_e)^{3/2}[1 + \frac{\pi^2(kT/\mu_e)^2}{(1 + 2m_e/\mu_e)} + \frac{\pi^2}{2} \frac{(m_e/kT)^2}{(\mu_e/kT)^4(1 + \frac{2m_e}{\mu_e})^2}]$$

$$+ 3(m_e/kT)^2 e^{-(\mu_e + m_e)/kT} K_2(m_e/kT) = 0$$

The values of the electron chemical potential μ_e calculated from the above expression when compared to those obtained in the ultrarelativistic and extreme degeneracy limit are found to be substantially different at moderate temperatures and densities.

5. RATES FOR NUCLEI IN THE MIDDLE OF THE fp-SHELL

To test the predictions of the model and to fix the parameter σ_N, we calculate the halflives of a number of nuclei in the fp shell with $A > 60$. The free decay is the $(\rho \to 0, T \to 0)$ limit. For the 13 nuclei considered i.e., ^{69}Cu, ^{68}Cu, ^{67}Ni, ^{66}Cu, ^{65}Ni, ^{65}Co, ^{64}Co, ^{63}Co, ^{63}Fe, ^{62}Co, ^{62}Fe, ^{62}Mn and ^{61}Fe the best fit to experimental value is obtained with $\sigma_N = 6.3$ MeV. Table 1 compares the free decay halflives thus calculated with the experimental values. It also gives the predictions of microscopic QRPA calculations of Klapdor *et al* (1984,1990) as well as those of the gross theory (Takahashi *et al* 1969,1973). For ^{64}Co, an Edgeworth expansion of the GT strength distribution with negative skewness of -0.3 and a $\sigma_N = 7.5$ gives the predicted half life to be 0.907 s (whereas the experimental value is 0.3 sec; note also the different calculated value reported in Table 1 which is for standard σ_N and symmetric Gaussian).

Table 1

Comparisons of calculated and experimental half lives

Nucleus	Experimental	Half life $\tau_{1/2}$(sec)		
			Calculated	
		This work	Gross theory	QRPA
^{69}Cu	180	251.4		
^{68}Cu	31	22.0		
^{66}Cu	306	305.5		
^{67}Ni	21	47.0	94	23
^{65}Ni	9072	589.0		
^{62}Co	90	15.1		
^{63}Co	27.4	52.1		
^{64}Co	0.3	3.53		10.0
^{65}Co	1.25	5.66	8	8.59
^{61}Fe	360	34.5		
^{62}Fe	68	183.4		
^{63}Fe	4.9	3.49	10	14.8
^{62}Mn	0.88	1.05	2	0.773

The finite temperature and density results for all the nuclei show the general trend that the rates decrease with increasing density. This is because increasing the density increases the chemical potential of electrons outside the nuclei impeding the decay process. The phase space factor f is very strongly dependent on the Q value - for free decays it goes as the fifth power of the Q-value if the Coulomb correction factor $F(Z,E)$ = 1. As the excitation energy in the daughter nucleus increases, the Q-value decreases and so that the phase space factor goes down. But on the other hand the GT strength rises because one goes higher up in energy towards the centroid from the tail region. Thus there is a competition between two opposing effects. The inclusion of excited states can increase the rate by a factor as large as 11.4 as seen for ^{69}Cu at the high density of of $\log \rho_{10} = -0.5, T = 6 \times 10^9$ K and Y_e =0.50. This contribution from the excited states is more for higher temperatures as also for higher densities.

Table 2 compares the β^- decay rates from the ground states for four isotopes of cobalt ^{62}Co, ^{63}Co, ^{64}Co and ^{65}Co with Y_e=0.47. For these isotopes the GT strength sum values are very close but the main difference is due to differing Q-values. Table

3 gives the electron capture rates of ^{62}Co. Rates for all 13 nuclei, reported at the conference cannot be reproduced here due to space restrictions but will be published elsewhere. The grid reported at the conference had Y_e = 0.42, 0.44, 0.47 and 0.50 while $\log\rho_{10}$ = -0.5, -1.0, -1.5, -2.0, -2.5 and T (in ^{o} K) = 2, 3, 4, 5, and 6 $\times 10^9$.

Table 2

β^- decay rates from the ground state for four isotopes of Cobalt for Y_e = 0.47.

$\log \rho_{10}$	Temperature (in ^oK)	Rates for the nucleus			
		^{62}Co	^{63}Co	^{64}Co	^{65}Co
−2.0	3×10^9	3.83×10^{-2}	9.31×10^{-3}	8.04×10^{-1}	1.06×10^{-1}
	4×10^9	3.86×10^{-2}	9.60×10^{-3}	8.54×10^{-1}	1.07×10^{-1}
	5×10^9	3.90×10^{-2}	9.89×10^{-3}	9.01×10^{-1}	1.07×10^{-1}
−1.0	3×10^9	3.64×10^{-3}	6.21×10^{-5}	2.77×10^{-1}	1.66×10^{-2}
	4×10^9	4.17×10^{-3}	1.40×10^{-4}	3.08×10^{-1}	1.79×10^{-2}
	5×10^9	9.79×10^{-3}	8.42×10^{-4}	4.71×10^{-1}	3.46×10^{-2}

Table 3

Electron capture rates for ^{62}Co

Electron Fraction	Temperature T (in ^oK)	Rates for $\log \rho_{10}$ -2.0	-3.0
0.5	3×10^9	9.25×10^{-7}	1.81×10^{-8}
	4×10^9	6.17×10^{-6}	3.25×10^{-7}
	5×10^9	2.42×10^{-5}	2.31×10^{-6}

Acknowledgments

A. R. thanks the Copernicus Astronomical Centre, Warsaw for hospitality. Travel and local support for him in Warsaw came from the NSF grant no INT87-15411-A01-TIFR and KBN grant no 2-1244-91-01 respectively. At Tata Institute, this research is part of the 8th Five Year Plan Project 8P-45.

References

Bloom, S.D. and Fuller, G.M. 1985, Nucl. Phys. A, **440**, 511.

Brown, B.A. and Wildenthal, B.H. 1985, At. Data Nucl. Data Tables. **33**, 347.

French, J.B., Kota, V.K.B., Pandey, A. and Tomsovic, S. 1988, Ann. Phys. (N.Y.) **181**, 235.

French, J.B. and Kota, V.K.B.1982, Ann. Rev. Nucl. Part. Sci. **32**, 35.

Fuller, G.M., Fowler, W.A. and Newman, M.J. 1982, Ap.J. Sup. **48**, 279.

Kar, K., Sarkar, S. and Ray, A. 1991, Phys. Lett. B., **261**, 217; 1992, Phys. Lett. B. **277**, 528.

Klapdor, H.V., Metzinger, J. and Oda, T. 1990, At. Data Nucl. Data Tables., **44**, 73.

———1984, At. Data Nucl. Data Tables, **34**, 81.

Kota, V.K.B. and Kar, K. 1989, Pramana, **32**, 647.

Moller, P. and Randrup, J. 1990, Nucl. Phys. A, **514**, 1.

Morita, M. 1973, Beta decay and muon capture (W.A. Benjamin, Inc).

Nakayama, K., Pio Galeao, A., and Krmpotic, F. 1982, Phys. Lett. B, **114**, 217.

Nomoto, K., Shigeyama, T., Kumagai, S. and Yamaoka,H. 1991, in "Supernovae and Stellar Evolution" eds. A. Ray and T. Velusamy (World Scientific, Singapore) p. 116.

Schenter, G.K. and Vogel, P. 1983, Nucl. Sci. Engg. **83**, 393.

Takahashi, K., Yamada, J.M. and Kondoh, J.T. 1973, At. Data Nucl. Data Tables, **12**, 101.

Takahashi, K. and Yamada, J.M. 1969, Prog. Theor. Phys. **41**, 1470.

Microscopic models for nuclear reaction rates

P. Descouvemont

Physique Nucléaire Théorique et Physique Mathématique, CP229, Université Libre de Bruxelles, B1050 Bruxelles, Belgium.

Abstract. We report on a microscopic description of low-energy nuclear reactions, in the framework of the Generator Coordinate Method. The model is briefly presented and illustrated by applications to the ^{7}Li(n,γ)^{8}Li , ^{7}Be(p,γ)^{8}B , and ^{12}C(α,γ)^{16}O capture reactions and to the ^{8}Li(α,n)^{11}B transfer reaction.

1. Introduction

Current studies of stellar evolution require an ever increasing number of nuclear reaction rates [1]. A special attention has recently been paid to new scenarios such as the hot CNO cycle [2], the inhomogeneous Big Bang model [3] or possible hot pp chains in low-metallicity massive stars [4]. At typical stellar temperatures, the astrophysically important energies between charged particles are so low with respect to the Coulomb barrier (typically 10%) that the cross sections are minute. For example, the ^{16}O(α, γ)^{20}Ne cross section at astrophysical energies is lower than 10^{-20} barns. In addition, several interesting scenarios involve short-lifetime nuclei, which can not be used as targets. Among these " exotic" reactions, we mention the ^{13}N(p,γ)^{14}O reaction which determines the frontier between the cold and the hot CNO cycles [2], and the ^{8}Li(α,n)^{11}B reaction which probably is the main source for $A > 12$ isotopes in an inhomogeneous Big-Bang hypothesis [3]. The smallness of the cross sections and the radioactive character of some nuclei make it experiments in laboratories very difficult. In spite of new detection techniques [1], and of the availability of radioactive beams [5], a theoretical support is often necessary to derive nuclear reaction rates at stellar temperatures.

Theoretical models can be roughly classified into three categories: (i) Models containing adjustable parameters, like the R-matrix [6] or the K-matrix [7] methods; in this case, parameters are fitted to available experimental data and are used to extrapolate the cross sections at astrophysical energies, usually inaccessible to experiment. (ii) Statistical models, where the relevant cross sections are calculated from the level properties in the compound nucleus [8]. (iii) "Ab initio" models where the cross sections are determined from the wave functions of the system.

The main drawback of (i) is the need of experimental data, ruling out most of reactions involving a radioactive nucleus. In addition, some effects, important at low energies, can be partly hidden in the energy range covered by experiment. This problem is well exemplified by the ^{12}C(α, γ)^{16}O cross section [1] where the low energy part is essentially given by the properties of two subthreshold states. Statistical models (*ii*) provide estimates of the cross section if the level density in the compound nucleus near the reaction threshold is high; consequently, most of the reactions involving light nuclei can not be investigated. This limitation is rather restrictive since it applies to many reactions involved in the CNO cycle, and to nuclei close to the stability line which are crucial in explosive burning. Models (*iii*) essentially concern potential models [9] for capture reactions or Distorted Wave Born Approximation (DWBA) models [10] for transfer reactions, where the wave functions of the system are determined from a nucleus-nucleus potential, and microscopic models [11, 12] where the cross sections are deduced from a nucleon-nucleon interaction and from fully antisymmetrized wave functions. Between them, a semi-microscopic approach, known as the Orthogonality Condition Model and taking antisymmetrization effects partly into account has been applied to astrophysical reactions a few years ago [13].

In this talk, we present a microscopic model, called the Resonating Group Method (RGM), equivalent to the Generator Coordinate Method (GCM). This cluster model offers a number of important advantages for the description of low-energy reactions. (*i*) It takes antisymmetrization exactly into account; (*ii*) Good quantum numbers are treated without approximation; (*iii*) Bound and scattering states are described in an unified way. This last property is extremely useful to test the validity of an experimentally unknown cross section from spectroscopic properties of the unified nucleus, which are generally better known. The heaviness of microscopic models is widely compensated by their predictive power since they only depend on a nucleon-nucleon interaction, and on a few assumptions concerning the cluster structure of the nuclei. Let us add that, if this talk focusses on astrophysical applications of the GCM, this method can be applied in a broad application field, like β decay, elastic and inelastic scattering, nucleus-nucleus bremsstrahlung, or nucleus-nucleus interaction for example (see refs in ref [12]).

2. The microscopic model

2.1 The Generator Coordinate Method

In a microscopic model [11, 12], the whole information concerning a A-nucleon system is deduced from the hamiltonian

$$ H \;=\; \sum_i^A T_i \;+\; \sum_{i<j}^A V_{ij} \tag{1} $$

where T_i is the kinetic energy of nucleon i and V_{ij} the two-body nucleon-nucleon interaction involving the Coulomb and nuclear contributions. In most of our applications, the central part of the nuclear term is chosen as a Volkov interaction, which is given by a superposition of gaussian terms, and a spin-orbit force; up to now, the tensor force has not been investigated.

The Schrödinger equation associated to hamiltonian (1) can not be solved exactly as soon as A is larger than three. Accordingly, some approximations must be done for the determination of the wave functions. In this paper, we briefly present the main properties of the GCM-RGM, but we send the reader to refs [11,12,14] for details.

The A nucleons of the system are assumed to be divided into two clusters with A_1 and A_2 nucleons, spins and parities $(I_1\pi_1)$ and $(I_2\pi_2)$ and internal wave functions $\phi^{I_1\pi_1}$

and $\phi^{I_2\pi_2}$ respectively. In RGM notations, the total wave functions of the system, with spin and parity $(J\pi)$, read

$$\Psi^{JM\pi} = \sum_{\ell I} \mathcal{A}[[\phi^{I_1\pi_1} \otimes \phi^{I_2\pi_2}]^I \otimes Y_\ell(\hat{\rho})]^{JM} g_{\ell I}^{J\pi}(\rho) \tag{2}$$

where ρ is the relative coordinate between the clusters, and $\mathcal{A}$ the A-body antisymmetrizor. The sum in (2) runs over all channel spins I and angular momenta ℓ. In the RGM, relative wave functions $g_{\ell I}^{J\pi}$ are obtained from a set of coupled integro-differential equations [11], whose expression is rather heavy and not systematic. In practice, this problem limits the application of the RGM to reactions involving a small number of nucleons. The GCM unables one to overcome this restriction by expanding the relative functions $g_{\ell I}^{J\pi}$ over a set of projected gaussian functions, centred at different generator coordinates R_n. It is then easy to show that (2) can be rewritten as

$$\Psi^{JM\pi} \simeq \sum_{\ell I} \sum_n f_{\ell I}^{J\pi}(R_n)\, \Phi_{\ell I}^{JM\pi}(R_n) \tag{3}$$

where $\Phi_{\ell I}^{JM\pi}(R_n)$ are projected Slater determinants, and $f_{\ell I}^{J\pi}(R_n)$ the generator function. This function is determined from the matrix elements of the overlap and of the hamiltonian between projected Slater determinants. Their calculation is quite systematic when going from one system to another one, and can therefore be applied to reactions much more complicated than in the RGM.

One of the colliding nuclei can be itself defined in a two-cluster [15] or even in a three-cluster [16] model. This extension of the "natural" two-cluster approach to a more refined multi-cluster model allows one to study reactions involving a deformed nucleus, such as ^{8}Be or ^{12}C for example. However, the number of clusters is limited by computer times since, because of projection on good quantum numbers, the calculation of matrix elements between GCM basis functions requires a numeral integration whose dimensions increase with the number of clusters [15].

A drawback of the GCM is the asymptotic behaviour of the wave functions (3). Because of the gaussian expansion of the radial functions, their asymptotic part is not physical neither for bound nor for scattering states. We solve this problem by using the Microscopic R-matrix Method (MRM) which is detailed in refs.[17, 18, 12].

2.2 Importance of antisymmetrization

Determining whether antisymmetrization effects in low-energy reactions are important or not is a delicate problem. Briefly speaking, one should say that in transfer reactions, antisymmetrization between the colliding nuclei is always important since the transfer process arises from the short-ranged nuclear force. For non-resonant radiative capture, the answer strongly depends on the considered reaction. A rough estimate of antisymmetrization effects can be given by comparing the sum R_{12} of the nuclear radii plus 1.5 fm (which can be considered as an upper limit of the antisymmetrization range) with a typical distance ρ_m encountered in the calculation of capture cross sections. It is shown in ref.[19], that the integrand involved in electromagnetic matrix elements presents a maximum near ρ_m, which can be evaluated from the asymptotic part of the wave functions [19]. Some typical values are given in table 1.

Table 1. Typical distances in radiative capture reactions; ℓ_i and ℓ_f are the initial and final angular momenta, and E_B the binding energy of the final state.

reaction	ℓ_i	ℓ_f	E_B (MeV)	ρ_m (fm)	R_{12} (fm)
$^3\text{He}(\alpha,\gamma)^7\text{Be}$	0	1	-1.59	6.5	5.0
$^7\text{Be}(\text{p},\gamma)^8\text{B}$	0	1	-0.136	38.0	4.8
$^{16}\text{O}(\text{p},\gamma)^{17}\text{F}$	1	2	-0.601	15.0	5.0
$^{12}\text{C}(\alpha,\gamma)^{16}\text{O}$	0	2	-0.245	35.0	5.6
$^{16}\text{O}(\alpha,\gamma)^{20}\text{Ne}$	2	0	-4.73	6.3	5.9

Table 1 shows that, for the $^3\text{He}(\alpha,\gamma)^7\text{Be}$ and $^{16}\text{O}(\alpha,\gamma)^{20}\text{Ne}$ reactions ρ_m is close to R_{12}; antisymmetrization effects are therefore expected to be important. The $^7\text{Be}(\text{p},\gamma)^8\text{B}$, $^{16}\text{O}(\text{p},\gamma)^{17}\text{F}$ and $^{12}\text{C}(\alpha,\gamma)^{16}\text{O}$ reactions which proceed to a weakly bound state are less affected by antisymmetrization effects since ρ_m is much larger than R_{12}. It should be mentionned, however, that in this case, the low-energy cross section is related to the asymptotic normalization of the final-state wave function and therefore might indirectly depend on antisymmetrization effects.

3. Applications

3.1 The $^7Li(n,\gamma)^8Li$ and $^7Be(p,\gamma)^8B$ reactions in a three-cluster model

3.1.1 Introduction

Since the early measurement of Imhof *et al.* [20], the $^7\text{Li}(\text{n},\gamma)^8\text{Li}$ reaction has been recently reinvestigated by Wiescher *et al.* [21], and by Nagai *et al.* [22]. The data of Imhof *et al.* were in good agreement with a microscopic three-cluster calculation [23], involving the $^7\text{Li}(3/2^-,1/2^-) + \text{n}$ configurations. This agreement is now questioned by the new data of Wiescher *et al.* , who find a cross section lower by a factor of 2 than the previous measurement. However, the Japanese group recently obtained a thermal cross section consistent with the data of Imhof *et al.*

The $^7\text{Be}(\text{p},\gamma)^8\text{B}$ reaction is primordial in the well known Solar-neutrino problem [24]. The current uncertainty on the zero-energy S-factor remains too large. Several potential model (see [25] and refs. therein) and microscopic [23,26] calculations have been performed, but the $S(0)$ values range from 0.0155 keV-barns to 0.030 keV-barns.

The present study generalizes our first calculation [23] by adding to the configuration space the $^7\text{Li}(7/2^-,5/2^-)+\text{n}$ and $^5\text{He}(3/2^-,1/2^-)+\text{t}$ configurations. This extension is expected to improve the wave functions for low-lying states and to give a description of some additional ^8Li or ^8B excited states. Furthermore, we test here the sensitivity of the cross sections by considering four Volkov forces (V1, V2, V3, V4). The ^7Li (^7Be) nucleus is defined in the $\alpha + \text{t}$ ($\alpha + {}^3\text{He}$) cluster model with a generator coordinate equal to 3.7 fm; for ^5He (^5Li), we choose 1 fm. The spin-orbit interaction is adjusted in order to reproduce the experimental energy difference between the ground and first excited states of ^7Li (^7Be). The Majorana parameter gives the correct ^8Li (^8B) binding energy. Details are given in ref.[27].

The ^{8}Li and ^{8}B spectra obtained with the V2 interaction are given in fig.1 and compared with experiment. The other Volkov forces provide nearly identical results. It turns out that the present configuration space allows one to describe most of experimentally known states, but the excitation energy of the first 1^+ state is somewhat underestimated. The fair description of the ^{8}Li and ^{8}B nuclei is supported by the spectroscopic properties given in table 2. Again, we only present the V2 results which differ from the other ones by a few percent only.

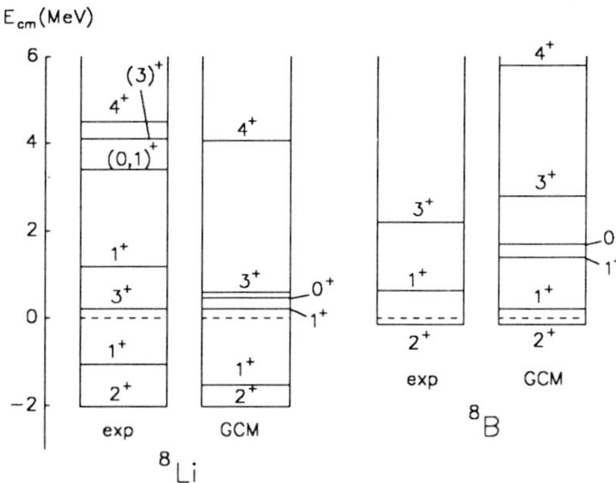

Fig.1 ^{8}Li and ^{8}B GCM spectra obtained with the V2 interaction. The ^{7}Li+n and ^{7}Be+p thresholds are represented by dashed lines.

Table 2. Spectroscopic properties of ^{8}Li and of ^{8}B. Experimental data are taken from ref.[28] except a), from ref.[29].

	^{8}Li		^{8}B	
	GCM	exp	GCM	exp
$Q(2^+)$ $(e.\text{fm}^2)$	3.1	2.4±0.2	7.3	
$\mu(2^+)$ (μ_N)	1.39	1.65	1.45	1.03
$B(\text{M1}, 1^+ \to 2^+)$ (Wu)	3.9	2.8±0.9	3.5	5.1±2.5
$B(\text{E2}, 1^+ \to 2^+)$ (Wu)	3.6	96±26$^{a)}$	13.7	
$B(\text{M1}, 3^+ \to 2^+)$ (Wu)	0.19	0.29±0.13	0.16	

The agreement between GCM and experiment is quite satisfactory except for the $B(\text{E2}, 1^+ \to 2^+)$ in ^{8}Li, whose experimental value is surprisingly high. This discrepancy is discussed in ref.[30].

3.1.2 The $^7Li(n,\gamma)^8Li$ reaction

The cross sections obtained with the different nucleon-nucleon forces are displayed in fig.2, together with the experimental data of Imhof *et al.* [20], and of Wiescher *et al.* [21]. The GCM results are close to the Imhof *et al.* experiment; at the typical energy of 25 keV, we get $\sigma = 54.2$, 54.9, 46.0 and 44.5 μb for V1, V2, V3 and V4 respectively. In this case, the sensitivity to the interaction is larger than for the spectroscopic properties. With respect to our earlier study [23], we get a very similar shape ($\sigma(25$ keV)$= 54.8$ μb in [23]), but the 3^+ resonance at 222 keV is now reproduced, with the introduction of excited configurations. The present study does not support the data of Wiescher *et al.*, whose overall normalization is smaller by a factor of 2.

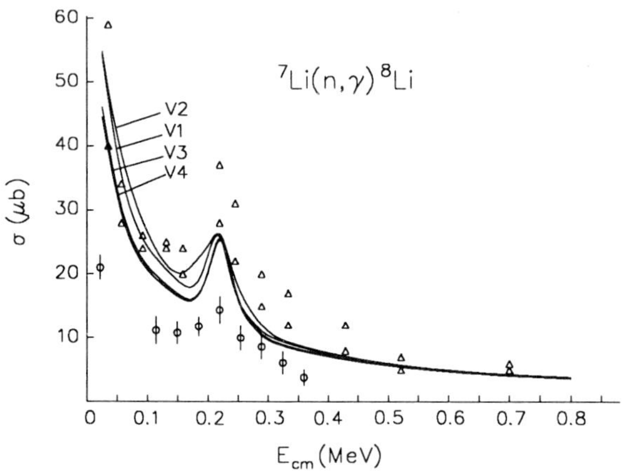

Fig.2 ^{7}Li(n,γ)^{8}Li cross sections for the different Volkov forces. The experimental data are taken from ref.[20] ($\triangle$) and ref.[21] ($\circ$)

3.1.3 The $^7Be(p,\gamma)^8B$ reaction

We present in fig.3 the GCM S-factors, together with experimental data. For the 1^+ partial wave which contains a resonance at 0.64 MeV, we have readjusted the Majorana parameter, in order to get the correct excitation energy of this resonance (see fig.1). The S-factors corresponding to the different interactions are almost proportional. At zero energy, extrapolation of our results gives $S(0)= 0.029$, 0.030, 0.028 and 0.027 keV-barns for V1, V2, V3 and V4 respectively. As for the ^{7}Li(n,γ)^{8}Li reaction, the present V2 S-factor is identical to our previous result [23]. The ground-state and low-energy scattering wave functions are therefore little modified by excited configurations.

At this stage, it is difficult to choose between the different nucleon-nucleon interactions. Each of them provides almost identical spectroscopic properties of ^{8}Li and ^{8}B, and the ^{7}Li(n,γ)^{8}Li data are not accurate enough to discriminate. Anyway, the present $S(0)$ values are much larger than those proposed by Barker and Spear [34] (0.017 keV-barns), and by Mukhamedzhanov and Timofeyuk [35] (0.0155 keV-barns).

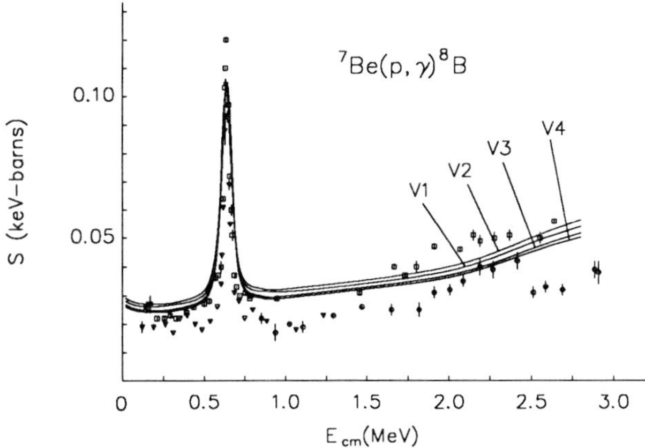

Fig.3 ^{7}Be(p,γ)^{8}B S-factors for the different Volkov forces. Experimental data are taken from ref.[31] ($\square$), ref.[32] ($\circ$) and ref.[33] ($\triangledown$).

3.2 The $^{12}C(\alpha,\gamma)^{16}O$ E2 S-factor in a four α model

In spite of its importance in stellar evolution, the $^{12}C(\alpha,\gamma)^{16}O$ cross section at low energies remains badly known [1]. Many experiments (see [36] and refs therein) have been devoted to this reaction, but their theoretical extrapolation down to astrophysical energies (typically 300 keV) remains controversial. The theoretical investigation of the $^{12}C(\alpha,\gamma)^{16}O$ reaction is very delicate since, at low energies, the S-factor is expected to involve an E1 component as well as an E2 component; in addition, they depend on spectroscopic properties of two weakly ^{16}O bound states (1_1^- at 7.12 MeV and 2_1^+ at 6.92 MeV). Theoretical approaches arise from different models: R-matrix [36,37,38], K-matrix [7], semi-microscopic [39,40], or microscopic [41,42] methods have tried to reduce uncertainties on the low-energy cross section.

We use here a microscopic four α model, where the ^{12}C nucleus is described by three α particles located on the apexes of an equilateral triangle. In ref.[16], we focussed on clustering effects in $\alpha + {}^{12}C$ scattering, and shown that the ^{12}C description is strongly improved with respect to the one-centre model. With an oscillator parameter $b=1.36$ fm, the minimum of the binding energy is obtained for a side of the triangle $R_C=2.8$ fm. This binding energy is larger by about 25 MeV that the value corresponding to $R_C=0$. The quality of the ^{12}C wave functions in the three-α model is illustrated in table 3 by some spectroscopic properties.

Table 3. Spectroscopic properties of ^{12}C for different R_C values.

	0.4 fm	1.6 fm	2.8 fm	4.0 fm	mixed	exp
$\sqrt{<r^2>_{0^+}}$ (fm)	2.11	2.21	2.47	2.99	2.58	2.48
$Q(2^+)$ $(e.\text{fm}^2)$	3.0	3.7	5.7	9.7	6.4	6 ± 3
$B(\text{E2}, 2^+ \to 0^+)$ $(e^2.\text{fm}^4)$	2.0	3.2	8.2	23.7	10.8	7.8 ± 0.4
$E(2^+) - E(0^+)$ (MeV)	1.28	1.51	2.75	3.52	2.92	4.44

We apply here the model to the E2 component of the $^{12}C(\alpha,\gamma)^{16}O$ cross section. The ^{12}C nucleus is described by a mixing of 4 configurations, corresponding to R_C=0.4, 1.6, 2.8 and 4.0 fm; the $\alpha+^{12}C(0^+,2^+)$ channels are taken into account. For a given channel, diagonalization of the 4×4 matrix leads to the ground (or first excited) ^{12}C state, and to three additional unphysical states, introducing well known distortion effects [43], and allowing ^{12}C to be deformed during the collision. The Majorana parameter of the V2 force (M=0.6305) is fitted to the 2_1^+ experimental energy. For the ^{16}O ground state we use M=0.6666, which represents a difference with the other partial waves much lower than in one-centre descriptions of ^{12}C [41,42]. We obtain, in ^{16}O :

$$\sqrt{<r^2>_{\mathrm{gs}}} = 2.64 \text{ fm} ; \quad B(\mathrm{E2}, 2_1^+ \rightarrow \mathrm{gs}) = 9.1 \; e^2.\mathrm{fm}^4 \tag{4}$$

whereas the experimental counterparts are 2.71 ± 0.02 fm and 7.8 ± 0.4 $e^2.\mathrm{fm}^4$ [44]. In the following, a small effective charge $\delta e = -0.08e$ will be used to take account of the slight difference between GCM and experimental $B(\mathrm{E2})$.

At low energies, the $\alpha + {}^{12}C$ phase shifts and the $^{12}C(\alpha,\gamma)^{16}O$ S-factors are affected by the location of several resonances in the ^{16}O spectrum. The 0_2^+ (-1.11 MeV) and 0_3^+ (4.89 MeV) states, the 2_2^+ (2.68 MeV) and 2_3^+ (4.36 MeV) states, and the 4_1^+ (3.19 MeV) and 4_2^+ (3.94 MeV) states in the 0^+, 2^+ and 4^+ partial waves respectively, are reproduced by the model (energies, α and γ are in reasonable agreement with experiment), but their properties do not exactly fit experiment. Since an accurate determination of the S-factor requires a very precise description of energies and widths, we have slightly modified the GCM properties to reproduce them. This method, which combines experimental information with microscopic results can be applied in a very consistent way within the R-matrix formalism [42]. Details are given in ref. [45].

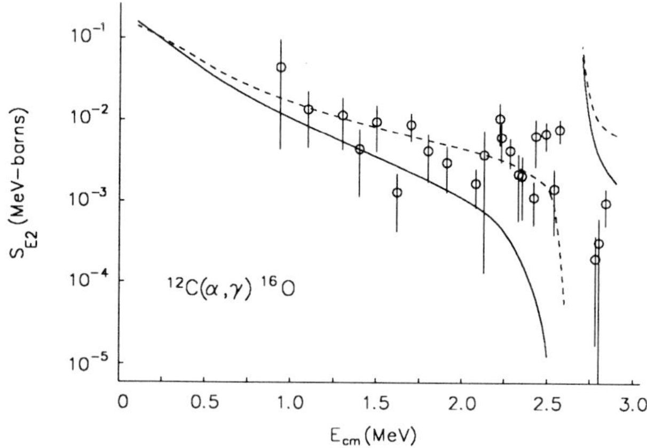

Fig.4 E2 $^{12}C(\alpha,\gamma)^{16}O$ S-factor relative to the ^{16}O ground state. The full line corresponds the full description of ^{12}C, and the dashed line to the single value $R_C = 2.8$ fm. The experimental data are taken from ref.[36]

The E2 S-factor, relative to the ^{16}O ground state, is compared to experiment [36] in fig.4. A simplified model, where a single R_C value (2.8 fm) is used, provides the dashed lines. At the lowest experimental energies, there is a good agreement between GCM and the data; near 2.7 MeV, the S-factor is strongly affected by the 2_2^+ resonance. At 300 keV, we get:

$$S_{E2}(300 \text{ keV}) \;=\; 0.09 \text{ MeV} - \text{barns} \tag{5}$$

This result is consistent with our previous calculations (0.10 MeV-barns [41] and 0.07 MeV-barns [42]) where stronger corrections had to be done on the energy spectrum and on the effective charges. Our S_{E2} value is much larger than the K-matrix extrapolation of Filippone *et al.* ($0.007^{+0.024}_{-0.005}$ MeV-barns [7]), but in agreement with the recent R-matrix fit of Barker and Kajino (0.05-0.18 MeV-barns [38]), who constraint their parameters through cascade-transition data.

Fig.5 presents the ^{12}C$(\alpha,\gamma)^{16}$O(2^+) S-factor. In the experimental energy range, the GCM underestimates the data by a factor of 2. This might be an indication that E1 capture, due to p wave contribution, is not negligible. Anyhow, this cascade transition does not significantly affect the S factor at astrophysical energies.

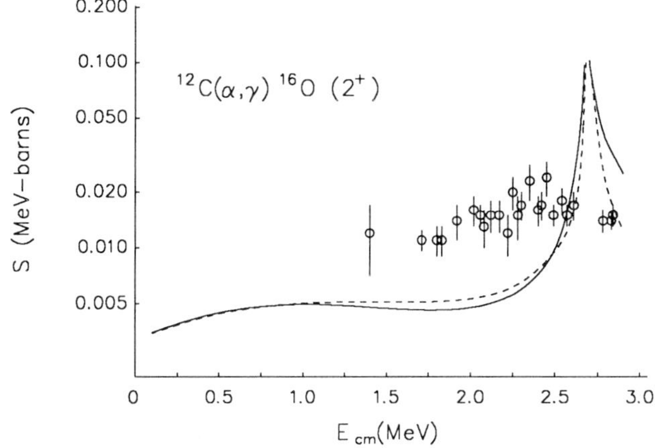

Fig.5 E2 ^{12}C$(\alpha,\gamma)^{16}$O S-factor relative to the ^{16}O 2^+ state. See caption to Fig.4.

The present calculation might be extended to E1 capture, but should require the introduction of $T = 1$ components in the wave functions. In ref. [42], we have shown that the ^{15}N+p and ^{15}O+n configurations give a satisfactory description of $T = 1$ mixing rates in ^{16}O. However, one should remind that a four cluster model requires the numerical evaluation of 7 dimension integrals [16] for the computation of GCM matrix elements. The occurence of s orbitals only makes it the vectorization of the code very efficient, and keeps computer times in reasonable limits. The introduction of ^{15}N+p and ^{15}O+n channels would require, either p orbitals if ^{15}N and ^{15}O are defined in the one centre model, or the extension to a five-cluster model. This latter possibility is convenient for the simplicity of the codes, but requires computer times which are currently prohibitive.

3.3 The $^8Li(\alpha,n)^{11}B$ reaction

This transfer reaction is important in the inhomogeneous Big-Bang theory [3]. According to this model, $A > 12$ isotopes might be produced through the chains

$$^1\text{H}(n,\gamma)^2\text{H}(n,\gamma)^3\text{H}(d,n)^4\text{He}(^3\text{H},\gamma)^7\text{Li}(n,\gamma)^8\text{Li} \tag{6}$$

and

$$^8\text{Li}(\alpha,n)^{11}\text{B}(n,\gamma)^{12}\text{B}(\beta)^{12}\text{C}(n,\gamma)^{13}\text{C}(n,\gamma)^{14}\text{C} \tag{7}$$

where the ^{8}Li$(\alpha,n)^{11}$B reaction plays a crucial role. Since the ^{8}Li nucleus has a short lifetime (0.85 s), the first measurement was carried out through the ^{11}B$(n,\alpha)^8$Li reverse reaction [46]. The use of a ^{11}B beam did however not allow to measure the contribution of ^{8}Li$(\alpha,n)^{11}$B* channels, which was later determined by Boyd *et al.* [47], who employed a radioactive ^{8}Li beam.

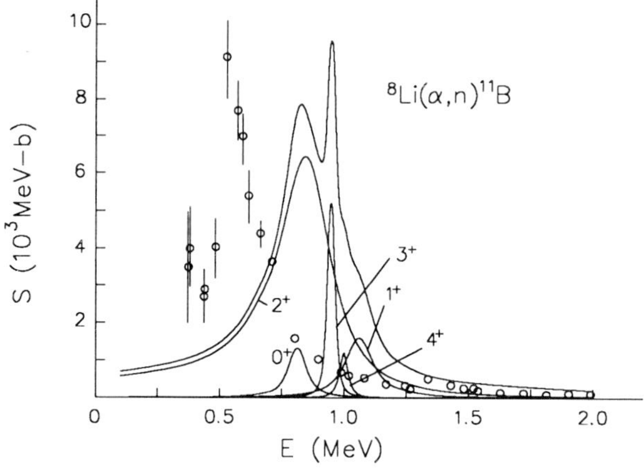

Fig.6 ^{8}Li$(\alpha,n)^{11}$B S-factor with a partial-wave decomposition. The experimental data are taken from ref.[46].

Here, we would like to briefly report on a microscopic approach [48], which was done *before* the publication of experimental data. We only compare to the data of Paradellis *et al.* [46] since our model does not contain any n + ^{11}B* excited configuration. The GCM calculation provides a good description of spectroscopic properties in ^{12}B low-lying states, and of the ^{11}B$(n,n)^{11}$B elastic cross section. It predicts the existence of a positive-parity "α-molecular" band located close to the α + ^{8}Li threshold, and whose states present a marked α + ^{8}Li structure. These resonances are clearly present in the

GCM S-factor displayed in fig.6, where it is decomposed into the different positive-parity partial waves. Negative-parity partial waves are completely negligible. Our S-factor is dominated by the 2^+ resonance which corresponds to an s wave in the entrance channel. This is consistent with the data of Paradellis *et al.* who observe a strong peak near 0.53 MeV (without spin assignment). The slight disagreement between the experimental and theoretical (0.89 MeV) energies is however remarkable if one keeps in mind that the nucleon-nucleon interaction has been fitted to ^{12}B low-lying states only. In addition, the shape of the resonance is very close to experiment, indicating that the partial alpha and neutron widths are realistic.

4. Conclusion

Owing to its basic grounds, a microscopic model is a powerful way to derive nuclear cross sections. The amount of necessary experimental information is restricted to some spectroscopic properties, which are in practice well known. The main limitations (which are partly correlated) are the description of the internal wave functions, and the extension of computer times. Some nuclei, (and unfortunately most of nuclei interesting in astrophysical scenarios!) can not be accurately defined in the one-centre harmonic oscillator model. Consequently, we have developed multi-cluster models [15,16], which enlarge the application field of the GCM to reactions involving a deformed nucleus. In this approach, the colliding nuclei are described by a structure composed of a closed shell nucleus (α or ^{16}O) and s shell particles, for which the harmonic-oscillator assumption is well justified.

An other way to improve the internal wave functions is to mix one-centre states corresponding to different oscillator parameters [43] which, here, are assumed to be equal for each cluster. This method raises enormous numerical and analytical problems because of the center of mass projection. The common limitations to both methods mainly rely on computer times, which necessitate a special attention to the efficiency of the codes. However, the importance of nuclear reaction rates in many astrophysical applications certainly deserves such an effort.

References

[1] Rolfs C and Rodney W S 1988, "Cauldrons in the Cosmos", (Chicago, London)

[2] Wallace R H and Woosley S E 1981, Astrophys. J. Suppl. **45** 309

[3] Malaney R A and Fowler W A 1988, in "Origin and Distribution of the Elements", ed. G.J. Mathews, World Scientific, p.76

[4] Mitalas R 1985, Astrophys. J. **290** 273

[5] Verviers J 1991, in "Nuclei in the Cosmos", ed. H. Oberhummer *et al.* , Springer-Verlag Berlin, p.9

[6] Lane A M and Thomas R G 1958, Rev. Mod. Phys. **30** 257

[7] Filippone B W , Humblet J and Langanke K 1989, Phys. Rev. **C40** 515

[8] Thielemann F K , Arnould M and Truran J W 1987, in "Advances in Nuclear Astrophysics", eds. Vangioni-Flam *et al.* ,525

[9] Tombrello T A 1965, Nucl Phys. **71** 459

[10] Oberhummer H and Staudt G 1991, in "Nuclei in the Cosmos", ed. H. Oberhummer *et al.* , Springer-Verlag Berlin, p.29

[11] Wildermuth K and Tang Y C 1977, "A Unified Theory of the Nucleus", Vieweg, Braunschweig

[12] Baye D and Descouvemont P 1989, Proc. 5th Int. Conf. on Clustering Aspects in Nuclear and Subnuclear Systems, Kyoto, Japan, 1988, J. Phys. Soc. Jpn. **58**, Suppl. p.103

[13] Langanke K and Friedrich H 1986, Adv. in Nuclear Physics **17**, ed. by J.W. Negele and E. Vogt (Plenum, New-York)

[14] Tang Y C 1981, in Topics in Nuclear Physics II, Lecture Notes in Physics, Springer, Berlin, **145** 572

[15] Baye D and Descouvemont P 1989, Proc. Sapporo Int. Symp. on Developments of Nuclear Cluster Dynamics, Sapporo, Japan, 1988, ed. Y. Akaishi *et al.* , World Scientific, p.6

[16] Descouvemont P 1991, Phys. Rev. **C44** 306

[17] Baye D , Heenen P H and Libert-Heinemann M 1977, Nucl. Phys. **A291** 230

[18] Baye D and Descouvemont P 1983, Nucl. Phys. **A407** 77

[19] Baye D and Descouvemont P 1985, Ann. Phys. (N.Y.) **165** 115

[20] Imhof W L, Johnson R G, Vaughn F J and Walt M 1959, Phys. Rev. **114** 1037

[21] Wiescher M, Steininger R and Käppeler F 1989, Astrophys. J. **A344** 464

[22] Nagai Y *et al.* 1991, Astrophys. J. **A381** 444

[23] Descouvemont P and Baye D 1988, Nucl. Phys. **A487** 420

[24] Filippone B W 1986, Ann. Rev. Nucl. Part. Sci. **36** 717

[25] Kim K H, Park M H and Kim B T 1987, Phys. Rev. **C35** 363

[26] Kolbe E, Langanke K and Assenbaum H J 1988, Phys. Lett. **214B** 169

[27] Descouvemont P and Baye D 1992, to be published

[28] Ajzenberg-Selove F 1988, Nucl. Phys. **A490** 1

[29] Brown J A *et al.* 1991, Phys. Rev. Lett. **66** 2452

[30] Descouvemont P and Baye D 1992, to be published

[31] Kavanagh R W *et al.* 1969, Bull. Am. Phys. Soc. **14** 1209

[32] Vaughn F J *et al.* 1970, Phys. Rev. **C2** 1657

[33] Filippone B W *et al.* 1983, Phys. Rev. **C28** 2222

[34] Barker F C and Spear R H 1986, Astrophys. J. **307** 847

[35] Mukhamedzhanov A M and Timofeyuk N K 1990, JETP Lett. **51** 283

[36] Redder A *et al.* 1987, Nucl. Phys. **A462** 385

[37] Barker F C 1987, Aust. J. Phys. **40** 25

[38] Barker F C and Kajino T 1991, to be published

[39] Langanke K and Koonin S E 1985, Nucl. Phys. **A439** 384

[40] Funck C, Langanke K and Weiguny A 1985, Phys. Lett. **152B** 11

[41] Descouvemont P, Baye D and Heenen P H 1984, Nucl. Phys. **A430** 426

[42] Descouvemont P and Baye D 1987, Phys. Rev. **C36** 1249

[43] Baye D and Kruglanski M 1992, Phys. Rev. **C45** 1321

[44] Ajzenberg-Selove F 1986, Nucl. Phys. **A460** 1

[45] Descouvemont P 1992, to be published

[46] Paradellis T *et al.* 1990, Z. Phys. **A337** 211

[47] Boyd R N *et al.* 1992, Phys. Rev. Lett. **68** 1283

[48] Descouvemont P and Baraffe I 1990, Nucl. Phys. **A514** 66

Properties of nuclei in astrophysical environments—binding energies, weak-interaction rates

K Takahashi and W Hillebrandt

Max-Planck-Institut für Astrophysik, W-8046 Garching bei München, FRG

Abstract. Some of the problems related to the evaluation of binding energies (masses) and weak-interaction rates in nuclei of astrophysical interest are briefly discussed by way of a few selected examples of those arising in stellar evolution and nucleosynthesis studies.

1. Introduction

The acquisition of precise nuclear data is a primary requisite to success for stellar evolution and nucleosynthesis studies. Only when that is accomplished, can one with some confidence decipher the elemental and isotopic abundances in Nature and thereby derive meaningful constraints on the associated astrophysical models.

Summaries of recent progress in nuclear data acquisition for astrophysical purposes can be found elsewhere [1-2]. Undoubtedly, one of the tasks of upmost importance is to improve the precision of thermonuclear reaction rates on light nuclei that control not only nucleosynthesis flows but also the energetics of stellar evolution. The continued, earnest efforts to that goal are reviewed also in this volume from both experimental [3] and theoretical [4] view-points. In this review, we beef up the discussions on a couple of other nuclear properties of astrophysical importance, namely, the binding energy and the weak-interaction rates.

2. Binding energies

2.1. In the nuclear ground states

2.1.1. Preliminaries

The binding energy, B, or the ground-state mass, M, is the most basic property of the nucleus. For a nucleus with Z protons and N neutrons, they are related by the definition $B(Z, N) = [m_n N + m_p Z - M(Z, N)]c^2$, where m_n and m_p are the neutron and proton masses, respectively.

As the atoms in the laboratory are usually neutral or nearly so, the tabulated experimental masses [5] refer to those of the neutral atoms: $M_0(Z, N) = M(Z, N) + m_e Z - B_e(Z)/c^2$, where m_e is the electron rest mass and $B_e(Z)$ ($\sim 0.014 \times Z^{2.39}$ keV [6]) is the total binding energy of the orbital electrons. It is also customary to introduce the atomic mass-excess $M_{ex}(Z, N) = M_0(Z, N) - [\text{amu}]\, A$, where the mass number $A = N + Z$ and the atomic mass unit $[\text{amu}] = M_0(^{12}\text{C})/12 = 931.501$ MeV/c^2, with which one can avoid large numbers.

2.1.2. Data needs

The table of experimentally known atomic masses [5] covers well those required in most stellar evolution and nucleosynthesis studies. A conspicuous exception to this concerns a large number of very neutron rich nuclei which are involved in a stellar nucleosynthesis process via rapid captures of neutrons, the r-process. This process, together with a slow neutron-capture process (the s-process), is thought to be responsible for the bulk abundance of naturally-occurring heavy elements. Despite the vigorous search ever since the pioneering works of B^{2}FH [7] and Cameron [8], the true astrophysical site(s) for the r-process has yet to be found (see dedicated reviews [9] and more recent attempts [10-11]). Nevertheless, it is imperative to have reliable nuclear data, such as masses, ready for use in r-process model calculations. Furthermore, one could then gain some hope, a dim one though it might be, for the possibility of segregating one astrophysical scenario from the other in a comparison of the calculated and observed compositions of r-process product nuclei (see e.g. [12]).

Remarkable progress has recently been made in direct mass measurement techniques (e.g. [13]), notably with the use of a recoil spectrometer on products of target- or projectile-fragmentation reactions at high energies, or by way of determining the cyclotron frequency of ions stored in a trap. Further progress in this field is expected to lead to a rapid improvement, in quality as well as in quantity, of mass determinations on nuclei far from stability. With this prospect apart, the available mass data are hardly sufficient as far as r-process calculations are concerned, this offering a justification for continued pursuit of improved mass predictions.

2.1.3. Models

We first recall the celebrated Bethe Weizsäcker (or the liquid-drop) mass formula which reads $B(Z,N) \sim a_v A - a_s A^{2/3} - a_i (N-Z)^2/A - a_c Z^2/A^{1/3} + \delta$. The adjustable, positive parameters $a_{v,s,i,c}$ are introduced for the volume, surface, symmetry and the Coulomb terms, respectively. The "even-odd" (or "pairing") term, δ, is canonically defined as nil for odd-A such that it gets positive for even-Z and negative for odd-Z in even-A cases.

While that simple formula describes observed nuclear masses remarkably well on the average, there remain some distinct deviations that can be attributed to nuclear shell structure. Among many models developed so far to deal with the residual "shell correction" [14,15], a purely empirical approach as initiated by Garvey *et al* [16] is to find a mass relationship, valid locally in Z, N and A, in such a way that the fluctuating parts largely cancel. It is no surprise that such mass relationships give very good fits to exisiting mass data (to an accuracy of 200 ~ 400 keV; see e.g. [17]), but also that their predictive powers are quite limited. The second approach is to add shell correction terms to the liquid drop formula (or the droplet formula [18]). If this is done empirically with many parameters (e.g. [19]), the goodness of the fit can be comparably high as in mass relationships. It is reduced considerably if the shell correction is obtained semiempirically with a bit of physics, with a resulting mean deviation of ~ 800 keV when the total number of parameters is limited to about 50. The mass formula of Hilf *et al* [20], which has been frequently used in r-process calculations because of its simplicity, belongs to this category. An alternative way is to deduce the shell correction from single-particle energies for a given nuclear potential by way of the so-called Strutinsky re-normalization procedure. Mass formulae thus obtained may be referred to as macro-micro formulae. An example of this type is one by Möller and Nix [21], which leads to a similarly good fit as the above but with about a half the number of parameters.

Despite some success, predictive powers of those semiempirical or macro-micro formulae are questionable as well. This probably reflects the fact that the handy separation of microscopic and macroscopic terms is insubstantial. One extreme method of describing the mass of the nucleus without that separation is to introduce local parameters for correlations possibly imaginable within the framework of the nuclear shell model as has been pursued by Zeldes and his collaborators (e.g. [22]). Although the fits are very good (within 300 keV on average), the results have rarely been used in r-process calculations, primarily because of the demerit that the method cannot be applied to unknown regions in the chart of nuclides.

It is impossible to describe the nucleus from first principles. Also, given the required accuracy in mass predictions, it is hardly imaginable that any fully microscopic methods starting from fundamental nucleon-nucleon interactions can be successful. In practice, however, one may try to find a compromise between the closeness to such a method and the goodness of the fit in order to reduce the number of parameters. In this respect, the "energy-density formalism", which is a kin to the Hartree-Fock method but allows force parameters to be varied, has turned out to be very promising. Tondeur [23] has demonstrated that it is possible to reproduce experimental masses within 1 MeV on the average by adjusting less than 10 parameters. With the self-consistency between the macro- and microscopic aspects being guaranteed, its predictive power is presumably higher than in other mass formulae. A drawback is that it requires much computational time, particularly for deformed nuclei. This reality has led to the development of a compromised (but to a certain extent still self-consistent) method what the authors originally refer to as the "extended Thomas-Fermi method with Strutinsky integrals" (e.g. [24]), and which apparently mimics the results of the energy-density formalism. Its predictive power has yet to be substantiated.

We add here that the validity of a "physical" mass formula is to be checked with regard also to other nuclear properties such as radii, deformations and fission barriers.

2.2. At finite temperatures

2.2.1. Nuclear excited states

At finite temperatures in stellar interiors, nuclear excited states are thermally populated according to a Boltzmann distribution $g_i = (2J_i + 1)\exp(-E_i/k_B T)$, where J_i and E_i are the spin and the excitation energy of a level i, k_B the Boltzmann constant and T is the stellar temperature. The importance of the "nuclear partition function", $g = \Sigma_i g_i$, becomes more as T increases. At the late stages of stellar evolution, for instance, temperatures exceed 5×10^9 K. Accordingly, all the nuclear reactions via strong and electromagnetic interactions proceed extremely fast, so that all the nuclear species can be considered in nuclear statistical equilibrium (NSE). The NSE (or Saha) equations can be solved for a given set of three variables: the temperature, density and the neutron-to-proton ratio. The last quantiy is often expressed in terms of electron concentration, Y_e, which is equal to the number fraction of protons, either free or bound to nuclei. To compute the abundances of nuclei in NSE requires proper evaluations of g in addition to precise values of ground-state binding energies. A difficulty stems from a lack of information on nuclear levels in beween the relatively well-known regions near the ground state and near the neutron separation energy of about 8 MeV.

As a consequence, one has to introduce a theory on nuclear level density. For most astrophysical applications, a handy "back-shifted Fermi-gas model" may be used with parameter values derived somehow empirically (e.g. [25]) or from theory such as the one based on a Hartree-Fock single-particle model coupled to the BCS pairing formalism (e.g. [26]). We shall come back in section 2.3 to some problems related to nuclear partition functions.

2.2.2. Ionization

A careful consideration of the changes in atomic binding energies along with increased degrees of ionization at high temperatures is due in a few cases. This concerns the calculations of β-decay Q-values between nuclei with very low atomic mass differences as will be disussed in section 3.2.2.

2.3. In high-density matter: nuclear equation of state

One of the most important ingredients for both neutron star and supernova models is the equation of state (EOS), namely the energy density and the pressure as functions of temperature T, density ϱ and the composition (see e.g. [27] for a review). Consequently, major fractions of the uncertainties in those models result from our incomplete knowledge of the EOS at ϱ from about 10^{-2} to 10 times nuclear matter density $\varrho_0 \simeq 3 \times 10^{14} \mathrm{g/cm^3}$. In particular, what the EOSs of supernova and early neutron star matter have in common are that the entropy per nucleon S is quite low (less than a few times k_B), and that the lepton fraction, or Y_e to be more specific (see section 2.2.1), is significantly lower than in normal matter. These are in contrast to the situation in heavy ion collisions, where S is a few orders of magnitude higher because of high excitation energies and only a moderate neutron excess can be achieved. Therefore, there is little hope that we can learn enough about the EOSs from laboratory experiments such as heavy ion collisions, and one has to rely on theoretical models and, possibly, on interpretations of astrophysical observations.

At rather low densities, $\varrho \lesssim 10^{-2}\varrho_0$, the EOS can in principle be calculated easily from a Boltzmann gas approximation for nucleons and nuclei in NSE, provided nuclear binding energies and partition functions are known. This, however, is not the case. Most of the nuclei present in the interior of a collapsing stellar core or in the outer layers of a neutron star would be very neutron-rich and thus short-lived under laboratory conditions, and many of them have not even been synthesized yet by experiments. As a result, one has to extrapolate the properties of stable and "mildly" unstable nuclei to unknown ones. Moreover, certain collective effects that are characteristic only at high matter density may become crucial.

The nuclear partition function, g as defined in section 2.2.1, may serve as an example to illustrate the difficulty mentioned above. It is well known that the concentration of free protons in a collapsing star more or less determines the effective electron-capture rate (see section 3.3) and thus the timescale of collapse, the mass of the homologous core and so on. Under NSE conditions, in turn, the proton fraction depends strongly on, among others, g because nuclei will have high excitation energies at the associated very high temperatures. The difficulty of obtaining reliable g is clearly observed in significantly different results of various calculations which are mainly based on the back-shifted Fermi-gas model. In particular, it is non-trivial a question how to decide which states should be counted as "bound" states still belonging to a nucleus above the neutron separation energy (e.g. [28]).

Abundances of nuclei in NSE are also very sensitive to the ground-state binding energies B because they enter the NSE abundance equations as $\exp(B/k_BT)$. The difficulty of determining precise B values is similarly clear.

At higher densities, the Boltzmann-gas approach to the nuclear part of the EOS is no longer valid. This happens if the nuclear radius R_N becomes comparable to the Coulomb interaction radius R_c. Phenomenologically one finds $R_N \sim A^{1/3}$ and R_c being roughly equal to the mean separation of adjacent nuclei, which is proportional to $(A/\varrho)^{1/3}$, leading to $R_N/R_c \simeq 10^{-5}\varrho^{1/3}$. If we impose $R_N/R_c \ll 1$, then ϱ has to be smaller than $10^{12}\mathrm{g/cm^3}$. At densities above this value, therefore, a self-consistent treatment becomes necessary in order to include at least the Coulomb correlation. At even higher densities $\varrho \gtrsim 0.1\varrho_0$, similar arguments indicate that also interactions among nucleons have to be included self-consistently.

The most advanced method that has been applied to the supernova problem so far is the temperature-dependent Hartree-Fock method [29-30], but little progress has been made in the last several years. In that method, one starts from a model Hamiltonian incorporating the nucleon-nucleon interactions in terms of an effective potential that is tailored to reproduce certain properties of finite nuclei and of cold nuclear matter as implied from the extrapolation of a mass formula. It is obvious that the resulting EOS will depend on the particular choice of the effective interaction, but we have to live with that uncertainty. Temperature effects are included in a statistical way. Although the concept of the above approach is clear and simple, it raises several problems in practice and in principle. Firstly, nuclei in the deep interior of a collapsing star will form a Coulomb lattice and, therefore, the Hartree-Fock equations have to be solved for a unit cell with periodic boundary conditions. This immediately leads to a three-dimensional problem which cannot be solved easily for the whole range of T, ϱ, and Y_e realized in a supernova core or the interior of a neutron star. One therefore generally assumes spherical symmetry and applies the Wigner-Seitz method to include Coulomb-lattice effects. This then means that one cannot impose periodic boundary conditions on the single-particle wave functions, resulting in some arbitrariness in the boundary conditions and consequently in the solutions. It is likely that some of the ongoing debates over the geometrical structure of matter near the transitional density to homogeneous nuclear matter stem from these ambiguities.

Secondly, one has to admit that, whereas in reality matter will be a mixture of various types of nuclei and free nucleons in NSE, those self-consistent models always assume that only the configuration with the lowest free energy (or free enthalpy) is present. It has been demonstrated [31] that the free energy per baryon versus A curve typically exhibits several distict minima. The lowest state could be a "nucleus" with several hundred nucleons (as opposed to lower A values familiar to ordinary nuclei), but the barrier between the various minima is often less than k_BT. This means that a mixture of very many "nuclei" with rather similar abundances will be present in equilibrium. Obviously, this effect will have some impact on the EOS.

Finally, the Hartree-Fock method is a single-particle approach and does not treat correlations in a self-consistent way. It therefore will gradually loose reliability with increased ϱ. In addition, the results will be sensitive to the choice of the effective interactions [30]. Moreover, it should be noted that in most supernova simulations even further simplifications of the method outlined above are used or the EOS is parametrized in a rather arbitrary way. Numerical results, therefore, should be taken with some care.

One might wonder why astrophysical models are to be so sensitive to the nuclear part of the EOS despite the fact that in most cases of interest at sub-nuclear densities the pressure is dominated by leptons. The main reason for that sensitivity is that for relativistic leptons the adiabatic index is equal to 4/3, the critical value for gravitational instability. Even a tiny deviation from this value as a result of nuclear effects is therefore often crucial. In addition, the density at which, for example, the core of a collapsing star bounces is determined by the nuclear part of the EOS, and so is in part the energy available to the shock.

We conclude this section with a few remarks on the nuclear EOS beyond nuclear saturation density ϱ_0. Inside a collapsing stellar core as well as in the deep interior of a neutron star, "nuclei" dissolve into a homogeneous fluid of free neutrons and protons once the density exceeds ϱ_0. Consequently, relativistic leptons are no longer the main contributor to the EOS, but the contributions from nucleon-nucleon interactions in a Fermi fluid become dominant. Phenomenologically determined nucleon-nucleon forces will gradually lose reliabilty as ϱ increases. It is therefore not surprising that the EOS at densities above, say, twice the nuclear saturation density remains subject to considerable dispute (see e.g. [32]). For supernova models, however, the situation does not seem to be that serious since most of realistic EOS calculations predict an adiabatic index of the order of 3 near ϱ_0 and thus a collapsing core will, in general, not overshoot this density by much. The situation is different during the early cooling phase of a nascent neutron star and in cold neutron stars. Depending on the mass of the neutron star, the density may exceed $10\varrho_0$ and the EOS at high ϱ becomes very important. It is obvious that phenomenological approaches at such high ϱ are not very reliable if they lack self-consistent treatments of, for example, enough correlations and hyperonic degrees of freedom.

For $\varrho \approx \varrho_0$, on the other hand, one might hope to deduce very good estimates of the nuclear EOS just from the (in)compressibilty of finite nuclei as determined by the measurements of breathing mode, i.e., the giant isoscalar monopole resonance (see e.g. [33]). However, that is not as simple as it sounds. First of all, in almost all cases of astrophysical interest, high density matter is neutron-rich and thus experimental data on, say ^{208}Pb, cannot be utilized in a straightforward way. Secondly, temperature effects are significant in general, and they cannot be mimicked by experiments. In conclusion, as far as astrophysical applications are concerned, the hot discussions (see e.g. [34]) about the true value of κ_0 of cold, symmetric nuclear matter appears to be a little academic.

3. Weak interaction rates

3.1. In the nuclear ground states

3.1.1. Preliminaries

The β-decay rate ($\equiv ln2/t_{\beta,1/2}$ with $t_{\beta,1/2}$ being the halflife) can be expressed as $\lambda_\beta = \Sigma(G^2/2\pi^3)|M|^2 f \approx \int(G^2/2\pi^3)S_\beta f \; \mathrm{d}E$, where G is either the vector or axial-vector coupling constant of the weak interaction, $|M|^2$ is the nuclear matrix element squared, and the known function f describes the lepton $(e^-\bar{\nu}_e, e^+\nu_e)$ phase volume (see e.g. [35]), while the summation runs over energetically possible final states. It is customary to consider the product $f \times t_{\beta,1/2}$ (or the ft-value) $\sim 1/\,|M|^2$ so as to illuminate nuclear structure effects more clearly. If the density of the final states is high, the β-strength function S_β can be defined as in the above.

In fact, the knowledge of S_β in a wider range of energy than the Q_β-value window is essential both in the understanding of nuclear β-decays and in many applications to astrophysical processes via weak interaction. The general behavior of S_β in energy and through the nuclear chart depends on the nuclear β-decay operator sandwiched in $|M|^2$, which is to match the changes of nuclear spins, J, and parities, π, of the initial and final states. Fast Fermi transitions ($|\Delta J| = 0, \Delta \pi = +$) with $\log_{10} ft \sim 3$ occur only in $N \approx Z$ nuclei where the isobaric analog state of a given initial state may fall in the Q_β window. Similarly because of a sumrule for $|M|^2$, Gamow-Teller transitions ($|\Delta J| \leq 1, \Delta \pi = +$, but excluding $0^+ \to 0^+$) proceed relatively slow with typical $\log_{10} ft = 4 \sim 6$ in a heavy nucleus where the Q_β window is off the range of the existing giant resonance which exhausts most of the strength. If $|\Delta J| \leq 1$ but $\Delta \pi = -$ (non-unique first-forbidden), $\log_{10} ft$ values are in the $5 \sim 8$ range.

3.1.2. Data Needs

In r-process models, β^- decays (i.e. electron emissions) on very neutron rich nuclei are crucial as they not only determine the abundance pattern but give a measure for its plausible duration timescale. The r-process is expected to occur at high temperatures ($\sim 10^9$ K), so that the decay halflives might be different from the terrestrial ones. Conventional wisdom, however, dictates that the possible consequence of temperature effects is minor since Q_β-values for such neutron-rich nuclei are so high that favorable Gamow-Teller channels are widely open (cf. section 3.2.1).

More important, studies of Gamow-Teller strength distribution in particular give basic information required for formulating weak interaction processes on heavy nuclei in astrophysical environments. A series of (p,n) reaction cross-section measurements at forward angle with a few hundred MeV protons (see e.g. [36] for an early review) have provided us with a clear picture of the global behavior of β^- strength functions, revealing the existence of Gamow-Teller giant resonances just as had been theoretically conjectured long since. Similar measurements of (n,p) cross-sections, as have been performed in limited cases so far (e.g. [37]), would certainly enhance our capability of reliably evaluating weak interaction rates under astrophysical conditions as will be discussed in section 3.3.

In passing, we note that in marginal cases the ft-values of highly-forbidden transitions are of some astrophysical interest. This concerns, for example, the case of ^{54}Mn as a clock for cosmic rays [38].

3.1.3. Models

Collective phenomena such as the Gamow-Teller giant resonance cannot be dealt by an independent-particle shell model of the nucleus. Probably the most convincing way of describing β-strength functions would be to perform a shell model calculation with realistic forces taking into account enough correlations among nucleons and thus in large enough a configuration space. In practice, such calculations are feasible only for nuclei with N and Z less than 20, i.e., up to s-d shell nuclei. Even there, one has some difficulties in explaining the whimsical nature with which nuclear β-decays are infamously associated (e.g. [39]).

Among several attempts at studying β-decays in heavy nuclei in a systematic way, the earliest one was the so-called gross theory of β-decay (e.g. [40]). This statistical theory has been used to predict β-decay properties $t_{\beta,1/2}$ in particular, by describing allowed and first-forbidden transitions essentially with one parameter (i.e. the width of the Gamow-Teller giant resonance which was still unknown) common for all nuclei. Its predictive power has in the last two decades had enough chance to be scrutinized. An apparent shortcoming of the prediction is said to be in its general tendency of considerably overestimating $t_{\beta,1/2}$ of neutron-rich nuclei when compared with accumulated experimental data. A part of the discrepancies can be reduced with the use of a more modern mass formula to obtain Q_β-values though. Apart from its lack of microscopic physics, the gross theory has the advantages of being quite stable with respect to the variation of parameter values and of having a reasonably high predictive power despite its extreme simplicity. For these reasons, it has been frequently used in r-process model calculations. Recently, some improvements of the theory have been made to consider the sumrules in a more consistent manner (e.g. [41]).

From microscopic view-points, there are two major ways of handling β decays in heavy nuclei. One is to perform a shell model calculation in a truncated (but as large as possible) configuration space. The outcome of sporadic case studies, however, implies that it is probably not worth the efforts of huge numerical calculations. The more pragmatic way is to assume apriori a Gamow-Teller residual interaction in a parametrized form such that at least the main feature of the Gamow-Teller strength distribution can be efficiently embodied. Once this is done, the question is how to describe the initial and final states with the use of certain approximations. The simplest case is referred to as a schematic shell model that treats particle-hole excitations in the so-called Tamm-Dankoff approximation (TDA). A step forward, the random phase approximation (RPA) allows to include part-time excitations in the initial ground state ("ground-state correlations"). Further, the BCS pairing can be embedded with the "quasi-particle" random-phase approximation (QRPA). Details on those approximations, including the more rigorous definitions, can be found elsewhere (e.g. [42]).

Along the above vein, the early predictions of $t_{\beta,1/2}$ of neutron-rich nuclei had relied on TDA or RPA before QRPA has become employed [43-44]. The model of Staudt *et al* [43] has reproduced existing halflife data quite well but by allowing for a different value of the residual interaction strength for each isotopic chain. It is imaginable that this model will have predicted right numbers for $t_{\beta,1/2}$ of unknown but nearby nuclei. But, it might well be that some potentially valuable information on hidden physics was lost in the fitting procedure such that long-range extrapolations were at risk, just as one had learned in a hard way during the course of developing mass formulae. The work of Möller and Randrup [44] has, on the other hand, sticked to a global set of parameters at the expense of less impressive fit. This model has an edge in that its background single-particle model is consistent with the one used to compute Q_β-values. It has recently been suggested [45] that some of the discrepancies of the calculated and experimental β-strengths may be reduced by the use of more adequately chosen nuclear deformation parameters than the model predicted.

One may question the validity of the QRPA. In fact, it has been demonstrated [46] that the QRPA is not necessarily good enough for simulating the results of realistic sd-shell model calculations. The study has confirmed also that the QRPA results are extremely sensitive to the assumed strength of particle-particle correlation. This type of correlation, additional to the particle-hole correlation responsible for the Gamow-Teller giant resonance, had been introduced in order to account for certain β^+ decays, particularly those involved in double β-decays (e.g. [47]). To go beyond the standard

QRPA is apparently a trying task [48].

An interesting alternative to the QRPA approach to β-decays in medium and heavy nuclei is the so-called "theory of finite Fermi systems" (or Migdal theory [49]), which can circumambulate some of the shortcomings of the QRPA (see e.g. [50]). So far, systematic calculations based on the Migdal theory have been performed only for β^- decay halflives with a then-justifiable neglect of particle-particle correlation [51].

3.2. At finite temperatures

3.2.1. Nuclear excited states

The effective β-decay rate of a nucleus at finite-temperature environments can be expressed as $\lambda_{eff} = \Sigma_i \lambda_i g_i / \Sigma_i g_i$ in terms of the population probability g_i as defined in section 2.2.1 and β-decay rate of i-th level under thermal equilibrium. Even at relatively low temperatures, the contributions from low-lying excited states could lead to an λ_{eff} value considerably different from the ground-state decay rate. This is the case when the latter rate is observed to be very low whereas much faster transitions are expected from low-lying excited states, with ^{99}Tc being a well-known example. The importance of excited-state β-decays under s-process conditions may serve us to make the point.

In contrast to the case of the r-process, the astrophysical sites for the s-process nucleosynthesis have become somewhat well known: central helium burning phase in massive ($\gtrsim 10$ M$_\odot$) stars produces s-process nuclei in the $60 \lesssim A \lesssim 90$ range, whereas during the thermal pulses developed in helium burning shell in low- and intermediate-mass ($1 \sim 5$ M$_\odot$) stars the heavier s-process nuclei can be synthesized, at typical temperatures T $\sim 10^8$ K and neutron densities n$_n$ $\sim 10^8$/cm^3. A driving force in the search for s-process sites has been the detailed analyses of the solar abundances in connection with temperature-enhanced β-decay rates on otherwise neutron-capturing stable or near-stable nuclei (see [52] for a review). Furthermore, case studies such as of ^{99}Tc$\cdot\beta$-decays at s-process temperatures can lead some clues to the "dredge-up" mechanism by which s-process products in the interiors of low- or intermediate-mass stars can be brought up to the stellar surface to be observed.

Unfortunately, there is no reliable way to evaluate ft-values of unknown β-transitions from excited states, and one has to rely utmost on semiempirical systematics or often even on a guess work [53]. For example, a large-scale (but truncated) shell-model calculation for ^{99}Tc turned out to be infertile [54].

3.2.2. Ionization

In stellar interiors, heavy elements are partially or fully ionized. When compared with the atomic mass, the mass of an ion increases along with the degree of ionization by loosing binding energies of orbital electrons. As a consequence, it may happen that β^- decays as observed in neutral atoms with very low Q-values ($\equiv Q_0$) become energetically forbidden. The Q-value in the case of full ionization is roughly $Q_0 - dB_e(Z)/dZ$ (see section 2.1.1). A catch is, however, β decay does occur by creating an electron in previously unoccupied orbit and thus gaining relatively large binding. This so-called bound-state β-decay, an inverse process of orbital electron capture, has been discussed theoretically for more than four decades (since [55]), often in connection with astrophysics (see e.g. [56-57]; also [58] for further references). Most interesting bound-state β-decays have indeed slightly negative Q_0-values: ^{163}Dy $\rightarrow$ ^{163}Ho (ground state) in s-process studies [57], ^{187}Re $\rightarrow$ ^{187}Os (9.75 keV excited state) in nucleo-cosmochronology [57,59], and ^{205}Tl $\rightarrow$ ^{205}Pb (2.3 keV excited state) which may also have some chronometric virtue [60].

The first confirmation of the bound-state β-decay process has just been reported for fully stripped ^{163}Dy by a storage-ring experiment at GSI, with the resulting halflife of about 47 days [61]. The good agreement of this and a theoretical value of 50 days [58] (aside from a factor of 7/8 with due consideration of hyper-fine structures [62]) confirms, among others, that the ft-value for that transition had been correctly estimated [63] from a Nilsson plus pairing model in this highly deformed region. Similar experiments on ^{205}Tl would be able to determine the ft-value for the transition to the first excited state in ^{205}Pb, which is in quest for designing a neutrino detector, while its theoretical (shell-model) estimates are very uncertain [64].

3.3. In high-density matter: electron captures

At high densities, electrons can have such high Fermi energies that their captures can occur even on stable nuclei. As mentioned briefly in section 2.3, electron captures on free protons play decisive roles on the fate of a core-collapsing (pre)supernova. This is particularly true in massive ($\gtrsim 10$ M$_\odot$) stars. While the evaluation of the capture rates on protons is quite easy, the trouble starts when electron-capture rates on heavy nuclei such as Fe-group nuclei or even on lighter ones are required as typically in scenarios on supernovae in less massive stars. Let us conclude this section with a few remarks on electron-capture rates on nuclei.

The capture rates of Fuller *et al* (e.g. [65]), which have been used widely in the literature, are largely based on a guess work of unknown ft-values. Better information on the required β^+ strength functions can be obtained from such (n,p) experiments as mentioned in section 3.1.2. The mere use of (n,p) data on ^{54}Fe [37] implies that some of those rates might be subject to alteration [66]. While one expect further (n,p) data in the future, one can also hope to have a theory that can reproduce the experimental data without too much of a twist. In that respect, simultaneous analyses of (n,p) and (p,n) data, such as the one [50] within the framework of Migdal theory, are encouraged. On the other hand, full-space sd-shell model calculations with the "best" available interactions have been carried out to evaluate the capture rates on lighter nuclei such as ^{20}Ne, ^{24}Mg and so on [67-68]. The detailed comparisons [68] of various models and generic (n,p) data obtained from the mirror (p,n) experiments imply some shortcomings of previous predictions.

4. Summary

We have discussed some problems related to nuclear binding energies and weak inter-action rates of astrophysical importance. As we have shown with some examples, the acquisition of nuclear data required in stellar evolution and nucleosynthesis studies has yet to rely largely on uncertain nuclear models.

Judging from remarkable progress made recently in developing novel experimental techniques, one can expect that some of the data in quest will become available in the near future. This particularly concerns masses and β-decay halflives of neutron-rich nuclei involved in the r-process nucleosynthesis. Also, further accumulation of (n,p) re-action cross-sections in relation to β^+ strength functions will be extremely valuable for studies of electron captures which play crucial roles in supernovae. However, whether laboratory experiments alone can sufficiently provide us with required information is an open issue. Indeed, they will hardly serve the astrophysical needs in cases where, for instance, nuclear excited states would come into play.

It is obvious that the aquisition of precise nuclear data of astrophysical interest has to be pursued along the line of interplay between experiment and theory. As we have outlined in this review, the development of convincing theoretical models encounters great difficulties of its own. In particular, the determination of the nuclear equation of state at high densities in supernova environments requires improved (self-consistent) theoretical treatments as experimental information will remain quite limited.

References

[1] Arnould M 1992 *Nucl. Phys.* **A538** 493c
[2] Takahashi K 1992 *Comm. Astrophys.* in press
[3] Trautvetter H P 1992 *J. Phys.* **G** this volume
[4] Descouvemont P 1992 *J. Phys.* **G** this volume
[5] Wapstra A H, Audi G and Hoekstra R 1988 *At. Data Nucl. Data Tables* **39** 281
[6] Foldy L L 1951 *Phys. Rev.* **83** 397
[7] Burbidge E M, Burbidge G R, Fowler W A and Hoyle F 1957 *Rev. Mod. Phys.* **29** 547
[8] Cameron A G W 1957 *Chalk River Report CRL-41* unpublished
[9] Hillebrandt W 1978 *Space Sci. Rev.* **21** 639
   Schramm D N 1982 *Essay in Nuclear Astrophysics* eds. C A Barnes *et al* (Cambridge Univ. Press, Cambridge) p. 325
   Mathews G J and Ward R A 1985 *Rep. Prog. Phys.* **48** 1371
   Cowan J J, Thielemann F-K and Truran J W 1991 *Phys. Rep.* **208** 267
[10] Woosley S E and Hoffman R 1992 *Astrophys.J.* submitted
   Meyer B S *et al* 1992 *Clemson University Preprint*
   Meyer B S 1992 *J.Phys.* **G** this volume
[11] Witti J, Janka H-Th, Takahashi K and Hillebrandt W 1992 *Proc. Int. Conf. "Nuclei in the Cosmos" II, Karlsruhe, 1992* to appear
   Takahashi K, Janka H-Th, Witti J and Hillebrandt W 1992 *Proc. 6th Int. Conf. on Nuclei far from Stability, Bernkastel-Kues, 1992* to appear
[12] Kratz K-L *et al* 1992 *Proc. 6th Int. Conf. on Nuclei far from Stability, Bernkastel-Kues, 1992* to appear
[13] Van Dyck R S Jr, Farnham D L, Bare J and Schwinberg P B 1992 *Proc. 9th Int. Conf. on Atomic Masses and Fundamnetal Constants, Bernkastel-Kues, 1992* to appear
   Bollen G *et al* 1992 *ibid*
   Wouters J M *et al* 1992 *ibid*
   Sharma K S *et al* 1992 *ibid*
[14] Maripuu S and Way K (special editors) 1976 *At. Data Nucl. Data Tables* **17** No.5-6 i
[15] Haustein P E (special editor) 1988 *At. Data Nucl. Data Tables* **39** 185
[16] Garvey *et al* 1969 *Rev. Mod. Phys.* **41** S1
[17] Jänecke J and Masson P J 1988 *in Ref.* [15] p. 265
[18] Meyers W D 1977 *Droplet Model of Atomic Nuclei* (IFI/Plenum Data Co., New York)
[19] Tachibana T, Uno M, Yamada M and Yamada S 1988 *in Ref.* [15] p. 251
[20] Hilf E R, von Groote H and Takahashi K *CERN Report 76-13* p. 142
[21] Möller P and Nix J R 1988 *in Ref.* [15] p. 213
[22] Liran S and Zeldes N 1976 *in Ref.* [14] p.431
[23] Tondeur F 1981 *CERN Report 81-09* p. 81
[24] Pearson J M *et al* 1991 *Nucl. Phys.* **A528** 1
[25] Gilbert A and Cameron A G W 1965 *Can. J. Phys.* **43** 1248
   Egidy T von, Behkami A N and Schmidt H H 1986 *Nucl. Phys.* **A454** 109
[26] Arnould M and Tondeur F 1981 *CERN Report 81-09* p. 229

[27] Hillebrandt W 1991 *Proc. Int. School of Physics "Enrico Fermi", Course 113, Varenna, 1989* p. 399

[28] Tubbs D L and Koonin S E 1979 *Astrophys. J.* **232** L59
El Eid M F and Hillebrandt W 1980 *Astron. Astrophys. Suppl.* **42** 215
Engelbrecht C A and Engelbrecht J D 1990 *Ann. Phys.* in press

[29] Bonche P and Vautherin D 1981 *Nucl.Phys.* **A372** 496

[30] Hillebrandt W and Wolff R G 1985 *Nucleosynthesis: Challenge and New Developments* ed. by W D Arnett and J W Truran (Univ. Chicago Press, Chicago) p. 131

[31] Hillebrandt W, Nomoto K and Wolff R G 1984 *Astron. Astrophys.* **133** 175

[32] Baron E and Cooperstein J 1990 *Astrophys. J.* **353** 597

[33] Shrama M M 1989 *The Nuclear Equation of State* eds. W Greiner and H Stöcker (Plenum Press, New York), p. 661

[34] Kahana S H 1989 *The Nuclear Equation of State* op. cit., p.721

[35] Gove N B and Martin M J 1971 *Nucl. Data Tables* **10** 205

[36] Goodman C D 1981 *Comm. Nucl. Part. Phys.* **10** 117

[37] Vetterli *et al* 1987 *Phys. Rev. Lett.* **59** 439

[38] Tarlé G, Ahlen S P and Cartwright B G 1979 *Astrophys. J.* **230** 607
Sur B *el al* 1989 *Phys. Rev.* **C39** 1511

[39] Muto K, Bender E and Oda T 1991 *Phys. Rev.* **C43** 1487

[40] Takahashi K, Yamada M and Kondoh T 1973 *At. Data Nucl. Data Tables* **12** 101

[41] Tachibana T, Yamada M and Yoshida Y 1990 *Prog. Theor. Phys.* **84** 641

[42] Eisenberg J M and Greiner W 1972 *Microscopic Theory of the Nucleus* (North Holland Pub., Amsterdam)

[43] Staudt A, Bender E, Muto K and Klapdor-Kleingrothaus H V 1990 *At. Data Nucl. Data Tables* **44** 79

[44] Möller P and Randrup J 1990 *Nucl. Phys.* **A514** 1

[45] Kratz K-L 1992 *private communication*

[46] Lauritzen B 1988 *Nucl. Phys.* **A489** 237

[47] Vogel P and Zirnbauer M R 1986 *Phys. Rev. Lett.* **57** 3148
Suhonen J, Taigel T and Faessler A 1988 *Nucl. Phys.* **A486** 91
Kuzmin V A and Soloviev V G 1988 *Nucl. Phys.* **A486** 118

[48] Zhao L and Sustich A 1992 *Ann. Phys.* **213** 378

[49] Midgal A B 1965 *Theory of Finite Fermi Systems and the Properties of Atomic Nuclei* (Nauka, Moscow; transl.: Interscience Pub., New York, 1967)

[50] Borzov I N, Fayans S A and Trykov E L 1990 *FEI Obninsk Preprint 2069*

[51] Lytostanskii Yu S and Panov I V 1988 *Sov. Astron. Lett.* **14** 70

[52] Käppeler, Beer H and Wisshak K 1989 *Rep. Prog. Phys.* **52** 945

[53] Takahashi K and Yokoi K 1987 *At. Data Nucl. Data Tables* **36** 375

[54] Takahashi K, Mathews G J and Bloom S D 1986 *Phys. Rev.* **C33** 296

[55] Daudel R, Benoist P, Jacques R and Jean M 1947 *C. R. Acad. Sci. (Paris)* **224** 1427

[56] Bahcall J N 1961 *Phys. Rev.* **124** 495
Pyper N C and Harston M R 1988 *Proc. R. Soc. Lond.* **A420** 277

[57] Takahashi K and Yokoi K 1983 *Nucl. Phys.* **A404** 578

[58] Takahashi K, Boyd R N, Mathews G J and Yokoi K 1987 *Phys. Rev.* **C36** 1522

[59] Yokoi K, Takahashi K and Arnould M 1983 *Astron. Astrophys.* **117** 65
Takahashi K and Baraffe I 1992 *Proc. 3rd. Int. Symp. on Weak and Electromagnetic Interactions in Nuclei, Dubna, 1992* to appear

[60] Yokoi K, Takahashi K and Arnould M 1985 *Astron. Astrophys.* **145** 339

[61] Jung M *et al* 1992 *Phys. Rev. Lett.* submitted

[62] Schiffer J P 1992 *private communication*

[63] Jonson B *et al* 1983 *Nucl. Phys.* **A396** 479c
Żylicz J, Hansen P G. Nielsen H L and Wilsky K 1966 *Arkiv. Fyz.* **36** 643

[64] Ogawa K 1992 *private communication*

[65] Fuller G M, Fowler W A and Newman M J 1982 *Astrophy. J. Suppl.* **48** 279

[66] Aufderheide M B 1991 *Nucl. Phys.* **A526** 161

[67] Kajino T *et al* 1988 *Nucl. Phys.* **A480** 175

[68] Takahara M *et al* 1989 *Nucl. Phys.* **A504** 167

Potential model and folding procedure: a new method of calculating astrophysical S-factors

H Krauss, K Grün, H Herndl, M. Muri and H Oberhummer†
R Zwiebel, H Abele, P Mohr and G Staudt‡
A Denker, W Hammer and G Wolf§

† Institut für Kernphysik, Technische Universität Vienna,
Wiedner Hauptstr. 8–10, A–1040 Vienna, Austria

‡ Physikalisches Institut, Universität Tübingen
Auf der Morgenstelle, W–7440 Tübingen, Germany

§ Institut für Strahlenphysik, Universität Stuttgart,
Allmandring 3, W–7000 Stuttgart, Germany

Abstract. The reactions ^{3}He$(\alpha,\gamma)^7$Be and ^{3}He$(^3$He,2p$)^4$He have been analyzed in the subCoulomb energy range using the potential model. In order to determine the optical folding potentials in the entrance channels and the bound–state potential for the ^{7}Be cluster states, elastic ^{3}He–^{4}He and ^{3}He–^{3}He scattering cross sections have been measured in the range 1–3 MeV. The absolute magnitude as well as the energy dependence of the astrophysical S–factor for both reactions calculated in the potential model shows excellent agreement with the experimental data.

1. Introduction

The potential model assumes that nuclear reactions proceed via the direct–reaction mechanism in a single–step process. For projectile energies well below the Coulomb and/or centrifugal barrier these models are often well suited to describe the experimental data for the astrophysical S–factors and to extrapolate this quantity to thermonuclear energies. A second important ingredient in our method is the folding procedure where the real parts of the optical and the bound–state potentials are determined by folding the density distributions of the involved nuclei with an effective nucleon–nucleon potential. A brief description of the potential model and the folding procedure is given in section 2. In section 3 we describe our study of elastic ^{3}He–^{4}He and ^{3}He–^{3}He scattering at energies below 3 MeV. From these experiments the optical folding-potentials are determined. In section 4 we calculate the astrophysical S–factors for the reactions ^{3}He$(\alpha,\gamma)^7$Be and ^{3}He$(^3$He,2p$)^4$He and compare them with the experimental data.

2. Potential Model and Folding Procedure

Potential models are based on the description of the dynamics of the reactions by a Schrödinger equation with local optical potentials in the entrance and/or exit channels. Such models are the "Distorted Wave Born Approximation" (DWBA) for transfer or the "Direct Capture Model" (DC) for radiative capture. The corresponding formulae for the cross sections in DWBA and DC can be found in [1] and references therein.

The most important ingredients in the potential models are the wave functions for the scattering and bound states in the entrance and exit channels. These are determined from potentials using the folding procedure. In this approach the nuclear densities taken from nuclear charge distributions [2] are folded with an energy and density dependent NN interaction v_{eff} [1, 3]:

$$V(R) = \lambda V_F(R) = \lambda \int \int \rho_a(\vec{r_1})\rho_A(\vec{r_2})v_{\mathrm{eff}}(E, \rho_a + \rho_A, \vec{s})d\vec{r_1}d\vec{r_2} \quad .$$

The variable $\vec{s}$ in the NN interaction term is given by

$$\vec{s} = \vec{R} + \vec{r_2} - \vec{r_1}$$

with $\vec{R}$ being the separation of the centers of mass of the two colliding nuclei. The normalization factor λ accounts for Pauli repulsion effects and dispersive parts in the potential $V(R)$ which are not included in the folding potential $V_F(R)$. This parameter can be adjusted to elastic scattering data and/or to bound and resonance state energies of nuclear cluster states. At the low energies considered in the nucleosynthesis often the imaginary term in the potential can be neglected. Therefore, in the potential model combined with the folding procedure for the potential the reaction cross sections can be calculated in many cases without any free parameter.

3. Elastic ^{3}He–^{4}He and ^{3}He–^{3}He Scattering

At incident energies $E_{\mathrm{Lab}}(^{3}\mathrm{He}) \approx 3 - 10\,\mathrm{MeV}$ differential cross sections for the elastic ^{3}He–^{4}He and ^{3}He–^{3}He scattering are well known [4, 5, 6, 7, 8]. In order to obtain cross section data at very low energies we have measured these scattering processes at 20 energies in the range from 1 to $3.3\,\mathrm{MeV}$ using the windowless, differential pumped and recirculating gas target system RHINOCEROS installed at the Stuttgart Dynamitron accelerator. In our experiment we used the jet configuration in which a supersonic jet produced by a laval nozzle serves as a nearly pointlike target zone with high density. The small and fixed size of this zone allows a good determination of angular distributions.

The target zone is centered in a 50 cm–diameter scattering chamber. The detector system consisted of ten surface-barrier detectors mounted at fixed positions. ^{3}He–beam intensities between 5 and $30\,\mu\mathrm{A}$ were used. In order to normalize the data (assuming that the ^{3}He–^{20}Ne scattering is dominated in this energy range by Rutherford interaction), a small quantity of ^{20}Ne gas was admixed to the respective ^{4}He and ^{3}He gas in the jet.

The differential cross sections were analyzed together with the known results at higher energies in terms of the optical model using double–folded potentials as mentioned above. The fit to the ^{3}He–^{4}He cross section data and to the energies of the $L = 1$ bound state and the $L = 3$ resonance-state doublet of the nucleus $^{7}\mathrm{Be} = {}^{4}\mathrm{He} \otimes {}^{3}\mathrm{He}$ results in a strong parity dependence of the potential: $\lambda_{\mathrm{odd}} = 1.85$; $\lambda_{\mathrm{even}} = 1.45$. Since the $L = 3$ doublet states mix slightly with higher non–cluster states, a somewhat deeper LS–potential is obtained for the $L = 3$ partial waves than for the $L = 1$ and $L = 2$ ones. Good agreement is found between the experimental [5] and calculated elastic phase shifts up to $E_{\mathrm{Lab}}(^3\mathrm{He}) \approx 12\,\mathrm{MeV}$.

The fit to the $^3\mathrm{He}$–$^4\mathrm{He}$ cross section data results in a potential for the spin–singulet partial waves (even L) with $\lambda_s = 1.802$ and in a potential for the spin–triplet waves (odd L) with $\lambda_t = 1.400$. The potentials describe very well the angular distributions and the elastic phase shifts calculated in the RGM model [9] up to an energy $E_{\mathrm{Lab}}(^3\mathrm{He}) \approx 30\,\mathrm{MeV}$.

4. Results for Reactions of the pp–Chain

The reactions $^3\mathrm{He}(^3\mathrm{He},2p)^4\mathrm{He}$, $^3\mathrm{He}(\alpha,\gamma)^7\mathrm{Be}$ and $^7\mathrm{Be}(p,\gamma)^8\mathrm{B}$ determine the branching ratios between the ppI, ppII and ppIII chains in hydrogen burning of main–sequence stars. The magnitude of their reaction cross sections is of special interest for the solar neutrino problem. The reason for this is that in the $^{37}\mathrm{Cl}$ experiment 14% and 77%, in the KamiokandeII experiment 0% and 100% and in the gallium experiment 26% and 11% of the detected neutrino flux stem from the high–energy neutrinos emitted in the ppII and ppIII chain, respectively [10]. Therefore, the reaction rates of the above nuclear processes determine the high–energy neutrino flux. In this work we investigate the first two reactions; the results for the last reaction $^7\mathrm{Be}(p,\gamma)^8\mathrm{B}$ will be published elsewhere [11].

$^3He(\alpha,\gamma)^7Be$

The reaction $^3\mathrm{He}(\alpha,\gamma)^7\mathrm{Be}$ was analyzed up to now in the DC model using a hard sphere potential [12] or by using more or less phenomenological parametrizations of the potentials [13, 14]. Furthermore many calculations exist within the resonating group theory [15, 16, 17, 18, 19, 20]. Our numerical calculations in the DI model were performed using the direct–capture code TEDCA [21]. The optical potential in the entrance channel as well as the potential for the bound state $^7\mathrm{Be} = {}^4\mathrm{He} \otimes {}^3\mathrm{He}$ was obtained as discussed in section 3. The spectroscopic factors were obtained from shell–model calculations giving $C^2S = 1.174$ (ground state) and $C^2S = 1.175$ (first excited state) [22].

The results of the DC calculation for the sum of the astrophysical S–factors for the transitions to the ground and first excited state of $^7\mathrm{Be}$ are compared with the experimental data in fig. 1. As can be seen from the figure there is excellent agreement with the experimental data. For low energies the S–factor can be parametrized as
$$S(E) = (0.520 - 3.61 \cdot 10^{-4}E)\,\mathrm{keV\,b}\ (E\ \mathrm{in\ MeV}).$$
The values extrapolated from original experimental papers including their 3σ errors give [10]
$$S(E) = (0.54 \pm 0.03 - 3.1 \cdot 10^{-4}E)\,\mathrm{keV\,b}\ (E\ \mathrm{in\ MeV}).$$
The astrophysical S–factor at the effective mean energy E_0 for the temperature of the sun $T_6 = 15$ is given in our model by $S(E_0 = 22.4\,\mathrm{keV}) = 0.512\,\mathrm{keV\,b}$. This value lies within the experimental error of the experimental S–factor $S(E_0 = 22.4\,\mathrm{keV}) = 0.53 \pm 0.03\,\mathrm{keV\,b}$ obtained from the above cited data.

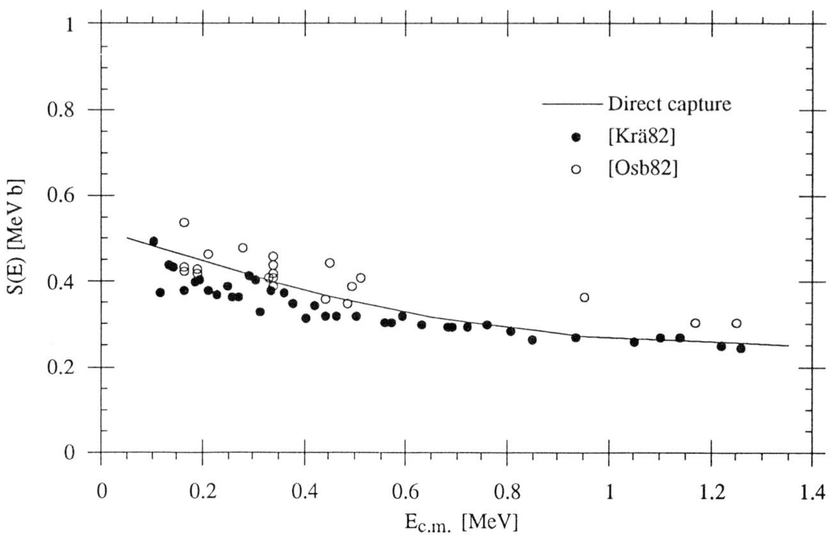

Figure 1. Astrophysical S–factor for the reaction ^{3}He$(\alpha,\gamma)^7$Be up to 1.3 MeV. The solid line is the result of the DC calculation. The experimental data are taken from [23].

$^3He(^3He,2p)^4He$

The reaction ^{3}He$(^3$He,2p$)^4$He was analyzed up to now using the R–matrix [24] or the resonating group theory [25]. Our numerical calculations in the DI model were performed using the finite–range DWBA code FREDICA [26]. The optical potential in the entrance channel was obtained as discussed in section 3. Since for the exit channel obviously there exists no density distribution of the diproton system we chose a Fourier–Bessel charge–distribution with a cut–off radius of 5 fm. The normalization constant $\lambda = 0.499$ was adjusted to reproduce the experimental reaction cross section at 150 keV. For the bound states ^{4}He=^{3}He+n and ^{3}He=2p+n we also used the folding procedure. In this case the depth of the folding potential λ was adjusted to reproduce the correct separation energies of $E = 20.578$ MeV and $E = 7.7184$ MeV for the first and second bound state, respectively. The spectroscopic factors for the first and second bound state are given by $C^2\mathcal{S} = 2$ and $C^2\mathcal{S} = 0.6667$.

The result of the DWBA calculation in finite range for the astrophysical S–factor is compared to the experimental data in fig. 2. As can be seen from the figure, there is excellent agreement with the experimental data. For low energies the S–factor can be parametrized as
$S(E) = (4.854 - 1.328E + 0.123E^2)\,\mathrm{MeV\,b}\ (E\ \mathrm{in\ MeV}).$
The values extrapolated from orginal experimental papers including their 3σ errors give [10]
$S(E) = 5.15 \pm 0.88 - 0.9E\ \mathrm{MeV\,b}\ (E\ \mathrm{in\ MeV}).$
The astrophysical S–factor at the effective mean energy E_0 for the temperature of the sun

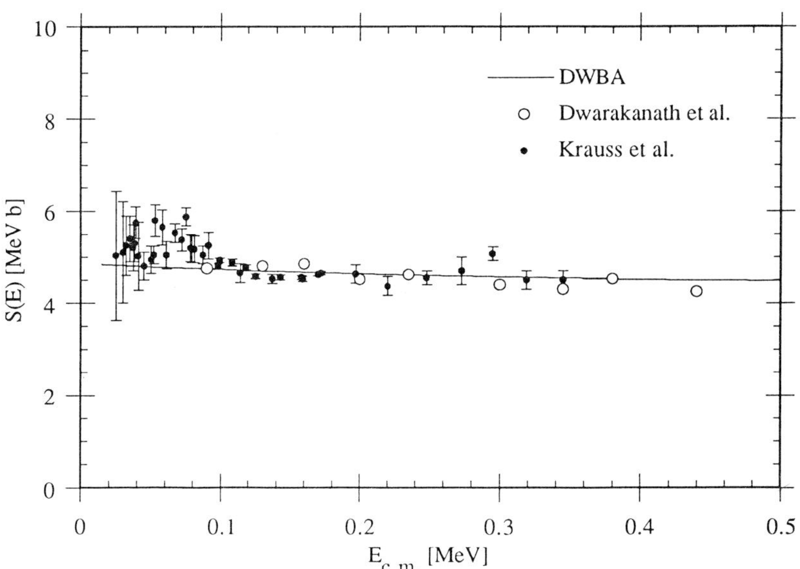

Figure 2. Astrophysical S-factor for the reaction $^3\mathrm{He}(^3\mathrm{He},2p)^4\mathrm{He}$ up to 0.5 MeV. The solid line is the result of the DWBA calculation. The experimental data are taken from [27, 28].

$T_6 = 15$ is given in our model by $S(E = 21.4\,\mathrm{keV}) = 4.63\,\mathrm{MeV\,b}$. This value lies within the experimental error of the S-factor $S(E = 21.4\,\mathrm{keV}) = 5.13 \pm 0.88\,\mathrm{MeV\,b}$ obtained from the above cited data.

5. Summary

The reactions $^3\mathrm{He}(\alpha,\gamma)^7\mathrm{Be}$ and $^3\mathrm{He}(^3\mathrm{He},2p)^4\mathrm{He}$ have been analyzed in terms of simple potential models, the direct capture model and the DWBA. In order to determine the optical potentials in the entrance channels and the bound state potential for the $^7\mathrm{Be}$ cluster states, elastic $^3\mathrm{He}$–$^4\mathrm{He}$ and $^3\mathrm{He}$–$^3\mathrm{He}$ scattering cross sections have been measured in the range 1–3 MeV. The shape of these potentials is determined by double–folding calculations. The diproton–$^4\mathrm{He}$ potential is adjusted to the experimental reaction cross section at E=150 keV. The absolute magnitude as well as the energy dependence of the astrophysical S–factor for both reactions calculated in the potential model shows excellent agreement with the experimental data. A similar result was obtained for the reaction $^7\mathrm{Be}(\mathrm{p},\gamma)^8\mathrm{B}$ [11].

Acknowledgements

We want to thank the Österreichische Nationalbank (project 3924), the FWF (project P7838-TEC) and the DFG (project Sta290/2-1) for their support.

References

[1] Oberhummer H and Staudt G 1991 *Nuclei in the Cosmos* ed H Oberhummer (Heidelberg: Springer Verlag) p 29

[2] de Vries H, de Jager C W and de Vries C 1987 *At. Data and Nucl. Data Tables* **36** 495

[3] Kobos A M, Brown B A, Lindsay R and Satchler R 1984 *Nucl. Phys.* *A* **425** 205

[4] Barnard A C L, Jones C M and Phillips G C 1964 *Nucl. Phys.* **50** 629

[5] Spiger R J and Tombrello T A 1967 *Phys. Rev.* **63** 964

[6] Chuang L S 1971 *Nucl. Phys.* *A* **174** 399

[7] Boykin W R, Baker S D and Hardy D M 1972 *Nucl. Phys.* *A* **195** 241

[8] Tombrello T A and Bacher A D 1963 *Phys. Rev.* **130** 1108

[9] Thompson D R and Tang Y C 1967 *Phys. Rev.* **159** 806

[10] Bahcall J N 1989 *Neutrino Astrophysics* (Cambridge: Cambridge University Press)

[11] Krauss H, Grün K, Rauscher T and Oberhummer H 1992 *Ann. d. Phys.* submitted

[12] Tombrello T A and Parker P D 1963 *Phys. Rev.* **131** 2582

[13] Kim B T, Izumoto T and Nagatani K 1983 *Phys. Rev.* *C* **23** 1

[14] Buck B, Baldock R A and Rubio J A 1985 *Journ. Phys.* *G* **11** L11

[15] Walliser H, Kanada H and Tang Y C 1984 *Nucl. Phys.* *A* **419** 133

[16] Kajino T and Arima A 1984 *Phys. Rev. Lett.* 739

[17] Langanke K 1986 *Nucl. Phys.* *A* **457** 351

[18] Mertelmeier T and Hofmann H M 1986 *Nucl. Phys.* *A* **459** 387

[19] Kajino T 1986 *Nucl. Phys.* *A* **460** 559

[20] Liu Q K K, Kanada H and Tang Y C 1986 *Phys. Rev.* *C* **33** 1561

[21] Krauss H 1990 *code TEDCA* not published

[22] Cohen D and Millener D J 1975 *Nucl. Phys.* *A* **238** 269

[23] Osborne J L, Barnes C A, Kavanagh R W, Kremer R M, Mathews G J, Zyskind J L, Parker P D and Howard A J 1984 *Nucl. Phys.* *A* **419** 115

[24] Jarmie N 1981 *Nucl. Sci. Eng.* **78** 404

[25] Typel S, Blüge G, Langangke K and Fowler W A 1991 *Z. Phys.* *A* **339** 249

[26] Grün K 1992 *code FREDICA* not published

[27] Dwarakanath R M and Winkler H 1971 *Phys. Rev.* *C* **4** 1532

[28] Krauss A, Becker H W, Trautvettter H P and Rolfs C 1987 *Nucl. Phys.* *A* **467** 273

Interpretation of rapidly rotating pulsars

F Weber Institute for Theoretical Physics, University of Munich, Theresien-str. 37/III, W-8000 Munich 2, FRG

N K Glendenning Nuclear Science Division, Lawrence Berkeley Laboratory, University of California, Berkeley, California 94720, USA

Abstract. The minimum possible rotational period of pulsars, which are interpreted as rotating neutron stars, is determined by applying a representative collection of realistic nuclear equations of state. It is found that none of the selected equations of state allows for neutron star rotation at periods below $0.8 - 0.9$ ms. Thus, this work strongly supports the suggestion that if pulsars with shorter rotational periods were found, these are likely to be strange-quark-matter stars. The conclusion that the confined hadronic phase of nucleons and nuclei is only metastable would then be almost inescapable, and the plausible ground-state in that event is the deconfined phase of (3-flavor) strange-quark-matter.

1 Introduction

The hypothesis that strange quark matter may be the absolute ground state of the strong interaction (not ^{54}Fe) has been raised by Witten in 1984 [1]. If the hypothesis is true, then a separate class of compact stars could exist, which are called strange stars. They form a distinct and disconnected branch of compact stars and are not part of the continuum of equilibrium configurations that include white dwarfs and neutron stars. In principle both strange and neutron stars could exist. However if strange stars exist, the galaxy is likely to be contaminated by strange quark nuggets which would convert all neutron stars that they come into contact with to strange stars [2, 3, 4]. This in turn means that the objects known to astronomers as pulsars are probably rotating *strange matter* stars, not neutron matter stars as is usually assumed. Unfortunately the bulk properties of models of neutron and strange stars of masses that are typical for neutron stars, $1.1 \lesssim M/M_\odot \lesssim 1.8$, are relatively similar and therefore do not allow to distinguish between the two possible pictures. The situation changes however as to the ability of fast rotation of strange stars, which has its origin in the completely different mass-radius relations of neutron and strange stars [5]. As a consequence of this the entire familiy of strange stars can rotate rapidly, not just those near the limit of gravitational collapse to a black hole as is the case for neutron stars. It is the concern of this paper to determine the minimum rotational period of a neutron star. possessing a mass of 1.45 $M_\odot$, below

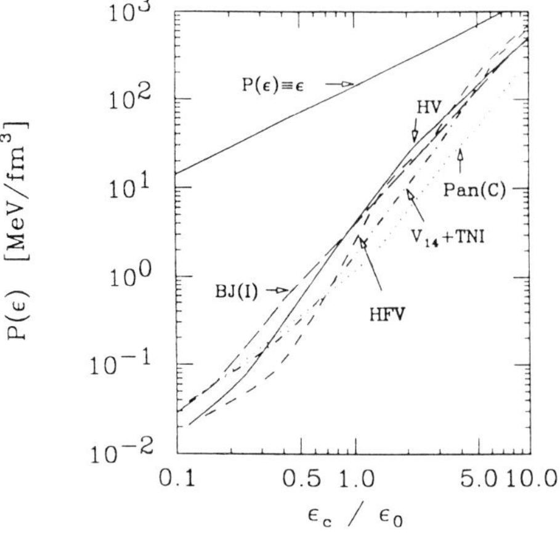

Figure 1: Graphical illustration of the equations of state BJ(I), Pan(C), FP(V_{14} + TNI), HV, and HFV.

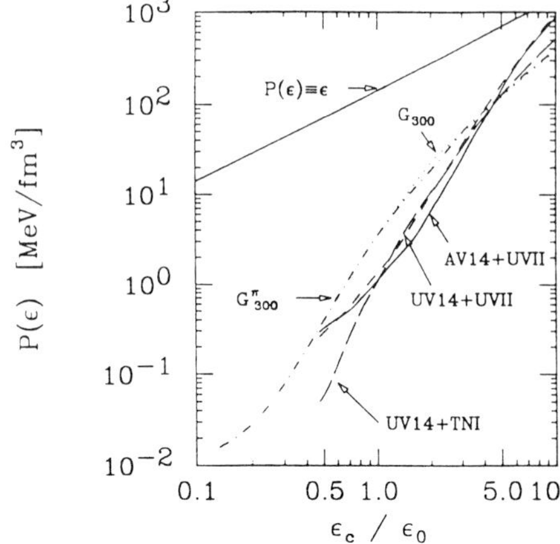

Figure 2: Same as Fig. 1, but for the equations of state WFF(AV_{14} + UVII), WFF(UV_{14} + UVII), WFF(UV_{14} + TNI), G_{300}, and G^{π}_{300}.

which it cannot rotate stably. Knowledge of that period is of decisive importance for the interpretation of the possible future detection of extremely rapidly rotating pulsars [2]. For this purpose general relativistic, rotating neutron star models are constructed by applying a representative collection of seventeen nuclear equations of state (Sec. 2), which serve as an input for solving the Einstein equations for a rotating object. For more details we refer to [6].

2   Collection of selected nuclear equations of state

The collection of nuclear equations of state applied for the construction of models of general relativistic rotating neutron stars is listed in Table 1. The equations of state are divided into two categories: (1) non-relativisitic potential model equations of state, and (2) relativistic equations of state which are determined in the framework of relativistic nuclear field theory (i.e., relativistic Hartree (entries 1 through 6, 8, 9), Hartree-Fock (entry 10), and T-matrix (entries 7, 11) approximations). An inherent feature of the latter is that they do not violate causality, i.e. the velocity of sound is smaller than the velocity of light at all densities, which is not the case for the potential models. Among the latter only the WFF(UV_{14} + TNI) equation of state does not violate causality up to densities relevant for the construction of models of neutron stars form it. The equations of state denoted G^{DCM1}_{225}, G^{DCM2}_{265}, G^{DCM1}_{B180}, and G^{DCM2}_{B180} have only recently been calculated [7] for electrically charge neutral neutron star matter in β-equilibrium for the derivative coupling Lagrangian of Zimanyi and Moszkowski [8]. The possibility of a phase transition of the dense core to 3-flavor quark matter is taken into account in equations of state G^{DCM1}_{B180} and G^{DCM2}_{B180}. Here a bag constant of $B^{1/4} = 180$ MeV has been used for the determination of the phase transition of baryon matter into quark matter, which places the energy per baryon of strange matter at 1100 MeV, well above the energy per nucleon in ^{56}Fe ($\approx$ 930 MeV). Not all equations of state of our collection account for neutron matter in β-equilibrium (entries 13 through 16). These models treat neutron star matter

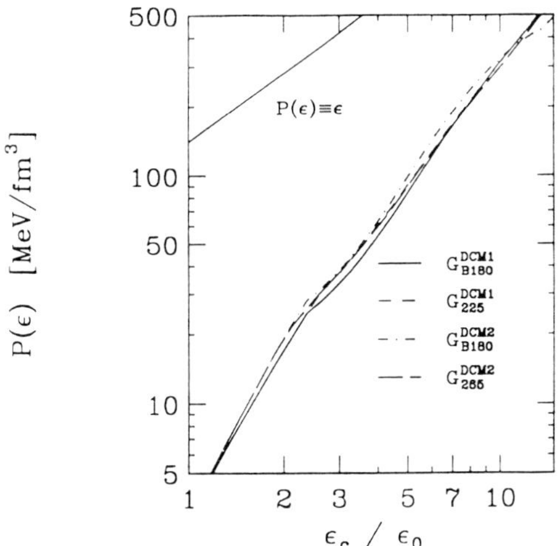

Figure 3: Same as Fig. 1, but for the equations of state G_{B180}^{DCM1}, G_{225}^{DCM1}, G_{B180}^{DCM2}, G_{265}^{DCM2}.

Table 1: Nuclear equations of state applied for the construction of models of general relativistic rotating neutron star models.

Label	EOS	Description[†]	Reference
	Relativistic field theoretical equations of state		
1	G_{300}	H, $K=300$	[9]
2	HV	H, $K=285$	[10, 11]
3	G_{B180}^{DCM2}	Q, $K=265$, $B^{1/4}=180$	[7]
4	G_{265}^{DCM2}	H, $K=265$	[7]
5	G_{300}^{π}	H, π, $K=300$	[9]
6	G_{200}^{π}	H, π, $K=200$	[12]
7	Λ_{Bonn}^{00} + HV	H, $K=186$	[13]
8	G_{225}^{DCM1}	H, $K=225$	[7]
9	G_{B180}^{DCM1}	Q, $K=225$, $B^{1/4}=180$	[7]
10	HFV	H, Δ, $K=376$	[11]
11	Λ_{HEA}^{00} + HFV	H, Δ, $K=115$	[13]
	Non-relativistic potential model equations of state		
12	BJ(I)	H, Δ	[14]
13	WFF(UV$_{14}$+TNI)	NP, $K=261$	[15]
14	FP(V$_{14}$+TNI)	N, $K=240$	[16]
15	WFF(UV$_{14}$+UVII)	NP, $K=202$	[15]
16	WFF(AV$_{14}$+UVII)	NP, $K=209$	[15]
17	Pan(C)	H, Δ, $K=60$	[17]

[†] The following abbreviations are used: N = pure neutron; NP = n, p, leptons; π = pion condensation; H = composed of n, p, hyperons ($\Sigma^{\pm,0}, \Lambda, \Xi^{0,-}$), leptons; $\Delta = \Delta_{1232}$-resonance; Q = quark hybrid composition, i.e. n, p, hyperons, u, d, s-quarks, leptons; K = incompressibility in MeV; $B^{1/4}$ = bag constant in MeV.

as being composed of only neutrons (label 14), or neutrons and protons in equilibrium with leptons (labels 13, 15, 16), which is however *not the true ground-state of neutron star matter predicted by theory* [10, 14, 17].

The equations of state, i.e. pressure P as a function of energy density ϵ (in units of normal nuclear matter density, $\epsilon_0 = 140$ MeV/fm^3), are graphically exhibited in Figs. 1 - 3. From Fig. 1 one sees the extremely soft behavior of the non-relativistic Pan(C) equation of state. Specifically, it is considerably softer than the other two non-relativistic equations of state shown in this figure, i.e. BJ(I) and FP(V_{14} + TNI). The BJ(I) and Pan(C) models account for baryon population in neutron star matter which leads to a slight flattening of the pressure curves at densities larger than respectively two and four times normal nuclear matter density. Two relativistic equations of state, HV and HFV, are shown too for the purpose of comparison. For $\epsilon \stackrel{>}{\sim} 3\epsilon_0$, the Hartree-Fock HFV equation of state behaves stiffer than the Hartree HV equation of state. The reason for this lies in the exchange contribution that is contained in the former equation of state. The kinks contained in the $P(\epsilon)$ curves of HFV and HV at densities of respectively $\epsilon \approx 1.6\,\epsilon_0$ and $\epsilon \approx 2\epsilon_0$ are caused by the onset of hyperon population. Figure 2 compares the non-relativistic equations of state of Wiringa, Fiks, and Fabrocini (WFF) with two relativistic ones. We recall that only the latter two describe neutron star matter composed of baryons in β-equilibrium with leptons (Table 1). Equation of state G^{π}_{300} additionally to baryon population also accounts for pion condensation in neutron star matter. According to this equation of state, condensation is predicted to set in at $\epsilon \approx 1.5\,\epsilon_0$ (dash-dotted curve in Fig. 2). At densities $\epsilon \stackrel{>}{\sim} 4\,\epsilon_0$ the WFF equations of state behave stiffer than the relativistic ones, violating causality at densities that are smaller than twice that value. The WFF(AV_{14} + UVII) and WFF(UV_{14} + UVII) equations of state are rather similar at sub-nuclear densities, which is not the case for the third WFF model (WFF(UV_{14} + TNI)) because of the different three-body-force model (TNI) in the latter case. The equations of state G^{DCM1}_{225}, G^{DCM2}_{265}, G^{DCM1}_{B180}, and G^{DCM2}_{B180} are graphically depicted in Fig. 3. The transition of confined hadronic matter into quark matter, which is taken into account in G^{DCM1}_{B180} and G^{DCM2}_{B180}, sets in at densities $\epsilon \stackrel{>}{\sim} 2.3\,\epsilon_0$. It lowers the matter's pressure relative to the confined phase. The mixed phase of hadrons and quarks ends (and the pure quark phase begins) at $\epsilon \approx 15\,\epsilon_0$, which is larger than the maximum density encountered in the cores of star models constructed from these equations of state.

3   Bounds on the properties of fast pulsars

In the following we present the results obtained for the bulk properties of a fast pulsar model that rotates at (1) its general relativistic Kepler period, and (2) that period at which the gravitational radiation reaction-driven instability sets in [13]. The latter sets a more stringent limit on stable rotation. According to Table 2, any observed, newly born pulsar created in a supernova and possessing a mass of typically 1.45 $M_\odot$ that rotates at a period below ≈ 1 ms would be in contradiction to our analysis. This can bee seen from Table 2 where the minimum possible rotational periods of hot (temperature $T = 10^{10}$ K) and cold ($T = 10^6$ K) pulsars are listed. (The dependence on temperature arises due to the viscosity dependence of the gravitational radiation reaction-driven instability [13].) Consequently, newly born pulsars observed in supernova explosions are predicted to have stable rotational periods $\stackrel{>}{\sim} 1$ ms as long as their masses are close to the above cited value, which is supported by supernova calculations [18]. As one sees, half-millisecond periods, for example. are completely excluded for pulsars made of baryon matter. Therefore, the

Table 2: Lower and upper bounds on the properties of a pulsar of $M \approx 1.45\,M_\odot$, calculated for the collection of equations of state of Table 1 (except for Pan(C)). The listed properties are: period at which the gravitational radiation-reaction instability sets in, $P(T_x)$ (in ms; T_x refers to the star's temperature in units of 10^x K; Kepler period, $P_{\rm K}$ (in ms); central density, ϵ_c (in units of the density of normal nuclear matter); moment of inertia, I (in g cm^2); redshifts of photons emitted at the star's equator in backward ($z_{\rm B}$) and forward ($z_{\rm F}$) direction.

	$P(T_6 = 1)$	$P(T_{10} = 1)$	$P_{\rm K}$	ϵ_c/ϵ_0	$\log I$	$z_{\rm B}$	$z_{\rm F}$	$z_{\rm p}$
upper bound	1.1	1.5	1	5	45.19	1.05	-0.18	0.45
lower bound	0.8	1.1	0.7	2	44.95	0.59	-0.21	0.23

possible future discovery of a single sub-millisecond pulsar, say 0.5 ms, would give a strong hint that such an object is a rotating strange star, not a neutron star, and that 3-flavor strange quark matter is the true ground-state of the strong interaction [2].

The upper and lower bounds on the minimum possible rotational period of an old neutron star are shifted, relative to the minimum period of a hot pulsar, toward smaller values. We find that the gravitational instability is damped by viscosity as long as the star's rotational period is larger than $0.8 \lesssim P/{\rm ms} \lesssim 1.1$, depending on the equation of state. Specifically an old pulsar of $T = 10^6$ K and mass $M \approx 1.45\,M_\odot$ cannot be spun up to stable rotational periods smaller than ≈ 0.8 ms. We note that the two fastest yet observed pulsars, rotating at 1.6 ms, are compatible with the periods in Table 2, provided their masses are larger than 1 $M_\odot$ [6]. The Kepler period, beyond which mass shedding at the star's equator sets in, sets an absolute lower limit on stable rotation. It might play a role in an old and cold pulsar whose rotation is stabilized by its large viscosity value.

4 Summary

The indication of this work is that the gravitational radiation-reaction instability sets a lower limit on stable rotation of a little more than $P \approx 1$ *ms for young and hot, and* $P \approx 0.8$ *ms for old and cold pulsars* having a typical mass of $M \approx 1.45\,M_\odot$. This has possibly very important implications for the nature of any pulsar that is found to have a shorter period, say below $P \lesssim 0.5$ ms. If pulsars with periods below that value were found, the conclusion that the confined hadronic phase of nucleons and nuclei is only *metastable* would be almost inescapable. The plausible ground-state state in that event is the deconfined phase of (3-flavor) strange quark matter. From the QCD energy scale this is as likely a ground-state as the confined phase. The possibility that the ground-state of baryonic matter at zero pressure is strange quark matter and that ordinary nuclei may only be metastable has important consequences for laboratory nuclear physics, the early universe, and astrophysical compact objects.

Acknowledgement: This work was supported by a grant of the Deutsche Forschungs-gemeinschaft.

References

[1] E. Witten, Phys. Rev. D **30** (1984) 272.

[2] N. K. Glendenning, Mod. Phys. Lett. **A5** (1990) 2197.

[3] J. Madsen and M. L. Olesen, Phys. Rev. D **43** (1991) 1069, ibid., **44**, 4150 (erratum).

[4] R. R. Caldwell and J. L. Friedman, Phys. Lett. **264B** (1991) 143.

[5] N. K. Glendenning, *Supernovae, Compact Stars and Nuclear Physics*, invited paper in Proc. of 1989 Int. Nucl. Phys. Conf., Sao Paulo, Brasil, Vol. 2, ed. by M. S. Hussein et al., World Scientific, Singapore, 1990.

[6] F. Weber and N. K. Glendenning, *Impact of the Nuclear Equation of State on Models of Rotating Neutron Stars*, Proc. of the Int. Workshop on Unstable Nuclei in Astrophysics, Tokyo, Japan, June 7-8, 1991, Eds. S. Kubono and T. Kajino, World Scientific, 1992.

[7] N. K. Glendenning, F. Weber, and S. A. Moszkowski, Phys. Rev. C **45** (1992) 844, (LBL-30296).

[8] J. Zimanyi and S. A. Moszkowski, Phys. Rev. C **42** (1990) 1416.

[9] N. K. Glendenning, Nucl. Phys. **A493** (1989) 521.

[10] N. K. Glendenning, Astrophys. J. **293** (1985) 470.

[11] F. Weber and M. K. Weigel, Nucl. Phys. **A505** (1989) 779, and references contained therein.

[12] N. K. Glendenning, Phys. Rev. Lett. **57** (1986) 1120.

[13] F. Weber, N. K. Glendenning, and M. K. Weigel, Astrophys. J. **373** (1991) 579.

[14] H. A. Bethe and M. Johnson, Nucl. Phys. **A230** (1974) 1.

[15] R. B. Wiringa, V. Fiks, and A. Fabrocini, Phys. Rev. C **38** (1988) 1010.

[16] B. Friedman and V. R. Pandharipande, Nucl. Phys. **A361** (1981) 502.

[17] V. R. Pandharipande, Nucl. Phys. **A178** (1971) 123.

[18] E. Müller, J. Phys. G, **16** (1990) 1571.

Extrapability of nuclear mass models

P Möller[1], J R Nix[1] and K-L Kratz[2]

[1]Theoretical Division, LANL, Los Alamos, NM 87545, USA
[2]Institut für Kernchemie, Universität Mainz, Federal Republic of Germany

1 Introduction

As one means of comparing different mass models, we investigate the results of various approaches when applied to new regions of nuclei that were not considered when the theory was formulated or its parameters determined. We also compare the ability of these approaches to provide new physical insight.

Global nuclear mass calculations that include a quantal treatment of the nuclear interaction are usually based on the macroscopic-microscopic method, with the microscopic correction determined from calculated single-particle levels by use of Strutinsky's method [1,2]. Reviews of early work may be found in refs. [3–5]. Commonly used potentials are the folded-Yukawa [6–8], Woods-Saxon [9], modified-oscillator [10] and two-center oscillator [11] single-particle potentials. In some nuclear mass calculations the microscopic terms are given by analytical expressions with adjustable parameters. As discussed later, these approaches, along with other non-microscopic nuclear mass models, diverge severely when applied to new regions of nuclei. They also suffer from the further limitation that the only nuclear-structure quantity they predict is the ground-state nuclear mass.

Over the years a large number of nuclear mass models have been developed and numerous nuclear mass calculations have been undertaken, as witnessed by the AMCO series of conferences and the issues of Atomic Data and Nuclear Data Tables being devoted to these items. Our own work on nuclear mass models has now resulted in a preferred formulation based on the folded-Yukawa single-particle potential and the finite-range droplet model. This model has its origin in a 1981 nuclear mass model [8] which utilized the folded-Yukawa single-particle potential developed in 1972 [6,12].

There were two major unresolved issues in our earlier calculations related to the pairing calculations and higher multipole effects. Extensive investigations of pairing models and their parameters have now been completed and resulted in an improved formulation of the pairing model [13]. We have now also minimized the potential energy with respect to ϵ_3 and ϵ_6 shape degrees of freedom. An overview of the results of this calculation has been given in a paper on Coulomb redistribution effects [14]. The theoretical error is 0.778 MeV for the region of nuclei from ^{16}O and above but much smaller for the heavier region. For example, in the region above $N = 65$ the theoretical error is only 0.465 MeV. This latest nuclear mass calculation, with the finite-range droplet model as the macroscopic model, is designated FRDM. Calculations based on the finite-range liquid-drop model are designated FRLDM.

Figure 1 compares our most recent results to those calculated with the ETFSI-1 model [15,16]. It is the only other recent global mass calculation based on a quantal treatment of the nucleon interaction that we are aware of. In the graph we show the difference between measured masses and calculated masses for the two models. For the FRDM we have limited the plot to $A \geq 36$, which is the region considered in the ETFSI-1 calculation.

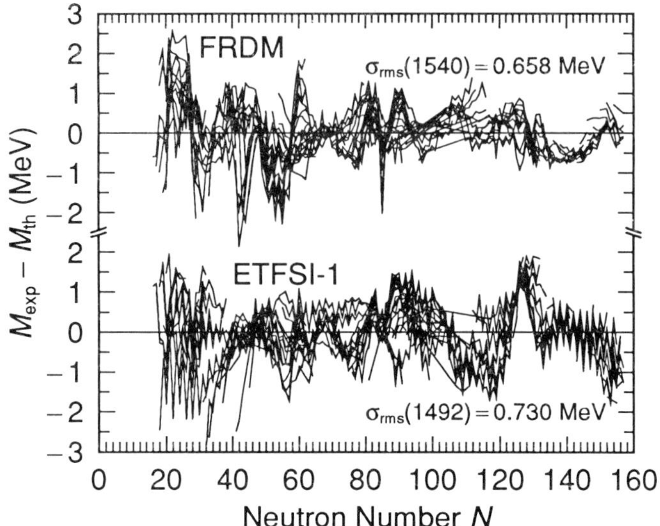

Figure 1: Comparison of discrepancies between measured and calculated masses for two models. The ETFSI-1 model shows a strong odd-even staggering, indicating a problem in the pairing model. the FRDM gives better agreement with measured masses, especially in the heavy region.

2   Extrapability of nuclear mass models

One test of the reliability of a nuclear mass model is to compare deviations between measured and calculated masses in new regions of nuclei that were not considered when the model parameters were determined to deviations in the original region. This type of analysis was used earlier by Haustein [17]. However, we here considerably modify his analysis. In addition to examining the raw differences between measured and calculated masses, we use these differences to determine the *model* mean discrepancy μ_{th} from the true masses and the *model* standard deviation σ_{th} around this mean.

Mass models based on postulated shell-correction terms and a fairly large number of parameters normally diverge outside the region where the parameters were determined. As an example of such behavior, we show in fig. 2 the error of the von Groote et al. [18] mass calculation in a region of newly measured masses. The square root of the second central moment is 1.15 MeV, or 72% larger than the 0.67 MeV rms error in the region where the parameters were adjusted. From this figure one also gets the intuitive feeling that the mean in the error grows with increasing distance from β-stability.

To study more quantitatively how the error depends upon distance from β-stability, we introduce bins in the error plot sufficiently wide to contain about 10–20 points and calculate the mean error and standard deviation about the mean for each of these bins. The results for the model shown in fig. 2 and for six other models are displayed in figs. 3 and 4. For each model the central, light-gray band representing the original error region extends one standard deviation on each side of zero. The solid dots connected by a thick black line represent the mean of the error for nuclei that were not considered when the model parameters were determined. The dark-gray area extends one standard deviation on each side of this line. The properties of the seven models displayed in figs. 3 and 4, as well as those of a recent neural-network calculation [19], are summarized in table 1.

It is of interest to note that for the three models that are based on a quantal treatment of the nuclear interaction, namely the three models in the lower part of fig. 4, *only one* of the points representing the mean deviation falls outside the original error region. Also, the full error of the

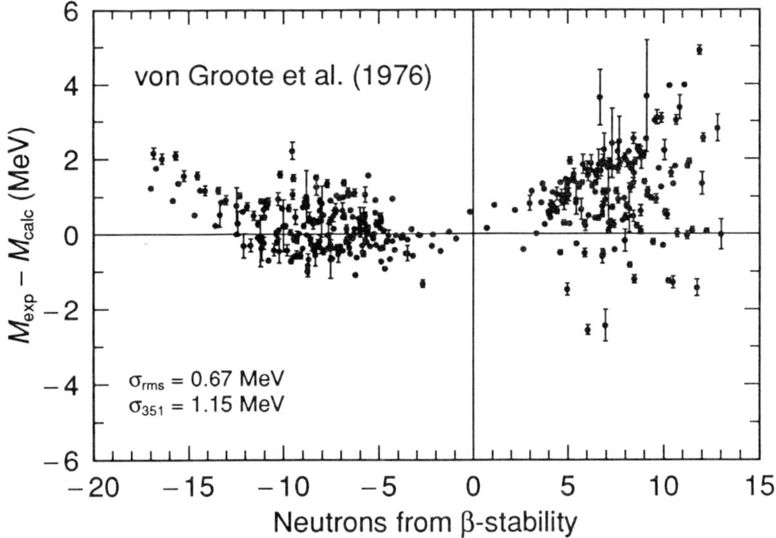

Figure 2: Study of the deviation between newly measured masses and calculated masses.

three models for new nuclei usually falls inside the error region corresponding to the original data set. Moreover, there are no systematic increases of the error with increasing distance from β-stability, with the possible exception of the Seeger-Howard model on the neutron-rich side. This model is based on a Nilsson modified-oscillator single-particle potential. The spin-orbit and pseudo-diffuseness parameters of this potential vary rather dramatically over the periodic system, in contrast to the behavior of these parameters in the folded-Yukawa single-particle potential used in the FRLDM and FRDM calculations. Therefore, even though they are based on a quantal treatment of the nuclear interaction, the Seeger-Howard results may be less reliable

Table 1: Comparison of errors of different mass calculations. The errors are tabulated both for the region in which the parameters were originally adjusted and for a set of new nuclei that were not taken into account in the determination of the parameters of the mass models. The error ratio is the ratio between the numbers in columns 9 and 3, except for the last line, where column 6 is used instead of column 9. It would have been preferable to use the error in column 4 instead of that in column 3, since σ_{th} does not contain contributions from the experimental errors. However, as can be seen in the table, the difference between the rms error and σ_{th} is small in the original region, where masses can be measured with smaller experimental errors than is possible far from stability. The quantity $\sigma_{\mathrm{th};\mu=0}$ is similar to a root-mean-square error but without contributions from experimental errors.

| Model | N_{par} | Original nuclei | | New nuclei | | | | | Error |
		σ_{rms} (MeV)	σ_{th} (MeV)	N_{nuc}	σ_{rms} (MeV)	μ_{th} (MeV)	σ_{th} (MeV)	$\sigma_{\mathrm{th};\mu=0}$ (MeV)	ratio
J. (G.-K.)	~ 500	0.118		337	1.461	-0.278	1.428	1.455	12.33
v. G. et al.	~ 50	0.67		351	1.193	0.612	0.978	1.154	1.72
H. et al.	~ 50	0.66		351	1.271	0.519	1.124	1.237	1.87
L.-Z.	178	0.276		346	0.912	-0.044	0.736	0.738	2.67
S.-H.	9	0.704		309	0.976	0.289	0.910	0.956	1.36
FRLDM	9	0.835	0.831	351	0.911	-0.321	0.826	0.884	1.06
FRDM	11	0.768	0.765	351	0.884	-0.049	0.847	0.848	1.10
Neural net	421	0.828		351	5.981				7.22

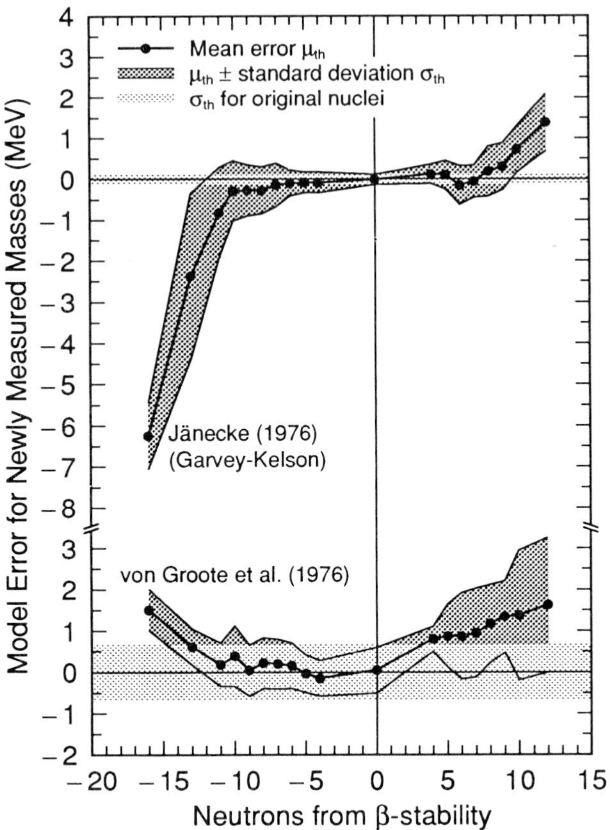

Figure 3: Comparison of error behavior for two models applied to new nuclei versus distance from stability. Compare this figure to fig. 4, which is plotted to the same scale.

for new regions of nuclei than calculations based on folded-Yukawa or Woods-Saxon potentials. In summary, we feel that at least the FRLDM and FRDM show substantial promise of being reliable as the proton and neutron drip lines are approached.

In contrast, it is clear that the four remaining models that are not based on a quantal treatment of the nuclear interaction quickly diverge when applied to nuclei outside the region where their parameters were originally adjusted. One can expect that they would become even more unreliable when applied even further from stability.

The fermion dynamical symmetry model (FDSM) [20], a truncated shell model, has so far only been applied to the calculation of masses heavier that Pb. One can raise many serious objections to this model. First, it has obviously been adjusted not only to measured masses but also to masses given by Wapstra systematics. Second, the peculiar shell-correction method employed is contrary to the ideas introduced by Strutinsky. Third, since a liquid-drop model expression is one part of the final expression for calculating nuclear masses, the parameters of this expression should also be counted as parameters adjusted to nuclear masses. Thus, the number of parameters adjusted to masses in the model exceeds 24, which is considerably greater than 13, the number claimed by the authors. This number should be compared to 11 adjustable parameters in the FRDM. When the FRDM is adjusted only to nuclei above Pb an rms deviation of 0.24 MeV is obtained, compared to 0.22 MeV for the FDSM model.

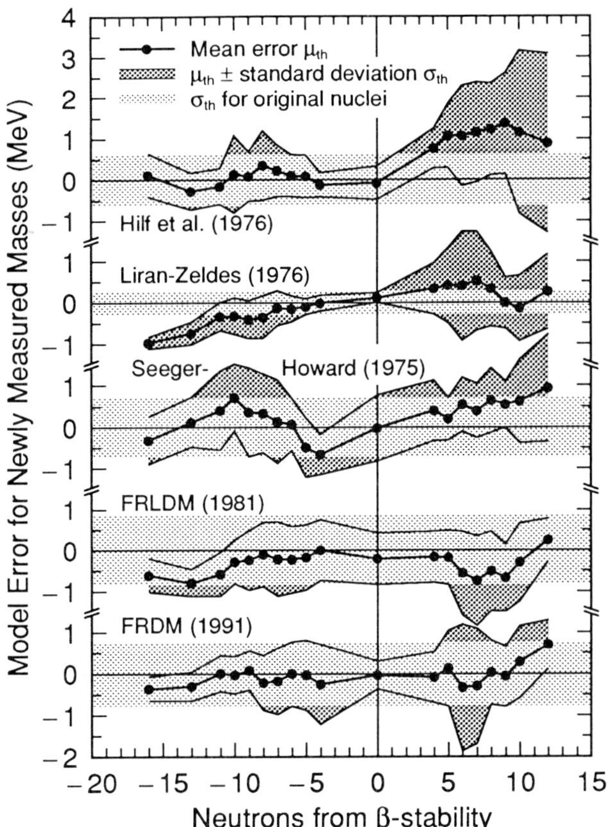

Figure 4: Comparison of error behavior for five mass models applied to new nuclei versus distance from stability. Compare this figure to fig. 3, which is plotted to the same scale.

3 Conclusions

Although mass models based on postulated microscopic corrections with a large number of adjustable parameters have small errors in the region where the parameters were determined, they diverge severely when applied outside this region. The error is larger by amounts ranging from 72% to 1133% in the new region than in the region where the model parameters were determined. In addition, the errors in most cases grow with distance from stability. In contrast, models based on a quantal treatment of the nuclear interaction show remarkable stability when applied to new nuclei that were not considered when the models were initially formulated and their parameters determined. The two most modern models in fig. 4 have errors in the new regions of nuclei that are only 6% and 10% larger than in the region where the parameters were adjusted. The reliability of the mass calculations based on quantal treatments has important consequences. As will be discussed in another contribution [21] to this conference, calculations based on such microscopic nuclear-structure models now make it possible to put new constraints on the stellar site and astrophysical conditions for the r-process, and also provide detailed insight into the isotopic r-abundance distribution.

From the many successes obtained by actually calculating microscopic effects, one can draw the following conclusions:

- Mass models based on postulated microscopic corrections and a large number of parameters are no longer worthwhile, and a disproportionate amount of effort on such models should be avoided. Instead, the focus should be to develop further the microscopic models that have provided so much insight into nuclear structure.

- A corollary to the above point is that it is meaningless to "optimize" the calculations to a small region of nuclei by adjusting some parameters in a limited region such as the actinide region with the hope of being able to predict the masses of close-lying nuclei better than in a global approach. As mentioned above, if we limit adjustment in our own model to the region above Pb we obtain an error of only 0.24 MeV. However, in such a limited study we would not be aware of the Coulomb redistribution effect, which lowers the mass of ^{272}110 by about 3 MeV relative to models that do not take Coulomb redistribution into account. *To be physically interesting a nuclear-structure calculation should normally use the same parameter set across several deformed and spherical regions.*

We are grateful to K. A. Gernoth and J. W. Clark for providing us the results of their neural-network calculations and to J. M. Pearson for providing us the results of his ETFSI calculations prior to publication. This work was supported by the U. S. Department of Energy.

References

[1] V. M. Strutinsky, Nucl. Phys. **A95** (1967) 420.

[2] V. M. Strutinsky, Nucl. Phys. **A122** (1968) 1.

[3] M. Brack, J. Damgaard, A. S. Jensen, H. C. Pauli, V. M. Strutinsky, and C. Y. Wong, Rev. Mod. Phys. **44** (1972) 185.

[4] J. R. Nix, Ann. Rev. Nucl. Sci. **22** (1972) 65.

[5] P. Möller and J. R. Nix, Proc. Third IAEA Symp. on the physics and chemistry of fission, Rochester, 1973, vol. I (IAEA, Vienna, 1974) p. 103.

[6] M. Bolsterli, E. O. Fiset, J. R. Nix, and J. L. Norton, Phys. Rev. **C5** (1972) 1050.

[7] P. Möller, S. G. Nilsson, and J. R. Nix, Nucl. Phys. **A229** (1974) 292.

[8] P. Möller and J. R. Nix, Nucl. Phys. **A361** (1981) 117.

[9] J. Dudek, Z. Szymanski, T. Werner, A. Faessler, and C. Lima, Phys. Rev. **C26** (1982) 1712.

[10] S. G. Nilsson, C. F. Tsang, A. Sobiczewski, Z. Szymański, S. Wycech, C. Gustafson, I.-L. Lamm, P. Möller, and B. Nilsson, Nucl. Phys. **A131** (1969) 1.

[11] U. Mosel and H. W. Schmitt, Phys. Rev. **C4** (1971) 2185.

[12] P. Möller and J. R. Nix, Nucl. Phys. **A229** (1974) 269.

[13] P. Möller and J. R. Nix, Nucl. Phys. **A536** (1992) 20.

[14] P. Möller, J. R. Nix, W. D. Myers, and W. J. Swiatecki, Nucl. Phys. **A536** (1992) 61.

[15] J. M. Pearson, Y. A. Aboussir, A. K. Dutta, R. C. Navak, M. Farine, and F. Tondeur, Nucl. Phys. **A528** (1991) 1.

[16] Y. Aboussir, J. M. Pearson, A. K. Dutta, and F. Tondeur, Nucl. Phys. **A** (1992) to be published.

[17] P. E. Haustein, Proc. 7th Int. Conf. on nuclear masses and fundamental constants (AMCO-7), Darmstadt-Seeheim, 1984 (Lehrdruckerei, Darmstadt, 1984) p. 413.

[18] H. von Groote, E. R. Hilf, and K. Takahashi, Atomic Data Nucl. Data Tables **17** (1976) 418.

[19] K. A. Gernoth, J. W. Clark, J. S. Prater, and H. Bohr (1992) to be published.

[20] X.-L. Han, C.-L. Wu, D. H. Feng, and M. W. Guidry, Phys. Rev. **C45** (1992) 1127.

[21] K. -L. Kratz, P. Möller, B. Pfeiffer, F. -K. Thielemann, A. Wöhr, and the ISOLDE collaboration, these proceedings (1992).

Enhancement of nuclear level density due to collective excitations

A Mengoni and G Maino

ENEA-INN.SVIL, Nuclear Data and Codes Laboratory
V.le G B Ercolani 8, 40138 Bologna (Italy)

Abstract. The contribution of collective degrees of freedom to
the nuclear level density for transitional class nuclei has been
calculated within the framework of the Interacting Boson Model
of Arima and Iachello. Following a previous work based on IBA-1,
the present calculation has been made within the IBA-2 model,
where the neutron and proton degrees of freedom are explicitly
taken into account. The level density enhancement factors for Sm
and Os isotopes belonging to different transitional classes have
been calculated and the results show that while for Os isotopes
a practically constant factor is obtained, for the Sm nuclei a
strong increase is observed as the axially symmetric deformed
shape is approached.

1. Introduction

Nuclear level densities are required for all the various
applications of the statistical theory of nuclear reactions.

Several models have been developed for the representation
of the different excitation modes of nuclei, for energies up
to several MeV. Most of them rely on the model assumption of
the nucleus as a Fermi gas of independent nucleons and then
apply corrections for the presence of effects such as the
nucleon pairing correlation and shell inhomogeneities.

Collective motion has also been considered using various
phenomenological descriptions of the nuclear spectra, usually
based on geometrical representations of the nuclear shape.

Following these approaches however, the collective degrees
of freedom need to be separated into a vibrational and a
rotational part, releasing the character of uniformity in the
description and leaving open the question of the treatment of

transitional nuclei where vibrations and rotations are not completely separable.

Recently [1], we have proposed to use the Interacting Boson Model of Arima and Iachello [2] to describe the collective properties of the nuclear motion and we have derived the proper enhancement factor for the nuclear level density as a function of the nuclear temperature.

This previous work was done for a special class of nuclei: those exhibiting dynamic symmetries, namely, those nuclei corresponding to spherical, axisymmetric deformed or γ-unstable shapes, respectively, in a geometrical picture.

In this work we have extended the application to nuclei in the transitional classes. To do this, the Interacting Boson Model-2 (IBA-2 for: Interacting Boson Approximation-2) has been applied.

2. Collective spectra in the Interacting Boson Models

In order to calculate the collective spectrum of a given nucleus the corresponding Hamiltonian has to be diagonalized.

Following the formalism of Ref. [2] we have adopted the usual IBA-2 Hamiltonian

$$H = E_0 + \epsilon(n_{d\nu}+n_{d\pi}) + \kappa Q_\nu^{(2)} \cdot Q_\pi^{(2)} + V_{\nu\nu} + V_{\pi\pi} + H_M. \qquad (1)$$

All the constants and operators in (1) have been defined in Refs. [2,3] and they will not be reported here. A complete description of the IBA-2 model, with a list of cases where it has been applied, can be found in Ref [2].

We only remind here that, in IBA-2, the neutron (ν) and proton (π) degrees of freedom are introduced and a microscopic interpretation of the bosons as correlated nucleon pairs is possible. Also, the transitional classes of IBA-2 are related to those of IBA-1, maintaining the possibility of a geometrical interpretation of the dynamic symmetries.

Moreover, it has been shown [2,3] that the parameters of the Hamiltonian (1) are either constant or vary smoothly and continuously within a transitional class. It is therefore possible, within this model, to predict the spectra of unmeasured nuclei.

3. The nuclear level density

Under the assumption of decoupling between intrinsic (noncollective) and collective degrees of freedom, the nuclear level density as a function of excitation energy, U, can be expressed by

$$\rho(U) = \rho_i(U) \cdot Z_{coll}(U) \qquad (2)$$

where

$$Z_{coll}(U) = \sum_\mu (2J_\mu + 1) \exp(-\beta E_\mu) \tag{3}$$

is the canonical partition function for the collective degrees of freedom of the system and β is the inverse of the nuclear temperature. The sum is over the whole collective energy spectrum.

In (2), the intrinsic level density ρ_i can be calculated from the Fermi gas expression, properly corrected for pairing correlation and shell effects as proposed, for example, in Ref. [4]. Z_{coll} acts as an enhancement factor to be applied to ρ_i in order to include collective effects.

We would like to stress here that in this description no distinction is made between the rotational and the vibrational degrees of freedom. This characteristic is maintained in the frame of the Interacting Boson Model and there appears to be no need for an a-priori classification of the nucleus into a specific geometrical specie.

4. The collective enhancement factor for Sm and Os

We have applied the IBA-2 model to the calculation of the energy spectra of several even-even Sm and Os isotopes in two transitional classes.

For the Sm case, the isotopes with A=148, 150 and 152 corresponding to the transitional class going from the spherical (A=148) to the axially symmetric deformed shape (A=152 or A=154) have been considered. The calculation of the energy spectra has been made diagonalizing the Hamiltonian (1), using the parameters of Ref. [5]. The code used is a version of NPBOS [6], modified in order to calculate the complete eigenvalues set.

Table 1. Calculated and experimental [7] values of the ratio between the yrast $J^\pi = 4^+$ and $J^\pi = 2^+$ levels for the Sm and Os isotopes.

	^{148}Sm	^{150}Sm	^{152}Sm		
Calculated	2.22	2.36	3.02		
Experimental	2.23	2.31	3.00		

	^{186}Os	^{188}Os	^{190}Os	^{192}Os	^{194}Os
Calculated	3.28	3.22	3.05	2.89	2.31
Experimental	3.17	3.08	2.93	2.82	

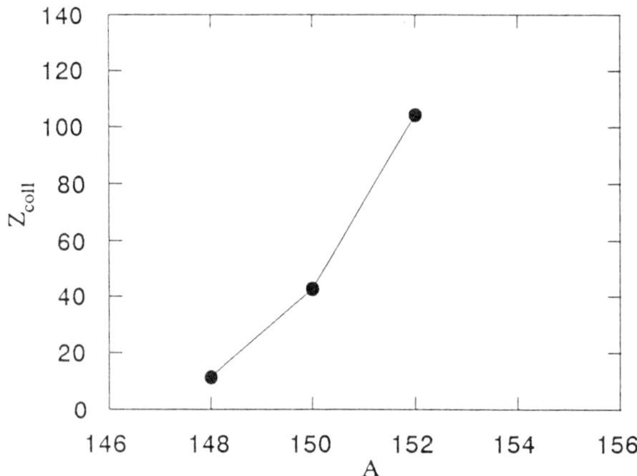

Figure 1. Collective enhancement factor for Sm isotopes.

The calculated spectra consist of several thousands of energy levels and cannot be reproduced here. As an example of the results obtained for the energy spectra, the values of the ratio between the yrast $J^{\pi}=4^{+}$ and $J^{\pi}=2^{+}$ energy levels are given in Table 1. The comparison is made with the experimental values of Ref. [7].

This ratio should be equal to 2 for spherical vibrators and about 3.3 for axially symmetric rigid rotators.

Once the energy spectrum is obtained, the calculation of Z_{coll} is straightforward using (3). For the three Samarium isotopes we have used the same value for the nuclear temperature, T=0.675 MeV, fairly well approximating the excitation energy of the neutron binding.

The results of the calculations are shown in Figure 1. An enhancement ranging from about 12 up to about 100 for A=152 is obtained.

The same kind of calculations was done for five Os isotopes with mass number, A, ranging from 186 up to 194. The results are shown in Figure 2. In this chain they represent the transition between the axially deformed nucleus with A=186 to the γ-unstable isotope with A=196, corresponding to the O(6) dynamic symmetry of IBA-2 [3]. In this case the nuclear temperature corresponding to the neutron separation energy is T=0.596 MeV.

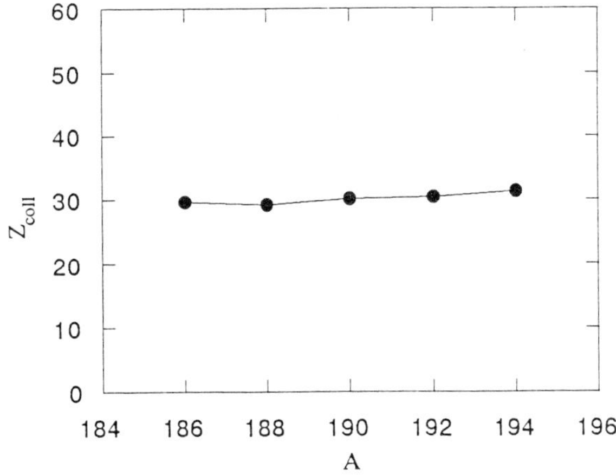

Figure 2. Collective enhancement factor for Os isotopes.

The results for Z_{coll} are here very different as compared to the Sm case. We observe a practically constant enhancement factor in the whole isotope chain.

5. Comments and concluding remarks

Our calculation show that the enhancement of the level density due to collective excitations can be substantial (up to a factor of 100) at excitation energies corresponding to neutron binding. This result confirms the calculations made using geometrical models [8].

The evaluation of the collective contribution to the level density for transitional class nuclei is important in order to estimate this quantity for the cases where experimental information is not available and the use of nuclear models and parameter systematics is necessary.

For instance, the systematics of the parameters of the Fermi gas model are usually evaluated including pairing correlations and shell effects. The inclusion of collective effects could improve the reliability of such a systematics and consequently the results of calculations based on these models.

Finally, we would like to mention here the necessity of studying the energy (or temperature) dependence of the collective effects. In fact, only in this way a more realistic extension of the applicability to the statistical model for the calculation of nuclear reaction cross sections could be achieved.

References

[1] Maino G, Mengoni A and Ventura A 1990, Phys. Rev. C **42** 988-92.
[2] Iachello F and Arima A 1987, The Interacting Boson Model (Cambridge: The University Press).
[3] Bijker R, Dieperink A E L, Scholten O and Spanhoff R 1980, Nucl. Phys. **A344** 207-32.
[4] Ignatyuk A V, Smirenkin G N and Tishin A S 1975, Sov. J. Nucl. Phys. **21** 255.
[5] Scholten O 1980, The Interacting Boson Model and Some Applications, PhD. Thesis, University of Groningen, The Netherlands.
[6] Otsuka T 1977, Computer program NPBOS, The University of Tokyo, Japan.
[7] Lederer C M and Shirley V S 1978, Table of Isotopes, 7th edition (New York: Wiley & Sons).
[8] Rastopchin E M, Svirin M I and Smirenkin G N 1990, Sov. J. Nucl. Phys. **52** 799-807.

Calculating charge change cross sections

P.Doll and H.J.Crawford [+]

Kernforschungszentrum Karlsruhe, Institut für Kernphysik 1,
P.O. Box 3640, WD-7500 Karlsruhe 1
[+] Space Science Laboratory, University of California, Berkeley, California 94720

Abstract. Investigations of the charge changing fragmentation cross sections for nuclei ranging from ^{12}C to ^{58}Ni and energy / nucleon from 300 to 1600 MeV on hydrogen emphasize the role of nucleon isobar excitations.

1. Introduction

The analysis and interpretation of the elemental and isotopic composition of cosmic-ray nuclei provide information on the sources and acceleration of cosmic-rays, and their propagation through interstellar matter. This interpretation requires the knowledge of precise cross section data [1,2]. The preliminary analysis of the heavy-ion fragmentation data, taken at the Lawrence Berkeley Laboratory HISS facility [2] and carried out by the 'transport collaboration', has yielded total charge changing cross sections $\sigma_{\Delta Z \geq 1}$ of beams from ^{22}Ne to ^{58}Ni on a liquid hydrogen target, over an energy range from 400 to 910 MeV/nucleon.

2. Model calculation

Our model calculation [3] was triggered by the observation that these charge changing cross sections showed a steep rise with increasing bombarding energy. Based on the optical model we describe in the present calculation [3] the absorption coefficient for a nucleon inside a nucleus by weighted cross sections for nucleon-nucleon collisions.

The use of the free nucleon-nucleon cross sections σ_{pp} and σ_{pn} for calculating the mean-free-path is probably justified because of the reduced nuclear density in the outer hemisphere of the nucleus, where most of the fragmentation takes place. Below 1GeV/c the proton-neutron cross sections σ_{pn} are almost three times larger than the proton-proton σ_{pp} or neutron-neutron cross sections which is due to the $^3S_1$ or $^1S_0$ spin and isospin structure of the nucleon-nucleon systems, respectively.

The strong increase of these cross sections above 1 GeV/c means an increasing pion and delta-isobar production during collisions in the outer boundary of the nucleus, depositing there an appreciable amount of excitation energy in the nucleons. Therefore, the question must be pursued, to which extent this excitation mode in the nucleons is instantaneously coupled to the fragmentation channels.

In fig. 1 a comparison is carried out with the new HISS data [2] for the nuclei ^{22}Ne, ^{26}Mg, ^{32}S and ^{40}Ca, including also some data from Webber et al. [4] for ^{32}S and ^{40}Ca. With respect to the semi-empirical calculations by Letaw et al.[5] given by dashed curves, our calculations are in most cases below the data.

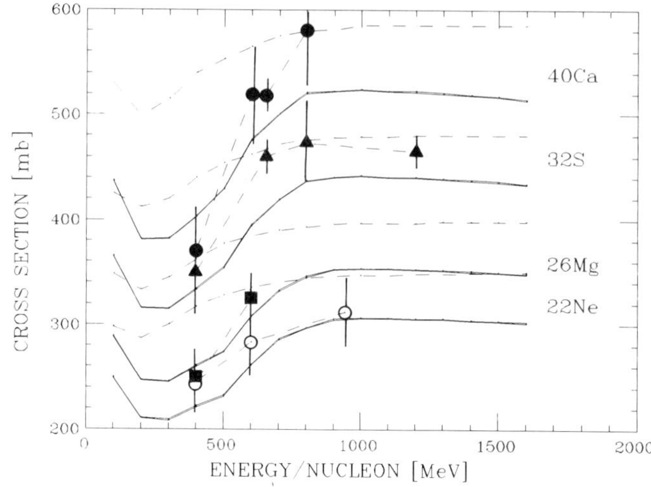

Fig. 1:   Calculations and data for the total charge changing cross sections.

The striking feature of the analysis presented above is the strong rising cross section by means of excited baryon states in the nucleons. This lends support to another collective mode in a nucleus which has a large parentage to the $S = 3/2$ and $I = 3/2$ spin and isospin Δ-excitation of the nucleons. The giant Gamow-Teller (GT) resonance (for $L=0$) as a spin and isospin excitation of the total nucleus is known to be partly quenched [6] by the nucleon isobar excitations. We asked for the modification of the absorptive part in the nuclear interaction, when this coupling to the GT-state in the corresponding nucleus is considered.

The calculation of partial cross sections for secondaries fragmentation is less straightforward [7]. We assume [3] that the partial cross sections follow from a deterministic ansatz which considers the target nucleus A^0 excited around the GT resonance. It may decay into fragments $A^0 \rightarrow A^i + x$. Therefore, the fragmentation cross section becomes proportional to $\exp(-Q^{0i})$, where Q^{0i} corresponds to the binding energy difference between A^0 and A^i. Preliminary results on partial cross sections support this ansatz.

3. Conclusion

This work investigates the strong energy dependence of the charge changing fragmentation cross sections for nuclei ranging from ^{12}C to ^{58}Ni and energy/nucleon from 300 to 1600 MeV. The calculation emphasizes the influence of the $\Delta(1232)$ excitations of the nucleon on the fragmentation cross sections.

References

[1]   M. Garcia-Munoz, J.A. Simpson, T.G. Guzik, J.P. Wefel, S.H. Margolis, *Astrophys. Jour. Suppl. Series* **64,269** (1987)

[2]   C.-X.Chen et al., *Contr.22 Int.*, Cosmic Ray Conf., Dublin, Ireland 1991 OG8.3-6

[3]   P. Doll et al. to be published.

[4]   W.R. Webber, J.C. Kish an D.A. Schrier, *Phys. Rev.* **41C,520** (1990)

[5]   J.R. Letaw, R. Silberberg and C.H. Tsao, *Astrophys. Jour. Suppl. Series* **51,271** (1983)

[6]   G.E. Brown and M. Rho, *Nucl. Phys.* **A 372,397** (1981)

[7]   P.Ferrando, W.R. Webber, P. Goret, J.C. Kish, D.A. Schrier, A. Soutoul, O. Testard, *Phys. Rev.* **37C, 1490** (1988)

The influence of one neighbour ion on a reacting pair in dense stellar matter

M Sahrling

Uppsala Astronomical Observatory, Box 515, 751 20 Uppsala, Sweden

Abstract. The rate of nuclear reactions, in dense matter, is strongly influenced by ions close to a reacting pair. This effect is normally treated by using an average distribution of "spectator" ions from the beginning, and using perturbation theory to calculate certain correction terms. This paper investigates the validity of this procedure, assuming one neighbour. It carefully calculates the reaction rate and compares it with the one obtained using the normal procedure. It turns out that this procedure gives a slightly higher value of the reaction rate than the more accurate one, when one ignores the correction terms. Including the correction terms the conventional rate becomes too low. It is therefore recommended that one uses the normal procedure, but ignores the correction terms.

1. Introduction

An important concept in astrophysics is the fusion rate for nuclear reactions. For the evolution of stars the reaction rate in matter of density higher than 10^6 g/cm^3 is important. High density situations that immediately come to mind are the carbon ignition in the late stages of evolution [1], and for white dwarfs in close binaries, the interplay between nuclear runaway and accretion that may determine the fate of the system. The characteristic feature of these reactions is that the surrounding ions exert a pressure on a fusing pair, thus forcing them together, increasing the bare Gamow reaction rate.

In the past 20 years reaction rates have been calculated with the assumption that "spectator ions" can be treated by using an average distribution from the beginning [2], [3], [4], [5], [6], [7]. Below, this assumption is referred to as the "mean potential method". The tunneling is, however, instantaneous. There might be certain configurations of spectators that, at the time of tunneling of the reacting pair, affects the reaction rate in a way that cannot be treated using mean potentials. Instead of using a mean potential one should in principle "average at tunneling", that is: calculate the reaction rate for each possible configuration of "spectators", then average the reaction rate over such configurations. This has been recognized by some authors [3] that used a perturbation theory approach to compensate for this effect. The validity of this perturbation approach has been questioned [8]. Here, a model is described to correctly account for a "spectator" that is the closest one to a reacting pair in a simplified 3D Coulomb plasma.

The paper is divided as follows: In section 2 the basic definitions are discussed. Section 3 describes the simplified 3D Coulomb plasma. Section 4 contains the results, and, finally section 5 discusses the conclusions.

2. Basic definitions

The one-component plasma (OCP) approximation is assumed. The ions are completely stripped of their electrons and move in a "jelly"-like sea of electrons. The properties of an OCP are described by two dimensionless parameters. The plasma parameter, Γ, describes the degree of coupling in the system and is defined as the ratio of the Coulomb and thermal energies. The second parameter is the Gamow exponent, τ, of the uncoupled fusion reaction.

The enhancement of the reaction rate, E, is defined as the ratio of the reaction rate for the coupled plasma to the uncoupled plasma. For the uncoupled plasma the reaction rate is proportional to $e^{-\tau}$ [3].

3. A simplified 3D Coulomb plasma.

This section describes a simplification that makes it possible to calculate the error one gets when using the mean potential method. It is too crude to be of use for calculating the true reaction rate in a stellar plasma.

Assume that there is only one neighbour ion close to the reacting pair and that all the other ions are so far away that they only feel one particle of charge 3Ze and not three distinct ones. The reaction system then consists of three particles. One can define the coordinate system such that the origin is at the center-of-mass of the reacting pair. Let r be the distance between them. The position of the neighbour ion is defined as $\bar{r}_3$. The mean potential, $w(r)$, defined by [3] can now be calculated.

The enhancement, E, in the conventional picture [3] becomes:

$$E_w = exp\left(-\frac{2}{\hbar}\sqrt{2m}\int_0^b \{[(Ze)^2/r + w(r) - (Ze)^2/b - w(b)]\}^{1/2}dr - \beta[(Ze)^2/b + w(b)] + \tau\right). \tag{1}$$

Above, m is the reduced mass of the reacting pair, b is the classical turning point for the reacting pair, Z is the charge of the ions and β is the inverse temperature in energy units.

If one instead performs the averaging at tunneling, the enhancement becomes:

$$E_{ex} = \int_V E(\bar{r}_3)d\bar{r}_3 \tag{2}$$

where

$$E(\bar{r}_3) = exp(-\frac{\sqrt{8m}}{\hbar} \int_0^b \{[(Ze)^2/r + W(r,\bar{r}_3) - (Ze)^2/b - W(b,\bar{r}_3)]\}^{1/2}dr - \beta[(Ze)^2/b + W(b,\bar{r}_3)] + \tau).$$

Here the potential $W(r,\bar{r}_3)$ is the potential energy for the reaction system when the neighbour is at position $\bar{r}_3$ and the distance between the reacting pair is r. It also includes the interaction with the electrons and all other ions. Here we follow the treatment in [3] and use only the value of b that gives the highest E.

To find the corrections of [3] in this model, one needs to calculate 2nd and higher order derivatives of the difference $< W > -w$.

4. Results

In the figure the difference between the expression for E using the mean potential method (1) and the more precise expression (2) is shown. The behavior is the same for all temperatures in the range $10^6 K < T < 10^{10} K$. The effect of using the mean potential is generally to raise the reaction rate. The reason for the small difference between the two averaging procedures when Γ is large is due to cancellation between two competing effects during tunneling. For small Γ's the classical turning point for the reacting pair is much smaller than the mean distance. The pair does not feel the third particle so much and the averaging procedure is not important.

All the effects above are small and will affect astrophysically interesting phenomena, such as the carbon ignition curve, only marginally. In the figure the corrections described in [3] are shown to 3rd order. It is interesting that the corrections generally give a too low value for the reaction rate.

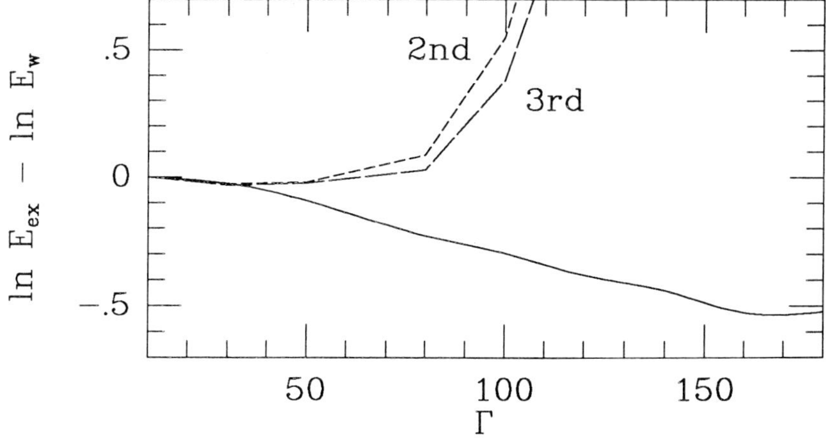

Figure. ln E_{ex}/E_w as a function of Γ for $\tau = 181.3$ corresponding to $T = 10^8 K$. The dashed curves represent the correction terms according to [3].

5. Conclusions

We have investigated the effect of accurately treating one neighbour ion to a reacting pair in dense stellar matter and found an astrophysically insignificant discrepancy compared to earlier averaging methods. We have used a simplified version of a 3D plasma to isolate the effect of one neighbour. Although this does not prove that the effect of many neighbours is not important in dense stellar matter, it suggests so. The conclusion is that when calculating the reaction rate in dense stellar matter one should only use the mean potential defined by [3] and ignore their correction terms. These results are different from an earlier one-dimensional approach [8] in the sense that the effect is generally not as big as there.

6. Acknowledgments

I would like to thank W.L. Slattery and H.E. DeWitt for stimulating discussions of my work. I am indebted to Christopher Pethick for many hours of stimulating discussions of physics in Copenhagen and Urbana. I am also grateful to Bengt Gustafsson, for his support during this work. I am especially grateful to Magnus Jändel who introduced me to this field. Finally, I would like to thank the scientific staff at the Uppsala observatory and the high energy astrophysics group at Urbana, Illinois.

This work has been supported by grants from the Foundation to the memory of Lars Hierta and The Royal Swedish Academy of Science.

References

[1]   Iben I and Renzini A 1983 *ARA&A* **21**   271

[2]   DeWitt H E, Graboske H C and Cooper M S 1973 *ApJ* **181** 439

[3]   Alastuey A and Jancovici B 1978 *ApJ* **226** 1034

[4]   Ichimaru S 1982 *Rev. Mod. Physics* **54** 1017

[5]   Itoh N, Totsuji H, Ichimaru S and DeWitt H E 1979 *ApJ* **234** 1079

[6]   Itoh N, Kuwashima F and Munakata H 1989 *ApJ* **362** 620

[7]   Ogata S, Iyetomi H and Ichimaru S 1991 *ApJ* **372** 259

[8]   Jändel M and Sahrling M 1992 *ApJ* **393**

Alpha-alpha bremsstrahlung in a microscopic model

M. Kruglanski[1] ,D. Baye and P. Descouvemont[2]

Physique Nucléaire Théorique et Physique Mathématique, CP229,
Université Libre de Bruxelles, B1050 Bruxelles, Belgium.

Abstract. Nucleus-nucleus bremsstrahlung is explored in the $\alpha + \alpha$ system with microscopic and potential models. Calculated cross sections show that experimental conditions exist where the deep-, shallow-potential and microscopic models would easily be discriminated.

1. Introduction

Nuclear reactions such as radiative capture play an important role in the understanding of stellar evolution. Since energies of astrophysical interest are small with respect to the Coulomb barrier, capture cross sections are weak and difficult to be measured. Therefore evaluation by theoretical models is necessary. In order to establish the validity of these models, it is important to have experimental tests of the quality of the wave functions.

Nucleus-nucleus bremsstrahlung offers a possibility of probing the inner part of scattering wave functions. It corresponds to a transition between continuum states, the two-particle relative motion being slowed down through a photon emission. This kind of reactions can be studied by microscopic[1,2] or potential[3,4] models. In principle, experiment should provide a way of testing them and choosing the most efficient ones. However, experimental data are very scarce, with rather large error bars. All the models agree reasonably well with these data[5] provided they correctly reproduce the phase shifts. Further experimental data are necessary to encourage progress in the theoretical description. But, one needs to know which type of data, and with which accuracy, might help discriminating between models. Here, we attempt to answer these questions by performing a comparison between microscopic and potential models for $\alpha + \alpha$ scattering.

The resonating-group method (RGM)[6] with the help of the generator-coordinate method (GCM)[7] provides an accurate microscopic description of elastic scattering which reproduces experimental phase shifts. The wave functions are fully antisymmetric and derived from a Schrödinger equation involving an effective nucleon-nucleon interaction and the exact Coulomb force. We shall try to find experimental conditions where the differences between models are enhanced, as much as possible.

---

[1]Boursier IRSIA
[2]Chercheur qualifié FNRS

2. Microscopic model for bremsstrahlung

A general RGM form of $\alpha + \alpha$ wave functions with monopolar distortion of the clusters reads, in partial wave ℓ,[8]

$$\Psi^{\ell m} = \sum_{\omega_1 \omega_2} \mathcal{A} \phi_1^{\omega_1} \phi_2^{\omega_2} Y_\ell^m(\Omega_\rho) g_{\omega_1 \omega_2}^\ell(\rho) \tag{1}$$

where $\mathcal{A}$ is the antisymmetrization projector over eight nucleons, $\boldsymbol{\rho} = (\rho, \Omega_\rho)$ is the relative coordinate between the clusters, and $\phi_i^{\omega_i}$ represents an internal wave function of cluster i in state ω_i of its monopolar excitation. The internal wave functions are obtained by a variational calculation involving a four-nucleon microscopic Hamiltonian and N_b selected translation-invariant antisymmetric basis functions $\phi_i(b)$, defined in the harmonic-oscillator (HO) model with N_b different oscillator parameters:

$$\phi_i^\omega = \sum_{k=1}^{N_b} c_{ik}^\omega \phi_i(b_k) \tag{2}$$

The basis function $\phi_i(b)$ is the internal part of a normalized Slater determinant $\Phi_i(b, \boldsymbol{S})$ defined in the HO shell model centred at $\boldsymbol{S}$ with size parameter b.

In the GCM, $\Psi^{\ell m}$ is expanded as a finite linear combination of projected Slater determinants in the two-centre HO-model context. Such determinants can be written as

$$\Phi(\boldsymbol{S}) = \mathcal{A} \Phi_1(b_1, -\tfrac{1}{2}\boldsymbol{S}) \Phi_2(b_2, \tfrac{1}{2}\boldsymbol{S}) \tag{3}$$

The distance $\boldsymbol{S}$ between the centres plays the role of a generator coordinate. The restoration of translation, rotation and reflection symmetries requires several steps[8] before providing the appropriate functions for (1). The first step removes the c.m. dependence with a traditional Peierls-Yoccoz projection on zero momentum. A second step makes uniform the widths of the relative parts of the functions with an integral transformation. Finally, basis functions $\phi^{\ell m}(R)$ are obtained with an angular momentum projection. The complete set of operations reads

$$\phi^{\ell m}(R) = (4\pi)^{-13/4} \left(\frac{128 b^2}{b_1^2 b_2^2 \beta^4} \right)^{3/4} \int d\Omega_R Y_\ell^m(\Omega_R)$$

$$\times \int d\boldsymbol{S} \exp((\boldsymbol{R} - \boldsymbol{S})^2/\beta^2) \int d\boldsymbol{a} \exp(-i\boldsymbol{a} \cdot \boldsymbol{P}_{cm}/\hbar) \Phi(\boldsymbol{S}) \tag{4}$$

where $\beta^2 = b^2 - \frac{1}{2}(b_1^2 + b_2^2)$, and the size parameter b may be chosen arbitrarily to fix the width of the relative gaussian function in $\phi^{\ell m}$. Basis functions $\phi_{\omega_1 \omega_2}^{\ell m}(R_n)$ involving distorted cluster wave functions can be obtained by combining $\phi^{\ell m}(R_n)$ with different oscillator parameters.

The wave function (1) is approximated as

$$\Psi^{\ell m} = \sum_{\omega_1 \omega_2} \sum_{n=1}^{N} f_{\omega_1 \omega_2}^\ell(R_n) \phi_{\omega_1 \omega_2}^{\ell m}(R_n) \tag{5}$$

after selection of a finite set of N values R_n of the generator coordinate. The coefficients $f_{\omega_1 \omega_2}^\ell(R_n)$ are determined from an 8-body Schrödinger equation. The microscopic R-matrix method (MRM) is employed to connect (5) to the correct asymptotic scattering behaviour[9].

Two α particles collide with relative momentum $\boldsymbol{p}_i$ in the z direction and energy $E_i = p_i^2/2\mu$ where μ is their reduced mass. After emission of a photon with energy E_γ and momentum $\boldsymbol{p}_\gamma = \hbar\boldsymbol{k}_\gamma$ in the direction $\Omega_\gamma = (\theta_\gamma, \varphi_\gamma)$, the final momentum of the system is $\boldsymbol{p}_f$ in the direction $\Omega_f = (\theta_f, \varphi_f)$. The final energy $E_f = p_f^2/2\mu$ satisfies $E_f = E_i - E_\gamma$ up to small recoil corrections.

Only even-ℓ partial waves $\Psi^{\ell m}$ are allowed. Hence odd-parity multipoles are forbidden. Since M1 transitions are also forbidden in the long-wavelength approximation, the E2 contribution dominates. The expression of the c.m. differential cross section $d^3\sigma/dE_\gamma d\Omega_f d\Omega_\gamma$ includes final-state phase shifts and reduced electric-quadrupole-operator matrix elements between the scattering partial waves: $\langle \Psi^{\ell_f} \| \mathcal{M}_2^E \| \Psi^{\ell_i} \rangle$. The slow convergence of the Coulomb part of this expression requires a special treatment which is explained in ref.[4]. Moreover the long-wavelength approximation and Siegert theorem cannot be used to simplify $\mathcal{M}_{2\mu}^E$. But by dividing the configuration space in two parts, the MRM restricts convergence problems to the external part where antisymmetrization can be neglected. The problem is therefore reduced to the calculation of matrix elements, between $\phi^{\ell m}(R)$ functions, of the operator

$$\mathcal{M}_{2\mu}^E \propto \sum_{j=1}^{8} \left[\boldsymbol{\nabla}(r^2 Y_2^\mu(\Omega_r)) \right]_{\boldsymbol{r}=\boldsymbol{r}_j - \boldsymbol{R}_{cm}} \cdot (\boldsymbol{p}_j - \tfrac{1}{8}\boldsymbol{P}_{cm}) \tag{6}$$

In the coplanar Harvard geometry, the α particles are detected in the direction $\Omega_1 = (\theta_1, 0)$ and $\Omega_2 = (\theta_2, 0)$ and the photon remains undetected. The Harvard cross section $d^2\sigma/d\Omega_1 d\Omega_2$ needs the same ingredients as the c.m. one.[1]

3. Comparison of cross sections in different models

Our goal is to analyze the validity of bremsstrahlung as a probe of scattering wave functions. A meaningful comparison of different models should rely on wave functions providing identical phase shifts. In addition, an experimental discrimination between the models requires that these phase shifts reproduce as accurately as possible the experimental ones.

A first model involves the deep Gaussian potential of Buck *et al*[10]. With only two parameters, it reproduces the experimental phase shifts in a very satisfactory way up to 15 MeV. Above 15 MeV, the experimental situation is less clear and several resonances are observed. Except in the vicinity of resonances, the deep-potential model (DPM) should remain valid up to at least 20 MeV. However, since the p+^{7}Li channel opens at 17 MeV, the single-channel description becomes less accurate beyond that energy.

A second potential model reproduces exactly the phase shifts of the DPM. Indeed, supersymmetry transformations allow one to construct potentials which are exactly phase-equivalent to a given potential, but with different numbers of bound states. However, the supersymmetric potential model (SPM)[11] provides potentials which must be singular in order to satisfy the generalized Levinson theorem. In the following, we employ the SPM to construct shallow potentials, where the two unphysical bound states for $\ell = 0$ and the other one for $\ell = 2$ are removed from the DPM. For $\ell \geq 4$, both potentials are identical.

The microscopic model summarized in sect.2 has already been applied, with the Minnesota force, to a description of $\alpha + \alpha$ scattering involving monopolar distortion of the α particles. A ten-channel approach was adopted corresponding to four α states: the ground state, the giant monopole resonance and two pseudostates. In order to have phase shifts as close as possible to the DPM and SPM ones, we adjust the force parameter u for each partial wave. With $u = 0.9413$ for $\ell = 0$, $u = 0.9506$ for $\ell = 2$ and $u = 0.9600$ for $\ell \geq 4$, the GCM phase shifts agree with an accuracy better than 3°. The ^{8}Be ground-state energy is well reproduced.

A fourth model is also employed because of the length of the 10-channel calculation. A single-channel GCM with distorted α clusters, where the interaction parameter is readjusted ($u = 0.977$, 0.966 and 0.969) provides phase shifts which, although not identical, are in reasonable agreement.

All experiments performed until now correspond to laboratory cross section in the Harvard geometry where the angles θ_1 and θ_2 are chosen equal. In the experiment realized by Peyer *et al*[5] ($\theta_1 = \theta_2 = 35°$) the experimental points are obtained below $E_i = 10$ MeV. In this energy range all theoretical calculations, including our models, reproduce reasonably well those data (except the last point which may be too low). No discrimination is possible there. But, as showed in fig.1, beyond 13 MeV, the situation is more cheering. The SPM cross section is smaller than the DPM one by a factor of two. The single-channel GCM is represented by the dashed line, and the crosses correspond to the 10-channel GCM calculation. The DPM and GCM provide different results probably due to antisymmetrization effects. However, the differences are not large enough and would require improved statistics for the experimental points.

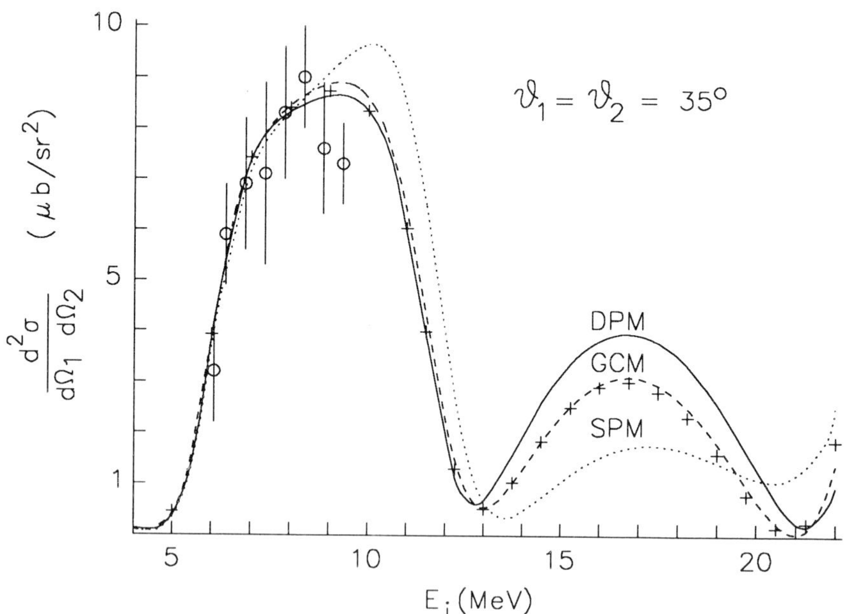

Fig.1 *Laboratory differential cross sections in the Harvard geometry*

It is possible to increase the differences between the models by exploiting phase-shift resonances. For instance, when $\theta_1 = \theta_2 = 27°$ the cross section, around $E_i = 11$ MeV, is enhanced: the transition corresponds to both the initial 4^+ to the final 2^+ resonances.

Laboratory cross section are difficult to interpret, and therefore c.m. cross sections are preferable. For two angles (30° and 90°), the c.m. cross sections, where the initial energy is fixed at $E_i = 16$ MeV, are displayed in fig.2 as a function of E_f. The 4^+ final resonance is visible as a broad bump in the 30° curves but not in the 90° curves. The 2^+ final resonance is apparent in both cases but is more important at 30°. At small photon energies, the cross sections are large but the differences between the models are small. Differences become significant for $E_\gamma \geq 8$ MeV. They are not directly related to the existence of a resonant behaviour but resonances are useful to increase the cross sections. The differences between the DPM and GCM are now very important and these models would clearly be discriminated by experiment.

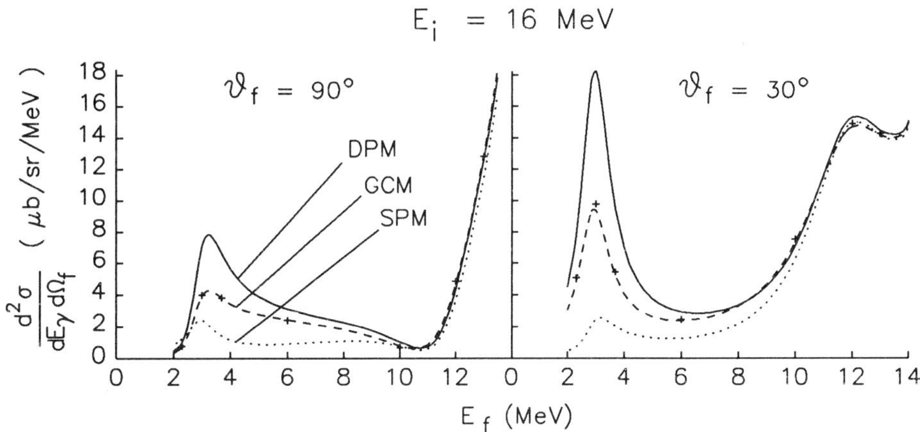

Fig.2 *C.m. differential cross sections at two angles θ_f*

4. Conclusion

Nucleus-nucleus bremsstrahlung should be a useful tool for probing scattering wave functions, with photon energies larger than about 10 MeV. Experiments in the Harvard geometry would be enough for discriminating between models. Resonant states play the role of amplifiers.

Since the SPM provides very different results from the other models, we think that the shallow potential model will be in contradiction with experiment. The differences between the DPM and GCM are evidence for antisymmetrization effects in scattering wave functions. The two versions of the GCM remain very similar at all energies.

References

[1] D. Baye and P. Descouvemont, Nucl. Phys. **A443** (1985) 302

[2] Q.K.K. Liu, Y.C. Tang and H. Kanada, Few-body Problems (1992) in press

[3] K. Langanke, Phys. Lett. **B174** (1986) 27

[4] D. Baye, C.Sauwens, P.Descouvemont and S. Keller, Nucl. Phys. **A529** (1991) 467

[5] U. Peyer, J. Hall, R. Müller, M. Suter and W. Wölfli, Phys. Lett **41B** (1972) 151

[6] K. Wildermuth and Y.C. Tang, "A Unified Theory of the Nucleus", Vieweg, Braunschweig (1977)

[7] H. Horiuchi, Prog. Theor. Phys. Suppl. **62** (1977) 90

[8] D. Baye and M. Kruglanski, Phys. Rev. **C45** (1992) 1321

[9] D. Baye, P.-H. Heenen and M. Libert-Heinemann, Nucl. Phys. **A291** (1977) 230

[10] B.Buck, H. Friedrich and C. Wheatley, Nucl. Phys. **A275** (1977) 246

[11] D. Baye, Phys. Rev. Lett. **58** (1987) 2738

New limits $T_{1/2}$ for 2β-decay of ^{116}Cd and $0\nu\beta^+$/electron capture of ^{106}Cd

F.A. Danevich, V.V. Kobychev, V.N. Kouts, V.I. Tretyak and Yu. Zdesenko

Institute for Nuclear Research of the Ukrainian Academy of Sciences, 252028 Kiev, Ukraine

Abstract.
The experiment (7000h) with three scintillation crystals ^{116}CdWO$_4$ (19; 14 and 13 cm^3) enriched by ^{116}Cd to 83% and natural one CdWO$_4$ (57 cm^3) has been made at the Solotvina Underground Laboratory. The limits $T_{1/2}=10^{22}$y have been set for the neutrinoless double beta decay of ^{116}Cd. Other results have been obtained for the 2β decay of ^{116}Cd to ground and excited states of ^{116}Sn and $0\nu\beta^+$/electron capture of ^{106}Cd.

The energy released in the 2β transition ^{116}Cd-^{116}Sn is 2802(4) keV. To search this process ^{116}CdWO$_4$ scintillators enriched by ^{116}Cd to 83% has been used [1]. Samples of metallic ^{116}Cd were obtained at the State Foundation (619g; 91%). A cadmium wolframite single crystal with a full mass of 510 g has been grown by the Czochralski technique from a composite material prepared by sintering a mixture of the oxides WO$_4$ and ^{116}CdO. The crystal was cleaved into five samples and the best from them (19; 14 and 13 cm^3) were used in the measurements. The number of ^{116}Cd nuclei in these samples is 2.18×10^{23}; 1.61×10^{23} and 1.42×10^{23} respectively. The energy resolution of the crystals with photomultiplier tube (PMT) XP2412 were 12 - 13% for γ-line of the ^{137}Cs with energy of 0.661MeV.

All searches have been performed in the Underground Laboratory of the INR which was built in the operating Solotvina salt mine. The Laboratory is situated 430 m deep from the earth surface that corresponds of 1000 m water-equivalent overburden which reduces the muon flux by a factor of 10^4. Due to low radioactive contamination of the salt the natural γ-background in Solotvina Laboratory is 10 - 100 times lower than in other Underground Laboratories [2].

The experiments were carried out with three kinds of the installation. In first set up two crystals (enriched - 19 cm^3 and natural - 57 cm^3) were coupled to PMT's through the light pipes 9 cm long. The active shielding consisted of four well-type CsI(Tl) crystals with diameter 15 cm and full length 40 cm. The ^{116}CdWO$_4$, CdWO$_4$ and CsI(Tl) scintillators were surrounded by the high purity mercury (8 cm thick), lead (23 cm) and polyethylene (24 cm). The plastic scintillator ($110\times110\times6$ cm^3) was operated as antimuon shielding. The measurements were made during more than 2600 h [3, 4].

The next set up consisted two enriched crystals ^{116}CdWO$_4$ (19 and 14 cm^3) which were coupled to PMT's (XP2412) through the quartz light pipes 25 cm long. The detectors were shielded by the OFHC copper (3.5 cm thick), mercury (8 cm), lead (23 cm) and polyethylene (24 cm). The background rate in these conditions was little less than early. Therefore it is possible to sum the last spectra which were collected during about 3000 h to the previous ones [3, 4]. The results of the adding are shown in Fig. 1, 2 and give for enriched crystals 850 hxkg (measuring time multiplied by mass), for natural one - 1392 h$\times$kg. The peaks presented in the spectra are caused mainly by ^{40}K and ^{208}Tl from PMT's and ^{137}Cs. There are two differences in these spectra: increase in counts below 320 keV for natural crystal and the small peak at the energy about 800 keV for enriched crystals. The first difference can be attributed to the fourth-forbidden beta decay of ^{113}Cd which is contained in the natural crystal (half life about 10^{16} y). Second one can be caused by α-particles from intrinsic radioactive contamination of the enriched scintillators with the level of impurities of about 5×10^{-11}g/g(^{235}U) or 3×10^{-10}g/g(^{238}U).

Using the data accumulated by enriched crystals during 850 h$\times$kg the half life limits for double β decay of ^{116}Cd to ground state of ^{116}Sn have been set:

$$0\nu2\beta \quad (g.s. - g.s.) \quad - 7 \times 10^{21}y \quad (90\%\ CL)\ ;$$
$$0\nu2\beta, M^\circ \quad (g.s. - g.s.) \quad - 1 \times 10^{21}y \quad (90\%\ CL)\ ;$$
$$2\nu2\beta \quad (g.s. - g.s.) \quad - 1 \times 10^{19}y \quad (99\%\ CL)\ .$$

Since in the measurements with an active shielding the background spectra have been registered in coincidence with the CsI(Tl) counters as well as in anticoincidence it is possible to search the 2β decay of ^{116}Cd to the excited states of ^{116}Sn and β^+/electron capture of ^{106}Cd which must be followed by γ-rays with energies of 1294 keV ($0^+ - 2^+$), 463 and 1294 keV ($0^+ - 0^+$) and 511 keV (β^+/E capture). Taking into account the numbers of nuclei, time of the measurements and background rates in the corresponding energy regions of the coincidence spectra the limits $T_{1/2}$ were deduced for the 2β decay of ^{116}Cd to the excited states of ^{116}Sn and for $0\nu\beta^+$/electron capture of ^{106}Cd:

$$0\nu2\beta \quad (0^+ - 2^+) \quad - 5 \times 10^{21}y \quad (90\%\ CL)\ ;$$
$$0\nu2\beta \quad (0^+ - 0^+) \quad - 1 \times 10^{20}y \quad (90\%\ CL)\ ;$$
$$2\nu2\beta \quad (0^+ - 2^+) \quad - 1 \times 10^{19}y \quad (99\%\ CL)\ ;$$
$$2\nu2\beta \quad (0^+ - 0^+) \quad - 6 \times 10^{18}y \quad (99\%\ CL)\ ;$$
$$0\nu\beta^+/EC \quad\quad - 1 \times 10^{19}y \quad (90\%\ CL)\ .$$

Comparing our experimental result for $0\nu2\beta$ decay of ^{116}Cd with theory [5] we find the limit of the neutrino mass less than 8 eV. The experiment is still running

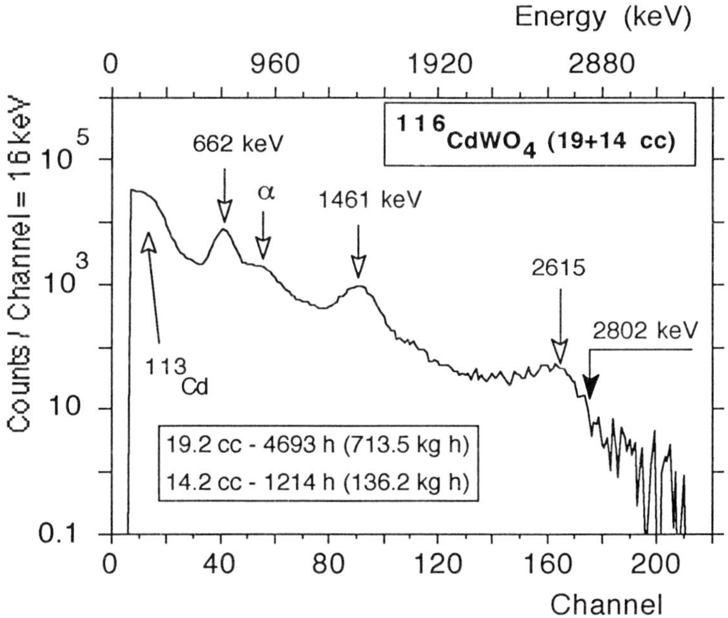

Fig. 1. Background spectrum of the enriched crystals

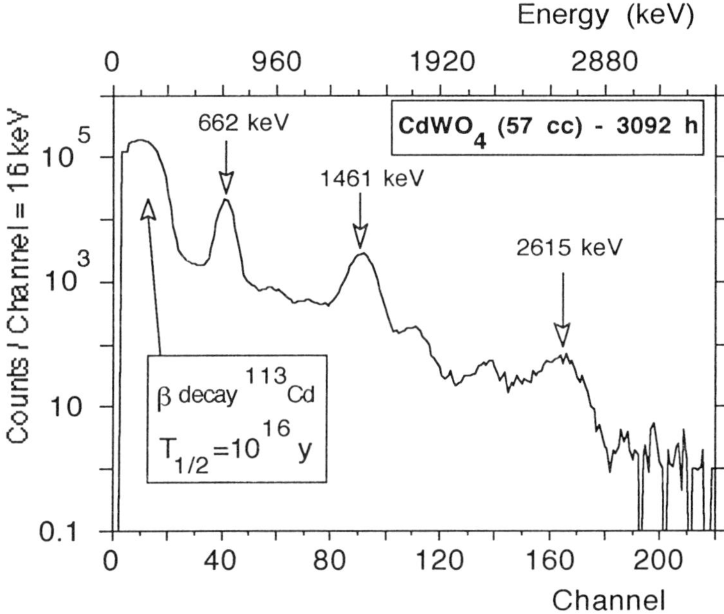

Fig. 2. Background spectrum of the natural crystal

now with three enriched crystals (19; 14 and 13 cm^3) and after the replacement of the PMT's XP2412 by the PMT's with the lower radioactive contamination (FEU-110). The energy resolution of the crystals is about 7.8% at 2.6 MeV and background rate around 2.8 MeV (within FWHM energy interval) is equal 2.3 counts/y/keV/kg (1.5 - for last set up with active and passive shielding). With 540 hxkg of a data taking the present limit $T_{1/2}$ is 10^{22}y (neutrino mass less than 7 eV). After one year of measurements it is possible to improve this limit till 3×10^{22}y (m_ν less than 4 eV). Now we are preparing the big scale experiment with ^{116}CdWO$_4$ crystals (20 kg of the ^{116}Cd) in which for $0\nu2\beta$ decay of the ^{116}Cd the limit $T_{1/2} > 10^{25}$y can be reached ($m_\nu < 0.2$ eV).

References

[1] F.A.Danevich et al. *Lett. JETP* **49** (1989) 417

[2] Yu.G. Zdesenko et al. *Proc. of Int. the Symp. on Underground Physics*, Baksan Valley, 1987 (Moscow: Nauka, 1988) **291**

[3] F.A. Danevich et al. *Abstact of the 14th EPS Conf. Rare nuclear decays and fundamental processes*, Bratislava, 1990 (Frankfurt: EPS) **141** (1990) 29

[4] Yu.G. Zdesenko *J. Phys. G: Nucl. Phys.* **17** (1991) S243

[5] A. Staudt, K. Muto and H.V. Klapdor-Kleingrothaus *Europhysics Letters* **13** (1990) 535

Section 3: The Early Universe and Galactic Evolution

Primordial nucleosynthesis and beryllium

Robert A. Malaney

Canadian Institute for Theoretical Astrophysics, University of Toronto, Toronto, ON, Canada M5S 1A7.

Abstract.   A review of the current status of primordial nucleosynthesis is presented. In particular, the consequences of a recent Monte Carlo analysis of standard big bang nucleosynthesis are discussed. Such a study addresses the validity of the standard model, and the limits imposed by it on the baryonic density of the universe. Inhomogeneous primordial nucleosynthesis models are reviewed in light of the recent observations of beryllium in very metal poor halo stars. Beryllium production in the very early galaxy by a new mechanism, namely neutrino nucleosynthesis, is also discussed.

1. Standard Big Bang Nucleosynthesis

In a recent and comprehensive study of standard big bang nucleosynthesis (SBBN) (Smith, Kawano and Malaney 1992) each of the 12 major big bang reaction rates were represented by a gaussian distribution centered on a mean rate value. In any particular computer run, a random number generator selected a number for each of the 12 reactions, which was then used to determine a reaction rate whose distribution over many runs would give a gaussian around the mean rate with a width equal to the $1\text{–}\sigma$ uncertainty assigned to the reaction. This feature was incorporated into the latest adaptation of Wagoner's code (Kawano 1992) and the primordial light element abundances were computed. For each value of the baryon-to-photon ratio η, 1000 runs of the computer program was carried out. This enabled the results to be represented by a mean value and a standard deviation for the abundance of each of the four light elements.

Uncertainties placed on inferred abundance ranges cannot be made statistically significant in the same sense as the numerical determinations of the abundance yields. However, the use of conservative abundance limits, coupled with the inclusion of nuclear uncertainties in a consistent fashion, allow the permitted range of η, and the validity of SBBN to be inferred with somewhat increased confidence.

From the inferred constraints on the primordial abundances, and from the limits of the abundances predicted by the Monte Carlo program, an allowed range of $2.86 \leq \eta_{10} \leq 3.77$ ($\eta_{10} = \eta \times 10^{10}$) was determined (Smith $et\ al.$ 1992). This range for η_{10} is bounded from below by the lower limit inferred on $D+^3He$, and bounded from above by the upper limit inferred on ^4He. In order to convert this allowed range of η_{10} into limits on the present baryon density, we require the present number density, n_γ, of microwave photons. This latter quantity can be readily obtained in terms of the present temperature of the microwave background, T_γ, through

$$n_\gamma = 20.3 T_\gamma^3 \text{ cm}^{-3} \quad , \tag{1}$$

where T_γ is in units of $^\circ K$. The most recent results from ground based as well as COBE data yield $2.75 \leq T \leq 2.79$ K (2σ) according to Walker $et\ al.$ (1991). It is convenient to express the baryon mass density, ρ_b, in terms of the critical mass density

$$\rho_c = \frac{3H_o^2}{8\pi G} \quad , \tag{2}$$

where $H_o = 100h$ km s^{-1} Mpc^{-1} is the present value of the Hubble parameter, and $0.4 \leq h \leq 1$ is a parameter which allows for the present uncertainty in the value of H_o as inferred from observation (for review see van den Bergh 1989). Defining $\Omega_b = \rho_b/\rho_c$ and combining eqs. (1) and (2) we find

$$\Omega_b h^2 = 3.6 \times 10^{-3} \left(\frac{T_\gamma}{2.75}\right)^3 \eta_{10} \quad . \tag{3}$$

Accounting for the uncertainties of T_γ and the allowed range of η, we find the permitted range of $\Omega_b h^2$ is given by

$$0.01 \leq \Omega_b h^2 \leq 0.014 \quad . \tag{4}$$

Allowing for the range $0.4 \leq h \leq 1$ this transforms into

$$0.01 \leq \Omega_b \leq 0.09 \quad . \tag{5}$$

There are several immediate inferences that can be drawn from the limits given by eq. (5). First, it is clear that the baryonic density falls far short of that required to close the universe. Second, since the amount of luminous matter in the universe is constrained to be $\Omega_{lum} \lesssim 0.007$ (eg. Pagel 1990) then baryonic dark matter is required. However, the argument for the necessity of non-baryonic dark matter is less persuasive. Although there is tentative evidence that Ω as inferred from galaxy cluster data (eg. $\Omega \approx 0.2$ for the Coma cluster - Hughes 1989) would seem to imply the existence of some form of non-baryonic matter, uncertainties could yet conspire to place Ω as determined from cluster data close to our inferred upper limit on Ω_b. If, however, larger values for Ω, such as those tentatively indicated by large scale density fluctuations and peculiar velocities ($\Omega = 0.75 - 1.15$ assuming the bias factor for the IRAS galaxies is close to unity - Yahil 1990; Kaiser *et al.* 1991), become well established then the constraints imposed by SBBN would imply the existence of some form of non-baryonic matter.

The SBBN model can account, within a narrow range of η, for all of the primordial abundances as inferred from observation. However, there has been much interest recently in extending the range of η beyond that allowed for by SBBN through the introduction of some new physics, or by the relaxation of some of the assumptions built into the SBBN model (for a review of non-standard primordial nucleosynthesis models see Malaney and Mathews 1992). Using only the inferred primordial abundances as a barometer, it was determined that deviations from the SBBN model will be *demanded* if the range of Y_p is determined to be outwith the range $0.237 \leq Y_p \leq 0.247$, assuming the constraints on D, ^{3}He and ^{7}Li remain firm. Although there is tentative evidence which suggest that the observed value of Y_p lies below this range (Fuller *et al.* 1991), the systematic uncertainties do not yet allow for a strong case (Pagel 1992). Any firm conclusion, however, that SBBN requires some modification will most likely come from this quarter.

2. Primordial Beryllium

Two recent detections (Gilmore, Edvardsson and Nissen 1991; Ryan *et al.* 1991) of ^{9}Be in the metal-poor star HD 140283 ([Fe/H]≈ -2.6) at the level Be/H$\sim 10^{-13}$ by number, raised the spectre of either primordial production of ^{9}Be in the early universe, or of some overlooked mechanism which dominated the production of ^{9}Be in the very early galaxy. This is especially so since there is tentative evidence which suggests that the ^{9}Be measured in HD 140283 cannot be accounted for by traditional models of cosmic ray spallation (see later discussion). We briefly review here the calculations which suggested the possibility of primordial beryllium production. For a more complete history on this subject the reader should consult the reviews of Reeves (1991) and Malaney and Mathews (1992).

Following the early two-zone calculations of inhomogeneous nucleosynthesis (eg. Applegate, Hogan and Scherrer 1987), it was shown that the diffusion of neutrons from one region to the other *during* the nucleosynthesis epoch played a more important role than hitherto believed (Malaney and Fowler 1988). There subsequently appeared a large number of investigations which utilized more sophisticated treatments of neutron diffusion (eg. multi-zoned) during the nucleosynthesis epoch (eg. Kurki-Suonio and Matzner 1989). Due to overproduction of ^{4}He in these more sophisticated codes the original hope of these models, namely a universe closed by baryons, became less compelling. It is worth noting, however, that late-time dissipation effects may occur when the photon mean free path becomes larger than the scale of the high-density inhomogeneity (Alcock *et al.* 1990). If these effects occur prior to the completion of nucleosynthesis they can have profound effects on the isotopic yields, the details of which are still unknown.

Inhomogeneous nucleosynthesis models must include a large number of nuclear reactions not included in the standard model networks. It was recognized early on that heavy (atomic mass $A > 7$) isotopic production, orders of magnitude larger than predicted by SBBN, could result as a consequence of these new reactions (Applegate and Hogan 1985). At high values of Ω_b the following reaction sequence, occurring in the neutron-rich regions, leads to the production of boron and carbon (Malaney and Fowler 1987; Applegate *et al.* 1988)

$$^7\text{Li}(n,\gamma)^8\text{Li}(\alpha,n)^{11}\text{B}(n,\gamma)^{12}\text{B}(\beta^-,\nu)^{12}\text{C} \ .$$

A key reaction with regard to Be production, however, which had been previously neglected was ^{7}Li$(t,n)^9$Be (Boyd and Kajino 1989). The inclusion of this latter reaction into the inhomogeneous nucleosynthesis codes led to a production of ^{9}Be orders of magnitude larger than that predicted by SBBN (Boyd and Kajino 1989) . Although it was subsequently determined that this was not always the case as the Be/H yield was found to be a sensitive function of the parameter space (Malaney and Fowler 1989), the possibility of producing observable quantities of primordial Be remained open.

The predictions of high Be yields were based on two-zone model calculations, however, in which the diffusion of neutrons during the nucleosynthesis was not accounted for. Using more sophisticated multi-zone calculations it was subsequently determined that an accurate treatment of neutron diffusion during nucleosynthesis could have a significant influence on the Be yield (Terasawa and Sato 1990). For example, in *all* of the parameter space investigated in the study of Terasawa and Sato (1990), it was found that the refinment of multi-zoning led to Be yields at least two orders of magnitude below the level of Be/H$\sim 10^{-13}$ recently determined in HD 140283. However, the importance of the diffusion coefficients and the length scale over which the neutrons diffuse could play a critical role in the final Be yield (eg. Applegate 1988, Malaney

1992a). In particular, large separation scales ($\gtrsim 10^8$ cm) between the high-density regions at the onset of nucleosynthesis, could drastically alter the efficiency of the neutron back-diffusion, resulting in high yields of Be. It is due to these theoretical caveats that a continued search for the *unique* observational signature of inhomogeneity in the early universe, namely a metallicity-independent "Be floor", should be given the highest priority.

3. Beryllium in the Early Galaxy

It is clear that SBBN cannot account for the ^{9}Be detected in HD 140283, and that inhomogeneous models do not yet make a clear prediction. Therefore new scenarios for ^{9}Be production in the early galaxy should also be considered. In fact, it has been recently proposed that the neutrino-induced production of ^{9}Be via ^{7}Li(t,n)^{9}Be in the helium shells of early generation supernovae may be of importance.

The neutrino-induced production of light isotopes ($\nu-$process) is not a new idea (eg. Domogatskii and Nadyozhin 1977; Woosley 1977; Epstein, Colgate and Haxton 1988; Woosley *et al.* 1990). However, the primary emphasis of earlier calculations was the reproduction of nuclear isotopes in solar-system proportions, and in the investigation of a possible r-process site. As such, these calculations are not directly applicable to the early galaxy. In particular, the initial metallicities of the progenitor supernova are too high ([Fe/H] > -2) to be of relevance to $\nu-$process production of ^{9}Be at [Fe/H]$\lesssim -3$, the range of interest with regard to ^{9}Be in HD 140283. Also, although it was found that significant yields of ^{11}B could be produced in the helium shell as a consequence of the neutrino burst, relatively little, if no, ^{9}Be was produced. However, it has been pointed out (Malaney 1992b) that at the low metallicities relevant to HD 140283 the ^{7}Li(t,n)^{9}Be nuclear reaction, not previously included in $\nu-$process nucleosynthesis, becomes of paramount importance. Indeed, if the isotopic yields in the helium shell adopted for the calculations of Malaney (1992b) are typical of the yields associated with a supernova event in the early galaxy (i.e. [Fe/H]$\lesssim -3$), then dilution factors of order 10^{-4} can readily to account for the ^{9}Be observed in HD 140283.

The yield of ^{9}Be from a given presupernova configuration in the neutrino-nucleosynthesis scenario is approximately metallicity independent (any such dependence would have to come through stellar structure), but the total yield in the early galaxy will depend on the supernovae rate. This primary nature of the ^{9}Be production would be evident through a linear relationship of its abundance with that of some other element that is primarily produced in a supernova. One such element would be oxygen. Indeed, it has been recently reported that the Be abundance in very metal-poor stars is directly proportional to the oxygen abundance (Gilmore *et al.* 1992). This behaviour is in direct contrast to that anticipated from traditional cosmic ray spallation reactions, which hitherto have been considered the likely source of all Be production in the galaxy (eg. Reeves and Meyer 1978; Walker, Mathews and Viola 1985). In the models invoked to account for light isotope yields at solar metallicities, the proton spallation on ambient CNO nuclei predicts a secondary relationship in that the Be abundance scales as the square of the O abundance. However, it should be noted that several authors have have attempted to accommodate the linear relationship of the low-metallicity Be and O data by arguing that modifications of the cosmic-ray proton spectrum in the early galaxy are justified, and that flatter proton spectrums allow for better agreement with the data (Gilmore *et al.* 1992; Duncan, Lambert and Lemke 1992; Prantzos *et al.* 1992; Walker *et al.* 1992). As discussed below, future observations will be able to discriminate between the different scenarios, including the primordial ^{9}Be production scheme, the latter of which would be observable as a constant ^{9}Be abundance with decreasing [Fe/H].

A recent determination of B/Be$= 10^{+5}_{-4}$ has been reported for HD 140283, with similar ratios reported for HD 19445 and HD 201891 (Duncan *et al.* 1992). The B/Be ratio in HD 140283 resulting from the $\nu-$process will be very sensitive to the stellar model and the details of the explosion thermodynamics. This is not only a consequence of sensitivity to the value of the shock temperature (the subsequent passage of the shock front can lead to Be destruction and B production), but also from the very large ^{11}B production that can occur within the carbon shell (Woosley *et al.* 1990) via direct neutrino spallation on ^{12}C. The amount of ^{11}B in the atmosphere of HD 140283 produced within the carbon shell will again be sensitive to the presupernova structure. Because of this sensitivity to initial conditions the B/Be ratio predicted by the early galactic ν-process is not well defined. The situation is further clouded by the study of Woosley and Weaver (1988) who show that various effects such as Coulomb corrections, semi-convection, convective overshoot, and the rate of the ^{12}C$(\alpha,\gamma)^{16}$O reaction, can reduce the C/O ratio within the carbon shell. Since the B production is directly related to the C abundance such effects (each of which can introduce an uncertainty of order 2) can be of significance. Even allowing for such effects it would seem that the lowest B/Be allowed for by the $\nu-$process would still be a factor of ~ 3 higher than that observed in HD 140283. But again we caution that hitherto there has been little motivation for the study of very low-metallicity objects. The relevant details of very metal poor stellar models, which would be required before a more accurate assessment of the B/Be ratio induced by the low-metallicity $\nu-$process was forthcoming, do not presently exist in the literature. Future study in this area would be of great value.

It is important to note, however, that galactic cosmic ray spallations on ambient interstellar CNO isotopes in a population II environment make a model independent prediction of $7 \lesssim$ B/Be $\lesssim 17$ (Walker *et al.* 1992). Observation in metal-poor stars (of either lower or higher metallicity than HD 140283) of a B/Be ratio outwith this range would be incompatible with the cosmic ray spallation hypothesis, but possibly compatible with the ν-process. Steigman and Walker (1992), and Walker *et al.* (1992) also point out the importance of future observation of lithium in metal-poor stars as a possible discriminate against cosmic ray spallation production, especially observations of ^{6}Li in hot halo stars.

It is of considerable interest to note the large quantities of ^{7}Li which are also produced in the ν-process. It is worthy to point out, however, that essentially no ^{6}Li is produced. The dilution factors required to account for the ^{9}Be observed in HD 140283 can simultaneously account for its measured lithium abundance (Rebolo *et al.* 1988), namely Li/H$= 1.1 \times 10^{-10}$. A verification of the ν-process described here, through continued spectroscopic study of very metal poor stars, may require a revision of the inferred primordial lithium abundance.

Acknowledgments

I would like to thank M. Smith and L. Kawano who were my collaborators on the Monte Carlo study, key results of which were presented in section 1.

References

Alcock, C. R., Dearborn, D. S., Fuller, G. M. , Mathews, G. J. and Meyer, B. S., 1990, *Phys. Rev. Lett.*, **64** 2607.

Applegate, J. H., 1988, *Phys. Rep.*, **163** 141.

Applegate, J. H. and Hogan, C., 1985, *Phys. Rev.* **D30** 3037.

Applegate, J. H., Hogan, C. J. and Scherrer, R. J., 1987, *Phys. Rev.*, **D35** 1151.

Applegate, J. H., Hogan, C. J. and Scherrer, R. J., 1988, *Ap. J.*, **329** 572.

Boyd, R. N. and Kajino, T., 1989, *Ap. J. Lett.*, **336** L55.

Domogatskii, G. V. and Nadyozhin, D. K., 1977, *M. N. R. A. S.*, **178**, 33p.

Duncan, D. K., Lambert, D. L., and Lemke, M., 1992, preprint.

Epstein, R. I., Colgate, S. A. and Haxton, W. C., 1988, *Phys. Rev. Lett.*, **61**, 2038.

Fuller, G. M., Boyd, R. N. and Kalen, J. D. 1991, *Ap. J. Letts.*, **371**, L11.

Gilmore, G., Edvardsson , B. and Nissen, P. E. , 1991, *Ap. J.*, **378** 17.

Gilmore, G., Gustafsson, B., Edvardsson, B. and Nissen, P. E., 1992, *Nature* **357**, 379.

Hughes, J. P., 1989, *Ap. J.*, **337**, 21.

Kaiser, N., Efstathiou, G., Ellis, R., Frenk, C., Lawrence A., Rowan-Robinson, M., and Saunders, W., 1991, *Mon. Not. R. astr. Soc.*, **252**, 1.

Kawano, L. H., 1992, Caltech/Kellog preprint OAP-714.

Kurki-Suonio, H. and Matzner, R. A., 1989, *Phys. Rev.* **D39** 1047.

Malaney, R. A., 1992a, in *Nuclear Astrophysics Symposium in Honor of W. A. Fowler's 80th Birthday*, to appear in *Physics Reports*.

Malaney, R. A., 1992b, CITA Report 92/5.

Malaney, R. A., and Fowler, W. A., 1987, in "The Origin and Distribution of the Elements", p76, ed. G. J. Mathews (Singapore: World Scientific).

Malaney, R. A., and Fowler, W. A., 1988, *Ap. J.* , **333** 14.

Malaney , R. A. and Fowler, W. A., 1989, *Ap. J. Lett.*, **345** L5.

Malaney, R. A. and Mathews, G. J., 1992, *Phys. Rep.* in press.

Pagel, B. E. J., 1990, in *Baryonic Dark Matter*, ed. D. Lynden-Bell and G. Gilmore (Kluwer: Netherlands) p237.

Pagel, B. E. J., 1992, *M. N. R. A. S.*, in press.

Prantzos, N., Casse, M. and Vangioni-Flam, E., 1992, *Ap. J.*, submitted.

Rebolo, R., Molaro, P. and Beckman, J. E., *Astr. Ap.*, 1988, **192**, 192.

Reeves, H., 1991, *Phys. Rep.*, **210**, 335.

Reeves, H., and Meyer, J. P., 1978, *Ap. J*, **226**, 613.

Ryan, S. G., Norris, J. E., Bessell, M. S. and Deliyannis, C. P., 1992, *Ap. J.* in press.

Smith, M. S., Kawano, L. H. and Malaney, R. A., 1992, *Ap. J. Supp.*, submitted.

Steigman, G., and Walker, T. P., 1992, *Ap. J. Lett.*, **385** L13.

Terasawa, N. and Sato, K., 1990, *Ap. J. Lett.*, **362** L47.

Van den Bergh, S., 1989, *Astron. Astrophys. Rev.*, **1**, 111.

Walker, T. P., Mathews , G. J. and Viola, V. E. , 1985, *Ap. J.*, **229** 745.

Walker, T. P., Steigman, G., Schramm, D. N., Olive, K. A. and Kang, H., 1991, *Ap. J.*, **376**, 51.

Walker, T. P., Steigman, G., Schramm, D. N., Olive, K. A., and Fields, B., 1992, *Ap. J.* submitted.

Woosley, S. E., 1977, *Nature*, **269**, 42.

Woosley, S. E. and Weaver, T. A., 1988, *Physics Reports*, **163**, 79.

Woosley, S. E. , Hartmann, D. H., Hoffman, R. D. and Haxton, W. C., 1990, *Ap. J.*, **356**, 272.

Yahil, A., 1990, in *Particle Astrophysics*, ed. J.-M. Alimi, A. Blanchard, A. Bouquet, F. M. de Volnay, J. T. T. Van (Editions Frontiers: France) p483.

Inhomogeneous universes from lattice QCD with dynamical quarks

W. Bürger, M. Faber, M. Hackel, H. Markum, M. Müller

Institut für Kernphysik, Technische Universität Wien,
A–1040 Vienna, Austria

Abstract. We investigate the free energy distribution across the interface of coexisting quark and hadron matter in the framework of lattice QCD. We calculate the interface tension α in the presence of dynamical quarks with four flavors. It turns out that α is very small and compatible with zero. The chiral condensate indicates the same width of the domain wall as the Polyakov loop distribution and the other thermodynamical observables under consideration.

1. Introduction

QCD thermodynamics on space-time lattices has demonstrated that matter can exist in two phases distinguishing between the confining hadron phase and the free quark-gluon plasma-phase. In pure gluonic QCD and in full QCD with four light dynamical fermions most numerical simulations support the fact that the phase transition is of first order [1] which implies that the two different phases can coexist at the critical temperature. This opens up the possibility of the creation of an inhomogeneous universe which can now be studied from the first principles of lattice QCD. The most important observable under consideration is the surface energy α between the quark-gluon plasma-state and the hadronic bubbles. The numerical value of α is a fundamental quantity for the inhomogeneity of the universe and represents an input parameter for the probability of nucleation and the average distance between nucleation centers, and further effects the nucleosynthesis of light elements [2]. The thickness of the domain wall of a coexisting two-phase system can also be associated with the skin of the fireball of a quark-gluon plasma which is currently under investigation in ultrarelativistic heavy-ion experiments.

To extract α one has to evaluate thermodynamical expressions demanding to differentiate the partition function with respect to the temperature, volume and interface area. This can be realized on the lattice by directly summing over plaquettes at fixed couplings which is called differential method. Lattice simulations of pure gluonic QCD on four-dimensional hypercubes of sizes $N_x \times N_y \times N_z \times 2$ have led to a non-vanishing

[1] Supported by "Fonds zur Förderung der wissenschaftlichen Forschung" under Contract No. P7510

$\alpha/T_c^3 \approx 0.24$ [3]. We perform an analysis for $N_t = 4$ in the pure gluonic QCD and in the presence of dynamical quark fields for which we choose the number of flavors $n_f = 4$ and the mass $m = 0.05$. This opens up the possibility to explore the domain wall between a two-phase system with spontaneously broken and restored chiral symmetry. In addition to the surface tension and other thermodynamical observables we calculate the distribution of the chiral condensate across the interface.

2. Theory

Starting from the relation for the free energy $F(T, V, A) = -T \ln Z(T, V, A)$ we express the partition function Z

$$Z(T, V, A) = \int \prod dU_{x\mu} \prod d\overline{\chi}_x \prod d\chi_x \exp(-S(U, \overline{\chi}, \chi)) \tag{1}$$

by a path integral over the lattice action $S(U, \overline{\chi}, \chi)$ [4]

$$
\begin{aligned}
S(U, \overline{\chi}, \chi) &= S_G(U) + S_F(U, \overline{\chi}, \chi) \\
&= \sum_x \left[\frac{1}{g_1^2} \frac{a_3}{a_0} (P_{01} + P_{02}) + \frac{1}{g_2^2} \frac{a_0}{a_3} (P_{13} + P_{23}) + \frac{1}{g_3^2} \frac{a_T^2}{a_0 a_3} P_{03} + \frac{1}{g_4^2} \frac{a_0 a_3}{a_T^2} P_{12} \right] \\
&\quad + \frac{n_f}{4} \left[\sum_x m \overline{\chi}_x \chi_x + \frac{1}{2} \sum_{x,\mu} \overline{\chi}_x \frac{\eta_\mu}{a_\mu} \left(U_{x\mu} \chi_{x+\mu} - U_{x-\mu,\mu}^{\dagger} \chi_{x-\mu} \right) \right] .
\end{aligned}
\tag{2}
$$

In the gluonic part of the action, the plaquettes $P_{\mu\nu}$ are defined as

$$P_{\mu\nu} = P_{\mu\nu}(x) = \mathrm{tr}[2 - U_{x\mu} U_{x+\mu,\nu} U_{x+\nu,\mu}^{\dagger} U_{x\nu}^{\dagger} - \mathrm{h.c.}] \tag{3}$$

and $U_{x\mu}$ means the gauge field on a link of a hypercubic lattice. The factors $1/g_i^2$ denote the couplings for an anisotropic lattice with the so-called Karsch coefficients c_s, c_t, $\beta_0 = c_s + c_t$ entering the renormalization group equation [4]. The anisotropic couplings determine the corresponding lattice spacings a_i. In the fermionic part of the action, according to the Kogut-Susskind formulation the fermion fields are represented by single component Grassmann fields $\overline{\chi}_x, \chi_x$ at the sites of the lattice [4]. The fermionic action describes n_f flavors with mass m and the Dirac matrices reduce to phase factors $\eta_\mu = (-1)^{x_1 + \ldots + x_{\mu-1}}$.

To perform QCD lattice-thermodynamics on a two-phase lattice of size $N_x \times N_y \times N_z \times N_t$ we place the interface in the (x,y) plane (see Fig. 1). To be able to take partial derivatives with respect to one variable while keeping two other variables constant we choose three different lattice constants a_0, $a_1 = a_2 = a_T$, a_3 which are all set equal after the derivation of the observables. The temperature, volume and interface size are given by $T = 1/N_t a_0$, $V = N_T^2 N_3 a_T^2 a_3$, $A = N_T^2 a_T^2$. Now we can proceed straightforward to derive the gluonic thermodynamical observables we are interested in, *i.e.* the surface energy α_G, energy ϵ_G, pressure p_G and entropy s_G [3]

$$
\begin{aligned}
\alpha_G \frac{A}{T} &= \left\langle A \frac{\partial S_G}{\partial A} \Big|_{T,V} \right\rangle \\
&= \left\langle \sum_x \left(\frac{1}{g^2} + \frac{1}{2}(c_t - c_s) \right) (2P_{03} - P_{01} - P_{02} - 2P_{12} + P_{13} + P_{23}) \right\rangle, \tag{4}
\end{aligned}
$$

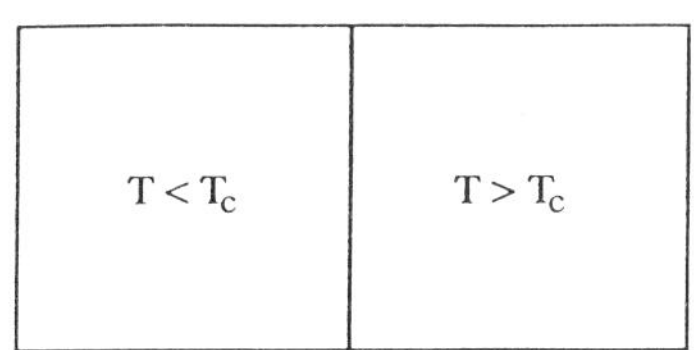

Figure 1. Realization of a two-phase system with an interface at T_c. The phase transition occurs in the z direction at $n_z = 8$ and $n_z = 16$ due to periodic boundary conditions.

which all reduce to sums over plaquettes. To relate the energy and pressure to the $T = 0$ case we have to subtract the vacuum contribution given by the average plaquette $P_{av}(g^2)$ on a symmetric lattice, $\epsilon_{vac} = -p_{vac} = -3\beta_0 P_{av}$. A further observable of interest is the Polyakov loop which on one hand represents the propagator in periodic time direction for a static quark

$$\langle L \rangle = \langle \frac{1}{3V} \sum_{\vec{x}} \mathrm{tr} \prod_{n_t=1}^{N_t} U_{x,\mu=0} \rangle \tag{5}$$

and on the other hand acts as an order parameter.

Similarly, we treat the fermionic part of the thermodynamical observables. After integration over the fermionic fields one obtains the fermion determinant

$$\exp(-S_F^{\mathrm{eff}}) = \frac{n_f}{4} \det[D(U) + m], \tag{6}$$

with the covariant derivative

$$D_{\mu,xy}(U) = \frac{\eta_\mu}{2a_\mu}[U_{x\mu}\delta_{x+\mu,y} - U^\dagger_{x-\mu,\mu}\delta_{x-\mu,y}]. \tag{7}$$

Performing the thermodynamical differentations we get the fermionic parts of the thermodynamical observables, *i.e.* the surface energy α_F, energy ϵ_F, pressure p_F and entropy s_F. The derivation was given for the first time for Wilson fermions in Ref. [3] and is formulated here for Kogut-Susskind fermions

$$\alpha_F \frac{A}{T} = \frac{n_f}{8} \langle \mathrm{tr}[(D_1 + D_2 - 2D_3^{\bullet})(D + m)^{-1}] \rangle. \tag{8}$$

In the fermionic system the chiral order parameter appears which is related to spontaneous chiral symmetry breaking and is a measure for the virtual quark density

$$\langle\langle \overline{\chi}_x \chi_x \rangle\rangle = \frac{n_f}{4V} \langle \mathrm{tr}(D + m)^{-1}_{xx} \rangle. \tag{9}$$

Single brackets mean path integration over the gauge field after fermionic integration whereas double brackets denote additional fermionic integration. The total expectation value of a thermodynamical observable $O = O_G + O_F$ consists of the gluonic and fermionic parts.

The simulations are realized on an $8 \times 8 \times 16 \times 4$ lattice with one half in the hadron phase at an inverse gluon coupling $\beta = \beta_c - \Delta\beta$ and the other half in the quark phase at $\beta = \beta_c + \Delta\beta$ (see Fig. 1). The partition wall is set to the critical coupling $\beta_c = 6/g_c^2$. Thus, the interface is forced by construction and is not created dynamically. To obtain the physical expectation value for a coexisting two-phase system, one has to extrapolate the observables to the critical point.

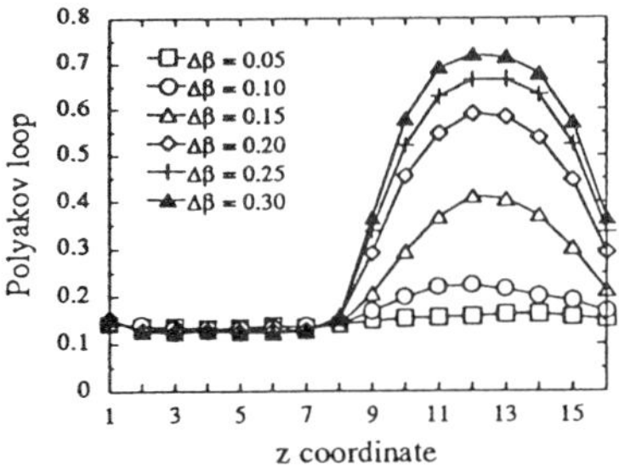
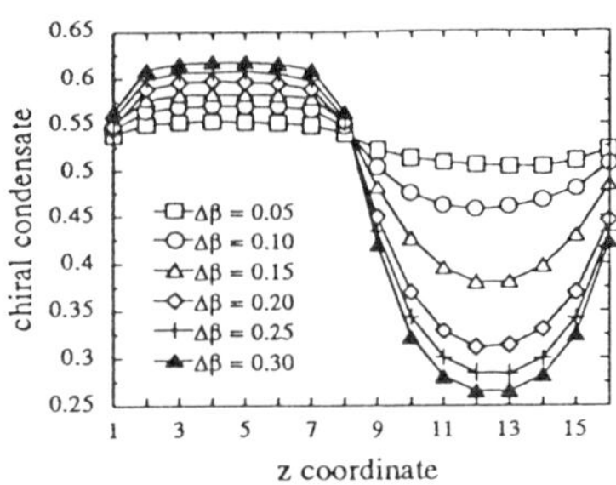

Figure 2. Profiles of the Polyakov loop and the chiral condensate for full QCD with four flavors and for various two-phase systems.

3. Results

We approximated the path integral by 5000 Monte Carlo iterations using the pseudo-fermionic algorithm with 50 fermionic steps and scanned several values of $\Delta\beta = 0.05, 0.10, 0.15, 0.20, 0.25, 0.30$. The dynamical quark field has flavor number $n_f = 4$ and mass $m = 0.05$ [5]. In Fig. 2 we present the Polyakov loop and the chiral order parameter. A clear change is seen at the transition point from confinement to deconfinement. The Polyakov loop has a non-vanishing expectation value due to the broken Z_3 symmetry from the fermions in the confining phase. The chiral symmetry is broken in the confinement and restored in the deconfinement. For the chiral order parameter profile crossing the interface at $n_z = 8$ we find an increase of the wall thickness when we approach the coexisting phases at $\Delta\beta \to 0$. For $\Delta\beta = 0.3$ the width is about 5 lattice spacings a which corresponds to roughly 1 fm. It turns out that the chiral condensate has the same width as the Polyakov loop distribution. The order parameter with dynamical quarks has not reached a plateau like in pure gluonic studies on larger lattices indicating a width of the domain wall of 2.5 fm [3]. The lattice results can be compared with a study of the σ model which predicts a width of about 4.5 fm [6].

Next we discuss the gluonic and fermionic contributions to the total thermodynamical observables in Fig. 3. We start with the distribution of the energy. Its gluonic part exhibits discretization effects at the transition point from the vacuum contribution. The fermionic part has a smooth behavior and clearly shows both phases, confinement and deconfinement. The vacuum corrections have been determined from a consistent fermionic simulation of an 8^4 lattice with the same parameters. The total energy as a sum of the gluonic and fermionic parts has discretization effects. In the deconfining phase the energy is in accordance with the Stefan-Boltzmann limit of an ideal gas with a tendency of overshooting.

We turn to the z component of the pressure in Fig. 3. Discretization effects at the phase transition are clearly visible. At high temperatures the ideal gas relation $\epsilon = 3p$ holds. The next plot presents the profile of the entropy. The discretization effects are partially compensated because pressure and surface energy enter into the entropy with different signs. For all thermodynamical observables it is found that the gluonic and fermionic contributions are of the same size.

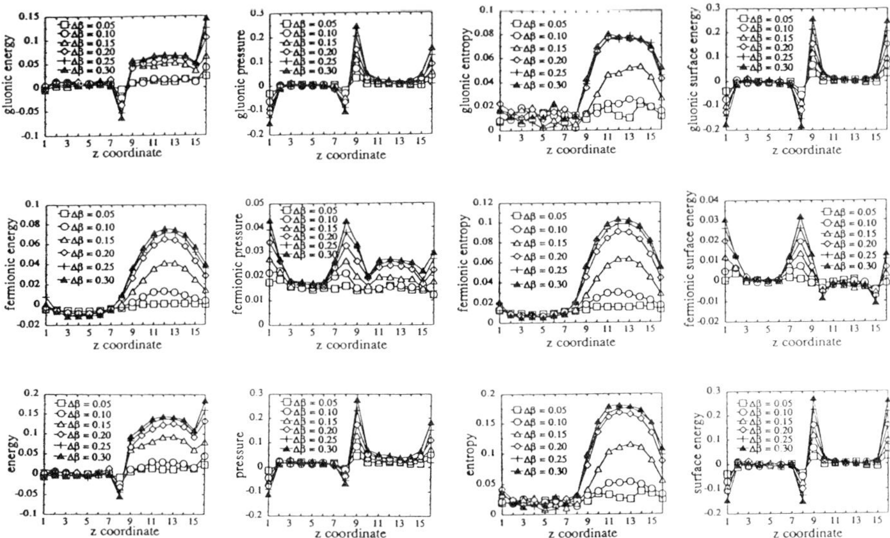

Figure 3. Thermodynamical observables for QCD with four dynamical quarks as a function of the z coordinate for different temperature gradients. The first row of the plots displays the gluonic contribution and the second row the fermionic part while the third row gives the sum (see text).

Now we look at the profile of the surface energy in Fig. 3. In the region of the phase transition the surface energy has a non-vanishing value. In comparison to the gluonic part having a positive and a negative peak, only a single positive peak is detected for the fermionic contribution. Thus, the total surface energy is stabilized by the fermionic contribution.

The physical expectation value of the surface energy is the sum of its profile in z direction. The left plot in Fig. 4 shows the surface energy normalized to T_c^3 in the presence of dynamical quarks as a function of the coupling gradient $\Delta\beta$. One finds that the fermionic part is smaller than the gluonic one. The extrapolated surface energy for a coexisting two-phase system is hard to extract and compatible with zero. In the right plot we compare our $N_t = 4$ computations with and without dynamical fermions. We find that the surface energy has a small numerical value decreasing with increasing time elongation and is lower than in the pure gluonic case, $\alpha/T_c^3 < 0.1$.

4. Summary

This study contains the first trial to extract the interface tension in the presence of dynamical quarks. We studied the fermionic behavior of different thermodynamical observables together with the order parameters of confinement and chiral symmetry. A crude estimate of the thickness of the domain wall gives for $\Delta\beta = 0.3$ about 5 lattice spacings which corresponds to roughly 1 fm. The wall thickness increases towards the

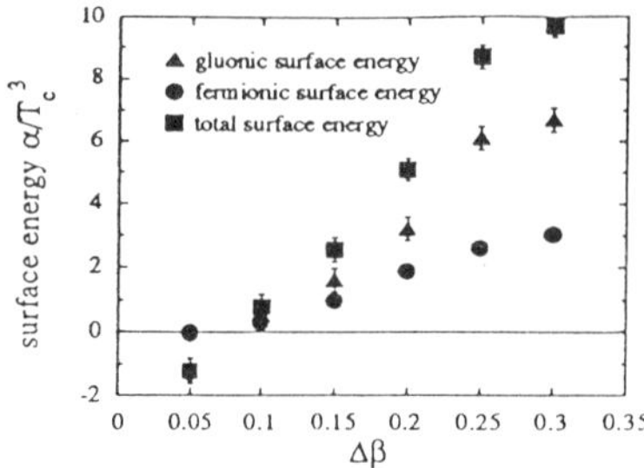

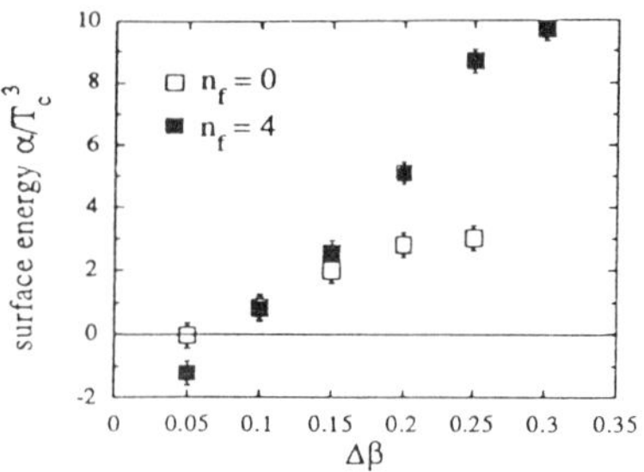

Figure 4. Total interface tension α/T_c^3 in the presence of dynamical quarks with $n_f = 4$ and its gluonic and fermionic contribution (left plot). Comparison of the interface in full QCD with the pure gluonic case (right plot). Error bars denote mean standard deviation.

coexisting two-phase system which might give some first principle information for heavy-ion experiments. For dynamical fermions with four flavors of equal mass we found that the gluonic and fermionic contributions to the interface tension are of comparable size but difficult to extrapolate to $T \to T_c$. Important for astrophysics, we can predict $\alpha/T_c^3 \approx 0.1$ as an upper bound for the interface tension. Since in the case of two light up and down quarks and a heavy strange quark the phase transition becomes even weaker, we can conclude that lattice calculations favor a rather homogeneous universe.

References

[1] Okawa M 1990 *Nucl. Phys.* B (Proc. Suppl.) **16** 562
     Bacilieri P et al. (APE Collaboration) 1988 *Phys. Rev. Lett.* **61** 1545
     Gottlieb S 1991 *Nucl. Phys.* B (Proc. Suppl.) **20** 247

[2] Applegate J H, Hogan C J and Scherrer R J 1987 *Phys. Rev.* D **35** 1151
     Fuller G M, Mathews G J and Alcock C R 1988 *Phys. Rev.* D **37** 1380

[3] Kajantie K, Kärkkäinen L and Rummukainen K 1990 *Nucl. Phys.* B **333** 100
     Huang S, Potvin J, Rebbi C and Sanielevici S 1990 *Phys. Rev.* D **42** 2864;
     1991 *Phys. Rev.* D **43** 2056

[4] Karsch F 1982 *Nucl. Phys.* B **205** 285
     Kajantie K and Kärkkäinen L 1988 *Phys. Lett.* B **214** 595
     Heller U and Karsch F 1985 *Nucl. Phys.* B **258** 29

[5] Hackel M 1991 *Diplomarbeit* (Vienna: Technical University)
     Sakuler W, Bürger W, Faber M, Markum H and Müller M 1992 *Phys. Lett.* B **276** 155

[6] Frei Z and Patkos A 1990 *Phys. Lett.* B **247** 381

The chemical evolution of galaxies

A Burkert

MPI für Astrophysik, Karl-Schwarzschild-Straße 1, W-8046 Garching, FRG

Abstract. The chemical evolution of galaxies is discussed starting with the simple box model. It is shown that this model with the additional assumption of metal-rich outflow provides a first useful insight into the chemical evolution of different stellar populations. More detailed models of galactic nucleosynthesis are then investigated which take into account galactic dynamics, non-instantaneous recycling and incomplete mixing. The implications of non-instantaneous recycling on the evolution of the element abundances in the galactic halo are discussed and constraints on the evolutionary time-scales of the halo are derived. Finally a model for the formation of the galactic halo is presented which takes into account local enrichment and incomplete mixing between different gas phases.

1. Introduction

Stars form in molecular clouds, evolve and synthesize heavy elements in their interior and eject metal-enriched gas into the interstellar medium at the end of their lifetime. The ejecta cool and condense into molecular clouds again. A new population of stars forms from this gas, contaminated by the ashes of an earlier stellar generation. This is the basic cycle of galactic nucleosynthesis which is studied by models of galactic chemical evolution.

In general, galaxies should not be considered as isolated systems. Mergers of spiral galaxies can form elliptical galaxies (Barnes 1989). Galaxy–galaxy encounters exchange metal-enriched gas and stars between different galaxies and affect their star formation rates. Thus galactic chemistry depends strongly on the interaction between galaxies. In addition we might have to extend the chemical cycle to scales of clusters of galaxies (Henry et al 1992). Imagine a first population of galaxies forming from primordial cluster gas, evolving on a short timescale and ejecting metal-enriched gas through galactic winds into the intergalactic medium again. Later generations of pre-enriched galaxies could then form through accretion of this gas by dark matter halos. Thus the chemical evolution of a galaxy might be strongly coupled with the evolution of the whole cluster. In this paper we will neglect the interaction with the environment and consider galaxies as isolated systems.

2. The simple box model

As a zero-order approximation of galactic chemical evolution the simple model describes galaxies as homogeneous, well mixed boxes, filled with two components: gas and stars. Stars form from the gas with a star formation rate Ψ and instantaneously return a mass fraction R of gas as well as a mass fraction y (the yield) of newly processed metals to the interstellar medium. Given these assumptions the chemical cycle can be described by the following set of differential equations:

$$\frac{dM_g}{dt} = -\Psi(1-R) - \left(\frac{dM_g}{dt}\right)_{outflow}$$

$$\frac{dM_*}{dt} = \Psi(1-R) \tag{1}$$

$$\frac{dM_{z,g}}{dt} = \frac{d}{dt}(Z_g \cdot M_g) = (y - Z_g)\frac{dM_*}{dt} - Z_g\left(\frac{dM_g}{dt}\right)_{outflow}.$$

Here M_g and M_* is the total mass of the gas and the stars, respectively. $M_{z,g}$ describes the total mass of metals in the gas, $Z_g = M_{z,g}/M_g$ is the gaseous metallicity and we allowed for gaseous outflow from the box with a rate $(dM_g/dt)_{outflow}$.

Neglecting outflow, the equations (1) can be integrated analytically and lead to a relation between Z_g and the gas fraction in the closed box (Searle and Sargent 1972):

$$Z_g = y \ln\left(\frac{M_{tot}}{M_g}\right) \tag{2}$$

where $M_{tot} = M_g + M_*$ is the total mass of the box and it was assumed that the initial gaseous metallicity $Z_g(M_g = M_{tot}) = 0$. With equation (2) one now can calculate the accumulative metallicity distribution of the stellar component $M_*(Z_* \leq Z)$ which determines the total mass of stars with a metallicity smaller than Z

$$M_*(Z_* \leq Z) = M_{tot}\left(1 - \exp\left(-\frac{Z}{y}\right)\right) \tag{3}$$

and the differential stellar metallicity distribution

$$\frac{dM_*}{dZ} = \frac{M_{tot}}{y}\exp\left(-\frac{Z}{y}\right) \tag{4}$$

In order to derive the yield y for a observed stellar population one often discusses the logarithmical metallicity distribution

$$\frac{dM_*}{d\log Z} \sim Z\exp\left(-\frac{Z}{y}\right) \tag{5}$$

which in the simple closed box model has a maximum at $Z = y$. Note also that the mass-weighted mean metallicity of the stellar component

$$\langle Z \rangle = y\left(1 - \frac{\ln\left(M_{tot}/M_g\right)}{M_{tot}/M_g - 1}\right) \tag{6}$$

cannot exceed the yield and in the limit $M_g ==> 0$ is equal to y. It was shown by Edmunds (1990) that equation (6) holds even with gas flows in the system and unenriched inflow.

According to the equations (2) to (4) the chemical enrichment does not depend on the star formation rate Ψ. Therefore the simple closed box approximation still represents one of the most consistent models of galactic chemical evolution as long as no theory of star formation exists. In more sophisticated chemical models (e.g. Matteucci and Francois 1989) an observed stellar abundance distribution is fitted by adjusting the star formation rate. Thus the star formation history of a stellar population can be predicted, assuming that the chosen model provides a good description of its evolution. However as long as it is not known whether such a star formation history is reasonable one cannot decide whether the chemical model itself is right.

2.1. The chemical evolution of the galactic bulge

The galactic bulge represents an ideal place for the simple closed box model because it is located in the center of our Milky Way and might be massive enough to retain most of its gas. Note however that any low-angular momentum debris from the galactic halo and disk is accreted by the bulge. Metal-enriched infall might therefore be important.

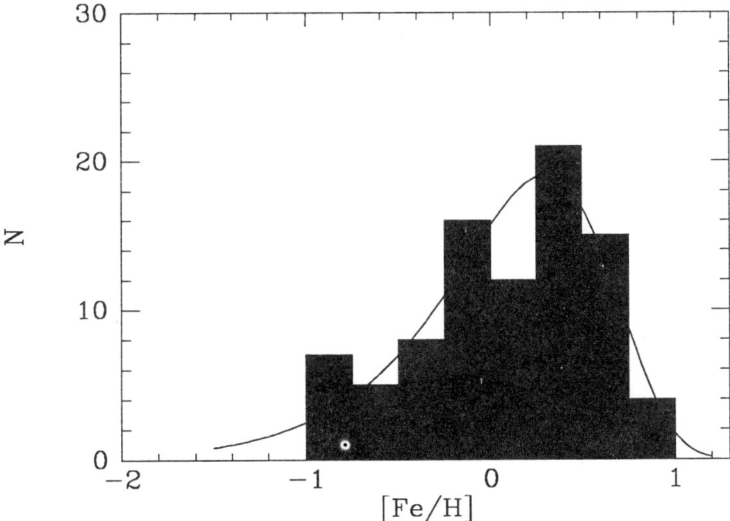

Figure 1. The iron abundance distribution of K giants in the galactic bulge is shown. The solid line indicates the predicted distribution of the simple box model with a yield of twice solar abundance.

Figure 1 shows the remarkable good fit of the simple closed box model to the differential logarithmical abundance distribution of the galactic bulge population (Rich 1988). The iron yield which is derived from the peak in the distribution is twice solar. Note that this peak is an artifact of plotting the number of stars as a function of their logarithmical metallicity (equation 5). The number of stars as a function of their iron metallicity $Z_{Fe} \sim 10^{[Fe/H]}$ peaks at $Z_{Fe} = 0$ and decreases exponentially with increasing Z_{Fe} (equation 4).

Iron is not a good metallicity indicator as it is produced by SN II and SN Ia which evolve on different timescales. Thus it depends on the evolutionary timescale of the bulge with respect to the timescale on which SN Ia explode whether the observed iron yield represents the true yield or only the fraction of iron which is ejected by high-mass stars. A better metallicity indicator is oxygen which represents the largest mass fraction of metals and is produced almost exclusively by high-mass stars which evolve on timescales of less than 10^7 yrs. Unfortunately the oxygen abundance is less accurately measurable than iron and not well determined for the bulge stars. If we adopt an $[O/Fe] = 0.5$ (Matteucci 1992) which is similar to the oxygen abundance in metal poor halo stars the true yield becomes $y_{eff} = 0.12$. This requires an initial mass function which is strongly peaked towards high-mass stars. It therefore seems more reasonable to assume that $[O/Fe] \approx 0$ for most of the bulge stars which leads to a true yield of order 0.04.

2.2. The galactic halo

Figure 2 shows the differential metallicity distribution of the Milky Way's halo globular cluster system (Webbink 1985; Zinn 1985). The abundance distribution can be fitted again by a closed box model with an effective yield which is equal to 3% of the solar iron metallicity. Note that a similar metallicity distribution with the same effective iron yield is found for the halo field star component (Pagel 1991) indicating that the globular clusters and field stars had a similar if not the same nucleosynthesis origin. Adopting an $[O/Fe] \approx 0.5$ the true effective yield of the galactic halo is $y_{eff} = 0.002$ and thus is a factor of twenty smaller than in the galactic bulge.

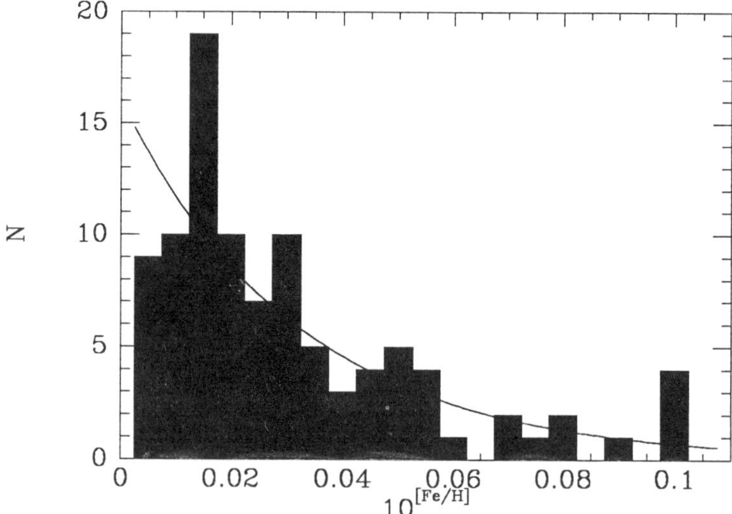

Figure 2. The differential metallicity distribution of the halo globular cluster system is compared with the prediction of the simple model (solid line) adopting a yield of 3% solar abundance.

Such a small yield might result from gaseous outflow. In fact, contrary to the bulge one expects the gas which was metal-enriched in the halo to settle into the galactic disk as a result of dissipational and radiative energy loss. Hartwick (1976) and Edmunds (1990) have shown that the stellar metallicity distribution of a box with outflow will only show the typical exponential decline which is characteristic for the simple model if one assumes that the outflow rate is proportional to the star formation rate:

$$\left(\frac{dM_g}{dt}\right)_{outflow} = \eta \cdot (1-R)\, \Psi. \tag{7}$$

With the assumption (7) the equations (1) lead to the following differential stellar metallicity distribution:

$$\frac{dM_*}{dZ} = \frac{M_{tot}}{y_{eff}} \exp\left(-\frac{Z}{y_{eff}}\right)$$

$$y_{eff} = \frac{y}{1+\eta}. \tag{8}$$

If the derived yield $y = 0.04$ of the galactic bulge represents the true yield and the initial mass functions of the bulge and halo stars did not differ significantly the observed effective halo yield predicts a halo outflow rate $\eta = 19$. Adopting a halo mass of order $3 \cdot 10^9\, M_\odot$ the total mass of gas which settled into the equatorial plane was $M_{disk} = \eta\, M_{halo} \approx 6 \cdot 10^{10}\, M_\odot$, in good agreement with the predicted disk mass of the Milky Way.

2.3. The galactic disk

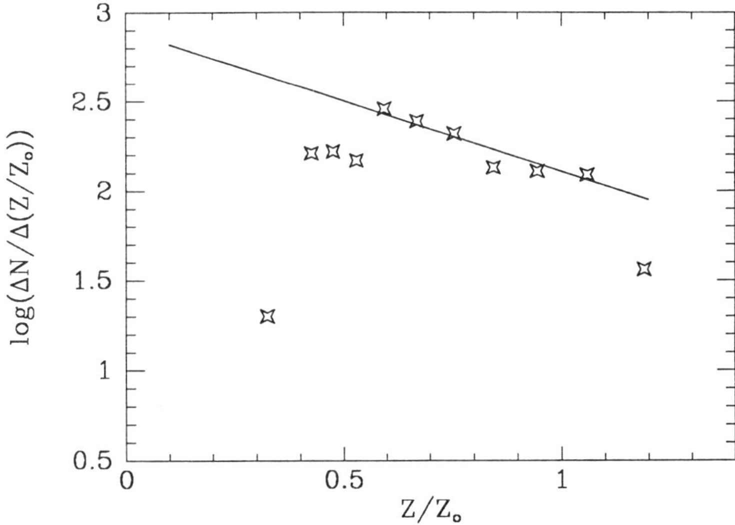

Figure 3. The oxygen abundance distribution of G dwarfs in the solar cylinder is compared with the prediction of the simple model (solid line) adopting a yield of 0.4 solar abundance.

The differential metallicity distribution for a complete sample of g-dwarfs in the solar neighbourhood according to Pagel and Patchett (1975) is shown in figure 3. Above $Z \approx 0.6\,Z_\odot$ the distribution can be well fitted by the simple closed box model with an effective yield $y_{eff} = 0.4\,Z_\odot$. There exists however a lack of metal poor g-dwarfs with respect to the simple model. Note that one has to correct the g-dwarf distribution for the fact, that the old galactic disk has a larger scale height than the young disk by this hiding metal poor, old g-dwarfs at larger vertical heights above the disk. Sommer—Larsen (1991) has however shown that this correction cannot solve the g-dwarf problem.

2.4. Late type galaxies

The simple model predicts a relation between the stellar mass fraction and the metallicity of the interstellar medium (equation 3). Figure 4 shows the observed stellar mass fraction of galaxies as a function of their gas metallicity (Matteucci and Chiosi 1983). The big spread indicates that in general the chemical evolution of galaxies is more complicated than the simple model. There exists however an upper metallicity limit $Z_{max}(M_* / M_{tot})$ which can be well described by the simple closed box assumption with a yield $y = 0.003$. This upper metallicity limit is explained by Edmunds (1990) who showed that almost all processes which affect the chemical evolution of galaxies will lead to metallicities which are smaller than in the simple model. We do not understand however why the derived yield for this limiting box model is much smaller than the true yield of the galactic bulge population.

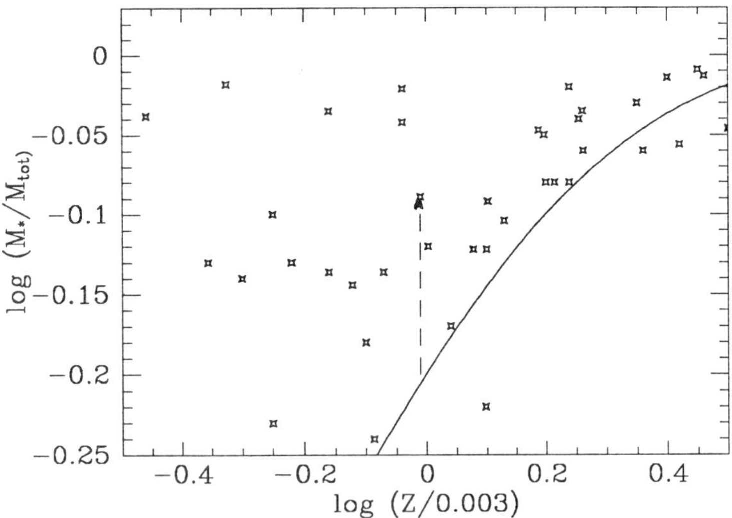

Figure 4. The stellar mass fraction of late type galaxies is plotted as a function of their gaseous oxygen abundance Z. The solid line shows the prediction of the simple model with a yield of 0.003. The dashed line indicates the influence of a strong galactic wind which removes 20% of the total baryonic mass.

Mass loss has been suggested as one of the reasons why galaxies have metallicities which deviate from the predictions of the closed box model. As is indicated by the dashed line in figure 4 a strong galactic wind, which throws out most of the gas on a short timescale, would not change the metallicity of the interstellar medium, it would however increase the stellar mass fraction. On the other hand this wind model requires that up to 90% of the baryonic matter had to be removed in some dwarf galaxies in order to explain their small metallicities. Then the systems could only remain gravitational bound if they were confined by a massive dark halo component which dominated the potential. In fact, high M/L - ratios are observed in dwarf galaxies.

3. Delayed recycling and inefficient mixing in galaxies

The simple model provides a first insight into the chemical evolution of galaxies. It is however not sophisticated enough in order to give a detailed description of the element enrichment in galaxies. Obviously the galactic dynamical evolution can strongly affect the chemical evolution as was already discussed in section 2. Detailed chemodynamical models of galactic evolution have been computed recently in order to understand the coupled chemical and dynamical evolution of dwarf galaxies, elliptical galaxies and galactic disks (Burkert and Hensler 1989; Theis et al 1992; Burkert et al 1992). A discussion of these numerical calculations is beyond the scope of this paper. Instead we will focuse on a few important aspects of galactic chemodynamics which are neglected in the simple chemical models.

3.1. Delayed recycling in the galactic halo

The rate P_i at which stars with an initial mass m and an initial metallicity Z eject an element i into the interstellar medium depends on their lifetime $\tau (m, Z)$ through the relation:

$$P_i (t) = \int_{m_{low}}^{m_{high}} m_i (m) \cdot \Phi(m) \cdot \Psi (t - \tau (m, Z)) \, dm \qquad (9)$$

where m_i is the ejected mass of element i, Φ is the initial mass function and Ψ is the star formation rate. Only in the limit where $|d\Psi / dt| \ll \Psi / \tau$ will the instantaneous recycling approximation be valid:

$$P_i (t) = \Psi (t) \cdot \langle y_i \rangle$$
$$\langle y_i \rangle = \int m_i \Phi \, dm \qquad (10)$$

In general however the history of the element ratios depends strongly on the formation and evolution of an observed stellar population and provides important information on the star formation rate as the primary sites for the production of different elements have different evolutionary timescales.

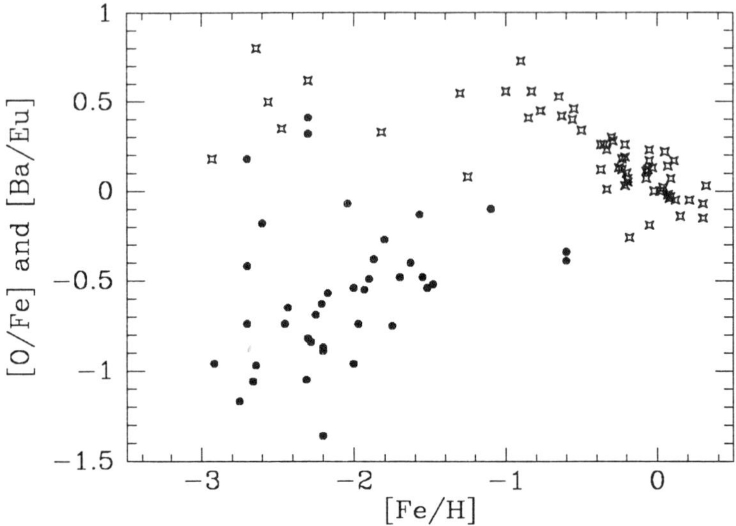

Figure 5. The observed $\lfloor O/Fe \rfloor$ abundance (four-pointed stars) of halo and disk stars is plotted as a function of $[Fe/H]$ with data taken from Sneden et al (1979), Clegg et al (1981), Lambert and Ries (1981) and Peterson et al (1990). Observations of $[Ba/Eu]$ are plotted as filled circles (Luck and Bond 1985; Gilroy et al 1988).

For example, equation (10) is a good approximation for oxygen which is produced by high mass stars with lifetimes of order 10^7 yrs; it is however not valid for elements like iron which are believed to originate primarily from SN Ia with evolution timescales ranging between 10^8 yrs and $3 \cdot 10^9$ yrs (Nomoto et al 1984).

Figure 5 shows the observed $[O/Fe]$ and $[Ba/Eu]$ - ratio versus $[Fe/H]$ for the galactic halo and disk population. The high $[O/Fe] \approx 0.5$ for metal poor halo stars with $[Fe/H] \lesssim -1$ is a typical signature of pure high mass star contamination indicating that the galactic halo formed before supernovae Ia could contribute to the iron enrichment.

Unfortunately the evolutionary timescale for SN Ia progenitors is not well determined as it depends critically on the orbital parameters of the binary system from which this supernova type is assumed to form. On the other hand the age spread of halo field stars and halo globular clusters (VandenBerg et al 1990; Schuster and Nissen 1989) indicates a formation timescale of the galactic halo of the order of $3 \cdot 10^9$ yrs which is consistent with recent calculations of the evolutionary timescale of SN Ia (Mathews et al 1992).

In contrast to $[O/Fe]$ which shows no significant trend for low iron abundances the $[Ba/Eu]$ - ratio grows on average by at least an order of magnitude between $-2.5 \lesssim [Fe/H] \lesssim -1.5$. Eu is a pure r - process element, associated with high mass stars whereas the s - process element Ba is assumed to originate from intermediate mass ($m \sim 1 - 7\,M_\odot$) AGB stars. According to Mathews et al (1992) the increase of the s/r - process ratio at $[Fe/H] = -2.5$ corresponds to an evolutionary timescale of the galactic halo of order 10^8 yrs. The $[Ba/Eu]$ - ratio achieved almost solar abundance ratios

$3 \cdot 10^8$ yrs after the onset of star formation in the galactic halo at $[Fe/H] = -1.5$.

3.2. Inefficient Mixing

Another important assumption of the simple model is the efficient mixing of metals into the star forming gas phase. Thus new stars form with a metallicity which is given by the mean mass fraction of metals in the gas. On the other hand local enrichment and inefficient mixing between different gas phases can significantly change the chemical evolution of a stellar population.

3.2.1. Inefficient mixing between gas phases

Imagine a situation where stars form from a cloudy component (CM) and return metal-enriched gas to a hot intercloud medium (ICM). If the mixing between the ICM and the CM is inefficient almost all ejected metals are stored in a hot gas phase and the metallicity of the stellar component will not change significantly on a timescale which is shorter than the mixing timescale between the two gas phases. If the hot gas phase results primarily from the stellar ejecta its mass M_{ICM} and its metallicity Z_{ICM} can be approximated by ($R \approx 0.2$)

$$
M_{ICM} = R \cdot M_* \approx 0.2 \cdot M_*
$$
$$
Z_{ICM} = \frac{y}{R} \approx 5 \cdot y \quad .
$$

$$(11)$$

Thus the hot component has a very high metallicity. It contains however only a small mass fraction of the whole galaxy. If most of the ICM is lost through a galactic wind at the end of the lifetime of the galaxy most of the metals but only a small amount of mass would be removed from the system. This process might explain why dwarf galaxies can loose most of their metals and still remain gravitational bound. It would also indicate that different gas phases in galaxies can have different metallicities and abundance distributions.

3.2.2. Local Enrichment

Local enrichment is an incomplete mixing process which plays an important role in a low-metallicity environment where the ejecta of a local OB-association can significantly affect the abundance distribution of the surrounding gas.

Let us consider a system of N high mass stars, each of which introducing an energy E and a mass of metals m_Z into the surrounding medium. A fraction f_e of the released energy will be converted into gas motion and will enrich a region with a given total gas mass M. With the relation

$$
0.5 \, M \, c_S^2 = f_e \, N \, E
$$

$$(12)$$

where c_S is the sound velocity of the surrounding gas which is swept up by the ejecta, one can estimate the metallicity change of the region

$$\Delta Z = \frac{N\, m_z}{M} = f_e^{-1}\left(0.5\, m_z\, c_s^2\, E^{-1}\right) \tag{13}$$

which according to Brown et al. (1991) is in the range of $10^{-2} - 10^{-1}\, Z_\odot$ for galactic conditions. A stellar population which forms in this locally enriched environment will then have a metallicity which can differ significantly from the mean metallicity $\langle Z \rangle$ of the interstellar medium as long as $\langle Z \rangle < 0.1\, Z_\odot$. Thus the galactic halo globular clusters might have formed from locally enriched primordial gas clouds (Brown et al, 1992). Note furthermore that the dynamical timescale for forming a cool, locally enriched gas phase in which a second generation of stars can form is only of the order of 10^7 yrs (Brown et al, 1991). Thus a metallicity spread of the order of $0.1\, Z_\odot$ in a sample of stars with small age spreads can be understood as a result of local enrichment.

3.2.3. Incomplete mixing in the galactic halo

The small age spread of most of the halo globular clusters (VandenBerg et al 1990) indicates that the bulk of the galactic halo population formed on a short timescale of $1 - 2 \cdot 10^8$ yrs which is smaller than the dynamical timescale of the protogalaxy. Thus incomplete mixing must have played an important role during this early star formation epoch of our Milky Way.

Figure 6 compares the metallicity distribution of the metal-poor halo globular clusters with the simple model. There exists a deficiency of metal−poor clusters with $[Fe/H] \leq -2.5$. Whereas according to the simple model we would expect 22 globulars in this metallicity range only 3 have been found up to now. This is a clear signature of self−enrichment which predicts a lower metallicity limit of this order (see § 3.2.2).

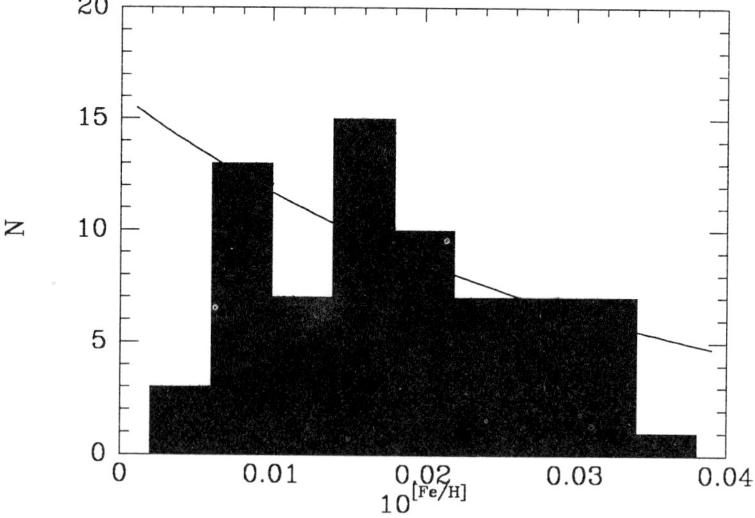

Figure 6. The metallicity distribution of metal poor halo globulars is compared with the predictions of the simple model, discussed in § 2.2.

Brown et al (1991, 1992) therefore constructed a model where all halo globular clusters formed from locally enriched, primordial gas clouds. Note that this self-enrichment scenario requires two distinct stellar generations: a first generation which provides the metals for a second generation which forms in the locally enriched environment. The globular clusters represent those second stellar populations which remained gravitational bound. The halo field stars on the other hand are a mixture of first generation stars and those second generation stars which resulted from failed, that is disrupted globular clusters.

Burkert and Truran (1992) show that the self-enrichment model predicts a star formation timescale for the galactic halo of the order of 10^8 yrs. As this timescale is much shorter than the dynamical timescale, the protogalaxy did not collapse and spin up significantly. Thus all the halo stars are slowly rotating. In addition self-enrichment predicts that high-metallicity halo stars with $[Fe/H] \approx$ -1 formed at the same time as stars with metallicities of the order of $[Fe/H] \approx$ -3. We therefore should not find an age-metallicity relationship in this halo population which formed during the first star burst.

The short star burst in the galactic halo destroyed all the primordial molecular clouds and heated the protogalaxy to temperatures of the order of 10^7 K by high—mass stars and supernova explosions. During the subsequent cooling phase which lasted $3 \cdot 10^9$ yrs the metals, ejected by SN II and SN Ia could efficiently be mixed throughout the whole extended protogalaxy leading to a mean gaseous iron abundance of $[Fe/H] = -1.5$. While this gas settled into the equatorial plane and started to form a thick disk component (Burkert et al 1992) a few younger halo globulars and field stars with $[Fe/H] \geq -1.5$ could form. This explains the observed age spread in the metal-rich halo component.

SN Ia never could contribute significantly to the halo enrichment even if their evolutionary timescale was shorter than 10^9 yrs. SN II exploded within molecular clouds and by this efficiently injected their metals into a high-density environment where self-enriched second generation stars formed. SN Ia on the other hand have longer lifetimes. So their progenitors were located in the hot diffuse intercloud medium and their ejecta were stored in the hot gas phase which did not condense into molecular clouds before it had settled in the equatorial plane or in the central region of our Galaxy. This scenario explains the intriguing coincidence that the SN Ia contribution is seen in the first disk stars however not in the halo component although there exists no physical connection between the evolutionary lifetime of SN Ia progenitors and the dynamical collapse time of our Milky Way.

References

Barnes J.E. 1989 *Nature* **338** 123

Brown J.H., Burkert A. and Truran J.W. 1991 *Astrophys. J.* **376** 115

Brown J.H., Burkert A. and Truran J.W. 1992 submitted to *Astrophys. J.*

Burkert A. and Hensler G. 1989 in *Evolutionary Phenomena in Galaxies* ed. J. Beckman, B.E.J. Pagel (Cambridge: Cambridge University Press) p. 230

Burkert A. and Truran J.W. 1992, in preparation

Burkert A., Truran J.W. and Hensler G. 1992 *Astrophys. J.* **391** 651

Clegg R.E.S., Lambert D.L. and Tomkin J. 1981 *Astrophys. J.* **250** 262

Edmunds M.G. 1990 *Mon. Not. R. Astr. Soc.* **246** 678

Gilroy K.K., Sneden C., Pilachowski C.A. and Cowan J.J. 1988 *Astrophys. J.* **327** 298

Hartwick F.D.A. 1976 *Astrophys. J.* **209** 418

Henry R.B.C., Pagel B.E.J., Lasseter D.F., Chincarini G.L. 1992 *Mon. Not. R. Astr. Soc.* in press

Lambert D.L. and Ries L.M. 1981 *Astrophys. J.* **248** 228

Luck R.E. and Bond H.E. 1985 *Astrophys. J.* **292** 559

Mathews G.J., Bazan G., Cowan J.J. 1992 *Astrophys. J.* **391** 719

Matteucci F. 1992 to appear in the proceedings of the workshop on *New Results on Standard Candles* ed. F. Caputo et al (Mem. Soc. Astron. It.)

Matteucci F. and Chiosi C. 1983 *Astron. Astrophys.* **123** 121

Matteucci F. and Francois P. 1989 *Mon. Not. R. Astr. Soc.* **239** 885

Nomoto K., Thielemann F.-K. and Yokoi K, 1984 *Astrophys. J.* **286** 644

Pagel B.E.J. 1991 in *Nuclei in the Cosmos* ed. H. Oberhummer (Berlin: Springer Verlag) p 89

Pagel B.E.J. and Patchett B. 1975 *Mon. Not. R. Astr. Soc.* **172** 13

Peterson R.C., Kurucz R.L. and Carney B.W. 1990 *Astrophys. J.* **350** 173

Rich R.M. 1988 *Astron. J.* **95** 828

Schuster W.J. and Nissen P.E. 1989 *Astron. Astrophys.* **222** 69

Searle L. and Sargent W.L.W. 1972 *Astrophys. J.* **173** 25

Sneden C., Lambert B.L. and Whitaker R.W. 1979 *Astrophys. J.* **234** 964

Sommer-Larsen J. 1991 *Mon. Not. R. Astr. Soc.* **249** 368

Theis C., Burkert A. and Hensler G. 1992 submitted to *Astron. Astrophys.*

VandenBerg D.A., Bolte M. and Stetson P. 1990 *Astron. J.* **100** 445

Webbink R.F. 1985 in *Dynamics of Star Clusters* ed. J. Goodman and P. Hut (Dordrecht: Reidel) p 541

Zinn R 1985 *Astrophys. J.* **298** 18

Effects of incomplete mixing in the Interstellar Medium on the secondary elements

M Wilmes† and J Köppen‡

† Institut für Theoretische Astrophysik, Im Neuenheimer Feld 561, D-6900 Heidelberg

‡ Institut für Theoretische Physik und Sternwarte, Olshausenstr. 40, D-2300 Kiel

Abstract. In this paper we show that the observed flat relation between nitrogen and oxygen does not neccessarely indicate that nitrogen is produced primarely. From our calculations we conclude that chemical evolution in an inhomogenous Interstellar Medium can mimic a primary production of nitrogen even if nitrogen is a pure secondary element.

1. Introduction

Conventional models for the chemical evolution of galaxies which assume that the Interstellar Medium (ISM) is always perfectly well mixed are known to fail in reproducing the observed relation between nitrogen and oxygen. Whereas oxygen is a primary element, i.e. produced from hydrogen within one stellar generation, nitrogen is expected to be made from carbon and oxygen in the CNO cycle. The conventional models for chemical evolution predict in this case $N \propto O^2$ which is much steeper than the observed relation (eg. Edmunds 1989).

One possible solution to explain the discrepancy between models and observations is the primary production of nitrogen during the Asymptotic Giant Branch (AGB) phase of stellar evolution when freshly produced carbon and oxygen from the core are mixed to outer layers of the star where hydrogen burning still goes on. It remains still uncertain if this so called "third dredge up" really occurs (Renzine and Fusi Pecci, 1988).

Our calculations show that chemical evolution with incomplete mixing of the ISM, i.e. where inhomogenities in the ISM are allowed to form and evolve, can provide an alternative explanation. Chapter 2 describes briefly the main features of an inhomogenous model while in chapter 3 we discuss some results of our calculations.

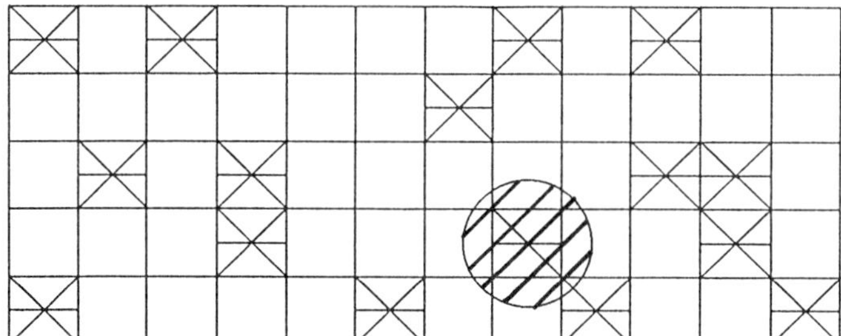

Figure 1. Example of a test volume divided into 60 subvolumes. The crosses mark subvolumes in which star formation occurs and the hatching shows the area polluted by the ejecta of one subvolume.

2. An inhomogenous model

We consider the test volume to be subdivided into a number (375) of chemically homogenous subvolumes (Fig.1) in which we follow the chemical evolution separately.

It is well known that star formation can be localized in distinct regions of the galaxie. We take this observation into account by concentrating star formation onto a small number of randomely choosen subvolumes. The probability for each subvolume to become "active" is given by a threshold T which is taken as a model parameter. In Fig.1 crosses mark those active subvolumes.

Mixing of newly synthesized elements to the surrounding ISM occurs instantly and immediately after stars are formed. These elements are dispersed over an area with finite pollution radius l. Only ejecta from stars can flow between separate subvolumes, we do not consider any other mixing processes in the gas. The hatched area in Fig.1 shows the polluted area by the ejecta of one subvolume for a pollution radius $l = 0.9$.

The Star Formation Rate (SFR) in an active subvolume is proportional to a power of the gas density: $\psi \propto \rho^n$, where the power law index n is another model parameter. The Initial Mass Function (IMF) is taken to be constant. Infall of gas from outside the test volume or outflows are not taken into account.

3. Results

On the following pages two sets of plots show the influence of the model parameters pollution radius l and power law index n on the relation between nitrogen and oxygen. The actual value for the variied parameter is given in every diagram. The crosses are the abundance values for each subvolume. The solid line is a fit to the data points we get from our inhomogenous model while the dashed line shows the relation predicted by conventional homogenous models.

The results are obtained from a model for the solar vicinity after an evolution time of 10 Gyr. The test volume was subdivided into 375 subvolumes. Note that in our model at every time we get a distribution of abundances while homogenous models give only one value for every abundance.

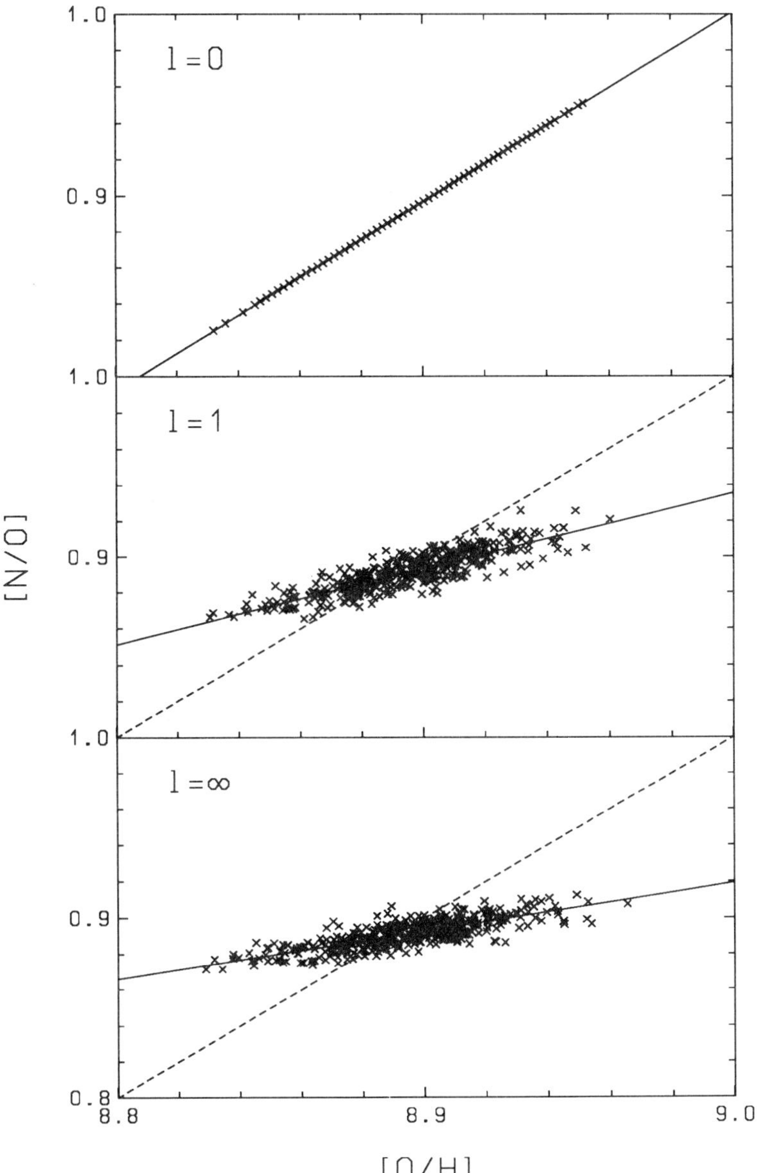

Figure 2. Abundances for different values of the pollution radius l. The solid line indicates the relation predicted by homogenous models while the dashed line is the relation we get from the inhomogenous model.

3.1. Varying the pollution radius

Fig. 2 shows the distribution of abundance values for pollution radii of $l = 0$, $l = 1$ and $l = \infty$ respectively. The star formation threshold T is set to 0.25 and the SFR is proportional to the gas density.

The plots clearly show that with increasing l the N-O-relation flattens. For $l = 0$ all ejecta are kept in the active subvolume and the single subvolumes act as homogenous closed box models with different evolution states. In this case we must obtain the relation predicted by homogenous models and as a consequence all values lay on a straight line with slope 1.

When l increases, inflow of newly synthesized material from outside the subvolume becomes more and more important. The N-O-relation flattens and we get a scatter in nitrogen abundances for a single oxygen abundance. This means that the nitrogen abundance is no longer determined by the abundance of the primary seed nuclei alone and the effective yield for nitrogen variies.

The minimum slope of approximately 0.3 is reached for $l = \infty$, i.e when the ejecta of an active subvolume are dispersed uniformly over the whole test volume. Then the amount of newly synthesized elements which are kept in an active subvolume is a minimum and the quantity of nucleosynthesis products entering a subvolume from outside is the same for all subvolumes. Note that even in this case localized star formaton leads to inhomogenities.

As a result we find that the amount of ejecta kept in an active subvolume tends to give a steep relation while material entering the subvolume from outside flattens the relation.

3.2. Varying the power law index

The second model parameter to be variied is the power law index n. Fig. 3 shows the results we get from the inhomogenous model for $n = 0.5$, $n = 1$ and $n = 2$ respectively.

All three plots are calculated for a infinite pollution radius. The diagrams show that the minimum value we can reach for the slope of the fitting line is determined by the power law index. For $n = 0.5$ the slope is approximately 0.5 while for $n = 2$ we almost get a constant. Increasing n to values even larger then 2 eventually leads to a negative slope, i.e $N \propto O^x$ with $x < 1$.

Varying n does not only change the minimum slope but also influences the time evolution of the N-O-relation. For $n < 1.5$ the relation is flat at early times but steepens when the evolution goes on. For larger values of n the relation becomes flatter when the evolution time increases.

So varying n may explain the observed for variations of the N-O-relation between different galaxies.

3.3. Varying the star formation threshold

As a third parameter we can vary the star formation threshold T. We do not show plots with different values for T but will briefly discuss the influence of T on the results obtained from our model.

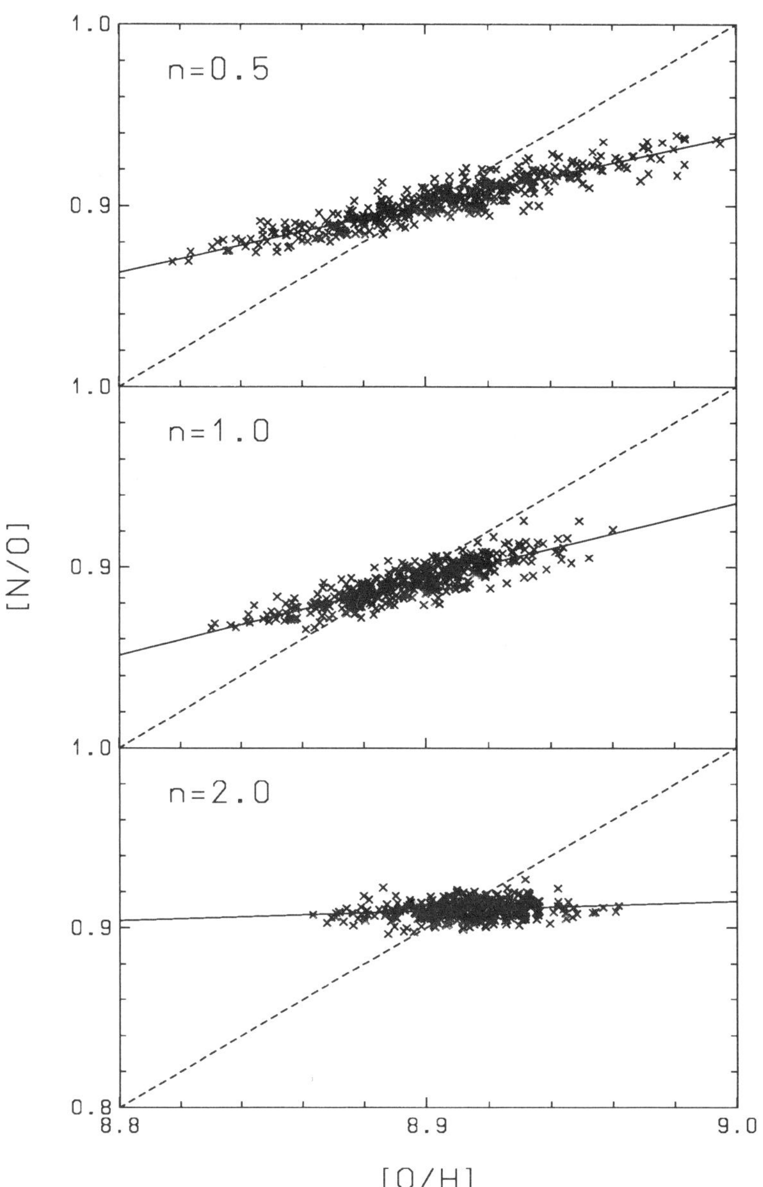

Figure 3. Abundances for different values of the power law index n.

As T is responsible for the "patchyness" of star formation, this parameter regulates the scatter of abundance values in the whole test volume: For great values of T nearly all subvolumes are active. If T is small, star formation is strongly localized. This means in our model that the star formation rate in an active subvolume becomes greater and so the change in abundances due to a star formation process rises.

On the other side we find that T has no influence on the steepness of the N-O-relation at all: Varying T while keeping all other parameters constant does not change the slope of the fitting line in any way.

This is important when we want to compare our results with observations: Due to the difficult estimation of the errors made in observing abundances, it is very uncertain if there is a real scatter in abundance values in the galaxie. Our model now tells us that even if there is no observable scatter very small deviations from inhomogenity can lead to a flat relation between nitrogen and oxygen.

4. Conclusion and outlook

We found that, while the average value of the abundance distribution we get from the inhomogeous model is the same as the value obtained from a similar homogenous model, the relation between nitrogen and oxygen is very sensitive to small deviations from homogenity. So we can say that especially to explain this relation inhomogenities in the ISM have to be taken into account by models for the cemical evolution of galaxies.

From observations we know that supernova explosions lead to superbubbles and cavities in the ISM. The observations and models describing superbubbles can help us to model the mixing of newly synthesized elements to the Interstellar Medium.

There are several processes which can eventually homogenize an inhomogenous ISM and may therefore influence our results. In our model we want to examine if diffusion due to turbulences in the ISM, radial gas flows and gas flows due to the differential rotation do change our results.

Acknowledgments

This work was supported by the Deutsche Forschungsgemeinschaft (DFG), Sonderforschungsbereich 328.

References

Edmunds M G 1989 *Summer School on Evolutionary Phenomena in Galaxies, Puerto de la Cruz (Spain), 4-15 Jul 1988* p. 356

Renzini A and Fusi Pecci F 1988 *Ann. Rev. Astron. Astrophys.* **26** 299

Galactic abundance gradients: application of the simple model of chemical evolution

W J Maciel

IAGUSP, Av. Miguel Stefano 4200, 04301-904, São Paulo SP, Brazil

Abstract. Determinations of the oxygen radial abundance gradient from galactic planetary nebulae and HII regions are reviewed. The application of the simple model for the chemical evolution of the Galaxy is shown to account for the observed gradients for type II planetary nebulae, which are disk objects with intermediate mass progenitors.

1. Introduction

Radial abundance gradients are observed in the Galaxy for oxygen (O/H) and other heavy element ratios (S/H, Ne/H, ...), both from HII regions and planetary nebulae (Pagel and Edmunds 1981, Shaver et al. 1983, Faúndez-Abans and Maciel 1986, Maciel 1992a). The gradients amount to:

$$\frac{d\log(\mathrm{O/H})}{dR} \cong -(0.05 - 0.08) \;\; \mathrm{dex/kpc} \tag{1}$$

with a typical value of -0.07 dex/kpc for HII regions.

Several models have been proposed to explain these gradients (see for example, Pagel 1992a,b, Köppen 1992), which generally involve complex assumptions often difficult to assess. Here we apply the so-called "simple model" for the chemical evolution of the Galaxy (cf. Pagel 1979, Tinsley 1980, Maciel 1992b) to the radial oxygen gradient derived from planetary nebulae.

2. Oxygen gradients from planetary nebulae

The galactic system of planetary nebulae includes a mixture of objects (Peimbert 1978, Maciel 1989, 1992a): type I nebulae, which belong to the young disk and have progenitors of masses 2–8 $M_\odot$; type II, intermediate disk objects (1–2 $M_\odot$); type III, old disk (0.8–1 $M_\odot$); type IV, halo nebulae, and type V, which are bulge objects.

The gradients derived from disk nebulae (types I-III) are slightly different from each other, possibly indicating a time evolution of the interstellar abundances. Table 1 gives the average gradients for each type of nebulae for the elements O, S, Ne, and Ar (Maciel and Köppen, 1992). As discussed earlier (Faúndez-Abans and Maciel 1986, Maciel, 1992a), type II nebulae better represent the interstellar conditions: they are relatively young, associated to the galactic disk both spatially and kinematically (Maciel and Dutra 1992), and their progenitor stars are intermediate mass stars, which do not enrich the interstellar medium with primary elements. Therefore, their oxygen abundances reflect interstellar abundances at the times the central stars were formed. The gradients for nebulae of types IIa and IIb are shown in figure 1, and the average gradient for type II is:

$$\log(O/H) + 12 \cong -0.06R + 9.12 \tag{2}$$

where R is the galactocentric distance to the planetary nebula, projected onto the disk.

Table 1: Abundance gradients from planetary nebulae [dlog(X/H)/dR dex/kpc]

Type	O	S	Ne	Ar
I	−0.001	−0.069	−0.008	−0.042
IIa	−0.045	−0.071	−0.060	−0.075
II	−0.056	−0.064	−0.058	−0.058
IIb	−0.058	−0.048	−0.041	−0.017
III	−0.059	−0.058	−0.053	−0.042

3. Application of the simple model

According to the simple model (cf. Maciel 1992b), we consider an interstellar medium of uniform composition, with no infall or mass flows. The stellar mass loss timescale is assumed to be much shorter than the age of Galaxy, and the metal production depends on the stellar mass only. The instantaneous recycling approximation (IRA) is adopted, and the initial mass function and chemical yields are assumed to be constants. In this case, the heavy element abundance Z is simply given by:

$$Z \cong y \ln(1 + \frac{M_s}{M_g}) \tag{3}$$

where y is the yield, that is, the mass of metals ejected to the interstellar gas relative to the mass that remains in the stars. M_s is the mass locked up in stars, and M_g is the gas mass. From a recent analysis of the abundances of helium, oxygen, nitrogen

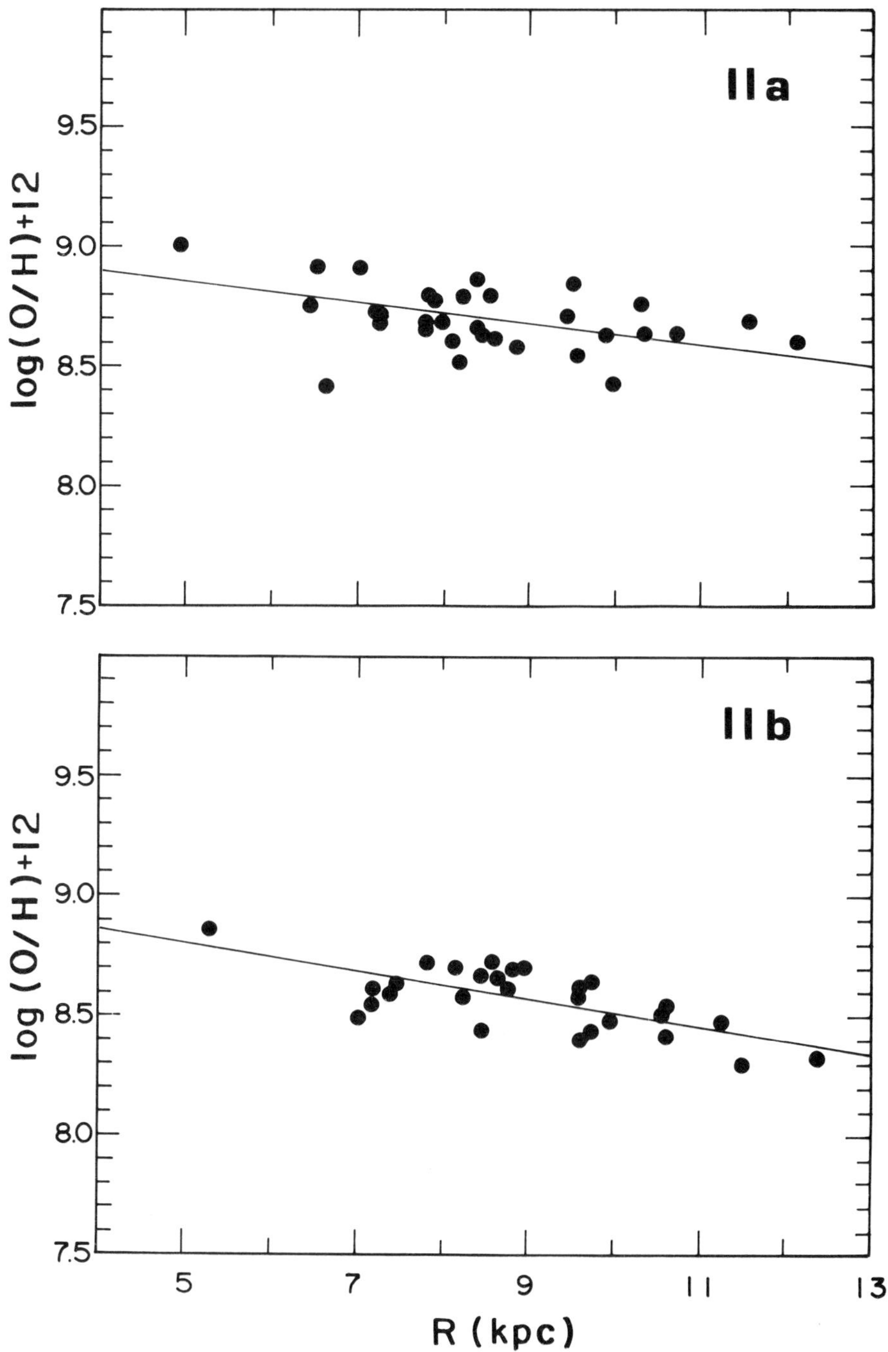

Figure 1: Oxygen abundance by number as a function of the galactocentric distance for planetary nebulae of types IIa and IIb (Maciel and Köppen 1992).

and carbon in planetary nebulae, Chiappini and Maciel (1992) obtained $Z \cong 26 \, O/H$, so that the oxygen abundance can be written in terms of the astration parameter $\log[\ln(1 + M_s/M_g)]$ as

$$\log\left(\frac{O}{H}\right) \cong \log\left(\frac{y}{26}\right) + \log\left[\ln\left(1 + \frac{M_s}{M_g}\right)\right] \tag{4}$$

The astration parameter as a function of position can be estimated from $(1 + M_s/M_g) \cong C \exp(-\alpha R)$, where α and C are constants (Pagel 1981, 1979, Shields and Searle 1978, Searle and Sargent 1972, Freeman 1970). Taking the following values at the sun's position ($R_0 = 8.5$ kpc): $\log[\ln(1+M_s/M_g)] \cong 0.32$ and $\alpha \cong 0.3$, we have $\ln C \cong 4.6$ (Pagel, 1992b, 1979).

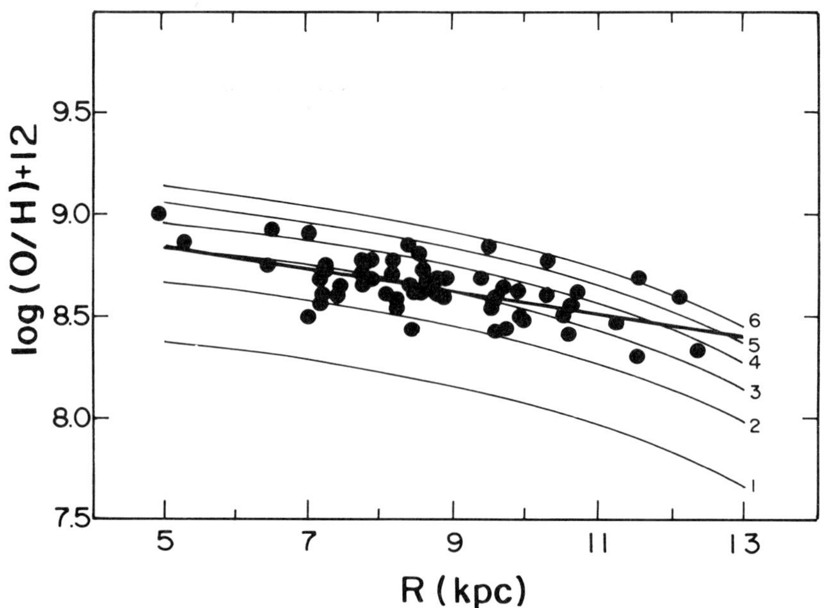

Figure 2: Oxygen data for type II nebulae, and calculated models. Curve 1: $y = 0.002$; 2: $y = 0.004$; 3: $y = 0.006$; 4: $y = 0.008$; 5: $y = 0.010$; 6: $y = 0.012$.

The calculated models are shown in figure 2, for different values of the yield. The curves are labelled 1 to 6 according to the adopted yield $y = 0.002, 0.004, 0.006, 0.008, 0.010,$ and 0.012. The data for type II nebulae are also shown. As can be seen from the figure, a good agreement is obtained for $0.004 < y < 0.010$. For $y = 0.006$ the least squares straight line is reproduced up to $R \cong 10$ kpc. This is close to the "canonical" value $y = 0.010$ (Talbot and Arnett 1973, Pagel 1979). At larger radii the hypothesis of a constant y seems to break up, but it should be noticed that uncertainties in O/H may reach 0.1 dex. These results are consistent with recent calculations of the yield (Köppen and Arimoto 1991), and from the calculated models, it can be concluded that the stellar masses above which black hole formation occurs are 30–40 $M_\odot$ (cf. Peimbert 1986a,b).

The origin of the gradients in this model is therefore linked to the mass distribution of gas and stars, and confirms that the main physical reason for the existence of gradients lies in the fact that a larger fraction of gas turned into stars in the inner regions than in the outer regions (Pagel, 1979).

Acknowledgements. This paper was partially supported by CNPq and FAPESP.

References

Chiappini C M L and Maciel W J 1992 (in preparation)

Faúndez-Abans M and Maciel W J 1986 *Astron. Astrophys.* **158** 228

Freeman K C 1970 *Astrophys. J.* **160** 811

Köppen J 1992 (private communication)

Köppen J and Arimoto N 1991 *Astron. Astrophys. Suppl.* **87** 109 (erratum: **89** 420)

Maciel W J 1989 *IAU Symposium 131* (Dordrecht: Kluwer) p 73

Maciel W J 1992a *Elements and the cosmos* (Cambridge: Cambridge University Press) (in press)

Maciel W J 1992b *Astrophys Space Sci.* (in press)

Maciel W J and Dutra C M 1992 *Astron. Astrophys.* (in press)

Maciel W J and Köppen J 1992 (in preparation)

Pagel B E J 1979 *Stars and Star Systems* (Dordrecht: Reidel) p 17

Pagel B E J 1981 *The structure and evolution od normal galaxies* (Cambridge: Cambridge University Press) p 211

Pagel B E J 1992a *IAU Symposium 149* (Dordrecht: Kluwer) (in press)

Pagel B E J 1992b *IUPAP Conference on primordial nucleosynthesis and evolution of early universe* (Dordrecht: Kluwer) (in press)

Pagel B E J and Edmunds M G 1981 *Ann. Rev. Astron. Astrophys.* **19** 77

Peimbert M 1978 *IAU Symposium 76* (Kluwer: Reidel) p. 215

Peimbert M 1986a *Star forming dwarf galaxies and related objects* (Paris: Frontières) p 403

Peimbert M 1986b *Publ. Astron. Soc. Pacific* **98** 1057

Searle L and Sargent W L W 1972 *Astrophys. J.* **173** 25

Shaver P A, McGee, R X, Newton L M, Danks A C, and Pottasch S R 1983 *Monthly Not. Roy. Astron. Soc.* **204** 53

Shields G A and Searle L 1978 *Astrophys. J.* **222** 821

Talbot R J and Arnett W D 1973 *Astrophys. J.* **186** 51

Tinsley B M 1980 *Fund. Cosm. Phys.* **5** 287

LiBeB production by cosmic rays

N. Prantzos

Institut d' Astrophysique de Paris-CNRS, and
Centre d' Etudes de Saclay-CEA, FRANCE

Abstract. Several models proposed recently in order to explain the observations of Be and B in old halo stars are presented and discussed here. It is concluded that some modifications are needed to the "standard" scenario of LiBeB production by cosmic rays, in order to explain the data.

1. Introduction

The recent observations of Be (Gilmore et al. 1991, 1992; Ryan et al. 1992) and B (Duncan et al. 1992) in low metallicity stars show that those light elements were present in the early galaxy in amounts larger than "conventional" nucleosynthesis by galactic cosmic rays (GCR) can account for (e.g. Vangioni-Flam et al. 1990). The absence of a Be (or B) "plateau" at low metallicity, however, "cooled" the initial excitement about a possible cosmological origin for those elements (Schramm, 1992; Malaney, this volume). Still, the observed linearity between the abundances of Be,B and Fe (the galactic "clock") is at variance with the expectations of the standard GCR scenario, which predicts a "secondary" behaviour for those light elements, i.e. a quadratic rather than a linear relationship (Prantzos, Cassé and Vangioni-Flam 1992b; hereafter PCFb). The observed B/Be ratio ($\sim$10 at all metallicities and 10$\pm$4 in the star HD140283) is, however, quite compatible with a GCR origin!

In the following we present a new model for the confinement of GCR in the early Galaxy that satisfies the observational constraints (developed in detail in Prantzos, Cassé and Vangioni-Flam 1992a; hereafter PCFa) and we discuss some recent relevant ideas concerning, in particular, the origin of GCR and the production site of LiBeB.

2. An Efficient Cosmic Ray Confinement in the Early Galaxy (?)

LiBeB isotopes are produced by GCR protons and alphas impinging on the CNO nuclei of the interstellar medium (ISM) and by GCR CNO nuclei impinging on ISM protons and alphas (hereafter components A- and B-, respectively). The contribution of the latter component to the LiBeB production is evaluated today to $\sim$20 % of the former one (Meneguzzi and Reeves 1975), and this must also have been the case in the past *if the composition of cosmic rays evolves like the one of the ISM*. This assumption is implied by current ideas on the origin of GCR (Meyer 1985; Ferrando 1992; see,

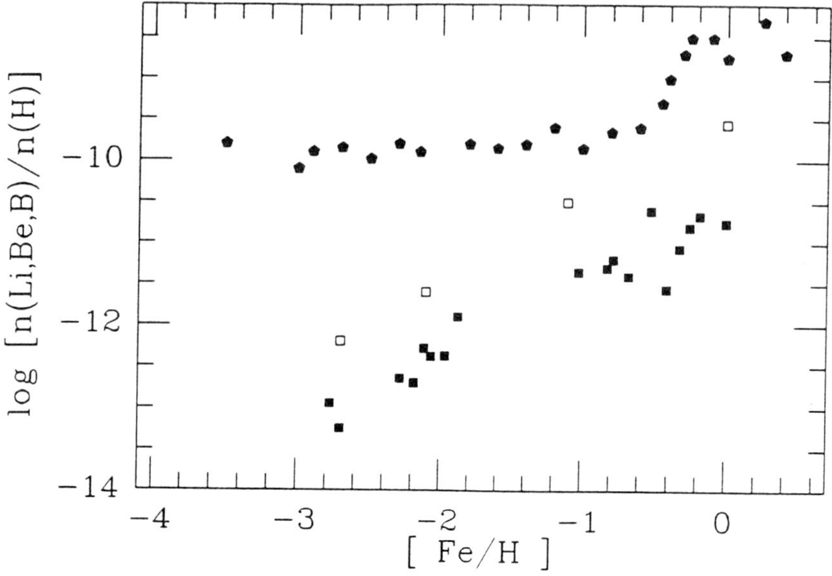

FIGURE 1: Li (pentagons), Be(black squares) and B(open squares) abundances as a function of metallicity. Only the upper envelope of the Li abundances (Spite 1991) is given. Solar system values are uncertain by a factor of ~1.5-2.

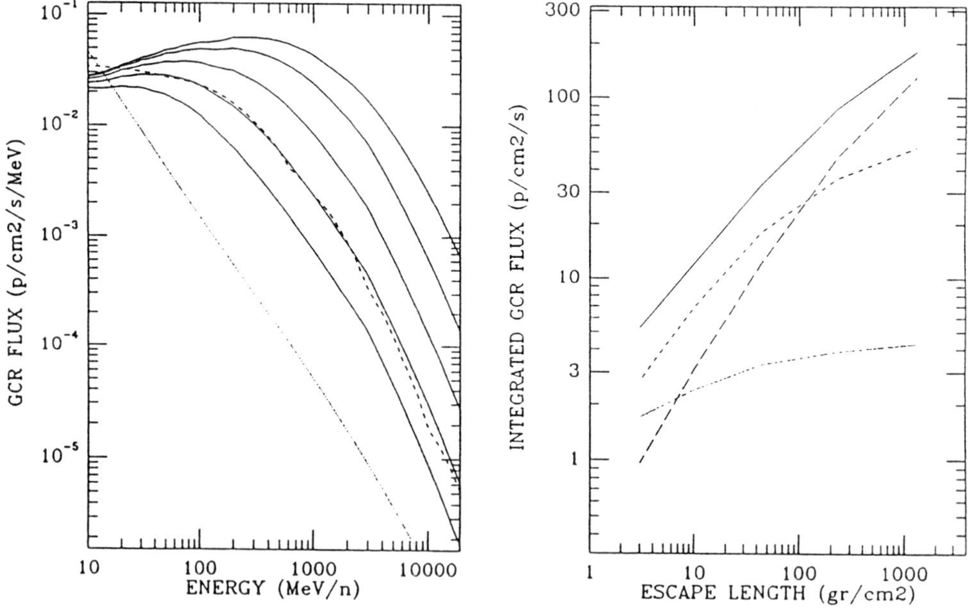

FIGURE 2: (a) Interstellar GCR spectra as a function of the escape length Λ (solid lines); larger fluxes correspond to larger Λ (3, 10, 40, 200 and 1200 g/cm^2, respectively). The dashed line represents a "realistic" current interstellar GCR spectrum and the dotted line a realistic source spectrum (see PCFa); (b) Dependence of GCR flux on Λ. *Solid line*: total flux; *long dashes*: flux in the 1000-10000 MeV/n range; *short dashes*: flux in the 100-1000 MeV/n range; *dotted line*: flux in the 10-100 MeV/n range.

however, Sec. 3), suggesting that the bulk of GCR particles originate mainly in solar-type stars. The production rate of the light (L) nuclei is given by:

$$\frac{dY_L}{dt} = <\sigma F_{p\alpha}^{GCR}> Y_{CNO}^{ISM} + <\sigma F_{CNO}^{GCR}\zeta> Y_{p\alpha}^{ISM} + (<\sigma F_{\alpha}^{GCR}> Y_{\alpha}^{ISM}) \quad (1)$$

where Y are the abundances (by number) of the various species, σ and F are the relevant cross-sections and GCR fluxes respectively (folded with the proper energy spectrum), ζ a decceleration term to account for the LiBeB produced by the CNO in flight (B-component), and the last term (C-component, in parenthesis) describes the production of ^{6,7}Li nuclei by fusion reactions between GCR alpha particles and ISM ones.

Assuming that the GCR flux F^{GCR} does not vary much and that B-comp. = 0.2 A comp. *always*, one obtains in the framework of a simple model for galactic chemical evolution (see PCFb): $Y_L \propto Z^2$ ($Z \sim Y_{CNO}$ being the metallicity of the ISM) i.e. a quadratic relationship between a secondary and a primary element. As emphasized in the Introduction, that relationship is not satisfied by the recent observations of Be and B and leads to theoretical abundances $\sim$30 times lower than the observed ones at [Fe/H]$\sim$-3.

One way to solve this problem is to assume very high (and rapidly decreasing) GCR fluxes in the early Galaxy, so as to compensate the large increase in Y_{CNO}^{ISM} in Eq. (1) and obtain a $\sim$ *const.* production rate for Be and B. Notice that this assumption should not be associated to a large supernova frequency (supernovae being thought to accelerate GCR), since large amounts of Fe would be produced at the same time. A "natural" way to obtain such high fluxes is proposed in the model of PCF, which is briefly described here.

It is currently thought that GCR are accelerated by SN shock waves and propagate in the Galaxy by diffusing on the irregularities of the galactic magnetic field and suffering losses from ionisation, nuclear reactions and leakage, according to the "leaky box" model for the galactic disk (e.g. Cesarsky and Soutoul 1992 for a review). The *escape length* Λ from the galactic disk, is evaluated today to $\Lambda_o \sim 10$ gr cm^{-2}. Notice that GCR particles are escaping from the galactic plane, i.e. the box is considered to be "leaky" only in the direction vertical to the plane and "closed" along the other two directions. The corresponding geometrical length, i.e. the thickness of the disk, is $H_o \sim 1$ kpc.

In the early Galaxy the escape length might have been quite different, because of the different geometry (larger vertical dimension H) and the larger gas fraction σ_g. The difficulties to evaluate it are underlined in PCF, where it is shown that under some plausible assumptions (in particular, assuming that the diffusion coefficient of GCR remains constant during galactic evolution) one is lead to $\Lambda \propto \sigma_g H$. This means that Λ might have been much larger in the halo phase, by factors as high as $\sim$150: indeed, σ_g (halo/today) $\sim$ 10, and H (halo/today) $\sim$ 15, since we can assume that GCR were confined during the halo phase up to a height of 15 kpc (i.e. the current disk radius).

A larger Λ means that GCR particles spend more time in the ISM before escaping from the Galaxy. In PCF it is shown that, since more and more particles are kept inside the galactic "box" (which becomes less "leaky"), *the total flux increases* approximately as $\sqrt{\Lambda}$, but the low-energy (10-100 MeV/n) flux is almost independent of Λ; moreover, since GCR particles are more subject to ionisation losses before leaving the Galaxy, *the equilibrium spectra become flatter up to higher energies*. These features appear clearly in Figs. 2a,b. The larger GCR flux favours greatly the production of the LiBeB isotopes in the early Galaxy (and particularly the production of the Be and B isotopes). The flatter GCR spectra help to keep relatively low the Li/Be production ratio at that epoch, when Li production was extremely favoured by the $\alpha + \alpha$ reactions. This helps to avoid an overproduction of Li whenever Be is significantly produced at low metallicities (see Steigman and Walker 1992).

An application of those prescriptions to a detailed model of galactic chemical evolution that reproduces well most of the observational features of the solar neighborhood (see PCFa), leads to the results presented in Figs. 3a,b,c,d. Now the agreement between theory and observations is satisfactory for all the LiBeB isotopes (this is not quite true for ^{11}B; all "standard" GCR scenarii give ^{11}B/^{10}B $\sim$2.5, instead of the observed value of $\sim$4). Due to the flatter GCR spectra in the early Galaxy, the B/Be ratio at [Fe/H]=-2.7 is found to be $\sim$11, to compare with the value of B/Be$\sim$13, obtained with current GCR spectra. This value is quite consistent with the value of B/Be for the star HD140283 (10$\pm$4). Notice, however, that the resulting Be vs. Fe relationship is still *not a linear one during the halo phase* (although the obtained slope is smaller than 2). Besides, several of the assumptions of the model (early vs. current confinement volume of GCR, $\sim$const. diffusion coefficient) should be thoroughly investigated before any definite conclusions can be drawn (see PCFa for a discussion).

3. Where do Cosmic Rays come from and where is LiBeB produced?

In Sec. 2 it was explicitly assumed that the second component of GCR (CNO nuclei on ISM protons and alphas) evolves as the first one. This is obviously true *if* GCR originate from the surfaces of F-K main sequence stars (see Meyer 1985); their abundances reflect then the one of the ISM at the moment of the formation of the star. In that case component B is completely symmetrical to component A, but *always* smaller (B$\sim$0.2 A).

Duncan et al. (1992) suggested that, during the halo phase, component B was more important than component A; this should happen *if the "metallicity" of GCR does not vary with time*, and an obvious way to obtain this, is to assume that SN *accelerate their own ejecta*. Then component B in Eq. (1) is $\sim$ *constant*, i.e. slowly decreasing with SFR but not affected by the large increase in ISM metallicity, and dominates Be and B production up to $Z \sim 0.2 \, Z_\odot$ (Fig. 4). A *linear* relationship between Be and Fe is naturally obtained then, without any need to invoke large GCR fluxes in the early Galaxy (see PCFb).

This interesting suggestion brings, once again, the problem of the *origin of GCR*. The *source composition* of GCR (i.e. derived from the observed one after taking into

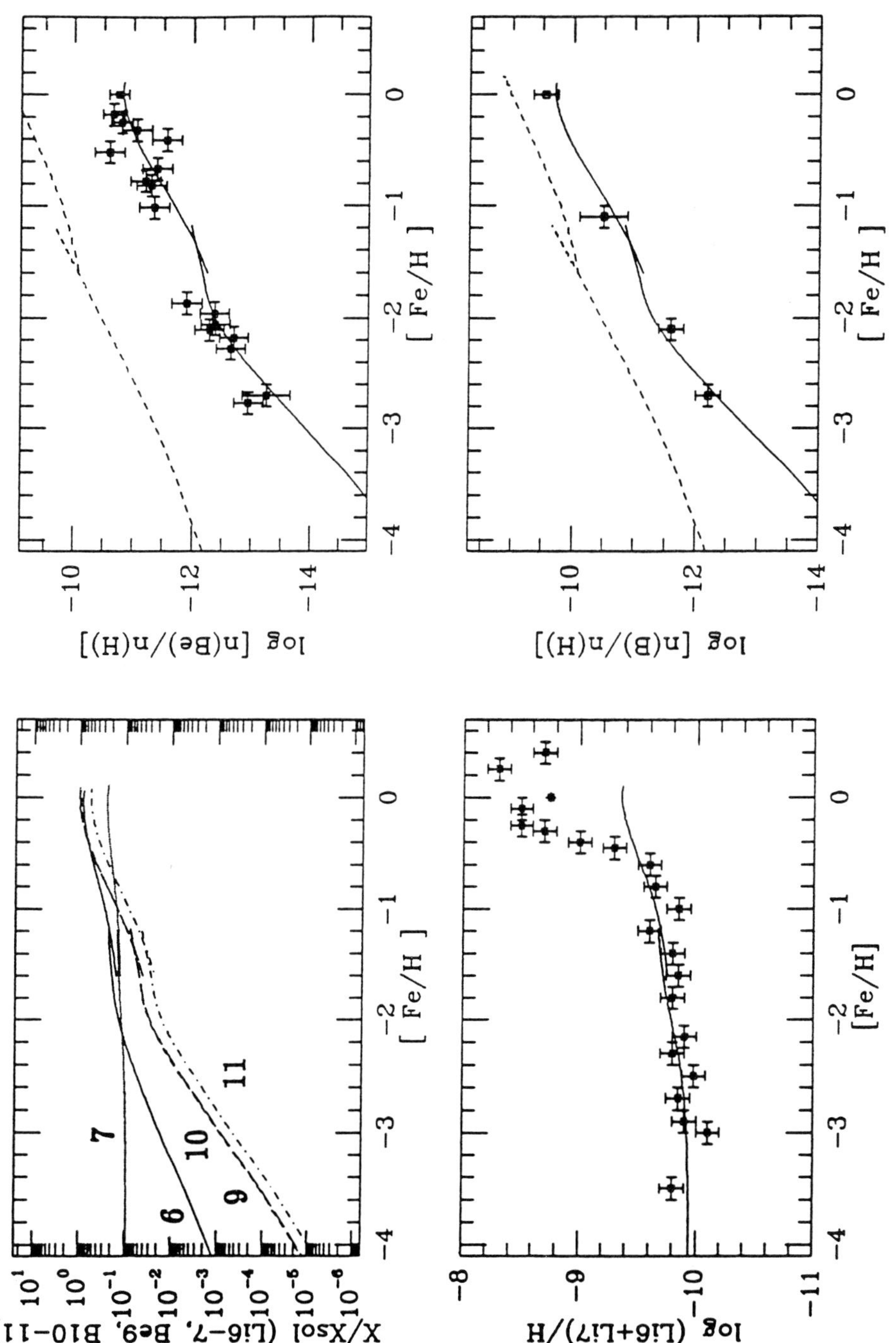

FIGURE 3: Evolution of LiBeB in the framework of a two-zone galactic evolution model with outflow during the halo phase (PCF). (a) Evolution of $X/X_\odot$ for all the isotopes (no stellar source for ^{7}Li is assumed); (b) Evolution of Li/H; (c) Evolution of Be/H; (d) Evolution of B/H.

account the propagation effects) is remarquably similar to the solar composition, with several important exceptions, however (see, e.g. Ferrando 1992, and Fig. 5):
a) a large depletion in H, He, N (by factors $\sim$10-30);
b) a smaller depletion in C, O, Ne, S, Ar (by factors $\sim$3-5);
c) an anomalous isotopic ratio ^{22}Ne/^{20}Ne$\sim$3 times solar.

It is now clear that the effect of the First Ionisation Potential (FIP; accounting for the moderate depletion of S, Ar, O, Ne) can only partially explain this pattern, and at least two other effects have to be introduced (Silberberg and Tsao 1990):
- a *rigidity* dependent depletion for H, He, N;
- an admixture of $\sim$2% material from WR stars (of the WC type) to account for the C/O ratio [(C/O)$_{GCR}$ >1, contrary to (C/O)$_{\odot}$ <1] and the ^{22}Ne/^{20}Ne ratio.

Could a SN origin explain that pattern with, at least, the same success? Obviously, the SNI and SNII ejecta are depleted in H, He, N (see Fig. 5) and feature (a) above is naturally explained. However, SNII ejecta are not particularly depleted in Ne, S, or Ar, so one has still to invoke a FIP bias for those species, and this time in a rather "unnatural" environment; notice that WR stars are still needed in order to "bring" C and ^{22}Ne in this model. Another important difficulty for the SN scenario is that no supernova (either SNI or SNII) produces heavy s-elements like Ba; those species should then be depleted in GCR, which is not the case, however (e.g. Ferrando 1992). Thus, we feel that the attractive (as far as the Be vs. Fe observations are concerned) idea of a SN origin of GCR, is not supported by our current knowledge on GCR source composition. It cannot be excluded, however, that some part of GCR come directly from SN, and that this component played an important role in the LiBeB production in the early Galaxy.

Another idea has been recently put forward by Feltzing and Gustafsson (1992): LiBeB production by GCR takes mostly place (at least during the halo phase) not on the diluted CNO of the ISM, but inside the CNO enriched supernova remnants. Obviously, one should replace in that case Y_{CNO}^{ISM} by dY_{CNO}^{ISM}/dt in Eq. (1), and a linear relationship between Be and metallicity is naturally obtained.

This idea, however, also meets with several serious difficulties. First, the CR spectra inside the supernova remnants should be quite steep (since the corresponding escape length would then be <1 g/cm^2), leading to unusual elemental and isotopic CR abundance ratios. Second, and perhaps most important, very high CR fluxes would be needed locally, in order to produce sizeable amounts of LiBeB during the $\sim$10^3 year period before the supenova remnant be diluted in the ISM; such high fluxes imply quite strong magnetic fields to confine them, and would produce intense emissions of high energy gamma-rays, that have not been observed in the Galaxy today (by e.g. COS-B). Despite those difficulties, this hypothesis needs some more study, since it also offers a "natural" linearity between Be,B and Fe abundances.

4. Conclusion

The recent observations of Be and B in low metallicity halo stars gave a new impetus to the study of the production of LiBeB by GCR. A new scenario for the propagation

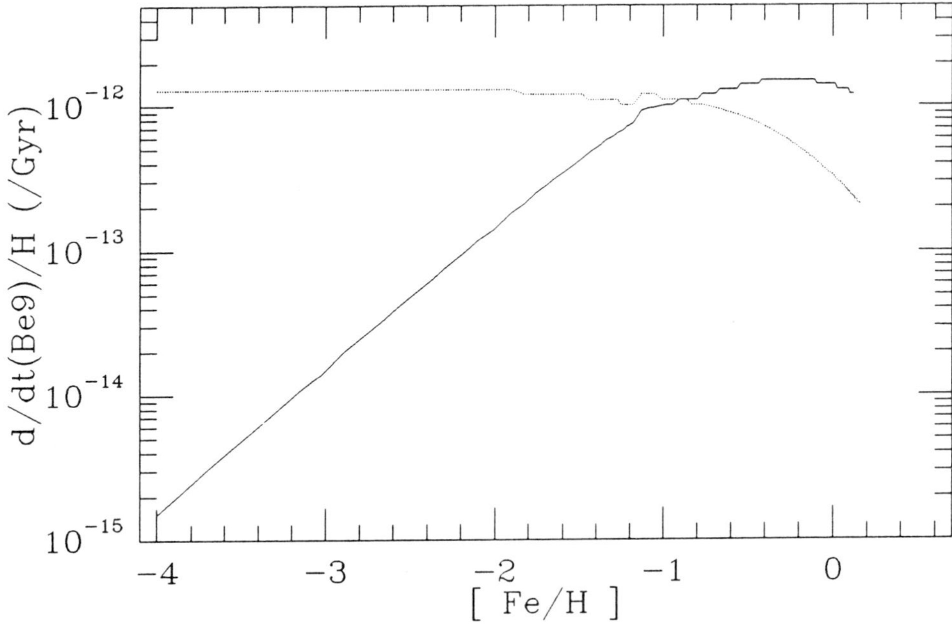

FIGURE 4: Evolution of the production rate of Be/H (in Gyr^{-1}) by components A (solid line) and B (dashed line), respectively (see text) *if GCR originate from supernovae*. Such an assumption leads naturally to a linear relationship between Be and Fe.

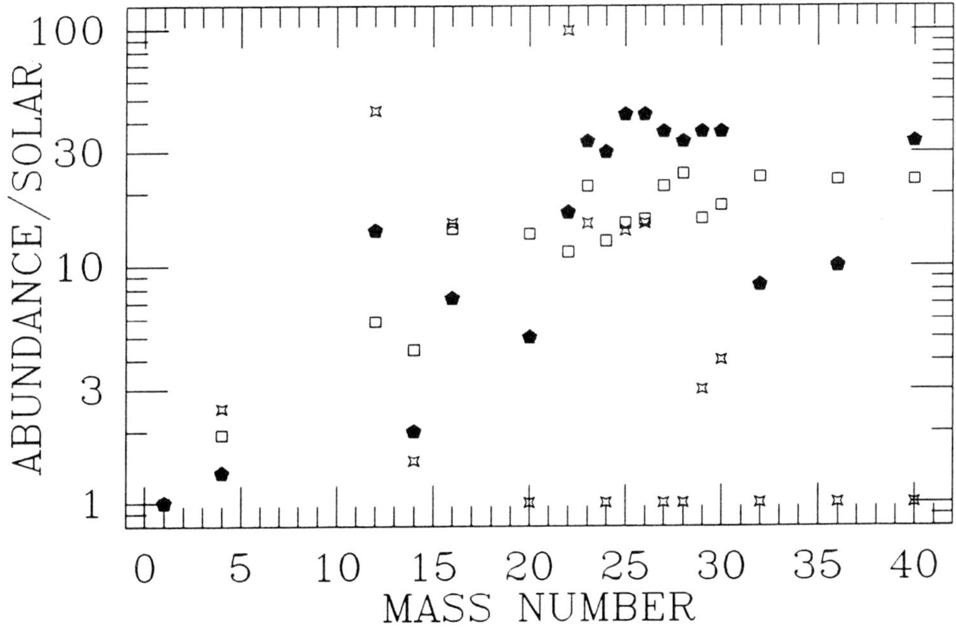

FIGURE 5: Composition of: GCR (from Silberberg and Tsao 1990; full pentagons); SNII ejecta (from Weaver and Woosley 1992; open squares); WR ejecta (from Prantzos et al. 1986; asterisks). $X/X_\odot$ values are given and the three sets of abundances are normalized to H=1.

of GCR in the early Galaxy is proposed by PCF(a,b) suggesting that, during the halo phase, GCR were more efficiently confined than today and had flatter spectra at low energy. Application of this scenario to a consistent model for the chemical evolution of the Galaxy leads to a satisfactory agreement between theory and observations. Still, the obtained Be vs. Fe relationship is not quite a linear one. A much better fit to the data is obtained if it is assumed that GCR originate from supernova, or if LiBeB is preferentially produced inside the CNO enriched supernova remnants, before their dilution in the ISM. Our current knowledge, however, on GCR source composition or CR fluxes inside supernova remnants does not support for the present those ideas.

In any case, we think that the observed Be and B abundances in low metallicity stars are best explained in terms of GCR production, with no need for a less conventional mechanism, like e.g. inhomogeneous big bang nucleosynthesis; the observed B/Be ratio ($\sim$10-12 at all metallicities, i.e. compatible with a GCR origin) is the strongest argument for that hypothesis. However, some of the "classical" ideas on GCR, concerning either their source composition or their confinement in the Galaxy, should be revised in order to account for those recent (and exciting!) observations.

References

Cesarsky C., Soutoul A. (1992) in *Origin and Evolution of the Elements*, Eds. N. Prantzos, E. Vangioni-Flam and M. Cassé (Cambridge Univ. Press), in press

Duncan D., Lambert D., Lemke D. (1992), *Ap. J.*, in press

Feltzing S., Gustafsson B. (1992) *Ap. J.*, submitted

Ferrando Ph. (1992), in *Nuclei in the Cosmos*, Ed. F. Kappeler, in press

Gilmore G., Edvardsson B., Nissen P. (1991), *Ap. J.*, **378**, 17

Gilmore G., Gustafsson B., Edvardsson B., Nissen P. (1992), *Nature* , submitted

Meneguzzi M., Audouze J., Reeves H. (1971), *A. A.*, **15**, 337

Meneguzzi M., Reeves H. (1975), *A. A.*, **40**, 110

Meyer J. P. (1985), *Ap. J.*, **57**, 173

Prantzos N., Doom C., Arnould M., de Loore C. (1986), *Ap. J.*, **304**, 695

Prantzos N., Cassé M., Vangioni-Flam E. (1992a), *Ap. J.*, in press

Prantzos N., Cassé M., Vangioni-Flam E. (1992b), in *Origin and Evolution of the Elements*, Eds. N. Prantzos, E. Vangioni-Flam and M. Cassé (Cambridge Univ. Press), in press

Ryan S., Norris J., Bessell M., Deliyannis C. (1992), *Ap. J.*, **388**, 184

Schramm D. (1992), in *Origin and Evolution of the Elements*, Eds. N. Prantzos, E. Vangioni-Flam and M. Cassé (Cambridge Univ. Press), in press

Silberberg R., Tsao C. (1990), *Ap. J. Let.*, **352**, L49

Spite F. (1991), in Proceedings of the Workshop "The problem of Lithium", Ed. F. D'Antona, *Mem. S. A. It.*, 61

Steigman G., Walker T. (1992), *Ap. J. Letters*, **385**, L13

Vangioni-Flam E., Cassé M., Audouze J., Oberto Y. (1990), *Ap. J.*, **364**, 568

Weaver T., Woosley S. (1992), in *Phys. Rep*, in press

The chemical evolution and abundance gradients in galaxy disks

T. Tsujimoto[1], K. Nomoto[1], T. Shigeyama[1], H. Saio[2], and Y. Yoshii[3]

[1] Department of Astronomy, Faculty of Science, University of Tokyo, Bunkyo-ku, Tokyo, Japan
[2] Department of Astronomy, Tohoku University, Sendai, Japan
[3] National Astronomical Observatory, Mitaka, Tokyo, Japan

Abstract.
By applying new supernova yields, the chemical evolution of star forming viscous disks is calculated. The resulting abundance gradients are compared with observations. Relation with the abundance distribution function of the bulge stars is discussed.

1. Introduction

The galactic chemical evolution models are evolving from the local model for the solar neighborhood to the models for global and dynamical properties of galaxies. For example, several attempts have been made to reproduce the abundance gradient in galaxies (e.g., Matteucci & Francois 1989). We calculate abundance gradients in the chemical evolution models of star forming viscous disks by applying the new yields of both type Ia and type II supernovae (Hashimoto et al. 1992; Tsujimoto et al. 1992). We discuss the relation of the chemical properties between the disk and the bulge.

2. Model

The equations to describe the dynamical and chemical evolution of star forming viscous disks are as follows (Lin & Pringle 1987; Yoshii & Sommer-Larsen 1989; Sommer-Larsen & Yoshii 1989, 1990; Clarke 1989):

$$\frac{\partial \Sigma}{\partial t} + \frac{\partial}{r \partial r}(\Sigma r v_{\mathrm{r}}) = -\frac{\Sigma}{t_*(t)} + f \quad ,$$

$$\frac{\partial}{\partial t}(\Sigma r^2 \Omega) + \frac{\partial}{r \partial r}(\Sigma r^3 v_{\mathrm{r}} \Omega) = r^2 \Omega (f - \frac{\Sigma}{t_*(t)}) - \frac{1}{2\pi}\frac{\partial}{r \partial r}(-2\pi \nu \Sigma r^3 \frac{\partial \Omega}{\partial r}) \quad ,$$

$$\frac{\partial}{\partial t}(Z\Sigma) = -\frac{\partial}{r \partial r}(Z\Sigma r v_{\mathrm{r}}) + \frac{\partial}{r \partial r}(\nu_{\mathrm{D}}\Sigma r \frac{\partial Z}{\partial r}) - Z\frac{\Sigma}{t_*(t)} + \Sigma(1-Z)\int dM \frac{y(M)\phi(M)}{t_*(t-\tau(M))} \quad .$$

Here Σ denotes the disk surface density of gas, v_{r} the gas radial velocity, f the rate of matter infall onto the disk, Z the mass fraction of each heavy element in the gas, ν_{D} the diffusive coefficient, $y(M)$ the yield from the star with mass M including both type Ia and type II supernovae, $\tau(M)$ the life time of stars, $\phi(M)$ the initial mass function.

The functional forms of the angular velocity of the disk Ω and the coefficient of kinematic viscosity ν are respectively (Lin & Pringle 1987):

$$\Omega = \Omega_0 \left(\frac{r}{r_0}\right)^{-\beta} \quad \left(0 < \beta \leq \frac{3}{2}\right), \quad \nu = \nu_0 \left(\frac{\Sigma}{\Sigma_0}\right)^a \left(\frac{r}{r_0}\right)^b \quad .$$

Here we assume flat rotation curve ($\beta = 1$). For the coefficient of kinematic viscosity, we consider the cases with linear viscosity ($a = 0$, $b = 1$) and the non-linear viscosity($a = 2$, $b = 3$).

The infall rate is assumed to have a form: $f = p \cdot \exp\left(-t/\alpha - r/l\right)$, where α denotes the time scale of infall, l the scale length. The infall is introduced to resolve the G-dwarf problem in the chemical evolution models. This is also required to prevent the disk gas from being consumed within a timescale shorter than the age of the Galaxy. The infall from halo is a natural result of galaxy formation, because the Galaxy contracted with a free fall time scale and then underwent quasi-static contraction.

The time scale of star formation t_* is assumed to be comparable to the viscous time scale t_ν as $t_* \sim t_\nu \sim r^2/\nu$. The star formation rate $\dot{\Sigma}_*$ is then given as Σ/t_*. The linear viscosity and the non-linear viscosity correspond to the star formation rate being proportional to Σ^1 and Σ^3, respectively.

Following Sommer-Larsen and Yoshii (1989), we assume $\nu_D \sim \nu$. We neglect the metallicity in the infalling gas (Lacey and Fall 1985). The mean mass fraction of each heavy element in stars Z_* is defined as

$$Z_*(r,t) = \left(\int_0^t dt\, t_*^{-1} Z(r,t)\Sigma(r,t)\right) \bigg/ \left(\int_0^t dt\, t_*^{-1} \Sigma(r,t)\right) \quad .$$

We integrate the first three equations with the initial condition that there is no matter in the disk.

3. Abundance gradients

The viscous evolution of the star forming disk forms an exponential disk as seen from the present distribution of the surface density of stars (Figure 1). This model naturally results in the formation of abundance gradients in stars and gas along the disk. Figures 2 and 3 show present radial distribution of the iron abundance [Fe/H] in stars and the oxygen abundance (O/H) in the interstellar gas, respectively.

These abundance gradients are formed by radial gas flows, thereby depending on the spatial distribution of the infalling gas. If the infall of matter is spatially uniform, the radial gas flow is always inward and produces a rather steep abundance gradient toward the center (dashed lines in Figs. 2 and 3). If the infall has a spatial gradient with $l = 4$ kpc, the radial flow is divided into outflow and inflow (Fig. 4) for the first ~ 3 Gyr, which makes the abundance gradient less steep (solid lines in Figs. 2 and 3).

These abundance gradients are shown to be consistent with the observational data by Neese and Yoss (1988) in Figure 2 and by Shaver et al. (1983) and Mezger et al. (1979) in Figure 3. The gradients of other elements are basically similar to Figures 2 and 3 due to the radial gas flow, which will be presented elsewhere.

4. Relation to the galactic bulge

Figure 2 shows that the metallicity in the star forming viscous disk model increases steeply toward the central region. It is interesting to see whether the galactic bulge can be regarded as an extension of the exponential disk (Weinberg 1992) in view of chemical abundances.

First let us examine the abundance distribution function of stars within 2 kpc from the center of the viscous disk. Figure 5 shows the comparison between the calculated abundance distribution function (solid line) and that of bulge K giants (dashed line; Rich 1990). The calculated distribution function has a strong peak around [Fe/H] $\sim$ +0.3 due to the inflow of metal-rich gas into the bulge region. However, it does not contain the observed super metal-rich stars as [Fe/H] = 0.5 – 1.

To produce such super metal-rich stars, we assume that viscosity ν is enhanced by a factor of $\sim$ 10 – 100 for a period of 0.5 – 1 Gyr during the age of 0.5 – 5 Gyr. It then enhances gas inflow to the central region and induces a burst of star formation/supernova explosion in the central region. Because of the enhanced enrichment of heavy elements, the resultant abundance distribution function of stars within 2 kpc from the center (solid line in Fig. 6) is in good agreement with that of bulge K giants (dashed).

In this scenario, type Ia supernovae have made a significant contribution to the increase in the metallicity [Fe/H]. However, recent observations suggest that the ratio of r-process elements (Eu) to Fe [Eu/Fe] in the bulge K giants and that of α-elements to Fe in the bulge M giants are significantly higher than the solar ratios (Rich 1992). These ratios, if confirmed, may indicate that contributions of SNe II dominate the chemical evolution of the bulge, which might be against the model with enhanced inflow of the disk material. In other words, heavy element enrichment in the bulge stars might have taken place in a short time ($\lesssim$ 1 Gyr).

The abundance ratios of the bulge K and M giants would favor the formation of the bulge in the early phase of galaxy collapse (Arimoto & Yoshii 1987; Matteucci & Brocato 1990). We are modeling this scenario including bursts of star formation/supernova explosions with SPH simulations, which will be reported elsewhere (Tsujimoto et al. 1992b).

References

Arimoto N., Yoshii, Y., 1987, A&A, 173, 23
Clarke C.J., 1989, MNRAS, 238, 283
Hashimoto M., Nomoto K., Tsujimoto T., Thielemann F.-K., 1992, in this volume
Lacey C.G., Fall S.M., 1985, ApJ, 290, 154
Lin D.N.C., Pringle J.E., 1987, ApJ, 320, L87
Matteucci F., Brocato E., 1990, ApJ, 365, 539
Matteucci F., Francois P., 1989, MNRAS, 239, 885
Mezger P.G., Pankonin V., Schmid-Burkg J., Thum C., Wink J., 1979, A&A, 80, L3
Neese C.L., Yoss K.M., 1988, ApJ, 95, 463
Rich R.M., 1990, ApJ, 362, 606
Rich R.M., 1992, in IAU Symposium 149, The Stellar Populations of Galaxies (Kluwer), p. 29
Shaver P.A., McGee R.X., Newton L.M., Danks A.C., Pottasch S.R., 1983, MNRAS, 204, 53
Sommer-Larsen J., Yoshii Y., 1989, MNRAS, 238, 133
Sommer-Larsen J., Yoshii Y., 1990, MNRAS, 243, 468
Tsujimoto T., Nomoto K., Shigeyama T., Saio H., Yoshii Y., 1992a, in this volume
Tsujimoto T., Nomoto K., Shigeyama T., 1992b, in IAU Symposium 153, Galactic Bulges, ed. H. Habing (Kluwer)
Yoshii Y., Sommer-Larsen J., 1989, MNRAS, 236, 779
Weinberg M.D., 1992, ApJ, 392, L67

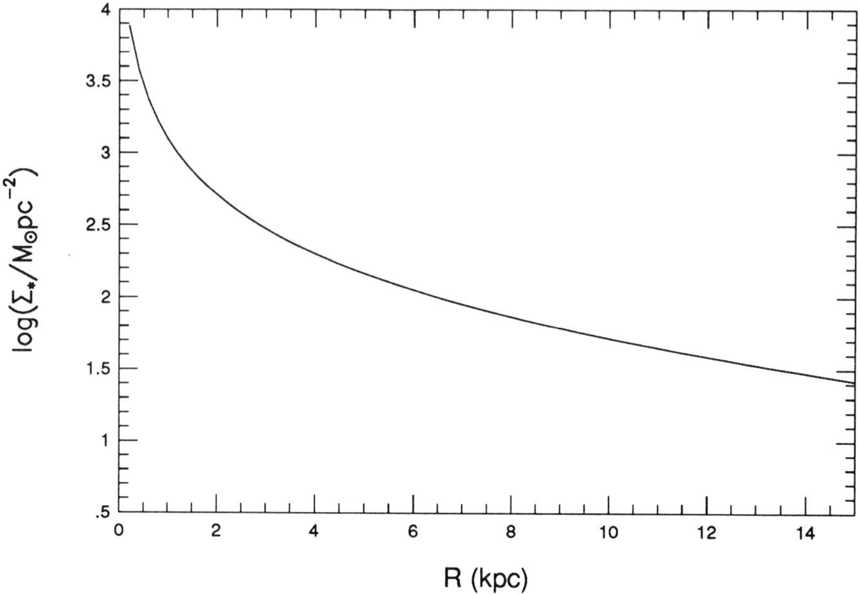

Figure 1.   Present distribution of the surface density of stars as a result of viscous evolution of the star forming disk.

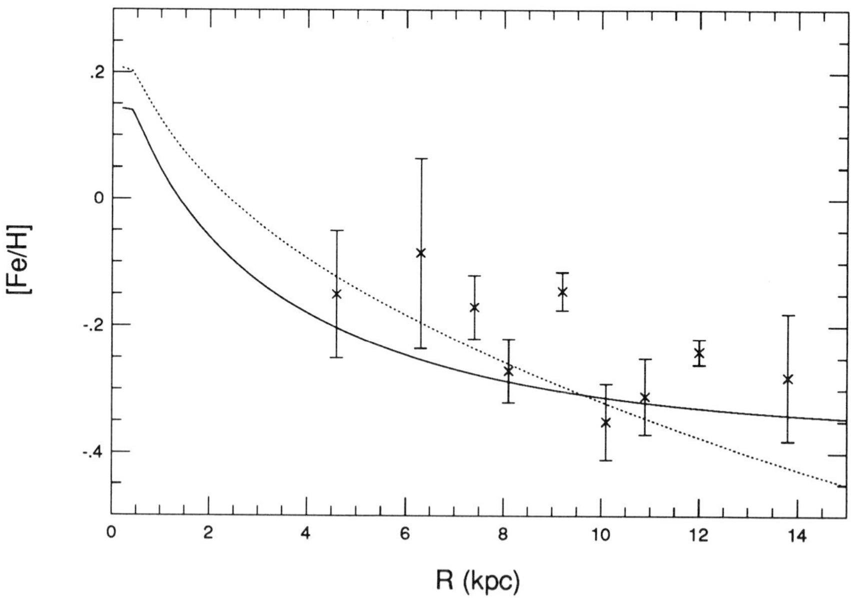

Figure 2.   Present radial distribution of the iron abundance [Fe/H] in stars as a result of viscous evolution of the star forming disk as compared with the observational data (Neese and Yoss 1988). The solid and the dotted lines assume the infall of matter with spatial distribution and the uniform infall, respectively.

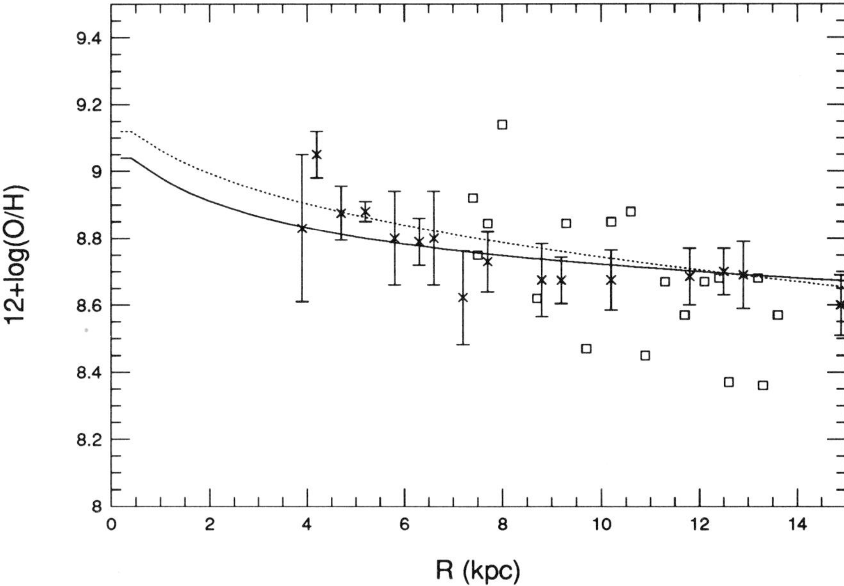

Figure 3. Present radial distribution of the oxygen abundance (O/H) in the interstellar matter as a result of viscous evolution of the star forming disk as compared with the observational data (Shaver et al. 1983; Mezger et al. 1979). The solid and the dotted lines are the same as in Figure 2.

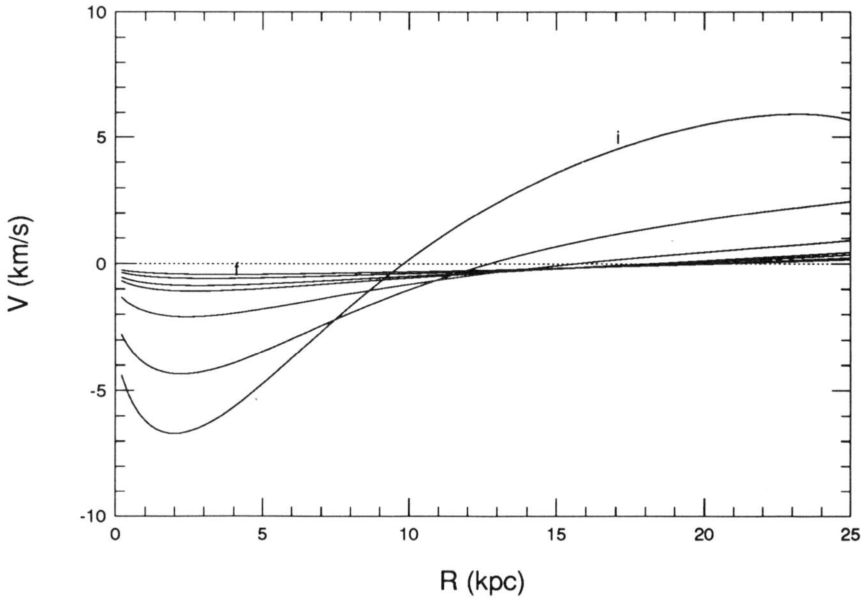

Figure 4. Velocity distribution of radial flows as a function of time from i (1.1×10^9 y) to f (14.9×10^9 y) for the infall of matter with a spatial gradient.

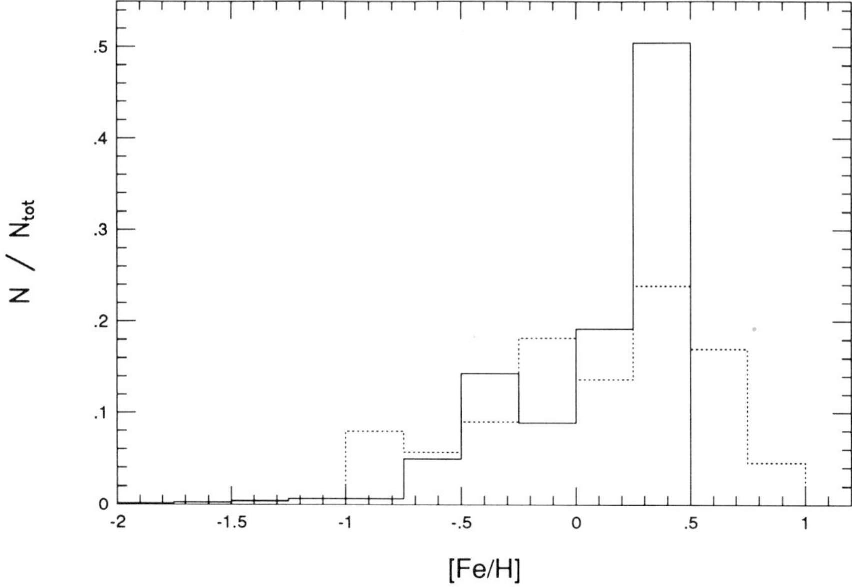

Figure 5. Abundance distribution function of stars within 2 kpc from the center of the viscous disk (solid) as compared with that of bulge K giants (Rich 1990).

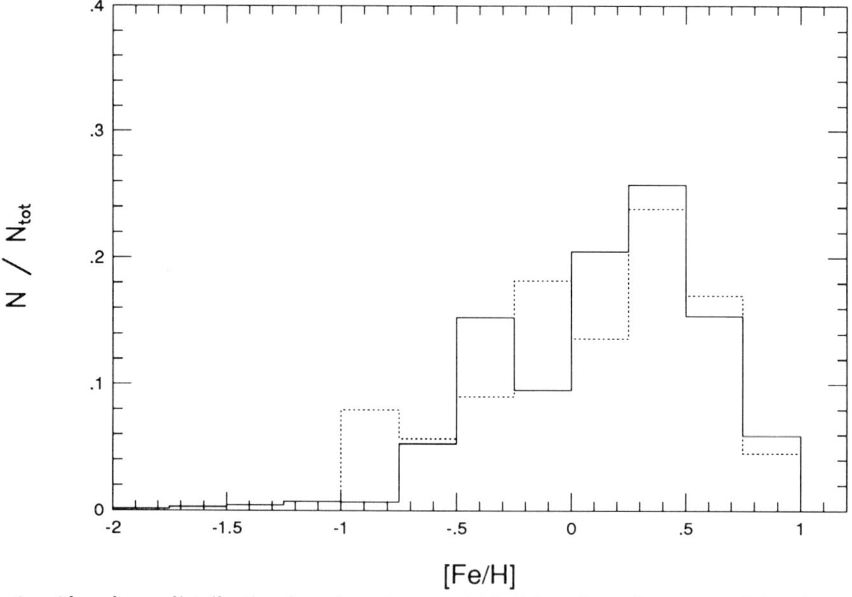

Figure 6. Abundance distribution function of stars within 2 kpc from the center of the viscous disk (solid) if the burst of gas inflow and associated star formation occurs. The dashed line is that of bulge K giants (Rich 1990).

Section 4: Stellar Models and Nucleosynthesis

Influence of new rates for the main s-process neutron sources in massive stars

G Meynet[1] and M Arnould[2]

1) Geneva Observatory, 51 ch. des Maillettes, CH-1290 Sauverny, Switzerland
2) Institut d'Astronomie et d'Astrophysique, C.P.165, Université Libre de Bruxelles, 50 av. F.D. Roosevelt, B-1050 Brussels, Belgium

Abstract. The two main neutron producing reactions in stellar He burning, $^{13}C(\alpha,n)^{16}O$ and $^{22}Ne(\alpha,n)^{25}Mg$, along with the $^{22}Ne(\alpha,\gamma)^{26}Mg$ competing channel, have been intensively studied recently, leading to new rate determinations. This work examines the impact of those new rates on the neutron production that is predicted to accompany central He burning in $M \geq 20 M_\odot$ stars with metallicity $Z = 0.02$. When the new rates are used, the corresponding $20 M_\odot$ star time integrated neutron flux is enhanced by about 30% with respect to the value obtained with the rates of Caughlan and Fowler (1988). The situation is qualitatively similar in more massive stars.

1. Introduction

Much experimental and/or theoretical effort has been devoted recently to $^{13}C(\alpha,n)^{16}O$ (Descouvemont 1987 ; Drotleff et al. this volume), $^{22}Ne(\alpha,n)^{25}Mg$ (Drotleff et al. this volume) and $^{22}Ne(\alpha,\gamma)^{26}Mg$ (Wolke et al. 1989), those reactions being considered as pivotal in the production of neutrons in stellar interiors. It results that the rates of those reactions differ significantly from those reported in the compilation by Caughlan and Fowler (1988 ; CF88), particularly in conditions appropriate to non-explosive He burning.

This work is devoted to an examination of the impact of those nuclear changes on the s-process that is predicted to develop during central He burning in massive stars. More specifically, two complete evolutionary sequences are computed for the central He burning stage of a $20 M_\odot$ star with metallicity $Z = 0.02$. The first one makes use of the new rates mentioned above, while the second one adopts the corresponding CF88 rates, all the other physical ingredients being identical. The new rates are also used to compute evolutionary sequences for 60 and $120 M_\odot$ stars.

The basic physical input of the computed stellar models is summarized in Sect. 2, with special emphasis on the ^{13}C and ^{22}Ne α-capture rates. Section 3 is devoted to a description of some of the results of our evolutionary computations, and to a discussion of the changes implied by the newly proposed α-capture rates on the neutrons produced during core He burning. Some conclusions are drawn in Sect. 4.

2. The physical ingredients of the stellar evolutionary models

Apart from the nuclear physics input, the stellar evolutionary models are computed as described by Schaller et al. (1992). In particular, the new radiative opacities of Rogers and Iglesias (1992) are used, the mass loss rates are derived from the empirical formula of de Jager et al. (1988), and a moderate core overshoot (dover/Hp = 0.2) is adopted. The nuclear reaction network of Schaller et al. is extended in order to include all the important neutron sources and sinks up to ^{27}Al. The neutron poisoning by the heavier nuclides is approximated by an effective neutron sink (see e.g. Jorissen and Arnould 1989). The neutron capture rates are taken from Beer and Voss (1991), while the charged particle capture rates are adopted from CF88, except, as mentioned above, for the α-particle captures by ^{13}C and ^{22}Ne in one of the computed evolutionary sequences, for which a set of rates referred to as "new rates" is selected. This set comprises the $^{13}C(\alpha,n)^{16}O$ rate calculated by Descouvemont (1987), the "best evaluation" for the $^{22}Ne(\alpha,n)^{25}Mg$ rate derived by Drotleff et al. (this volume), and the geometrical mean of the upper and lower bounds to the $^{22}Ne(\alpha,\gamma)^{26}Mg$ rate obtained by Wolke et al. (1989).

The new rates differ in several respects from the CF88 ones. Without going into details, let us simply notice that the new $^{13}C(\alpha,n)^{16}O$ rate is roughly three times more rapid than the CF88 one at temperatures (about 10^8 K) that are typical of the center of massive stars at the beginning of their core He burning, when ^{13}C is expected to burn. This enhancement cannot affect the neutron capture process in the massive stars under consideration, ^{13}C being just depleted earlier in the He burning phase than predicted with the use of the CF88 rate.

In contrast, ^{22}Ne is known to burn significantly only in the late stages of the core He burning phase, when the temperature reaches values in excess of about $3\ 10^8$ K. In those conditions, the new $^{22}Ne(\alpha,n)^{25}Mg$ rate is about twice the CF88 one, and is roughly equal to the new rate of the competing $^{22}Ne(\alpha,\gamma)^{26}Mg$ reaction. At the same temperatures, the (α,n) channel is predicted to be about four times more rapid than the (α,γ) one when use is made of the CF88 rates. The impact on the neutron budget of the speeding up of both the (α,n) and (α,γ) channels implied by the new rates can be evaluated quantitatively with the aid of our computed detailed evolutionary models.

3. Some results of the evolutionary computations

Figure 1 displays the time evolution of the ^4He, ^{13}C and ^{22}Ne mass fractions during the core He burning phase of the considered $20M_\odot$ star when the set of new rates is adopted. Figure 2 presents the corresponding mean neutron flux

$$< N_n(t)v_T(t) > = \frac{1}{M_{cc}} \int_0^{M_{cc}} N_n(t,m_r)v_T(t,m_r)dm_r \qquad (1)$$

where $N_n(t,m_r)$ and $v_T(t,m_r)$ are the neutron concentration and mean thermal velocity at time t and mass coordinate m_r, and M_{cc} is the mass of the convective He burning core.

At the beginning of He burning, ^{13}C is rapidly destroyed through $^{13}C(\alpha,n)^{16}O$, which produces a first burst of neutrons. Thereafter, ^{14}N left in large abundance in the stellar core by the preceding CNO burning of hydrogen is transformed into ^{22}Ne by the well known reaction sequence $^{14}N(\alpha,\gamma)^{18}F(e^+\nu)^{18}O(\alpha,\gamma)^{22}Ne$. That ^{22}Ne keeps a roughly constant abundance (mass fraction of the order of 0.01) during most of the He burning phase, and is destroyed by α-capture close to He exhaustion only, where a second burst of neutrons produced by $^{22}Ne(\alpha,n)^{25}Mg$ is observed. That burst is

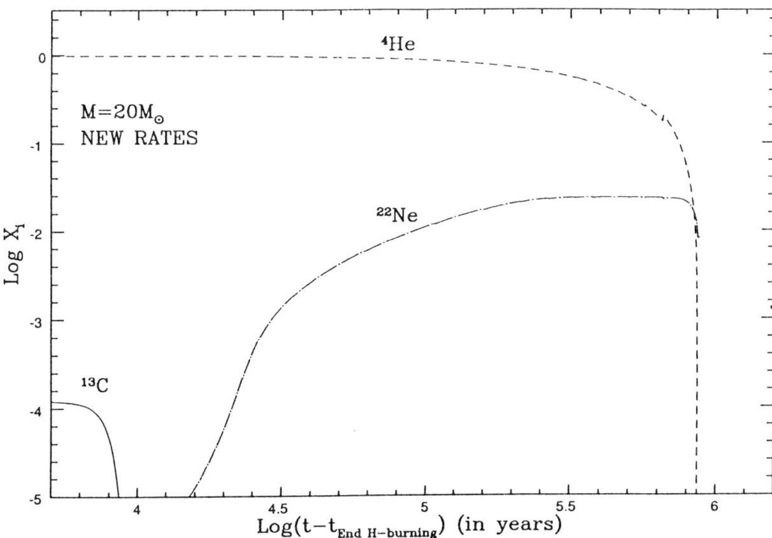

Figure 1 : *Time evolution of ^4He, ^{13}C and ^{22}Ne mass fractions during the core He burning phase of our $20M_\odot$ model when the new rates are used. The time $t_{End\ H-burning}$ denotes the moment at which the central abundance of hydrogen falls to zero.*

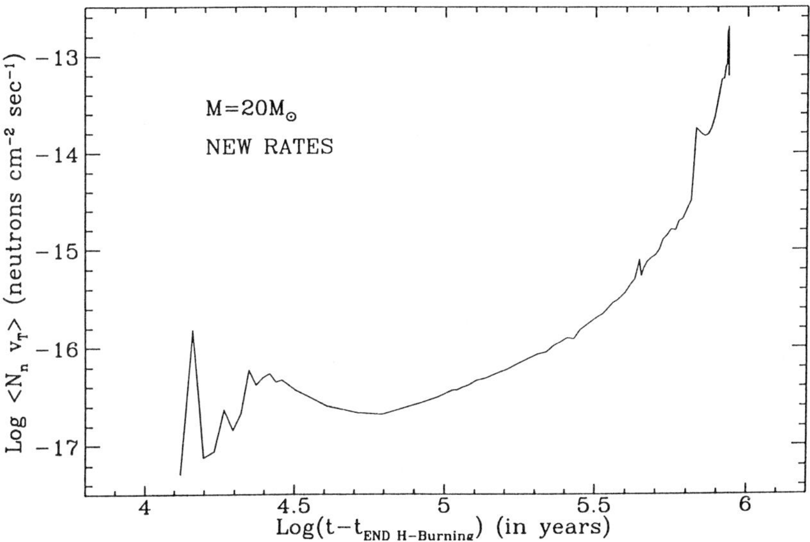

Figure 2 : *Time evolution of the mean neutron flux during the core He burning phase in our $20M_\odot$ model when the new rates are used.*

much more important than the ^{13}C one, the available ^{22}Ne being about two orders of magnitude more abundant than the ^{13}C emerging from the CNO burning. On the other hand, it is strongly peaked near the end of central He burning; in fact, the last percent of the total duration $t_{He} = 860000$ yr of that phase is responsible for 80% of the total neutron exposure, defined as the mean neutron flux (Eq. 1) integrated over t_{He}.

We now want to analyze the extent to which the results just reported would be affected by the use of the CF88 rates. Figure 3 compares the time evolution of the ^{13}C mass fraction predicted with the new $^{13}C(\alpha, n)^{16}O$ rate and with the CF88 one. As expected, the ^{13}C neutron burst develops slightly earlier (by about 500 yr in our $20M_\odot$ model) with the new rate. Figure 4 presents the variations of the ^{22}Ne mass fraction versus the core He mass fraction close to the end of He burning. The new rates predict an increase by 50% of the burned ^{22}Ne at He exhaustion. Correspondingly, the total neutron exposure rises by 30% only, this being due to the competing effect of the $^{22}Ne(\alpha, \gamma)^{26}Mg$ channel. It has to be noted that the neutron exposure increase is due entirely to the portion of the He burning phase displayed in Fig. 4. During the earlier evolutionary stages, the mean neutron flux obtained with the CF88 rates is in fact larger than the one computed with the new rates. This simply results from the fact that the new $^{22}Ne(\alpha, n)^{25}Mg$ rate is smaller than the CF88 one for temperatures $T < 2.4\ 10^8$ K.

In addition to the calculations for the $20M_\odot$ star reported above, the evolution of $Z = 0.02\ 60$ and $120M_\odot$ stars has also been computed with the new rates, the corresponding models using the CF88 rates being just taken from Schaller et al. (1992). Table 1 compares the quantities of ^{22}Ne burnt at the end of the He burning phase, $\Delta^{22}Ne$, that are derived for the three considered massive stars when the new and CF88 rates are used.

Table 1

$M/M_\odot$	$\Delta^{22}Ne$(CF88)	$\Delta^{22}Ne$(new)
20	0.009	0.014
60	0.011	0.016
120	0.011	0.016

The variations in $\Delta^{22}Ne$ are quite similar for all three stars so that we can expect similar enhancements of the neutron exposure when the new rates are used.

As stated above, the adopted new rates for the α-captures by ^{22}Ne are intermediate between experimentally determined lower and upper bounds (Wolke et al. 1989; Drotleff et al. this volume). A rough analysis has been performed in order to evaluate the impact of the remaining rate uncertainties on the quantity n of neutrons liberated by $^{22}Ne(\alpha, n)^{25}Mg$ in typical core He burning conditions. At $T = 3\ 10^8$ K, and independent of the uncertainties in the $^{22}Ne(\alpha, \gamma)^{26}Mg$ rate, we obtain $n(upper)/n(CF88) = 3.2$, $n(best)/n(CF88) = 2.4$ and $n(lower)/n(CF88) = 1.5$, where the parentheses indicate that the upper, best, lower (Drotleff et al. this volume) or CF88 rates are used for $^{22}Ne(\alpha, n)^{25}Mg$. It thus appears that the newly measured ^{22}Ne α-capture rates favor in any case the He-burning s-processing. However, due to the uncertainties in the $^{22}Ne(\alpha, n)^{25}Mg$ rate, the amount of neutrons liberated during core He burning cannot be determined reliably to better than a factor of about 2, this result being insensitive to the $^{22}Ne(\alpha, \gamma)^{26}Mg$ rate uncertainties. It has to be noted

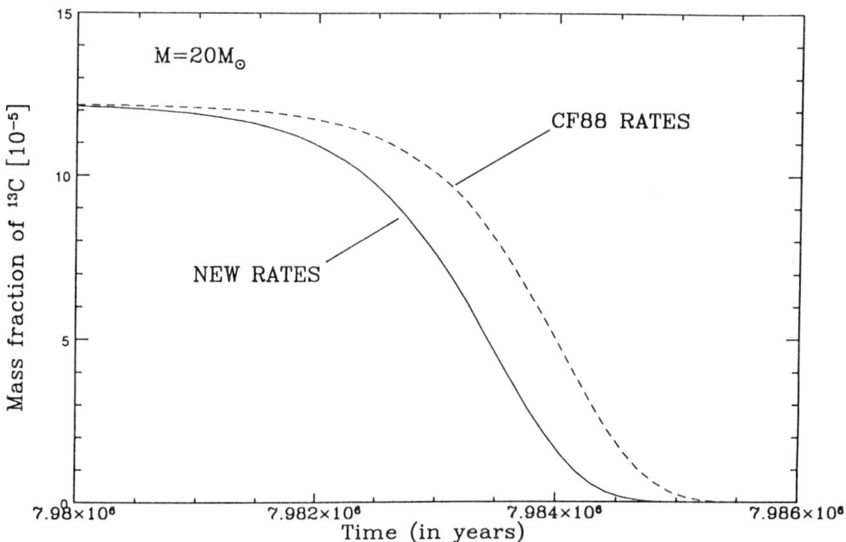

Figure 3 : *Comparison of the time evolutions of the ^{13}C mass fraction predicted with the new $^{13}C(\alpha,n)^{16}O$ rate and with the CF88 one.*

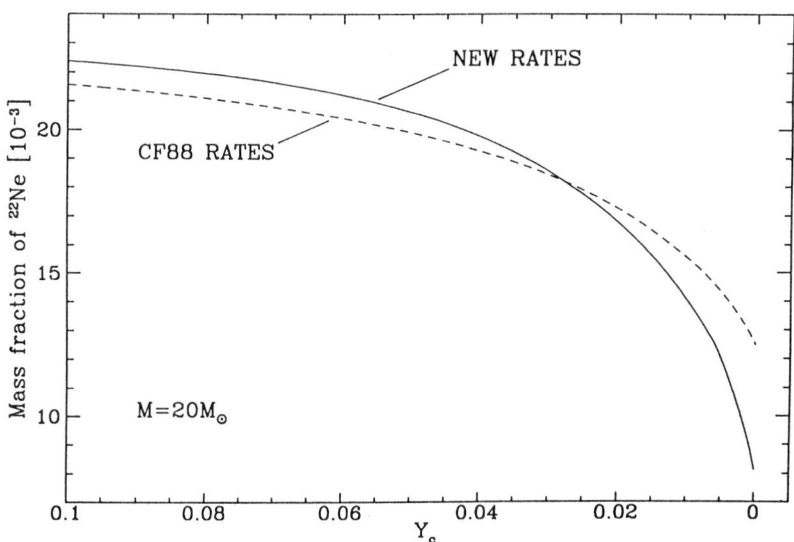

Figure 4 : *Variations of the ^{22}Ne mass fraction versus the core He mass fraction, Y_c, close to the end of He burning for the two sets of rates.*

that the values of the neutron ratios mentioned above, as well as their attached uncertainties, are lower for higher temperatures. For example, $n(upper)/n(CF88) = 1.7$, $n(best)/n(CF88) = 1.4$ and $n(lower)/n(CF88) = 1.0$ at $T = 7\ 10^8$ K, which could be typical of carbon or of explosive He burning.

4. Conclusions

The newly determined rates for the α-captures by ^{13}C and ^{22}Ne do not drastically affect the overall picture of the s-processing during core He burning in massive stars, as it emerges from many computations performed with the CF88 rates (for the most recent calculations of this type, see e.g. Baraffe et al., this volume). However, they lead to a predicted strenghtening of the neutron capture nucleosynthesis. For example, the use of the new ("best estimate") ^{22}Ne α-capture rates instead of the CF88 ones lead to a neutron exposure increase by about 30% during core He burning in a $20M_\odot$ model star. The situation is qualitatively similar in higher mass stars.

The neutron exposure associated to core He burning cannot be predicted with an accuracy that is better than a factor of about 2. This is due to the uncertainties remaining in the $^{22}Ne(\alpha,n)^{25}Mg$ rate, those still affecting the competing (α,γ) channel having no significant influence in that respect. Without further nuclear physics progress, it is thus quite premature to discuss in all details the subtleties of the massive star core He burning neutron capture nucleosynthesis in relation with the presumably associated solar-system s-process weak component.

Our results also imply that the new ^{22}Ne α-capture rates could likely increase the strength of the shell He burning s-process predicted on grounds of the CF88 rates (Arcoragi et al. 1991), but would in contrast reduce the efficiency of the already very weak s-processing accompanying carbon burning (Arcoragi et al. 1991 ; Raiteri et al. 1992), as most of the neutrons are made available by the destruction of ^{22}Ne left over from core He burning.

Acknowledgements. It is a pleasure to thank W. Hammer and his collaborators for having made available their data on the ^{13}C and ^{22}Ne α-capture rates prior to publication. This work has been supported in part by the SCIENCE Program SC1-0065. M.A. is Chercheur Qualifié F.N.R.S. (Belgium).

References

Arcoragi, J.-P., Langer, N., Arnould, M., 1991, A&A 249, 134

Beer, H., Voss, F., 1991, preprint

Caughlan, G.R., Fowler, W.A., 1988, Atomic Data Nuc. Data Tables 40, 283

de Jager, C., Nieuwenhuijzen, H., van der Hucht, K.A., 1988, A&A 173, 293

Descouvemont, P., 1987, Phys. Rev. C36, 2206

Jorissen, A., Arnould, M., 1989, A&A 221, 161

Raiteri, C.M., Gallino, R., Busso, M., 1992, ApJ 387, 263

Rogers, F.J., Iglesias, C.A., 1992, ApJS 79, 507

Schaller, G., Schaerer, D., Meynet, G., Maeder, A., 1992, A&AS in press

Wolke, K., Harms, V., Becker, H.W., Hammer, J.W., Kratz, K.L., Rolfs, C., Schröder, U., Trautvetter, H.P., Wiescher, M., Wöhr, A., 1989, Z.Phys. A334, 491

s-Process nucleosynthesis in massive stars as a function of the metallicity

I. Baraffe[1,2], M. F. El Eid[1] and N. Prantzos[3]

[1] Universitäts-Sternwarte, Geismarlandstr. 11, W-3400 Göttingen, Federal Republic of Germany
[2] Max-Planck-Institut für Astrophysik, Karl-Schwarzschild-Str. 1, W-8046 Garching, Federal Republic of Germany
[3] Service d'Astrophysique, Centre d'études de Saclay, F-91191 Gif sur Yvette, France

Abstract. We present the results of s-process nucleosynthesis in massive stars as a function of the metallicity Z. This neutron capture process produces elements ranging from Cu to Zr. It is found that the final production is a complex function of the metallicity. At intermediate metallicities, the s-process efficiency is higher than the one implied by a linear relationship, whereas at low Z, it drops below. Stellar yields, which are required quantities for chemical evolution models of the Galaxy are given for a range of metallicities $10^{-4} \leq Z \leq 0.02$.

1. Introduction

The slow neutron capture process, the so-called s-process, is found to be operative in the helium burning core of massive stars, triggered by the $^{22}Ne(\alpha,n)^{25}Mg$ neutron source. Elements with mass number $60 \leq A \leq 90$ are produced with an efficiency which depends on the initial metallicity Z of the star (Prantzos et al. 1990). In the following, we will summarize the effects of Z on the s-process as investigated by Baraffe et al. (1992). The understanding of such dependence is necessary to follow the evolution in the Galaxy of elements produced by this process.

2. S-yields as a function of the metallicity

The s-process calculation is performed during the helium burning phase of 15, 20 and 30 $M_\odot$ stars with initial metallicity $10^{-4} \leq Z \leq 0.02$. Details of input physics can be found in Baraffe et al. (1992).

The s-process efficiency depends on two main effects. First, it is influenced by the initial abundance of O and Fe in the star. According to observations (Barbuy 1988; Gratton 1990; Spite and Spite 1991) the O/Fe ratio is higher at lower metallicity compared to solar one. An increase of the O/Fe ratio increases the efficiency of the s-process for the following reasons.

The production of neutrons in the helium convective core is due to the $^{22}Ne(\alpha,n)^{25}Mg$ reaction which is efficient near the end of core helium burning. When neutrons are released, they are captured by Fe nuclei initially present in the star, giving rise to a significant production of heavier elements up to A $\sim$ 90.

The abundances of ^{22}Ne in the core depends on the initial abundance of CNO. Indeed, those nuclei are essentially transformed in ^{14}N during core hydrogen burning, whereas ^{14}N is converted into ^{22}Ne through $^{14}N(\alpha,\gamma)^{18}F(\beta^+\nu)^{18}O(\alpha,\gamma)^{22}Ne$. Therefore, an increase of O relative to Fe results in a increase of the neutron number per Fe seed nuclei. This shifts the production peak toward heavier elements in the mass range $60 \leq A \leq 90$. The heaviest elements in this mass range which can be produced (Sr, Y, Zr) are very sensitive to this effect (cf. Fig.1), whereas the lightest (Cu, Zn) are rather unaffected.

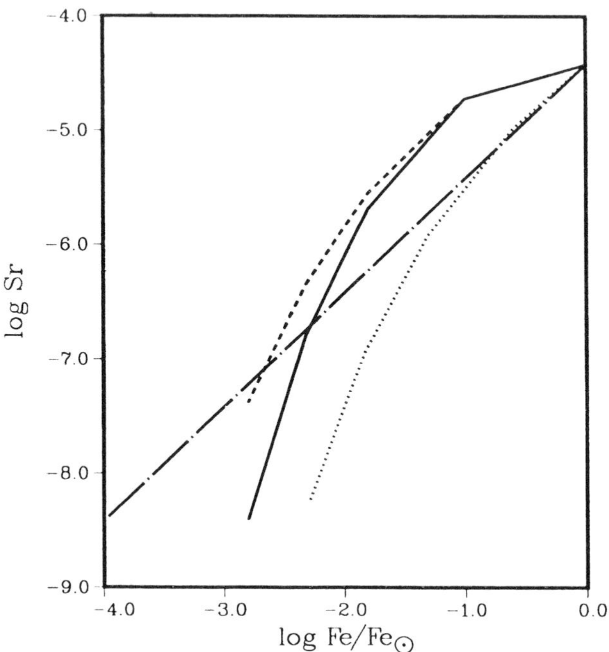

Figure 1. Final abundance in mass of Sr produced by s-process in the helium convective core of a 30 $M_\odot$ star as a function of the metallicity. The nucleosynthesis calculation is made under different conditions. The dotted line corresponds to $[O/Fe] = 0$ for all metallicities (with the usual definition $[O/Fe] = \log (O/Fe)_\star - \log (O/Fe)_\odot$). The solid and dashed lines are obtained in the case of an initial O/Fe ratio such that $[O/Fe] = 0.5$ at subsolar metallicities. The solid line is obtained for the $^{17}O(\alpha,\gamma)$ rate of Caughlan and Fowler (1988), whereas the dashed line corresponds to a rate equal to zero. A comparison is also made with the linear case where the production of Sr is assumed to be proportional to the initial Fe (—— - ——).

The s-process efficiency depends also on the effect of light nuclei (C, O, Ne) produced during helium burning and which absorb neutrons i.e. the so-called primary poisons. This effect is negative as it reduces the number of neutrons which may be captured by Fe. As the abundance of primary poisons in the helium convective core is rather unsensitive to a change of Z, whereas the neutron production decreases with Z, the poison effect increases as the metallicity decreases. This means that the s-process efficiency tends the decrease with the metallicity.

In stars with initial metallicity $Z \leq 10^{-3}$, ^{16}O is the dominant neutron poison via $^{16}O(n,\gamma)^{17}O$. Its poison effect is very sensitive to the competition between the reactions $^{17}O(\alpha,n)^{20}Ne$ and $^{17}O(\alpha,\gamma)^{21}Ne$. According to Caughlan and Fowler (1988), the (α,γ) rate is a factor ~ 10 smaller than the (α,n) rate at helium burning temperatures. This estimation may however be an upper limit of this rate which is still very uncertain. A decrease of the (α,γ) rate relative to the (α,n) rate would reduce the poison effect of ^{16}O as it favors the restitution of neutrons. We can see such an effect in Fig. 1 where both cases with the rate of Caughlan and Fowler (1988) and a rate set to null are taken into account. In the case of the lowest metallicity presented ($Z = 10^{-4}$), the production of Sr increases with a factor ~ 10 when the rate is set to null, as shown in Fig. 1.

Finally, we give in Table 1 the production of elements from Cu to Zr from s-process in massive stars of variable metallicity. The tabulated values refer to the stellar yield as derived in Baraffe et al (1992) i.e. the mass of an element synthesized in a star and ejected in the interstellar medium. This quantity is useful for galactic chemical evolution studies.

Table 1. Stellar yields in $M_\odot$ of several elements significantly produced in massive stars of different mass M (in $M_\odot$) and metallicity Z. Those results are obtained for an initial O/Fe ratio such that $[O/Fe] = 0.5$ at subsolar metallicities and with the $^{17}O(\alpha,\gamma)$ rate of Caughlan and Fowler (1988).

M	Z	Ni	Cu	Zn	Ga	Ge	Kr	Sr	Y	Zr
15	0.02	$1.9\ 10^{-4}$	$5.1\ 10^{-5}$	$9.3\ 10^{-5}$	$7.2\ 10^{-6}$	$1.6\ 10^{-5}$	$3.6\ 10^{-6}$	$1.4\ 10^{-6}$	$1.0\ 10^{-7}$	$9.5\ 10^{-8}$
	$5\ 10^{-3}$	$4.1\ 10^{-5}$	$1.5\ 10^{-5}$	$4.5\ 10^{-5}$	$5.7\ 10^{-6}$	$1.6\ 10^{-5}$	$8.3\ 10^{-6}$	$5.7\ 10^{-6}$	$4.8\ 10^{-7}$	$3.8\ 10^{-7}$
	10^{-3}	$6.9\ 10^{-6}$	$2.4\ 10^{-6}$	$6.6\ 10^{-6}$	$7.9\ 10^{-7}$	$2.2\ 10^{-6}$	$9.8\ 10^{-7}$	$6.1\ 10^{-7}$	$4.8\ 10^{-8}$	$3.7\ 10^{-8}$
	$3\ 10^{-4}$	$2.3\ 10^{-6}$	$6.5\ 10^{-7}$	$1.3\ 10^{-6}$	$1.2\ 10^{-7}$	$2.8\ 10^{-7}$	$7.9\ 10^{-8}$	$3.5\ 10^{-8}$	$2.4\ 10^{-9}$	$2.0\ 10^{-9}$
	10^{-4}	$4.0\ 10^{-7}$	$7.3\ 10^{-8}$	$9.2\ 10^{-8}$	$4.3\ 10^{-9}$	$7.9\ 10^{-9}$	$1.2\ 10^{-9}$	$7.5\ 10^{-10}$	$7.3\ 10^{-11}$	$6.7\ 10^{-11}$
20	0.02	$6.7\ 10^{-4}$	$1.8\ 10^{-4}$	$3.5\ 10^{-4}$	$2.7\ 10^{-5}$	$6.2\ 10^{-5}$	$1.4\ 10^{-5}$	$5.7\ 10^{-6}$	$4.0\ 10^{-7}$	$3.6\ 10^{-7}$
	$5\ 10^{-3}$	$5.5\ 10^{-5}$	$2.1\ 10^{-5}$	$8.0\ 10^{-5}$	$1.2\ 10^{-5}$	$4.0\ 10^{-5}$	$2.9\ 10^{-5}$	$2.9\ 10^{-5}$	$3.1\ 10^{-6}$	$3.0\ 10^{-6}$
	10^{-3}	$9.9\ 10^{-6}$	$3.7\ 10^{-6}$	$1.3\ 10^{-5}$	$1.8\ 10^{-6}$	$5.6\ 10^{-6}$	$3.5\ 10^{-6}$	$3.1\ 10^{-6}$	$3.0\ 10^{-7}$	$2.7\ 10^{-7}$
	$3\ 10^{-4}$	$3.8\ 10^{-6}$	$1.2\ 10^{-6}$	$3.2\ 10^{-6}$	$3.5\ 10^{-7}$	$9.2\ 10^{-7}$	$3.5\ 10^{-7}$	$2.0\ 10^{-7}$	$1.5\ 10^{-8}$	$1.1\ 10^{-8}$
	10^{-4}	$1.3\ 10^{-6}$	$3.0\ 10^{-7}$	$4.7\ 10^{-7}$	$3.0\ 10^{-8}$	$6.2\ 10^{-8}$	$1.1\ 10^{-8}$	$4.7\ 10^{-9}$	$3.9\ 10^{-10}$	$3.8\ 10^{-10}$
30	0.02	$2.1\ 10^{-3}$	$7.5\ 10^{-4}$	$2.2\ 10^{-3}$	$2.6\ 10^{-4}$	$7.3\ 10^{-4}$	$3.3\ 10^{-4}$	$2.1\ 10^{-4}$	$1.6\ 10^{-5}$	$1.3\ 10^{-5}$
	$5\ 10^{-3}$	$1.5\ 10^{-4}$	$6.0\ 10^{-5}$	$2.4\ 10^{-4}$	$3.8\ 10^{-5}$	$1.3\ 10^{-4}$	$1.0\ 10^{-4}$	$1.1\ 10^{-4}$	$1.3\ 10^{-5}$	$1.3\ 10^{-5}$
	10^{-3}	$2.7\ 10^{-5}$	$1.1\ 10^{-5}$	$3.9\ 10^{-5}$	$5.8\ 10^{-6}$	$1.9\ 10^{-5}$	$1.3\ 10^{-5}$	$1.3\ 10^{-5}$	$1.3\ 10^{-6}$	$1.2\ 10^{-6}$
	$3\ 10^{-4}$	$1.1\ 10^{-5}$	$3.8\ 10^{-6}$	$1.1\ 10^{-5}$	$1.2\ 10^{-6}$	$3.5\ 10^{-6}$	$1.5\ 10^{-6}$	$9.6\ 10^{-7}$	$7.5\ 10^{-8}$	$5.8\ 10^{-8}$
	10^{-4}	$3.7\ 10^{-6}$	$9.7\ 10^{-7}$	$1.7\ 10^{-6}$	$1.3\ 10^{-7}$	$2.8\ 10^{-7}$	$5.8\ 10^{-8}$	$2.4\ 10^{-8}$	$1.7\ 10^{-9}$	$1.6\ 10^{-9}$

3. Concluding remarks

The results of s-process analysis in massive stars as a function of metallicity can be summarized as follows. The lightest elements which can be produced by this process (Cu, Zn) show a linear behavior with the initial iron abundance. The production of heavier elements (Ga - Zr) is ruled by the competition between the positive effect of the initial O/Fe ratio and the negative effect of the primary poisons. At intermediate metallicity ($10^{-3} \leq Z < 0.02$), the production is higher than the one corresponding to a linear relationship. This is due to the increase of the O/Fe ratio relative to the solar one. At lower metallicities, the s-process efficiency drops below the linear relationship because of the dominant effect of the primary poisons.

Using those results in a model of chemical evolution of the Galaxy (Baraffe and Takahashi 1992), the contribution of s-process in massive stars is found to be too low to account for the observed abundances of elements as Cu, Sr, Y and Zr relative to Fe at low metallicities.

Those conclusions are based on the current knowledge of nuclear physics data and stellar evolution theory, which still have uncertainties. The question is whether those uncertainties could modify the results such as to increase the s-process efficiency in massive stars with low Z.

In the present work, we only take into account the effects which may mostly affect the results of s-process, as the O/Fe ratio and the $^{17}O(\alpha,\gamma)$ rate. If modifications of reaction rates involving light nuclei, which may influence the poison effects at low metallicity, are to be found in the future, the present results may need to be reconsidered.

Fortunately, the new estimation of the $^{22}Ne(\alpha,n)^{25}Mg$ reaction rate (see Hammer et al. 1992, this meeting) has little influence on the s-process production in massive stars, as shown by Meynet and Arnould (1992, this meeting).

Concerning uncertainties of stellar models, we have checked that the new set of opacities of the group of Livermore (Iglesias and Rogers 1991) does not affect our results. Indeed, those input data modify significantly the atmosphere properties of a star, compared to models obtained with the Los Alamos opacities (Huebner et al. 1977). But they do not alter the inner structure such as to modify the production of s-process in massive stars.

On the other hand, treatment of convection can modify the central conditions of a star. The choice of the criterion for the onset of convection is still controversial. Our stellar models were computed using the Schwarzschild criterion for convection. As alternative, the Ledoux criterion which is more restrictive causes a reduction of the size of the helium convective core and consequently of the central temperature. This has even the effect of lowering the s-process production. We then do not expect that this type of uncertainty in the stellar models will help resolving the problem of efficiency drop of the s-process at low metallicity.

References

Baraffe I., El Eid M. F., Prantzos N., 1992, A&A 258, 357

Baraffe I., Takahashi, K., 1992, in preparation

Barbuy B., 1988, A&A 191, 121

Caughlan G. R., Fowler W. A. 1988, Atom. Data Nucl. Data Tables 40, 283

Gratton R. G. 1990, in: Evolution of Stars: the Photospheric Abundance Connection. Proc IAU Symp. 145, Golden Sands, Bulgaria

Huebner, W. F., Merts, H. L., Magee, H. N., Jr., Argo, M. F. 1977, Los Alamos Sci. Lab. Report LA 6760 MA

Iglesias, C. A., Rogers F. J., 1991, ApJ 371, L73

Prantzos N., Hashimoto M., Nomoto K., 1990, A&A 234, 211

Spite M., Spite F., 1991, A&A 252, 689

Constraints to the weak *s*-component from nucleosynthesis in massive stars

C. M. Raiteri[†][1], R. Gallino[‡], M. Busso[†] and F. Käppeler[§]

† Osservatorio Astronomico di Torino, Strada Osservatorio 20, 10025 Pino Torinese, Italy

‡ Istituto di Fisica Generale dell'Università, via P. Giuria 1, 10125 Torino, Italy

§ Kernforschungszentrum Karlsruhe, Institut für Kernphysik, P.O. Box 3640, D-7500 Karlsruhe, Germany

Abstract.
New results of *s*–nucleosynthesis calculations from a generation of massive stars are presented, using the compilation of neutron capture cross sections by Beer, Voß & Winters (1992): the temperature-dependence of $< \sigma_{n,\gamma} >$ is taken into account, which is of particular importance when dealing with shell C-burning conditions. Proper extrapolations of beta-decay rates of the unstable isotopes at high temperature are also included. The results confirm that the weak *s*-component in the solar system is reproduced once standard prescriptions on the chemical evolution of the Galaxy are employed. A 20% higher solar abundance for Kr is derived with respect to that quoted by Anders & Grevesse (1989). The behaviour of the neutron flow at relevant branching points and the comparison with the phenomenological approach are discussed.

1. Introduction

A number of works have been devoted to the analysis of the *s*–process in massive stars with the aim of explaining the origin of the bulk *s*–nuclei from iron to strontium in the solar system, that from the phenomenological point of view belong to the *weak* component. The neutron source is the reaction $^{22}\mathrm{Ne}(\alpha,\mathrm{n})^{25}\mathrm{Mg}$, which is activated when the temperature exceeds 2.5×10^8 K. In the range of atomic mass considered, also the *main* component is playing an important role, and so in order to explain the solar abundances of the light *s*–isotopes we need a self–consistent description of both the two components. Raiteri et al. (1991a,b) have shown that when taking into account the *s*–nucleosynthesis

---

[1] E-mail: RAITERI@ASTTO2.INFN.IT (BITNET).

occurring during both core He and shell C burning in a whole generation of massive stars, the weak component can be reproduced. The contribution by the main component was derived from a best fit exponential distribution of neutron exposures, as calculated from the s-nucleosynthesis occurring in low mass stars suffering Thermal Pulses on the Asymptotic Giant Branch (TP–AGB) phase (Käppeler et al. 1990).

We want here to take into account the effect of important improvements in the nuclear physics on the s–nucleosynthesis in stellar models and to make a comparison with the results of the phenomenological approach to the weak component.

Since the compilation of the recommended set of maxwellian averaged neutron-capture cross sections at 30 keV by Bao & Käppeler (1987), many new experimental determinations and theoretical estimates have been made. They have been recently analyzed by Beer, Voß & Winters (1992; hereafter BVW92), that also made a careful estimate of the $< \sigma_{n,\gamma} >$ temperature dependence. Actually, relevant cross sections in the s–path show important departures from the classical $1/v$ trend. While this fact in massive stars does not change much the situation in He burning conditions, where $KT{\sim}30$ keV, it has important consequences when dealing with shell C burning, which occurs at ${\sim}91$ keV. As for unstable nuclei, the β–decay rates by Takahashi & Yokoi (1987) have been extrapolated by Neuberger (1992) to the high temperatures suitable for shell C burning. The temperature dependence of the β–decay rate of ^{79}Se has been taken from Klay & Käppeler (1988). The solar abundances are the meteoritic data by Anders & Grevesse (1989).

2. The Reproduction of the Solar s–Composition

The s–process nucleosynthesis in massive stars was calculated following stellar models by Nomoto & Hashimoto (1988). These models are characterized by the choice of the Schwarzschild's criterion for the treatment of convection and by the $^{12}C(\alpha,\gamma)^{16}O$ reaction rate estimated by Caughlan et al. (1985); after He burning a large convective C shell is developed, affecting most of the C-O core.

Table 1 shows the contribution to the solar system abundances of the s–processing in stellar models for the relevant isotopes; all but ^{58}Fe, ^{86}Kr and ^{87}Rb are s–only nuclei. The contributions from massive stars (MS) were obtained by integrating over a whole stellar generation in the framework of a simple model for the chemical evolution of the Galaxy (see Raiteri et al. 1991b for a discussion). The s–contributions from low mass stars (LMS) are also indicated (Gallino, Busso & Raiteri 1992); they have been calculated after updating the corresponding nuclear network according to BVW92. An appropriate treatment of the difficult branching at ^{85}Kr was adopted working inside the present uncertainties of the neutron capture cross sections of Kr isotopes, in order to overcome the longstanding problem of a ^{86}Kr–overproduction by the main component. For both MS and LMS we chose the lower limit for the $< \sigma_{n,\gamma} >$ of ^{84,85}Kr and the upper one for $< \sigma_{n,\gamma} >$ of ^{86}Kr. In LMS, during the major neutron irradiation released by the $^{13}C(\alpha,n)^{16}O$ reaction at the low temperature of $KT{\sim}12$ keV, the isomeric ratio of ^{84}Kr is increased by ${\sim}25\%$, thus diminishing the flow to ^{86}Kr. Furtheron we added an extra–contribution of 1 mb to the $< \sigma_{n,\gamma} >$ of ^{86}Kr in the hypothesis of a consistent direct capture contribution at low temperatures.

Table 1. Contributions to solar–system abundances

	MS (weak)	LMS (main)	TOT
^{58}Fe	0.33	0.02	0.35
^{70}Ge	0.66	0.16	0.82
^{76}Se	0.66	0.31	0.97
^{80}Kr	0.91	0.15	1.06
^{82}Kr	0.56	0.64	1.20
^{86}Kr	0.15	1.04	1.19
^{87}Rb	0.14	0.97	1.11
^{86}Sr	0.25	0.89	1.14
^{87}Sr	0.17	0.78	0.95

A global inspection of Table 1 shows the degree of reproduction of the solar s–abundances for $60<A<90$. In more detail, some considerations can be drawn. First af all, about 1/3 of ^{58}Fe is accounted for by the s–process. The solar abundance of ^{70}Ge can be better reproduced by recalling that such isotope has an estimated p–contribution of about 10% (Beer 1985). Analogously, a 15% p–correction should also be applied to solar ^{80}Kr. The reproduction of Kr isotopes can then be obtained if we remind that the solar Kr abundance cannot be experimentally derived, but only estimated through the s–systematics. Consequently, our results indicate a 20% higher solar Kr than in Anders & Grevesse (1989). The abundances of ^{87}Rb and ^{86}Sr are inside the errors coming from the determination of their cross sections and solar abundances. Furthermore, one has to consider that ^{87}Rb is a long–lived isotope that suffers partial decay into ^{87}Sr after being ejected into the interstellar medium and before the formation of the solar system.

3. Branching analysis in stellar models

There are only two branchings of the neutron flow in the atomic mass range dominated by the weak component: those at ^{79}Se and at ^{85}Kr. The first one is affected by the heavy dependence on temperature of the ^{79}Se β–decay rate, while the second one is only sensitive to the neutron density.

3.1. The Branching at ^{79}Se

In massive stars, the low neutron density during core He burning ($n_n \sim 10^6$ cm^{-3}) favours the decay of ^{79}Se and thus the production of ^{80}Kr with respect to ^{82}Kr. On the other hand, shell C burning is characterized by a quite higher neutron density, starting from 10^{11} cm^{-3} and then decaying nearly exponentially down to a few 10^9 cm^{-3}. This implies that ^{79}Se suffers neutron captures building up ^{80}Se, so that the production of ^{80}Kr (and ^{82}Kr) is delayed until when the neutron density has consistently decreased.

A different behaviour of the neutron fluence at the ^{79}Se branching is experienced by low mass TP–AGB stars. Here, the branching is not operating when the major fraction of neutrons is produced by the ^{13}C source. The small contribution of 15% to ^{80}Kr comes only from the secondary neutron peak released at about 23 keV by the marginal activation of the ^{22}Ne source.

3.2. The Branching at ^{85}Kr

Massive stars are not critical in the synthesis of ^{86}Kr, since the low neutron density in core He burning conditions makes the ^{85}Kr branching to completely bypass ^{86}Kr, while producing ^{86}Sr and ^{87}Sr. The situation is different in the shell C burning: here the isomeric state of ^{85}Kr has been considered as thermalized, and the high average neutron density forces the s–path to mainly proceed towards the neutron capture channel. Consequently, a consistent production of ^{86}Kr and ^{87}Rb is built. The isotopes ^{86}Sr and ^{87}Sr are not fed, except in the final phase when n_n falls. One has also to take into account a partial β^+–decay of ^{87}Sr which occurs at such high temperatures.

In TP–AGB stars the final production of ^{86}Kr depends on the peak neutron density released by the major ^{13}C neutron source. The results presented here are determined by the adopted choice of neutron capture cross sections on Kr isotopes, and wait for an experimental confirmation.

4. Comparison between the phenomenological approach and stellar models

The phenomenological study solves the problem of a ^{86}Kr (and ^{87}Rb) overproduction by assuming a pulsed neutron exposure for the main component with a pulse duration of <10 yr and a low constant neutron density (Beer & Macklin 1988). Unfortunately, this effect is not efficient enough during thermal pulses in TP–AGB stars.

The weak component is more difficult to be inspected by the phenomenological approach, since both the branchings at ^{79}Se and at ^{85}Kr, from which effective temperature and neutron density might be inferred, are affected by the way the main is formulated. Hence, no firm conclusions can be drawn as for the neutron density, the temperature, as well as for the weak neutron exposure. Beer, Walter & Käppeler (1992; hereafter BWK92) have recently investigated possible scenarios for the weak component. Starting from their prescriptions on the main component, in order to reproduce the solar abundance ratio $(^{80}\mathrm{Kr}/^{82}\mathrm{Kr})_\odot$ they had to formulate the weak s–process as the superposition of two single neutron irradiations: one due to "He burning" and the other to "shell C burning" processing. Assuming the neutron exposure during the second irradiation to be negligible as a first approximation, and only affecting the branching–dependent isotopes, they found that roughly 82% of the s–abundances processed by the weak component must have experienced both the neutron irradiations.

The s–contributions by BWK92 are summarized in Table 2. Here too ^{58}Fe is accounted only for 1/3, and ^{86}Kr is overproduced by 10%. The results shown in Table 2 can be compared with the predictions from stellar models (Table 1). Different contributions are found for ^{80}Kr: while from the phenomenological point of view this isotope is consistently produced by both the main and the weak component, in the astrophysical scenario its synthesis must be ascribed almost entirely to massive stars. The analytical treatment of the main component introduces an effective higher temperature (and a lower neutron density) with respect to the s–process occurring in TP–AGB stars, so that the

Table 2. Main and weak s–contributions according to the phenomenological description by BWK92

	weak ("He+C")	main (pulsed)	TOT
^{58}Fe	0.28	0.03	0.31
^{70}Ge	0.85	0.15	1.00
^{76}Se	0.58	0.29	0.87
^{80}Kr	0.33	0.51	0.84
^{82}Kr	0.40	0.58	0.98
^{86}Kr	0.41	0.69	1.10
^{87}Rb	0.05	0.65	0.70
^{86}Sr	0.04	0.93	0.97
^{87}Sr	0.01	1.01	1.02
$n_n(10^8$ cm$^{-3})$	0.01(He),80(C)	2.0	
KT(keV)	25(He),86(C)	25	

production of ^{80}Kr at the ^{79}Se branching is favoured. As for the weak component, the model of BWK92 gives a greater weight to the high–temperature, high–neutron density phase (which corresponds to shell C burning), during which ^{80}Kr is partly bypassed.

The abundances of ^{86}Sr, ^{87}Sr, ^{86}Kr and ^{87}Rb depend on the different conditions that play at the ^{85}Kr branching. In the TP–AGB stars a higher average neutron density for the main component implies that the neutron flow feeds ^{86}Kr and ^{87}Rb, avoiding an excessive production of ^{86}Sr and ^{87}Sr. We have already recalled how the problem of ^{86}Kr overproduction is overcome by the phenomenological analysis assuming a pulsed model for the main component. On the other side, the important "C burning" irradiation at high neutron density gives rise to a large contribution to this isotope (and a low one for ^{86}Sr and ^{87}Sr).

5. Conclusions

The s–processing in a whole generation of massive stars has been investigated with an updated input nuclear physics, that takes into account the temperature–dependence of neutron capture cross sections and proper extrapolations of beta–decay rates at high temperature. When massive stars s–contributions to the solar composition are added to those coming from TP–AGB stars of low mass the weak component is reproduced. In order not to overproduce ^{86}Kr by the main component it is necessary to slightly increase the neutron capture isomeric ratio to ^{85}Kr. This prescription is inside the present experimental uncertainty. Our results indicate a 20% increase of the solar Kr abundance, for which only theoretical estimates exist. The difficulties of the phenomenological treatment of the weak component are briefly discussed, and its results are compared to the predictions by stellar models.

References

Anders, E., & Grevesse, N. 1989, *Geochim. Cosmochim. Acta*, **53**, 197

Bao, Z. Y., & Käppeler, F. 1987, *Atomic Data & Nucl. Data Tables*, **36**, 411

Beer, H. 1985, *Kernforschungszentrum Karlsruhe*, Internal Report

Beer, H., & Macklin, R. L. 1988, *ApJ*, **331**, 1047

Beer, H., Voß, F. & Winters, R. R. 1992, *ApJS*, **80**, 403 (BVW92)

Beer, H., Walter, G., & Käppeler, F. 1992, *ApJ*, **389**, 784

Caughlan, G. R., Fowler, W. A., Harris, M. J., & Zimmerman, B. A. 1985, *Atomic Data and Nucl. Data Tables*, **32**, 197

Gallino, R., Busso, M., & Raiteri, C. M. 1992, this meeting

Holmes, J. A., Woosley, S. E., Fowler, W. A., & Zimmerman, B. A. 1976, *Atomic Data and Nucl. Data Tables*, **18**, 305

Käppeler, F., Gallino, R., Busso, M., Picchio, G., & Raiteri, C. M. 1990, *ApJ*, **354**, 630

Klay, N., & Käppeler, F. 1988, *Phys. Rev. C*, **38**, 295

Neuberger, D. 1992, in preparation

Nomoto, K., & Hashimoto, M. 1988, *Phys. Reports*, 163, 13

Raiteri, C. M., Busso, M., Gallino, R., Picchio, G. & Pulone, L. 1991a, *ApJ*, **367**, 228

Raiteri, C. M., Busso, M., Gallino, R., & Picchio, G. 1991b, *ApJ*, **371**, 665

Takahashi, K., & Yokoi, K. 1987, *Atomic Data & Nucl. Data Tables*, **36**, 375

Woosley, S. E., Fowler, W. A., Holmes, J. A., & Zimmerman, B. A. 1978, *Atomic Data and Nucl. Data Tables*, **22**, 371

Synthesis of ^{19}F in the He-burning zones of massive stars

G Meynet[1] and M Arnould[2]

1) Geneva Observatory, 51 ch. des Maillettes, CH-1290 Sauverny, Switzerland
2) Institut d'Astronomie et d'Astrophysique, C.P.165, Université Libre de Bruxelles, 50 av. F.D. Roosevelt, B-1050 Brussels, Belgium

Abstract. The mechanisms for the synthesis of ^{19}F are not well known, even if many scenarios have been proposed. Here we explore the ability of the He-burning zones in massive stars to produce significant amounts of ^{19}F. To this end we have computed detailed evolutionary sequences for models of 20, 60 and $120M_\odot$ from the ZAMS to the end of the carbon-burning phase ($Y = 0.30$, $Z = 0.02$). Some preliminary results have also been obtained for a 9 and $12M_\odot$ model. We find that the He-burning shell at work in those of the considered stars that experience core carbon burning contributes in a modest way to the synthesis of ^{19}F. In contrast, the ^{19}F abundance in the winds of the 60 and $120M_\odot$ Wolf-Rayet stars during their WC stage are enhanced by factors between 7 and 50, depending on the initial mass and metallicity. Wolf-Rayet stars could thus contribute substantially to the enrichment in ^{19}F of the interstellar medium.

1. Introduction

Among the elements ranging from carbon to calcium, fluorine (the only stable isotope of which is ^{19}F) is the least abundant in the solar system, its mass fraction being $4\,10^{-7}$ only (Anders and Grevesse, 1989). As reviewed by Goriely et al. (1989), different sites have been proposed for the synthesis of ^{19}F. The possibility of production of that nuclide in low and intermediate mass stars during their Asymptotic Giant Branch (AGB) phase has been studied with the aid of detailed evolutionary models by Forestini et al. (1992), or on qualitative grounds by Jorissen et al. (1992). On the other hand, the ^{19}F yields from hydrostatic and explosive nucleosynthesis for initial masses comprised between 12 and $40M_\odot$, taking into account the "neutrino process" have been computed recently by Weaver and Woosley (1992). In the present work, and in order to investigate quantitatively a suggestion by Goriely et al. (1989), we examine specifically the ability of the He burning shell of massive stars to produce ^{19}F. This is done with the aid of detailed models of stars with masses $M \leq 20M_\odot$ up to the end of central C burning. In addition, we calculate the extent to which the winds of Wolf-Rayet (WR) stars could contribute to the ^{19}F enrichment of the interstellar medium (ISM), a possibility that has not been explored yet. The evolutionary models are computed as described by Schaller et al. (1992). However, the adopted nuclear reaction network is extended, especially in order to include the reactions involved in the production and destruction of ^{19}F (see also Meynet and Arnould, this volume, for the description of other changes). Those reactions are briefly presented in Sect. 2. Section 3 describes our results, and some conclusions are drawn in Sect. 4.

2. The reaction network for the ^{19}F synthesis

2.1 H-burning

The abundance of ^{19}F is followed through a weak branching in the CNO cycle

$$^{14}N(p,\gamma)^{15}O(\beta^+)^{15}N(p,\gamma)^{16}O(p,\gamma)^{17}F(\beta^+)^{17}O(p,\gamma)^{18}F(\beta^+)^{18}O(p,\gamma)^{19}F$$

$$^{19}F(p,\alpha)^{16}O$$

The adopted $^{19}F(p,\alpha)^{16}O$ rate is the geometrical means of the lower and upper limits to that rate proposed by Kious (1990).

2.2 He-burning

We use the same nuclear reaction network as the one briefly presented by Meynet and Arnould (this volume). Three main reactions chains are considered :

$$^{14}N(\alpha,\gamma)^{18}F\begin{cases}(\beta^+)^{18}O(p,\alpha)^{15}N(\alpha,\gamma)^{19}F(\alpha,p)^{22}Ne\\(n,p)^{18}O(p,\alpha)^{15}N(\alpha,\gamma)^{19}F(\alpha,p)^{22}Ne\\(n,\alpha)^{15}N(\alpha,\gamma)^{19}F(\alpha,p)^{22}Ne\end{cases}$$

The synthesis of ^{19}F thus requires the availability of neutrons and protons. They are mainly produced by $^{13}C(\alpha,n)^{16}O$ and $^{14}N(n,p)^{14}C$. The ^{13}C α-capture rate is taken from Descouvemont (1987), whose theoretical prediction of an increase of the astrophysical S-factor at low energies is confirmed by recent experimental data (Drotleff et al., this volume). The $^{14}N(n,p)^{14}C$ rate is taken from Brehm et al. (1988)(*). It has to be noted that the first branching is much more important that the other ones in the massive star conditions of relevance here.

3. Results

3.1 H-burning phase

During the main sequence, ^{19}F is essentially destroyed. For example, the ^{19}F mass fraction drops from $4 \ 10^{-7}$ to $0.01 \ 10^{-7}$ between the beginning and the end of H burning in the core of the $20M_\odot$ star.

3.2 Core He-burning phase

As can be seen in Figure 1a, ^{19}F is built up through

$$^{14}N(\alpha,\gamma)^{18}F(\beta^+)^{18}O(p,\alpha)^{15}N(\alpha,\gamma)^{19}F$$

at the beginning of He-burning, while it is destroyed by $^{19}F(\alpha,p)^{22}Ne$ at the end of that phase (see Fig. 1b). Thus material experiencing the whole He-burning episode cannot be ^{19}F enriched. At least two ways are conceivable in order to avoid that

---

(*) That rate is a factor of two lower that the one proposed by Koehler and O'Brien (1989). As such, it leads to a lower limit of the calculated ^{19}F yields.

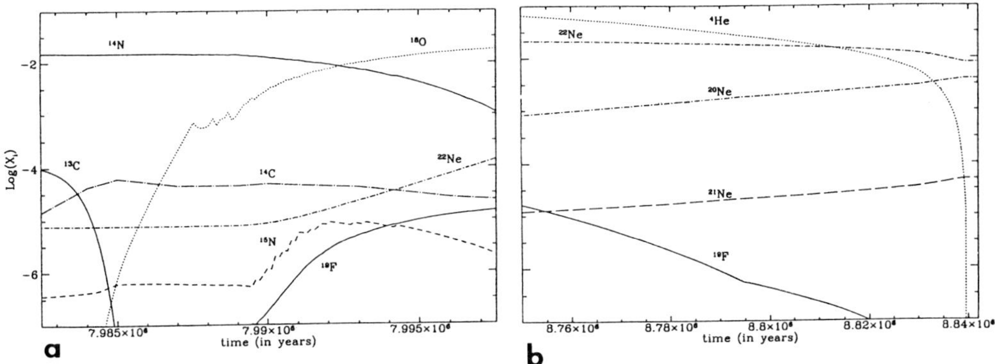

Figure 1 : *(a) Evolution of the central mass fractions of various nuclides at the beginning of the He-burning phase in the $20M_\odot$ star model; (b) same as (a), but at the end of the He-burning phase, ~ 750000 years later.*

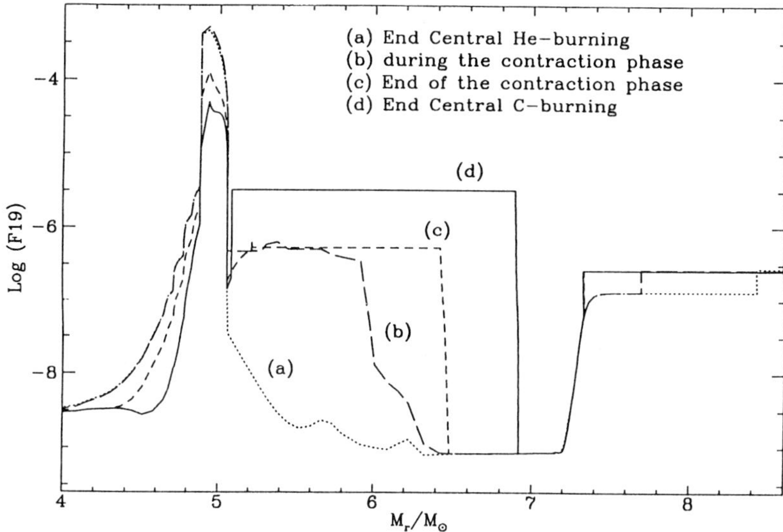

Figure 2 : *Evolution with time of the profile of the ^{19}F mass fraction in the $20M_\odot$ model. In the outermost layers, ^{19}F is not affected.*

depletion : (1) ^{19}F is ejected from the star before completion of the He-burning phase. This can occur in WR stars; (2) the burning of He is only partial and some ^{19}F can survive up to the presupernova stage. In massive stars, this situation may be found either at the periphery of the convective He-burning core, or in the He-burning shell.

3.3 Contribution of the He-burning shell

Figure 2 displays the ^{19}F profile inside our $20M_\odot$ star model at various remarkable epochs in its evolution, while Figure 3 presents the profiles of some nuclides at the end of C burning. Two ^{19}F-enriched zones can be identified : a well marked peak around mass coordinate $M_r = 5M_\odot$, and a plateau that becomes wider and wider as time passes. The peak results from the shrinkage of the convective core close to the end

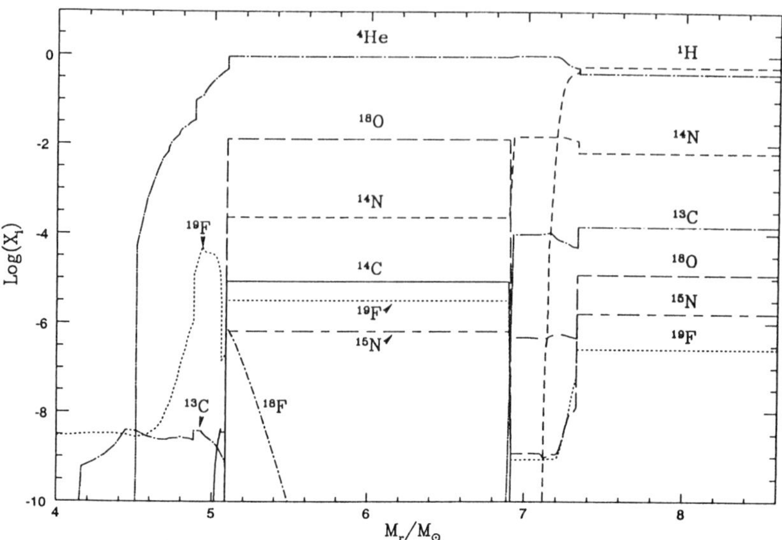

Figure 3 : *The mass fractions of some elements taking part in the synthesis of ^{19}F are represented as a function of the mass (in $M_\odot$) inside the $20M_\odot$ model at the end of the carbon-burning phase.*

of He burning, leaving some ^{19}F unburned at the edge of the core. This ^{19}F is then partly destroyed during the carbon-burning phase. The second zone contains the ^{19}F produced by the He-burning shell.

If all the ^{19}F present at the end of the carbon burning phase is ejected into the ISM at the supernova stage (clearly a rough estimate), then

$$<^{19}F(ejecta)> /^{19}F(ISM) = \sim 2$$

for our $20M_\odot$ star model, where $<^{19}F(ejecta)>$ and $^{19}F(ISM)$ are the ^{19}F mean abundance in the ejecta and the solar system abundance, respectively. This enhancement factor is well in the range of values found in the presupernovae models of Weaver and Woosley (1992, see in their tables 2 and 3, the columns corresponding to an enhancement factor of 3 for the $^{12}C(\alpha,\gamma)^{16}O$ reaction rate). For lower initial masses, we obtain a lower production factor (~ 1.4 for our $9M_\odot$ model). In higher mass models, which become WR stars, the entire He-burning shell has been wind ejected by the end of the C-burning phase, so that no contribution can come from that shell. From these results we can conclude that the contribution of the He-burning shell in massive stars is quite modest.

3.4 Contribution of the wind of the WR stars

Figure 4 shows the evolution of the abundances of various nuclides at the surface of our $60M_\odot$ star versus the remaining mass (or equivalently, time). The initial surface abundances start being altered when the products of hydrogen burning are uncovered by mass loss. As a consequence, the ^{19}F abundance decreases (see Sect. 3.1). In contrast, the ^{19}F amount increases when the products of He burning appear at the surface during the WC phase. The following results have been obtained :

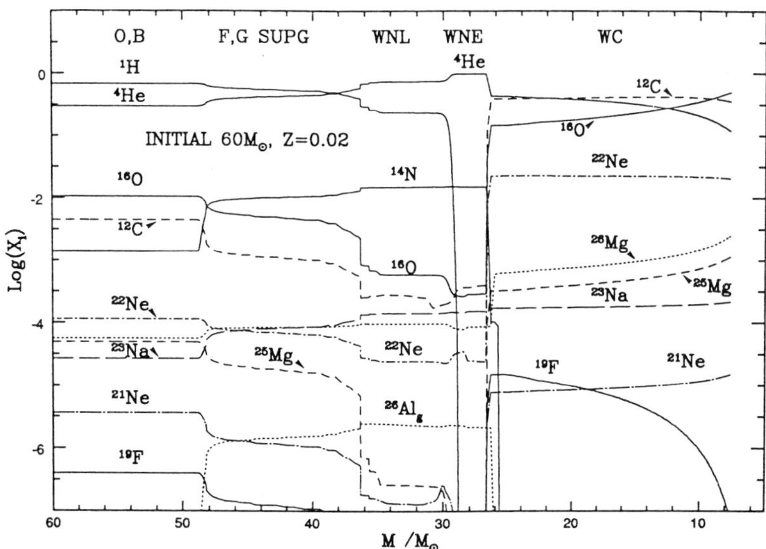

Figure 4 : *Evolution of the surface abundances (in mass fraction) versus the remnant mass M for the $60M_\odot$ model. During the WC phase, ^{19}F is enhanced by a factor 18 on average.*

$$60M_\odot \ , \ Z = 0.020, <^{19} F(WC) > /^{19}F(ISM) =\sim 18$$

$$60M_\odot \ , \ Z = 0.030, <^{19} F(WC) > /^{19}F(ISM) =\sim 51$$

$$120M_\odot, \ Z = 0.020, <^{19} F(WC) > /^{19}F(ISM) =\sim 7,$$

where $<^{19} F(WC) >$ is the mean abundance of ^{19}F in the wind of WC stars. Let us estimate the importance of this production for the ^{19}F enrichment of the ISM :

• The net rate of ^{19}F enrichment of the ISM by the wind of WR stars, $\dot{M}_{19}$, is

$$\dot{M}_{19} = (<^{19} F(WN) > -^{19}F(ISM))\dot{M}_{WN} + (<^{19} F(WC) > -^{19}F(ISM))\dot{M}_{WC} \quad (1)$$

where $\dot{M}_{WN}$, $\dot{M}_{WC}$ are the rates of mass input into the ISM through stellar winds by WN, WC stars, respectively, and $<^{19} F(WN) >$ is the mean ^{19}F abundance in the wind of WN stars. According to Abbott (1982), $\dot{M}_{WN} = 2.0 \ 10^{-5}$ and $\dot{M}_{WC} = 2.9 \ 10^{-5} M_\odot$ per *year* and kpc^2. If the values of $<^{19} F(WN) >$ and $<^{19} F(WC) >$ derived for our $Z = 0.02 \ 60M_\odot$ star are inserted into Eq. (1),

$$\dot{M}_{19} \sim 2 \ 10^{-10} M_\odot \ per \ year \ and \ kpc^2.$$

• The contributions of the winds to the net yields may be written $y_{19} = \frac{\dot{M}_{19}}{\Psi_1(1-R)}$, where $R = 0.17$ is the fraction of stellar gas returned to the ISM, and Ψ_1 is the star formation rate, which we adopt equal to $\sim 10M_\odot$ per *Gyr* and pc^2 (Tinsley, 1980). Thus the net yields becomes

$$y_{19} = 2.4 \ 10^{-8}$$

• In the closed model of chemical galactic evolution, $^{19}F(WR) = y_{19} \ln \mu^{-1}$ for an initially enriched gas, μ being the mass fraction of the gas in the Galaxy. If $\mu \sim 5-10\%$ (values for the solar neighbourhood),

$$^{19}F(WR)/^{19}F(ISM) =\sim 0.15.$$

If the results for the $60M_\odot$ at $Z = 0.03$ are used, we obtain

$$^{19}F(WR)/^{19}F(ISM) =\sim 0.46.$$

These results indicate that the winds of WR stars can contribute non-negligeably to the ^{19}F galactic content.

4. Conclusion

Detailed evolutionary models indicate that the He-burning shells of $M \geq 9M_\odot$ stars cannot be important contributors to the galactic ^{19}F. In contrast, we show that ^{19}F is sufficiently overabundant in the winds of WC stars for them to contribute more or less significantly to the ^{19}F enrichment of the ISM, this contribution increasing with metallicity. More models, especially of low metallicity stars, are needed in order to perform more quantitative evaluations of the role played by WR stars in the ^{19}F enrichment of galaxies. Finally, let us stress that it would be of high interest to confront the predictions of our calculations with the spectroscopic determination of the fluorine abundance in WR stars. It remains to be seen if such observations are really feasible.

Acknowledgements : We thank P. Aguer, M. Kious and J.P. Thibaud for providing us with their experimental data concerning $^{19}F(p,\alpha)^{16}O$ prior to publication, as well as for discussions concerning other reactions involved in the synthesis of ^{19}F. This work has been support in part by the SCIENCE Program SC1-0065. M.A. is Chercheur Qualifié F.N.R.S. (Belgium).

References

Anders, E., Grevesse, N., 1989, Geochim. et Cosmochim. Acta 53, 197

Brehm, K., Becker, H.W., Rolfs, C., Trautvetter, H.P., Käppeler, F., Tatynski, W., 1988, Z. Phys. A330, 167

Descouvemont, P., 1987, Phys. Rev. C36, 2206

Forestini, M., Goriely, S., Jorissen, A., Arnould, M., 1992, A&A, to appear

Goriely, S., Jorissen, A., Arnould, M., 1989, in *Proc. 5th Workshop on Nuclear Astrophys.*, eds. W. Hillebrandt & E. Müller, Max Planck Inst. für Astrophys. Rep., p. 60

Jorissen, A., Smith, V., Lambert, D., 1992, A&A, to appear

Kious, M., 1990, Ph. D. Thesis (C.S.N.S.M. Orsay), unpublished

Koehler, P.E., O'Brien, H.A., 1989, Phys. Rev. C39, 1655

Schaller, G., Schaerer, D., Meynet, G., Maeder, A., A&AS, to appear

Tinsley, B., 1980, Fundamentals of Cosmic Physics 5, 287

Weaver, T.A., Woosley, S.E., 1992, preprint

Analysis of the double neutron pulse model in TP–AGB stars

R. Gallino†[1], M. Busso‡, C.M. Raiteri‡

† Istituto di Fisica Generale dell'Universita' di Torino, Via P. Giuria 1, 10125 Torino, Italy

‡ Osservatorio Astronomico di Torino, Strada Osservatorio 20, 10025 Pino Torinese, Italy

Abstract.
The astrophysical conditions for the activation of the double neutron pulse in low mass stars (M=1–3 $M_\odot$) suffering recurrent convective He instabilities on the asymptotic giant branch are considered. The development of thermal pulses is followed taking into account the decrease in time of the intershell mass, of the overlapping factor between pulses and of the interpulse period, according to recent stellar models. The s-nucleosynthesis is calculated by adopting the new recommended values for neutron capture cross sections by the Karlsruhe Group, including their dependency on temperature. This improved scenario confirms that an asymptotic distribution of s-abundances is eventually reached, which is essentially determined by the asymptotic neutron exposure and by the "freeze-out" conditions of the neutron density. The problems concerning the activation of the ^{13}C neutron source are briefly examined.

1. Introduction.

Many observational data provide information on the galactic history of the s-isotopes. The solar system distribution of chemical abundances (Anders & Grevesse 1989) is the most important one: it derives from a composite process of nucleosynthesis that occurred in previous generation of stars, requiring three distinct s-components: a main, a weak and a strong one. The main component accounts for the bulk of the heavy s-isotopes between Y and Pb, giving minor contributions to isotopes with atomic mass number A < 90; the weak component works only for A < 90; the strong component has been introduced to explain the s-contribution to ^{208}Pb.

[1] E-mail: GALLINO@TO.INFN.IT (BITNET).

Other constraints come from high-resolution spectroscopic observations of dwarf and giant stars of different populations, revealing an unexpected trend for several elements that, in the galactic disc, are mainly contributed by the s-process. Among them, there are Y and Ba: the logarithmic ratios [Y/Fe] and [Ba/Fe] in the interstellar medium remain constant for stars with metallicity [Fe/H] > -1.5 (see Mathews, Bazan & Cowan, 1992 and references therein). This is typical of "primary" species, and is at odd with the classical expectation of a "secondary" nature for the s-process, according to which the above logarithmic ratios should instead increase linearly with [Fe/H].

Moreover, among evolved stars, a number of peculiar red giants of low mass, the MS, S, C (N-type), CH and Ba stars, show correlated ^{12}C and s-enhancements: many of them show also the presence of the unstable isotope ^{99}Tc, hence indicating that the s-nucleosynthesis is an ongoing process in the He burning zones of these stars (Smith & Lambert 1990; Busso et al. 1992a).

Finally, the analysis of the isotopic composition of SiC microcrystals embedded in pristine meteorites reveals that many elements in these stardusts are enriched in their s–isotopes (Ott et al. 1988; Lewis et al. 1990; Virag et al. 1992; Prombo et al. 1992; Richter et al. 1992). Actually, the s-signature is not solar, but rather corresponds to a spread of mean neutron exposures, lower than that responsible for the main component. The condensation of SiC grains requires C/O > 0.8, and it is natural to infer that most of them form in the mass-losing envelopes of C-stars (Lewis et al. 1990; Gallino et al. 1990).

2. The main component and the phenomenological analysis.

The phenomenological analysis is a useful tool in order to understand the origin of the s-elements in the solar system (Mathews & Ward 1985; Käppeler, Beer & Wisshak 1989), being independent of stellar models and neutron sources. It has revealed that the main component can be reproduced through an exponential distribution of neutron exposures, and is characterized by a mean exposure $\tau_0 = (0.306 \pm 0.010)$ [kT (keV)/30]$^{1/2}$ mb^{-1}, by an irradiated fraction G $= (0.057 \pm 0.004)$ % of solar Fe, and by a number of neutrons captured per iron seed $n_c = 10.7 \pm 0.7$ (Käppeler et al. 1990). Moreover, the analysis of the various branchings in the s-path allows one to estimate the effective neutron density and temperature. Indeed, some s-only isotopes such as ^{80}Kr, ^{152}Gd and ^{164}Er are "thermometers" of the s-process, their production depending on brachings sensitive to the temperature. Other s-only isotopes, such as ^{148,150}Sm and ^{186}Os depend on branchings that are mainly controlled by the neutron density. The presently accepted values of the above parameters are (Käppeler et al. 1990): T$=(3.3 \pm 0.5)\times10^8$ K, $n_n =(3.4 \pm 1.1)$ cm^{-3}. These results are important for characterizing the astrophysical sites of the s–nucleosynthesis of heavy elements.

3. The s-processing in thermal pulses

Low mass thermally pulsing asymptotic giant branch (TP-AGB) stars appear as good candidates for the production of the main component. The ^{22}Ne neutron source here is only marginally activated, while the alternative reaction $^{13}C(\alpha,n)^{16}O$ may efficiently operate (Gallino et al. 1988; Hollowell & Iben 1989; Gallino 1989; Käppeler et al. 1990). As a result, two episodes of neutron irradiations occur, giving rise to the double neutron pulse scenario. The ^{13}C neutron source requires a primary production of ^{13}C during the intershell period, through the mixing of a small amount of protons from the envelope into the ^{12}C-rich zone. A ^{13}C-enriched pocket containing a few 10^{-6} $M_\odot$ of ^{13}C is thus formed and is ingested by the next convective instability, releasing neutrons at kT=12 keV (Iben & Renzini 1982; Hollowell & Iben, 1988, 1989). The efficiency of this mechanism is still a matter of debate, especially for galactic disc stars. Actually, during the III dredge-up phase, that occurs after each instability once thermal pulses have reached full amplitude, the external convection penetrates inwards, and a sharp discontinuity in chemical composition results between the hydrogen-rich matter of the envelope and the carbon-rich matter of the He shell. In these conditions, a region of marginal instability is expected (Busso et al. 1992a), favouring the required proton penetration. The occurrence of such an instability still needs to be investigated.

In past years, modelling the s-process nucleosynthesis during TP-AGB phases has been followed using thermal pulse average conditions. Also in our models (Käppeler et al. 1990) we considered a constant amount of ^{13}C ingested per pulse and adopted typical fixed prescriptions for pulse shapes, thermal conditions, and dilution factor D (ratio between the shell mass at ^{13}C burning and that at maximum convective expansion).

Actually, evolutionary calculations of TP-AGB phases by different authors (Iben, 1977, 1982; Packzynski, 1975; Schonberner, 1979; Boothroyd & Sackmann, 1988) have shown that the main characteristics of the models (convective shell mass, interpulse duration, overlapping between adjacent pulses) are monotonically decreasing functions of a unique parameter, the core mass M_H. Their variation in time may have significant effects on the way in which the neutron exposed material is mixed with fresh iron seed matter during the growth of each instability and on the neutron density, which increases pulse after pulse. For these reasons, while still assuming a constant amount of ^{13}C burnt per pulse (that for the moment has to be considered as a free parameter), we have modified our models according to the above scenario.

We have followed in particular the results by Chieffi, Limongi & Straniero (1992; see also Busso et al. 1992b), who computed new thermal pulses for stars of solar metallicity and mass M = 1.5, 3, 5 and 7 $M_\odot$ with the FRANEC evolutionary code and assuming standard mass loss rates. The main features of these models are: i) the III dredge-up is naturally found, even for core masses below 0.7–0.8 $M_\odot$; ii) after dredge-up is started, the evolutionary track is modified and the strength of the pulses is enhanced; iii) the amount of dredge-up matter increases in time, from $\simeq 10^{-4}$ $M_\odot$ to $\simeq 10^{-3}$ $M_\odot$.

Using the above models, and our network including about 500 nuclides, we recomputed the s-process nucleosynthesis adopting the most recent prescriptions for cross sections and β-decay rates. Concerning cross sections, the compilation by Beer, Voß & Winters (1992) also includes the $< \sigma_{n,\gamma} >$ temperature dependence. In stellar conditions, correction factors had to be applied to take into account captures from low-lying excited levels and unthermalized isomeric states. This is particularly important for ^{85}Kr. As for the beta-decay rates of heavy unstable isotopes at stellar temperatures, they are based on the estimates by Takahashi & Yokoi (1987) and more recent updatings.

The results obtained with these revised models confirm our previous finding that the average neutron density has a dynamical history and that its maximum is determined by the rate of ingestion of the ^{13}C-pocket by the growing convective instability. Moreover they point out that, for any amount of ^{13}C burnt, the decrease in the convective shell mass leads to increasing neutron exposures per pulse $\Delta\tau$. In the meantime, the pulse shape also changes: in particular, the dilution factor decreases monotonically. The combined effects of increasing $\Delta\tau$ and of decreasing dilution cause the *effective* neutron exposure to reach an asymptotic limit.

The final abundance of most isotopes is "freezed-out" when the neutron density drops at the quenching of the instability (Käppeler et al. 1990). The peak neutron densities reached in the pulses influence only the final abundances of a few neutron-rich isotopes such as ^{86}Kr, ^{87}Rb, and ^{96}Zr; it is necessary however that n_n does not exceed a few 10^9 cm^{-3}, in order not to overproduce them. Of importance is also the temperature dependence of cross sections for relevant isotopes, since the bulk of the neutron exposure comes from ^{13}C burning at about 12 keV, and only a minor contribution derives from the ^{22}Ne source at about 26 keV. In this respect, the enhancement by about 40 % of $\sigma(^{138}$Ba$)$ at 12 keV recently found by Beer et al. (1992) confirms a major point in the analysis by Gallino et al. (1992) on the main component and on the Ba isotopic anomalies in meteoritic SiC grains.

The overabundances of s-nuclei in the He-shell corresponding to the mean neutron exposure that best fits the main component are shown in Fig. 1. The value of this mean exposure has slightly decreased with respect to previous analyses (see above) due to the adoption of the new cross sections, and is now $\tau_0 \simeq 0.28$ mb^{-1}. In our models, it corresponds to stars with metallicities of about 1/3 of the solar one.

As a byproduct, our study also predicts the distribution of r-process contributions to non s-only nuclei, which are shown in Fig. 2.

4. Conclusions

New calculations of neutron-capture nucleosynthesis in TP-AGB stars of low mass have been performed, in which average prescriptions for the double neutron pulse model are relaxed, and the temporal evolution of the main parameters involved is considered. This improved scenario does not modify the general results we have obtained so far; in particular an asymptotic distribution of s-abundances is eventually reached. Moreover, we confirm that thermal pulses in low mass TP–AGB stars are a suitable astrophysical site that can simultaneously account for the *main* s-process component in the solar system, for the spectroscopic observations of s-enriched AGB stars, for the primary-like behaviour of the s–elements in disc stars and for the s-process isotopic anomalies in meteoritic SiC grains.

References

Anders, E., & Grevesse, N. 1989, *Geochim. Cosmochim. Acta*, 53, 197.

Beer, H., Corvi, F., Mauri, A. & Athanassopulos K., 1992, this Volume.

Beer, H., Voß, F. & Winters, R. R. 1992, *ApJSuppl.*, 80, 403.

Boothroyd, A. I., & Sackmann I.-J. 1988, *ApJ*, , 328, 653.

Busso, M., Gallino, R., Lambert, D.L., Raiteri, C.M., & Smith, V.V. 1992a, *ApJ*, 399, in press.

Busso, M., Chieffi, A., Gallino, R., Limongi, M., Raiteri, C.M., & Straniero, O. 1992b,in *Planetary Nebulae*, R. Weinberg and A. Acker (eds.), (Dordrecht: Kluwer), in press.

Chieffi, A., Limongi, M. & Straniero. O. 1992, in preparation

Gallino R. 1989, in *Evolution of Peculiar Red Giant Stars*, (ed. H. R. Johnson & B. Zuckerman), (Cambridge Univ. Press), 176.

Gallino, R., Busso, M., Picchio, G., Raiteri, C. M. & Renzini, A. 1988, *ApJ*, 334, L45.

Gallino, R., Busso, M., Picchio, G., & Raiteri, C.M. 1990, *Nature*, 348, 298.

Gallino, R., Raiteri, C.M., & Busso, M. 1992, *ApJ*, in press.

Hollowell, D. E. & Iben, I. Jr. 1988, *ApJ*, 333, L25.

Hollowell, D. E. & Iben, I. Jr. 1989, *ApJ*, 340, 966.

Käppeler, F., Beer, H., & Wisshak, K. 1989, in *Report Progr. Phys.*, 52, 945.

Käppeler, F. Gallino, R., Busso, M., Picchio, G., & Raiteri, C.M. 1990, *ApJ*, 354, 630.

Iben, I. Jr. 1977, *ApJ*, 217, 788.

Iben, I. Jr. 1982, *ApJ*, 260, 821.

Iben, I. Jr., & Renzini, A. 1982, *ApJ*, 263, L23.

Lewis, R. S., Amari, S., & Anders, E. 1990, *Nature*, 348, 293.

Mathews, G., Bazan, G., & Cowan, J. J. 1992, *ApJ*, 391, 719.

Mathews, G. J., & Ward, R. A., 1985, *Report Progr. Phys.*, 48, 1371.

Ott, U., Begemann, F., Yang, J. & Epstein, S. 1988, *Nature*, 332, 700.

Paczynski, B. 1975, *ApJ*, 214, 812

Prombo, C. A., Podosek, F. A., Amari, S. & Lewis, R. S. 1992, *ApJ*, in press.

Richter, S., Ott, U., & Begemann, F., 1992, this Volume.

Schonberner, D. 1979, *A&A*, 79, 108.

Smith, V. V. & Lambert, D. L. 1990 *ApJSuppl*, 72, 387.

Takahashi, K., & Yokoi, K. 1987, *Atomic Data & Nucl. Data Tables*, 36, 375.

Virag, A., Wopenka, B., Amari, S., Zinner, E. K., Anders, E., & Lewis, R. S. 1992, *Geochim. Cosmochim. Acta*, 56, 1715.

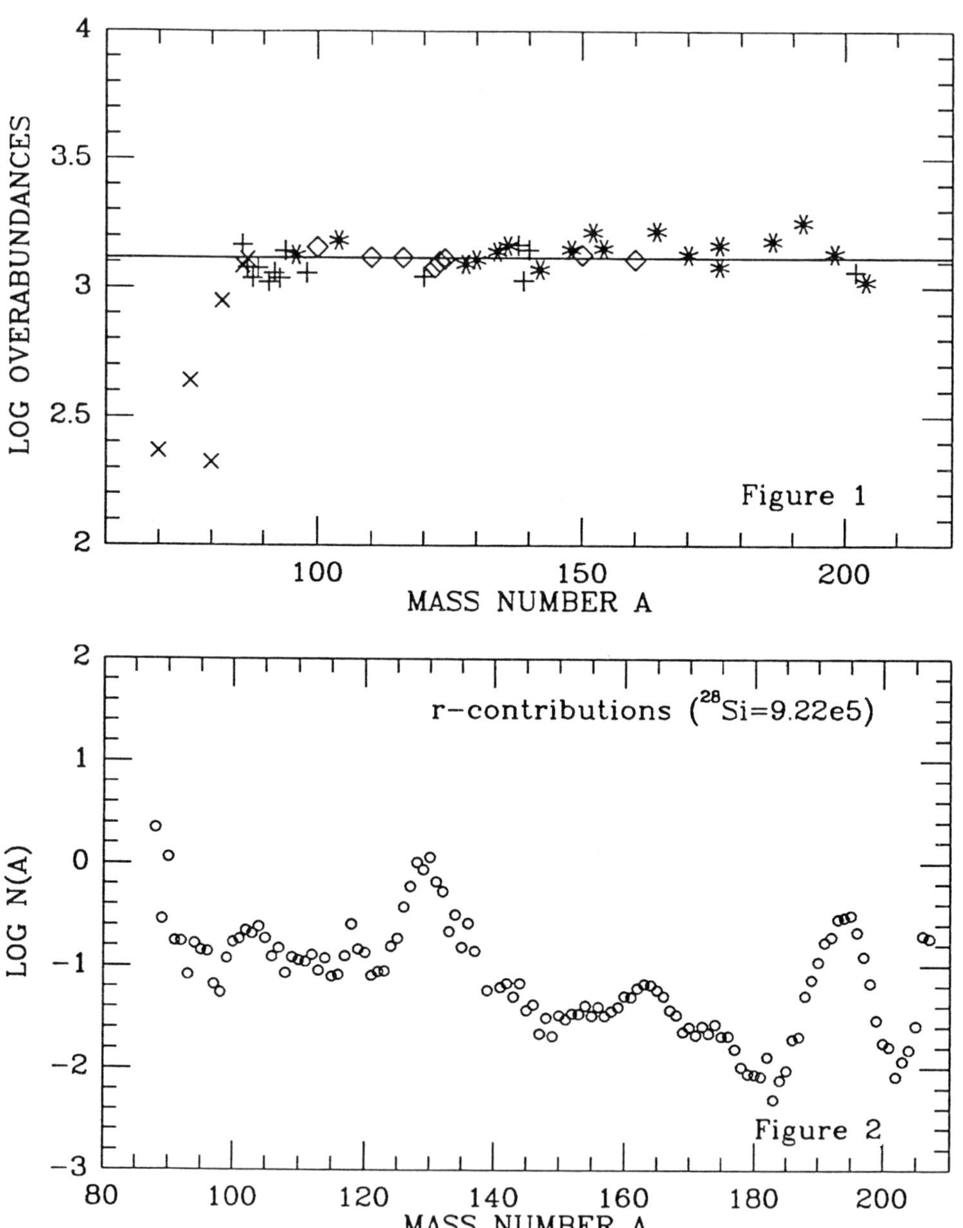

LOG OVERABUNDANCES
MASS NUMBER A
Figure 1
r-contributions (^{28}Si=9.22e5)
LOG N(A)
MASS NUMBER A
Figure 2

Nucleosynthesis and mixing in C and Ba stars

M. Busso†[1], L. Beglio† R. Gallino‡ D.L. Lambert§, C. M. Raiteri†and V.V. Smith§

† Osservatorio Astronomico di Torino, 10025 Pino Torinese, Italy

‡ IFG, Universita' di Torino, via P. Giuria 1, 10125 Torino, Italy

§ Department of Astronomy, University of Texas, 15398 Austin, Texas

Abstract.

New nucleosynthesis models for thermally pulsing asymptotic giant branch stars of low mass are used to predict the enrichment in s–process nuclei and in light species at the surface of carbon and barium stars. We show that the observed distributions can be reproduced by neutron capture models based on the ^{13}C neutron source attaining neutron exposures τ_0 in the range 0.2 – 0.4 mb^{-1}. This range is the same found for MS and S giants, of which C and Ba stars can be interpreted as being evolved descendants. Rather stringent constraints on the amount of ^{13}C required are derived. We also confirm that the Rb/Sr ratio in classical Ba stars definitely excludes the possibility that ^{22}Ne be the main neutron source.

1. Introduction

It has been known for twenty years (Ulrich, 1973) that thermally pulsing asymptotic giant branch (TP–AGB) stars are an astrophysical site suitable to produce neutron exposures of the exponential form required to account for the solar–system *main s-process component* (i.e. s–nuclei with A $\geq$ 80–85).

The first quantitative models computed within this scenario were based on stars of intermediate mass ($3 \leq$ M/M$_\odot \leq 8$) which, during thermal instabilities, develop rather hot (T$_{bottom} \simeq 3.5\ 10^8$ K) convective He–burning shells and can hence activate the neutron source ^{22}Ne$(\alpha,$n$)^{25}$Mg (Truran and Iben, 1977). These models, however, have been subsequently questioned: indeed, on one side they predict high luminosities (M$_{bol} \leq -6.5$), while most s–enriched stars in the Galaxy and in the Magellanic Clouds are fainter (M$_{bol} \leq -4$: Blanco et al., 1980; Iben and Renzini, 1983). On the other

---

[1] E-mail: BUSSO@ASTTO2.INFN.IT (BITNET).

hand, the distribution of s–elements induced by the ^{22}Ne$(\alpha,n)^{25}$Mg reaction cannot resemble that of the solar system (Busso et al., 1988).

An alternative picture has been outlined recently (Hollowell and Iben, 1988; Gallino et al., 1988), according to which neutrons can be released mainly by the ^{13}C$(\alpha,n)^{16}$O neutron source, that is activated at lower temperatures and hence is efficient in low mass stars (M $\leq$ 3 M$_\odot$), evolving at lower luminosities. Under this hypothesis both the solar system *main* s–process component and other constraints coming from isotopic anomalies in meteorites were accounted for (Käppeler et al., 1990; Gallino et al., 1990). However, in this second scenario the production of the neutron source ^{13}C relies on subtle partial mixing processes, necessary to introduce some protons from the convective envelope into the ^{12}C–rich region. Attempts to model this process were so far successful only for population II stars (Hollowell and Iben, 1988). Therefore it has become important to derive external constraints on the amount of ^{13}C that is required to explain the known s–process distributions in various galactic environments (solar system, unevolved stars, evolved red giants); such an analysis can then be used as a guideline for improved stellar models looking for ^{13}C production.

Among these constraints, of particular importance are the observations of red giants showing enhanced s–element abundances due to dredge–up of material from internal regions. Following Lambert (1991)'s definition we shall divide them in two classes:

i) *intrinsic* TP-AGB stars: they include MS, S and C (N-type) stars showing the unstable nucleus ^{99}Tc ($\tau_{1/2} = 2\ 10^5$ yr) as a probe that they are *presently* undergoing nucleosynthetic processes.

ii) *extrinsic* TP-AGB stars: they include the various classes of G and K-type Ba stars and the cooler S stars not showing Tc, which are thought of as being former Ba stars evolved up to the early AGB phase.

In a previous work (Busso et al., 1992) we analyzed MS and S giants (both with and without Tc), showing how they can be well fitted by models assuming the reaction ^{13}C$(\alpha,n)^{16}$O as the main origin for neutrons. We want to perform here a similar study for the more evolved C (N-type) giants and for their decendants Ba stars, which are believed to be produced by mass transfer within a binary system, from an AGB primary component to a dwarf or giant secondary.

2. The Sample Stars and The Nucleosynthesis Models

Our knowledge of the CNO and s-process photospheric abundances of intrinsic and extrinsic TP-AGB stars is summarized in a number of recent reviews (see e.g. Lambert, 1991). Concerning C-stars (N-type), detailed spectroscopic observations of light element abundances were given by Lambert et al. (1986). On the contrary, the distribution of s-process nuclei is still rather poorly known, relying only on the data by Utsumi (1985), having large uncertainties (0.4 dex at least, see Gustafsson, 1989). The abundances observed in the various classes of Ba stars were reviewed by Lambert (1985). Observational uncertainties are smaller than in C stars: typical values are around 0.2 dex. Concerning MS and S stars without Tc which, due to their having a white dwarf companion (Brown et al., 1990), should be evolved representatives of the

Ba star class, a thorough review of their composition is given in Smith and Lambert (1990). The data accuracy is 0.2 – 0.3 dex.

In order to compare observed distributions to model predictions we performed a series of TP–AGB nucleosynthesis calculations, assuming that some mixing at the base of the convective envelope produces ^{13}C and parametrizing its amount in mass. We followed $\alpha-$, n– and p–captures in the He–shell through a network of 450 nuclei, using cross sections by Beer, Voss and Winters (1992, hereafter BVW). For ^{138}Ba, the new measurement at low temperature by Beer et al. (1992, hereafter BCMA) was adopted (see below). Standard prescriptions for the stellar evolution along the AGB were followed. In particular, the relations giving the convective shell mass (ΔM_{csh}), the interpulse period (Δt_{ip}) and the overlapping factor between adjacent pulses (r) as a function of the core mass M_H were taken from Schonberner (1979) and Chieffi et al. (1992: private communication). Mass loss was taken into account, using the Reimers (1975) parametrization with variable η. We then followed the evolution of surface abundances as long as material from the He-shell is dredged to the surface during the interpulse periods, using simple descriptions for the dredge-up process and assuming pre-AGB photospheric CNO compositions from observations of normal giants (see e.g. Lambert, 1991). The initial stellar mass was taken to be 1.5 $M_\odot$.

The outcome of this work is a number of atmospheric evolutionary sequences whose predicted abundances depend on the amount of ^{13}C burnt per cycle and on the metallicity Z. These two parameters control the neutron exposure, since they give the number of neutrons produced and the abundance of seeds, respectively. For each case, up to 40 pulses were computed, the actual number being limited by the adopted mass loss parameter η.

3. Results and Discussion

Figure 1 (solid lines) shows a sample of surface enrichment sequences: one moves from one line to another by changing the neutron exposure τ_0 and along a particular line by varying the dilution of the He-shell material into the envelope. Also reproduced are the loci occupied by the observations of MS, S, C, CH, mild-Ba and Ba II stars. One can see that C–star compositions are characterized by the same τ_0 values of MS and S star; they however need higher concentration of He–shell material, since they show higher absolute abundances. This is consistent with their being more evolved than normal S stars, having seen further mass loss and mixing. Concerning the various classes of Ba stars (including S stars without Tc), their position in the plot is folded down again into the area of common S stars; this is what one expects by diluting a C–star envelope through a mass transfer onto a companion. Hence we can conclude that, within observational uncertainties, MS, S, C and Ba stars can be explained by a unique nucleosynthesis process, producing τ_0 values in a limited range (from 0.2 up to 0.4 – 0.5 mb^{-1}), the different absolute abundances being due to the combined operation of dredge–up, mass loss and mass transfer. For the metallicities observed in the sample stars, the τ_0 values we found correspond to the burning of a few 10^{-6} $M_\odot$ of ^{13}C per pulse.

Table 1. FIT OF UU Aur

Element	Observed Abundances (Utsumi, 1985); (Lambert et al., 1986)	Model Abundances A (new $\sigma(^{138}Ba)$) $M=1.5M_\odot$, $\tau_0=0.44$	Model Abundances B (old $\sigma(^{138}Ba)$) $M=1.5M_\odot$, $\tau_0=0.44$
12C/13C	52	55	55
C/O	1.06	1.16	1.16
[Fe/H]	0.20	0.20	0.20
[Y/Fe]	1.20	0.99	0.99
[Zr/Fe]	1.30	1.02	1.02
[Ba/Fe]	1.00	1.24	1.31
[La/Fe]	1.30	1.18	1.09
[Nd/Fe]	1.10	0.99	0.90
[Sm/Fe]	1.30	0.91	0.82

As an example of the quality of model predictions, Tables 1 and 2 show detailed fits to the abundances of the C–star UU Aur and of the Ba–star HR774. In the tables, columns labelled 'model A' adopt the new ^{138}Ba cross section by BCMA, while those labelled 'model B' use the previous data by BVW. Acceptable solutions for UU Aur are obtained when mixing of He–shell material into the envelope has reached a dilution factor of about 1/35. For HR774, this same envelope composition has to be further diluted by a factor 1/3 (as expected by mass transfer). In both cases a good agreement between predictions and observations is obtained; moreover we notice that, as far as the abundance ratios of Ba to heavier nuclei are concerned, results of model A are systematically better than those of model B. Actually, an increase in the ^{138}Ba cross section by around 30–40 % was found to be necessary both to reproduce the observed Ba/Nd ratios in MS and S stars (Busso et al., 1992) and to account for the measured Ba anomalies in meteoritic SiC grains (Gallino et al., 1992). The new measurement exactly provides the expected enhancement at the low temperature (T $\simeq$ 12 keV) at which the ^{13}C source operates.

Concerning Ba stars, we have to underline that, since they have higher temperatures and better known atmospheric structures than normal AGB stars, their abundances provide more stringent constraints on the nucleosynthesis process and hence on the neutron source. To show this, Figure 2 presents the average Rb/Sr ratio measured by Malaney and Lambert (1988) in Ba II giants (heavy dot), compared to our predictions by thermal pulses in low mass stars (LMS) at various neutron exposures τ_0. Also shown (arrow) is the area of intermediate mass star results (IMS) by Busso et al. (1988). Since Rb has one only measured line, the error bar in the observed data was assumed to be rather large (0.3 dex); despite this uncertainty, it is obvious that IMS models using the ^{22}Ne source are ruled out by observations.

Table 2. FIT OF HR 774

Element	Observed Abundances (Smith,1983); (Tomkin and Lambert,1983)	Model Abundances A (new $\sigma(^{138}Ba)$) $M=1.5M_\odot$, $\tau_0=0.44$	Model Abundances B (old $\sigma(^{138}Ba)$) $M=1.5M_\odot$, $\tau_0=0.44$
12C/13C	>23	24	24
C/O	0.95	1.00	1.00
[Fe/H]	-0.30	-0.30	-0.30
[Cu/Fe]	0.41	0.12	0.12
[Zn/Fe]	0.21	0.08	0.08
[Ge/Fe]	0.18	0.30	0.30
[Rb/Fe]	0.46	0.83	0.83
[Sr/Fe]	1.12	0.98	0.98
[Y/Fe]	1.11	1.04	1.04
[Zr/Fe]	1.05	1.06	1.06
[Nb/Fe]	1.18	1.05	1.05
[Mo/Fe]	1.05	0.95	0.95
[Ba/Fe]	1.19	1.30	1.37
[La/Fe]	1.17	1.24	1.15
[Nd/Fe]	1.23	1.04	0.95
[Eu/Fe]	0.50	0.42	0.37

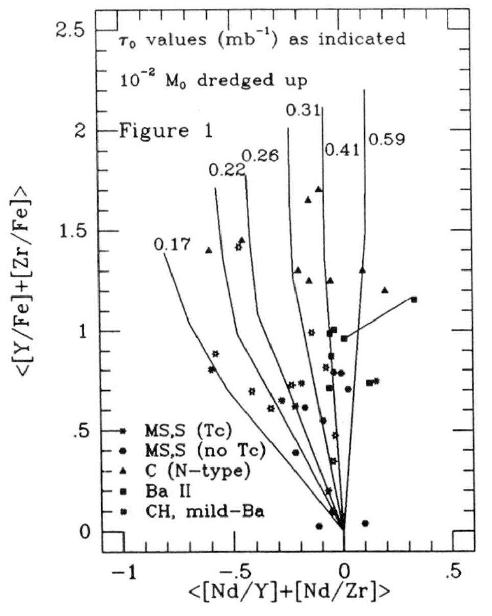

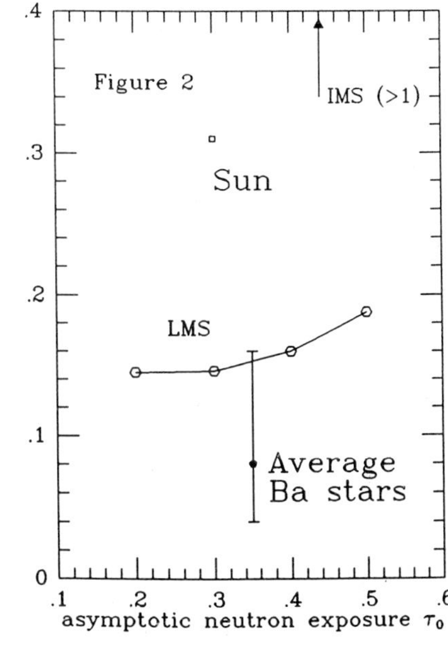

4. Conclusions

We have shown that distributions of s–process elements in C and Ba stars can be well fitted by models of thermal pulse nucleosynthesis in which ^{13}C is the main neutron source. The agreement between observations and models is further improved by recent results for the cross section of ^{138}Ba. C and Ba stars require the same neutron exposures typical of MS and S giants, and can be interpreted as being their descendants. IMS models based on ^{22}Ne burning are ruled out by observations of the ratio Rb/Sr in Ba II giants.

References

Beer, H., Voß, F. & Winters, R. R. 1992, *ApJ Suppl.* , **80**, 403.

Beer, H., Corvi, F., Mauri, A. & Athanassopoulos, K. 1992 (this conference).

Blanco, V.M., McCarthy, M.F., & Blanco, B.M. 1980 *ApJ* **242**, 938.

Brown, J.A., Smith, V.V., Lambert, D.L., Dutchover, E.Jr, Inkle, K.H., & Johnson, H.R., 1990, *A.J.* **99**, 1930.

Busso, M., Picchio, G., Gallino, R., & Chieffi A. 1988, *ApJ* , **326**, 196.

Busso, M., Gallino, R., Lambert, D.L., Raiteri, C.M., & Smith, V.V. 1992, *ApJ* **399** (in press).

Käppeler, F., Gallino, R., Busso, M., Picchio, G., & Raiteri, C. M. 1990, *ApJ* , **354**, 630

Gallino, R., Busso, M., Picchio, G., Raiteri, C. M., & Renzini, A. 1988, *ApJ* **334**, L45.

Gallino, R., Busso, M., Picchio, G., & Raiteri, C. M. 1990, *Nature* **348**, 298.

Gallino, R., Raiteri, C. M., & Busso, M. 1992, *ApJ* (in press)

Gustafsson, B. 1989, *Ann Rev. Astr. Ap.* **27**, 701.

Hollowell, D.E., & Iben, I.Jr 1988 *ApJ* **333**, L25.

Iben, I., Jr., & Renzini, A. 1983 *Ann. Rev. Astr. Ap.* **21**, 271.

Lambert, D.L., 1985, *Cool Stars with Excesses of heavy Elements*, ed. M. Jaschek and P.C. Keenan (Dordrecht: Reidel), p. 191.

Lambert, D.L., Gustaffson, B., Eriksson, K., Inkle, K.H. 1986, *ApJ Suppl.* **62**, 373.

Lambert, D.L. 1991, in *Evolution of Stars: The Photospheric Abundance Connection*, ed. G. Michaud and A. Tutukov (Dordrecht: Kluwer Acad. Press), p. 299.

Malaney, R.A., & Lambert, D.L., 1988, *MNRAS* **235**, 695.

Reimers, D. 1975, Mem. Soc. Roy. Liege VI ser. **8**, 369.

Schonberner, D. 1979, *A&A* **79**, 108.

Smith, V.V. and Lambert, D.L. 1990, *ApJ Suppl.* **72**, 387.

Tomkin J. and Lambert, D.L. 1983, *ApJ* **273**, 722.

Truran, J.W., & Iben, I, Jr 1977, *ApJ* **227** 209.

Ulrich, R.K. 1973, in *Explosive Nucleosynthesis*, ed. D.N. Schramm and W.D. Arnett (Univ. Texas), p. 139.

Utsumi, K 1985, in *Cool Stars with Excesses of heavy Elements*, ed. M. Jaschek and P.C. Keenan (Dordrecht: Reidel), p. 243.

Energy generation in convective shells of low mass, low metallicity AGB stars

Grant Bazan[1,2,4] **and John Lattanzio**[3,4,5]

[1]McDonald Observatory, University of Texas, Austin, TX 78712,
[2]Department of Astronomy, University of Illinois, Urbana, IL 61801,
[3]Department of Mathematics, Monash University, Australia,
[4]I.G.P.P., L-413, Lawrence Livermore National Laboratory, CA 94550,
[5]Institute of Astronomy, Cambridge, ENGLAND CB3 0HA.

Abstract. We investigate the details of the flash-driven convection zone in low mass AGB stars. As this convection moves outward in mass and ingests the previously deposited ^{13}C, energy is released by the subsequent alpha captures on the ^{13}C, together with the associated s-processing. We include, for the first time, this energy release. We find that the effect on nucleosynthesis is very small, but that there may be significant changes to the boundaries of the convective shell. These could have dramatic consequences for the nucleosynthesis and mixing in AGB stars.

1. Introduction

In low mass stars of low metallicity, a sizable amount of ^{13}C in a small region ($\Delta m \leq 5 \times 10^{-4}\ M_\odot$) can be made by semi-convection mixing proton-rich and carbon-rich material during interpulse periods (Iben and Renzini 1982). This material is ingested by an expanding convective shell and burned at high temperatures (Gallino *et al* 1988). In all previous stellar evolution calculations of thermally pulsing AGB evolution, the energy output of the neutron emission and capture reactions and beta decays has been ignored. We have calculated this energy and included it in AGB evolutionary models.

2. Calculations

Because of the impracticality of implementing a full neutron capture network into a stellar evolution code, we have iterated between calculations of the energy associated with neutron generation and captures and calculations of the stellar evolution resulting from the consideration of the energy. We derive time dependent energy generation rates through the solution of a charged particle/neutron capture nucleosynthesis network of 605 nuclei. The ^{13}C is initially assumed to lie outside the convective shell in a mass distribution similar to that of Hollowell (1988). The ingestion rate, which was shown by Hollowell and Iben (1990) to determine the magnitude of the neutron density, is (initially) taken from the fifteenth pulse of Hollowell (1988). The neutron density is assumed to take its equilibrium value. Initially, the temperature and density behaviors are also taken from Hollowell (1988). Temperature and density values in later iterations are taken from the new AGB models with mass dependencies according to the isentropic conditions in the convective shell. Nucleosynthesis via neutron irradiation is determined according to the usual equations (see Bazan and Lattanzio 1992 for details). The resulting energy generation is then employed in an AGB model of C-O core mass 0.59 $M_\odot$ and metallicity $Z = 0.001$ calculated from a ZAMS mass of 1.5 $M_\odot$ using the same code outlined in Lattanzio (1986).

3. Altered AGB Evolution

The integrated luminosity through the convective shell from neutron captures, beta decays, and the $^{13}C(\alpha, n)^{16}O$ reactions varies between 10^3 and $10^4 L_\odot$, according to the model parameters for the 15th thermal pulse from Hollowell (1988). For comparison, the total energy in the convective shell (radiative + convective) peaks at $\sim 1500 \ L_\odot$ at this time. This extra energy causes the upper convective boundary to be pushed out in mass (the lower convective boundary is also advanced in mass). The more rapid growth of the upper convective boundary means that the actual ^{13}C ingestion rate should be higher than described by evolution calculations not considering the extra energy, which from Hollowell and Iben (1990) means that the neutron generation and capture proceeds at a faster rate. The resultant energy generation should then evolve faster with a higher peak value.

From the nucleosynthesis code we determine the energy generation as a function of time for our assumed ingestion rate of the ^{13}C pocket. The evolution code then determines the new time dependence of the boundary of the convective shell and a new ^{13}C ingestion rate. We use this new ingestion rate in the nucleosynthesis code to determine the time dependence of the energy generation, which is then used to determine the new convective shell boundary behavior. The procedure 'converges' (as determined by the ingestion rate) after about 3 iterations. Figure 1 shows the detailed evolution of the shell due to the changes in the rate of ^{13}C ingestion, which begins at $t \sim 11540$ years in the figure. We show the changes in the convective boundaries for 3 iterations, with the solid line being the standard case. The first iteration is shown by the dashed line and gives the new evolution assuming added energy according to the ingestion rate of the Hollowell (1988) model. The dot-dashed line represents our 'converged' model, after we have performed two iterations of energy calculations and stellar modelling. As the rates of the last two iterations were very nearly equal, we stopped calculations here.

There are still differences between the runs, however, as shown clearly in figure. The shell is almost quenched due to the energy from neutron captures and assorted other reactions. In run '3', the maximum extent of the convection is still enough to engulf all the ^{13}C (which extends over $\Delta m = 5 \times 10^{-4} \ M_\odot$ only). But we note that the details of the processes leading to the ^{13}C pocket are not well understood, and if the pocket extended much further, we could have the situation where the temporary quenching of the convection and its subsequent growth and renewed ^{13}C ingestion could lead to mini-pulses of s-processing *within* a single thermal pulse. Details, of course, will depend on the dynamics of the convection and nuclear burning.

4. Nucleosynthesis Considerations

A larger ingestion rate leads to higher peak and average neutron densities experienced by material in the shell (Hollowell and Iben 1990). In the usual case we find a peak neutron density of $5 \times 10^9 \ cm^{-3}$, while the final iteration has a peak of $2 \times 10^{10} \ cm^{-3}$.

(The latter value would be higher still if not for the retreat of the convective shell and the corresponding decrease in temperature at the base of the shell.) Following the arguments of Hollowell and Iben (1990), the peak neutron density in run '3' should be roughly three times that in run '2', since the mass expansion rate in run '3' is roughly three times that in run '2'. The fact that the neutron densities differ by only about 0.05 dex is the result of the decrease in the base temperature. The magnitude of the neutron density in the converged model seems to conflict with previous work on parameters governing the s-process. Parameter studies of the s-process (Kappeler *et al* 1990, Bauer *et al* 1991) have hinted that the s-process average neutron density should be $\sim 10^8$ cm^{-3} in order to reproduce the relative abundances of the s-only isotopes. Another problem is the lack of any measurable ^{96}Zr in MS and S stars (Smith and Lambert 1985, 1986, 1990). This mandates that the neutron density be no more than a few times $\sim 10^8$ cm^{-3}.

Another descriptive parameter of the s-process, the integrated neutron flux or exposure, changes due to the new energy source. A simple model of s-processing in AGB stars relates exposures received by material in individual thermal pulses to the mean exposure of the 'classical' model exposure distribution (Ulrich 1973). Initially, our models give a per pulse exposure of 0.30 mb^{-1}, while in our final run it reaches only 0.19 mb^{-1}. Taking into account the convective shell dilution factor reduces this further, but does not lead to severe differences when compared to previous studies.

5. Conclusion

We have shown that the inclusion of energy from neutron production, capture, and nuclear decay reactions significantly alters the development of the flash-driven convective shell in low-mass, low-metallicity AGB stars. Our exploratory analysis indicates the possibility of mini-pulses of s-processing within a single thermal pulse. A significant change is the more rapid growth of the outer boundary of the convective shell, leading to a much more rapid ingestion of the ^{13}C pocket. This leads to a neutron density higher than cases where the extra energy is ignored. As a result of the altered evolution, mean exposure values fall, but there is only a small change in the expected s-process abundances.

Finally, a word of caution. Busso and Gallino (private communication) have pointed out that the results of this paper are very dependent on the temperature in the convective shell when the C^{13} is ingested. The difference in the duration of the convective shell in the present calculations and those of Hollowell (1988) is some 50 years (approximatly 250 years versus 300), and the details of how the two models are combined can cause significant changes in the importance of the energy source discussed in this paper. The larger effect is found when the C^{13} is ingested earlier in the life of the convective shell (i.e. at lower temperatures, so that its importance relative to the triple-alpha reactions is enhanced). Indeed, for the suggested mini-pulses to occur may require ingesting C^{13} even earlier than done in this paper (which may have been "too early" to be consistent with standard models). If these mini-pulses can be shown to provide the required nucleosynthesis occurring in these stars, then this may be used as a probe of the C^{13} distribution, and the semiconvective (or possibly other) mixing responsible.

6. Acknowledgements

The authors wish to thank Jim Truran, Grant Mathews, Michael Howard, and Stan

Woosley for their comments. Special thanks to Maurizzio Busso and Roberto Gallino for their detailed analysis of this work, as well as their stimulating and insightful comments !

This work was performed under the auspices of the U.S. Department of Energy by the Lawrence Livermore National Laboratory under contract No. W-7405-ENG-48, for the University of Illinois under NSF contract AST-8611500, and for the McDonald Observatory, University of Texas at Austin.

References

Bauer, R. W., Bazan, G., Becker, J. A., Howe, R. E., and Mathews, G. J. 1991, *Phys. Rev. C*, **43**, 2004.

Bazan, G., and Lattanzio, J. C., 1992, *Astrophys. J.*, submitted.

Gallino, R., Busso, M., Picchio, G., Raiteri, C. M., and Renzini, A. 1988, *Astrophys. J. Lett.*, **334**, L45.

Hollowell, D. E. 1988, *Ph. D. Thesis*, University of Illinois.

Hollowell, D. E. and Iben, I. Jr. 1990, *Astrophys. J.*, **349**, 208.

Iben, I. Jr. I. Jr. and Renzini, A. 1982, *Astrophys. J. Lett.*, **259**, L79.

Kappeler, F., Gallino, R., Busso, M., Picchio, G., and Raiteri, C. M. 1990, *Astrophys. J.*, **354**, 630.

Lattanzio, J. C., 1986, *Astrophys J.*, **311**, 708.

Smith, V. V. and Lambert, D. L. 1985, *Astrophys. J.*, **294**, 326.

Smith, V. V. and Lambert, D. L. 1986, *Astrophys. J.*, **311**, 843.

Smith, V. V. and Lambert, D. L. 1990, *Astrophys. J. Suppl.*, **72**, 387.

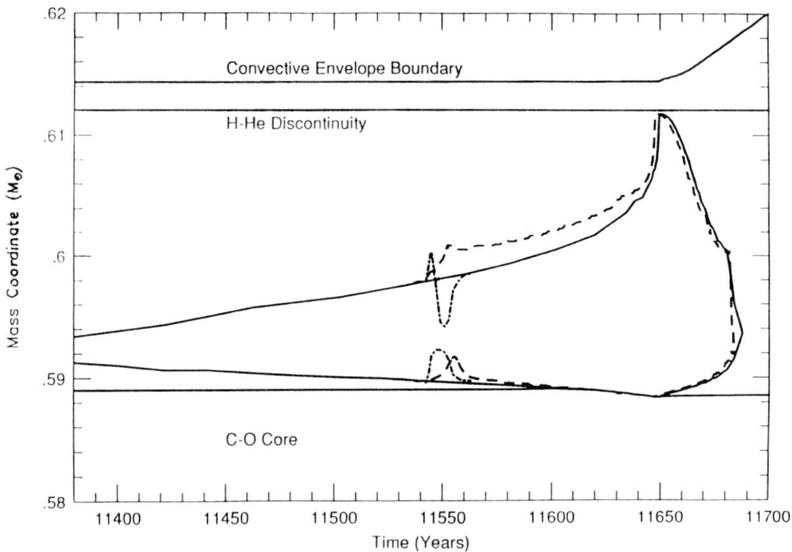

Figure 1: The convective shell behavior under the influence of added energy. The models shown include an unaltered model (solid line), run '1' (dashed line), and run '3' (dot-dashed line).

SiC particles from AGB stars: Mg burning and the s-process

Lawrence E Brown[1] **and Donald D. Clayton**

Department of Physics and Astronomy, Clemson University, Clemson, SC 29634-1911, USA

Abstract. The "new astronomy" of SiC particle isotope measurements provides many clues for the nuclear astrophysics of their production sites. We compare the measurements of some important lighter isotopes in the large SiC particles with the predictions of a $5.5 M_\odot$ asymptotic giant branch (AGB) star model. We find remarkable agreement for the isotopic abundances of C, Si, ^{22}Ne , and ^{26}Al .

1. Introduction

With the discovery of s-process Xe in SiC particles (Lewis, *et al* , 1990) it was quickly realized that AGB stars were a likely source for these particles and Gallino, *et al* (1990) calculated detailed predictions of s-process abundances for these particles. In this work we emphasize the lighter isotopes found in these particles. We treat the main constituents of the particle: Si and C, and the important radioactive decay products ^{22}Ne and ^{26}Mg . We have used standard AGB models and found that the Si measurements demand a high core mass (*ie* total stellar mass $\gtrsim 4 M_\odot$). We present a $5.5 M_\odot$ model which predicts the Si composition to a high degree. From this, we find reasonable agreement with carbon measurements. We find that the important diagnostic isotope ^{22}Ne must come from ^{22}Na in this model (see also Brown and Clayton, 1992b). We produce ^{26}Al which more than explains the observed abundance.

2. TP-AGB Model

Detailed descriptions of thermal pulse-AGB star evolution are available (*eg* Renzini and Voli, 1981). A full description of the particulars of our model is in preparation (Brown and Clayton, 1992c). Figure 1 gives an overview of our $5.5 M_\odot$ model. Our convective envelope hot bottom burning (HBB) uses mass averaged reaction rates (for now) and a hydrostatic envelope model. Our parametrized mass loss and core/shell prescriptions are from Bazan (1991).

[1] E-mail: elwin@gamma.phys.clemson.edu (INTERNET).

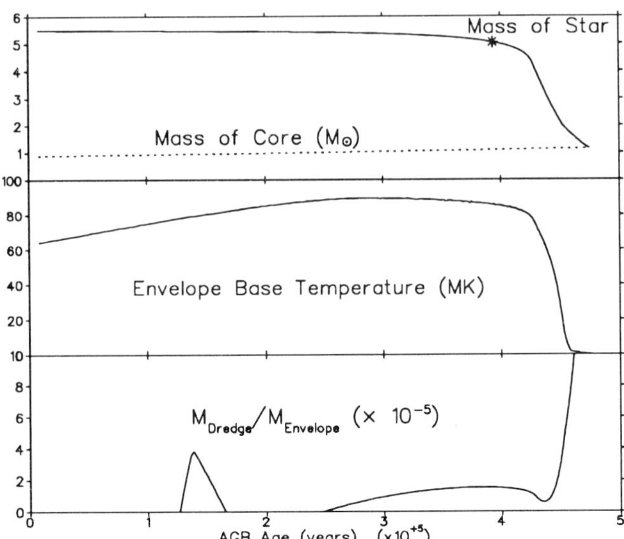

Figure 1. The total stellar mass, hydrogen depleted core mass, convective envelope base temperature, and ratio of dredge-up mass to total envelope mass as a function of TP-AGB lifetime. The asterisk in the first graph marks transition to a carbon star.

As the core mass grows, the He-shell flashes become hotter and produce more neutron irradiation and, eventually, Mg burning in stars of sufficient mass, such as this $5.5M_\odot$ model. As the envelope is lost in a "superwind" driven by pulsations, large amounts of carbonaceous dust form. The convective envelope base temperature determines the amount of CNO processing in the envelope and produces radioactive species ^{26}Al and ^{22}Na when seed material is available. The dredge-up mass in our model is a complicated function of core mass and star mass but it determines how much altered Si is released to the envelope and how much seed ^{25}Mg and ^{21}Ne are available for ^{26}Al and ^{22}Na production.

Models of AGB star evolution are notoriously uncertain in important features like dredge-up and mass loss because modeling is difficult. Certain parameters are often adjusted essentially freely. In this work we have decided to remain with a single standard model with the exception of shell temperature noted below. By not adjusting any other parameters (not that we weren't tempted) we hope to show just how likely it is that these particles, or at least a subset of them, came from an intermediate mass AGB star (IMS). In the future, it seems likely that the SiC particles will become a powerful tool for studying stellar structure and evolution; *ie* the particles can be used to determine the parameters.

3. Silicon

The silicon isotopic composition of most of the large SiC particles (Virag, *et al* , 1991) and all of the "platy" subset found by Stone, *et al* (1991), lies along a line of slope ≈ 1.4

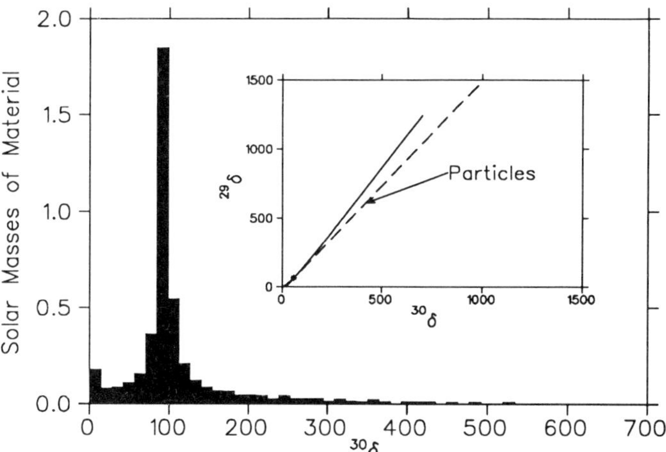

Figure 2. Inset shows the evolutionary track of the *average* surface silicon isotope composition for a $5.5M_\odot$ AGB star. Dashed line is the correlation line of observed particles (Stone, *et al* 1991; Virag, *et al* 1991). Histogram shows the amount of mass ejected vs the *average* surface $^{30}\delta$. Most of the particles are expected to have modest isotopic excesses near 100 parts per thousand, as observed in meteoritic SiC.

in the $^{29}\delta$-$^{30}\delta$ plane.[2] Brown and Clayton (1992a) have argued that compositions in this region can *only* be a product of Mg burning. In standard IMS models this *requires* a 10-20% increase in the base temperature of the convective He burning shell. For core masses above $0.96M_\odot$, these are based entirely on the three models of Iben (1977), so there is room for some uncertainty. In our $5.5M_\odot$ model, we increased the shell temperature by a constant 20%. The agreement shown in figure 2 is good. While the track in the inset reaches much higher δ values than measured in the particles, it is clear from the histogram that this is only in a small amount of mass at the end of the stellar lifetime. During the "superwind" mass loss phase, mixtures of shell and envelope material will produce exactly the correlation and magnitudes seen in the Si anomaly. Further, the overall abundance of Si is raised considerably at Mg burning temperatures (by α chain production of ^{28}Si) so that SiC (as opposed to graphite) becomes more favored. This may provide a "projection operator" to explain why few if any large SiC particles with standard s-process composition (along a line of slope ~ 0.5) are found.

4. Carbon

The carbon composition is a difficulty of a different type, because it is set primarily by HBB which is rapidly changing during dust formation. The Mg burning and envelope endmembers for Si are stationary while the envelope carbon endmember is not (see figure 3). In very loose language, this could be our prescription: a convective blob of Si rich

---

[2] $^{29}\delta = 1,000 \times [(^{29}Si/^{28}Si)/(^{29}Si/^{28}Si)_\odot - 1]$, $^{30}\delta$ and $^{13}\delta$ are defined similarily with ^{28}Si and ^{12}C in the denominators respectively.

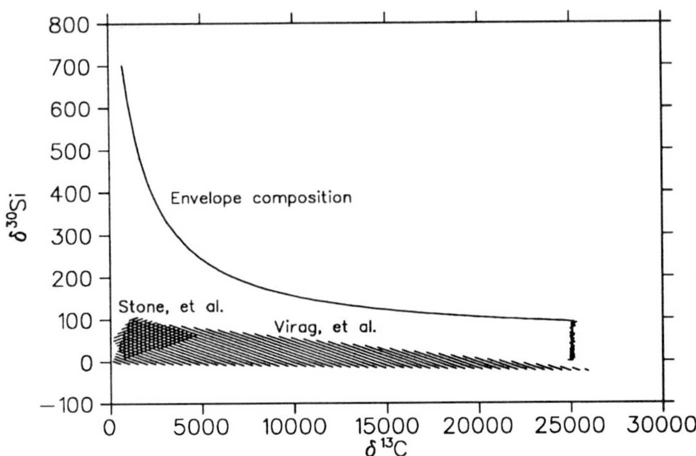

Figure 3. Envelope composition evolution curve for carbon and silicon. The shaded areas show measured ranges in a general sample of large SiC grains (Virag, *et al* , 1991) and a sample of "platy" particles (Stone, *et al* , 1991). Shell carbon composition is $Y(^{12}C) \approx 0.03$.

matter carrying pure ^{12}C carbon nucleates SiC in a dredged up sample of shell material. The grain picks up more mass with the average composition of the envelope as it grows. The size of the shell-material nucleus determines the size of the Si anomaly. All of the Si picked up from the envelope will move the overall anomaly along the same $^{29}\delta$-$^{30}\delta$ slope. The envelope carbon composition is rapidly changing, however, so the ^{12}C core, corresponding roughly to the size of our Si anomaly, is diluted by envelope material having $^{13}\delta$ anywhere in the range 30,000-700 depending on the mass and evolutionary state of the star. Particles should form with carbon compositions not along the envelope evolutionary track as in silicon, but to the left of it an undetermined amount. While we are qualitatively in agreement with the measurements in our model, quantitatively more work needs to be done. $^{13}\delta$ is strongly influenced by HBB envelope base temperature so this picture could change (see below under ^{26}Al). It is apparent that the detailed formation history of the SiC grains will become very important for this question.

5. ^{22}Ne

The presence of nearly pure ^{22}Ne (called Ne-E) was one of the main initial diagnostics of these particles. Lewis, *et al* (1990) and Gallino, *et al* (1990) hypothesized that this ^{22}Ne was implanted as ^{22}Ne into the particles by a wind from the exposed ^{22}Ne -rich He shell of a forming planetary nebula. This is exactly the scenario necessary in a low mass AGB star, which will not produce ^{22}Na . However, it fails to generate the Si isotopic correlation. If we require Mg burning to get the Si isotopic composition right, all of the ^{22}Ne in the shell must be burned away in each thermal pulse leaving none for implanting (see figure 4). Our HBB envelope produces 2.6yr ^{22}Na at abundance levels ^{22}Na /^{23}Na of up to 3×10^{-5}. This is adequate to explain observed concentrations of the SiC ^{22}Ne

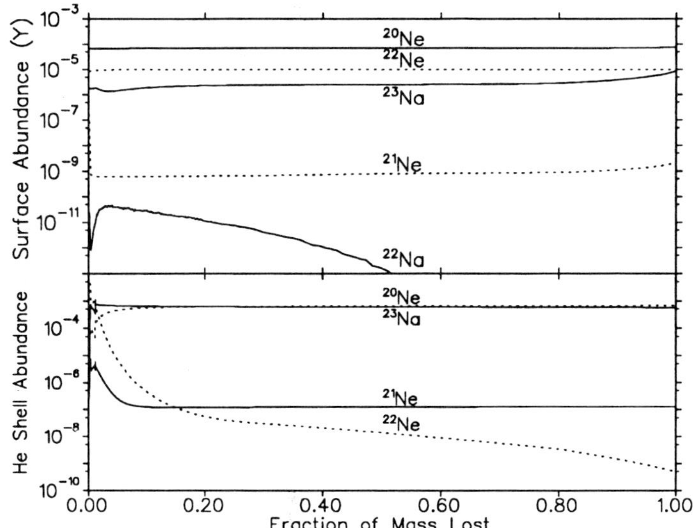

Figure 4. Envelope and Helium burning shell compositions for the Ne-Na cycle isotopes shown as a function of fraction of ejected mass. Note that the He shell can not be directly exposed for wind ejection, etc, until all mass is lost

abundance. Since the ^{22}Na drops off as the envelope is lost and is no longer able to support the high HBB temperatures necessary for ^{22}Na production, we would predict that only $\sim 10\%$ of the particles should carry Ne-E. This selectivity has recently been seen (Nichols, 1992). One of the main diagnostics of the history of these particles will probably be ^{22}Ne . Brown and Clayton (1992b) discuss these points in more detail.

6. ^{26}Al

The measured abundance of initial ^{26}Al content in these particles (inferred from the abundance of ^{26}Mg) ranges over ^{26}Al $/^{27}$Al $= 10^{-4} - 10^{-2}$. Our calculated ratio of ^{26}Al $/^{27}$Al is higher than observed at the "superwind" phase (see figure 5). However, this ratio is a strong function of convective-shell base temperature; moreover, for a slightly different mixing length to scale height ratio, $\alpha = 1.25$ as opposed to our standard $\alpha = 1.5$, ^{26}Al $/^{27}$Al $= 0.03$ which is more in line with observations (note that this does not strongly effect the ^{22}Na production since the Coulomb barrier is much lower for ^{21}Ne $(p, \gamma)^{22}$Na than for ^{25}Mg $(p, \gamma)^{26}$Al). ^{26}Al in SiC, in conjunction with γ-ray astronomical results for the diffuse ^{26}Al line emission, could become an important diagnostic of AGB envelope physics.

7. Conclusions

The Mg burning/IMS model is shown to be remarkably robust for the light elements. Much more detailed models of AGB stars are necessary before conclusions are definitive.

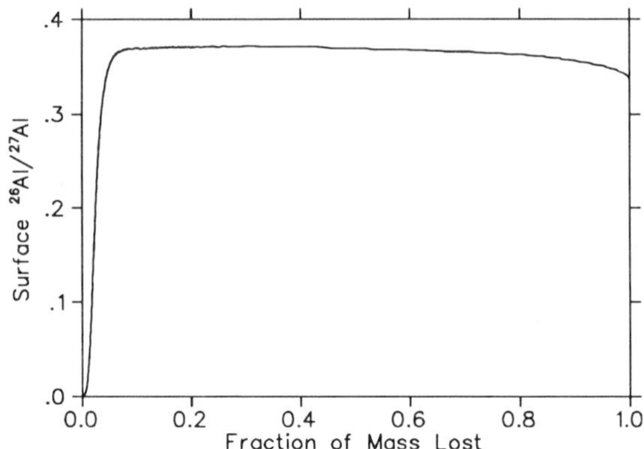

Figure 5. Surface Al isotope ratio as a function of mass lost.

It is still unclear whether most of these particles could have originated in a single star (Cameron, 1992). Particular points which strongly influence our results are: dredge-up models, the temperature of the convective He shell, a more complete s-process network, the convective and opacity prescriptions for the envelope, and our not quite justifiable use of averaged reaction rates in the envelope. We are working on all of these and more.

Acknowledgements

We thank Grant Bazan and Mounib El Eid for comments and suggestions and Icko Iben for the use of his hydrostatic envelope model code. This work was supported by NASA in the Origins of Solar Systems program and Planetary Materials and Geochemistry program.

References

Bazan G 1991 Ph.D. thesis, University of Illinois

Brown L E and Clayton D D 1992a *Ap. J. (Lett.)* **392** L79

——1992b submitted to *Science*

——1992c in preparation for *Ap. J.*

Cameron A G W 1992 in *Protostars and Planets III* (Univ. Arizona Press: Tuscon), in press

Gallino R, Busso M, Pichio G and Raiteri C M 1990 *Nature* **348** 298

Iben I 1977 *Ap. J.* **217** 788

Lewis R S, Amari S and Anders E 1990 *Nature* **348** 293

Nichols R, private communication

Renzini A and Voli M 1981 *Astron. Astrophys.* **94** 175

Stone J, Hutcheon I D, Epstein S and Wasserburg G J 1991 *Earth Planet. Sci. Lett.* **107** 570

Virag A, Wopenka B, Amari S, Zinner E, Anders E and Lewis R S 1992 *Geochim. Cosmochim. Acta* **56** 1715

Galactic chemical evolution: the intermediate mass elements

S. E. Woosley[1,2], **F. X. Timmes**[1], **and Thomas A. Weaver**[2]

[1]Board of Studies in Astronomy and Astrophysics, UCO/Lick Observatory, University of California at Santa Cruz, Santa Cruz, CA 95064

[2]General Studies Group, Physics Department, Lawrence Livermore National Laboratory, Livermore, CA 94550

Abstract. Nucleosynthesis in the intermediate mass range (carbon through nickel) has been calculated for a grid of stellar masses between 10 and 40 $M_\odot$ for solar metallicity, and 12 and 75 $M_\odot$ for zero metallicity, with a total of 26 stars evolved to the presupernova state. Explosion has been simulated in 13 of these and the final nucleosynthetic yields, including the ν-process, determined. Except for the products of the neutrino process (fluorine and boron), the presupernova abundances of isotopes lighter than about A = 40, closely resemble the final yields. These results, when incorporated into a model for Galactic chemical evolution that includes contributions from lower mass stars and Type Ia supernovae, give present day abundances that are in good agreement with those observed in the solar system when a particular choice is made for the $^{12}C(\alpha, \gamma)^{16}O$ reaction rate (S(300 keV) = 0.17 MeV barns). We find primary nitrogen is produced for zero metallicity stars whose mass is 30 $M_\odot$ or greater. Some preliminary predictions of the evolution of N, O, Na, Al, Mg, and Fe as a function of metallicity are presented.

1. INTRODUCTION

Nucleosynthesis in a massive star is a function of at least six physical variables: a) the mass of the star (and thus the initial mass function), b) its initial composition, c) its mass loss history, d) the set of nuclear physics employed in its study, e) the treatment of convection, and f) the "explosion mechanism" (by which we include a host of uncertainties such as the mass cut, explosion energy, reimplosion mass, etc.). Of these, convection, the rate for $^{12}C(\alpha,\gamma)^{16}O$, and the manner in which explosion is simulated are the greatest sources of uncertainty and cause of difference among the various groups who study this subject. Especially during advanced nuclear burning stages, one must use a time-dependent theory of mixing. Ours, as all others with which we are familiar, is based upon mixing length theory, a qualitative model at best. One must also choose between an instability criterion based upon temperature gradients (Schwarzschild) or a combination of density and composition gradients (Ledoux) and decide what to do when one is satisfied (Schwarzschild) and the other is not. This is the condition for semiconvection. As discussed elsewhere (Weaver, Zimmerman, and Woosley 1978; Woosley and Weaver 1988), we use a Ledoux-based model with semiconvection incorporated in a simple parametric way. The specific models discussed here use a diffusion coefficient calculated using $\alpha = 0.1$. Nomoto and his colleagues use the Schwarzschild criterion, which is approximately the $\alpha = $ infinity case.

Then there is the issue of convective overshooting. A convective region will not have a sharp boundary. How thick is the boundary and what is the diffusion coefficient within this region? Chiosi assumes a lot of overshoot mixing, Meader assumes less, we use an even smaller amount, and Nomoto, none. Overshoot mixing can affect many aspects of nucleosynthesis. It can make the helium core larger, which essentially recalibrates the mass scale for nucleosynthesis, and it may be instrumental in making primary nitrogen. For the present, we represent overshoot mixing using a single slowly mixing zone that bounds all convective regions on both top and bottom. This prescription allows a convective region that would grow, were it not for composition barriers, to slowly expand. However, it is not a very physically appealing prescription because it can give results that are sensitive to the numerical mesh adopted. We are in the process of changing this aspect of our calculations.

Once the convective physics is determined, for better or for worse, the nuclear physics, hydrodynamics, and thermodynamics coded, and rotation ignored, one can proceed with the calculation. Recent developments in RISC based workstations have allowed us to greatly accelerate the rate at which we calculate presupernova models. Given this numerical ability, we have embarked on a survey, still in its initial phases, to explore the effect of the some of the other physical variables mentioned above. Since the goal of the survey is to match the Galactic abundances at a particular time and place (solar system formation) the stellar models must be incorporated into at least a simple representation of Galactic chemical evolution (§4). This paper reports our first attempt to do so. For this study we have used models that, at the time of iron core collapse, had typically 500 mass shells (250% that of Weaver, Zimmerman, and Woosley 1978) and had taken 15,000 time steps. Each of these zones carries through its entire evolution a nuclear reaction network of 150 isotopes (used for nucleosynthesis calculations, though not for energy generation). We now feel that even this resolution was not wholly adequate, and the next phase, already in progress, typically uses with

850 zones and carries 200 isotopes in each. To repeat, however, results reported here used the smaller network and coarser zoning.

So far, only stars of zero and solar metallicity have been studied. The initial composition was either that of the Big Bang ("Z" series - by mass 76% H, 0.0091% ^{2}H, 0.0034% ^{3}He, 24% ^{4}He, and 8×10^{-8} % ^{7}Li) or taken from Anders and Grevesse (1989; "S" series). Stars of 12, 13, 15, 18, 20, 22, 25, 30, 35, 40, 50, and 75 $M_\odot$ were considered for the "Z" series. Stars of 11, 12, 13, 15, 18, 19, 20, 21, 22, 25, 27, 30, 35, and 40 $M_\odot$ were considered for the "S" series (sometimes, as in §2, a somewhat smaller grid was used for surveys in which key physics was varied). In the future other intermediate values of metallicity between "Z" and "S" need to be included, but then the results from this first set of calculations plus a Galactic chemical evolution model are necessary to estimate the starting compositions for the intermediate points.

The present calculations did not include mass loss. In collaboration with Norbert Langer, we are currently studying the effects of mass loss on our models (Woosley, Langer, and Weaver 1993ab). For the most part, the nucleosynthesis is not greatly altered from the constant mass case presented here unless the helium core is uncovered before helium burning is complete. If the helium core is uncovered early on, then one may get quite a different final state which we (and many others) believe is the origin of Type Ib supernovae.

2. THE RATE FOR ^{12}C$(\alpha, \gamma)^{16}$O

Figure 1, taken from Weaver and Woosley (1992) shows results from the "S" series (for variable values of the ^{12}C$(\alpha, \gamma)^{16}$O reaction rate) integrated over a simple initial stellar mass function. Fifty-seven presupernova models were calculated to prepare this graph. Explosive nucleosynthesis was not calculated. However, as we shall discuss later, explosion does not substantially modify the expected nucleosynthesis for A less than 40. The suggested value of the ^{12}C$(\alpha, \gamma)^{16}$O rate is 1.7 ($\pm$ 50%) times the Caughlan and Fowler (1988) value, i.e., 1.7 times 100 keV barns for the S-factor at 300 keV for all components of the cross section. The requisite error bar for an adequate determination of stellar nucleosynthesis would be 20%. Pending further experimental study (e.g., Buchman et al, this volume) we shall adopt this value in subsequent work.

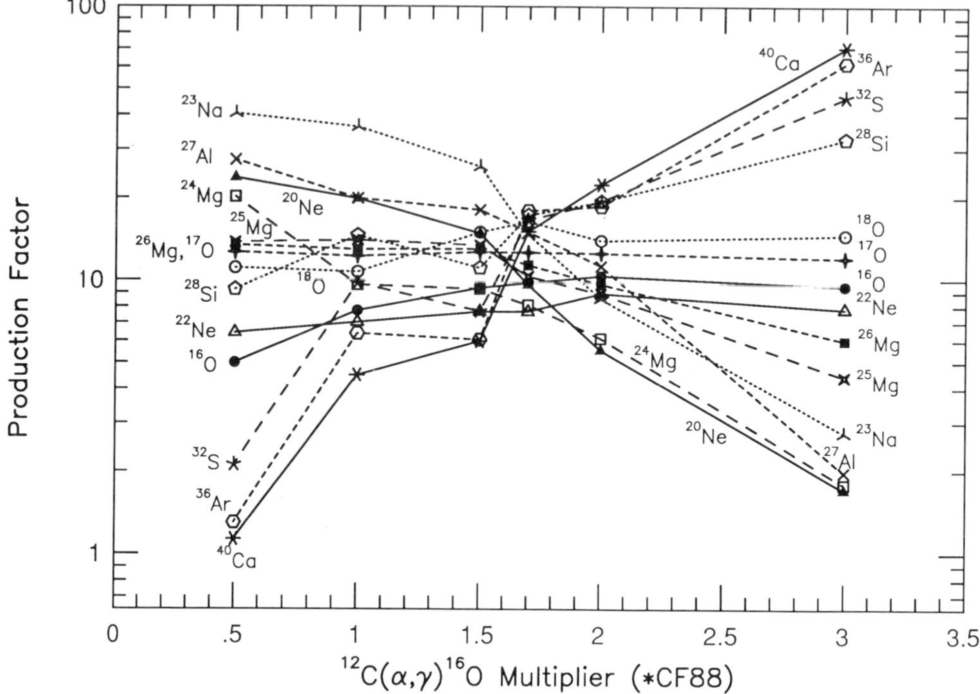

Fig. 1.— Production factor for 14 isotopes of elements from oxygen through calcium as a function of ^{12}C$(\alpha,\gamma)^{16}$O reaction rate for models having a "nominal" amount of semiconvection. Each point represents an average over an initial stellar mass function of a grid of stellar masses from 10 to 40 $M_\odot$ folded with an IMF of slope -2.3. This result is insensitive to the assumed slope of the IMF. See Weaver and Woosley (1992) for details.

3. SIMULATING THE EXPLOSION

Explosion was simulated in a subset of our stars by choosing a mass cut and using a piston (Weaver and Woosley 1982, 1992). The mass cut is one of the most critical and uncertain choices one must make (lacking a set of reliable quantitative explosion models) and especially affects the synthesis of the iron group. Iron nucleosynthesis is far more sensitive to this choice of the mass cut and the structure of the presupernova star than it is, for example, to the explosion energy (Woosley, Langer, and Weaver 1993a). Here we chose mass cuts (Table 1) outside the neutronized iron core and at the location of an abrupt entropy increase if one was nearby. An unknown amount of material will accrete onto the protoneutron star as the explosion develops and more may fall back several hours later when the reverse shock arrives back at the core. Our choice of mass cut here probably errs on the side of being too deep in and thus producing too much iron (SN 1987A, an approximately 20 $M_\odot$ star made 0.07 $M_\odot$ of ^{56}Ni). Other mass cuts should be explored in the future. The entry in Table 1 for the 20 $M_\odot$ "S" model (0.01 $M_\odot$) is not a misprint, but reflects a particularly steep density gradient at the edge of the iron core in that model. No great significance should be attached to the difference in iron yields or residual masses in the stars of differing metallicity. They are within the variation expected when the mass of the star is varied slightly or different choices made for the mass cut.

In each case the piston was first moved smoothly in from the initial radius of the chosen mass cut in the presupernova star to a radius of 500 km in a time of 0.45 s, then quickly moved out at initial velocities between 1.9 and 2.2 $\times 10^9$ cm s^{-1} and gradually slowed as if on a ballistic trajectory. The explosion is then parameterized in terms of the final kinetic energy imparted to the ejecta at infinity (Table 1) which includes any energy from nuclear processing.

TABLE 1

Explosion Parameters

Mass ($M_\odot$)	M_{baryon} ($M_\odot$)	KE_∞ (10^{51} erg)	$M(^{56}$Ni$)$ ($M_\odot$)	Mass ($M_\odot$)	M_{baryon} ($M_\odot$)	KE_∞ (10^{51} erg)	$M(^{56}$Ni$)$ ($M_\odot$)
15	1.62	1.01	0.08	15	1.57	0.98	0.19
20	1.42	0.96	0.12	18	1.64	1.18	0.19
25	1.90	1.49	0.45	20	1.54	1.67	0.01
30	1.87	1.51	0.30	22	1.83	1.71	0.34
35	2.40	1.28	0.23	25	1.67	1.10	0.22
40	1.51	1.37	0.31	30	1.99	1.71	0.40
				35	2.00	1.54	0.51

[a] The first set of columns is for the zero metallicity stars; the second set for those initially solar.

4. CHEMICAL EVOLUTION OF THE GALAXY

For a given set of stellar yields the Galactic chemical evolution was calculated using a multi-zone chemical evolution code capable of evolving an arbitrary number of isotopes. The instantaneous recycling approximation is not invoked; the full set of integro-differential equations that describe zone models of chemical evolution are solved. Input nucleosynthesis is allowed to depend on both the mass and metallicity. Stellar yields for values of metallicity other than zero and solar are interpolated linearly in Z. During testing, the code reproduced the results of several previous Galactic chemical evolution studies (Talbott & Arnett 1971, 1973ab, 1975; Chiosi 1980; Chiosi & Matteucci 1982; and Matteucci & Greggio 1986).

A novelty of our chemical evolution code is the adoption of a "solar masses in and solar masses out" formalism in lieu of the "production matrix" formalism for describing the nucleosynthesis. For example, Table 2 gives the final (post-explosive) yields in solar masses of our adopted 25 $M_\odot$ models of zero and solar metallicity. A total of 26 such stars stars have been evolved to the presupernova state and explosion has been simulated in 13 of them. The final yields include the ν-process contributions (Woosley et al. 1990).

TABLE 2

Explosive Yields for a Z7 and S7 Series 25 $M_\odot$ Star

^AZ	Z7 Yield	S7 Yield	^AZ	Z7 Yield	S7 Yield
^{1}H	9.32e+00	9.38e+00	^{40}Ar	1.46e-10	1.64e-05
^{2}H	8.82e-06	1.08e-04	^{39}K	5.48e-05	1.10e-03
^{3}He	9.52e-05	6.14e-04	^{40}K	2.29e-08	6.20e-06
^{4}He	9.43e+00	8.63e+00	^{41}K	4.93e-06	6.48e-05
^{7}Li	4.65e-07	3.47e-07	^{40}Ca	1.93e-02	1.59e-02
^{10}B	9.06e-12	2.57e-09	^{42}Ca	3.68e-06	2.41e-04
^{11}B	8.94e-07	1.26e-06	^{43}Ca	1.54e-06	1.02e-05
^{12}C	4.99e-01	3.92e-01	^{44}Ca	2.77e-04	2.04e-04
^{13}C	1.66e-05	1.48e-03	^{46}Ca	1.92e-13	8.12e-07
^{14}N	4.07e-06	7.95e-02	^{45}Sc	3.06e-07	6.06e-06
^{15}N	4.37e-05	1.27e-04	^{46}Ti	8.68e-07	7.38e-05
^{16}O	2.46e+00	3.25e+00	^{47}Ti	2.48e-06	1.14e-05
^{17}O	4.95e-08	9.62e-04	^{48}Ti	6.61e-04	4.44e-04
^{18}O	1.32e-06	4.34e-03	^{49}Ti	1.10e-05	2.26e-05
^{19}F	3.88e-05	1.21e-04	^{50}Ti	2.11e-11	1.66e-05
^{20}Ne	4.39e-01	6.64e-01	^{50}V	1.95e-09	1.73e-07
^{21}Ne	2.68e-05	4.76e-04	^{51}V	2.01e-05	4.74e-05
^{22}Ne	8.58e-06	4.71e-02	^{50}Cr	7.99e-06	1.83e-04
^{23}Na	1.58e-03	2.89e-02	^{52}Cr	5.30e-03	2.90e-03
^{24}Mg	1.54e-01	8.22e-02	^{53}Cr	2.04e-04	2.99e-04
^{25}Mg	7.34e-04	2.80e-02	^{54}Cr	1.05e-09	3.51e-05
^{26}Mg	6.81e-04	3.34e-02	^{55}Mn	6.49e-04	1.49e-03
^{27}Al	2.69e-03	2.76e-02	^{54}Fe	9.65e-04	1.26e-02
^{28}Si	2.65e-01	3.00e-01	^{56}Fe	4.55e-01	2.44e-01
^{29}Si	9.52e-04	5.83e-03	^{57}Fe	1.38e-02	1.01e-02
^{30}Si	3.39e-04	5.10e-03	^{58}Fe	5.58e-09	8.94e-04
^{31}P	2.58e-04	2.59e-03	^{59}Co	5.22e-04	1.07e-03
^{32}S	1.08e-01	1.44e-01	^{58}Ni	1.94e-02	2.26e-02
^{33}S	1.18e-04	9.26e-04	^{60}Ni	1.07e-02	6.65e-03
^{34}S	1.65e-04	7.34e-03	^{61}Ni	7.84e-04	7.57e-04
^{36}S	6.64e-09	4.40e-05	^{62}Ni	4.10e-03	4.48e-03
^{35}Cl	4.68e-05	9.97e-04	^{64}Ni	6.21e-12	3.09e-04
^{37}Cl	1.26e-05	2.95e-04	^{63}Cu	4.13e-05	4.48e-04
^{36}Ar	2.09e-02	2.30e-02	^{65}Cu	1.29e-16	9.19e-05
^{38}Ar	7.32e-05	7.68e-03	^{64}Zn	5.93e-06	2.68e-03

The nucleosynthesis of low and intermediate mass stars is taken from Renzini & Voli (1981), who derived yields for stars in the range 1 $M_\odot$ < M < 9 $M_\odot$. Renzini & Voli took into account the convective dredge-up mechanism of Iben & Truran (1978), the nuclear processing in the deepest layers of the convective envelope, and post main sequence mass loss. In particular, we have adopted their Z=0.04 and Z=$Z_\odot$ case A tables.

Deflagrations in carbon-oxygen white dwarfs, whose mass is near the Chandrasekhar limit, can reproduce the majority of the observed features in Type Ia supernova, which are very important contributors to enrichment of the interstellar gas in iron. Theoretical models computed by Nomoto et al. (1984), Woosley et al. (1984), Thielemann et al. (1986), and Woosley and Weaver (1992) have varying ratios of iron peak to intermediate mass elements due to different assumptions made about the velocity of the deflagration flame. We are presently studying the nucleosynthetic yields of a deflagration model using the flame propagation results of Timmes & Woosley (1992). However, for the present study we have adopted the popular W7 model of Nomoto et al. (1984) as representative of Type Ia nucleosynthesis.

Finally, we have adopted the radial and time dependent mass infall model of Chiosi (1980), a k=2 Schmidt star formation rate (Schmidt 1959), a stellar mass – lifetime homology relation (Talbott and Arnett 1971), and a Salpeter (1955) initial mass function. Additional details of the dynamical models and their parametric sensitivities will be presented in Timmes, Woosley, and Weaver (1993). While the dynamical models of the Galaxy presented in this work are certainly very simple, the chemical models are the most detailed to appear in the literature.

Figure 2 shows the resulting nucleosynthesis in the solar vicinity assuming

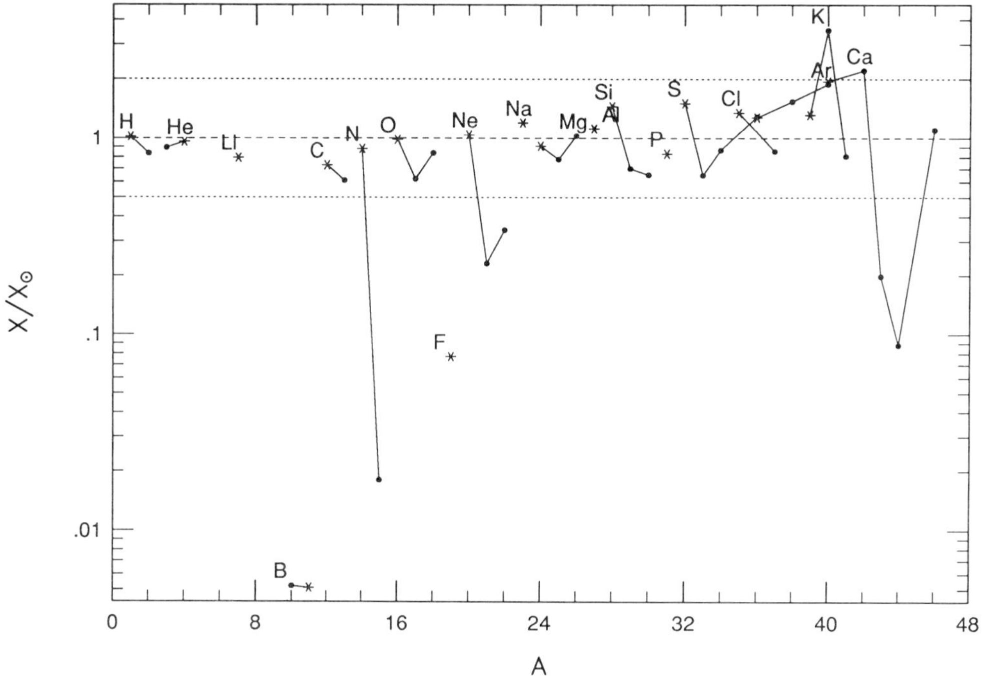

Fig. 2.— The computed isotopic composition of the Solar zone at the time when the Sun was born compared to the Anders and Grevesse (1988) mass fractions. Explosive Processing in massive stars and Type 1a supernova have not been included.

an n = -1.5 Salpeter IMF over the mass range 0.8 – 100 $M_\odot$, where the solar mass fractions are taken from Anders and Grevesse (1989). In this zone the total mass in the present epoch is 55 $M_\odot$ pc^{-2}, with the gas comprising 4.4 $M_\odot$ pc^{-2}, the mass accretion rate is 0.33 $M_\odot$ pc^{-2} Gyr^{-1}, the stellar birthrate is 1.2 $M_\odot$ pc^{-2} Gyr^{-1}, the Type II supernova rate integrated over the Galactic disk is 4.17 per century, the integrated Type I rate is 1.08 per century, the change of helium with respect to metals $\Delta Y/\Delta Z$ is 2.44, and the total metallicity of the gas at the time the Sun was born is 1.855e-2. Overall, the agreement of the computation with the observations is quite good.

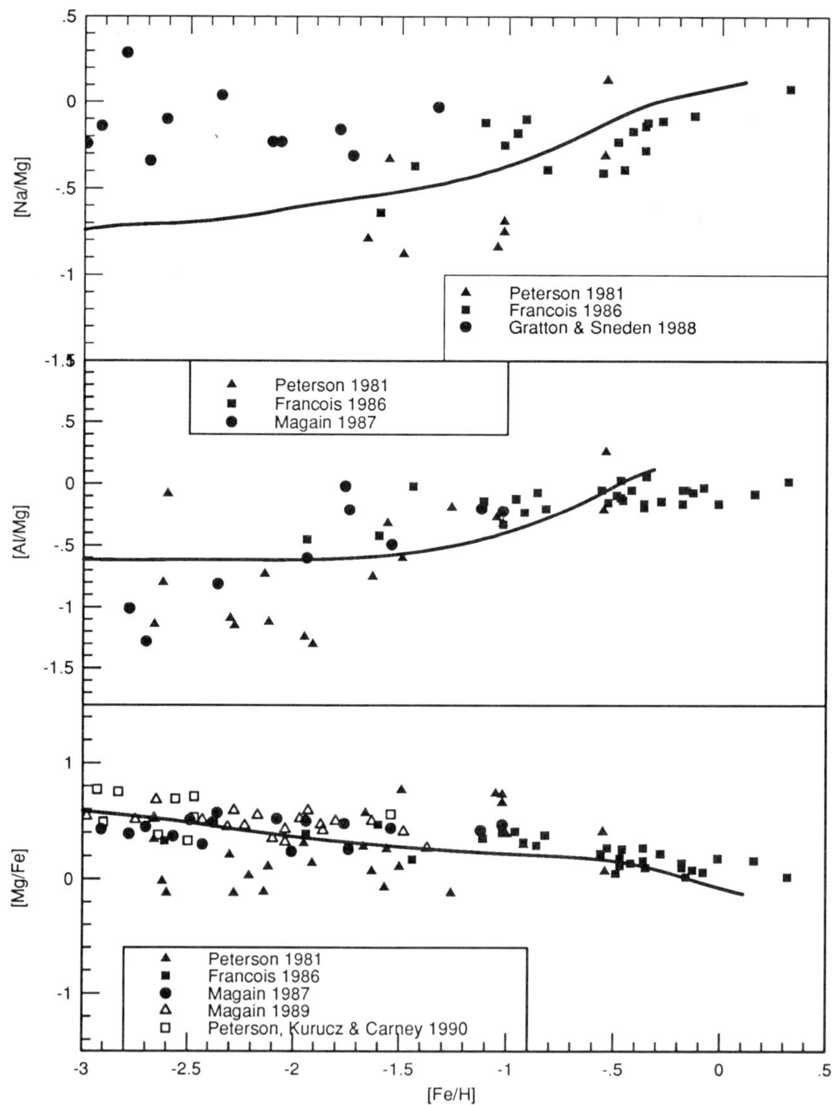

Fig. 3.— a) Sodium to magnesium, b) aluminum to magnesium, and c) magnesium to iron as a function of metallicity. Solid lines are the theoretical result.

Figure 3 shows the expected evolution Na/Mg, Al/Mg, and Mg/Fe as a function of metallicity (the evolution of Fe is discussed in the next section). The observations of [Na/Mg] and [Al/Mg] (Peterson 1981) for Z > -2.0 clearly show a relative deficiency of the light odd Z elements that is in reasonable accord with our calculations. These are very preliminary results. It is expected that use of a larger reaction network will give more neutronization during carbon burning and that Na and Al yields in zero and low metallicity stars may be increased in our future studies.

Primary nitrogen is produced in all our zero metallicity stars heavier than 30 $M_\odot$ as the helium convective shell penetrates into the hydrogen shell. The exact amount is not determined very well computationally as it is sensitive to how convection is treated. However, no such primary production is observed, using the same convective prescription, in stars of initially solar metallicity. The nitrogen is made as the helium convective shell penetrates into the hydrogen shell with violent, almost explosive, consequences. Figure 4 shows the computed and observed [N/Fe] as a function of [Fe/H] chronometer. Initially the [N/Fe] ratio is large owing to the production of primary nitrogen in the zero metal massive stars, then it decreases due to mass and metallicity effects that suppress the production of primary nitrogen, and finally rises again due to the production from the convective dredge-up mechanism of low mass stars. Caution is advised as these results are very uncertain and sensitive to zoning and the treatment of convection. They may also change significantly when stars of intermediate metallicity are calculated.

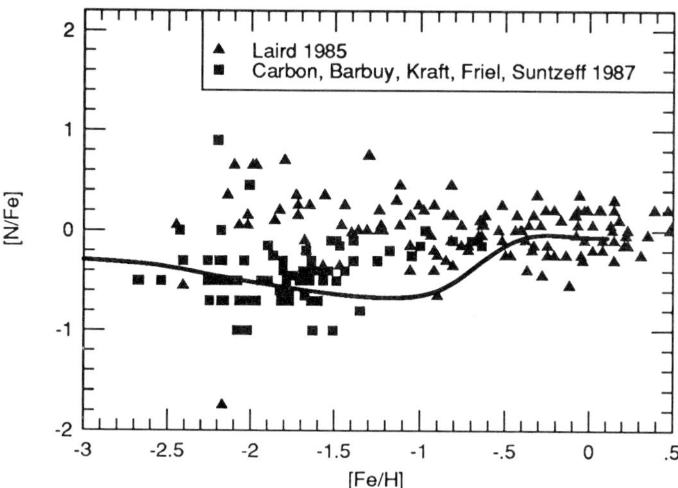

Fig. 4.— Observed and computed [N/Fe] as a function of the [Fe/H] chronometer. The initial value of [N/Fe] is due to the production of primary nitrogen in zero metal high mass stars. Mass and metallicity effects then cause the curve to decrease, and finally the contributions from the convective dredge-up mechanism of low mass stars cause the [N/Fe] ratio to approach its solar value. The low metallicity values given here from our model are very uncertain.

5. COMBINED EXPLOSIVE YIELDS - II-p and Ia

Figure 5 shows the evolution of oxygen and iron when nucleosynthesis from the exploded versions of the massive star models (Table 1) are combined with that from Type Ia (as typified by Nomoto's Model W7). Figure 6 gives the present calculated abundances at the solar circle from a) all stars except Type Ia and b) all sources. Since Figure 6b is our principal result it warrants some discussion.

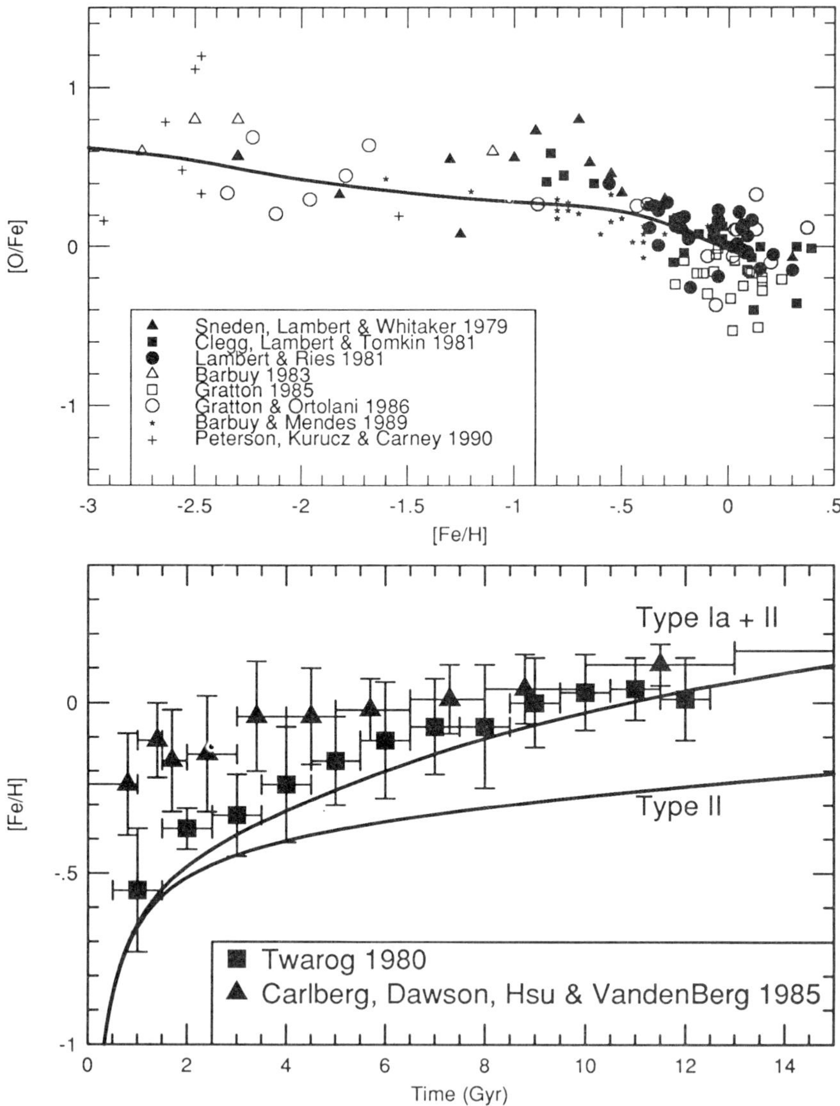

Fig. 5.— a) Oxygen to iron and b) iron to hydrogen as a function of metallicity. Iron is made here in the explosions of both Type Ia and Type II supernovae.

The light isotopes, hydrogen, helium, and lithium originate, as expected, in the Big Bang though there is a 5 - 10% contribution to ^{7}Li from the neutrino process. ^{11}B and ^{19}F are products of the neutrino process. A low value of μ- and τ-neutrino temperature was adopted here (6 MeV) compared to earlier studies (8 MeV; Woosley et al. 1990). Uncertainties in the models allow both; the production of Li, F, and B scales as T_ν^2 and could easily be a factor of two greater. ^{9}Be, ^{10}B, and ^{15}N need alternate nucleosynthetic sites. Other productions up to Ca are in excellent agreement with solar values except for ^{21,22}Ne, ^{40}K, and ^{40}Ar. The solar abundances of ^{40}K and ^{40}Ar are very uncertain. ^{21}Ne may be low because a value of ^{20}Ne$(n,\gamma)^{21}$Ne of 0.12 mb was adopted during helium burning, which is correct (Winters and Macklin 1988), *and in carbon burning* which is incorrect. A value twice as large is appropriate, will be included in future calculations, and may increase the ^{21}Ne yield (much of ^{21}Ne is made in carbon burning). The low yield of ^{22}Ne is perplexing and may indicate a problem in nuclear cross sections or convection.

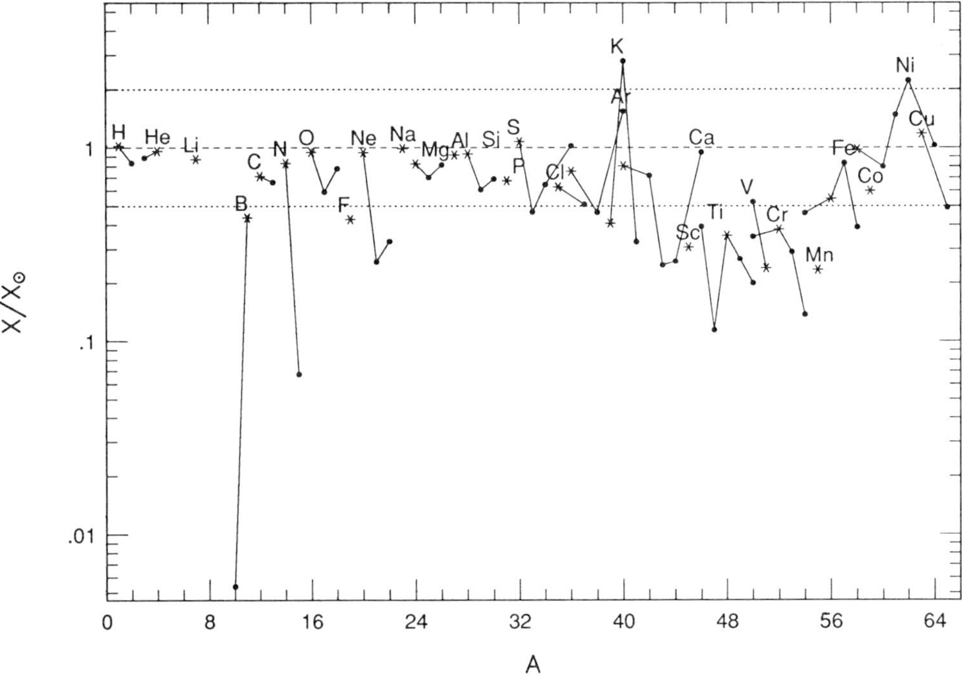

Fig. 6a.— Ratio of the final theoretical mass fractions at the solar circle compared with those observed in the sun (Anders and Grevesse 1988). Isotopes of the same element are connected by lines and the most abundant isotope indicated by an asterisk. The horizontal dashed band is a range of a factor of two about the solar value and points found in the band are regarded as successes (see text for discussion). Figure 6a includes only the yields from low mass stars and Type II supernovae.

Above calcium the calculated abundance ratios show a much larger spread reflecting the greater uncertainty associated with explosive models compared to hydrostatic ones. ^{43}Ca and ^{47}Ti have historically presented problems (Woosley 1986). ^{44}Ca is low reflecting an inadequate production as ^{44}Ti in the α-rich freeze out. All three of these isotopes could be made in helium detonation models of Type I (e.g.., Woosley, Taam, and Weaver 1986). ^{50}Ti, ^{54}Cr, and ^{48}Ca, all products of neutron-rich nse are low. They would be produced in a Type Ia deflagration model if the flame were given an initially slow speed (Woosley and Weaver 1992). Similarly ^{54}Fe and ^{58}Ni are high and this reflects a problem in carbon deflagration models that can be cured by a more realistic prescription for flame propagation (Timmes and Woosley 1992, Woosley and Weaver 1992). The high production of ^{62}Ni may be a consequence of a network that terminated at zinc, or perhaps a symptom of how the explosion was parameterized.

All in all it's not a bad fit.

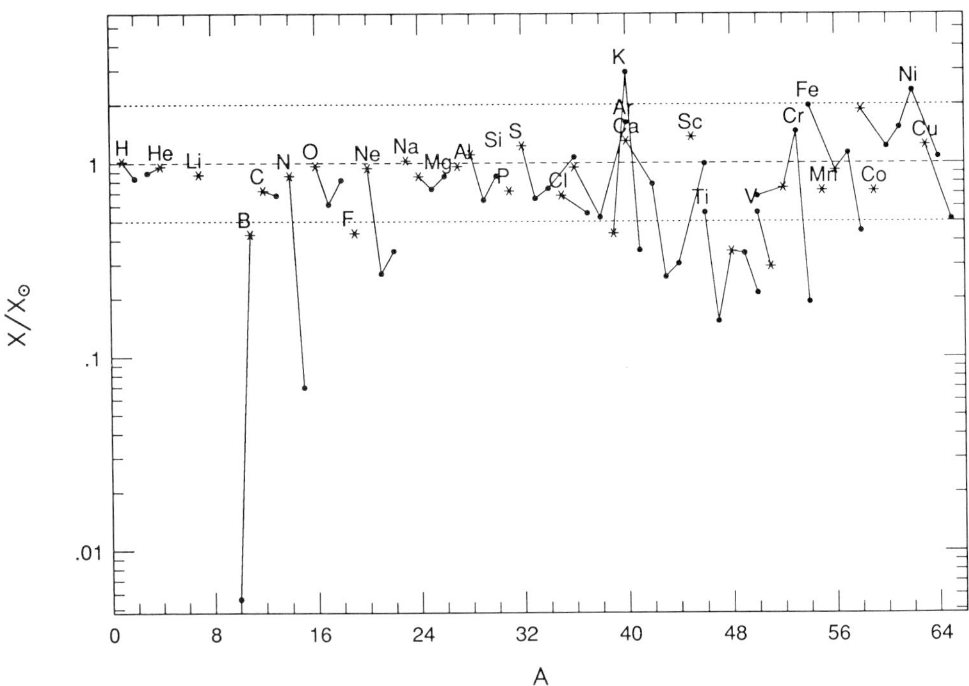

Fig. 6b.— Same as Figure 6a but adds the contribution from supernovae of Type Ia (carbon deflagration).

This work has been supported by the National Science Foundation (AST 91-15367, by the California Space Institute and, at Livermore, by the Department of Energy (W-7405-ENG-48). We are grateful to Rob Hoffman for help in keeping our nuclear reaction network current and valuable discussions of nucleosynthetic processes.

REFERENCES

Anders, E. & Grevesse, N. 1989, Geochim. Cosmochin. Acta, 53, 197

Barbuy, B. 1983, AA, 123, 1

Barbuy, B., & Mendes, M. E. 1989, AA, 214, 239

Carbon, D. F., Barbuy, B., Kraft, R. P., Friel, E. D. & Suntzeff, N. B. 1987, PASP, 99, 335

Carlberg, R. G., Dawson, P. C., Hsu, T., Vandenberg, D. A. 1985, ApJ, 294, 674

Caughlan, G. A., & Fowler, W. A. 1988, Atomic Data and Nuclear Data Tables, 40, 238

Chiosi, C. 1980, AA, 83, 206

Chiosi, C., & Matteucci, F. 1982, AA, 105, 140

Clegg, R. E. S., Lambert, D. L., & Tomkin, J. 1981, ApJ, 250, 262

Francois, P. 1986, AA, 160, 264

Gratton, R. G. 1983, AA, 148, 105

Gratton, R. G., & Ortolani, S. 1986, AA, 169, 201

Gratton, R. G. & Sneden, C. 1988, AA, 204, 193

Iben, I. Jr., & Truran, J. W. 1978, ApJ, 220, 980

Laird, J. B. 1985, ApJ, 289, 556

Lambert, D. L. & Ries, L. M. 1981, ApJ, 248, 228

Magain, P. 1987, AA, 179, 176

Magain, P. 1989, AA, 209, 211

Matteucci, F. & Greggio, L. 1986, AA, 154, 279

Nomoto, K. Thielemann, F. K., & Yokoi, Y. 1984, ApJ, 286, 644

Peterson, R. C. 1981, ApJ, 244, 989

Peterson, R. C., Kurucz, R. L., & Carney, B. W. 1990, ApJ, 350, 173

Renzini, A. & Voli, M. 1981, AA, 94, 175

Salpeter, E. E. 1955, ApJ, 121, 161

Schmidt, M. 1959, ApJ, 129, 243

Sneden, C., Lambert, D. L. & Whitaker, R. W. 1979, ApJ, 234, 964

Talbott, J. T. Jr., & Arnett, W. D. 1971, ApJ, 170, 409

Talbott, J. T. Jr., & Arnett, W. D. 1973a, ApJ, 186, 51

Talbott, J. T. Jr., & Arnett, W. D. 1973b, ApJ, 186, 69

Talbott, J. T. Jr., & Arnett, W. D. 1975, ApJ, 197, 551

Thielemann, F. K., Nomoto, K., & Yokoi, Y. 1986, AA, 158, 17

Timmes, F. X., & Woosley, S. E. 1992, ApJ, 397, 220

Timmes, F. X., Woosley, S. E., & T. A. Weaver 1993, in Proceedings of the VI Advanced School of Astrophysics in São Paulo, Brazil, ed. Barbuy, B., Freitas Pacheco, J. A. & Janot-Pacheo, E. (São Paulo: IAGUSP), 333

Twarog, B. A. 1980, ApJ, 242, 242

Weaver, T. A., Zimmerman, Z. B., & Woosley, S. E. 1978, ApJ, 225, 1021

Weaver, T. A., & Woosley, S. E. 1992, Physics Reports, in press

Winters, R. R., and Macklin, R. L. 1988, ApJ, 329, 943

Woosley, S. E. 1986, in Nucleosynthesis and Chemical Evolution, ed. B. Hauck, A. Meader, & G. Meynet, (Geneva: Swiss Society of Astrophysics and Astronomy, Geneva Observatory), 1

Woosley, S. E., & Weaver, T. A. 1982, in Essays in Nuclear Astrophysics, ed. C. A. Barnes, D. D. Clayton, and D. N. Schramm (Cambridge Univ. Press: Cambridge), 377

Woosley, S. E., Axelrod, T. S., & Weaver, T. A. 1984, in Stellar Nucleosynthesis, ed. Chiosi, C., & Renzini, A. (Dordrecht: Reidel), 263

Woosley, S. E., Taam, R. E., & Weaver, T. A. 1986, ApJ, 301, 601

Woosley, S. E., & Weaver, T. A. 1988, Physics Reports, 163, 179

Woosley, S. E., Hartmann, D., Hoffman, R., & Haxton, W. 1990, ApJ, 356, 272

Woosley, S. E., & Weaver, T. A. 1992, in Les Houches, Session LIV 1990: Supernovae, ed. J. Audouze, S. Bludman, R. Mochkovitch, J. Zinn-Justin (Elsevier Science Publishers: BV), 000

Woosley, S. E., Langer, N., & Weaver, T. A. 1993a, ApJ, submitted

Woosley, S. E., Langer, N., & Weaver, T. A. 1993b, ApJ, in preparation

The r-process in the hot bubble

Bradley S. Meyer[†][1]

† Department of Physics and Astronomy, Clemson University, Clemson, SC
29634-1911, USA

Abstract.   The high-entropy, evacuated region forming just outside the
proto-neutron star during a core-collapse supernova makes an excellent site
for the r-process. The high entropy per baryon allows this region to have
a high free neutron/seed nucleus ratio, a necessary requirement for the r-
process, even though the material is not particularly neutron rich. The
neutrino-driven wind blowing from the hot bubble lasts several seconds and
has a mass loss rate of roughly 10^{-5} $M_\odot$ s^{-1}. The total mass of r-process
matter produced during the supernova is thus $\sim 10^{-4}$ $M_\odot$, in good agreement
with arguments from galactic chemical evolution. The hot-bubble r-process
achieves local steady beta flow, in agreement with results from experimental
nuclear physics. Deficiencies in the present model may reflect in part failings of the nuclear mass extrapolation used. It may be possible to constrain
nuclear and neutrino physics through the r-process.

1. Introduction

Roughly one half of the nuclei in nature with mass number A greater than 70 were made
in the *rapid* neutron capture process, or r-process. In this process of nucleosynthesis,
neutron captures occur more rapidly than beta decays, so that an r-process path some
5-20 neutrons rich of the valley of beta stability is populated. When the neutron bath
disappears, the nuclei beta decay back to the beta stable nuclei and give rise to the
r-process nuclei in nature. (For a review of the r-process, see Cowan *et al* (1991)).

The temperature, neutron density, and timescale requirements for the r-process
suggest supernovae as the likely site for this process. Despite some thirty years of work,
however, no definitive supernova site for the r-process has been determined. Nevertheless,
recent developments in our understanding of delayed core-collapse supernovae have lead
to the identification of a promising site–the hot bubble. It is the purpose of this paper
to present a model of the r-process in this site and to discuss the model's strong and
weak points. We also present a discussion of related nuclear and particle physics issues.

---

[1] E-mail: brad@cosmo.phys.clemson.edu (Internet).

We begin with a description of a primary r-process in terms of freeze out from nuclear statistical equilibrium. A primary r-process is one in which the seed nuclei for neutron captures are produced during the r-process event itself. (This is to be contrasted with a secondary r-process in which the seed nuclei pre-exist.) We find that the photon-to-baryon ratio, or equivalently the entropy per baryon, provides a useful means of classifying r-processes. We then discuss where in supernovae we might locate high- and low-entropy r-processes. The hot bubble provides a natural site for a high-entropy r-process, and we give a detailed discussion of this site. We present a model for such a high-entropy r-process. In the last section, we discuss how the hot-bubble r-process does achieve steady beta flow and how steady beta flow suggests that the r-process that produced the solar-system nuclei was primary, how deficiencies in our model may be telling us something about nuclear physics, and how the r-process may constrain neutrino properties.

2. Freeze out from NSE

In a primary r-process, material begins at high temperature and density and then expands and cools. The material starts at sufficiently high temperature that all strong and electromagnetic nuclear reactions come into equilibrium with their reverse reactions. The material is then in nuclear statistical equilibrium (NSE) and the abundance per baryon of a nuclide of atomic number Z and mass number A is simply

$$Y(Z,A) = G(Z,A)[\zeta^{A-1}\pi^{(1-A)/2}2^{(3A-5)/2}]A^{3/2}(kT/m_N c^2)^{3(A-1)/2}\phi^{1-A}$$

$$\times X_p^Z X_n^{A-Z}\exp\left(B(Z,A)/kT\right), \tag{1}$$

where $G(Z,A)$ is the nuclear partition function, $\zeta(3)$ is the Riemann zeta function of argument 3, k is Boltzmann's constant, T is the temperature, m_N is the nucleon mass, c is the speed of light, ϕ is the photon-to-baryon ratio, X_p and X_n are the free proton and neutron mass fractions, respectively, and $B(Z,A)$ is the binding energy of nucleus (Z,A). In terms of the temperature and density, ϕ is

$$\phi = \frac{n_\gamma}{\rho N_A} = \frac{1}{\pi^2}\frac{g_\gamma}{(\hbar c^2)^3}\frac{\zeta(3)(kT)^3}{\rho N_A}, \tag{2}$$

where n_γ is the photon number density, ρ is the mass density, N_A is Avagadro's number, and g_γ is the spin factor for photons. We can write equation (2) in terms of the temperature T_9 (in units of $10^9\ K$) and density ρ_5 (in units of $10^5\ g\ cm^{-3}$) as

$$\phi = 0.34\frac{T_9^3}{\rho_5}. \tag{3}$$

From equation (1), we can see that the abundance $Y(Z,A)$ is strongly dependent on ϕ. For a given temperature, the larger ϕ is, the smaller will be the abundance of a given heavy nucleus.

As the material expands and cools, the nuclear reactions slow down and start to fall out of equilibrium. Eventually the reactions become much slower than the expansion and they effectively shut off or "freeze out". The first reactions to freeze out are the charged-particle reactions. At the time of freeze out of the charged-particle reactions, there are nuclei and free neutrons. The nuclei then capture the free neutrons and beta-decay. This is the r-process. Initially in this r-process the neutron capture, or (n, γ), reactions are in equilibrium with their reverse reactions, neutron disintegrations, or (γ, n), reactions. This is the "classical" or $(n, \gamma) - (\gamma, n)$ equilibrium r-process phase. Eventually $(n, \gamma) - (\gamma, n)$ equilibrium can no longer be maintained, and non-equilibrium processing occurs. Finally, the neutron capture and disintegration reactions also freeze out, and only beta decays occur until nuclei reach the valley of beta stablilty. A primary r-process is thus just one phase of the freeze out from NSE. It occurs when the material is sufficiently neutron rich that a significant number of free neutrons are present at the time of charged-particle reaction freeze out.

How neutron rich must the material be to undergo an r-process? The r-process produces uranium (A=238) from seed nuclei ($A \approx 50 - 100$), therefore, some 100 free neutrons are needed per seed nucleus. The required neutron richness thus depends on the abundance of seed nuclei.

If $\phi \lesssim 1$, NSE favors iron-group nuclei ($A \approx 60$). These nuclei are the seeds for the r-process once the charged-particle reactions have frozen out. Because the r-process needs $\gtrsim 100$ free neutrons per seed, a seed like ^{56}Fe requires total neutron-to-proton (n/p) ratio of roughly $(30+100)/26=5$. This is quite neutron-rich material.

If $\phi \gtrsim 10$, NSE favors ^{4}He at high temperature. As the temperature drops in the expanding material, NSE begins to favor heavier mass nuclei. For large ϕ, this occurs late in the expansion. At this late time, the two reaction sequences that begin the assembly of ^{4}He nuclei into heavier mass nuclei, $\alpha + \alpha + \alpha \rightarrow {}^{12}$C and $\alpha + \alpha + n \rightarrow {}^9Be(\alpha, n)^{12}$C, may be operating at a low level, or may have frozen out. Unlike these two reaction sequences, which are slow because they depend on three-body reactions, alpha captures on ^{12}C and heavier nuclei occur rapidly and build up heavy mass nuclei. At $A \approx 100$, disintegrations impede the further flow. Eventually the alpha-capture reactions freeze out. The material now consists of free neutrons, heavy seed nuclei, and many unassembled ^{4}He nuclei. At this point, the free neutrons begin to capture on the seed nuclei, but not on the ^{4}He nuclei. An r-process thus occurs. A typical composition at charged-particle reaction freeze out is 20% free neutrons by mass, 10% heavy seed nuclei, and 70% alpha particles. What n/p ratio do we have in this case and what is the free neutron/seed nucleus ratio? Suppose the seed is $Z = 35$, $A = 100$. The free neutron/seed nucleus ratio is in this case 200, more than sufficient for an r-process. The n/p ratio is $(.7/4 \times 2 + .1/100 \times 65 + .2)/(.7/4 \times 2 + .1/100 \times 35) = 1.6$. This is much less neutron rich than in the low-ϕ scenario.

3. Supernova sites

In the previous section, we classified a primary r-process according to whether the photon-to-baryon ratio ϕ was large or small. For small ϕ, all protons are bound up into seed nuclei at the beginning of the r-process. In such an r-process, quite neutron-rich matter is needed to produce actinide nuclei. For large ϕ, most protons remain bound up in alpha particles. The few seed nuclei present have the free neutrons all to themselves. In such a case, only moderately neutron-rich matter is required to synthesize the actinides. We turn now to the question of where in supernovae small-ϕ and large-ϕ r-processes might occur.

3.1. Small-ϕ r-process

A small-ϕ r-process requires n/p$\gtrsim$5. Such conditions would occur in the neutronized outer layers of the proto-neutron star. Hillebrandt, Takahashi, and Kodama (1976) calculated the r-process in such an environment and were able to obtain a quite good fit to the solar-system r-process distribution.

There is a fundamental problem with the small-ϕ r-process, however, and that is how to eject the correct amount of neutron-rich matter. There are 3×10^4 $M_\odot$ of r-process material in the galaxy (inferred from the data in Anders and Grevesse (1989) and Käppler, Beer, and Wisshak (1989)). The galaxy is roughly 10^{10} years old and the supernova rate is between 10^{-1} yr^{-1} and 10^{-2} yr^{-1}. If every supernova undergoes some r-process nucleosynthesis, then each supernova ejects $\sim 10^{-4}$ $M_\odot$ of r-process matter.

How is it that $\sim 10^{-4}$ $M_\odot$ of neutron-rich material is ejected from the outer neutronized layers of the proto-neutron star? Why is not some other mass ejected? In the detailed dynamical models of Hillebrandt, Takahashi, and Kodama (1976), 0.1 $M_\odot$ of neutron-rich material had to be ejected to get the n/p$\sim$ 7 matter required to producd actinide nuclei. In such a case, supernovae would greatly overproduce r-process matter in the galaxy. Of course, the bulk of this material could fall back on the proto-neutron star. However, this would require just the right mixing to get out the high n/p material, which seems like an extremely unlikely scenario. On the other hand, it may be that only some one in one thousand supernovae can eject neutron-rich matter into the interstellar medium. Here it may be that only a small fraction of supernovae have high rotation rates and magnetic fields that allow magnetohydrodynamical instabilities to eject r-process matter (LeBlanc and Wilson 1970; Meier *et al* 1976; Symbalisty 1984). This possibility is certainly plausible and deserves attention, but the current calculations seem to indicate that unnaturally large rotation rates and magnetic fields are required for ejection of neutron-rich matter from the proto-neutron star.

3.2. Large-ϕ r-process

A large-ϕ r-process requires n/p$\sim$ 1.6 This material is less neutron rich than that required for the small-ϕ r-process; therefore, this r-process could occur further out in the neutronized material ejected during a core-collapse supernova. It is imperative, however, that we find a site that has a large enough ϕ. Too small of a photon-to-baryon ratio leads to too much assembly of alpha particles into seed nuclei. This gives too many seed nuclei per free neutron and no r-process. In general, high temperatures are associated with high density in supernovae, so it is somewhat difficult to find a site where the temperature is high but the density is low. The high-entropy wind, or "hot bubble", that develops in a delayed-mechanism core-collapse supernova, which we discuss in the next section, is a high-temperature, low-density site, and it is at present the most promising site for the r-process. This site also offers a natural explanation for why $\sim 10^{-4}$ $M_\odot$ of r-process matter is ejected per supernova.

4. The hot-bubble site

The hot bubble that forms during a core-collapse supernova is a promising site for a large-ϕ r-process. What is this hot bubble and how does it develop? To answer these questions, we turn briefly to the details of core-collapse supernovae. For more information, the reader should consult the original papers (Bethe and Wilson (1985); Mayle and Wilson (1988); Mayle and Wilson (1991)).

Stars of mass 9 $M_\odot$ or larger evolve until they develop an iron core of mass roughly 1.4 $M_\odot$. The combined effects of electron capture, photo-disintegrations, and neutrino losses then remove the pressure support of the core. The inner ~ 0.7 $M_\odot$ collapses homologously until it reaches densities in excess of nuclear density. At this point the pressure rises dramatically and resists further collapse. The homologous core bounces and drives out a shock wave that works its way through the remainder of the initial iron core. The material in front of the shock has low entropy per baryon ($\sigma \approx 1$, where σ is the entropy per baryon in units of Boltzmann's constant) and consists predominantly of heavy nuclei. The shock disintegrates the material it passes through, leaving it with an entropy of $\sigma \approx 10$. The disintegration of the heavy nuclei absorbs kinetic energy from the shock. In most cases the shock stalls before it reaches the outer boundary of the initial iron core.

As the shock stalls, the material behind the shock accretes back onto the proto-neutron star. This evacuates the region between the stalled shock and the proto-neutron star. The gravitational binding energy released by the accreting material radiates away in the form of energetic neutrinos. These neutrinos deposit energy in the evacuated region. This leads to production of electron-positron pairs which themselves can scatter neutrinos. This leads to further energy deposition and further increase in the entropy. In this way the entropy can build up to $\sigma \approx 100 - 200$ or greater. The pressure in this high-entropy, or "hot", bubble also builds up and revives the stalled shock's outward motion. If the shock breaks through the remainder of the initial core, we have a successful, "delayed" supernova explosion.

Because the temperature is high and the density is low in the hot bubble, ϕ is large. Since most of the entropy is in relativistic particles,

$$\sigma \approx \frac{4}{3} \frac{a(kT)^3}{\rho N_A},\tag{4}$$

where a is the Stephan-Boltzmann constant

$$a = \frac{g\pi^2}{30} \frac{1}{(\hbar c)^3}.\tag{5}$$

Here g is a spin factor. The relativistic particles are photons, electrons, and positrons, so $g = g_\gamma + \frac{7}{8}(g_{elec} + g_{pos}) = 11/2$. From equation (2), we thus have

$$\phi \approx 0.10\sigma.\tag{6}$$

Thus, entropies of $\gtrsim 100$ give $\phi \gtrsim 10$. The entropy or ϕ is right for an r-process. What about the n/p ratio?

The n/p ratio in the hot bubble is set by the charged-current captures on nucleons: $\nu + n \rightarrow p + e^-$ and $\bar{\nu} + p \rightarrow n + e^+$. At early times, the material is slightly proton rich. This is because the neutrinosphere and anti-neutrionsphere are essentially at the same place in the proto-neutron star so that the fluxes of the ν's and $\bar{\nu}$'s are roughly the same. The lower mass for the proton favors neutrino capture on neutrons over antineutrino capture on protons. As the proto-neutron star cools, however, electron captures make

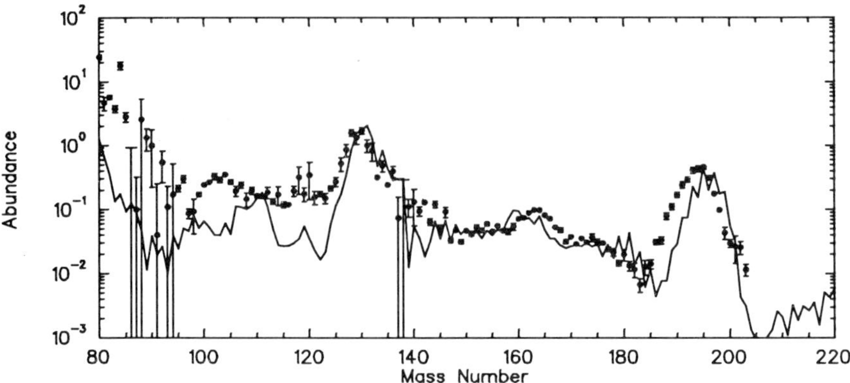

Figure 1. Weighted-sum r-process and solar r abundances

the proto-neutron star more and more neutron rich. Therefore, inside the proto-neutron star, the mean free path for electron neutrinos becomes shorter than that for electron anti-neutrinos. The anti-neutrinosphere moves inwards to smaller radius and higher temperature than the neutrinosphere. The $\bar{\nu}_e$'s will thus have a higher temperature. Anti-neutrino capture on protons will begin to dominate neutrino capture on neutrons in the hot bubble and thus will drive the hot-bubble matter neutron rich. The material in the hot bubble is continuously blowing out as a high-entropy, neutrino-driven wind that begins proton rich but become neutron rich with time. If the wind becomes neutron rich enough, it will undergo r-process nucleosynthesis. Interestingly, the mass loss rate for this high-entropy wind is roughly 10^{-5} $M_\odot$ s^{-1} (Duncan *et al* 1986; Woosley 1992); therefore, if this wind lasts for a few seconds, as indicated in the supernova models of Mayle and Wilson, $\sim 10^{-4}$ $M_\odot$ of r-process matter will be ejected. As we have seen above, this is just what is required to explain the galactic r-process mass fraction. We should finally note that a similar wind (and a similar r-process) can occur during the collapse of an accreting white dwarf (Woosley and Baron 1992).

5. R-process calculations

We have computed the r-process that results during a core-collapse supernova (Meyer *et al* (1992)). We used the conditions $\sigma = 400$ and initial η varying from 0.0 to 0.4, where η is the neutron-richness parameter defined as

$$\eta = \sum_i (N_i - Z_i) Y(Z_i, A_i). \tag{7}$$

Here the sum is over all nuclear species present and $N_i = A_i - Z_i$ is the neutron number of species i. Other conditions are as described in Meyer *et al* (1992).

No single run (*i.e.* of given initial η) gave a good representation of the solar-system r-process abundance curve. When we summed these runs with the weighting

$\exp(-3\eta)$, however, we found quite good agreement with the solar-system abundance curve. Our results are shown in figure 1. The solid curve represents the weighted sum of the individual runs with different initial η's. The filled circles with error bars are the solar-system r-process abundances, as determined by Käppler, Beer, and Wisshak (1989). Our results have the $A = 130$ and $A = 195$ peaks shifted to slightly too large mass. There are deficiencies around $A = 120$ and $A = 140$, both of which may be due to errors in the nuclear mass extrapolation used in the calculations. (See the next section for a discussion of these points.) The low-mass range ($A < 110$) is also deficient, but this is probably filled in by the α-process–that is, material with lower entropy and neutron richness that may come from an earlier phase of the wind blowing from hot bubble than the phase that makes the r-process(Woosley and Hoffman (1992)). Despite the deficiencies, however, the solar-system r-process abundance cure is quite well reproduced. The model matches the heights and widths of the peaks in the solar-system distribution well and does a good job between the peaks. In particular, there is a strong hint of the rare-earth peak around $A = 160$, which, since there is no fission in our model, is due to some nuclear structure effects at this mass number. In summary, the model is extremely successful.

How realistic are the conditions we have used? The wind probably does not get as neutron rich as $\eta = 0.40$. In fact, $\eta = 0.20$ is probably the most neutron rich the wind gets. To get a successful r-process with a lower η requires a higher entropy. $\sigma = 0.40$ is already large; however, recent results indicate that σ may get to be as large as 700 (Wilson 1992), or, in models with large-scale convection, as large as 1000 (Herant, Benz, and Colgate 1992). Furthermore, the expansion timescale may be five times less than the one we used (Woosley 1992). Faster expansion leads to less assembly of alpha particles and a higher free neutron/seed ratio.

Our calculations show that the hot bubble does provide a likely site for the r-process. Whether in fact the hot bubble provides the conditions necessary to synthesize all of the solar-system r-process nuclei is a question that awaits further work on the delayed-supernova mechanism for resolution. On the other hand, if the r-process does occur in the hot bubble, the r-process provides some important constraints on super-novae.

6. Nuclear physics

The recent advances in our understanding of the r-process in the context of the hot bubble have largely been due to the recent advances in supernova theory. It is important to realize that recent advances in nuclear physics are also improving our understanding of the r-process. And just as we may find that the r-process can provide us with important constraints on the supernova mechanism, the r-process may provide important constrains on nuclear and particle physics.

6.1. Steady beta flow

Let us define the total abundance per baryon of charge Z nuclei to be

$$Y_Z = \sum_N Y(Z, A),\tag{8}$$

where $N = A - Z$ is the neutron number and we fix Z in the summation. Let us also define λ_Z, the average beta-decay rate of charge Z nuclei, to be

$$\lambda_Z = \sum_N \lambda_\beta(Z, A) Y(Z, A)/Y_Z, \qquad (9)$$

where $\lambda_\beta(Z, A)$ is the total beta-decay rate of nucleus (Z, A). The rate of change of Y_Z in the r-process is

$$\frac{dY_Z}{dt} = -\lambda_Z Y_Z + \lambda_{Z-1} Y_{Z-1}. \qquad (10)$$

Here we see that the abundance of charge Z nuclei decreases due to beta flow out of chain Z but increases due to beta flow from chain $Z - 1$ into chain Z. If the r-process operates at roughly constant conditions for times $>> 1/\lambda_Z$, the beta flow may become steady. In this case, $dY_Z/dt \to 0$ and $\lambda_Z Y_Z \to constant$.

Kratz *et al* (1986) and Kratz *et al* (1988) (see also Kratz 1988) have measured the beta-decay lifetimes of several nuclei with $N = 50$ and $N = 82$ on the r-process path. These nuclei, because their neutron shells are closed, have long beta-decay lifetimes and are thus "waiting point" nuclei at which nuclei wait and the abundance builds up during the r-process. The abundance build up at these nuclei gives rise to the r-process peaks after beta-decay back to the valley of beta stability. The significance of the experimental results is that, when they are combined with the assumption of $(n, \gamma) - (\gamma, n)$ equilibrium and of steady beta flow, they give an excellent representation of the solar-system r-process abundances in the $A = 80$ and $A = 130$ peaks. Thus, from experimental work we have strong, albeit circumstantial, evidence that the r-process operated under the conditions of $(n, \gamma) - (\gamma, n)$ equilibrium and local steady beta flow. Such a condition probably requires that the r-process be primary, since only a primary r-process would achieve the temperatures and densities necessary for $(n, \gamma) - (\gamma, n)$ equilibrium and the long timescales necessary for steady beta flow.

The steady beta flow is not global, however. The constant $\lambda_Z Y_Z$ for nuclei on the $N = 50$ closed shell ($A \approx 80$ nuclei) is not the same constant for nuclei on the $N = 82$ closed shell ($A \approx 130$ nuclei). A possible implication of this is that the r-process cannot be fit by a single component (single value for initial η, σ, and expansion timescale) but rather requires several different components (Kratz *et al* 1992). We found this to be true in our model of the hot-bubble r-process, and in fact this point has long been appreciated by theoreticians (Seeger, Fowler, and Clayton 1965; Kodama and Takahashi 1975; Hillebrandt, Takahashi, and Kodama 1976). In the case of the hot bubble, we expect different components naturally to be coming in the high-entropy wind because ϕ and η are continuously changing in the hot bubble. The earlier phases of the wind will have lower ϕ and η (as discussed in section 4) and will predominantly make the $A = 80$ and $A = 130$ peaks in the r-process distribution. The later phases will have higher ϕ and η and will make the $A = 195$ peak.

That the hot-bubble naturally gives local steady beta flow may be seen in figure 2. In this plot the solid curve gives the beta flow versus atomic number for the $\eta=0.225$ component, which strongly contributes to the $A = 130$ peak, and the dashed curve gives the beta flow for $\eta=0.400$ component, which strongly contributes to the $A = 195$ peak. Each curve is for conditions shortly before falling out of $(n, \gamma) - (\gamma, n)$ equilibrium. In each component, the flow is not globally steady. This is because there is a limited number of nuclei working their way up the network. In small regions, however, especially between waiting point, the beta flow has time to go into an approximately locally steady state.

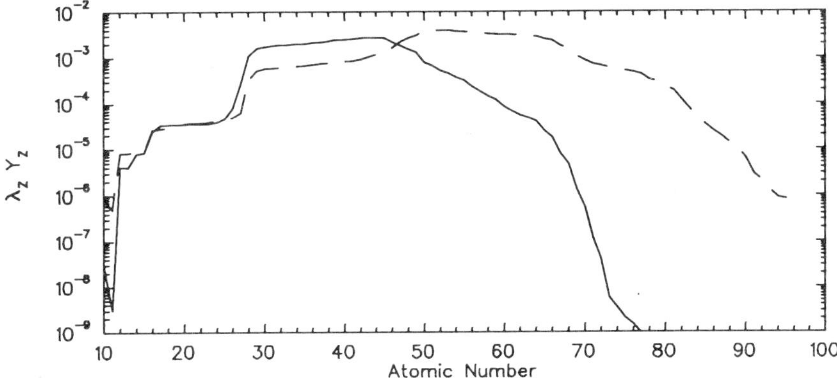

Figure 2. $\lambda_Z Y_Z$ for the $\eta=0.225$ component (solid curve) and for the $\eta=0.400$ component (dashed curve).

6.2. The $A \approx 120$ Dip

One of the deficiencies of our r-process model is the dip around $A = 120$. The calculated abundances are an order of magnitude below those of the solar system. Because this deficiency shows up in all of the η-components that are added together to make up figure 1, it seems to be the effect of an overly strong closed shell at $N = 82$ in the nuclear mass extrapolation we have used (Möller 1991). Because of the strong shell, the r-process path leaps over ten mass units from $A \approx 115$ to $A \approx 125$, bypassing these nuclei and giving rise to the dip. For a nuclear mass extrapolation with a weaker $N = 82$ closed shell, the approach to the $N = 82$ closed shell is less sudden, the r-process path does not leap over the $A \approx 120$ nuclei, and no dip forms. (See, for example, Kratz *et al* 1992.)

Such information will be invaluable to nuclear theorists in the future. Study of the r-process may be the only way to constrain the properties of many of these neutron-rich nuclei and to allow nuclear theorists to refine their models. However, we must proceed with caution. It may be that some η-component we have not computed does fill in the $A \approx 120$ dip. Neutrino inelastic scattering during the r-process may move some material from the $A = 130$ peak into the $A \approx 120$ dip. Or it may be that another process unrelated to the hot-bubble r-process, such as a weak neutron burst in the He-burning shell during a supernova, fills in the abundances at $A \approx 120$ (e. g. Cameron *et al* 1992). Before we definitively conclude that the $A \approx 120$ dip is due to an overly strong closed neutron shell in current mass extrapolations, we should rule out astrophysical sources for filling in this dip.

6.3. The $A = 139$ Dip

Our r-process model also proved deficient in the region around $A = 139$. We have traced at least some of this deficiency back to the nuclear mass extrapolation we have used (Möller 1991). Around $A \approx 139$, the r-process path passes through the region

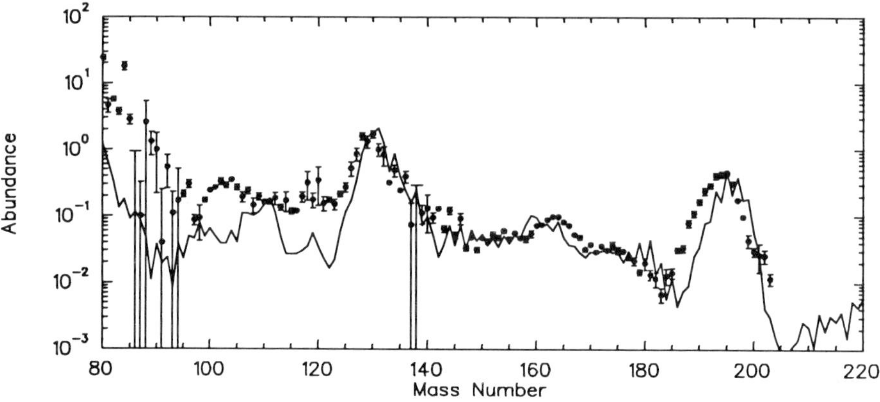

Figure 3. Same as figure 1 but with interpolated values for neutron-separation energies for certain $Z = 50 - 55$ isotopes (see text).

of transition from spherical to deformed nuclei in the Möller mass extrapolation. In particular, ^{139}Te is the first deformed Te isotope. Because of the nuclear-shape transition, the nuclear mass of ^{139} is less than the value one would interpolate from its odd-mass Te neighbors. This gives a small neutron binding energy for ^{139}Te and a weak abundance build up at $A = 139$. ^{141}Sn, ^{138}Sb, ^{138}I, ^{139}Xe, and ^{140}Cs all show a similar effect.

Figure 3 shows the sensitivity of our model to the effects of the nuclear-shape transition. In this calculation, the neutron binding energies of ^{141}Sn, ^{138}Sb, ^{139}Te, ^{138}I, ^{139}Xe, and ^{140}Cs were all set to the values interpolated from the binding energies of the neighboring isotopes. For example, the neutron binding energy of ^{138}Sb was set equal to the value interpolated from the binding energies of ^{136}Sb and ^{140}Sb, the neighboring even-mass isotopes of antimony. All other parameters were exactly as used to compute the curve in figure 1. From figure 3 we see that the $A = 139$ dip is largely filled in, although the model is still deficient around $A = 142$. Our calculations and the solar-system r-process abundance curve may be telling us that the nuclear-shape transition may be occurring in nature at different mass for the $Z = 50 - 55$ elements than in the Möller mass extrapolation. We should note that the above-mentioned I, Xe, and Cs isotopes all have experimentally measured masses. Each of these isotopes is the first to show a negative systematic deviation of the theoretical from experimental mass, which shows that the Möller mass formula has a systematic effect at these isotopes. From the r-process we may be gleaning important nuclear clues. However, the astrophysical caveats mentioned above also hold here.

6.4. ν-oscillations

If the ν_τ has a cosmological closure mass of $m_{\nu_\tau} \approx 30$ eV, it may experience matter-enhanced resonant conversion into ν_e between the surface of the proto-neutron star and the stalled shock (Fuller *et al* 1992). The tau neutrinosphere is deeper inside the proto-neutron star than the electron neutrinosphere; therefore, the emergent tau neutrinos have a greater energy than the electron neutrinos. Once the tau and electron neutrinos interconvert, the energy spectrum of the ν_e's is given by that of the initial ν_τ's. Thus, the initial heating of the hot bubble is greater than the case where there is no neutrino interconversion. In this way, these matter-enhanced ν-oscillations increase the entropy of the hot bubble and the explosion energy of the delayed supernova. This is significant because current supernova models, which do not include ν-oscillations, yield explosions that are too weak.

The neutrino conversion may also have profound effects on the r-process. The increase in entropy will help by increasing the neutron-to-seed ratio. On the other hand, the energy spectrum of the ν_e's after conversion from the ν_τ's is more energetic than that for the $\nu_{\bar{e}}$'s, even at late times. This means that, if the resonant conversion occurs, the hot bubble will not become neutron rich. In this case, the r-process cannot occur.

Such considerations may provide important constraints on the properties of neutrinos, of supernovae, or of the r-process. If there is ν_e-ν_τ interconversion as described above, the hot bubble may not be the site of the r-process. If, on the other hand, the r-process does occur in the hot bubble, it may be that the tau neutrino does not have a mass in the $m_{\nu_\tau} \approx 10 - 100$ eV range or that the $\nu_e - \nu_\tau$ mixing angle is $< 10^{-4}$ radians (Fuller *et al* 1992). This would provide important constraints on neutrinos and cosmology. Or it may be that the density drop off outside the proto-neutron star becomes so steep at late times that the ν-oscillation length becomes greater than the characteristic length scale for change in density. In this case, there would be no resonant conversion (Bethe 1986) at late times and the hot bubble would become neutron rich in time for the r-process to occur. The future for study of the interplay of neutrino physics, supernova physics, and nucleosynthesis is bright.

7. Conclusions

The hot bubble provides a high-entropy, neutron-rich environment which is ideal for an r-process. The r-process mass produced in such an environment is in agreement with that expected from simple galactic chemical evolution arguments. Whether the hot bubble is indeed the site for the r-process, however, is a question that will require more work on core-collapse supernovae. Supernova modellers must determine if the hot bubble has high enough entropy and neutron rich enough material to produce the actinide nuclei.

Nuclear physicists have told us that the r-process achieved local steady beta flow, which has profound implications for the site of the r-process, namely that it be primary and be result in a sum of components. If we know that the r-process happens in the hot bubble, can we then tell the nuclear physicists about the strength of the $N = 82$ closed neutron shell, the location of the nuclear-shape transition of $Z = 50 - 55$ elements, and the mass of the tau neutrino? These are questions to which we must turn our attention in the near future.

References

Anders E and Grevesse N 1989 *Geochim. Cosmochim. Acta* **53** 197

Bethe H A 1986 *Phys. Rev. Lett.* **56** 1305

Bethe H A and Wilson J R 1985 *Ap. J.* **295** 14

Cameron A G W, Thielemann F-K, and Cowan J J 1992 *Phys. Rept.* submitted

Cowan J J, Thielemann F-K, and Truran J W 1991 *Phys. Rept.* **208** 267

Duncan R C, Shapiro S L, and Wasserman I 1986 *Ap. J.* **309** 141

Fuller G M, Mayle R W, Meyer B S, and Wilson J R 1992 *Ap. J.* **389** 517

Herant M, Benz W, and Colgate S 1992 *Ap. J.* **395** 642

Hillebrandt W, Takahashi K, and Kodama T 1976 *Astron. Astrophys* **52** 63

Käppler F, Beer H, and Wisshak K 1989 *Rept. Progress Phys.* **52** 945

Kodama T and Takahashi K 1975 *Nucl. Phys.* **239** 489

Kratz K-L 1988 *Rev. Mod. Astron.* **1** 184

Kratz K-L *et al* 1986 *Z. Phys. A* **325** 483

Kratz K-L, Thielemann F-K, Hillebrandt W, Möller P, Harms V, Wöhr A, and Truran J W 1988 *J. Phys. G* **14** Suppl. 331

Kratz K-L, Bitouzet J-P, Thielemann F-K, Möller P, and Pfeiffer B 1992 *Ap. J.* submitted

LeBlanc J M and Wilson J R 1970 *Ap. J.* **161** 541

Mayle R W and Wilson J R 1988 *Ap. J.* **334** 909

Mayle R W and Wilson J R 1991 *Supernovae* (New York: Springer Verlag) p 333

Meier D L, Epstein R I, Arnett W D, and Schramm D N 1976 *Ap. J.* **204** 869

Meyer B S, Mathews G J, Howard W M, Woosley S E, and Hoffman R D *Ap. J.* **399** to be published

Möller P 1991 private communication

Seeger P A, Fowler W A, Clayton D D 1965 *Ap. J. Suppl.* **11** 121

Symbalisty, E M D 1984 *Ap. J.* **285** 729

Wilson J R 1992 private communication

Woosley S E 1992 in preparation

Woosley S E and Baron E 1992 *Ap. J.* **391** 228

Woosley S E and Hoffman R 1992 *Ap. J.* **395** 202

Gamma-ray emission from the late detonation model for the type Ia supernova 1991T

S. Kumagai[1], K. Nomoto[1], T. Shigeyama[1], and H. Yamaoka[2]

[1] Department of Astronomy, University of Tokyo, Bunkyo-ku, Tokyo 113, Japan
[2] Department of Physics, College of General Education, Kyushu University, Fukuoka 810, Japan

Abstract.
We predict the time change in X-ray and γ-ray fluxes from the recent unique type Ia supernova SN 1991T in NGC4527, based on the *late detonation* models. The model predictions are marginally consistent with the upper limit to the 847 keV line flux obtained by the OSSE observation. The maximum flux of the 812 keV line from the ^{56}Ni decay is sensitive to the ^{56}Ni distribution, so that the observations at an early epoch can distinguish the models.

1. Introduction

The recent type Ia supernova (SN Ia) 1991T in NGC4527 has shown quite a unique spectral evolution (Filippenko et al. 1992; Ruiz-Lapuente et al. 1992). The inferred composition structure of SN 1991T is as follows: (I) The outermost layer is composed of Ni and Fe with expansion velocities $v_{\mathrm{exp}} \gtrsim$ 13,000 km s^{-1}. (II) The intermediate layer is rich in Si/Ca with $v_{\mathrm{exp}} \sim$ 10,000 km s^{-1}. (III) The central layer is again dominated by Fe.

The composition structure of the inner layers (II) and (III) of SN 1991T can be well accounted for by the carbon deflagration model W7 (Nomoto et al. 1984) and its synthetic spectra (Jeffery et al. 1992). However, the presence of high velocity heavy elements is not consistent with W7, because the highest velocity of newly synthesized elements in W7 is $\sim$ 15,000 km s^{-1}.

To account for these features, Yamaoka et al. (1992) have presented *late detonation* models in which a transition from the convective deflagration to the detonation is assumed to take place in the outermost layer. The carbon deflagration burns bulk of the white dwarf matter and forms a central Fe/Co/Ni core whose ^{56}Ni mass is 0.58 $\sim$ 0.70$M_\odot$ and a Si/S/Ca layer; in the outermost layer the deflagration is transformed into a detonation which forms Fe/Co/Ni layer containing 0.04$\sim$0.13$M_\odot$ ^{56}Ni.

Such a composition structure with the Si/S/Ca layer sandwiched by the two Fe/Co/Ni layers is consistent with the optical spectral evolution of SN 1991T. Since these models produce some ^{56}Ni in the outermost layers, it is expected that more intense γ-rays and X- rays emerge in the early phase than from W7. If this is the case, observations of these emissions from SN 1991T and similar type of SNe Ia will be quite important. In this paper, we predict the time change of γ-rays and X-rays from this type of SNe Ia and discuss their detectabilities and differences from W7.

2. Explosion models

We adopt the late detonation models in Yamaoka et al. (1992). Characteristic features of these models are summerized in Table 1. The composition structure of model W7DT is shown in Figure 1, where the mass of the ejecta is 1.378$M_\odot$.

Using these models, we perform the Monte Carlo simulation of transport of X-ray and γ-ray photons which are emitted from the decaying ^{56}Ni and ^{56}Co in the ejecta (see Kumagai et al. 1988, 1989). Six lines from ^{56}Ni and 45 lines from ^{56}Co are included. We then obtain the line γ-ray flux emitted from the surface of the ejecta without suffering from Compton scattering or photo-electric absorption.

Model	E_{exp} (erg)	$M_{^{56}\mathrm{Ni}}$ ($M_\odot$)	$V_{^{56}\mathrm{Ni}}$ (km s^{-1})	M_{trans} ($M_\odot$)
W7	1.3×10^{51}	0.58	13,000	
W7DT	1.5×10^{51}	0.78	29,000	1.13
W7DN	1.4×10^{51}	0.63	31,000	1.20
W7DHE	1.7×10^{51}	0.73	40,000	1.20
W8DT	1.5×10^{51}	0.80	50,000	1.25

Table 1. The explosion energy, the mass of synthesized ^{56}Ni, the velocity of outermost ^{56}Ni and the mass at which the transition from deflagration to detonation takes place for five explosion models.

3. Line γ-ray light curves

Figures 2 and 3 show the light curves of the dominant γ-ray lines from the decays of ^{56}Ni and ^{56}Co at a distance of 10 Mpc for W7DT and W7, respectively. After 40 days, these light curves depend only on the total mass of ^{56}Ni synthesized at the explosion (see also Burrows et al 1991; Müller et al. 1991). In contrast, they show a significant model dependence in the early phase.

Figure 4 compares the light curves of the most dominant γ-ray lines from the decays of ^{56}Ni (812 keV) and ^{56}Co (847 keV) for the five models. The dates and fluxes at the peak of the light curves and the ratios of the peak fluxes for the five models are summarized in Table 2.

It is seen that the date of the 847 keV line flux maximum is insensitive to the model, and the maximum flux of this line depends only on the total mass of ^{56}Ni. In contrast, the 812 keV line flux maximum from the ^{56}Ni decay depends on the ^{56}Ni distrubution, thus being sensitive to the model. Table 2 shows the ratio of maximum intensities of 812 keV and 847 keV lines. Clearly the observations at the earliest epoch are crucial to distinguish the models.

Model	f_{812} (cm^{-2} s^{-1})	d_{812} (days)	f_{847} (cm^{-2} s^{-1})	d_{847} (days)	f_{812}/f_{847}
W7	4.5×10^{-6}	28	3.4×10^{-5}	85	0.13
W7DT	2.6×10^{-5}	14	5.1×10^{-5}	80	0.51
W7DN	9.2×10^{-6}	10	3.7×10^{-5}	85	0.25
W7DHE	5.1×10^{-5}	7	4.7×10^{-5}	75	1.09
W8DT	4.0×10^{-5}	6	5.2×10^{-5}	77	0.77

Table 2. The flux and date at the peaks of the light curves of 812 keV and 847 keV γ-ray lines, and the ratio of the peak flux of these two lines.

4. Comparison with the OSSE observation SN 1991T

If we apply W7DT to SN 1991T, the distance to this supernova can be estimated as follows: The bolometric luminosity at maximum brightness of W7DT is 1.15×10^{43} erg s^{-1}, which gives $M_{bol} = -18.95$ (Yamaoka et al. 1992). Taking $M_B - M_{bol} = 0.28$ and $m_B = 11.20$ for the blue maximum of SN 1991T (Branch and Tamman 1992), we get 12.2 Mpc as a distance to SN 1991T. Then the 847 keV line flux from the supernova is expected to be 3.5×10^{-5} photons cm^{-2} s^{-1} after 50 days.

The OSSE experiment on the COMPTON observatory has observed SN 1991T and obtained only the upper limit to the 847 keV γ-ray line flux as 4.4×10^{-5} photons cm^{-2} s^{-1} on $66 - 79$ days and 3.9×10^{-5} photons cm^{-2} s^{-1} on $176 - 190$ days (Lichti et al. 1992). This is marginally consistent with the predicted flux for W7DT, implying that the ^{56}Ni mass in SN 1991T would be smaller than $0.98 M_\odot$.

5. Detectability of γ-ray lines

Figure 5 compares the maximum flux of the 847 keV γ-ray lines for the five models as a function of the distance to the supernova. Figure 6 summarizes the maximum fluxes (at 10 Mpc) of the dominant γ-rays lines from the decays of ^{56}Ni and ^{56}Co as a function of the line energies. In these figures, the detection limits of GRO (Krufess 1988), DUET (Kamae 1992) and INTEGRAL (Durouchoux & Matteson 1989) are shown.

If a type Ia supernova occurs at a distance of $\lesssim 10$ Mpc, the 847 keV line for all the five models can be detected with the OSSE experiment, while the 812 keV line (^{56}Ni) is detectable only for W7DT, W7DHE and W8DT.

References

Branch D., Tammann G.A., 1992, Ann. Rev. Astr. Ap., in press
Burrows A., Shankar A., Van Riper K.A., 1991, ApJ 379, L7
Filippenko A.V., et al., 1992, ApJ 384, L15
Jeffery D.J., Leibundgut B., Kirshner R.P., Benetti S., Branch D., Sonneborn G., 1992, ApJ 397, 304
Kamae T., 1992, private communication
Kumagai S., Itoh M., Shigeyama T., Nomoto K., Nishimura J., 1988, A&A 197, L7
Kumagai S., Shigeyama T., Nomoto K., Itoh M., Nishimura J., Tsuruta S., 1989, ApJ 345, 412
Kurfess J.D., 1988, In: Nuclear Spectroscopy of Astrophysical Sources, eds. N. Gehrels and G. Share, AIP, New York
Lichti G., et al., 1992, A&AS, in press
Durouchoux P., Matteson J., 1989, In: Supernovae, ed. S.E. Woosley (Springer Verlag), p. 291
Müller E., Höflich P., Khokhlov A., 1991, A&A 249, L1
Nomoto K., Thielemann F.-K., Yokoi K., 1984, ApJ 286, 644
Ruiz-Lapuente P., Cappellaro E., Turatto M., Gouiffes C., Danziger I.J., Della Valle M., Lucy L.B., 1992, ApJ 387, L33
Yamaoka H., Nomoto K., Shigeyama T., Thielemann F.-K., 1992, ApJ 393, L55

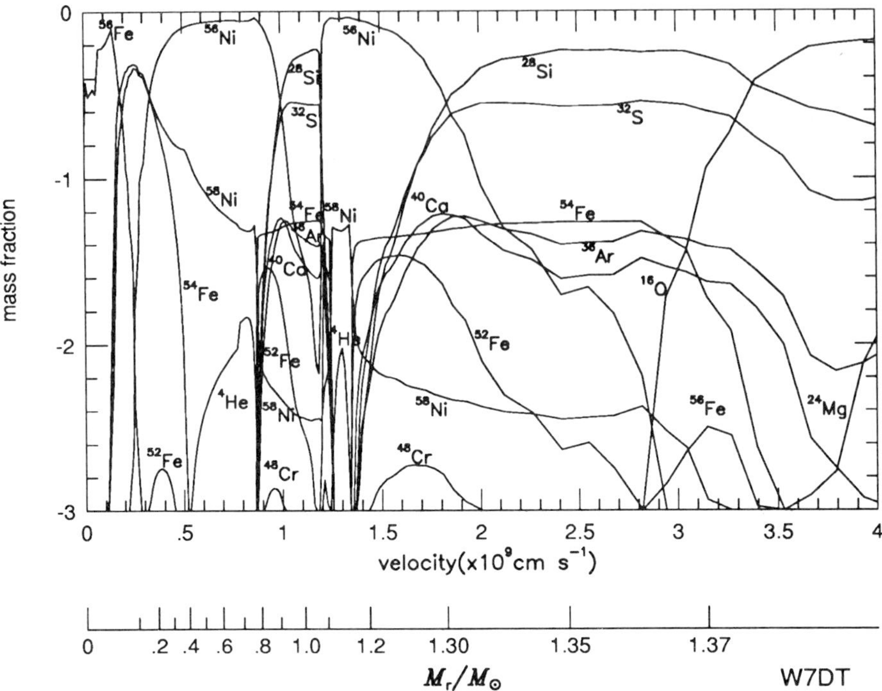

Figure 1. Composition of the late detonation model W7DT as a function of M_r, where the transition from deflagration to detonation takes place at $M_r = 1.13 \ M_\odot$.

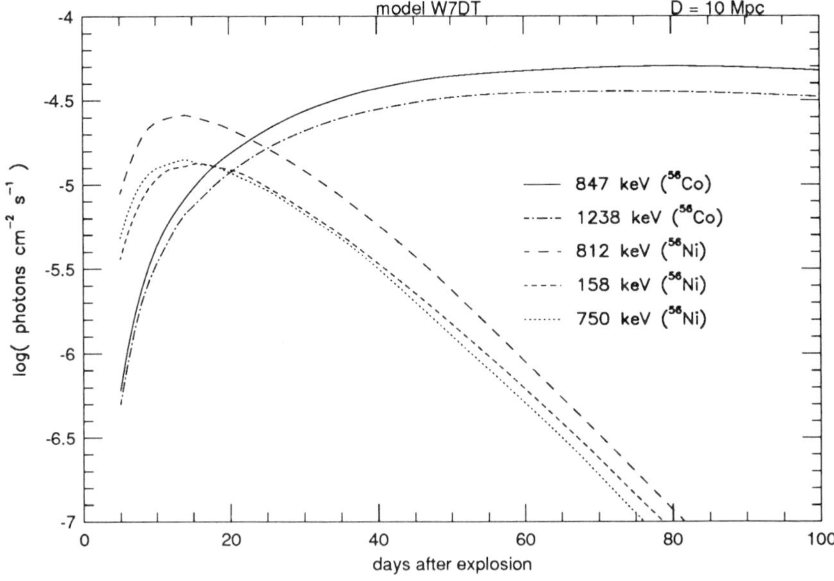

Figure 2. Light curves of γ-ray lines of 812 keV, 158 keV and 750 keV from the ^{56}Ni decay and the 847 keV and 1238 keV lines from the ^{56}Co decay for model W7DT at a distance of 10 Mpc.

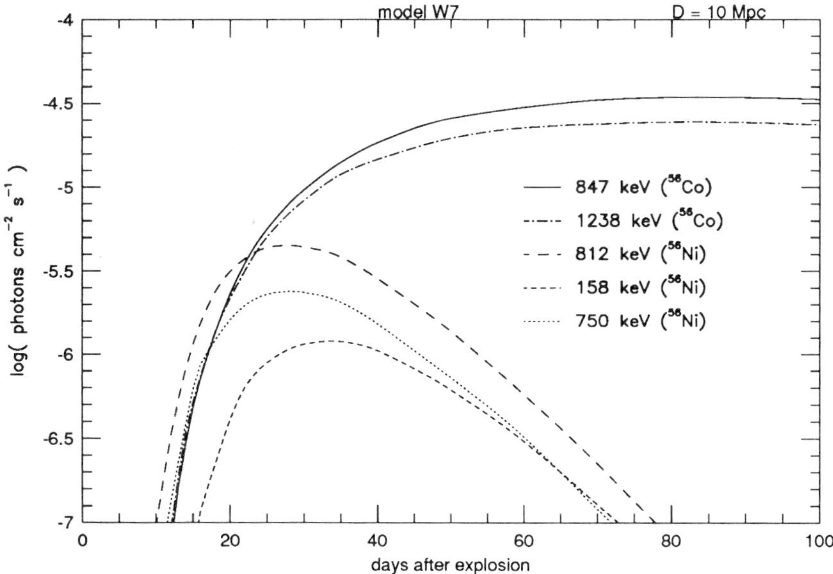

Figure 3. Same as Figure 2 but for model W7.

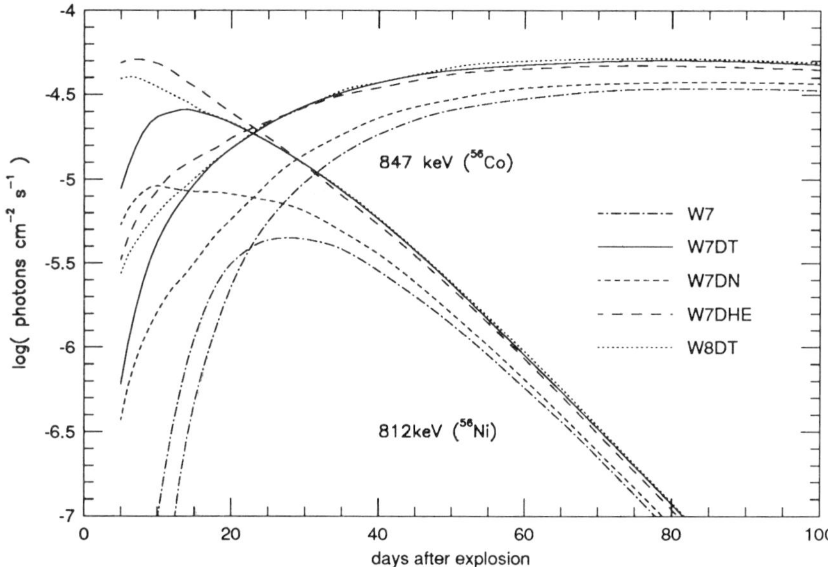

Figure 4. Light curves of γ-ray lines of 812 keV (^{56}Ni) and 847 keV (^{56}Co) for W7 (dash-dotted), W7DT (solid), W7DN (dashed), W7DHE (long-dashed) and W8DT (dotted) at a distance of 10 Mpc.

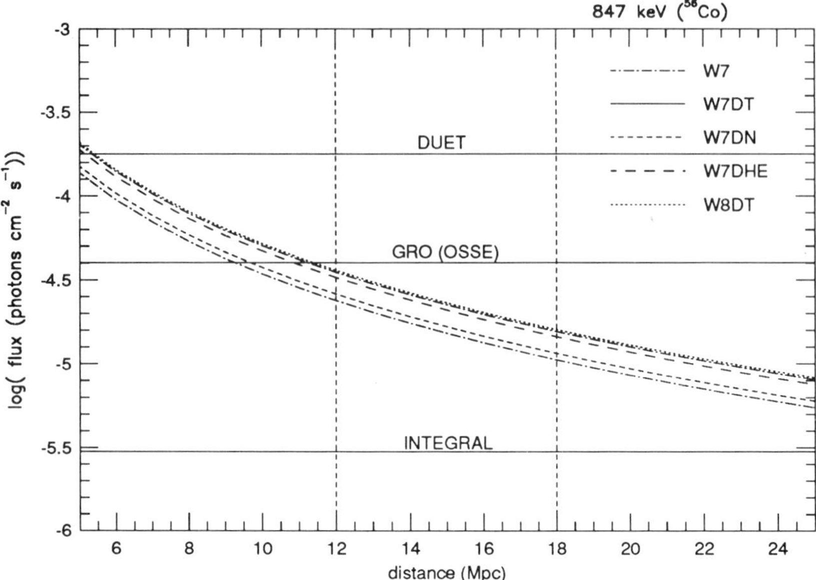

Figure 5. The maximum fluxes of the 847 keV line from the decay of ^{56}Co from models W7, W7DT, W7DN, W7DHE and W8DT as a function of the distance to the type Ia supernova.

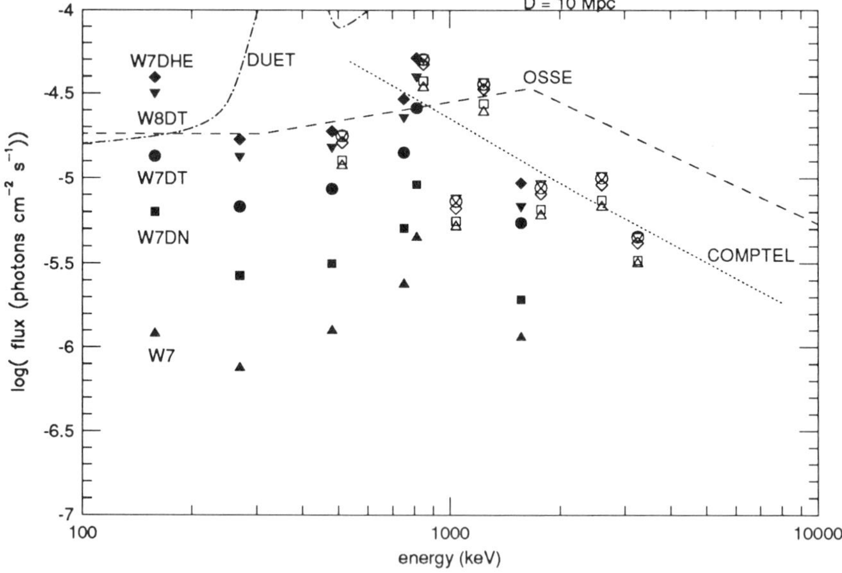

Figure 6. The maximum fluxes of the dominant γ-ray lines from the decays of ^{56}Ni and ^{56}Co for W7, W7DT, W7DN, W7DHE and W8DT at a distance of 10 Mpc as a function of the energy of lines. Filled and open symbols indicate the lines from ^{56}Ni and ^{56}Co, respectively. The line sensitivities of DUET (observing time 10^4 sec; dash-dotted), OSSE (10^6 sec; dashed) and COMPTEL (10^6 sec; dotted) are also shown.

Nucleosynthesis in Type Ia supernovae: effects of non-spherical detonations

Matthias Steinmetz and Ewald Müller

Max–Planck–Institut für Astrophysik, Karl-Schwarzschild-Straße 1,
W-8046 Garching b. München

Abstract. We have investigated the nucleosynthesis during a carbon detonation of a low mass white dwarf ($M \approx M_\odot$) with a central density of $3 \ 10^7$ g/cm^3, which is typical for a Type Ia-progenitor produced by the so-called merging scenario. Nucleosynthesis with an α-network and hydrodynamics are calculated simultaneously. One-dimensional models show, that some of the commonly used approximations for the equation of state, the reaction rates and the time discretization lead to errors in the final composition of the order of 50%. The models fit the observed abundances of intermediate mass elements like S, Si or Ca quite well, but the total mass of the white dwarf is too small. Rotation allows to increase the mass of the white dwarf at fixed density up to the Chandrasekhar mass with only moderate deformation of the star. The resulting relative abundances of the rotating models remain nearly unchanged. The yields increase linearly with the mass of the rotating models. Because the merging scenario implies the possibility of an off-center carbon ignition, we have also studied the propagation and nucleosynthesis of a detonation wave ignited off-center in a spherically symmetric white dwarf. We find an increase of the total production of intermediate mass elements of about 10%, the Ni-production being slightly reduced. The final configuration consists of an almost spherically symmetric Ni-core and a highly asymmetric envelope composed of intermediate mass elements. The asymmetric density, velocity and composition distribution of the ejecta will probably influence observable quantities like the linear polarization and may give rise to a direction dependent luminosity.

1. Introduction

The currently favoured model of Type Ia supernovae (SNIa) is a thermonuclear explosion of a near Chandrasekhar mass carbon oxygen white dwarf. Whether the explosion is mediated by a supersonic burning front as in the detonation model of Arnett (1969), by a subsonic flame as in the deflagration model of Nomoto et al. (1976), or by a combination of both in the so-called delayed detonation model proposed by Khokhlov (1991, see also Woosley 1992) is still a matter of controversial debates.

Among others the strongest argument against the detonation model is its failure to produce intermediate mass elements (IMEs) like Si, S and Ca which are seen in SNIa spectra (Branch et al. 1985). As shown by Imshennik and Khokhlov (1984), this failure is not caused by the fact the explosion being a detonation, because it is also possible to produce IMEs in detonations if the density of the fuel is less than a few 10^7 g/cm^3. However, in the C/O-detonation model of Arnett (1969), the SNIa-progenitor has typical central densities of a few times 10^9 g/cm^3, i.e. only a negligible amount of IMEs are produced. As shown in a recent study (Steinmetz et al. 1992, henceforth SMH92) the amount of mass in the low density regime can be increased drastically by rotation. On the other hand, the total mass of the star grows beyond 2 M$_\odot$, and all the mass will be burned during a detonation. To get a reliable amount of IMEs ($\approx 0.3 - 0.5$ M$_\odot$), an unacceptable large overproduction of ^{56}Ni would result.

In the following we consider the nucleosynthesis in a low mass white dwarf with a central density of a few 10^7g/cm^3, which is typical for SNIa progenitors in the so-called merging scenario. The results can also partially be applied to the detonation phase in the delayed detonation model.

2. Numerical method

As in SMH92 the hydrodynamic equation are integrated by the piecewise-parabolic-method (PPM) of Colella and Woodward (1984). To describe the transition between incomplete burning, which results in the production of IMEs, and complete burning into iron group elements (IGEs) an α-network is solved simultaneously. The reaction rates were taken from the library of Thielemann (Thielemann et al. 1986 and private communication), for more details, see Steinmetz and Müller (1992, in preparation). Solving the nuclear network the following problems must be addressed:

- If one considers only the (α, γ) reactions between the different α-nuclei, a white dwarf of about 1 M$_\odot$ produces $\approx$0.25 M$_\odot$of IGEs and $\approx$0.75 M$_\odot$ IMEs. However, the combination of an (α, p)-reaction followed by a (p, γ)-reaction is often faster than the (α, γ)-reaction. Taking this into account in the network, the production of IMEs reduces to about 0.3 M$_\odot$.

- A further problem is related to the effects of electron screening. We have simulated electron screening by increasing the reaction rates beyond Si by a factor of 6 and 30, respectively. The changes in the yields are less than 1%.

- The equation of state (EOS) used in our calculations includes contributions from 13 ideal Boltzmann gases, radiation and an arbitrarily degenerate and arbitrarily relativistic electron-positron gas. Because the evaluation of the EOS is quite expensive, some authors approximate the electron contribution by a completely degenerate gas (e.g. Benz et al. 1989). This approximation is quite good in the radiation dominated regime ($T > 7\ 10^9$ K) and for low temperatures ($T < 10^9$ K).

In the intermediate regime, the error in energy at a given temperature is of the order of 10%, which appears acceptable. However in a hydrodynamic calculation, temperature is determined from energy and density. Because the energy depends only weakly on temperature this 10% error in the energy or pressure determination results in a temperature error of 50% or even more. Because the reaction rates depend exponentially on temperature, a large error in the yields can occur. Indeed, a test calculation with the completely degenerate EOS gives only 0.11 $M_\odot$ of IMEs, i.e. a factor of three less than the correct value.

- The reaction rates depend exponentially on temperature and at least quadratic on composition. Therefore, during a time step changes in composition and temperature have to remain small. We find that a composition change of 5-10% during a timestep is acceptable. The changes in temperature must be less than 1% to get yields accurate within 1%. Allowing for a 3% change in temperature increases the error to more than 10%.

- All calculation were performed with a 1D-PPM version with 500 grid points. Calculations with 2000 grid points agree within less than 1%.

3. Spherically symmetric models

The initial hydrostatic models are constructed with a completely degenerate, arbitrarily relativistic electron gas for different central densities between 10^7 g/cm^3 and $1.2\ 10^8$ g/cm^3. The detonation is induced by raising the central temperature to $2\ 10^9$ K, which results in the formation of a detonation wave (see SMH92). The explosion is followed until homologeous expansion is reached and all burning processes are quenched. The yields of the different elements are shown in the following table, where mass and yields are given in $M_\odot$, the central densities ϱ_c in 10^7 g/cm^3.

ϱ_c	M_{tot}	He	C	O	Ne	Mg	Si	S	Ar	Ca	Fe	Ni
1	0.821	0	0.013	0.160	0.006	0.004	0.200	0.192	0.059	0.063	0.006	0.113
2	0.947	0.001	0.004	0.084	0.002	0.001	0.144	0.143	0.046	0.055	0.011	0.455
4	1.063	0.009	0.002	0.035	0.001	0	0.095	0.094	0.029	0.038	0.008	0.751
8	1.160	0.012	0.001	0.014	0	0	0.055	0.054	0.017	0.023	0.006	0.977
12	1.238	0.017	0	0.005	0	0	0.029	0.028	0.009	0.012	0.008	1.128

The yields for Ti and Cr are negligible. One can see that even the most massive model has a mass still considerably less than the Chandrasekhar mass. In the first model, a remarkable amount of unburned C and O is left over and some Ne and Mg is produced, while most of the mass ends up as IMEs from Si–Ca. As expected, the production of IGEs increases, that of IMEs decreases with increasing central density or total mass. Note, that with increasing mass the production of He increases, too, which is due to photodesintegration of Ni in the dense inner layers with temperatures beyond $5\ 10^9$ K. In the most massive model, nearly all carbon/oxygen is burned into IGEs. Observed SNIa spectra imply, that the velocity of IMEs is in the range 10000 to 25000 km/sec, which cannot be reproduced by the two more massive models, where the IMEs expand too fast. On the other side, the mass of the first three models is too small, i.e. the light curve rises too fast for masses below 1.2 $M_\odot$ (Khokhlov et al. 1992). Furthermore, the production of Ni in the first model is too small to power a SNIa light curve.

4. The influence of rotation

In the case of rotation and especially in the case of differential rotation, the total mass of a degenerate configuration can be increased for a fixed central density by up to 30% with only a moderate amount of deformation. Most of this additional mass is located in the low density outer regions of the star (SMH92). Therefore one would expect at least a linear increase of the yields of IMEs with the mass of the white dwarf.

From the dynamical point of view the behaviour of the detonation is similar to that in the more massive rotating white dwarfs investigated in SMH92: While the detonation wave is of spherical shape, until burning becomes quenched in the low density layers close to the surface ($\varrho < 10^5$ g/cm^3), the remaining shock wave shows remarkable asymmetries. This is in accordance with the theory of Chapman and Jouguet (see e.g. Courant and Friedrichs 1948); for a more detailed discussion of this aspect we refer to SMH92.

In the table below the relative abundances of the rotating (R) and the non-rotating model (S) are quite similar, i.e. the yields roughly scale with the mass of the star. When the total mass increases by 14%, the amount of synthesized IGEs, IMEs by and unburned matter increases by 15%, 13% and 7%, respectively. This shows again, that dynamics effect the nucleosynthesis only weakly, the yields being mainly determined by the density stratification of the initial model. However, as the deformation of the investigated models is only moderate, it is unclear whether this result also holds for highly deformed objects.

5. Off-center detonation

According to the merging scenario, the carbon fuel can ignite off-center (Mochkovitch and Livio 1989). In this case, part of the detonation wave has to propagate up the density gradient. Moreover, in the centrally ignited models an already well developed detonation wave propagates into the low density regime, whereas in the off-center case the detonation first must build up, i.e. one can expect that the wave is not so strong as in the centrally ignited case. Both effects do increase the production of IMEs. Although the merging scenario, in principle, requires 3D simulations, a simplified but still at least qualitatively correct picture of the dynamics may be obtained by considering an off-center detonation in a spherically symmetric white dwarf. Then, the axis through the stellar center and the ignition point defines a symmetry axis, i.e. the problem is reduced to a 2D one. The yields of an off-center detonation in a white dwarf with a central density of 3 10^7 g/cm^3 are shown in the table below (Model O). The radius of the ignition point is 1/3 stellar radius.

	M_{tot}	He	C	O	Ne	Mg	Si	S	Ar	Ca	Fe	Ni
S	1.023	0.006	0.003	0.050	0.001	0.001	0.111	0.112	0.036	0.046	0.011	0.645
R	1.162	0.008	0.002	0.053	0.001	0	0.129	0.127	0.040	0.050	0.012	0.740
O	1.023	0.006	0.004	0.047	0.002	0.001	0.114	0.116	0.039	0.052	0.011	0.630

As one can see, the results are only slightly different from the centrally ignited model (S), the yields of Ni, S and Si being changed only by a few percent. The maximum effect (12%) occurs for the elements Ar and Ca. With increasing radius of the ignition point, first Ar and Ca, then Si and S and finally C and O will be enhanced, because with increasing distance density and temperature behind the shock decrease, i.e. the burning becomes less and less complete.

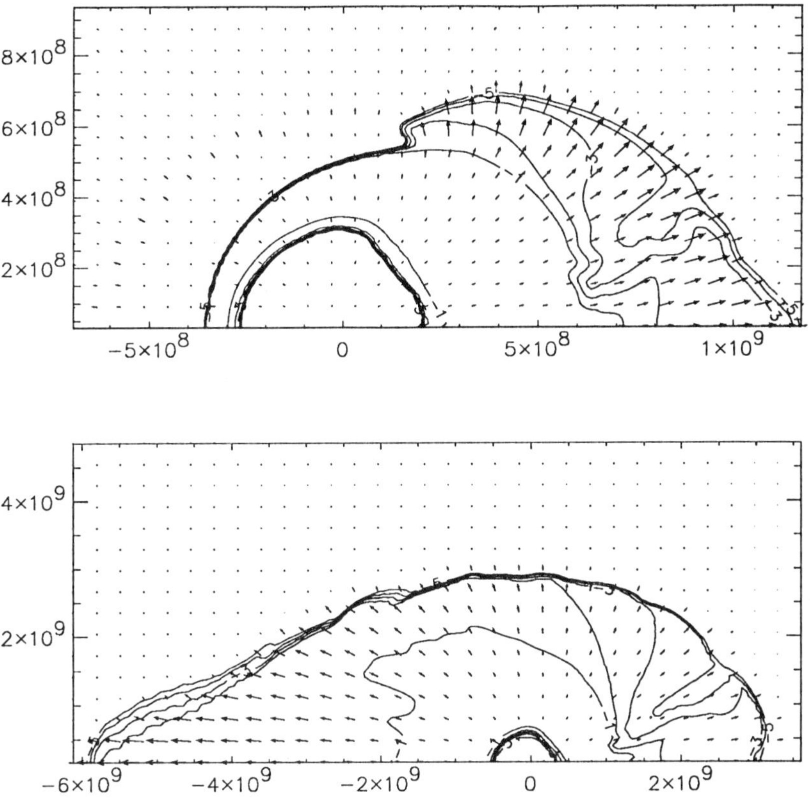

Figure 1.

Velocity field and distribution of the IMEs for an off-center carbon detonation. Snapshots are taken at 0.5 (top) and 2.3 sec (bottom). The contours correspond to a mass fraction of 10^{-5}, 10^{-4}, 10^{-3}, 10^{-2} and 10^{-1}, respectively.

Of more interest is the dynamical evolution of the off-center detonation and the spatial distribution of the elements. In the figure above, both contour plots show the composition of the IMEs together with the flow pattern. The first snapshot is taken about 0.5 sec after the ignition of the detonation, the second one at $t = 2.3$ sec. One can see that the detonation wave first reaches the surface in the hemisphere containing the ignition point. The shock waves propagating around the center of the star in both directions collide in the hemisphere opposite to the one where the detonation started. Due to energy and momentum conservation the material is then highly accelerated and ejected. For a more detailed discussion of this mechanism see Steinmetz and Höflich (1992).

The abundance distribution in the inner layers of the star is quite spherical, while the distribution in the outer layers is highly anisotropic. The density stratification shows a similar behaviour. Due to the geometrical dilution of the expanding ejecta, the early light curve of the supernovae is dominated by emission from matter in the highly anisotropic outer layers giving rise to a direction dependent luminosity and causing linear polarization. At late epochs, one looks deeper and deeper into the expanding object, i.e. the light curve is mainly due to emission from matter in the spherically symmetric central layers and the linear polarization vanishes (Höflich and Steinmetz 1991, Steinmetz and Höflich 1992).

6. Conclusions

We have performed 2D simulations of carbon detonations in white dwarfs of low central densities to study the effect of non-spherical detonations onto the nucleosynthesis in Type Ia supernovae. A couple of 1D calculations show that one has to check carefully the physical and numerical approximations used in the calculations. Because of the exponential dependence of the reaction rates on temperature small errors can lead to large errors in the yields. The investigation of detonations in rotating stars and of off-center detonations show that the nucleosynthesis is only slightly modified by the non-spherical dynamics. However, especially in the off-center case the distribution of the elements is highly anisotropic which probably has observable consequences, e.g. linear polarization or a direction dependent luminosity.

References

Arnett W D 1969 *Astrophys. Space Sci.* **5** 180

Benz W, Hills J G and Thielemann F K 1989 *Astrophys. J.* **348** 647

Branch D, Dogget J B, Nomoto K and Thielemann F K 1985 *Astrophus. J.* **294** 614

Colella P and Woodward P R 1984 *J. Comp. Physics* **54** 174

Courant R and Friedrichs K O 1948 *Supersonic Flow ans Shock Waves* (Interscience, New York)

Höflich P and Steinmetz M 1991 *Proceedings of the 6th Workshop on Nuclear Astrophysics, held at Ringberg, February 1991* (Proceedings MPA/P5) P 103

Imshennik V S and Khokhlov A M 1984 *Sov. Astron. Lett.* **10** 262

Khokhlov A M 1991 *Astron. Astrophys.* **245** 114

Khokhlov A M, Müller E and Höflich P 1992 Astron. Astrophys. **253** L9

Mochkovitch R and Livio M 1989 Astron. Astrophys. **209** 111

Nomoto K, Sugimoto D and Neo S 1976 *Astrophys. Space Sci.* **39** L37

Steinmetz M and Höflich P 1992 *Astron. Astrophys.* **257** 641

Steinmetz M, Müller E and Hillbrandt W 1992 *Astron. Astrophys.* **254** 177

Thielemann F K, Arnould M and Truran J W 1986 in *Advances in Nuclear Astrophysics* (eds. Vangioni-Flam E, Audouze J, Casse M, Chieze J P and Tran Than Van, Editions Frontiers, Gif sur Yvette, France) pp 525-540

Woosley S E 1992 in *Proceedings of the Les Houches Summer School on Supernovae* (eds. Audouze J, Bludman S, Mochkovitch R and Zinn-Justin J, Elsevier, in press)

Nucleosynthesis in Type Ia supernovae

Nucleosynthesis in type Ia supernovae

J. Isern[1,2], E. Bravo[3,2], R. Canal[4,2], J. Labay[4,2]

1) Centre d'Estudis Avançats Blanes (CSIC) Spain
2) Laboratori d'Astrofísica (IEC) Spain
3) Departament de Física i Enginyeria Nuclear (UPC) Spain
4) Departament d'Astronomia i Meteorologia (UB) Spain

Abstract. The thermonuclear deflagration of an accreting white dwarf in a close binary system is commonly thought to be at the origin of Type Ia supernovae. This model is characterized by the overproduction of species like ^{54}Fe, ^{58}Ni and ^{54}Cr which seem incompatible with the Solar System abundances. We show here that taking into account the detailed physical properties of white dwarf interiors during the cooling prior to the explosion and the contribution of Type II supernovae, this problem disappears or is strongly reduced.

1. Introduction

There is nowadays a wide consens that Type Ia supernovae (SNIa) are the result of the thermonuclear deflagration of a white dwarf star accreting matter from its companion in a binary system. In this scenario, the white dwarf made of C and O approaches to the Chandrasekhar's mass and ignites its fuel under degenerate conditions. The density at which the ignition happens, $2\,10^9 \leq \rho \leq 10^{10}$ g/cm^3 (Hernanz et al. 1988) depends on the accretion rates, chemical composition of the accreted matter and temperature and mass of the white dwarf when accretion starts. These parameters depend on the particular binary system considered (two white dwarfs, a white dwarf plus a helium star or a white dwarf plus a red giant star) which, in turn, depends on the masses and orbital parameters of the primeval binary system. All these binary systems have to face several objections (Canal 1992) and there is not a general agreement on which is the best one, although the merging of two white dwarfs seems to be the must widely accepted.

Once the thermonuclear runaway has started, it must propagate through a significant part of the star in order to obtain the energy necessary to power the explosion. The properties of the burning front are not known at present. On theoretical grounds and in the lack of 3-D numerical simulations it is impossible to predict the correct behavior of the flame and the only way of getting insight on the problem is to constrain it from the observation of the phenomenon (light curve, expansion velocity, spectrum...) and from its impact on the chemical evolution of the galaxy.

One of the last kind of constraints is the production of neutron rich isotopes in the region of iron peak elements. In the context of the standard deflagration model, the electron captures on the incinerated material plus the neutron content stored in ^{22}Ne nuclei lead to the overproduction of these isotopes in a quantity that is clearly incompatible with the solar abundances (Thieleman et al. 1986, Woosley 1992) of such elements.

Recently it has become clear that Type II and Type Ib also contribute to the iron content of the Galaxy. In the first case, SN1987A has proven the injection of $0.07 M_\odot$ of iron per event, and in the second case, SN1983N has proven the injection of $0.3 M_\odot$ per event (Graham 1985), quantities that compare to the $0.6 M_\odot$ of iron per event predicted by the best SNIa models. If it is accepted that the rates of SNIa are considerably smaller, roughly a factor 10 (van den Bergh & Tamman 1991), than the combined rates of SNII and SNIb/c, the contribution of the former ones to the present rate of increase of the iron content of the galaxy would only be 1/3. The purpose of this paper is to estimate the contribution of Type Ia supernovae to the galactic evolution of iron peak elements and to examine the constraints introduced to the SNIa models.

2. Theoretical models

2.1. SNIa models

In order to compute the yields of the iron peak elements, we have constructed several models of SNIa explosion. Our calculations have started from an isothermal degenerate core with $\rho_c = 4\,10^9$ g/cm^3, $T = 5.5\,10^7$ K and equal abundances of carbon and oxygen by mass and a small quantity ($X_{\mathrm{Ne}} = 0.01$) of ^{22}Ne. This ^{22}Ne was not uniformly distributed through all the star but concentrated in the central regions due to a process of sedimentation induced by crystallization (Bravo et al. 1992). The hydrodynamic equations have been solved in the way described by Bravo et al. (1992) and the propagation of the deflagration was computed adopting the prescriptions proposed by Sutherland and Wheeler (1984) with a value of the free parameter $\alpha = 0.7$ (models labelled R). Models labelled D were designed to allow the burning front to propagate as a deflagration in the central regions (at $v_{\mathrm{def}} = 0.03c$) and as a detonation for densities greater than $7\,10^7$ g/cm^3. Table 1 displays the yields, obtained for both kind of models, of the Fe-group elements.

Table 1. $[^{A}\mathrm{Z}/^{56}\mathrm{Fe}] \equiv log\left[X(^{A}\mathrm{Z})/X(^{56}\mathrm{Fe})\right] - log\left[X_\odot(^{A}\mathrm{Z})/X_\odot(^{56}\mathrm{Fe})\right]$

^{A}Z	R	D	^{A}Z	R	D	^{A}Z	R	D	^{A}Z	R	D
^{46}Ti	-2.38	-1.64	^{52}Cr	0.03	0.56	^{58}Fe	0.44	0.97	^{63}Cu	-1.07	-1.32
^{47}Ti	-3.10	-2.46	^{53}Cr	0.12	0.54	^{59}Co	0.09	-0.39	^{65}Cu	-1.45	-1.58
^{48}Ti	-0.78	0.01	^{54}Cr	1.33	1.78	^{58}Ni	0.73	0.20	^{64}Zn	-0.63	-1.78
^{49}Ti	-0.23	0.07	^{55}Mn	0.23	0.32	^{60}Ni	0.48	-0.09	^{66}Zn	0.34	-0.58
^{50}Ti	1.02	1.50	^{54}Fe	0.47	0.45	^{61}Ni	-0.15	-0.56	^{67}Zn	-0.91	-1.09
^{50}V	-1.45	-1.24	^{56}Fe	0.00	0.00	^{62}Ni	0.53	0.54	^{68}Zn	-1.12	-2.69
^{51}V	-0.30	0.31	^{57}Fe	0.21	-0.07	^{64}Ni	0.30	0.65	^{70}Zn	-0.99	-1.54
^{50}Cr	-0.37	0.14									

2.2. Chemical evolution models

The following approximations have been made (Abia et al. 1991): 1) Instantaneous recycling has not been considered. The star lifetimes, τ_M, were taken from Talbot & Arnett (1971). 2) The star formation rate $\Psi(t)$ is related to the surface gas density via $\Psi(t) = \alpha \sigma_g^n(t)$, with $\alpha = 0.6$ Gyr^{-1} and n=1. An exponentially decreasing unenriched infall with a e-folding time of 5 Gyr has been included in such a way that the present accretion rate would be $f \lesssim 1.0$ M$_o dot$pc^{-2}Gyr^{-1} (Mirabel & Morras 1984) and the present total surface mass density $\sigma_T \simeq 70$ M$_\odot$pc^{-2} (Bahcall & Soneira 1984). 3) The adopted Initial Mass Function (IMF), taken constant in space and time, was from Scalo (1986).

Concerning the supernova scenarios, we adopted the idea that SNIb/c and SNII are a consequence of the gravitational collapse of the iron core of massive stars (the former ones due to the explosion of Wolf-Rayet stars), and that Type Ia are a consequence of the merging of two white dwarfs. Of course, this scenario is not complete, but it is enough for our purposes. Supernova rates have been obtained from Van den Bergh & Tamman (1991) assuming that the Milky Way is a Sbc galaxy with $L_B = 2\,10^{10}$ L$_\odot$ and that $H_0 = 65$ km/s/Mpc. They are: $R_{II} = 3.32\,10^{-2}$ yr^{-1}, $R_{Ib/c} = 0.65\,10^{-2}$ yr^{-1}, $R_{Ia} = 0.41\,10^{-2}$ yr^{-1}. Since the properties of supernova explosions, both gravitational and thermonuclear, have not been yet settled, the yields of chemical elements are rather uncertain. It seems reasonable, however, to adopt a characteristic average of 0.1, 0.3 and 0.6 $M_\odot$ of iron per SNII, SNIb/c and SNIa per event respectively.

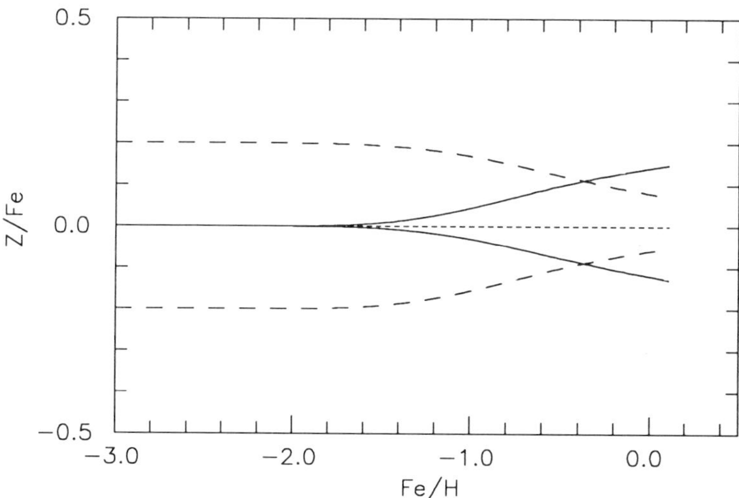

Figure 1. Evolution of [Z/Fe] vs. metallicity for different yields of SNIa and gravitational collapse SN. Solid lines: [Z/Fe] = -0.2 and 0.2 in SNIa, and 0.0 in SNIb and SNII. Dashed lines: [Z/Fe] = -0.2 and 0.2 in SNIb and SNII, and 0.0 in SNIa.

With these parameters the main characteristics (age-metallicity relationship, current fraction of gas, metallicity in G-dwarfs and galactic evolution of O and Fe) are well reproduced.

3. Results and discussion

Figure 1 shows the behavior of the solutions as a function of the adopted yields of iron peak isotopes. From this behavior it is clear that the abundances of iron peak elements in extremely metal poor stars are entirely due to gravitational supernovae. These family of solutions have to be compared with the isotopic abundances in the Solar System, the elemental abundances in the stellar atmospheres and to the yields provided by supernova theory.

Recently, Gratton & Sneden (1988, 1991) have shown that Cr follows the Fe abundances and that Co and Ni are slightly underabundant in extremely metal poor stars, i.e. $< [Cr/Fe] >= -0.04 \pm 0.05$, $< [Co/Fe] >= -0.12 \pm 0.05$, and $< [Ni/Fe] >= -0.04 \pm 0.08$ in the region $-3 \lesssim [Fe/H] \lesssim -1$. These values immediately suggest that gravitational supernovae underproduce cobalt and niquel and that, consequently, the thermonuclear ones must produce an excess of such elements. In the context of the model of chemical evolution of the galaxy proposed here, these observational data can be fitted at a 1σ level if the yields of gravitational and thermonuclear supernovae follow the values quoted in Table 2. Table 3 displays the same values obtained theoretically by several authors and in this paper. Before comparing Tables 2 and 3 it is necessary to take into account that the results concerning thermonuclear supernovae are strongly influenced by the uncertainties on the absolute yields and supernovae rates. For instance, if the iron yield of thermonuclear supernovae is in the range of 0.40 to 0.74 $M_\odot$ and that of gravitational supernovae in the range of 0.04 0.1 $M_\odot$, the fitted ratios for thermonuclear supernovae can take values on the intervals: $-0.01 \leq [Cr/Fe] \leq 0.47$, $-0.02 \leq [Mn/Fe] \leq 0.57$, $-0.54 \leq [Co/Fe] \leq 0.34$ and $0.05 \leq [Ni/Fe] \leq 0.61$, which cover almost all the figures quoted in Table 3 for thermonuclear models. The situation is very different for the case of gravitational supernovae since these factors only introduce a range of variation of ± 0.01. That implying that the average yield of iron peak elements in gravitational supernovae is strongly constrained by the observation of these elements in metal-poor stars, and no one of the models quoted in Table 3 can account for the observations.

Table 2. Abundance ratios necessary to fit the observed abundances of the iron peak elements

	gravitational SN	thermonuclear SN
[Cr/Fe]	-0.11 ± 0.01	0.08 ± 0.09
[Mn/Fe]	-0.30 ± 0.04	0.08 ± 0.07
[Co/Fe]	-0.09 ± 0.06	-0.06 ± 0.13
[Ni/Fe]	-0.07 ± 0.02	0.17 ± 0.10

Table 3. Abundances of iron peak elements compared to Fe for gravitational and thermonuclear supernovae

	gravitational SN					thermonuclear SN			
	a	b	c	d	e	f	g	h	i
[Cr/Fe]	-0.03	0.02	-0.11	0.04	-0.54	-0.23	0.43	0.12	0.61
[Mn/Fe]	-0.50	-0.39	-0.44	-0.52	-0.86	-0.05	-0.05	0.16	0.25
[Co/Fe]	-0.05	-0.32	-0.23	-0.15	-0.27	-0.41	-0.42	0.02	-0.45
[Ni/Fe]	0.62	0.25	0.01	-0.16	0.30	0.55	0.11	0.59	0.10

a) Thielemann et al. 1990 ($M_{cut} = 1.59\ M_\odot$)
b) Thielemann et al. 1990 ($M_{cut} = 1.63\ M_\odot$)
c) Woosley & Pinto 1988 ($M = 18\ M_\odot$)
d) Woosley & Pinto 1988 ($M = 20\ M_\odot$)
e) Woosley & Weaver 1982 ($M = 25\ M_\odot$)
f) Thielemann et al. 1986
g) Khokhlov 1991 (model N21)
h) This work, model R
i) This work, model D

4. Conclusions

A simple model of the chemical evolution of the galaxy has been constructed and the constraints imposed by observations of the chemical abundances in stars to the models of Type Ia supernovae have been examined. The results show that, within this context, all the models of Type Ia supernovae are compatible with those observations and that there is no reason based on nucleosynthesis to look for an ad hoc behaviour of the burning front to explain the observations. By contrast, if the model is correct, the current models of Type II supernovae do not provide satisfactory results and must be revised.

Acknowledgments

This work has been supported in part by the CICYT grants PB87-0304 and PB90-0912, by the PICS 114 of the CSIC, and by an EASI/CESCA project.

References

Abia C., Canal R., Isern J., 1991, ApJ 366, 198
Bahcall J.N., Soneira R.M., 1984, ApJS 55, 67
Bravo E., Isern J., Canal R., Labay J., 1992, A&A 257, 534
Canal R., 1992, in *Proc. Les Houches School on Supernovae*, in press

Graham J.R., Meikle W.P.S., Allen D.A., Longmore A.J., Williams P.M., 1985, MN-RAS 218, 93

Gratton R.G., Sneden C., 1988, A&A 204, 193

Gratton R.G., Sneden C., 1991, A&A 241, 501

Hernanz M., Isern J., Canal R., Labay J., Mochkovitch R., 1988, ApJ 324, 331

Khokhlov A.M., 1991, A&A 245, L25, and 245, 114 (g)

Mirabel I.F., Morras R., 1984, ApJ 279, 86

Scalo J.M., 1986, Fundamentals of Cosmic Physics 11, 1

Sutherland P.G., Wheeler J.C., 1984, ApJ 280, 282

Talbot R.J., Arnett W.D., 1971, ApJ 170, 409

Thielemann F.K., Nomoto K., Yokoi K., 1986, A&A 158, 17 (f)

Thielemann F.K., Nomoto K., Hashimoto M., 1990, in *Chemical and Dynamical Evolution of Galaxies*, eds. F. Ferrini, J. Franco, F. Matteucci (a,b)

Van den Bergh S., Tamman G.A., 1991, ARA&A 29, 363

Woosley S., 1992, in *Proc. Les Houches School on Supernovae*, in press

Woosley S.E., Weaver T.A., 1982, in *Essays in Nuclear Astrophysics*, eds. C.A. Barnes, D.D. Clayton, D.N. Schramm (Cambridge University Press: Cambridge), p. 377 (e)

Woosley S., Pinto P.A., 1988, Proc. A.S.A. 7, 4 (c,d)

The p-process and Type Ia supernovae

W. M. Howard[†] and B. S. Meyer[‡]

[†]Institut d'Astronomie et d'Astrophysique Université de Libre Bruxelles
Brussels, Belgium

and

[‡]Department of Physics and Astronomy Clemson University Clemson, SC 29634 USA

Abstract: We study p-process nucleosynthesis in the context of a delayed-detonation model for Type Ia supernovae. In particular, we use the model of Khokhlov to study the nucleosynthesis that occurs on the surface (the outer few hundredths of a solar mass) of a carbon-oxygen white dwarf that undergoes disruption by the delayed detonation mechanism. Under such circumstances, most of the mass of the white dwarf reaches nuclear statistical equilibrium (NSE) and produces elements near the iron peak. However, on the surface of the white dwarf the peak temperatures during the disruption are not high enough for the nuclei to attain NSE, although significant transmutation of heavy elements into p-process nuclei can occur. We find that the heavy element surface composition of the white dwarf before the explosion has important implications for the subsequent p-process nucleosynthesis. We also discuss the production of the interesting p-process chronometers ^{92}Nb, ^{97}Tc, ^{98}Tc and ^{146}Sm.

1   Introduction

The p-process is the conversion via photodisintegration reactions and proton capture reactions of s- and r-process nuclei into the proton-rich p-nuclei (Woosley and Howard, 1978; and recent review by Lambert, 1992). Unlike the r- and s-processes, which occur in very special astrophysical circumstances, the p-process occurs in many places in nature. For example, massive stars undergoing quiescent oxygen burning, convert heavy elements into p-nuclei through a series of photodisintegration reactions (Arnould, 1976). Almost all massive stars undergoing Type II supernovae explosions also produce many p-nuclei. In this paper, we will discuss the contribution of Type Ia supernovae to the Galactic content of p-nuclei. The p-process occurs at a rather low entropy (entropy per baryon of 1), when the astrophysical timescale is less than that required for the nuclei to attain nuclear statistical equilibrium (NSE) (Meyer, 1992). In other words the break-down by photodisintegration reactions of the s- and r-process nuclei into iron is not complete. In such a case nuclei such as ^{144}Sm and ^{124}Xe and ^{126}Xe are produced in great abundance.

The problem for the p-process is to identify those astrophysical sites that are major contributors to the Galactic abundance of the p-nuclei. For example, it is known that Type II supernovae can be major contributors to the bulk of the p-nuclei (Woosley and Howard, 1978; Rayet, Prantzos and Arnould, 1990; Prantzos, Hashimoto, Rayet and

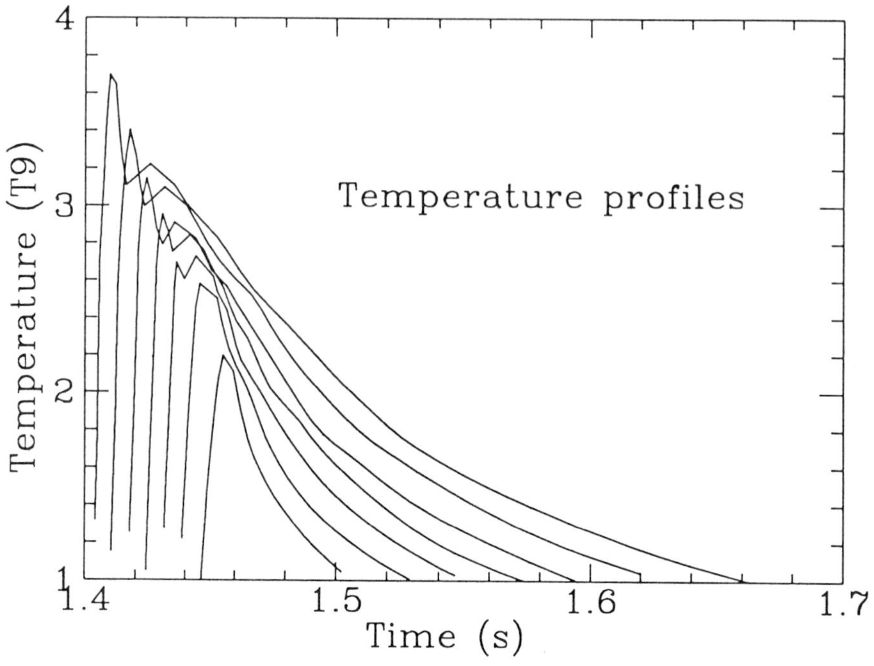

Figure 1: Temperature profiles in Khokhlov's model N21

Arnould, 1991). However, the major problem for these models is the underproduction of the low mass p-nuclei: ^{74}Se, ^{78}Kr, ^{84}Sr, ^{92}Mo, ^{94}Mo, ^{96}Ru and ^{98}Ru. Although these nuclei are rather few in number, they are the most abundant p-nuclei and pose a series challenge for p-process theory. As pointed out by Woosley and Howard (1978) there are insufficient seeds to produce ^{92}Mo by photodisintegration reactions alone on a solar-system like distribution. In addition, the proton fluences during Type II supernovae explosions are too low to induce significant proton captures which may contribute to the production of ^{92}Mo.

It is in this context that we consider Type Ia supernovae. Howard, Meyer and Woosley (1991, hereafter HMW) considered the p-process in Type Ia supernovae, but in a simple parametrized context. That is, they assumed an initial uniform density of $\rho = 10^7 g/cm^3$, an e-folding timescale of 0.6 s for the temperature and a uniform mass distribution in peak temperature. In this paper, we study p-process nucleosynthesis in a detailed supernova model. Khokhlov (1991) has calculated a series of delayed detonation models for Type Ia supernovae. In the outer zones of these models NSE is not attained and intermediate mass elements are produced. Moreover, intermediate mass elements are observed in the spectra of Type I supernovae. Figure 1 shows temperatures as a function of time in some of the outer 15 zones of model N21 by Khokhlov (1991). Time is measured from the time that the deflagration wave starts in the center of the star. These zones have the proper thermodynamic histories to produce p-process elements. We notice that the timescale for an e-folding of the temperature is approximately 0.2 s, which is much shorter than assumed by HMW. The outer 15 zones compries about 0.05 solar masses. The largest zone has a mass of 0.008 solar masses, while the smallest zone has a mass of 0.001 solar masses.

There are two reasons why we consider Type Ia supernovae to produce the low mass p-nuclei. First, it offers the possibility for high proton fluences (HMW) from carbon burning at relatively high densities in the outer zones to produce the p-isotopes of Se, Sr, Kr, Mo and Ru via proton-induced reactions. Second, the surface of the white dwarf has the possibility to be enriched in s-process seeds through accretion from an Asymptotic Giant Branch (AGB) companion, or through having produced s-process elements in thermal instabilities on its surface. The s-process composition on the surface of a white dwarf is rather uncertain, but both possibilities have important consequences for the p-process.

2   Results

We calculate the p-process yields by the method described earlier (HMW). For an initial solar-system like composition (Anders and Grevesse, 1989), the result for the p-process yields, mass-averaged over the outer 15 zones of Khokhlov's model is shown in Figure 2. We plot only the p-nuclei and the overproduction factors are relative to solar-system abundances. The overproduction trends with respect to mass number are similar to those observed in Type II model calculations (Prantzos, Rayet and Arnould, 1991), except that the yields of ^{92}Mo, ^{94}Mo, ^{96}Ru and ^{98}Ru are somewhat higher in our calculation. They are not high enough, however, to explain the solar-system abundances of these nuclei. Furthermore, the yields of ^{74}Se, ^{78}Sr and ^{84}Kr are also high, as well as the yield of ^{28}Si, that results from the carbon burning during the p-process nucleosynthesis. We assume that 1 percent by mass of ^{22}Ne is initially present in these layers. However, the neutrons released via the ^{22}Ne(α,n)^{25}Mg reaction have little effect upon the p-process yields, as demonstrated by a comparison calculation with an initial solar-system mass fraction of ^{22}Ne. The proton-capture reactions are less effective than in the parametrized model calculations of HMW because the density in the Khokhlov model is somewhat lower than $10^7 g/cm^3$, and more importantly, because the expansion timescale is much shorter than assumed by HMW. The yields of ^{138}La, ^{164}Er and ^{180}Ta are not produced in this calculation because of either relatively low binding energy (as in the case of ^{138}La and ^{180}Ta) or its relatively high abundance (^{164}Er). In addition there are significant yields of ^{74}Se, ^{78}Kr and ^{84}Sr, especially relative to Mo and Ru. However, as discussed in HMW, for the case of a solar-like composition, the overproduction factors are too small for a significant Galactic contribution to p-nuclei. This is because of the relatively small amount of mass that experiences the p-process in a Type Ia supernova explosion. This conclusion is independent of the explosion mechanism.

We now investigate what happens when we assume that the surface composition is enriched with a weak s-process irradiation. For example, we consider a single exposure of $\tau = 0.30 mb^{-1}$ on a timescale of 10^5 yr. Such an exposure produces a peak near the $N = 50$ closed shell, due to the small neutron capture cross section of ^{88}Sr. Although the overall trend in mass number is improved, there is a significant relative overproduction (by three orders of magnitude) of ^{74}Se, ^{78}Kr and ^{84}Sr. Again the overall yield of the p-nuclei is too small, except for ^{74}Se, ^{78}Kr and ^{84}Sr, to make a significant Galactic contributions. This simply reflects the overproduction of the initial seeds in this mass region. The point is that the p-process nucleosynthesis is extremely sensitive to the initial distribution of s- and r-process elements.

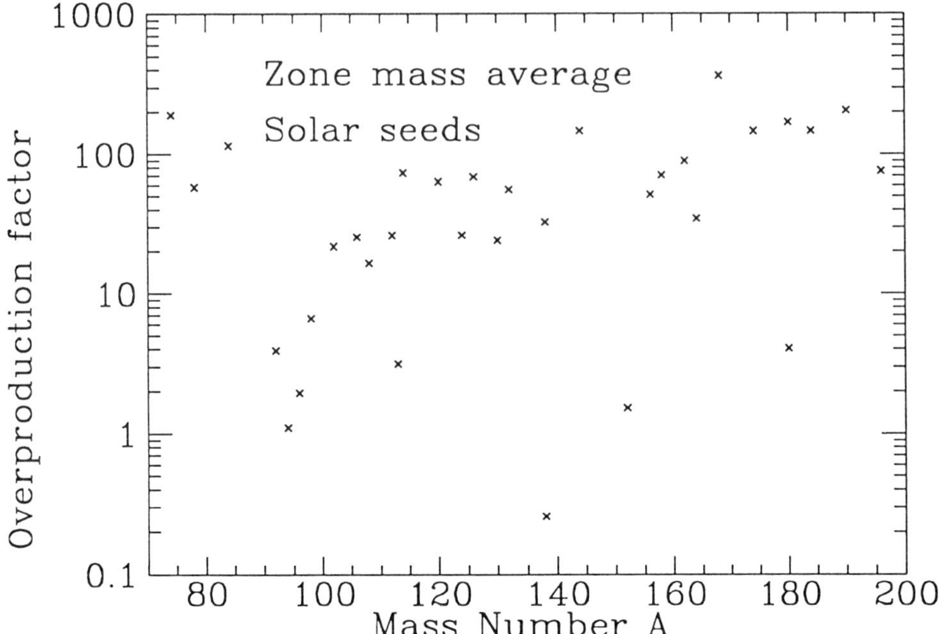

Figure 2: p-process yields from an initial solar-like composition

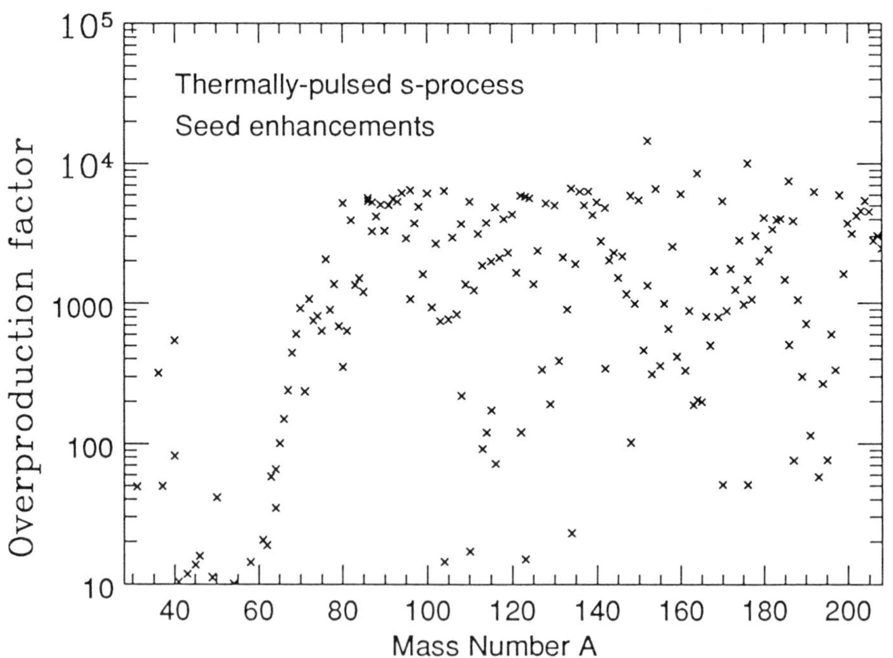

Figure 3: s-process yields from a pulsed neutron exposure

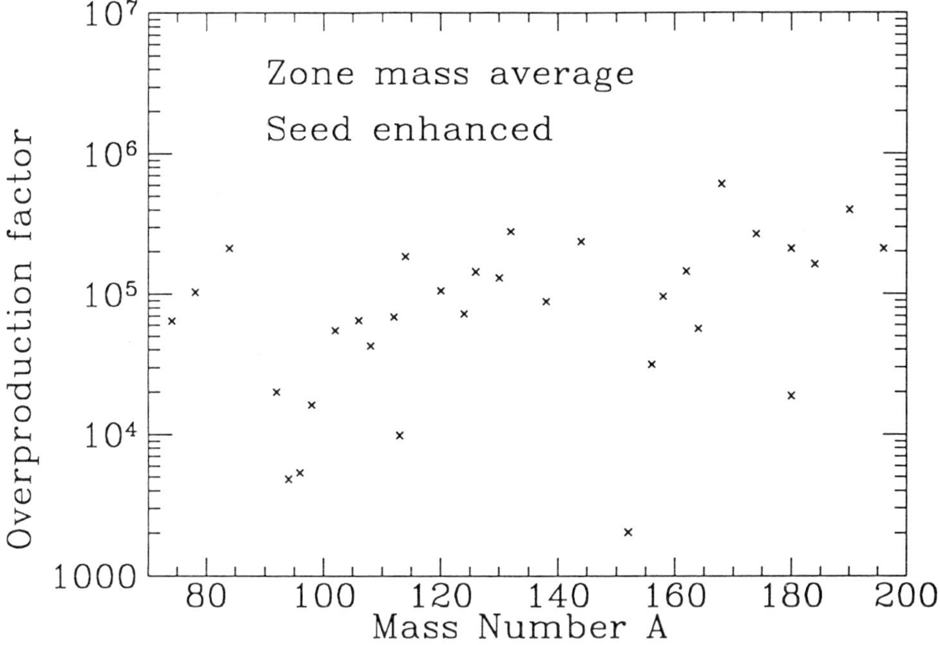

Figure 4: p-process yields from an initial seed-enhanced composition

To further emphasize this point we carry out one more calculation. For an initial s-process seed we take a distribution from a major or solar-system like component. We follow a thermally-pulsed s-process with an effective τ of 0.30 mb^{-1} for 10 pulses, after which the s-process nuclei obtain their asymptotic abundances. This initial seed distribution for stable nuclei, in terms of overproduction factors relative to solar, is shown in Figure 3. The seed distribution is quite different from that obtained from a weak component exposure in that all of the s-process nuclei up to ^{208}Pb are enriched. The result of the p-process nucleosynthesis from such an initial seed distribution in Khokhlov's model is shown in Figure 4. The effect is that almost all of the p-nuclei yields are enhanced by a factor of more than 10^4. As discussed by HMW these yields are enough to make a significant Galactic contribution to p-nuclei. The yields are rather uniform as a function of mass number. However, the yields of ^{92}Mo, ^{94}Mo, ^{96}Ru and ^{98}Ru are still underproducted by a factor 10 relative to the other p-nuclei. The ^{74}Se, ^{78}Kr and ^{84}Sr are not overproduced as they were in the calculations of HMW because of the the inefficiency of proton captures. The contribution to ^{180}Ta comes from one zone that reaches a peak temperature of $T_9 = 2.2$.

3   The p-process chronometers

The p-process has some interesting chronometers, including ^{146}Sm ($\tau_{1/2}$: 1.08×10^8 yr), ^{92}Nb ($\tau_{1/2}$: 3.6×10^7 yr), ^{97}Tc $\tau_{1/2}$: 2.6×10^6 yr) and ^{98}Tc ($\tau_{1/2}$: 4.2×10^6 yr). The production of these isotopes many pose interesting constraints on the p-process. Unfortunately, the nuclear physics related to the production of ^{146}Sm is uncertain, the production of ^{92}Nb may have others sources and the identification of ^{97}Tc and ^{98}Tc is only a possibility (Yin, Jagoutz and Wanke, 1992). We give the production ratio for these isotopes for our last calculation. The production ratios are similar to those obtained by HMW. Although ^{98}Tc has a longer half-life that ^{97}Tc, its production in the p-process is less than that of ^{97}Tc by several orders of magnitude. For our seed-enhanced calculation we obtain: ^{92}Nb/^{92}Mo $= 0.01$, ^{97}Tc/^{92}Mo $= 0.03$, ^{98}Tc/^{92}Mo $= 0.0001$, and ^{146}Sm/^{144}Sm $= 0.04$.

4   Conclusions

We demonstrate that the p-process operates effectively on the surface of a degenerate carbon-oxygen white dwarf when it undergoes a delayed detonation explosion to produce a Type Ia supernova. However, with an initial solar- system like composition of s- and r-process elements, there is probably an insufficient Galactic yield of p-nuclei. With an enriched s-process composition, the yields become more interesting, but there can be significant under productions of ^{92}Mo , ^{94}Mo, ^{96}Ru and ^{98}Ru. The character of the p-process yields are extremely sensitive to the s- and r-process enhancements. We encourage theoretical studies of the surface compositions of white dwarfs and the production of heavy elements through either accretion or through thermal instabilities on the surface. Contrary to the conclusions of Lambert (1992) the Galactic contribution to p-nuclei remains an open question. The outer layers of delayed detonations on white dwarfs remain an interesting site for the p-process. However, further studies of its surface composition are required.

We are grateful to A. M. Khokhlov for sharing details of his models. WMH is grateful for the hospitality of the Institut d'Astronomie et d'Astrophysique of the Universite Libre de Bruxelles, and for the financial support of the Fonds National Belge de la Recherche Scientifique (F.N.R.S.) during his sabbatical stay. He is also grateful to the Institut Medical Edith Cavell, where an unexpected visit allowed some time for the preparation of this manuscript. This work is also supported by the US Department of Energy under contract to the Lawrence Livermore National Laboratory.

REFERENCES

Arnould, M. 1976, *Astron. Astrophys.*, *46*, 117.
Anders, E. and Grevesse, N. 1989, *Geochim Cosmochim. Acta*, *53*, 197.
Howard, W. M., Meyer, B. S. and Woosley, S. E. 1991, *Astrophys. J. (Lett.)*, *373*, L5.
Khokhlov, A. M. 1991, *Astron. Astrophys.*, *245*, 114.
Lambert, D. L. 1992, *Astron. Astrophysics Rev.*, *3*, 201.
Meyer, B. S. 1992, Physics Reports, to be published.
Prantzos, N., Hashimoto, M., Rayet, M. and Arnould, M. 1990, *Astronom. Astrophys.*, *238*, 445.
Rayet, M., Prantzos, N. and Arnould, M. 1990, *Astronom. Astrophys.*, *227*, 271.
Woosley, S. E. and Howard, W. M. 1978, *Astrophys. J. Suppl.*, *36*, 285.
Yin, Q., Jagoutz, E. and Wanke, H. 1992 Research for extinct ^{99}Tc and ^{98}Tc in the early solar system, Proc. 55th Annual Meteoritical Society Meeting, Copenhagen, 26 July-1 Aug, 1992

Relative frequencies of type Ia and type II supernovae in the Galaxy, LMC, and SMC

T. Tsujimoto[1], K. Nomoto[1], M. Hashimoto[2], S. Yanagida[3], and F.-K. Thielemann[4]

[1] Department of Astronomy, Faculty of Science, University of Tokyo, Bunkyo-ku, Tokyo 113, Japan
[2] Department of Physics, College of General Education, Kyushu University, Fukuoka 810, Japan
[3] Department of Physics, Ibaraki University, Mito, Ibaraki 310, Japan
[4] Harvard Smithsonian Center for Astrophysics, 60 Garden Street, Cambridge MA 02138, USA

Abstract.
The yields from type II and type Ia supernovae are combined and compared with the solar isotopic abundances as well as the abundances of LMC and SMC. The occurrence frequencies of these two types of supernovae are derived. The relative frequency of type Ia supernovae is larger for the LMC and SMC than in the Galaxy.

1. Introduction

Supernovae are the sites of heavy element production in galaxies. Because of the short lifetimes of their massive progenitors, type II supernovae (SNe II) and type Ib/Ic supernovae (SNe Ib/Ic) contribute to the heavy element production in the early chemical evolution of galaxies. Type Ia supernovae (SNe Ia), in contrast, produce heavy elements with a longer time scale (~ 1 Gyr) in the late phase of the galactic evolution. Here we assume that stars more massive than initially $\sim 10\ M_\odot$ end up as SNe II if single stars and as SNe Ib/Ic if being in close binary systems, and that SNe Ia are due to thermonuclear explosions of accreting white dwarfs. See Branch et al. (1991) for more details of the progenitor–supernova connection.

The yields from SNe II and SNe Ia and thus their nucleosynthesis roles in the chemical evolution of galaxies are also distinctly different. Here we demonstrate from the comparison between the observed and theoretical abundance ratios that the relative roles of SNe II and SNe I are different among the Galaxy, LMC, and SMC.

2. Yields from type II and type Ia supernovae

For SNe II yields we use the elemental abundances after explosive burning in $13 - 70\ M_\odot$ stars with interpolation and extrapolation in stellar masses (Hashimoto et al. 1992). Figure 3 of Hashimoto et al. (1992) shows that the abundance ratios of O, Ne, Mg relative to Fe in SNe II are significantly larger than the solar ratio.

For SNe Ia yields we adopt nucleosynthesis products from the carbon deflagration model W7 (Nomoto et al. 1984) with a slight modification that the initial ^{22}Ne abundance is set to be zero in the C+O white dwarf (model W70). This corresponds to the assumption that most of the white dwarf progenitors either belong to Population II or undergo the separation of ^{22}Ne during solidification (Ogata et al. 1992). Figure 1 shows the abundances of W70 (Thielemann et al. 1992), where O/Fe is much smaller than the solar ratio. In other words, SNe Ia enhance mostly iron in the galactic evolution.

3. Solar abundances

To quantitatively demonstrate the relative nucleosynthesis roles of SNe Ia and SNe II, we sum up the nucleosynthesis products as $[rX_{i,\mathrm{Ia}} + (1-r)X_{i,\mathrm{II}}]$ and determine r by minimizing the deviation from the solar abundances (Yanagida et al. 1990; Nomoto et al. 1991). The best fit to the solar abundances (Anders & Grevesse 1989) is obtained for $r = 0.09$ as shown in Figure 2, where the deviation from the solar ratios is mostly less than a factor of ~ 2. (See Hashimoto et al. 1992 for discussion on the relatively large deviation for several elements.) The obtained ratio of SNe Ia/SNe II may be consistent with the recent observed estimate by van den Bergh & Tammann (1991).

4. Abundances in the LMC and SMC

The LMC and SMC abundances are different from the solar. They are metal deficient by a factor of 2 - 4 relative to the solar. The abundance patters are also different; for example, the O/Fe ratio in the LMC and SMC is smaller than the solar ratio. Such a difference can be understood as follows.

We obtained the ratio $r/(1-r)$ between SN Ia and SNe II by minimizing the deviation from the observed abundances (Russel and Dopita 1990). The obtained values are $r = 0.19$ for the LMC (Fig. 3) and 0.25 for the SMC (Fig. 4). Apparently, the frequency of SNe Ia in the LMC and SMC is significantly larger than in the Galaxy, being larger for more metal deficient galaxies.

5. Chemical evolution models for the solar neighborhood

To understand the source of the above difference in the relative frequency of SNe Ia in the Galaxy, LMC and SMC, we need to clarify what processes control the chemical evolution of galaxies. Let us start from the chemical evolution model for the Galaxy with a continuous star formation history. The basic equations to describe the evolution in terms of star formation, matter infall, etc. are given in Tsujimoto et al. (1992).

Figure 5 shows the evolutionary change in $[\mathrm{O/Fe}] = \log_{10} [(\mathrm{O/Fe})/(\mathrm{O/Fe})_\odot]$ as a function of metallicity [Fe/H]. (Observed points are taken from Andersen et al. 1988, Barbuy 1988, and Gratton & Ortolani 1986.) The [O/Fe] ratio in low metallicity stars is determined by SNe II, while the decrease in [O/Fe] at $[\mathrm{Fe/H}] \gtrsim -1$ is due to the iron production from SNe Ia whose lifetime is assumed to be ~ 1 Gyr.

The [O/Fe] ratio from SNe II depends on the lower mass bound M_{bh} to the stars that become black holes, where no heavy element ejection is assumed for stars with masses $M > M_{\mathrm{bh}}$. Since the ejected oxygen mass is larger for more massive stars while the iron mass is much less sensitive to the progenitor mass (Nomoto et al. 1991), the integrated [O/Fe] is larger for larger M_{bh}. To demonstrate this effect, the dashed lines in Figure 5 show the four cases with $M_{\mathrm{bh}} = 100\ M_\odot$ (case A), 55 $M_\odot$ (case B), 40 $M_\odot$ (case C), and 32 $M_\odot$ (case D), where the integration of the yield is made over 10 $M_\odot$ – M_{bh} with Salpeter's IMF. The resulting yields of SNe II give [O/Fe] = 0.70 (A), 0.50 (B), 0.40 (C), 0.26 (D) at metallicity of $[\mathrm{Fe/H}] \lesssim -1$. For comparison, the solid line shows the integration over 10 – 100 $M_\odot$ but oxygen yield is assumed to saturate at constant value for 40 – 100 $M_\odot$ due to larger mass loss from more massive stars.

To be consistent with the observed [O/Fe] = 0.41 $\pm$ 0.15 averaged over low metallicity stars of $[\mathrm{Fe/H}] \lesssim -0.8$ in the solar neighborhood (Gratton 1991), M_{bh} ranges from $\sim 32~M_\odot$ to $\sim 55~M_\odot$ with the best estimate of $40~M_\odot$.

Possible contributions of type II-BL and II-L supernovae to the chemical evolution would decrease [O/Fe], if the AGB star models for these types of supernovae and their iron production are assumed (see Nomoto et al. 1991 for more details).

6. Chemical evolution models for the LMC and SMC

The difference in the relative frequency of SNe Ia in the chemical evolution of the LMC and SMC compared with the Galaxy may be due to the differences of the initial mass function (Matteucci & Brocato 1990; Russel & Dopita 1992) or different star formation history (e.g., star bursts; Gilmore & Wyse 1991) in the LMC and SMC.

The SN II rate is directly related to the star formation rate. On the other hand, the SN Ia rate depends on the binary evolution. For example, if the accretion onto the white dwarf is due to the Roche lobe overflow of the companion star, the SNe Ia rate is determined by the lifetime of the low mass companion; the rate depends also on the initial separation of the double white dwarfs, if the merging scenario is adopted.

If the LMC and SMC have undergone burst of star formation, element production due to SNe II has occurred in a burst while that of SNe Ia has taken place continuously. Such a difference of the star formation history among the Galaxy and the LMC/SMC could cause the difference in the relative frequencies of SN II and SN Ia.

Figure 6 shows possible models for the chemical evolution of the LMC with star bursts (upper solid) and a different initial mass function (lower solid). The dashed curve shows the chemical evolution model for the Galaxy with continuous star formation and Salpeter's initial mass function. Further analysis is needed to clarify which effect dominates the evolution of the LMC and SMC as well as irregular galaxies.

References

Anders E., Grevesse N., 1989, Geochim. Cosmochim. Acta, 53, 197
Andersen J., Edvardsson B., Gustafsson B., Nissen P.E., 1988, in IAU Symposium 132, p. 43
Barbuy B., 1988, A&A, 191, 121
Branch D., Nomoto K., Filippenko A.V., 1991, Comments on Ap., 15, 221
Gilmore G., Wyse R.F.G., 1991, ApJ, 367, L55
Gratton R.G., 1991, in IAU Symposium 145, ed. G. Michaud et al. (University of Montreal), p. 27
Gratton R.G., Ortolani S., 1986, A&A, 169, 201
Hashimoto M., Nomoto K., Shigeyama T., 1989, A&A, 210, L5
Hashimoto M., Nomoto K., Tsujimoto T., Thielemann F.-K., 1992, in this volume
Matteucci F., Brocato E., 1990, ApJ, 365, 539
Nomoto K., Shigeyama T., Tsujimoto T., 1991, in IAU Symposium 145, Evolution of Stars: the Photospheric Abundance Connection, ed. G. Michaud (Kluwer), p. 21
Nomoto K., Shigeyama T., Yanagida S., Hayakawa S., Yasuda K., 1990, in Chemical and dynamical evolution of galaxies, eds. F. Ferrini et al. (ETS Editrice, Pisa), p. 366
Nomoto K., Thielemann F.-K., & Yokoi K., 1984, ApJ, 286, 644
Ogata, S., Iyetomi, H., Ichimaru, S., Van Horn, H.M., 1992, preprint
Russell S.C., Dopita M.A., 1990, ApJS, 74, 93
Russell S.C., Dopita M.A., 1992, ApJ, 384, 508
Thielemann F.-K., Nomoto K., Hashimoto M., 1992, in Origin and Evolution of the Elements, ed. N. Prantzos and E. Vangoni-Flam (Cambridge University Press), in press
Tsujimoto T., Nomoto K., Shigeyama T., Saio H., Yoshii Y., 1992, in this volume
van den Bergh S., Tammann G.A., 1991, ARA&A, 29, 363
Yanagida S., Nomoto K., Hayakawa S., 1990, in Proceedings of the 21st International Cosmic Ray Conference (Adelaide), 4, 44

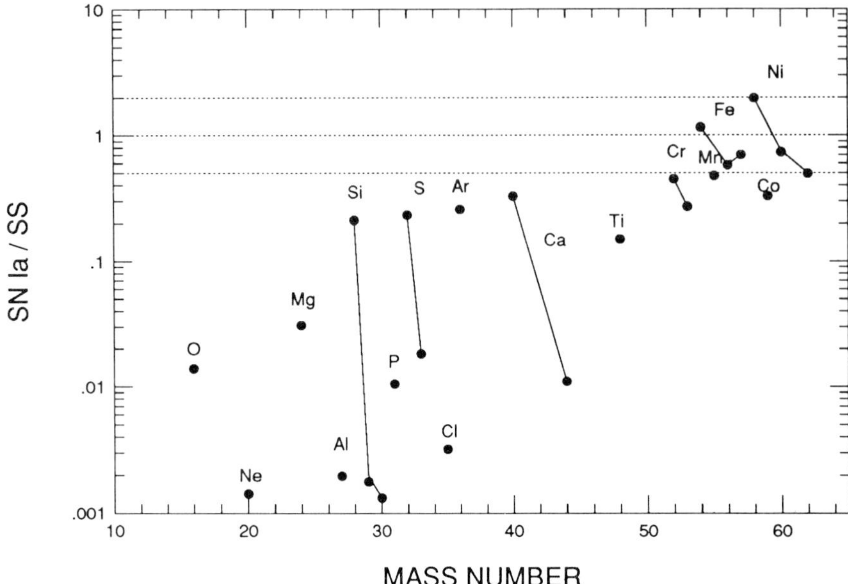

Figure 1. Nucleosynthesis products from the carbon deflagration model W70 for Type Ia supernovae (SNe Ia) as compared with the solar system abundances.

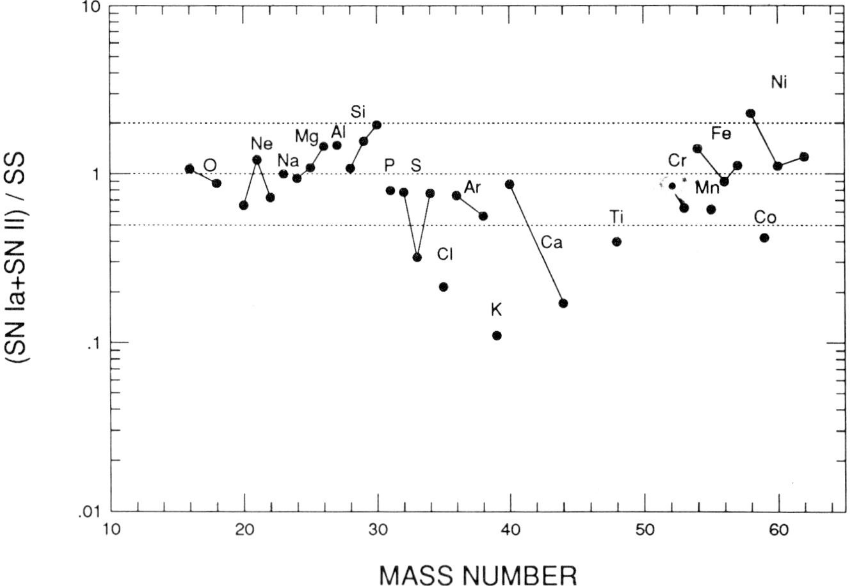

Figure 2. The combined nucelosynthesis products from SNe Ia and SNe II. The SNe Ia/SNe II ratio of 0.09/0.91 is obtained by minimizing the deviation from the solar abundances.

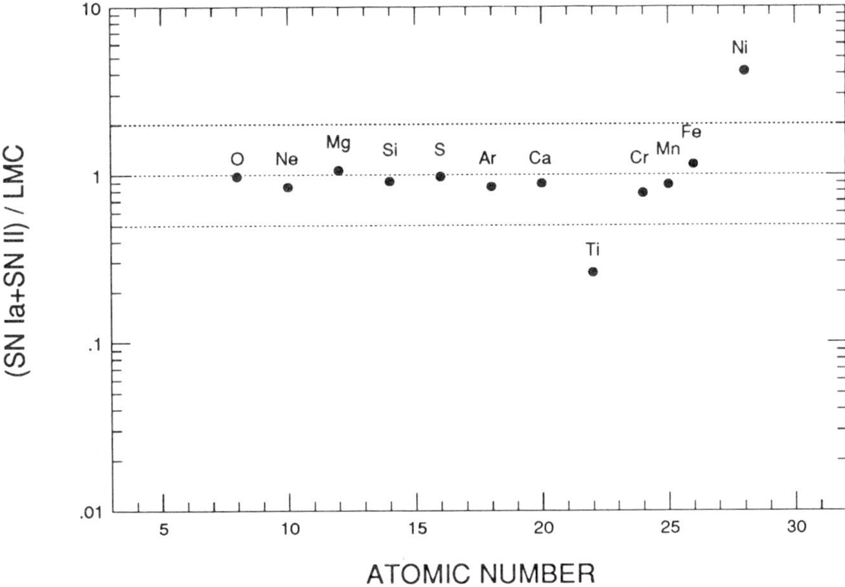

Figure 3. The combined nucelosynthesis products from SNe Ia and SNe II. The SNe Ia/SNe II ratio of 0.19/0.81 is obtained by minimizing the deviation from the LMC abundances.

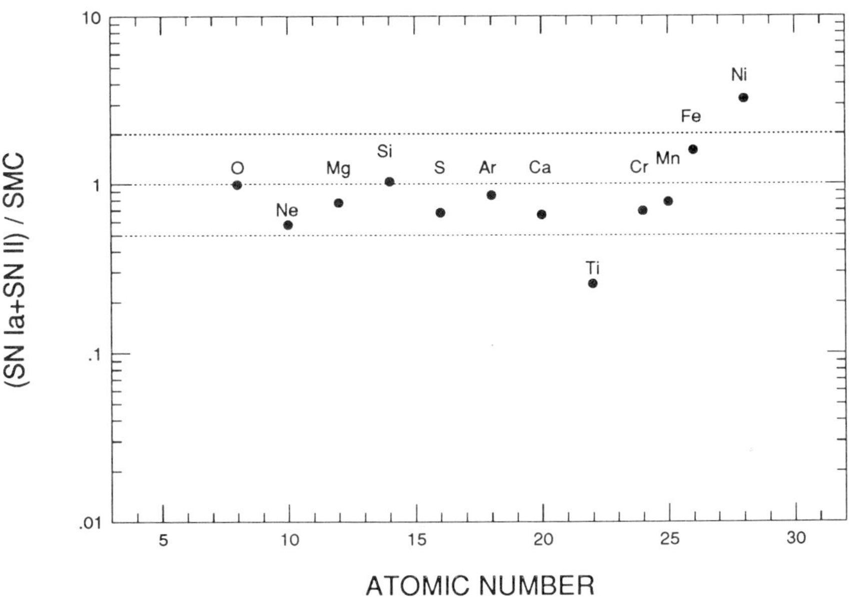

Figure 4. The combined nucelosynthesis products from SNe Ia and SNe II. The SNe Ia/SNe II ratio of 0.25/0.75 is obtained by minimizing the deviation from the SMC abundances.

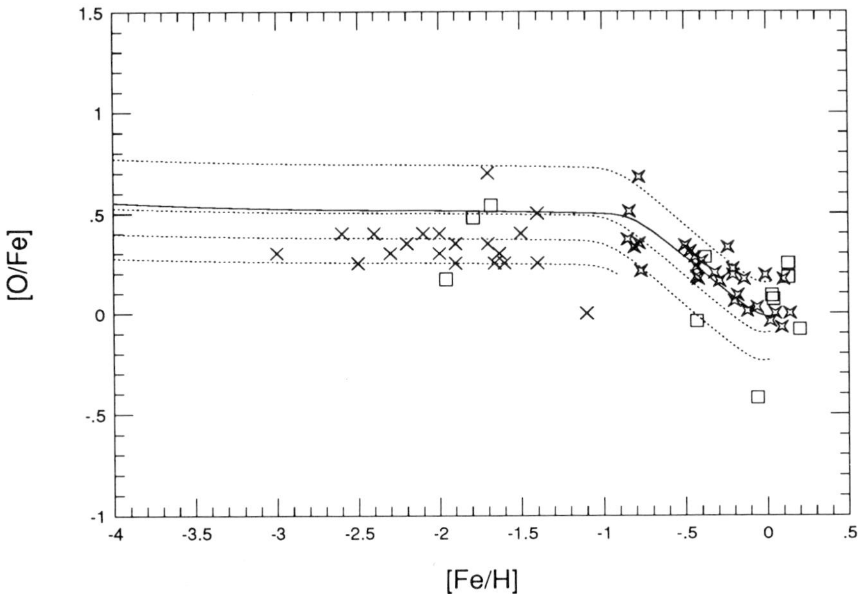

Figure 5.   The chemical evolution of the Galaxy, [O/Fe] ratio against metallicity [Fe/H]. The observations of low metallicity stars are compared with four different theoretical models (see text).

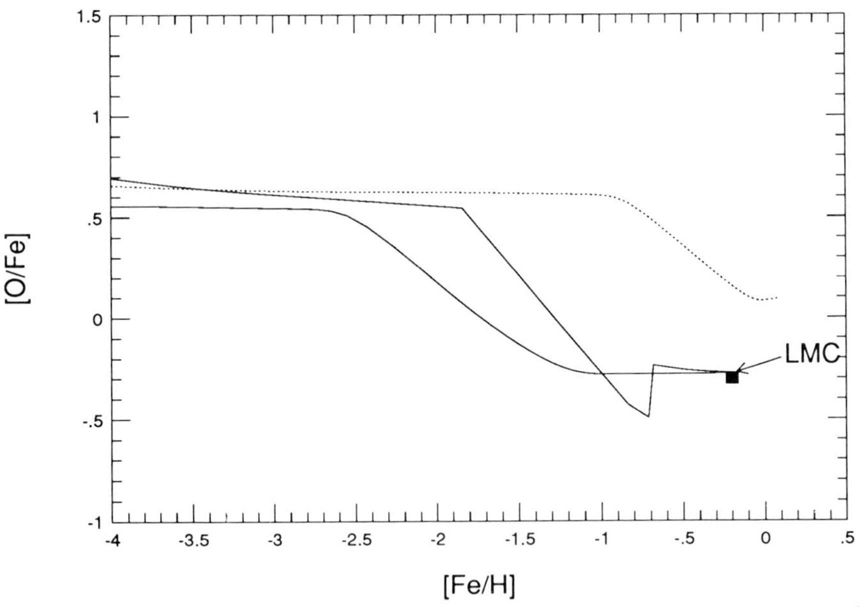

Figure 6.   Possible models for the chemical evolution of the LMC with star bursts (upper solid) and a different initial mass function (lower solid). The dashed curve shows the chemical evolution model for the Galaxy with continuous star formation and Salpeter's initial mass function.

Explosive nucleosynthesis in massive stars: 13–70 M. stars

M. Hashimoto[1], K. Nomoto[2], T. Tsujimoto[2], and F.-K Thielemann[3]

[1]Department of Physics, College of General Education, Kyushu University, Ropponmatsu, Fukuoka 810, Japan

[2]Department of Astronomy, University of Tokyo, Tokyo 113, Japan

[3]Harvard Smithsonian Center for Astrophysics, Cambridge MA 02138, USA

1. Introduction

Explosive nucleosynthesis in massive stars is one of the major sources of nuclei in the cosmos (e.g., Fowler 1984). It involves still relatively large uncertainties which stem from unknown nuclear reaction rates, such as the $^{12}C(\alpha, \gamma)^{16}O$ rate, and uncertain physical processes, such as convection and the supernova explosion mechanism. On the other hand, detailed study of nucleosynthesis can provide important clues to the understanding of these unknown processes and reaction rates. In this regard, comparison with supernova observations (abundance determinations, optical light curves of SN 1987A and type Ib/Ic supernovae, and γ-rays lines) has been a new tool to test the production of the radioactive nuclei ^{56}Ni, ^{57}Ni, ^{44}Ti (e.g., Kumagai et al. 1991). Further it provides the basic input data for the studies of chemical evolution of galaxies and the origin of the solar system abundances.

With this motivation, we have studied the presupernova evolution and explosive nucleosynthesis in massive stars on a dense grid of stellar masses (Hashimoto & Nomoto 1992). This is an extension of our previous studies given in Nomoto & Hashimoto (1988). The main-sequence star masses considered here range from 13 to 70 $M_\odot$. The systematic study of such a wide mass range enables us to understand how explosive nucleosynthesis depends on the presupernova stellar structure and thus on the progenitor mass. Then, the supernova yields can be integrated over the initial mass function in order to compare with the solar system or early galactic abundances.

2. Presupernova models and initial compositions

Presupernova models are obtained by evolving helium stars with masses $M_\alpha = 3.3$ $M_\odot$, 4.0 $M_\odot$, 5.0 $M_\odot$, 6.0 $M_\odot$, 8.0 $M_\odot$, 16.0 $M_\odot$, and 32.0 $M_\odot$ (Nomoto & Hashimoto 1988; Hashimoto & Nomoto 1992). We use the Schwartzschild criterion for convection and neglect overshooting. With this choice, these helium star masses approximately correspond to the main-sequence masses of $M_{\rm ms} = 13$, 15, 18, 20, 25, 40, and 70 $M_\odot$, respectively (Sugimoto & Nomoto 1980). We have evolved these helium stars from helium burning to the onset of the Fe core collapse. The initial compositions are: $X(^4He) = 0.9879$ and $X(^{14}N) = 0.0121$, where all the original CNO elements are assumed to be converted into ^{14}N during core hydrogen burning. Nuclear reaction rates are mostly taken from Caughlan & Fowler (1988) (see Nomoto & Hashimoto

1988 for more details); for the uncertain rate of $^{12}C(\alpha,\gamma)^{16}O$, we have used the rate by Caughlan et al. (1985), which is larger than the 1988 rate by a factor of ~ 2.4. As will be discussed in §5, this choice of the reaction rate leads to a better agreement between the isotopic abundances in the type II supernova ejecta and the solar abundances.

At the end of core helium burning, the formation and abundance of a carbon-oxygen core is largely influenced by the $^{12}C(\alpha,\gamma)^{16}O$ rate. Especially the C/O ratio affects the abundances of Ne, Mg, Al relative to O in the more evolved cores. (For silicon burning and the final size of the iron core, see Nomoto & Hashimoto 1988). Neutronization during oxygen/silicon shell burning affects the isotopic ratio of iron peak elements. The electron number per ion, Y_e ($= 0.5 - \eta/2$, where η is the neutron excess) ranges from 0.490 to 0.499 around the mass cut that divides the neutron star and the ejecta. It should be noted that the presupernova models and the distribution of neutron excess include uncertainties involved in the nuclear data and electron capture rates. Figure 1 shows the presupernova composition structure for $M_\alpha = 5\ M_\odot$.

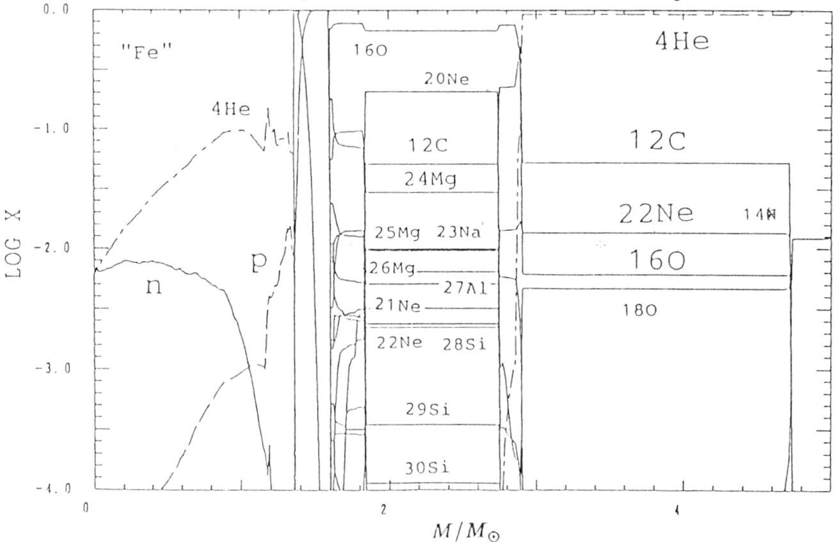

Fig. 1: Presupernova composition structure for $M_\alpha = 5\ M_\odot$.

3. Explosive nucleosynthesis

Hydrodynamical models of supernova explosions and associated explosive nucleosynthesis are calculated for the seven presupernova models as has been done for SN1987A ($6\ M_\alpha$; Shigeyama et al. 1988). For each mass zone of the explosion models explosive nucleosynthesis was calculated with the use of a nuclear reaction network (Hashimoto et al. 1989). The ejected amount of ^{56}Ni, whose decay powers the supernova light curve, depends on the explosion energy, location of the mass cut, and the structure of the presupernova models such as the distributions of density and Y_e.

Since the mechanism of the supernova explosion after core collapse is still not fully understood, the explosion energy and the mass cut are uncertain, except for SN 1987A. Since the explosion energy of SN 1987A has been estimated to be $\sim 10^{51}$ erg from the analysis of the light curve (e.g., Nomoto et al. 1991a for a review), the final kinetic energy of the explosion is assumed to be 10^{51} erg for all the initial models. A shock wave is generated by depositing thermal energy at the mass cut (see Aufderheide et al. 1991 for the uncertainties introduced by this approach).

The ejected mass of ^{56}Ni depends critically on the location of the mass cut. In the present study, the mass cut is chosen at $M_r \approx 1.3 M_\odot$ for $13 - 15$ $M_\odot$ stars to produce 0.15 $M_\odot$ ^{56}Ni and to explain the light curves of type Ib/Ic supernovae (Shigeyama et al. 1990) and at $M_r \approx 1.6 M_\odot$ for $18 - 20$ $M_\odot$ stars to produce 0.07 $M_\odot$ ^{56}Ni, which is consistent with the light curve of SN 1987A (Hashimoto et al. 1989). For stars more massive than 25 $M_\odot$, there is no clear observational constraint but we assume a production of ≤ 0.07 $M_\odot$ of ^{56}Ni in view of the absence of bright type Ib with slow decline (Shigeyama et al. 1990). The use of supernova light curves would be currently the only reasonable way to legitimate the location of the mass cut assumed for the massive stars considered here.

The general features of explosive nucleosynthesis in type II supernovae have been discussed by Woosley et al. (1973), Hashimoto et al. (1989), and Thielemann et al. (1990, 1992). The elements produced explosively depend on the peak temperature, T_p, attained through the passage of the shock. For example, at $T_p > 5.0$ 10^9 K, explosive silicon and oxygen burning produces important radioactive nuclei, like ^{56}Ni, ^{57}Ni, and ^{44}Ti depending on the distribution of neutron excess. For $2.6\ 10^9 K < T_p < 3.0\ 10^9$ K, where explosive neon burning takes place, the p-process occurs via the photodisintegration $(\gamma, p), (\gamma, n), (\gamma, \alpha)$ of nuclei (e.g. Arnould et al. 1991).

The inner part of the star, with $T_p > 2.0\ 10^9$ K, experiences explosive nucleosynthesis while the outer part, including most of the original oxygen-rich layer, is ejected relatively unprocessed by explosive nuclear burning. This leads to a mass dependence of the ratio between explosive and hydrostatic burning products.

4. Results

Figure 2 a–c shows the abundance distribution after explosive processing (upper) and the integrated abundances of the ejecta relative to the solar values (Anders & Grevesse 1989) (lower) for $M_\alpha = 3.3, 5$, and 8 $M_\odot$. Among the elements shown here, the ratio $[O/Fe] = \log_{10}[(O/Fe)/(O/Fe)_\odot]$ is crucial to determine the relative role of type Ia and type II supernovae for the chemical evolution of galaxies (Nomoto et al. 1991b). ^{56}Fe is produced from explosive nucleosynthesis via $(^{56}Ni \rightarrow ^{56}Co \rightarrow ^{56}Fe)$; its mass is in principle determined by the mass cut, but, in the present study, is given from the supernova light curve models as discussed earlier. On the other hand, ^{16}O is mostly produced by hydrostatic nuclear burning. Therefore, its yield depends strongly on the stellar mass, being much larger for massive mass. Figure 2 shows this strong dependence of $[O/Fe]$ as a function of stellar mass.

The major abundances resulting from hydrostatic and explosive nucleosynthesis are: ^{16,18}O, ^{20,21,22}Ne, ^{23}Na, ^{24,25,26}Mg, ^{27}Al, ^{28,29,30}Si, ^{31}P, ^{32,33,34}S, ^{35}Cl, ^{36,38}Ar, ^{39}K, ^{40,44}Ca, ^{48}Ti, ^{52,53}Cr, ^{54,56,57}Fe, ^{55}Mn, ^{59}Co, and ^{58,60,62}Ni.

5. Discussion

Figure 3 shows the isotopic abundances relative to their solar values (Anders & Grevesse 1989) after integrating over the mass range from 13 to 70 $M_\odot$ with an initial

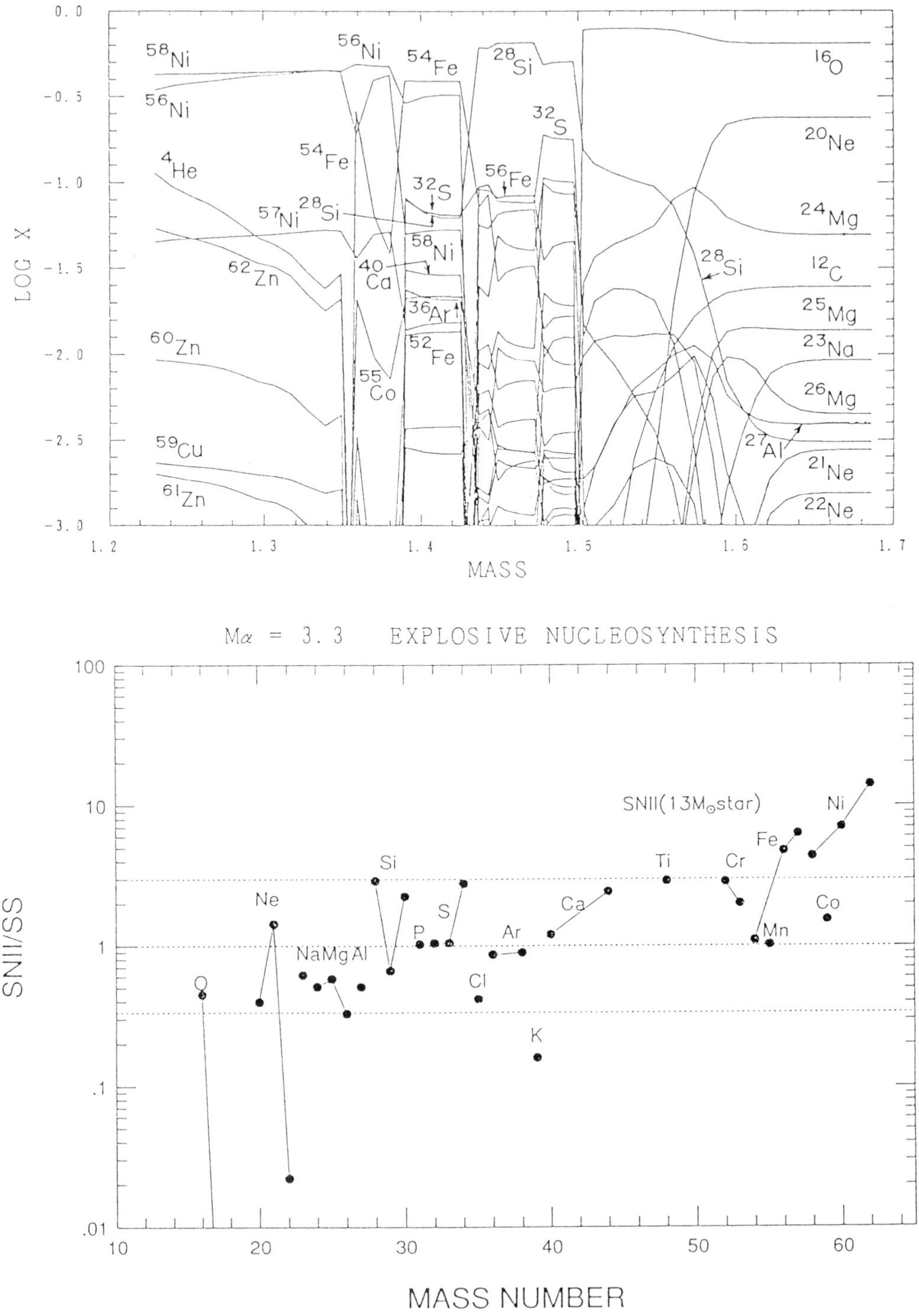

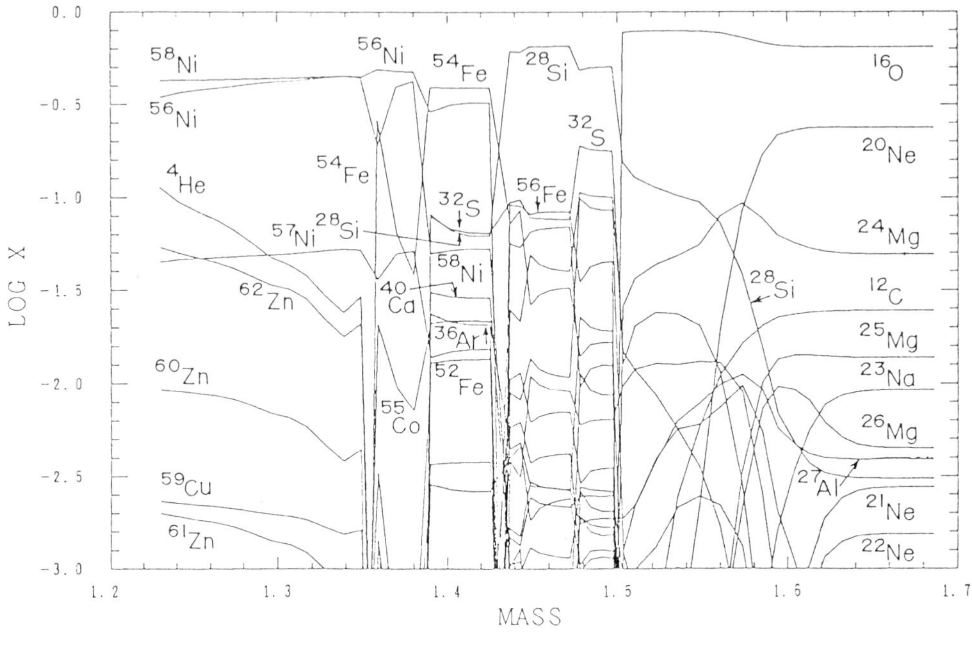

Fig. 2a: The abundance distribution after explosive nucleosynthesis (upper) and the integrated abundances of the ejecta relative to the solar values (Anders & Grevesse 1989) for $M_\alpha = 3.3\ M_\odot$.

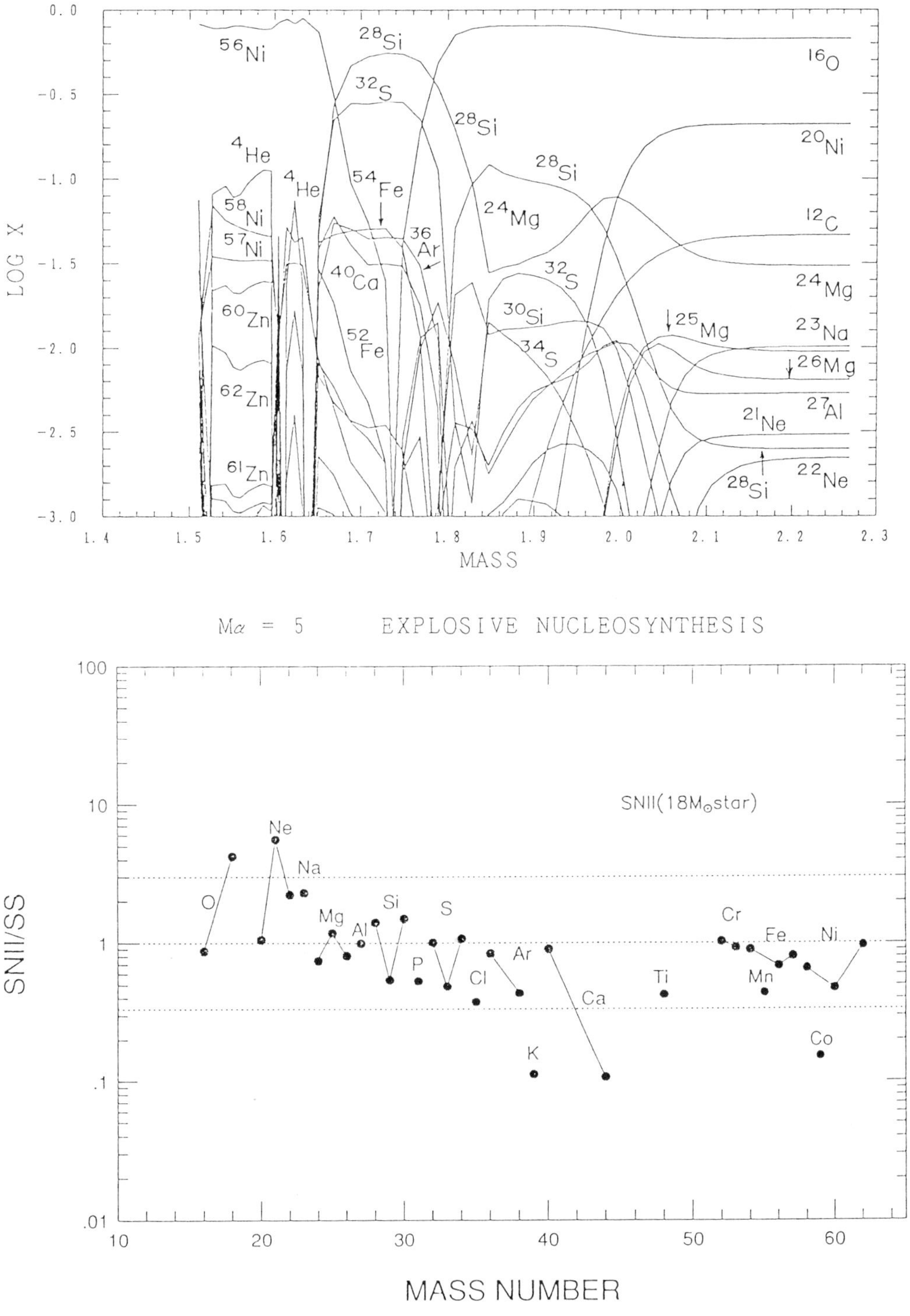

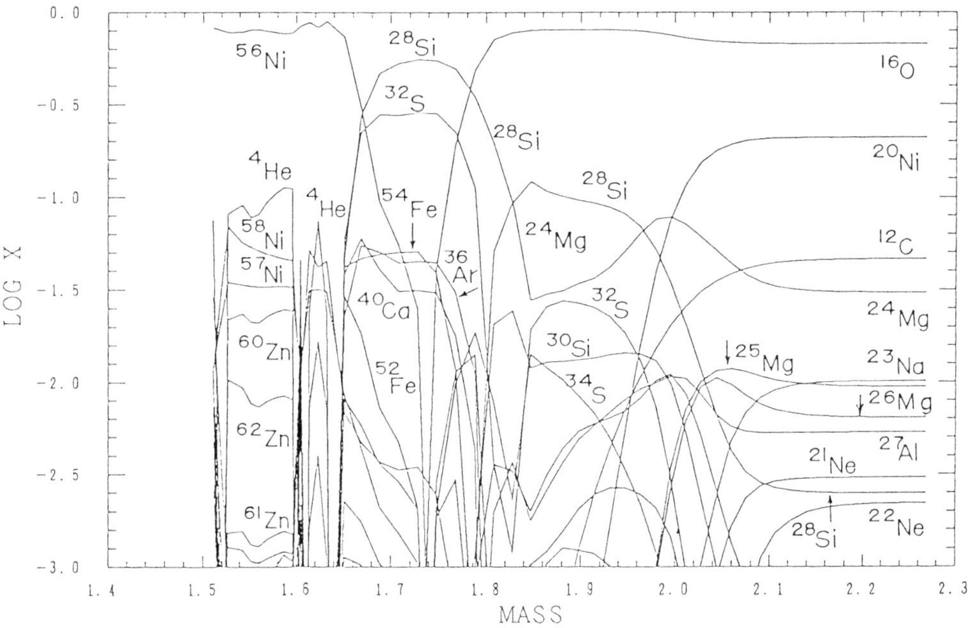

Fig. 2b: Same as Fig. 2a but for $M_\alpha = 5\ M_\odot$.

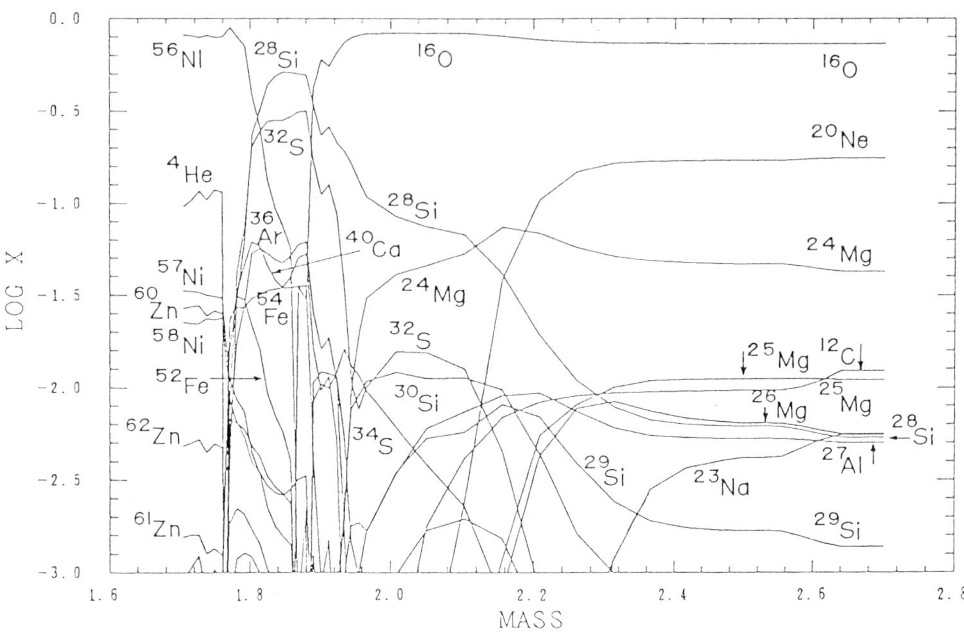

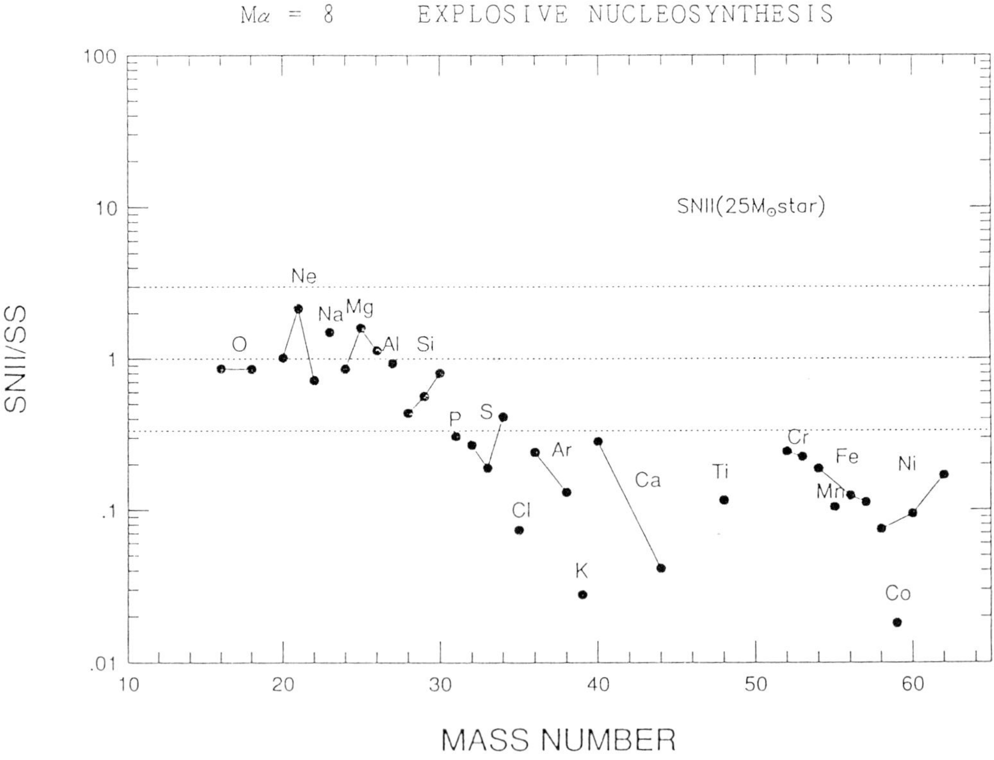

$M\alpha = 8$ EXPLOSIVE NUCLEOSYNTHESIS

Fig. 2c: Same as Fig. 2a but for $M_\alpha = 8\ M_\odot$.

mass function $\propto M^{-7/3}$. Because the 25 $M_\odot$ star gives the largest contribution to the oxygen production, [O/Fe] exceeds zero. For typical hydrostatic burning products below $A < 27$, however, the relative abundance ratios from massive stars are in good agreement with the solar ratios. Note that the isotopic ratios in Figures 2 & 3 depend also on overshooting. The good agreement suggests that the $^{12}C(\alpha, \gamma)^{16}O$ reaction rate from Caughlan et al. (1985), which is a factor of ~ 2.4 larger than that of Caughlan & Fowler (1988), is more consistent with the observations than the latter, if no convective overshooting is included. If overshooting at the core helium burning would reduce the C/O ratio, a smaller $^{12}C(\alpha, \gamma)$ rate would be a better choice.

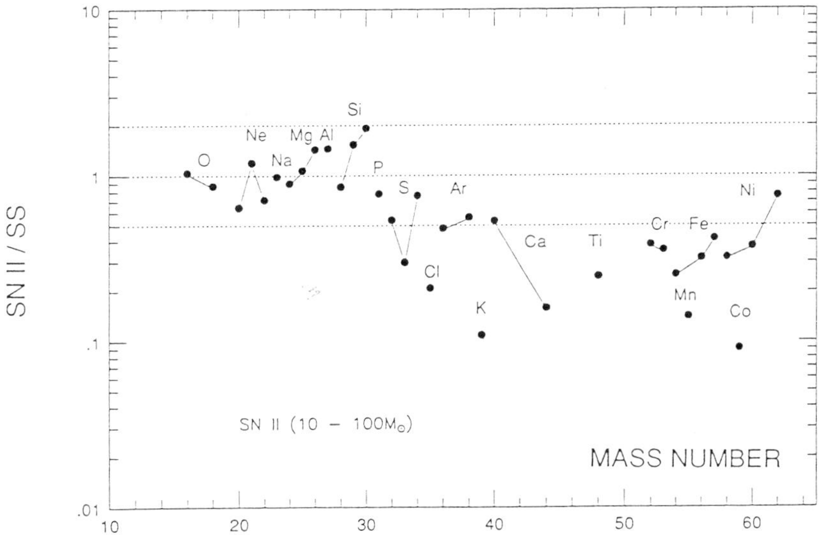

Fig. 3: The isotopic abundances relative to the solar values (Anders & Grevesse 1989), being integrated over the mass range from 13 to 70 $M_\odot$ with an initial mass function ($\propto M^{-7/3}$).

If we combine the contributions from both type I & type II supernovae with an appropriate ratio, the supernova abundances agree with the solar abundance ratios within a factor 2–3 for most species (see Tsujimoto et al. 1992 in this volume). Some species, ^{35}Cl, ^{39}K, ^{44}Ca, ^{48}Ti, and ^{59}Co, are underproduced and ^{62}Ni are overproduced. If we include the contribution from the s-process that occurs during core helium burning (Prantzos et al. 1990a), ^{48}Ti and ^{59}Co are enhanced appreciably compared with the seed (solar) abundances. However, ^{35}Cl, ^{39}K and ^{44}Ca are enhanced only a factor of $\sim$ 2. Since the produced s-nuclei in the central region will be destroyed during both the subsequent hydrostatic burning stage and the supernova explosion, these nuclei could have a strong contribution from helium and carbon shell burning.

On the other hand, ^{62}Ni is originally produced as ^{62}Zn and decays to ^{62}Ni ($^{62}Zn \rightarrow^{62} Cu \rightarrow^{62} Ni$). This nucleus, being rather neutron-rich ($Y_e = 0.484$ for ^{62}Zn), is produced in the neutron-rich layers near the mass cut region in type II supernovae as well as in type Ia supernovae. The type Ia contribution of neutron-rich species may depend on the age and accretion rate of white dwarfs (Nomoto et al. 1984; Thielemann et al. 1986; Yamaoka et al. 1992). The contribution of type II supernovae depends on the mass cut. The overproduction of ^{62}Ni would be smaller if, compared with the models adopted here, the mass of the neutron-rich layer above the mass cut is much smaller and a larger fraction of the iron peak elements is produced from oxygen-rich layer where $Y_e = 0.498 - 499$. This would imply that the explosion is probably due to delayed neutrino heating so that a significant amount of mass continues to contract after the collapse has started (Wilson and Mayle 1990). As a result of this delayed infall, material with large neutron excess would become part of the neutron star and the oxygen-rich layer would be located closer to the mass cut. Massive stars are also the sites for the origin of nuclei beyond iron-peak elements. For example, the s-process occurs to produce the weak component of nuclei $50 < A < 100$ during hydrostatic helium burning (e.g., Prantzos et al. 1990a). During the passage of the shock in the oxygen rich layer, the p-process occurs through the photodisintegration of nuclei and produces about 50 % of the p-nuclei (Prantzos et al. 1990b; Arnould et al. 1991). However, the production of s-nuclei and p-nuclei depends essentially on the seed composition. For example, if we use the LMC abundances as the seed ($Z \approx Z_\odot/4$), we cannot get an appreciable overproduction for s-nuclei and p-nuclei (Prantzos 1990a; Arnould et al. 1991). The effect of initial metallicity on the yields from hydrostatic and explosive nuclear burning needs further quantitative studies; comparison with observations (e.g., Gustafsson 1992) would provide important constraints on stellar nucleosynthesis models.

References

Anders E., Grevesse, N., 1989, Geochim. Cosmochim. Acta 53, 197

Arnould, M., Rayet, M., Hashimoto, M., 1992, in International Workshop on Unstable Nuclei in Astrophysics, ed. S. Kubono & T. Kajino (World Scientific), p23

Aufderheide, M., Baron, E., Thielemann, F.-K. 1991, ApJ 370, 630

Caughlan, G.R., Fowler, W.A., Harris, M.J., & Zimmerman, B.A. 1985, Atomic Data & Nuclear Data Tables, 32, 197

Caughlan, G.R., Fowler, W.A., 1988, Atomic Data & Nucl. Data Tables, 40, 283

Fowler, W.A. 1984, Rev. Mod. Phys., 56, 149

Gustafsson, B., 1992, this volume

Hashimoto, M., & Nomoto K. 1992, in preparation

Hashimoto, M., Nomoto K., & Shigeyama, T., 1989, A&A, 210, L5

Kumagai, S., Shigeyama, T., Hashimoto, M., Nomoto K., 1991, A&A, 243, L13

Nomoto, K., & Hashimoto, M. 1988, Phys. Rep. 163, 13

Nomoto, k., Shigeyama, T., Kumagai, S., & Yamaoka H. 1991a, in Supernovae and Stellar Evolution, ed. A. Ray & T. Velusamy (World Scientific), p. 116

Nomoto, K., Shigeyama, T., Tsujimoto, T. 1991b, in IAU Symposium 145, Evolution of Stars: the Photospheric Abundance Connection, ed. G. Michaud (Kluwer), p. 21

Nomoto, K., Thielemann, F.-K., & Yokoi, K. 1984, ApJ, 286, 644

Prantzos, N., Hashimoto, M., Nomoto, K., 1990a, A&A 234, 211

Prantzos, N., Hashimoto, M., Rayet, M., Arnould, M., 1990b, A&A 238, 455

Shigeyama, T., Nomoto, K., & Hashimoto, M. 1988, A&A, 196, 141

Shigeyama, T., Nomoto, K., Tsujimoto, T & Hashimoto, M. 1990, ApJ, 361, L23

Sugimoto, D., & Nomoto, K. 1980, Space Sci. Rev., 25, 155

Thielemann, F.-K, Nomoto, K. Yokoi, K., 1986, A&A 158, 17

Thielemann, F.-K., Hashimoto, M., & Nomoto, K. 1990, ApJ, 349, 222

Thielemann, F.-K., Nomoto, K., & Hashimoto, M. 1992, in Les Houches, Session LIV ed. J. Audouze, S. Bludman, R. Mochkovitch & J. Zinn-Justin, in press

Tsujimoto, T., Nomoto, K., Hashimoto, M., & Thielemann, F.-K. 1992, this volume

Woosley, S.E., 1992, this volume

Woosley, S.E., Arnett, W.D., Clayton, D.D., 1973, ApJ, Suppl. 26, 231

Yamaoka, H., Nomoto, K., Shigeyama, T., Thielemann, F.-K., 1992, ApJ, Lett. 393, L55

The r-process: at last some results with microscopic nuclear predictions

S. Goriely and V. Bouquelle

Institut d'Astronomie et d'Astrophysique, Université Libre de Bruxelles C.P. 165, Av. F.D. Roosevelt, 50, B-1050 Bruxelles, Belgium

Abstract.
The influence of nuclear mass models on the r-abundance distribution is investigated in the framework of the standard r-process model. Among the nuclear properties of interest for the r-process, nuclear masses clearly have the most decisive influence. All r-process calculations have so far made use of droplet-type mass formulae only, the reliability of which remains very uncertain along the r-process path (i.e. far off the experimentally known region). A new mass table based on the microscopic Extended Thomas-Fermi plus Strutinsky Integral method is now available and has been used to calculate the production of r-nuclei within the waiting point approximation. The r-abundance distribution obtained with microscopic masses shows significant differences from the one using a droplet-type mass formula. The impact of the nuclear mass model on the nuclear properties of interest for the r-process calculation, namely the neutron separation enegies and the β-decay rates, and on the final r-abundance profile is analysed in detail.

1. Introduction

The rapid neutron-capture process, or r-process, is known to be of fundamental importance for explaining the origin of approximately one half of the $A > 60$ stable nuclei observed in nature, as well as to be responsible for the production of long-lived nucleocosmochronometers used to evaluate the age of our Galaxy. For the last few decades, nuclear astrophysicists have built up more and more sophisticated r-process models, trying desperately to add new astrophysical or nuclear physics ingredients to reproduce the solar system composition in a more satisfactory way. As regards the nuclear physics requirements, the r-process calculations demand the knowledge of the nuclear properties of thousands of (mostly unknown) neutron-rich nuclei. Those include neutron capture cross sections, photodisintegration rates, β-decay half-lives, rates of β-delayed single and multiple neutron emission, α-decay half-lives and β-delayed, as well as spontaneous and neutron-induced fission probabilities. Unfortunately, although a great effort has been devoted to improve the nuclear physics theories required to calculate those properties, some uncertainties still remain on more basic nuclear ingredients, such as the nuclear binding energies. In particular, the nuclear masses determine the β-decay Q-values and the values of the neutron separation energy. Consequently, however powerful the nuclear theories will be to predict the neutron capture cross-sections, the β-rates or the fission probabilities, we cannot expect those calculations to be reliable, as long as the evaluation of the nuclear masses will not have reached a given degree of accuracy.

Recently, the canonical model of the r-process has been used by Goriely and Arnould (1992) in order to study the variations in the r-process abundance predictions implied by the adoption of nuclear masses calculated with a microscopic model of the nucleus, known as the Extended Thomas-Fermi plus Strutinsky Integral method (ETFSI), instead of masses derived from a droplet model (von Groote et al., 1976). The present work extends that first analysis by confronting the r-process yields calculated with the ETFSI masses and those predicted by the latest version of the macroscopic–microscopic mass formula, known as the finite range droplet model (FRDM) (Möller et al., 1992). In Sect. 2, the ETFSI and FRDM predictions of the nuclear properties of interest for the r-process are compared. The influence of the nuclear mass model on the r-abundances is shown and discussed in Sect. 3.

2. The nuclear mass models

The r-process requires the knowledge of the masses of several thousands of nuclei many of which are deformed. Mass predictions for these have until very recently only been obtained through droplet-type mass formulae, among which the so-called FRDM (Möller et al., 1992) constitutes the most sophisticated version. With some 25 adjustable parameters this formula gives a fit to the 1540 experimental masses with an rms error of only 658 keV.

Despite its increasing reliability, the FRDM approach has to face some criticisms, the main one being the lack of coherence between the macroscopic (i.e. the droplet model) and the microscopic (i.e. the shell and pairing corrections) parts. This inconsistent treatment is avoided when using more fundamental approaches, such as the extended Thomas-Fermi plus Strutinsky Integral method (ETFSI) (Dutta et al., 1986). This method is based on the extended Thomas-Fermi approximation to the HF method using a Skyrme-type effective interaction. Such a technique is characterized by a high degree of coherence between the macroscopic and microscopic terms, the unifying factor being the Skyrme force that underlies both parts (Pearson et al., 1991).

Compared with the FRDM, the ETFSI method gives comparable fits to the experimental data (Aboussir et al., 1992). Its rms deviation to the 1492 experimental masses amounts to 730 keV (such a fit has been achieved with only 9 parameters). However the two models extrapolate differently. Figure 1 shows the mass differences between the ETFSI model and the FRDM. The ETFSI method appears to overestimate the nuclear masses with respect to the FRDM for light and medium nuclei ($Z < 70$) while the reverse can be observed for heavier nuclei. Yet both mass models agree within ± 5 MeV in the prediction of the $Z < 80$ masses, while differences as large as 20 MeV were encountered between the droplet masses of von Groote et al. (1976) and the ETFSI masses (Goriely and Arnould, 1992).

Figure 1 also shows important systematic differences at the neutron magic numbers reflecting the differences in the prediction of the shell corrections. In particular, it can be shown that while both mass models globally agree on the strength of the shell corrections at the neutron magic number $N = 82$, the FRDM predicts a stronger $N = 50$ shell effect, while the ETFSI models overestimates, relative to the FRDM, the $N = 126$, $N = 184$ and $Z = 50$ shell corrections. Some differences between the mass models can also be found in the different treatment of the deformation, as will be emphasized in the next Section.

Figures 2 (a)–(b) compare how the two mass models extrapolate the neutron separation energy and the β-decay rate far away from the stability valley. The question now is to know if the differences between the two mass models (± 1 MeV on the neutron separation energy, and a factor of 5 on the β-decay rates) can affect significantly the r-process abundance predictions.

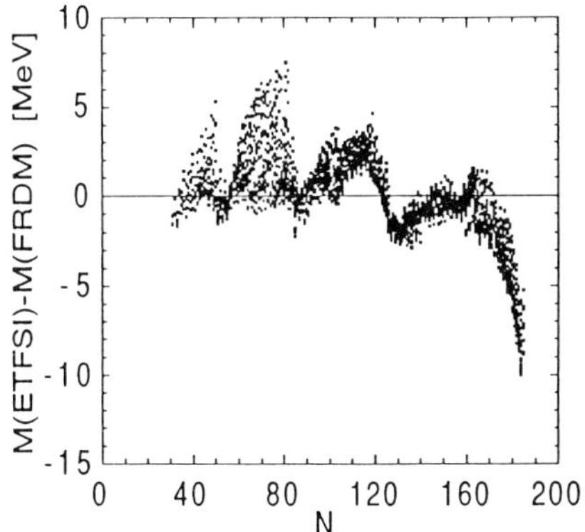

Figure 1. Differences as a function of the neutron number N between ETFSI masses (Aboussir *et al*, 1992) and FRDM masses (Möller et al., 1992) for all the nuclei $20 \leq Z \leq 100$. Only nuclei stable under neutron emission for both nuclear models are considered.

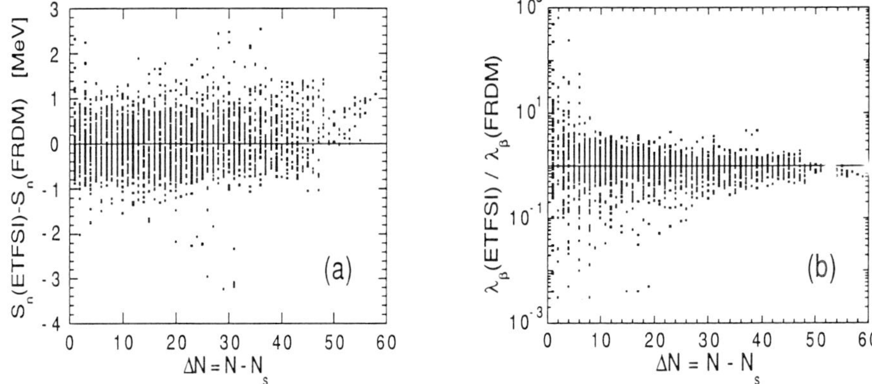

Figure 2. a) Differences of the neutron separation energy as a function of the neutron surplus $\Delta N = N - N_s$ (where N_s is the heaviest stable isotope of the considered element). b) Ratio of the β-decay rates as a function of the neutron surplus. The rates are derived from the gross theory.

3. Influence of the nuclear mass model on the r-process abundances

In order to answer that question, we adopt the canonical r-process model and input nuclear physics already used by Goriely and Arnould (1992).

Figure 3 shows the abundance patterns derived for the arbitrary set of astrophysical parameters $T_9 = 1$, $N_n = 10^{24}$ cm^{-3} (i.e. $S_a^0 = 2$ MeV[†]) and $\tau = 1$ s using both the FRDM and the ETFSI mass models. The two abundance distributions show significant differences due to different predictions of the S_n and Q_β-values, and more particularly of different shell effects and deformations.

The stronger FRDM shell corrections at $N = 50$ leads the r-process path to a region closer to the stability line, and consequently delays the synthesis of the heavy elements. In particular, the difference in the predicted masses of ^{78}Ni is responsible for the faster

---

[†] $S_a^0 = S_{2n}^0/2 = (34.075 - \log N_n + 3/2 \ \log T_9)T_9/5.04$ is defined as the even–odd averaged neutron separation energy and has been shown to be of great interest to characterize the r-process path (Goriely and Arnould, 1992)

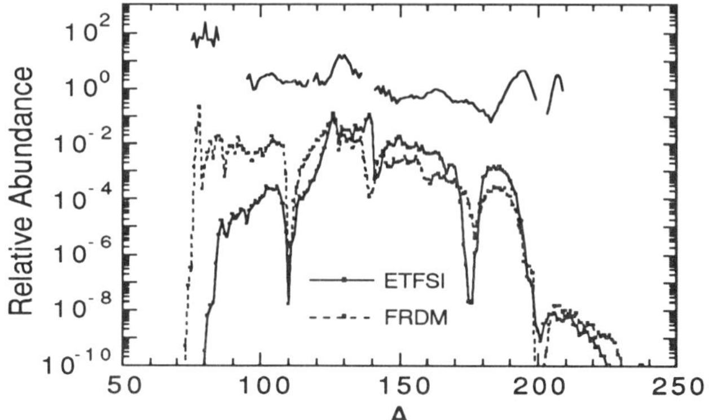

Figure 3. r-process yield curves obtained in the framework of the canonical model for the set of physical parameters $T_9 = 1$, $N_n = 10^{24}$ cm^{-3} and $\tau = 1$ s. The full curve has been obtained making use of the FRDM mass predictions, while the dashed curve corresponds to the use of the ETFSI masses. Both yields have been obtained with the initial abundance N(^{56}Fe)=1. The top curve corresponds to the solar system r-process abundances (Anders and Grevesse, 1989) arbitrarily normalized.

depletion of the light nuclei in the ETFSI case.

The more or less equally strong shell corrections at the neutron magic number $N = 82$ leads to the $A = 127$ peak with the same width and the same depth on the edges. In particular, it can be seen that the same dip on the left side of the abundance peak is present in both profiles. Yet differences can be observed in the $A = 190$ peak due to a stronger ETFSI shell effect, relative to the FRDM prediction, at the $N = 126$ magic number. Therefore, the ETFSI peak is slightly wider and shifted to higher masses. Moreover, it displays a much steeper edge on its left side. The strong $Z = 50$ ETFSI shell closure creates a bottleneck (through an increase of the Q_β) that hinders material from being driven to $Z > 50$. This explains the secondary peak at $A = 139$ which does not appear in the FRDM r-abundance curve.

The different treatment of the deformation in the two mass models also affects the predicted depletion of certain nuclei. In particular, the FRDM abundance depression around A=139 is mainly due to an important variation of the nuclear deformation when passing from the Z=49 to the Z=50 isotopic chain. A similar behaviour in the ETFSI abundance profile can be observed for slightly higher masses ($A = 141$). This effect results from the ETFSI prediction of deformations in the $Z = 51$ isotopic chain.

When other sets of astrophysical conditions with higher S_a^0 values are considered, the r-process paths move away from the neutron drip line and the abundance peaks are shifted to higher A-values. Effects resulting from different shell corrections and deformations predicted by the two mass models remain and still lead to the same characteristic profile.

Such modifications in the abundance distributions have important astrophysical consequences. In particular, fits to the observed abundances constrain the astrophysical parameters in domains that might strongly differ from one nuclear mass model to the other, especially if it is assumed—as is generally done (Kratz et al., 1992)—that the solar system r–abundances originate from a few single supernova events.

As there is no compelling evidence for such a limitation, we have tested the influence of the nuclear mass models by allowing the mixing of a large number of r-process events characterized by the astrophysical conditions: $1 \leq T_9 \leq 3$ (in steps of 0.2),

$10^{18} \leq N_n(\mathrm{cm}^{-3}) \leq 10^{28}$ (in steps of 10), and such that $1.5 \leq S_a^0(\mathrm{MeV}) \leq 6$. Moreover, each of those 89 constant neutron irradiations are assumed to take place during 15 different times τ. Three timescales t_{max}^p have been determined in order to produce the nuclides beyond the three observed peaks at $A^p = 80, 130, 195$. They are defined by

$$t_{max}^p = \sum_{Z=Z_0}^{Z^p} \tau_\beta(Z), \tag{1}$$

where $Z_0 = 26$ is the considered initial seed element, Z^p is the value of the chain for which the waiting point is located at $A = A^p$, and $\tau_\beta(Z)$ is the effective β-decay half-life of the Z–chain. An upper limit of $10\,\mathrm{s}$ is chosen for t_{max}^p. Finally, for each of these three timescales t_{max}^p, 5 equally distributed values $\tau = n/5\ t_{max}^p$ (where $n=1$ to 5) are considered. The r-abundance curve resulting from the superposition of these 1335 r-processes is shown in Fig. 4 for both the ETFSI and FRDM masses, Fig. 5 displaying the cumulative counts of events in each selected bin of thermodynamic conditions.

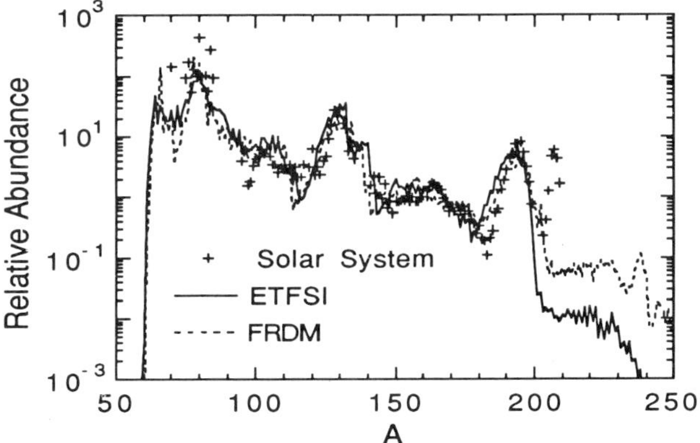

Figure 4. r-abundance distributions obtained, as explained in the text, with the ETFSI and FRDM masses and compared with the observed solar system distribution (crosses).

As can be seen, the fit to the solar system abundances is relatively good and does not depend significantly on the mass model. As a matter of fact, since a great number of r-process paths are considered, the differences in the mass predictions are averaged. However, as the two predicted r-abundance curves result from the superposition of the same sets (T_9, N_n), their temporal histories differ because of differences in the values of $\tau_\beta(Z)$.

4. Conclusion

At this point, it is impossible to tell if a small or a large number of supernova events have contributed to the solar system r-abundances. In the first case, this paper shows that the influence of the mass model on the abundance predictions is drastic. In contrast, the uncertainites related to the exact location of the r-process path are removed when the mixing of a large number of supernova events is considered. However, in the latter case, the timescales that could have led to the solar system abundances

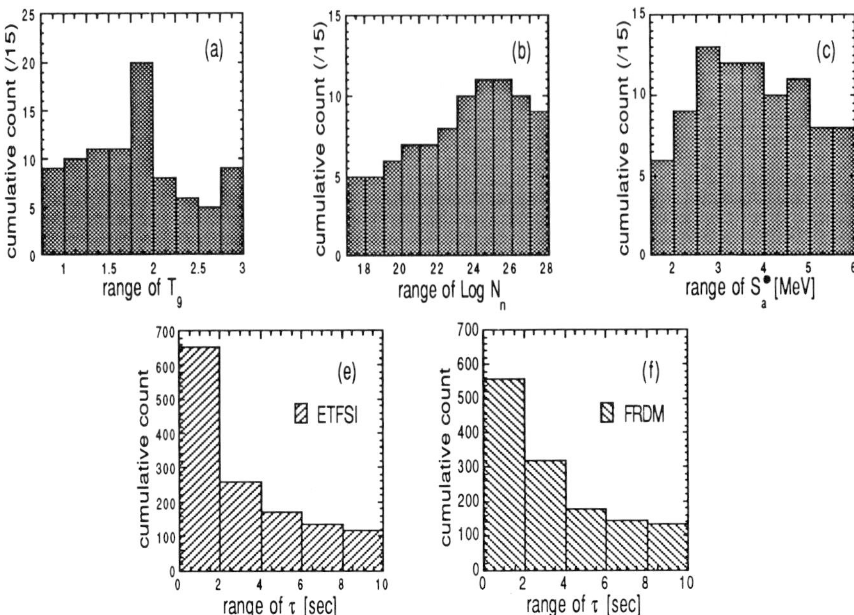

Figure 5. Thermodynamic histograms leading to the r-abundance curves in figure 4, i.e. the cumulative count in each of the ranges of temperature (a), neutron density (b), S_a^0-value, characterizing the r-process path (c) and time in the ETFSI case (d) and FRDM case (e).

still depend on the choice of the nuclear mass model. They are, obviously, also affected by other nuclear physics uncertainties, such as the β-decay theory.

In any case, with the numerous uncertainties remaining in the description of the astrophysical environments as well as in the theoretical estimate of the nuclear physics input, the reliability of the r-process calculations comes naturally to the mind. It is our opinion that these problems should call for a much deeper study of each of the input data before rushing into numerical results and, more particularly, into premature comparisons of these results with the observed r-abundances.

Acknowledgments

This work has been supported in part by the SCIENCE Program SCI–0065. S.G. is F.N.R.S. Research Assistant.

References

Aboussir, Y., Pearson J.M., Dutta, A.K., Tondeur, F.: 1992, *preprint*
Anders, E., Grevesse, N.:1989, *Geochim. Cosmochim. Acta* **53**, 197
Dutta, A.K., Arcoragi, J.-P., Pearson, J.M., Behrman, R., Tondeur, F.: 1986, *Nucl. Phys.* **A458**, 77
Goriely, S., Arnould, M.: 1992, to appear in *Astron. Astrophys.*
Kratz,K.-L., Möller, P., Pfeiffer, B., Thielemann F.-K., Wöhr, A.: 1992, *This Volume*
Möller, P., Nix, J.R., Myers, W.D., Swiatecki, W.J.: 1992, *At. Data Nucl. Data Tables* to be published
Pearson, J.M., Aboussir, Y., Dutta, A.K., Nayak, R.C., Farine, M., Tondeur, F.: 1991, *Nucl. Phys.* **A528**, 1
von Groote, H., Hilf, E.R., Takahashi, K.: 1976, *At. Data Nucl. Data Tables* **17**, 418

Nucleosynthesis in neutrino driven type-II supernovae

J.Witti, H.-Th. Janka, K.Takahashi, W. Hillebrandt

MPI für Astrophysik, Karl-Schwarzschild-Straße 1, W-8046 Garching, FRG

Abstract.   It has recently been suggested that the astrophysical r-process might take place in the so-called 'hot bubble' region around a newly formed neutron star. This extended region of low density and high entropy is a direct result of heating by neutrinos, which escape from the central compact remnant in a type-II supernova. We have further pursued that possibility by using data from a hydrodynamical simulation of the time evolution of such a high-entropy supernova bubble. In particular, we have investigated the α-rich freeze-out following the breakdown of nuclear statistical equilibrium in the expanding and cooling matter. Our results indicate that the amount and composition of processed material yields a good starting point for a subsequent r-process.

1. The hot bubble

Type-II supernovae are regarded as a very promising site for the production of heavy elements via the rapid neutron capture process. However, in previous scenarios that explained the explosion by a hydrodynamical 'prompt' mechanism on shock propagation time scales, the r-process was assumed to take place in matter with high density. Therefore, extreme neutron excesses were required to get enough neutrons per seed nucleus. Furthermore the amount of processed material per supernova event was found to be much too large to explain the abundances in the galaxy (see e.g. Hillebrandt [1]).

The delayed explosion mechanism reveals quite different conditions. According to current models for the explosion of massive stars of $8M_\odot \lesssim M \lesssim 25M_\odot$ in a type-II supernova, energy deposition by neutrinos just outside the newly formed neutron star causes a violent expansion of the overlaying material. On a time scale of about 1-2 seconds after core bounce (*'delayed explosion'*) this heating may transfer enough energy to drive a powerful shock into the mantle of the progenitor star and to explain the energetics of a type-II supernova event (see Fig. 1). Between the central neutron star and the overlaying stellar mantle an extended, rapidly expanding region of low density ($\rho \approx 10^5 g/cm^3$) and typical temperatures of about $6 \cdot 10^9 K$ is created (*'hot bubble'*) (see Figs. 1 and 2). Absorption of electron neutrinos ν_e on neutrons and electron antineutrinos $\bar{\nu}_e$ on protons contributes most to the heating in this region. Since capture of antineutrinos slightly dominates, these reactions also tend to produce a neutron excess, i.e. to drive the electron concentration Y_e to values below 0.5.

2. The α-rich freeze-out

As shown by Woosley & Hoffman [2] and by Meyer [3], the high entropies in the matter blown away from the neutron star enable the production of heavy nuclei during

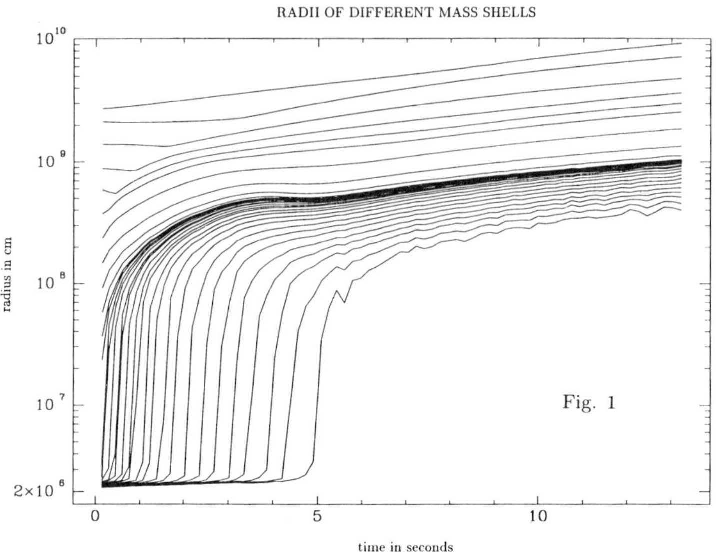

Figure 1. Above the newly formed neutron star in the center of a type-II supernova a region of low density and high entropy, the so-called *'hot bubble'*, is formed by neutrino heating. The matter inside this bubble is rapidly expanding, accompanied by additional material blown off from the neutron star surface by the action of neutrinos: *'neutrino wind'*.

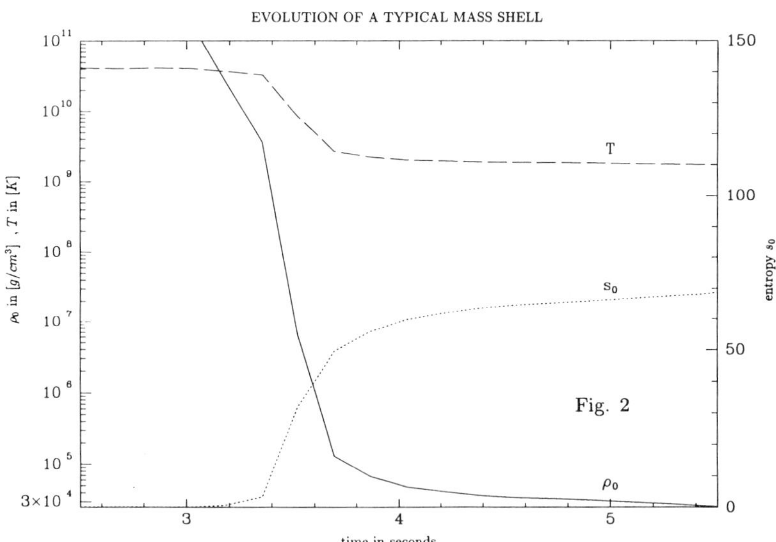

Figure 2. Different matter elements in the neutrino wind run through a similar evolution: Temperature and density drop quickly, while the radiation entropy per baryon ($s \propto T^3/\rho$) is raised significantly due to neutrino heating. Such conditions should lead to the α-rich freeze-out. The plotted evolution of a mass shell defines our standard background model.

an α-rich freeze-out phase which follows the cooling of the expanding matter and the breakdown of nuclear statistical equilibrium. The density drops very quickly, and so only a fraction of the initially unbound nucleons and α-particles can recombine to heavier elements. This is due to the fact that the crucial three-body reactions $\alpha + \alpha + n \rightarrow {}^9Be$ and $3\alpha \rightarrow {}^{12}C$ diminish rapidly with decreasing density. Thus one ends up with material containing mostly α-particles and neutrons, and only a small fraction of nuclei with mass numbers around 100 (see Fig. 4). This contamination with heavy nuclei serves as seed for a subsequent r-process, since under the described conditions sufficiently high neutron to seed ratios appear, even without very large neutron excesses.

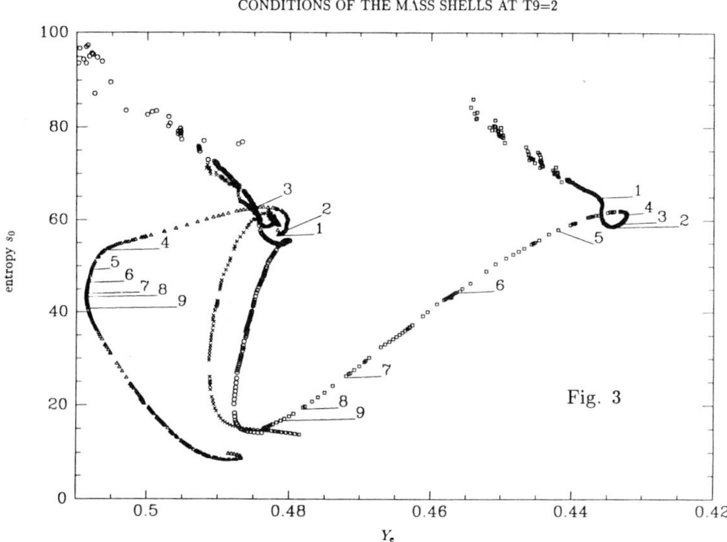

Figure 3. The critical parameters for the α-rich freeze-out are the entropy per baryon s (given in units of the Boltzmann constant), and the electron concentration Y_e, which determines the neutron excess. The symbols in the figure distinguish the conditions in samples of mass shells for four different hydrodynamical models of the hot bubble evolution. The numbers denote the integral mass content of the shells above the marked points in $10^{-3} M_\odot$. Mass elements with higher entropies are blown off the neutron star surface at later times.

3. Results

In the presented work we have studied the conditions and the outcome of the α-rich freeze-out (Figs. 4–7). Our investigations are based on, and interpreted in their meaning for, detailed hydrodynamical models of type-II supernova explosions (Janka [4]). The conditions of four models we examined so far are displayed in Fig. 3. Unfortunately the entropies are too low to allow for a successful freeze-out and r-process. We determined the required conditions by a parametric study (Fig. 5), where we varied Y_e in the limits of 0.49 to 0.33 and artificially increased the radiation entropy $s \propto T^3/\rho$ by scaling down the density along the evolution track of a representative mass shell (Fig. 2). Because about 100 free neutrons for each seed nucleus are needed to form elements

FREEZE-OUT PATH AND DISTRIBUTION OF THE MASS FRACTION
IN THE Z-A-PLAIN ($T_9 = 4.5$)

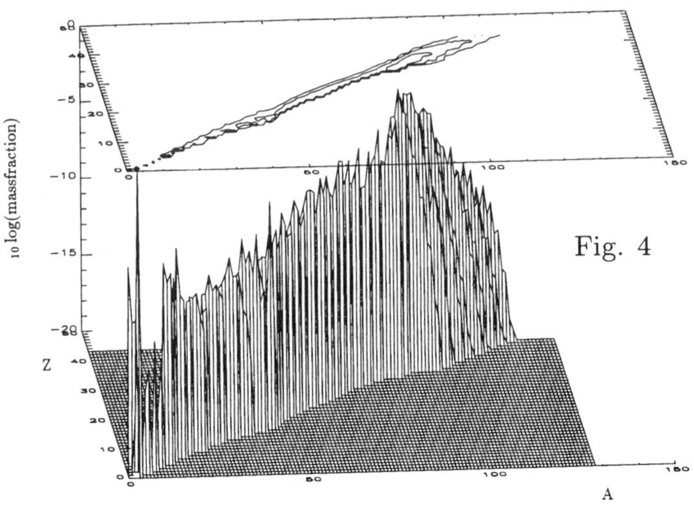

Figure 4. During the α-rich freeze-out (displayed results are for a calculation with $s = 10 \cdot s_0$ and $Y_e = 0.40$) α- and n-captures build up the nuclei along the valley of stability and shift them to high charge numbers Z and mass numbers A. When all charged particle reactions have ceased, the nuclei will move to the n-rich side, and the r-process will take over.

NEUTRONS TO SEED RATIO
shaded with produced mass fraction ($A > 50$)

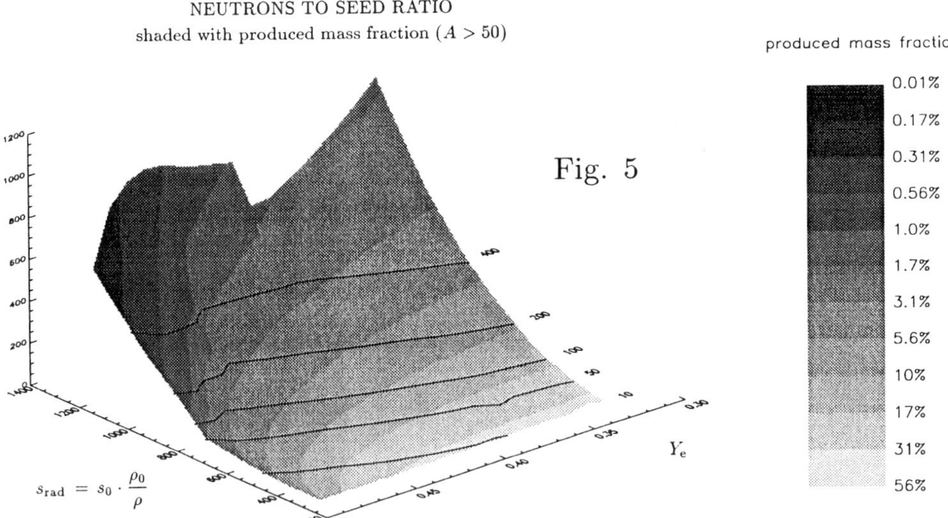

Figure 5. This figure shows the results for the α-rich freeze-out from a parametric study, where the quantities $s(t)$ and $Y_e(t)$ of our standard model (Fig. 2) were scaled up and down, respectively. With an abundance peak around $A = 100$ in a successful freeze-out one arrives at about 100 neutrons per seed nucleus and a mass fraction of about 10% in heavies for conditions $Y_e \approx 0.45$ and $s \approx 600k_B/$nucleon.

with mass numbers over 200, our results indicate that with Y_e between 0.48 and 0.40 entropies of about 600 k_B/nucleon (at $T = 2 \cdot 10^9 K$) are required for a promising r-process. Since the radiation entropy increases with time, this corresponds to a value of around 300 k_B/nucleon for the relevant freeze-out temperatures around $T = 5 \cdot 10^9 K$. For such conditions the mass fraction of produced seed nuclei, and therefore also of the final r-process nuclei, is about 10% (Fig. 6).

Finally, we have studied the dependence of the α-rich freeze-out on the time scale of the matter expansion (Fig. 7): In a faster freeze-out, nucleons and α-particles have less time to recombine. Therefore a smaller amount of heavies is produced, and the final neutron to seed ratio stays higher. This means that a faster expansion has the same effect as an increase of entropy.

As stated above, our hydrodynamical models show too low values for the radiation entropy in the hot bubble region when the α-rich freeze-out takes place. Supposed that in all relevant mass shells of our hydrodynamical models (Fig. 3) the density were lower by the appropriate factor of 6–10, the whole mass with suitable conditions would be less than about $10^{-3} M_\odot$. Roughly 10% of this material is consumed by the r-process, thus we get an estimate of the order of $10^{-4} M_\odot$ r-process material per supernova event. This is just what one would expect from the age of the galaxy, the supernova rate and the observed abundances of r-process elements.

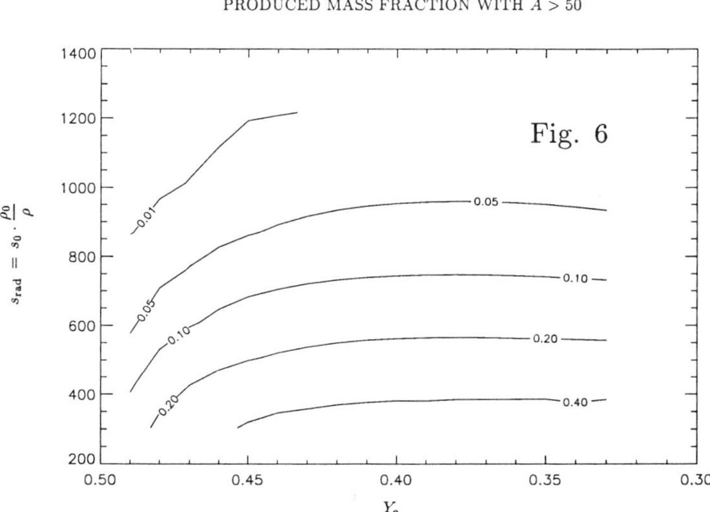

Figure 6. A mass fraction of 10–15% in heavies demands an entropy of 400 to 600 k_B/nucleon. This means that the density has to be lower by a factor of up to 10 compared with our standard model (Fig. 2). This yields, with 10% of the mass in heavy nuclei, around $10^{-4} M_\odot$ seed (see Fig. 3). These numbers are of the right order of magnitude to explain the content of r-process elements in our galaxy as originating from type-II supernova events.

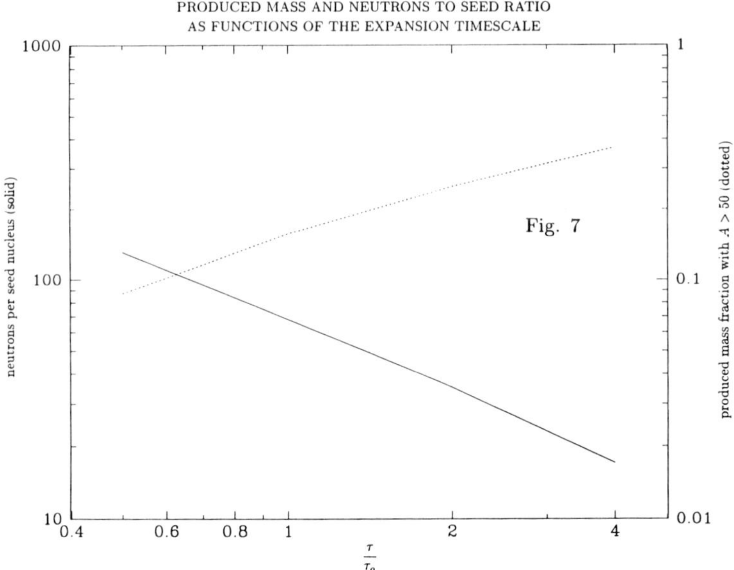

Figure 7. Reducing the time scale of the matter expansion has a similar effect as increasing the entropy, because the time for the α-particles to reassemble and for subsequent capture of neutrons is shorter. During a more rapid evolution a smaller amount of heavies is produced (dotted line) and more free neutrons remain (solid line), improving the conditions for the r-process.

4. Conclusions

The creation of heavy elements in a neutron-rich environment during the α-rich freeze-out allows a subsequent r-process to take place. This will be investigated in a continuation of the presented work. The conditions and the amount of material in the hot bubble region around a newly formed neutron star suggest type-II supernovae to be a good site for the formation of r-process elements.

Acknowledgments

H.-Th. Janka would like to thank the Höchstleistungsrechenzentrum HLRZ des Forschungszentrums Jülich (KFA) for grants of computer time on the Cray Y-MP.

References

[1] W. Hillebrandt, 1978, SSR,**21**,639
[2] S. E. Woosley and R. D. Hoffman, 1992, to appear in Astrophys. J.
[3] B. S. Meyer *et al.*, preprint (1992).
[4] H.-Th. Janka, in preparation.

Nuclear statistical equilibrium, neutron-rich α-rich freeze out and the r-process

W. M. Howard, S. Goriely, M.Rayet and M. Arnould

Institut d'Astronomie et d'Astrophysique
Université Libre de Bruxelles
Brussels, Belgium

*Abstract:*We study the high-entropy alpha-rich freeze out of neutron-rich nuclei as suggested by Woosley and Hoffman (1992) and the subsequent formation of r-process elements. In particular, we use a single dynamical reaction network to study the establishment of nuclear statistical equilibrium (NSE), the charged-particle freeze out and the subsequent neutron-captures and beta decays that produce r-process elements. We study the effect of using various nuclear models for the nuclear binding energies, and in particular, the extended Thomas-Fermi plus Strutinsky Integral method and the finite-range droplet model. We also use self-consistent beta decay rates calculated with the gross theory. We compare some of our results with those of other researchers.

1   Introduction

Recently much interest has been generated by the study of the neutron-rich alpha-rich freeze out that is expected to occur in the neutrino-energized wind that emerges from the nascent neutron star in a Type II supernova (Woosley and Hoffman, 1992, hereafter WH). Such a wind is expected to have high entropy and relatively low Y_e due to the interaction of electron neutrinos and anti-neutrinos with the protons and neutrons escaping in the wind. It was recognized by WH that such conditions might provide an attractive site for the astrophysical r-process, as well as contributing to the Galactic production of some intermediate mass elements that are usually ascribed to the astrophysical r- and p-processes. Meyer et al. (1992) demonstrated that a solar-system distribution of r-process nuclei may be produced for an interesting range of entropy and neutron excess.

2   Reaction Network

In view of the large uncertainties in the nuclear physics of the r-process, as well as the uncertainties of the modeling of the astrophysical site, we plan to study the formation of r-process elements with various nuclear models and for a wide range of entropy and Y_e conditions. We first construct a nuclear reaction network that can calculate the establishment of nuclear statistical equilibrium (NSE) at high temperature and density, as well as the subsequent freeze-out of charged-particle induced reactions and the ensuing r-process. Since the nuclear binding energies affect the establishment of NSE, as well as the r-process path and the beta decay rates, we supplement experimentally determined masses with the recent mass calculations based on the Extended Thomas-Fermi plus

Strutinsky Integral (ETFSI) approach (Pearson et al. 1991). We also plan to use in the future the finite-range droplet model masses of Möller et al. (1988, 1992). We have a proper transition from experimentally to theoretically determined masses by taking differences only between sets of experimental or theoretical masses to determine binding energies.

Apart from linking protons and neutrons with ^4He, our network includes nuclides up to $Z = 90$ ranging from the valley of β-stability up to $Y_e = 0.30$. Proton- and α-particle induced reactions are considered up to Sn and Sm, respectively. Their inclusion is necessary in order to follow properly the neutron-rich and α-rich freeze out. All the necessary neutron captures and β-decays, as well as reverse reactions, are also duly taken into account. The network is solved implicitly without linearization of the differential equations using the LSODE Solver (Hindmarsh, 1983).

The rates of the charged-particle reactions on targets up to Si are taken from Caughlan and Fowler (1988). The rates for heavier targets and for neutron captures are calculated with the Hauser-Feshbach model of Thielemann et al. (1986). Some neutron capture rates on light species are taken from more recent experimental data (Thielemann et al. 1991). The reverse reaction rates are calculated from detailed balance using the ETFSI masses when experimental data are not available, and temperature-dependent partition functions obtained from known nuclear levels and the level density formula of Thielemann et al. (1986). The experimentally determined β-decay rates are complemented with the predictions of Fuller et al. (1985) and of the gross theory (Takahashi et al. 1973). In the latter case, the rates are calculated with the ETFSI masses.

Our code has been tested by replicating some of the NSE calculations of Hartmann et al. (1985), many of the calculations of WH, and a r-process as calculated by Meyer et al. (1992). In this paper, we discuss two examples of these calculations. The first one is for a low-entropy freeze out, and the other one is for a high-entropy freeze out that establishes a r-process.

3   Results

We distinguish betweeen a low-entropy NSE freeze out and a relatively high entropy α-rich freeze out. At very high temperatures ($T > 6 \times 10^9 K$), NSE favors protons, neutrons and α-particles. As the matter cools the protons and neutrons recombine into α-particles. In turn these recombine into heavier nuclei through the triple-alpha reaction to ^{12}C, or through $^4He(\alpha n, \gamma)^9Be(\alpha, n)^{12}C$ in a neutron-rich ($Y_e < 0.45$) freeze out. Both of these reaction paths are extremely sensitive to the density, as well as of course to the temperature. Moreover, the reverse reactions are also important in determining the freeze out abundances. The final α-particle abundance is determined by the extent of their recombination in the cooling and expanding gas. For low Y_e and high entropy, we find that $^{12}C(n, \gamma)^{13}C$ is also important for determining the final α-particle abundance. For the $^{12}C(n, \gamma)^{13}C$ rate we use a $1/v$ extrapolation of the 30 keV cross section measured by Nagai et al. (1991). At high entropy only some of the α-particles may recombine to heavier nuclei. When only a few percent of the α-particles do so, the neutron density can remain high because the neutrons are not captured on α-particles. This ensures a high neutron to seed ratio and helps establish a r-process. For a low-entropy freeze out, the α-particles efficiently recombine to heavier nuclei.

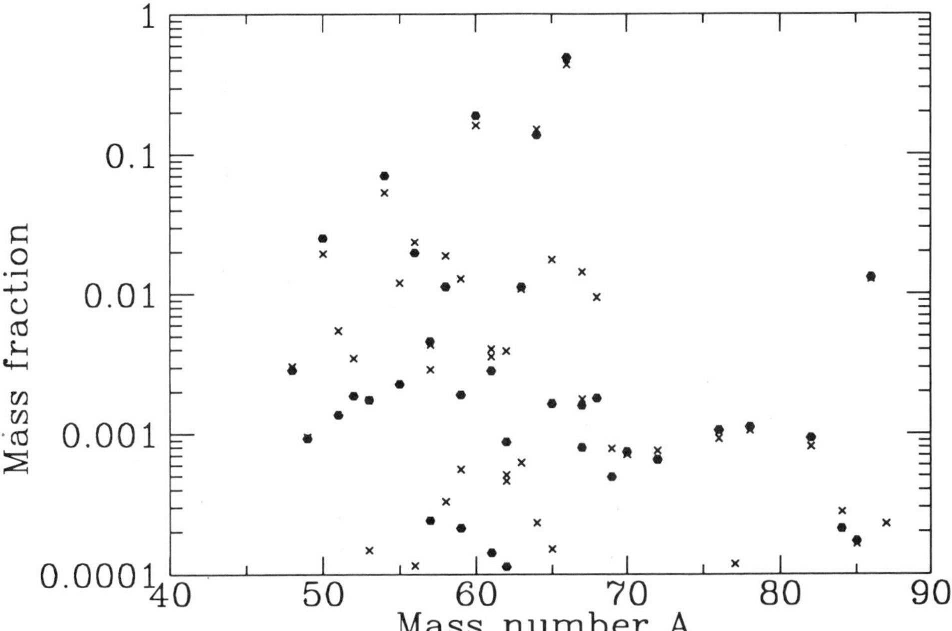

Figure 1: Abundances at freeze out from NSE for an entropy per baryon of 20 and $Y_e = 0.43$, for two values of the freeze-out temperature. The light crosses correspond to a freeze out temperature of $T_9 = 3.5$ and the dark circles to a freeze out temperature of $T_9 = 1.0$.

However, as pointed out by WH, the neutrons play a different role in this case. Almost all the matter forms heavy nuclei (near the iron group), so that all the neutrons are incorporated into these nuclei during freeze out, producing such isotopes as ^{54}Cr, ^{60}Fe and ^{66}Ni.

The first test calculation we perform concerns a relatively low entropy per baryon of 20, where an α-rich freeze out is not expected to occur (Hartmann et al. 1985). The initial temperature is taken as $T_9 = 10$ and the initial density as $\rho = 1.7 \times 10^7 g/cm^3$, with $Y_e = 0.43$. The expansion is assumed to be adiabatic on a hydrodynamical timescale. The result is shown in Fig. 1 for two values of the freeze-out temperature, that is the temperature at which the charged-particle induced reactions are stopped. This is a classical NSE freeze-out calculation, where the final α-particle mass fraction is approximately 4×10^{-5}. The main products are ^{66}Ni, ^{64}Ni, ^{60}Fe, ^{54}Cr and ^{50}Ti. At $T_9 = 3.5$, the nuclei are still in NSE, so the differences in abundances at the two temperatures are deviations from NSE due to charged-particle reactions. As one can see, the freeze-out effects are rather small for the most abundant nuclei, while the differences can be an order of magnitude or more for the rare ones such as ^{65}Ni, ^{67}Ni and ^{68}Ni. The ^{66}Ni decays to ^{66}Zn, which is the most abundantly produced isotope. If this process were the source of the Galactic ^{66}Zn, then it would also contribute 20 percent to the Galactic ^{64}Ni, ^{54}Cr and ^{50}Ti.

The second calculation is an α-rich freeze out starting at $T_9 = 10$, $\rho = 8.543 \times 10^5 g/cm^3$ and $Y_e = 0.35$. This corresponds to an entropy per baryon of approximately 400, and

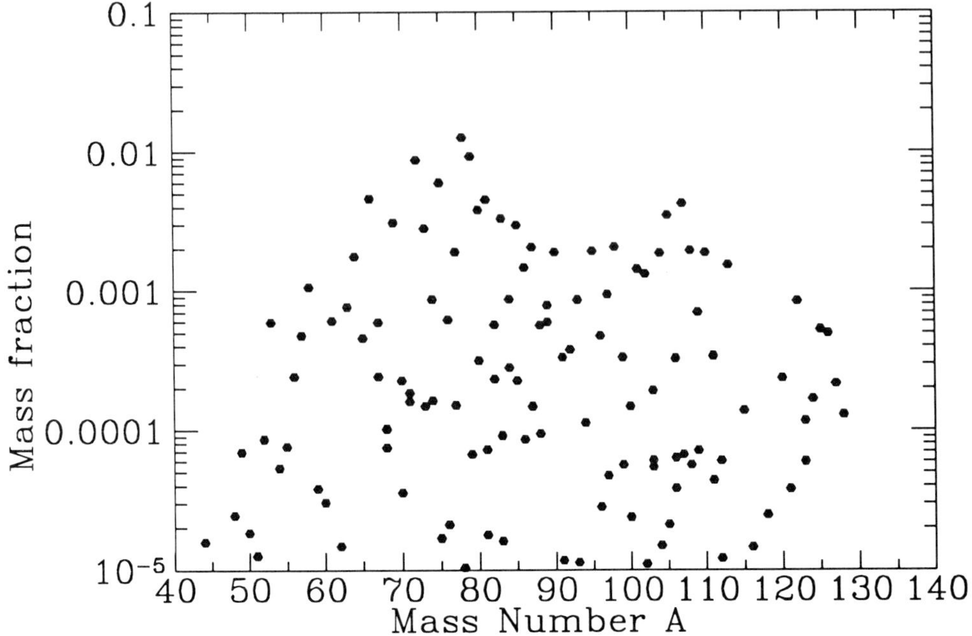

Figure 2: Abundances for a neutron-rich ($Y_e = 0.35$) α-rich freeze out at a time $t = 0.3257s$. The charged-particle reactions produced these seeds and with a large free neutron abundance, a r-process will ensue.

provides enough free neutrons for a subsequent r-process. We follow the temperature and density profile from WH for the case of $Y_e = 0.35$ and $\rho/3$ (Table 9 of WH). Figures 2 and 3 show distributions at times $t = 0.3257s$ and $t = 2.9s$, respectively. Time $t = 0.3257s$ corresponds to the point where WH give a detailed abundance distribution that is used by Meyer et al. (1992) to calculate the subsequent r-process. The seed nuclei (approximately 10 percent by mass) have started to capture neutrons and to make nuclei beyond the $N = 50$ closed shell. However, the $^4He(\alpha n, \gamma)^9Be(\alpha, n)^{12}C$ reaction sequence has not frozen out and continues to generate seeds at a significant rate until approximately $t = 0.8s$, at which point we have a neutron-to-seed ratio of approximately 91. The freeze out value of the α-particle mass fraction is 0.575. At $t = 0.3257s$, the abundance distribution is peaked near the $N = 50$ closed shell in the form of ^{78}Ni, ^{79}Cu and ^{80}Zn. There is a secondary peak beyond the $N = 50$ closed shell at ^{105}Y and ^{107}Y. The peak near ^{78}Ni agrees with the calculation of WH. However, we disagree with WH beyond $N = 50$, where they obtain a peak at ^{90}Se. We attribute this difference to the the fact that the experimentally unknown masses are predicted from different formulae. WH adopt the mass relation by Comay et al. (1988), while we use the ETFSI model, which should be more reliable for extrapolation to neutron-rich nuclei (Möller and Nix, 1992).

Figure 3 shows the distribution at time $t = 2.9s$ when the neutrons have just frozen out. At that point $T_9 = 0.8$ and the free neutron density is $n_n = 2.5 \times 10^{20} cm^{-3}$. After being produced at earlier times, the peak at the $N = 82$ closed shell is eroded and the peak at $N = 126$ is well established. Additional smoothing can be provided by β-delayed

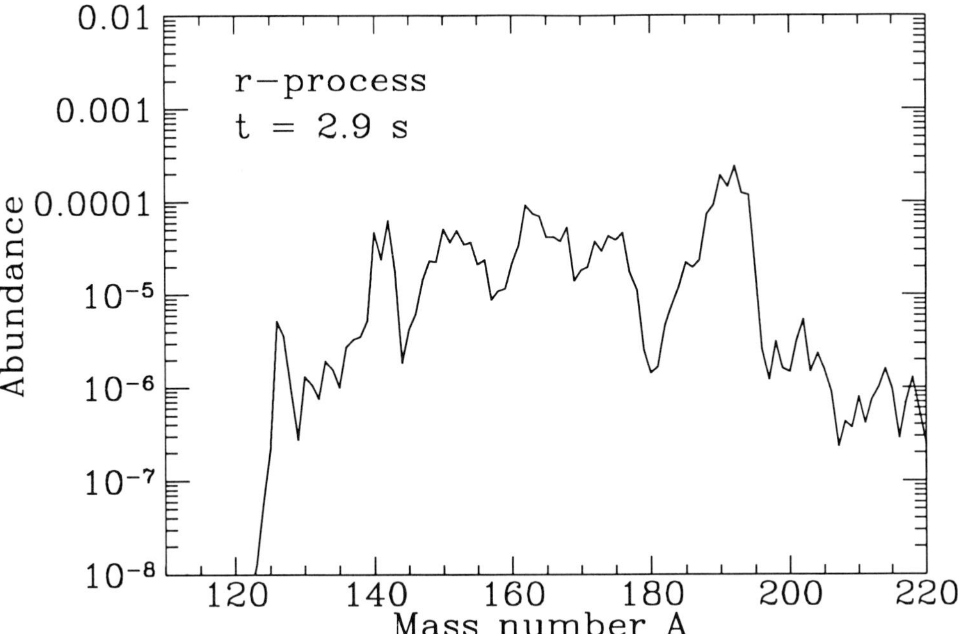

Figure 3: Abundance distribution at a time $t = 2.9s$.

neutron emission which is not included our calculation. We have the same ack of r-process nuclei near $A = 120$ that is observed in other calculations. We also predict a dramatic drop-off before the $N = 126$ closed shell that is probably due to our use of the ETFSI masses. Our results are in contrast to those of Meyer et al. (1992) who predict a freeze out of the neutron reactions at $t = 2.0s$. We find that the neutron freeze out is *extremely sensitive* to the generation of seed nuclei, which in turn depends drastically on the light charged-particle and neutron capture reactions below ^{20}Ne. In addition, as stated above, we obtain a significant seed production beyond $t = 0.3257s$. Moreover, the β-decay rates used in our calculation for the neutron-rich nuclei are slower than the rates of Klapdor et al.(1984) used by Meyer et al. (1992) and WH by almost an order of magnitude. Differences in the β-decay rates and the seed generation rates probably account for our later neutron freeze out, but we have not explored this issue in detail. We report this r-process calculation as a preliminary result. Obviously, much future work and exploration is necessary. A complete description of our code, along with detailed results, will be published elsewhere.

4 Conclusions

We have developed a code to follow the establishment of NSE in a high temperature and high density environment, the freeze-out of the charged-particle induced reactions, and the subsequent r-process in the case of a neutron-rich high-entropy environment. We have discussed some of the physics input of our code and have demonstrated two examples of its use. The study of NSE and the neutron-rich α-rich freeze out will be a fruitful topic for many years to come.

WMH would like to thank the Institut d'Astromonie et d'Astrophysique of the Université Libre de Bruxelles for its kind hospitality and for a generous allocation of computer time. He is also thankful for the financial support of the Fonds National Belge de la Recherche Scientifique (F.N.R.S.) during his sabbatical stay. This work is supported by the SCIENCE Programme SCI-0065 and the US Department of Energy through contract with the Lawrence Livermore National Laboratory. M. A. and M. R. are F.N.R.S. Research Associate and S. G. is F. N. R. S. Research Assistant.

REFERENCES

Caughlan, G. R. and Fowler, W. A. 1988, *Atomic Nucl. Data Tables*, *40*, 283.

Comay, E., Kelson, I. and Zidon, A. 1988, *Atomic Nucl. Data Tables*, *39*, 235.

Fuller, G. M., Fowler, W. A. and Newman, M. J. 1982, *Astrophys. J. Suppl.*, *42*, 447.

Hartmann, D., Woosley, S. E. and El Eid, M. 1985 *Astrophys. J.*, *297*, 837.

Hindmarsh, A. C. 1983, in *Scientific Computing*, eds. R. S. Stepleman et al., North Holland, Amsterdam, 55.

Klapdor, H. V., Metzinger, J. and Oda, T. 1984, *Atomic Data Nucl. Data Tables*, *31*, 81.

Meyer, B. S., Mathews, G. J., Howard, W. M., Woosley, S. E. and Hoffman, R. D. 1992, *Astrophys. J.*, in press.

Möller, P., Myers, W. D., Swiatecki, W. J. and Treiner, J. 1988 *Atomic Nucl. Data Tables*, *39*, 225; also P. Möller, P., Nix, J. R., Myers, W. D. and Swiatecki, W. J. 1992, *Atomic Data Nucl. Data Tables*, to be published.

Möller, P. and Nix, J. R. 1992, *Proc. 6th International Conference on Nuclei Far from Stability and 9th International Conference on Atomic Masses and Fundamental Constants*, Bernkastel-Kues, Germany, 19-24 July, 1992, to be published.

Nagai, Y., Igashira, M., Takeda, K., Mukai, N., Motoyama, S., Uesawa, F., Kitazawa, H. and Fukude, T. 1991, *Astrophys. J.*, *372*, 683.

Pearson, J. M., Aboussir, Y., Dutta A. K., Navak, R. C., Farine, M. and Tondeur, F. 1991, *Nucl. Phys.*, *A528*, 1.

Takahashi, K., Yamada, M. and Kondoh, T. 1973, *Atomic Data Nucl. Data Tables*, *12*, 101.

Thielemann, F.-K., Arnould, M. and Truran, J. W. 1986, in *Adances in Nuclear Astrophysics*, eds. E. Vangioni-Flam et al., Editions Frontiéres; Gif-sur-Yvette, 525.

Thielemann, F.-K., Applegate, J. H., Cowan, J. J. and Wiescher, M. 1991, in *Nuclei in the Cosmos*, eds. H. Oberhummer and C. Rolfs, Springer-Verlag, Berlin, 147.

Woosley, S. E. and Hoffman, R. D. 1992, *Astrophys. J.*, in press.

p-Process nucleosynthesis in pair creation supernovae

M. Rayet[1], M. El Eid[2], M. Arnould[1]

[1] Institut d'Astronomie et d'Astrophysique, Université Libre de Bruxelles, CP165,
50 av. F.D.Roosevelt, B-1050 Bruxelles, Belgium.
[2] Universitäts-Sternwarte Göttingen, Geismarlandstr. 11, W-3400 Göttingen, Germany

Abstract. This paper presents the first detailed calculation of the p-process in very massive stars leading to pair-creation supernovae. The specific stellar model considered here leads to a substantial production of the p-nuclides from Ba to Hg in relative proportions close to solar, the deficiency of the lighter p-nuclides resulting from their lock-up in a black hole remnant. The impact of a change of the remnant mass $M_{b.h.}$ on the predicted production of p-nuclides is also explored, different $M_{b.h.}$ values being possibly obtained for models that differ slightly from the one explored here. We also demonstrate for the first time that hydrostatic O/Ne burning can provide a dominant component to the total p-process yields from certain explosive events. The calculations described here will be extented to other models of very massive stars with different masses and metallicities.

1. Introduction

The so-called p-process is made responsible for the stellar synthesis of the few ($\approx$ 35) proton-rich nuclei heavier than Fe (p-nuclei), which are bypassed by the neutron capture processes.

As first proposed by Arnould (1976), that process could develop in the O/Ne layers of massive stars already at the pre-supernova stage, where temperatures in excess of 10^9 K can lead to the synthesis of the p-nuclides by the photodisintegration of seed s- and r-nuclei. Suitable conditions could also be obtained in those layers at the supernova stage, as explored by Woosley and Howard (1978), and as confirmed by detailed calculations based on realistic models for stars in the $M_{ZAMS} = 15-25\,M_\odot$ range exploding as type II supernovae (SNII) (Prantzos *et al* 1990, Arnould *et al* 1992). In both hydrostatic and explosive burning conditions, most of the p-nuclei can be produced in proportions close to solar. Conditions appropriate to the p-process could also be achieved in a very thin external zone of massive white dwarfs exploding as type Ia supernovae (SNIa) (Howard *et al* 1991). Contrary to some initial hope, this scenario does not appear able to cure some of the problems faced by the SNII models, like the underproduction of the Mo and Ru p-isotopes (Howard and Meyer, this Conference).

It has also been suggested that very massive stars ending their life as pair-creation supernovae (PCSN) are efficient producers of p-nuclei (Rayet *et al* 1988). The late evolution of stars with an initial mass of $\sim 100\,M_\odot$ or more consists of a quasi-hydrostatic contraction phase followed by the explosion of their massive oxygen core triggered by an electron-positron–pair instability (El Eid and Langer 1986). During the quasi-hydrostatic phase, as well as during the explosion, the temperature exceeds 10^9 K in a large fraction of the core, where p-nuclei are therefore expected to be produced. PCSN's thus offer a natural site for a p-process occuring both (quasi-)hydrostatically and explosively inside the same star. This paper describes the first quantitative study of such a unique scenario.

The adopted stellar model leading to a PCSN, the nuclear physics input, and the seed abundances are described in Sect. 2. The derived p-nuclei abundances are presented and discussed in Sect. 3. Some conclusions are drawn in Sect. 4.

2. The input physics

2.1. The stellar model

The evolution of a star with initial mass $M_{ZAMS} = 140\,M_\odot$ is computed from the main sequence up to the PCSN explosion. The convective zones are determined by the Schwarzschild criterion, and no overshooting is included. As far as mass loss rates are concerned, the values recommended by de Jager et al. (1988) are adopted during most of the main sequence phase. They are increased artificially close to the termination of that phase in order to prevent the model star to evolve to the red giant branch, where objects with the derived high luminosity are not observed. During the Wolf-Rayet phase, an average mass loss rate of $3 \times 10^{-5}\,M_\odot\,\mathrm{y}^{-1}$ is adopted, which lies in the range observed for those objects (Van der Hucht 1992). On the other hand, the Los Alamos opacities (Huebner *et al* 1977) are adopted (the new Livermore opacities were not yet available when our calculations have been performed). Finally, the energy production and the nucleosynthesis during all the evolutionary phases are calculated from a detailed nuclear reaction network (see Sect. 3).

With this input physics, the evolution of the star proceeds as follows. By the end of core helium burning, the model star resembles a WNL object. It is made of an oxygen core of $\sim 50\,M_\odot$ surrounded by an extended helium envelope still containing some hydrogen, and enriched in nitrogen.

The further evolution of the considered star proceeds through short phases of core carbon and neon burnings, followed by a core oxygen burning stage during which the conditions are suitable for the creation of $e^\pm$–pairs that lead to the collapse of the inner core, accompanied with the explosive burning of about 4 $M_\odot$ of oxygen on a time scale of ≈ 50 s (El Eid 1991). The associated release of energy is sufficient to reverse the collapse into an explosion with a kinetic energy of 4.4×10^{51} erg.

The time evolution of the temperature in various layers of the oxygen core during its quasi-hydrostatic contraction, collapse and explosion is shown in Fig. 1. It appears that some of those layers are well suited for the development of a p-process.

A full treatment of the radiative transport (Herzig 1992) after the explosive oxygen burning episode (temperatures lower than about 10^9 K) leads to the prediction that the most central 19 $M_\odot$ of material reimplode into a black hole remnant that traps some of the layers that are suited to the development of the p-process (see Fig. 1). It has to be stressed that the remnant mass is a very sensitive function of the peak temperatures achieved during the collapse phase. This in turn implies a high sensitivity to the total amount of consumed oxygen, or to the initial mass of the oxygen core, its value relative to M_{ZAMS} depending upon various ingredients of the model (metallicity, mass loss rate, description of the convection, extent of the overshooting,...). In order to examine qualitatively the impact of a change of the black hole mass $M_{b.h.}$ on the predicted p-process yields, the composition of the ejecta has been calculated for various remnant masses, the computed temperature and density profiles being assumed to remain unchanged.

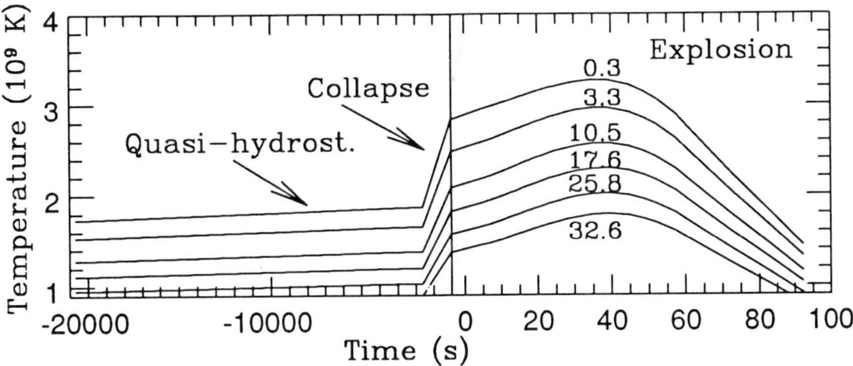

Figure 1. Time evolution of the temperature in various layers of the $50\,M_\odot$ oxygen core during its quasi-hydrostatic contraction, collapse and explosion (Note the change in the time scales). The various layers are labelled with the corresponding values of $M_r/M_\odot$

2.2. Nuclear physics input

The adopted nuclear reaction network is described in detail by Rayet et al. (1990). It includes 1040 stable, as well as unstable (mostly) neutron-deficient nuclides between H and Bi, and about 11000 reactions with neutrons, protons, α-particles, as well as photodisintegrations. In such conditions, all light particle abundances are calculated self-consistently.

Up to silicon, reaction rates are taken from the compilations of Caughlan and Fowler (1988) and Bao and Käppeler (1987). Above Si, the experimental information concerning reaction rates is extremely scarce, so that the necessary cross sections are calculated in the framework of a Hauser-Feshbach model described by Thielemann *et al* (1986).

In view of the rather lengthy phase of quasi-hydrostatic contraction prior to the PCSN (Fig. 1), β^+-decays may play a non-negligeable role, as already stressed by Arnould (1976). When experimental data are missing, β^+-decay rates are computed from the gross theory of Takahashi *et al* (1973). Note that e^--captures are always too slow to be of any practical significance in the astrophysical scenario considered here.

2.3. Seed abundances

The s-process that develops during the He-burning phase provides the seeds for the p-process. The seed abundance distribution is calculated as described by e.g. Prantzos *et al* (1987). As is typical of the s-process in massive stars, the resulting abundances drop abruptly, compared to the solar ones, at A in excess of about 90.

3. Results and discussion

Various layers of the massive oxygen core considered here are found, as expected from an inspection of Fig. 1, to be well suited for the development of the p-process.

The calculated yields of p-nuclides from the PCSN described in Sect. 2.1 are depicted in Fig. 2. The results obtained for $M_{b.h.} = 19\,M_\odot$ derived from our evolutionary computations indicate a deficiency of the p-nuclides lighter than Ba relative to the heavy ones. As can be seen from an analysis of the results of Fig. 2 obtained for lower values of $M_{b.h.}$, this is mainly due to the trapping in the remnant of the layers that are hot enough for producing significant amounts of light p-nuclides, while not contributing to the synthesis of the heavy ones. The existence of such zones results from the fact that each species is produced in a very narrow range of temperatures, this extreme sensitivity being in turn due to the key role played by photodisintegrations in the nuclear flow that leads to the synthesis of the p-nuclides. More specifically, and as already stressed in previous studies of the p-process in hydrostatic or explosive O/Ne burning (e.g. Arnould 1976, Arnould *et al* 1992, and references therein), the heavy p-nuclei are produced at lower temperatures than the light ones, which are more resistant to photodisintegrations. An increase of the remnant mass thus depletes preferentially the light p-nuclides relative to the heavy ones.

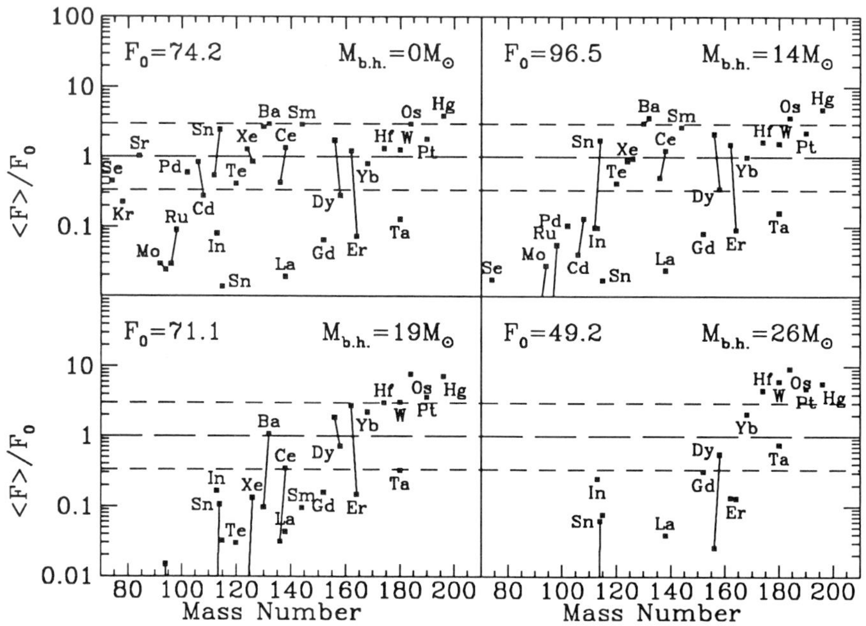

Figure 2. Yields of the p-nuclides derived from the PCSN model described in Sect. 2.1. They are expressed in terms of the normalized overproduction factor $< F >/F_0$, where $< F >$ is calculated from the total mass of each p-nuclide ejected by the PCSN divided by the corresponding mass if the PCSN had solar composition, while F_0 is the average of the $< F >$ values predicted for the 35 p-nuclides. So, $< F >/F_0$ would be 1 for all p-nuclei if the obtained abundance pattern happened to be solar. The yields are presented for various black hole masses $M_{b.h.}$, 19 $M_\odot$ being the value derived from the detailed evolutionary computations presented in Sect. 2.1

Figure 2 also indicates that the PCSN p-nuclide abundance pattern would qualitatively resemble the SNII abundances (e.g. Arnould *et al* 1992, and references therein) in absence of any remnant, a situation that could well be obtained in model stars that are slightly different from the one computed here, as already emphasized in Sect. 2.1. In particular, the majority of the p-nuclides would be overproduced within a factor of ≈ 3 of the mean overproduction F_0, while a severe underproduction would be obtained for the Mo and Ru p-isotopes. As in SNII calculations, some species that are not considered as pure p-nuclides (like ^{113}In, ^{115}Sn, ^{152}Gd, or ^{164}Er) would also be produced in too small amounts.

As far as the odd-odd neutron-deficient nuclides are concerned, ^{138}La is, as usual, largely underproduced, independent of the adopted remnant mass. In contrast, ^{180}Ta^m is nicely coproduced with the other heavy p-nuclides in relatively cool layers, at least for black hole masses in excess of about 19 $M_\odot$. A significant production of that nuclide is also obtained for the various SNII models considered up to now (Arnould *et al* 1992). As discussed in detail by Prantzos *et al* (1990), this results from a subtle and highly temperature-dependent balance between its destruction and its production from the relatively abundant ^{181}Ta.

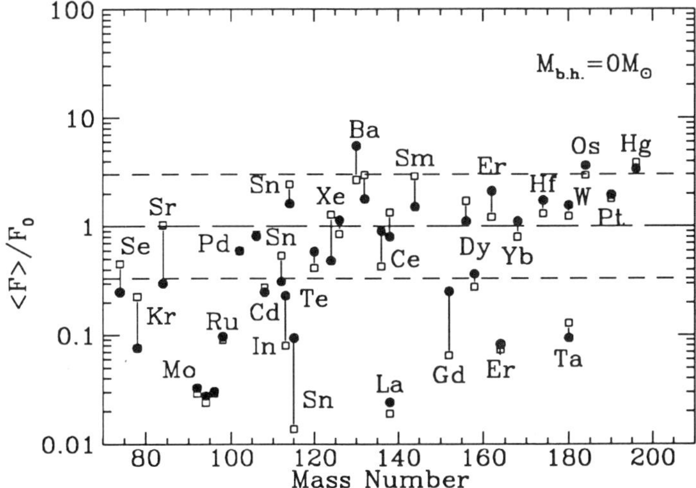

Figure 3. Comparison between the p-process yields obtained just prior (black dots) and after (open squares) the PCSN explosion (see Sect. 2.1), assumed here to leave no remnant. The relative abundance changes due to the explosion would not be qualitatively different for other $M_{b.h.}$ values

Finally, we want to stress that, as demonstrated in Fig. 3, the p-process yields presented in Fig. 2 result to a very large extent from the quasi-hydrostatic evolutionary phase prior to the explosion (see Fig. 1). More specifically, the heavy p-nuclides, which emerge from relatively cool regions, appear to be almost unaffected by the explosion. In contrast, the explosion produces a large fraction of the three lightest ones, whose synthesis requires a high temperature (and which would be locked in the remnant if $M_{b.h.}$ is larger than about 15 $M_\odot$). In the intermediate-mass range, the p-nuclides are either slightly produced or destroyed during the explosion, the largest destruction occuring in fact for low-abundance species, like ^{113}In, ^{115}Sn or ^{152}Gd. This work thus substantiates for the first time on quantitative grounds the early claim by Arnould (1976) that hydrostatic O/Ne burning in massive stars could be an adequate site for p-nuclide production.

4. Conclusions

This paper presents the first detailed calculation of the p-process in very massive stars leading to a PCSN, and complements similar computations performed in the framework of detailed SNII models. The specific PCSN model considered here leads to a substantial production of the p-nuclides from Ba to Hg in relative proportions close to solar, the deficiency of the lighter p-nuclides resulting from their lock-up in a $M_{b.h.} = 19\ M_\odot$ black hole remnant. The impact of a change of $M_{b.h.}$ on the predicted production of p-nuclides is also explored, different $M_{b.h.}$ values being possibly obtained for models that differ slightly from the one explored here. We also demonstrate for the first time that hydrostatic O/Ne burning can provide a dominant component to the total p-process yields from certain explosive events.

At this point, the question of course arises of the possible contribution of PCSN to the galactic content of p-nuclides. In order to try solving that problem, it is necessary to extend the calculations reported here to other very massive stars with different masses and metallicities. Another key aspect concerns the frequency of the PCSN events. Obviously, such stellar explosions cannot be very frequent, and despite the very large mass of processed material ejected per single event, their net yields may turn out not to be sufficient to account for the p-nuclei at the solar level if it is truly representative of the whole galactic disk, which is not known at the present time.

Acknowledgments

We thank the Brussels Free Universities Computing Center for a generous allocation of computer time on its Cray YMP. This work has been supported in part by the SCIENCE Program SC1-0065 of the European Economic Community. M.A. and M.R. are Chercheurs Qualifiés F.N.R.S. (Belgium).

References

Arnould, M. 1976, *Astron. Astrophys.*, **46**, 117

Arnould, M., Rayet, M. and Hashimoto, K. 1992, in *Unstable Nuclei in Astrophysics*, eds. S. Kubono and T. Kajino (World Scientific, Singapore), p. 23

Bao, Z.Y. and Käppeler, F. 1987, *Atomic Data Nuclear Data Tables*, **36**, 411

Caughlan, G.R. and Fowler, W.A. 1988, *Atomic Data Nuclear Data Tables*, **40**, 283

de Jager, C., Nieuwenhuijzen, H., Van der Hucht, K.A. 1988, *Astron. Astrophys. Suppl.*, **72**, 259

El Eid, M.F. 1991, in *Supernovae*, ed. S.E. Woosley (Springer-Verlag, Berlin), p. 568

El Eid, M.F. and Langer, N. 1986, *Astron. Astrophys.*, **167**, 274

Herzig, K. 1992, Ph. D. Thesis, University of Göttingen (unpublished)

Howard, W.M., Meyer, B.S. and Woosley, S.E. 1991, *Astrophys. J. (Lett.)*, **373**, L5

Huebner, W.F., Merts, A.L., Meggee, N.H. and Argo, M.F. 1977, Los Alamos Sci. Rep. LA-6760-M

Prantzos, N., Arcoragi, J.-P. and Arnould, M. 1987, *Astrophys. J.*, **315**, 209

Prantzos, N., Hashimoto, M., Rayet, M. and Arnould, M. 1990, *Astron. Astrophys.*, **238**, 455

Rayet, M., Prantzos, N. and Arnould, M. 1988, in *Origin and Distribution of the Elements*, ed. G.J. Mathews (World Scientific, Singapore), p. 625

Rayet, M., Prantzos, N. and Arnould, M. 1990, *Astron. Astrophys.*, **227**, 271

Takahashi, K., Yamada, M. and Kondo, T. 1973, *Atomic Data Nuclear Data Tables*, **12**, 101

Thielemann, F.-K., Arnould, M. and Truran, J.W. 1986, in *Advances in Nuclear Astrophysics*, eds. E. Vangioni-Flam, J. Audouze, M. Cassé, J.-P. Chièze and J. Tran Thanh Van (Editions Frontières, Gif-sur-Yvette), p. 525

Van der Hucht, K.A. 1992, *Astron. Astrophys. Review*, in press

Woosley, S.E. and Howard, W.M. 1978, *Astrophys. J. Suppl.*, **36**, 285

Nucleosynthesis in massive Thorne–Żytkow objects

Robert C. Cannon

Institute of Astronomy, Madingley Road, Cambridge, CB3 0HA, U.K.

Abstract. Calculations have been performed of the reaction paths in the convective envelope of a Thorne-Żytkow object. The proton-rich material at the base of the envelope, continuously linked to the outer regions by convection, provides an ideal site for the *rp*-process. In addition, material may repeatedly pass through the burning zone, with large intervals between each passage, allowing the formation of heavy nuclei up to $A \sim 150$. Evolutionary calculations of the structure and composition change show that in most cases a significant fraction, $\sim 5\%$, of the envelope is converted to metals beyond iron, and for masses in the upper range of consistent solutions, many of the *p*-nuclei are produced in abundances reminiscent of those found for the solar system. If Thorne-Żytkow objects are a later stage of the evolution of massive X-ray binaries after the components have coalesced, there would have been sufficient in the Galaxy comfortably to account for the observed abundances of many *p*-nuclei.

1. The *rp*-process in TŻOs

The structure of red giant stars with degenerate neutron cores (TŻOs) has been discussed by Thorne and Żytkow (1977). They found that stars of above about $11\,M_\odot$ generate most of their luminosity from nuclear burning in the convective envelope rather than accretion onto the core. They expressed reservations about their use of reaction rates appropriate to in situ burning to construct equilibrium models, but Biehle (1991) has since demonstrated that the *rp*-process is dominant in the envelope, and that self-consistent solutions do indeed exist.

The main difference between TŻOs and other sites for the *rp*-process (Wallace and Woosley, 1981) is the potential for material making repeated passages through the burning zone. If the extent of the reaction sequence in one passage is limited by long β decays ($\gtrsim 1\,\text{s}$) rather than by photodissociations then on a subsequent passage further proton additions may take place and a new range of elements is made accessible. The long periods spent outside the burning zone allow material to decay back towards the line of beta stability from which further proton additions are more favourable. In this case the limiting condition on the mass numbers reached is the rate of proton addition reactions on stable nuclei. Using the approximations to Hauser Feshbach theory of Woosley *et al* (1978) and integrating over the envelope from a temperature at the base of $T_9 \times 10^9\,\text{K}$ one may derive an approximate stopping condition:

$$A_{\text{max}} \approx 50 T_9 \ .\tag{1}$$

This relation breaks down at temperatures in excess of $\sim 3 \times 10^9\,\text{K}$ where photodissociations become important even for nuclei near stability.

2. A model of convective nucleosynthesis in TŻOs

The conventional treatment of convective mixing by a diffusion equation eliminates the possibility that the composition of rising and falling convective elements may differ.

Near the base of a TŻO however, the convective timescale is ~ 0.01 seconds – much shorter than many of the β decays – so the approximation of constant composition at a given radius breaks down. We use instead the scheme illustrated in figure 1. At each level there are two composition variables for each species - one for rising elements, and one for falling ones. Material is allowed to advect from above or below, and diffuse sideways. In the limit of very rapid lateral diffusion it reduces to a diffusion equation for the vertical composition profile. At the other limit, where no material is exchanged between columns, it represents a 'conveyor belt' scheme (e.g. Eich *et al* 1989).

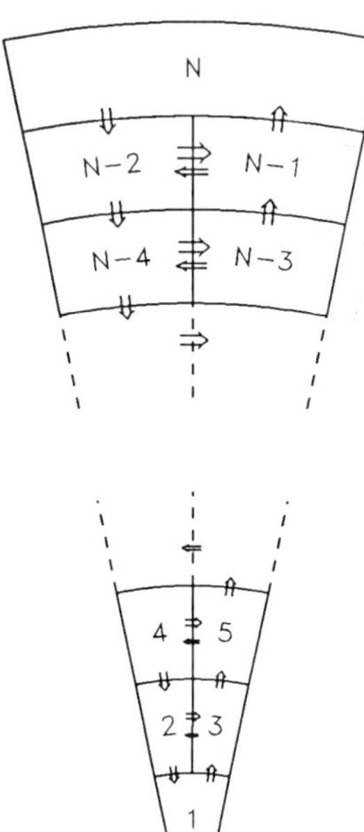

Figure 1. Schematic representation of the allowed flows in the nucleosynthesis model. With upstream differencing, the composition in each cell is only linked to those from which material flows into it. Thus, for even n, the n'th cell is linked to cells n+1, and n+2. For odd n, it is linked to cells n-2, and n-1. Either way, only groups of three consecutively numbered cells are linked; this is necessary in order to keep the resulting matrix as sparse as possible given the physical requirements.

The solution of a full network of several hundred species twice at each level, linked by advection and diffusion terms is not necessary since the rp-process conveniently divides across β decays or proton additions whose inverse is negligible. It may thus be solved as a series of smaller weakly linked blocks with only minor losses in accuracy.

3. Results

In view of the uncertainties in the actual values of the convective velocity and mixing length from standard mixing length theory, we have calculated a range of TŻO models in which various scalings have been applied to these quantities in the hopes of spanning the region of physical interest. The main effect of these changes is to alter the temperature in the burning region. The luminosity of the star hardly changes.

The yield of heavier elements is shown in figure 2. Since the line of beta stability is approached from the proton-rich side the envelope is very rich in p-nuclei. The

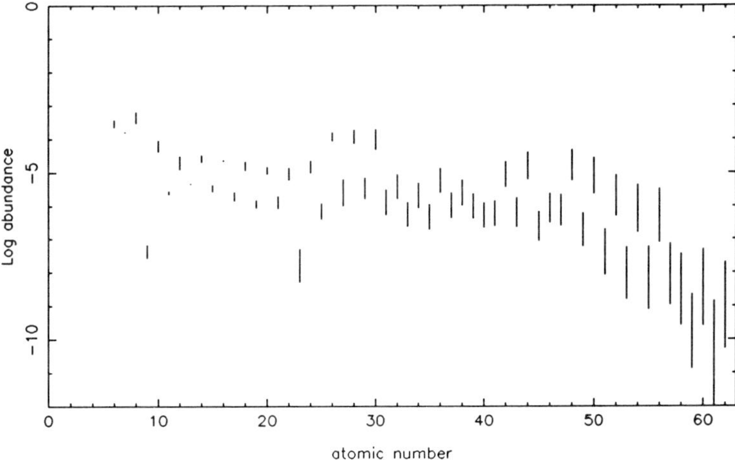

Figure 2. Envelope number abundances by element. Vertical lines show the change in abundance between 10^5 years and 4×10^5 years.

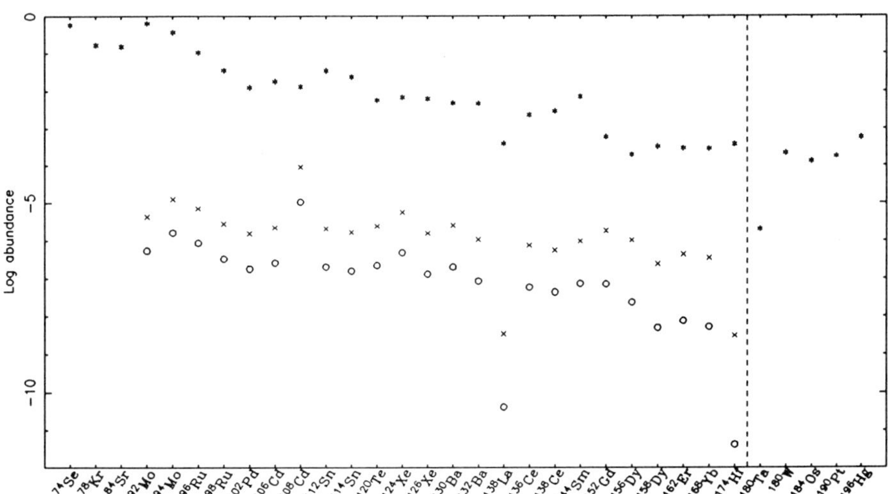

Figure 3. Envelope number abundances of p-process elements between molybdenum and hafnium. at 10^5 years (circles) and 4×10^5 years (crosses). Solar system abundances are shown by stars, with an arbitrary normalisation. The beginning of the network was modelled by a steady rate of supply, giving the smaller spread between abundances at 10^5, and 4×10^5 years. The dashed line shows the maximum extent of the network.

abundances of these nuclei are shown in figure 3, with those occurring in the solar system for comparison.

4. Discussion

Supposing TŻOs are a later stage in the evolution of short period (less than a few days) massive x-ray binaries, (MXRBs), after the components have coalesced, (Ray *et al* 1987) one may assess their significance as sites for the production of p-nuclei. The birthrate of MXRBs has been estimated at about one per thousand years (Meurs and van den Heuvel 1989), by assuming a lifetime of $\sim 2.5 \times 10^4$ years, with between 20 and 60 such sources in the Galaxy at present. Thus, over the last 10^{10} years, there should have been about 10^7 TŻOs. Each may yield up to 0.5% of it's envelope mass in p-nuclei, giving 0.1 $M_\odot$ each, or about 10^6 $M_\odot$ in the Galaxy. Observational data (Cameron 1973) suggest an abundance by mass of 5×10^{-9}, giving 5×10^3 $M_\odot$ in total for a galactic mass of 10^{12} $M_\odot$. The upper limit therefore represents an overproduction of p-nuclei by a factor of about a thousand. There are at least two reasons why this upper limit should not be realised:

1 The lifetime may be rather shorter than assumed in these calculations. Mass loss rates for red supergiants are very uncertain, but any rate in excess of $\sim 10^5$ $M_\odot$ yr^{-1} would reduce the length of the TŻO phase below the time required to exhaust the envelope of seed nuclei. Due to the non-linearity of the heavy element yields a factor of two in the lifetime could reduce the yield by an order of magnitude.

2 Only the hotter models produce p-nuclei in large quantities. The temperature in the burning zone is determined by the details of the convection to give the right luminosity, so whatever the correct physical description, on would expect some range of masses to correspond to the right temperature range. Since p-nuclei are produced over about the upper quarter of the range in $\log T$ corresponding to consistent solutions, depending in the mass distribution of MXRBs, this factor may reduce the yield by an order of magnitude or more.

In conclusion, more work is required on formation scenarios for massive TŻOs but if their birthrate turns out to be sufficiently high ($\gtrsim 1$ per 10^5 years) they may be the 'alternative exotic sites' of H-rich layers alluded to by Lambert (1992).

References

Biehle G T 1991 *Ap.J.* **380** 167

Cameron A G W 1973 *Space Sci. Rev.* **15** 121

Eich C, Zimmermann M E and Thorne K S 1989 *Ap.J.* **346** 277

Lambert D L 1992 *Astron. Astrophys. Rev.* **3** 201

Meurs E J A and van den Heuvel E P J 1989 *Astron. Astrophys.* **226** 88

Ray A, Kembhavi A K and Antia A 1987 *Astron. Astrophys.* **184** 164

Thorne K S and Żytkow A N 1977 *Ap.J.* **212** 832

Wallace R K and Woosley S E 1981 *Ap.J.* **45** 389

Woosley S E, Fowler W, Holmes J A and Zimmerman, B. A., 1978 *Orange Aid Preprint 422* Caltech

Nucleosynthesis from accreting white dwarfs

J. José[1,2,4], M. Rayet[2], M. Arnould[2], and M. Hernanz[3,4]

[1]Departament de Física i Enginyeria Nuclear, EUPVG, Universitat Politècnica de Catalunya, Avda. Víctor Balaguer, E-08800 Vilanova i la Geltrú (Barcelona), SPAIN.

[2]Institut d'Astronomie et d'Astrophysique, Université Libre de Bruxelles, CP 165, Av. F.D. Roosevelt 50, B-1050 Bruxelles, BELGIUM.

[3]Centre d'Estudis Avançats de Blanes, CSIC, Camí de Santa Bàrbara, E-17300 Blanes (Girona), SPAIN.

[4]Laboratori d'Astrofísica, Societat Catalana de Física, IEC, Carme 15, E-08001 Barcelona, SPAIN.

Abstract. We report on preliminary calculations of recurrent weak flashes that take place on the surface of massive accreting white dwarfs, for a range of mass accretion rates and stellar masses. A simple two-zone model in a plane-parallel approximation is assumed to simulate the periodic hydrogen shell flashes arising from the preliminary accretion of hydrogen-rich fuel. We study in detail the nucleosynthesis occuring during a whole sequence of such events, by following the evolution of 60 nuclei ranging from neutrons to ^{28}Si and linked by a fully updated network that includes more than 260 nuclear reactions. Some considerations concerning the eventual fate of such systems and their connection with astrophysical phenomenae of interest for nuclear astrophysics will be presented.

1. Introduction

Mass-accretion onto the surface of white dwarfs in close binary systems plays an important role for the explanation of a widely range of astrophysical phenomena. In particular, novae eruptions, type Ia supernova explosions and low-mass x-ray binaries (related with the origin of neutron stars in binary systems due to the accretion-induced collapse of a white dwarf) deal with such scenario.

All these astrophysical sites point out the crucial role played by the outer burning shells in the long-term evolution of accreting white dwarfs. Despite this interest, a detailed analysis of the whole white dwarf evolution, taking simultaneously into account the progress of hydrogen and helium burning in the accreted layers and the core evolution, is not attainable, because of very different evolutionary time-scales and prohibitive computer time requirements. In this sense, it is necessary to implement simplified models to study the external layers, in order to determine the *gross* properties of the double H-He shell burning proceeding through a large number of recurrent flashes. In particular, the accretion history of the white dwarf, will provide some important features that can be figured out from the analysis of the nucleosynthesis associated with the outbursts. Detailed calculations of the chemical composition of the outer layers will be of interest to model a pre-supernova progenitor, with strong implications on the composition of the ejecta.

2. Model & input physics

We follow the evolution of the external burning layers of a massive accreting white dwarf ($M \geq 1.2\ M_\odot$) by means of a simplified double-shell model (José et al. 1992), built as an extension of the one-zone model from Paczynski (1983). The model assumes a top shell with a solar-like composition accreted from the companion. Its helium-rich ashes, coming from the series of recurrent hydrogen flashes that take place in the top shell, build up a bottom layer from which helium flashes can also be modelled.

Within the framework of the previous model by Paczynski some simplificative assumptions are made. A plane-parallel geometry is used to account for the structure of both thin and few massive shells. Furthermore, changes in the highly degenerate interior are neglected since the characteristic time-scales for the evolution of the burning shells are much shorter than those for the inner core. Hence, one can describe the white dwarf core via two parameters: a constant surface gravity, g, related with its total mass and radius, and a heat flux exchanged through the border between the core and the external shells.

Regarding the input physics, an equation of state including ions, electrons and radiation, is adopted. Neutrino losses as well as convective energy transport have been neglected in the computations. With respect to opacity, both conductive and radiative terms have been taken into account.

The synthesis of nuclear species during the recurrent flashes is described following the time evolution of 60 nuclei, ranging from neutrons to ^{28}Si and linked by a fully updated network that includes more than 260 nuclear reactions, which can play a role during hydrogen or helium burning phases. Screening factors have been also included via the method described by DeWitt et al. (1973) and Graboske et al. (1973). The network also deals with a neutron sink, parametrized by means of an effective neutron capture cross section of 9.45 mb (Bao & Käppeler 1987, Jorissen & Arnould 1987). The time evolution of the nuclei abundances is followed by the application of Wagoner's two-step linearization technique to the corresponding set of differential equations (Wagoner 1969). This procedure provides a second-order approximation for the evaluation of the abundances which ensures conservation of the baryonic number to the 11th digit. Time step is chosen in such a way that relative abundance variations of nuclear species with $Y>10^{-14}$ remain below 15%, as a compromise between accuracy and computation time. Neutrons, deuterons and tritium are assumed to be in equilibrium.

Details regarding the method to solve the system of stellar equations as well as the progress of the successive shell flashes have been described elsewhere (José et al. 1992).

3. Results & conclusions

We report on preliminary calculations of the nucleosynthesis associated with recurrent weak flashes resulting from accretion of solar-like matter onto the surface of a 1.2 $M_\odot$ white dwarf. Appropiate values for the mass-accretion rate, leading to weak outbursts, have been considered. The corresponding range is limited, on one hand, by hydrogen-steady burning ($\dot{M} \simeq 2\times10^{-7}$ $M_\odot.yr^{-1}$, Iben 1982) as well as by the critical rate for nova eruptions ($\dot{M} \leq 10^{-9} - 10^{-10}$ $M_\odot.yr^{-1}$, MacDonald 1983), for which extremely violent outbursts with ejection of a part of the accreted envelope are expected.

Some general properties of shell flashes, such as the influence of the white dwarf mass and accretion rate on the violence of the outbursts and its recurrence time have been pointed out by several authors (Fujimoto 1982, Paczynski 1983, José et al. 1992). In addition, we shall focus on the interest of shell flashes from the nucleosynthesis viewpoint. The explosion of a massive white dwarf destabilized by accretion has been proposed as a realistic scenario for the origin of type Ia supernova explosions. In this case, the abundances of the ejecta will strongly depend on the previous accretion history of the white dwarf. Hence, the resulting chemical composition will be much different than the obtained during hydrostatic or hydrodynamic burning, as expected from different time-scales. In particular, we should stress the crucial role played by ^{13}C, ^{21}Ne and ^{22}Ne as neutron-sources during the long-term evolution of the system. In this framework, depending on neutron concentrations, the explosive nucleosynthesis accompanying a SNIa may result in the production of neutronized species (a *mini* s-process) or, on the other hand, in the photo-disintegration of the seed-nuclei, leading to the production of the neutron deficient stable isotopes known as p-nuclei.

Let's follow in detail the time-evolution of the accreted shells during a series of recurrent flashes (i.e., up to the fifth outburst), resulting from accretion of solar-like matter at a rate $\dot{M}=5\times10^{-8}$ $M_{\odot}$.yr^{-1} onto a 1.2 $M_{\odot}$ white dwarf. A full flash-cycle driven by accretion starts with an extremely short phase ($\tau \simeq 10$ sec), in which proton captures on very light nuclei destroy very efficiently ^{6}Li, ^{7}Li, ^{9}Be, ^{10}B and ^{11}B at a temperature of 3×10^7 K. In addition, ^{13}C is enhanced via ^{12}C(p,γ)^{13}N(β^+)^{13}C. As time goes on, the evolution of the burning layers is followed by a moderate increase of temperature, resulting from gravitational compression carried by the infalling matter. Due to the metal content (Z $\simeq$ 0.03), the *cold* CNO cycle (mainly by ^{12}C(p,γ)^{13}N(β^+)^{13}C(p,γ)^{14}N(p,γ)^{15}O(β^+)^{15}N(p,α)^{12}C) dominates the contribution played by the p-p chains during the first rise of temperature (T $\simeq$ 7×10^7 K). Hydrogen content is progresively shifted towards helium-rich isotopes. In addition, as expected from CNO cycle, ^{14}N becomes one of the main products of the evolution. The flash takes place 65 years from the end of the previous cycle, with a peak temperature of 1.5×10^8 K. At this point, there is a competition between *cold* and *hot* modes of CNO cycle, plus some traces of the Ne-Na and Mg-Al cycles. Some *leakage* to heavier isotopes like ^{29}Si is obtained, due to the partial break of the CNO cycle. The composition of the shell after the flash is given in table 1. The most abundant isotope is ^{4}He (98% by mass). Some representative species present in the helium-rich ashes are ^{14}N (roughly 1%) and, finally, ^{12}C, and ^{20}Ne. Regarding the evolution of the neutron-source isotopes, ^{13}C is enhanced up to 6×10^{-4} by mass, while neon isotopes (specially ^{22}Ne) are destroyed during the flash. Overproduction factors, relative to solar abundances, are also shown in figure 1.

Regarding the effect of the mass accretion rate on the calculations, longer interflash periods (i.e., time elapsed between two consecutive outbursts) as well as higher peak temperatures have been obtained, as expected, for lower accretion rates. Preliminary implications on the resulting nucleosynthesis can be also sketched: ^{21}Ne and ^{22}Ne are more efficiently destroyed while ^{13}C is less enhanced during the accretion-driven shell flashes that take place for higher accretion rates. On the other hand, abundances of ^{14}N, ^{20}Ne or ^{25}Mg increase under the same conditions (see table 1 for detailed comparison). The role played by the white dwarf mass is also shown in table 1. Longer interflash periods result from lighter degenerate dwarfs. In this case, as a consequence of a higher peak temperature, lighter white make more ^{13}C and destroy more ^{21}Ne and ^{22}Ne.

Table 1. Isotopic abundances at the end of the 5th flash for several models. Solar values are given for comparison.

Element	solar composition	1.2 M$_\odot$ 1×10^{-7} M$_\odot$.yr^{-1}	1.2 M$_\odot$ 5×10^{-8} M$_\odot$.yr^{-1}	1.2 M$_\odot$ 5×10^{-9} M$_\odot$.yr^{-1}	1.3 M$_\odot$ 5×10^{-8} M$_\odot$.yr^{-1}
^{3}He	2.94×10^{-05}	7.26×10^{-21}	1.32×10^{-19}	9.56×10^{-19}	1.12×10^{-21}
^{4}He	2.76×10^{-01}	9.84×10^{-01}	9.84×10^{-01}	9.84×10^{-01}	9.84×10^{-01}
^{6}Li	6.51×10^{-10}	2.57×10^{-19}	2.94×10^{-18}	1.74×10^{-17}	3.32×10^{-20}
^{7}Li	9.37×10^{-09}	7.70×10^{-16}	1.99×10^{-16}	2.27×10^{-16}	3.29×10^{-15}
^{9}Be	1.67×10^{-10}	1.87×10^{-20}	1.22×10^{-19}	6.29×10^{-19}	5.44×10^{-21}
^{10}B	1.07×10^{-09}	2.09×10^{-17}	2.80×10^{-17}	7.77×10^{-17}	5.50×10^{-17}
^{11}B	4.74×10^{-09}	3.00×10^{-15}	1.11×10^{-15}	1.31×10^{-15}	4.66×10^{-15}
^{12}C	3.04×10^{-03}	1.57×10^{-03}	1.74×10^{-03}	2.70×10^{-03}	2.29×10^{-03}
^{13}C	3.66×10^{-05}	4.90×10^{-04}	5.38×10^{-04}	6.66×10^{-04}	7.05×10^{-04}
^{14}N	1.11×10^{-03}	1.06×10^{-02}	1.04×10^{-02}	8.76×10^{-03}	9.51×10^{-03}
^{15}N	4.37×10^{-06}	6.23×10^{-07}	6.60×10^{-07}	7.91×10^{-07}	7.80×10^{-07}
^{16}O	9.61×10^{-03}	1.28×10^{-04}	1.47×10^{-04}	4.13×10^{-04}	1.85×10^{-04}
^{17}O	3.90×10^{-06}	2.90×10^{-06}	3.01×10^{-06}	3.25×10^{-06}	3.30×10^{-06}
^{18}O	2.17×10^{-05}	5.40×10^{-09}	5.44×10^{-09}	5.35×10^{-09}	8.25×10^{-09}
^{19}F	4.06×10^{-07}	1.58×10^{-09}	2.15×10^{-09}	2.28×10^{-08}	3.23×10^{-09}
^{20}Ne	1.62×10^{-03}	1.60×10^{-03}	1.59×10^{-03}	1.52×10^{-03}	1.55×10^{-03}
^{21}Ne	4.14×10^{-06}	1.64×10^{-07}	2.98×10^{-07}	3.38×10^{-06}	4.87×10^{-07}
^{22}Ne	1.30×10^{-04}	7.36×10^{-07}	7.93×10^{-07}	1.07×10^{-06}	7.98×10^{-07}
^{23}Na	2.35×10^{-05}	1.72×10^{-06}	1.56×10^{-06}	1.26×10^{-06}	1.19×10^{-06}
^{24}Mg	5.16×10^{-04}	5.80×10^{-08}	5.98×10^{-08}	9.91×10^{-08}	5.95×10^{-08}
^{25}Mg	6.78×10^{-05}	8.85×10^{-06}	8.33×10^{-06}	4.99×10^{-06}	6.65×10^{-06}
^{26}Mg	7.78×10^{-05}	4.74×10^{-07}	4.77×10^{-07}	9.72×10^{-07}	4.14×10^{-07}
^{27}Al	5.81×10^{-05}	4.43×10^{-06}	4.44×10^{-06}	4.13×10^{-06}	4.42×10^{-06}
^{28}Si	6.54×10^{-04}	1.68×10^{-03}	1.69×10^{-03}	1.66×10^{-03}	1.72×10^{-03}

As a summary, we shall conclude that the implementation of a two-zone model allowing the building up of a bottom shell from succesive series of H-flashes, has proven very useful to make a preliminary analysis of the resulting nucleosynthesis. Despite the simplifying assumptions used, better abundances than ones assumed neglecting the accretion history have been derived in this work. The determination of chemical abundances during the accretion phase can be crucial to describe the nucleosynthesis accompanying the eventual fate of the system. In particular, SNIa explosions are expected to lead to interesting processes requiring the knowledge of the detailed chemical composition of their progenitors, like, for instance, the p-process.

This work has been supported by the CICYT grant PB87-0304, by the PICS project *Origin and evolution of light elements* and by a CESCA grant.

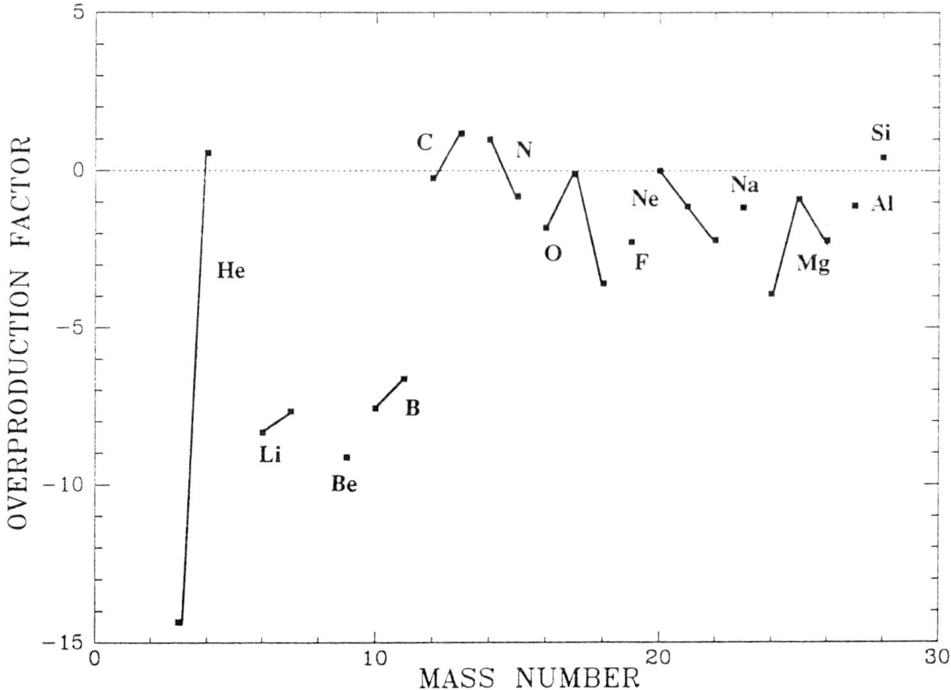

Figure 1. Overproduction factors relative to solar abundances versus mass number, for the 5th H-flash resulting from accretion at a rate $\dot{M}=5\times10^{-8}$ $M_\odot.yr^{-1}$ onto a 1.2 $M_\odot$ white dwarf. The dashed line corresponds to solar abundances.

One of us (J.J.) would like to thank M. Arnould and M. Rayet for kind hospitaly during his stay at the Institut d'Astronomie et d'Astrophysique, where part of this work was done.

References

Bao Z.Y. and Käppeler F. 1987 *Atomic Data Nucl. Data Tables* **36** 411

DeWitt H.E., Graboske H.C. and Cooper M.S. 1973 *Astrophys. J.* **181** 439

Fujimoto M.Y. 1982 *Astrophys. J.* **257** 752

Graboske H.C., DeWitt H.E., Grossman A.S. and Cooper M.S. 1973 *Astrophys. J.* **181** 457

Iben I. 1982 *Astrophys. J.* **259** 244

Jorissen A. and Arnould M. 1989 *Astron. Astrophys.* **221** 161

José J., Hernanz M. and Isern J. 1992 *Astron. Astrophys* (submitted)

MacDonald J. 1983 *Astrophys. J.* **267** 732

Paczynski B. 1983 *Astrophys. J.* **264** 282

Wagoner R.W. 1969 *Astrophys. J. Suppl.* **18** 247

Status of solar neutrino experiments

W. Hampel

Max-Planck-Institut für Kernphysik, P.O. Box 103980, 6900 Heidelberg 1, Germany

Abstract.   The present status of all four running solar neutrino experiments is reviewed. Updated results from the Chlorine, Kamiokande and SAGE detectors are presented along with the first result from the GALLEX detector [83 $\pm$ 19 (stat.) $\pm$ 8 (syst.) SNU]. This number, while consistent with the presence of the full pp neutrino flux as predicted by the Standard Solar Model, confirms the reduced fluxes of high energy solar neutrinos observed in the Chlorine and Kamiokande experiments. Possible scenarios (non-standard solar models, neutrino mixing, neutrino decay and the interaction of magnetic fields in the solar convective zone with a possible magnetic moment of the neutrino) are discussed in order to explain the deviations between the Standard Solar Model predictions and the actual experimental data.

1.   Introduction

Experiments intended to observe the neutrinos produced in the fusion reactions in the interior of the Sun provide the only direct way to test the theories of energy generation in stars. A second motivation arises from particle physics: these difficult and ambitious experiments may be the only way to obtain information on neutrino properties which reveal themselves only over astronomical distances between neutrino source and detector. For almost twenty years only one experiment of this kind has been operative. In the past few years this situation has changed dramatically. At this time three additional solar neutrino detectors have yielded experimental data.

At a recent conference, Neutrino-92 in Granada, Spain (June 8-12, 1992), new results from all four running solar neutrino experiments have been reported. The present article summarizes these new data, compares them to the expectations from so-called Standard Solar Model (SSM) calculations and finally discusses possible reasons for the observed discrepancies.

2.   Standard Solar Model

A new Standard Solar Model (SSM) has been calculated by Bahcall and Pinsonneault [1]. The most important changes in the input data compared to the Standard Solar Model of Bahcall and Ulrich [2] are new elemental abundances, new opacities and a revised S-factor for the ^{7}Be(p,γ) reaction. In addition, for the first time the diffusion of helium

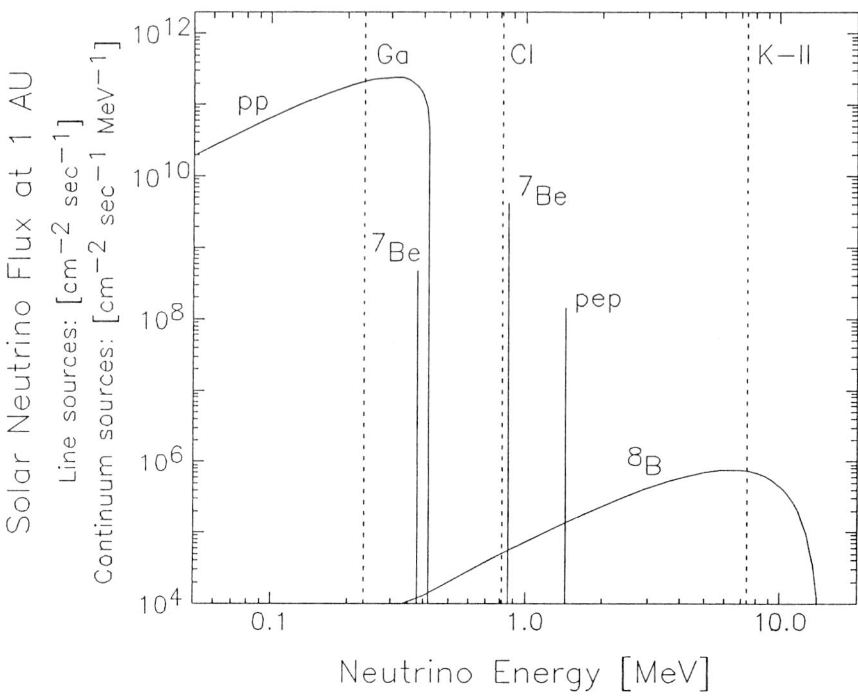

Fig. 1: Energy spectrum of solar neutrinos produced in the pp chain. The vertical dashed lines indicate the energy thresholds for the Gallium (Ga), Chlorine (Cl) and Kamiokande (K-II) experiments.

with respect to hydrogen was included in the model. The resulting production rates for the Chlorine, Kamiokande and Gallium experiments differ by less than 2% compared to the SSM of Bahcall and Ulrich [2]. Since most of the theoretical papers considered in section 4 refer to the SSM neutrino fluxes of Bahcall and Ulrich [2], we adopt these here for comparison with the experimental results.

Bahcall and Pinsonneault [1] also checked different independent solar model computer codes for consistency. The codes employed by Bahcall and Ulrich [2], Turck-Chièze et al. [3], Sienkiewicz et al. [4] and the Yale code used in [1] all agree to within 1.3% in the predicted Cl detector rate if the same input data are used.

The spectrum of solar neutrinos produced in the proton-proton reaction chain (except for the ^{3}He+p neutrinos) is displayed in Figure 1 along with the thresholds for the Gallium (Ga), Chlorine (Cl) and Kamiokande (K-II) experiments. Table 1 gives the fluxes of the different solar neutrino sources along with the corresponding production rates for the Cl and Ga experiments as predicted by the SSM of [2].

3. New experimental results

3.1 Chlorine experiment

The radiochemical Cl detector, located at the Homestake gold mine in Lead, South Dakota (USA), is based on the neutrino capture reaction ^{37}Cl$(\nu,e^-)^{37}$Ar ($T_{1/2} = 35$ d).

Table 1: Solar neutrino fluxes and capture rates for the Chlorine and Gallium detectors [2].

Neutrino source and energy [MeV]	Flux at earth $[10^{10}\mathrm{cm}^{-2}\mathrm{sec}^{-1}]$	Production rate [SNU]	
		^{37}Cl experiment	^{71}Ga experiment
pp ≤ 0.42	6.0	–	70.8
pe$^-$p 1.44	0.014	0.2	3.0
^{7}Be 0.38,0.86	0.47	1.1	34.3
^{8}B < 15	0.00058	6.1	14.0
^{3}He p ≤ 18.77	$7.6 \cdot 10^{-7}$	0.03	0.06
^{13}N ≤ 1.20	0.061	0.1	3.8
^{15}O ≤ 1.73	0.052	0.3	6.1
^{17}F ≤ 1.74	0.00052	0.004	0.06
Total	6.60	7.9	132

The ^{37}Ar atoms are subsequently extracted from the target liquid (615 tons C_2Cl_4) and counted by observing their radioactive decay in a proportional counter. The ^{37}Ar production rate averaged over 91 runs covering the time period from 1970 to the end of 1990 is 2.23 ± 0.22 SNU [5] (1 SNU = 1 neutrino reaction per second in 10^{36} target atoms), a value which is almost a factor of four lower than the SSM prediction of 7.9 SNU (see Table 1). This discrepancy is known since long as the Solar Neutrino Problem (SNP). There have been speculations that the Cl rate is not constant in time but varies in anticorrelation with the solar sun spot activity. We shall address this question in section 4.4.

3.2 Kamiokande experiment

Since 1987 there exist experimental data from a second experiment, the Kamiokande detector located in the Kamioka mine in Japan. This experiment measures the Cerenkov light emitted by relativistic electrons produced in elastic collisions between solar neutrinos and electrons in the inner fiducial 680 tons of a large tank filled with a total of 2180 tons of water. For background reduction, the threshold has to be set to a neutrino energy of 7.3 MeV, thus this detector is sensitive only to the upper end of the ^{8}B neutrino spectrum (see Figure 1). In 1990 data taking was interrupted for 9 month in order to replace dead photomultipliers, to implement new electronics and to add reflective mirrors for better light collection. Data taking was resumed in December 1990. Since then the detector monitored the solar ^{8}B neutrino flux for an additional 220 days [6]. The data of this period have been called Kamiokande-III.

Figure 2 depicts the new data along with the results of the previous measurements [7]. The measured signal of Kamiokande-III is (in units of the expected SSM value of reference [2]) 0.59 with a statistical error of $+0.11/-0.09$ and a systematic error of $\pm$ 0.06.

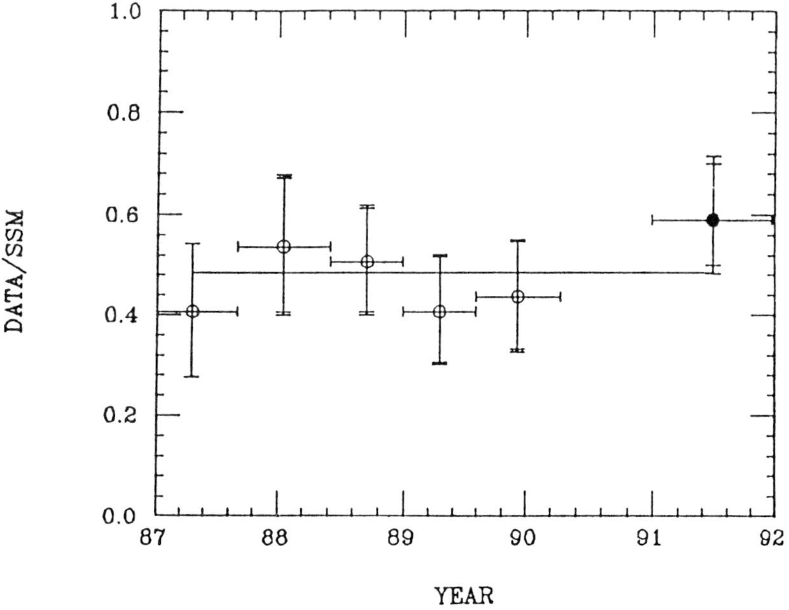

Fig. 2: Results of the Kamiokande detector for six different time periods [6][7].

When averaged with the Kamiokande-II data, the result is 0.48 ± 0.05 (stat.) ± 0.06 (syst.). A possible time dependence of the Kamiokande signal is discussed in section 4.4.

3.3 Gallium experiments

It has been realized for some time that experiments with an energy threshold low enough to observe the low-energy neutrinos generated in the primary fusion reaction in the Sun (pp-neutrinos, see Figure 1) should provide qualitatively new and different information. This is the reason why two independent groups, SAGE [8] and GALLEX [9] (see below) have been working on the realization of a new radiochemical solar neutrino detector based on the reaction $^{71}\mathrm{Ga}(\nu,e^-)^{71}\mathrm{Ge}$ ($T_{1/2}$ = 11.43 d). The largest fraction (56%) of the total production rate (132 SNU) predicted by the SSM [2] is caused by the pp and pep neutrinos, the contribution from ^{7}Be- and ^{8}B-neutrinos is 26% and 11%, respectively (see Table 1).

Apart from the first additional step employed in the chemical extraction for SAGE the procedure in both experiments is similar in principle. After an exposure time of three to four weeks, the solar neutrino produced ^{71}Ge atoms are extracted from the target and converted into the gas germane, GeH_4. The germane is then filled together with xenon into miniaturized low-level proportional counters where the ^{71}Ge electron capture decay is observed. The energy deposited by Auger electrons and X rays emitted in this decay results in two peaks, the L peak at about 1.2 keV and the K peak at 10.4 keV. Typical counting efficiencies for these two peaks achieved for the GALLEX counters are 31% for the L peak and 35% for the K peak [9].

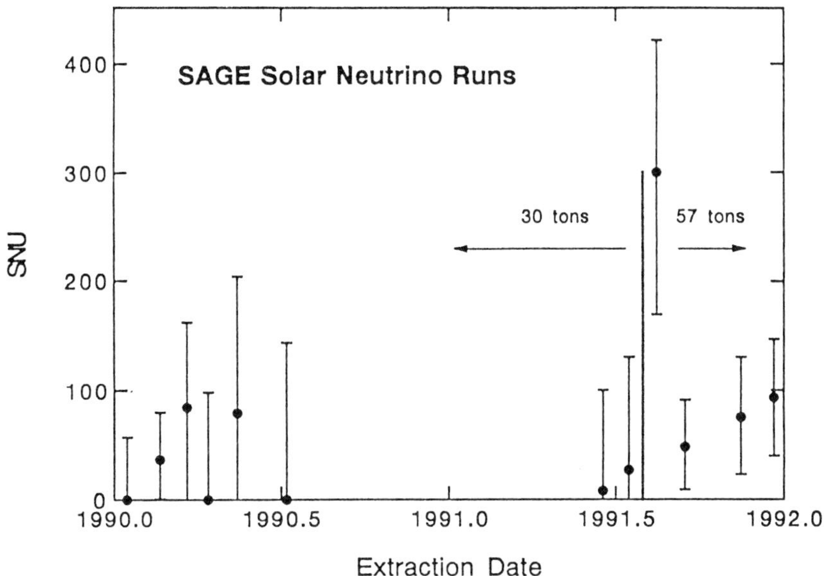

Fig. 3: Results of SAGE solar neutrino runs [10]. Five of the runs in 1990 (excluding the May run) constitute the SAGE data block published in 1991 [8].

3.4 *SAGE experiment*

SAGE (Sowjet-American-Gallium-Experiment) is a Russian-American collaboration. The detector uses gallium in its metallic form (total available amount 57 tons) and is located at the Baksan Underground Laboratory in the North Caucasus at a shielding depth of 4800 meter water equivalent [8]. Because the counter background in the L peak is too high, the SAGE collaboration so far can use only the K peak data for analysis.

The first result of SAGE, based on 5 single runs performed from January to July 1990 using an average amount of 27 tons of gallium gave a production rate of 20 SNU with a statistical error of +15/-20 SNU and a systematic error of $\pm$ 32 SNU (1σ) [8]. This is only about 15% of the SSM prediction (Table 1). After the GALLEX collaboration has published their first result (see below), SAGE has reported data from seven additional runs [10]. One of these (May 1990) belongs to the time period of the first SAGE data block. Figure 3 displays the results of all 12 SAGE runs. Since the investigation of systematic errors for the new runs was not completed, no new average value could be given. If one simply averages the 7 new runs unweighted, one arrives at a rate around 90 SNU. If all 12 SAGE runs were averaged that way, a central value close to 60 SNU is obtained. However, these numbers can serve only as a hint. As mentioned above, the final data analysis including evaluation of systematic errors is still pending [10].

3.5 *GALLEX experiment*

GALLEX is an international collaboration with members from institutes in Germany, France, Italy, Israel and the United States [9]. The target material of this detector consists of 30.3 tons of gallium in form of 101 tons of $GaCl_3$ solution, it is located at

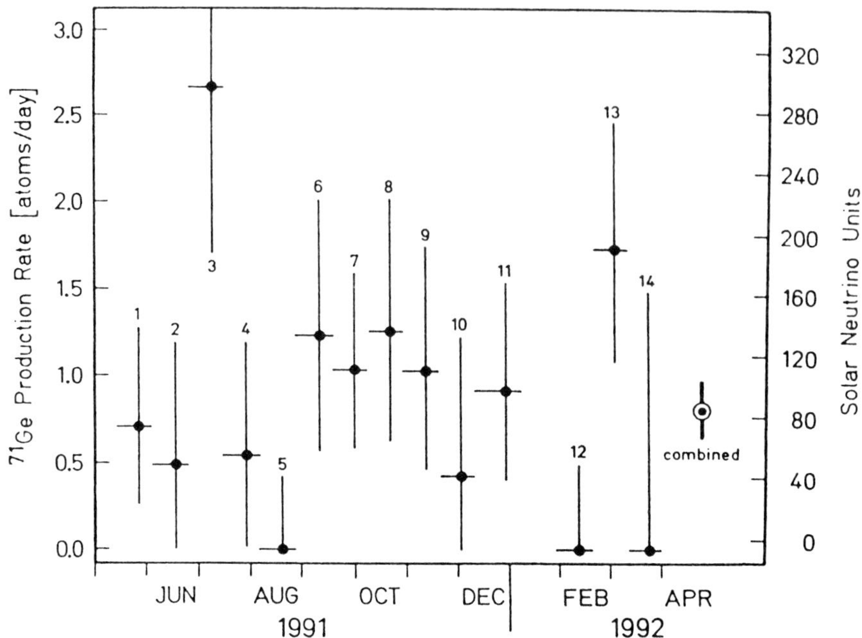

Fig. 4: Results of the first 11 months of solar neutrino recording by GALLEX. The error bars plotted are 1σ. The left hand scale is for the measured ^{71}Ge production rate in atoms per day, on the right hand scale the net solar neutrino production rate in SNU is given [9].

the Gran Sasso Underground Laboratory (shielding depth 3300 meter water equivalent) near L'Aquila, Italy. Counter backgrounds are low enough so that the information from both the L and K peaks are used in the data analysis.

The results of the first 14 solar neutrino runs covering the time period from May 1991 to March 1992 are displayed in Figure 4. After corrections for side reactions and other backgrounds, the observed average solar neutrino production rate in SNU is 83 $\pm$ 19 (stat.) $\pm$ 8 (syst.). This corresponds to 63% of the SSM prediction [2].

The Gallium detector result (like those of the other solar neutrino experiments) is based on the observation of very few events. In order to compare the capabilities of both Ga detectors, it is instructive to calculate the number of neutrinos which in principle could have been detected for a given solar neutrino production rate (here we shall assume 80 SNU). The main factors entering such a calculation are the number of runs performed, the amount of gallium used per run, the exposure time per run, the chemical yield and the counting efficiency. The result is 12 for the five SAGE runs published in 1991 [8], 37 for all 12 SAGE runs displayed in Figure 3 and 66 for all 14 GALLEX runs presented in Figure 4. Because of this better statistics of the GALLEX data and because of the above mentioned lack of a new SAGE average, we adopt the GALLEX value as representative of the gallium result.

4.   Possible interpretations of the experimental data

4.1   Solar models

The most important conclusion to be inferred from the new data is the observation of the pp neutrinos by GALLEX. There is no way to produce the measured signal (83 SNU) without having at least half of the 74 SNU predicted by the SSM for the pp and pep neutrinos (see Table 1). This holds for all scenarios which will be discussed below in order to explain the SNP. More than fifty years after Bethe [11] has suggested the proton-proton reaction as the starting reaction for fusion of hydrogen into helium, its operation in the sun has been experimentally verified. On the other hand, the gallium result confirms the solar neutrino deficit observed by the Chlorine and Kamiokande experiments (despite the fact that with a few percent probability the present errors of the GALLEX rate still allow a value close to the SSM prediction).

Astrophysical solutions to the SNP (i.e. so-called "non-standard solar models") have been proposed since the SNP has been revealed more than twenty years ago. Most of these models have been tailored in order to reduce the solar central temperature and thus lower the ^{8}B and (to a smaller extent) the ^{7}Be neutrino fluxes [12]. It has, however, been realized about two years ago (see for instance [13] [14]) that the different suppression factors observed for the Chlorine and Kamiokande experiments tend to exclude this solution to the SNP. The updated results corroborate this finding. This is demonstrated in Figure 5 where in the upper part the combined experimental Kamiokande and Chlorine data (solid square with 1σ error bars) are compared with the predictions from the SSM of [2] (filled circle) and from a variety of so-called non-standard solar models (open circles) which have been proposed over the last 25 years in order to explain the SNP (for references see for example [12]). Also shown are the predictions of a simplified model (dashed curve, see [15]) in which the fluxes of the different solar neutrino sources have been calculated as a function of the solar core temperature, using the relations given by Bahcall and Ulrich [2]. The ellipse around the experimental data point indicates the 95% confidence region which has been calculated from the statistical and systematic experimental errors including an assumed additional 10% error which accounts for the uncertainties in the neutrino reaction cross section for both experiments.

It is evident from Figure 5 that there is very little overlap between the combined Kamiokande-Chlorine experimental data and any non-standard solar model prediction. A proper calculation yields only 1.2% probability for such an overlap which implies that a reduction of the solar central temperature (for whatever reason) is no longer a likely solution of the SNP. This conclusion is supported by recent improvements in the field of helioseismology [16]. While the SSM fairly well matches the experimental helioseimological data, any departure from the SSM in order to diminish the SNP gives a worse fit to the observations.

The lower part of Figure 5 shows the analogous plot for the combined Gallium-Chlorine experiments. Though for models consistent with the Chlorine detector result the corresponding predictions for Gallium are above the central GALLEX value, the present error range of the GALLEX result does not exclude astrophysical solutions to the SNP on the basis of the Gallium-Chlorine data alone.

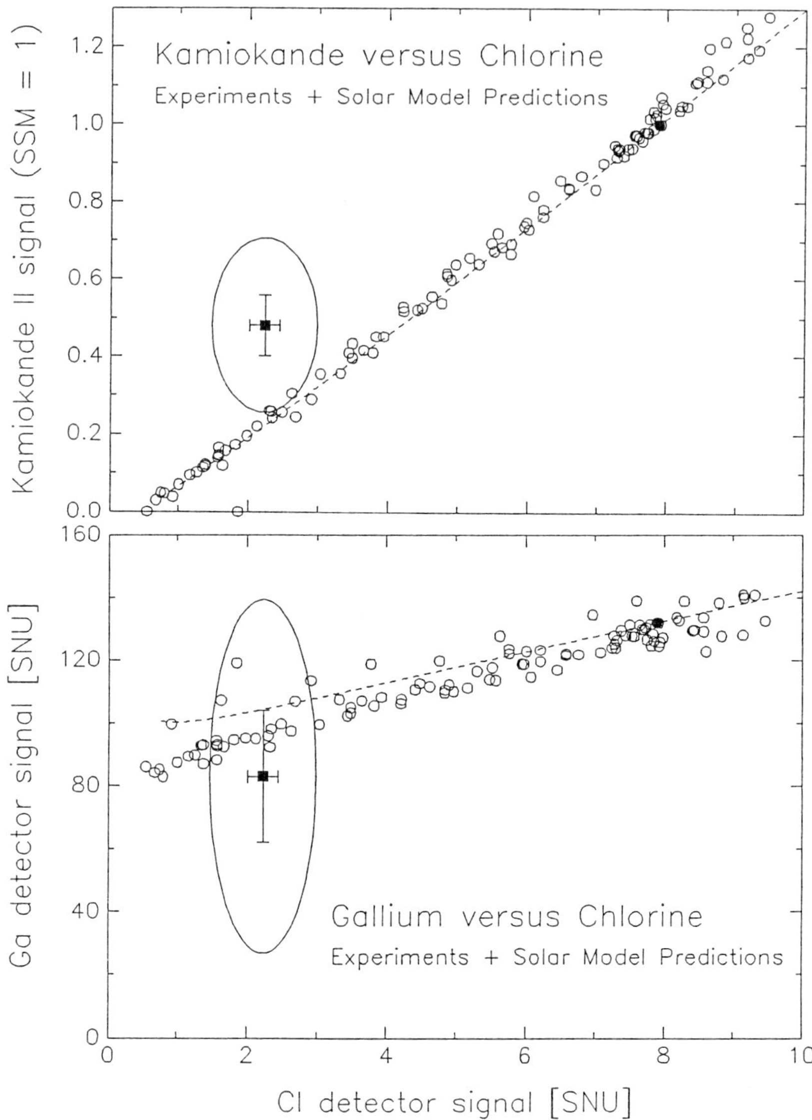

Fig. 5: Comparison of the combined Kamiokande-Chlorine measured value with the predictions of solar models (upper part, see text for details). The lower part shows the analogous plot for the combination Gallium-Chlorine.

4.2 *Neutrino mixing*

The overall conclusion from the previous section leaves the other widely discussed solution to the SNP, neutrino oscillations, as the more likely candidate. Here one considers especially the matter enhanced neutrino oscillations (MSW effect). If neutrinos do have a rest mass and if in neutrino interactions the different flavours are mixed (as in the

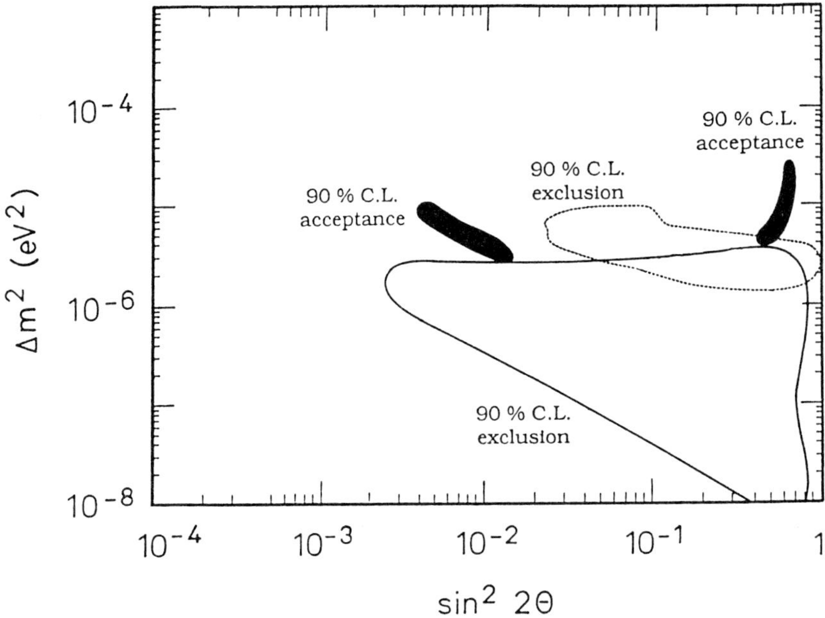

Fig. 6: Diagram of Δm^2 versus $\sin^2 2\theta$. Parameter values inside the black areas are allowed by the combined results of the Cl, Kamiokande and Ga experiments. The regions inside the dotted and solid lines are excluded by the Kamiokande and the GALLEX data (see text) [15].

case of weak interactions of quarks), then a certain fraction of the electron-type neutrinos (ν_e) produced in the solar core may be converted to a muon- or tau-type neutrino (ν_μ, ν_τ) during their passage through the solar matter [12]. Since the radiochemical Cl and Ga detectors are sensitive to ν_e only and since in the energy range relevant for the Kamiokande detector the $\nu_\mu e^-$ and $\nu_\tau e^-$ scattering cross sections are a factor of seven smaller than that for the $\nu_e e^-$ reaction, neutrino oscillations can lead to reduced signals for all three detectors. In contrast to the case for vacuum neutrino oscillations, the MSW effect can produce substantial flavour conversion even if the mixing angle is small. Also, because the extent of flavour conversion depends on the neutrino energy, quite different suppression factors for the three experiments may result.

If mixing is restricted to the two-neutrino case, then the expected suppression factors (relative to the SSM predictions) for all three experiments can be calculated as a function of two parameters, Δm^2 (the difference of the squared masses of the two mixing neutrino eigenstates ν_1 and ν_2) and $\sin^2 2\theta$ (where θ is the mixing angle). Figure 6 (taken from [15]) shows the result of such a combined analysis. There are only two small areas (marked in black) in the two-dimensional Δm^2 - $\sin^2 2\theta$ parameter space in which the suppression factors due to the MSW effect are for all three experiments in agreement (at the 90% confidence level) with the observations. Figure 6 shows in addition two regions which are not allowed by the data at the 90% confidence level. The area inside the dotted curve is excluded from the absence of a day-night effect in the Kamiokande real-time data [17]. The region inside the solid line is excluded by the GALLEX result, here the expected suppression would be larger than actually observed [15]. We should mention here that uncertainties in the SSM neutrino fluxes have been neglected in the calculations on which Figure 6 is based [15]. The above conclusions, however, do not substantially change if these uncertainties are taken into account (see for instance [18]).

For completeness it should finally be noted that there is also a solution for vacuum neutrino oscillations which can accomodate (within error limits) the results of all three experiments [19]. In this case Δm^2 must be close to 10^{-10} eV^2 (which corresponds to an oscillation length of 1.65 Astronomical Units for a 10 MeV neutrino) and mixing must be nearly maximal ($\sin^2 2\theta \simeq 1$).

4.3   *Neutrino decay*

Another principal solution to the SNP is neutrino decay. The most straightforward form, the decay of the ν_e produced in the sun into some sterile species, has been ruled out by the observation of the $\bar{\nu}_e$ signal from the Supernova 1987A, since these neutrinos have survived a possible decay over a distance about 10^{10} times larger than the sun-earth distance. A combination of neutrino decay and flavour mixing, however, can explain the observed Supernova 1987A $\bar{\nu}_e$ signal and still result in a Cl detector rate much below the SSM prediction [20].

A new detailed calculation of such a scenario in which the heavier mass eigenstate ν_2 decays into ν_1 (or $\bar{\nu}_1$ or both) and a Majoron has been performed by Berezhiani et al. [21]. These authors showed that for certain ranges of mixing angles θ and neutrino lifetimes it is possible to reproduce the observed reduction factors (relative to the SSM prediction of [2]) for both the Chlorine and Kamiokande detectors. The corresponding prediction for the Gallium detector yields values between 20 and 45 SNU. This is far away from the central value (83 SNU) measured by GALLEX, leaving only a few percent probability that neutrino decay is the correct explanation for the SNP.

4.4   *Time variations of the solar neutrino flux*

The rate given for the Cl detector in section 3.1 is a time-averaged rate, yet it has been speculated (see for example [22][23]) that the individual data are not randomly distributed in time but exhibit an anticorrelation with the eleven-year sun spot cycle. On the other hand, the Kamiokande-II data [7] did not show such a dependence. The latter is confirmed by the new Kamiokande-III result [6]. Figure 7 (taken from [6]) displays the K-II and K-III data (solid circle) as a function of the sun spot number. It is evident that (within error limits) the Kamiokande signal is constant in time, covering the time period from the solar activity minimum in 1987 to the maximum near 1991.

Spin precession for Dirac neutrinos [24] or spin-flavour conversion for Majorana neutrinos [25][26] with a large magnetic moment ($\mu_\nu > 10^{-11} \mu_{Bohr}$) in the time-dependent magnetic fields of the solar convective zone have been proposed as explanation for the observed time variation of the Cl detector signal. The latter mechanism (eventually combined with the MSW effect, see section 4.2) is preferred since it is in principle capable to produce a sufficiently large time variation for the Cl detector rate and at the same time restrict the corresponding change in the Kamiokande signal to a smaller amplitude. The spin-flavour conversion of Majorana neutrinos has the additional advantage that the

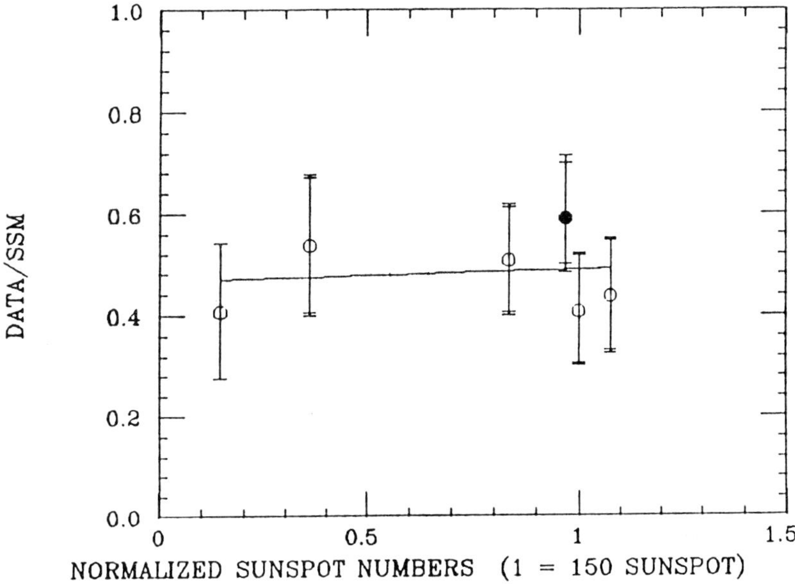

Fig. 7: Kamiokande solar neutrino data as a function of the sun spot number [6].

upper bound on μ_ν ($< 10^{-12}\mu_{Bohr}$) derived from the observation of neutrinos from the supernova 1987A does not apply [27].

Two specific models of the spin-flavour conversion scenario which are (within error limits) capable to reproduce the observed reduction factors as well as the time behaviour of the Chlorine and Kamiokande signals [28] [29] both predict Gallium detector rates below 40 SNU at the present near maximum of the solar cycle. This is in obvious contradiction to the measured GALLEX rate of 83 ± 21 SNU.

5.   Conclusions

GALLEX has observed the pp neutrinos generated in the primary proton-proton fusion reaction. This constitutes the first direct experimental evidence for the proton-proton cycle to be operative in the sun. On the other hand, GALLEX confirms the deficit of higher enery neutrinos observed by the Chlorine and Kamiokande experiments. The origin of this deficit is still not clear. However, the astrophysical solution (non-standard solar models having a lower central temperature than the SSM) is no longer very likely because of the quite different observed reduction factors for the Chlorine and Kamiokande detectors. This leaves neutrino oscillations (MSW effect) as the more plausible explanation for the SNP. In two small distinct regions of the Δm^2 - $\sin^2 2\theta$ parameter space the MSW prediction for all three experiments gives a good fit to the experimental data. Other possible explanations for the SNP like neutrino decay or two specific models of spin-flavour conversion are not supported by the GALLEX result.

References

[1] Bahcall J N and Pinsonneault M H 1992, preprint IASSNS-AST 92/10, to be published in *Reviews of Modern Physics*

[2] Bahcall J N and Ulrich R K 1988 *Rev. Mod. Phys.* **60** 297-372

[3] Turck-Chièze S, Cahen S and Cassé M 1988 *Astrophys. J.* **335** 415-424

[4] Sienkiewicz R, Bahcall J N and Paczynski B 1990 *Astrophys. J.* **349** 641

[5] Lande K, contribution to the *Neutrino '92 Conference, Granada, Spain (June 8-12, 1992)*, to be published in the Conference Proceedings

[6] Nakamura K, contribution to the *Neutrino '92 Conference, Granada, Spain (June 8-12, 1992)*, to be published in the Conference Proceedings

[7] Hirata K S et al. 1990 *Phys. Rev. Lett.* **65** 1297-1300

[8] Abazov A I et al. 1991 *Phys. Rev. Lett.* **67** 3332

[9] Anselmann P et al. 1992 *Phys. Lett.* **B285** 376-389

[10] Bowles T, contribution to the *Neutrino '92 Conference, Granada, Spain (June 8-12, 1992)*, to be published in the Conference Proceedings

[11] Bethe H A 1939 *Phys. Rev.* **55** 434

[12] Bahcall J N 1989, *Neutrino Astrophysics*, Cambridge University Press

[13] Hampel W 1990 *Physics World* **3** No. 9, 20-21

[14] Bahcall J N and Bethe H A 1990 *Phys. Rev. Lett.* **65** 2233-2235

[15] Anselmann P et al. 1992 *Phys. Lett.* **B285** 390-397

[16] Gough D and Toomre J 1991 *Ann. Rev. Astron. Astrophys.* **29** 627-684

[17] Hirata K S et al. 1991 *Phys. Rev. Lett.* **66** 9

[18] Bludman S A, Hata N, Kennedy D C and Langacker P G 1992, preprint UPR-0516T, University of Pennsylvania

[19] Barger V, Phillips R J N and Whisnant K 1990 *Phys. Rev. Lett.* **65** 3084-3087

[20] Frieman J A, Haber H E and Freese K 1988 *Phys. Lett.* **B200** 115-121

[21] Berezhiani Z G, Fiorentini G, Moretti M and Rossi A 1992 *Z. Phys.* **C54** 581-586

[22] Filippone B W and Vogel P 1990 *Phys. Lett.* **B246** 546-550

[23] Bahcall J N and Press W H 1991 *Astrophys. J.* **370** 730-742

[24] Voloshin M B, Vysotskii M I and Okun L B 1986 *Soviet Phys. JETP* **64** 446

[25] Lim C S and Marciano W J 1988 *Phys. Rev.* **D37** 1368

[26] Akhmedov E Kh 1988 *Phys. Lett.* **B213** 64

[27] Leurer M and Liu J 1989 *Phys. Lett.* **B219** 304-308

[28] Ono Y and Suematsu D 1991 *Phys. Lett.* **B271** 165-171

[29] Babu K S, Mohapatra N and Rothstein I Z 1991 *Phys. Rev.* **D44** 2265-2277

List of Participants

J. Andrzejewski	University of Lodz	Poland
C. Angulo	Ruhr-Universität Bochum	Germany
M. Arnould	Université Libre de Bruxelles	Belgium
I. Baraffe	MPl Garching	Germany
B. Barbuy	University of Sao Paulo	Brasil
F.C. Barker	The Australian National University	Australia
C. A. Barnes	California Institute of Technology	USA
G. Baur	Forschungszentrum Jülich	Germany
H. Beer	Kernforschungszentrum Karlsruhe	Germany
G. Bogaert	CSNSM Orsay	France
A.I. Boothroyd	University of Toronto	Canada
V. Bouquelle	Université Libre de Bruxelles	Belgium
L.E. Brown	Clemson University	USA
R. Bruß	Technische Universität Berlin	Germany
L. Buchmann	University of British Columbia	Canada
H. Bucka	Technische Universität Berlin	Germany
A. Burkert	MPI für Astrophysik Garching	Germany
M. Busso	Osservatorio Astronomico di Torino	Italy
R. Cannon	University of Cambridge	England
F. Corvi	CBNM EURATOM	Belgium
K. Cunha	University of Austin in Texas	USA
S. Czajkowski	CSNSM Orsay	France
K. Czerski	Technische Universität Berlin	Germany
P. Decrock	University of Leuven	Belgium
T. Delbar	Université Cath. de Louvain	Belgium
P. Descouvemont	Université Libre de Bruxelles	Belgium
D. Disdier	CRNS Strasbourg	France
P. Doll	Kernforschungszentrum Karlsruhe	Germany
S. Druyts	CBNM EURATOM	Belgium

642

M.F. El-Eid	Universität Göttingen	Germany
P. Engelmann	Kernforschungszentrum Karlsruhe	Germany
P. Ferrando	CEN Saclay	France
R. Gallino	Università di Torino	Italy
W. Galster	Université Cath. de Louvain	Belgium
J. Garrett	Oak Ridge National Laboratory	USA
M. Giller	Institute of Nuclear Studies	Poland
Yu.M. Gledenov	JINR Dubna	Russia
S. Goriely	Université Libre de Bruxelles	Belgium
C. Grama	Inst. of Atomic Physics Bucharest	Romania
U. Greife	Ruhr-Universität Bochum	Germany
E. Grosse	GSI Darmstadt	Germany
K. Grün	Technische Universität Wien	Austria
K. Guber	Kernforschungszentrum Karlsruhe	Germany
B. Gustafsson	Astr. Observatoriet Uppsala	Sweden
W. Hammer	Universität Stuttgart	Germany
W. Hampel	MPI für Kernphysik Heidelberg	Germany
S. Harissopulos	NRC Democritos	Greece
C. L. Harper	Harvard University Cambridge	USA
M. Hashimoto	Kyushu University Ropponmatsu	Japan
P. Heide	Technische Universität Berlin	Germany
A. Heße	Universität-Gesamthochschule Siegen	Germany
W. Hillebrandt	MPI für Astrophysik Garching	Germany
W. M. Howard	Lawrence Livermore National Laboratory	USA
J. Humblet	Université Liège	Belgium
Ch. Iliadis	University of Notre Dame	USA
J. Isern	Centre D'Estudis Avangats Blanes	Spain
S. Jaag	Kernforschungszentrum Karlsruhe	Germany
C.W. Johnson	California Institute of Technology	USA
A. Jorissen	ESO Garching	Germany
J. José	EUPVG-UPC	Spain
M. Junker	Ruhr-Universität Bochum	Germany
F. Käppeler	Kernforschungszentrum Karlsruhe	Germany
R.W. Kavanagh	California Institute of Technology	USA
A.M. Khoklov	MPI für Astrophysik Garching	Germany
I. Khubeis	University of Jordan	Jordan
J. Kiener	CSNSM Orsay	France
T. Kirsten	MPI für Kernphysik Heidelberg	Germany
F. Knape	Technische Universität Berlin	Germany
U. Koelmel	Kernforschungszentrum Karlsruhe	Germany
S. Kossionides	NRC Democritos	Greece

K.-L. Kratz	Universität Mainz	Germany
L. Kraus	CRNS Strasbourg	France
H. Krauss	Technische Universität Wien	Austria
M. Kruglanski	Université Libre de Bruxelles	Belgium
D.L. Lambert	University of Austin in Texas	USA
J. Lattanzio	Monash University Clayton	Australia
P. Leleux	Université Cath. de Louvain	Belgium
M. Lemoine	IAP Paris	France
I. Linck	CRNS Strasbourg	France
W. J. Maciel	University of Sao Paulo	Brasil
R. A. Malaney	University of Toronto	Canada
H. Markum	Technische Universität Wien	Austria
K. Marti	University of California San Diego	USA
P. Mazzali	ESO Garching	Germany
A. Mengoni	ENEA Bologna	Italy
B. Meyer	Clemson University	USA
G. Meynet	Geneva Observatory	Switzerland
M.F. Mohar	Lawrence Berkeley Laboratory	USA
P. Möller	Sci. Computing Los Alamos	USA
G. Molnar	Institute of Isotopes Budapest	Hungary
Y. Nagai	Tokio Institute of Technology	Japan
Zs. Németh	Institute of Isotopes Budapest	Hungary
R. Neubert	Technische Universität Berlin	Germany
K. Nomoto	University of Tokyo	Japan
H. Oberhummer	Technische Universität Wien	Austria
A. Omont	IAP Paris	France
U. Ott	MPI für Chemie Mainz	Germany
B. Pagel	NORDITA Copenhagen	Denmark
T. Paradellis	NRC Democritos	Greece
Yu. Popov	JINR Dubna	Russia
N. Prantzos	CEN Saclay	France
G. Raimann	Ruhr-Universität Bochum	Germany
W. Ratynski	Soltan Institute for Nucl. Studies, Swierk	Poland
A.K. Ray	Tata Institute of Fundamental Res.	India
M. Rayet	Université Libre de Bruxelles	Belgium
H.-G. Rebel	Kernforschungszentrum Karlsruhe	Germany
Ch. Reineke	Technische Universität Berlin	Germany
S. Richter	MPI für Chemie Mainz	Germany
C. Rolfs	Ruhr-Universität Bochum	Germany
G. Roters	Ruhr-Universität Bochum	Germany
M. Sahrling	Astronomical Observatory Uppsala	Sweden

G. Schatz	Kernforschungszentrum Karlsruhe	Germany
H. Schatz	Kernforschungszentrum Karlsruhe	Germany
V. Schönfelder	MPI für Extraterr. Physik Garching	Germany
C. Schwee	Universität Karlsruhe	Germany
T. Shima	Tokio Institute of Technology	Japan
A.C. Shotter	University of Edinburgh	Great Britain
M. Simon	Universität-Gesamthochschule Siegen	Germany
V.V. Smith	University of Austin in Texas	USA
E. Somorjai	ATOMKI Debrecen	Hungary
O. Sorlin	Université de Paris	France
G. Staudt	Universität Tübingen	Germany
R. Steininger	Universität Karlsruhe	Germany
M. Steinmetz	MPI für Astrophysik Garching	Germany
H. Stoll	Kernforschungszentrum Karlsruhe	Germany
G. Strazzulla	Città Universitaria Catania	Italy
K. Takahashi	MPI für Astrophysik Garching	Germany
Ch. Theis	Kernforschungszentrum Karlsruhe	Germany
F.-K. Thielemann	Harvard University Cambridge	USA
K. Toukan	University of Jordan	Jordan
H.-P. Trautvetter	Ruhr-Universität Bochum	Germany
N. Tsoglin	Ukrainian Academy of Sciences	Ukraina
Yu. Tsoglin	Ukrainian Academy of Sciences	Ukraina
S. Typel	Universität Münster Germany	
M. Uhl	Universität Wien	Austria
F. Voss	Kernforschungszentrum Karlsruhe	Germany
C. Wagemans	University of Gent	Belgium
W. Walters	University of Maryland	USA
R. A. Ward	Lawrence Livermore Natl. Lab.	USA
G.J. Wasserburg	California Institute of Technology	USA
F. Weber	Universität München	Germany
N. Weber	Kernforschungszentrum Karlsruhe	Germany
M. Werner	MPI für Chemie Mainz	Germany
M. Wiescher	University of Notre Dame	USA
M. Wilmes	Universität Heidelberg	Germany
K. Wisshak	Kernforschungszentrum Karlsruhe	Germany
A. Wöhr	Universität Mainz	Germany
S.E. Woosley	University of California Santa Cruz	USA
D. Zahnow	Ruhr-Universität Bochum	Germany

Author Index